Figures available in three downloadable sizes (resolutions)

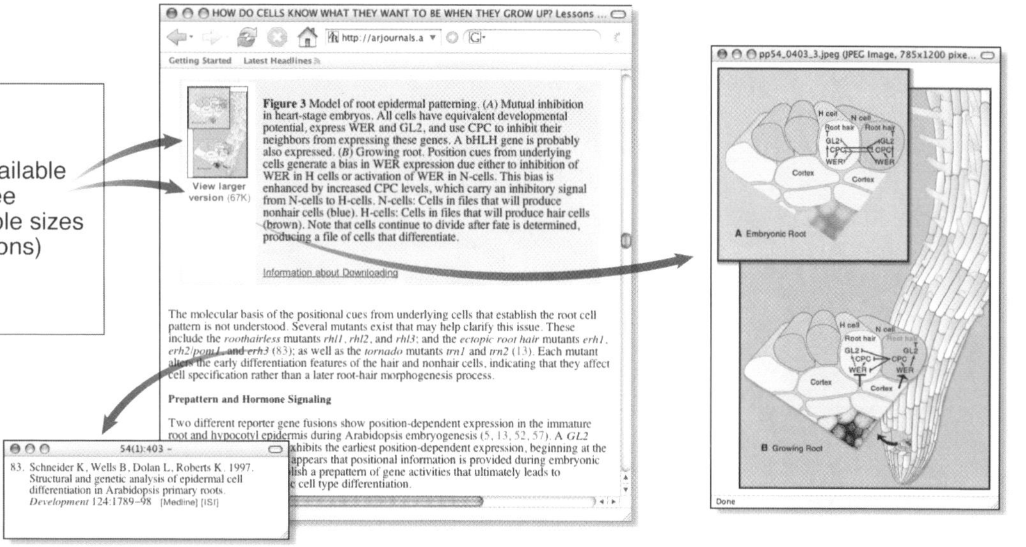

Citations in text link to references in bibliography

References in Annual Reviews chapter bibliography link out to sources of cited articles online

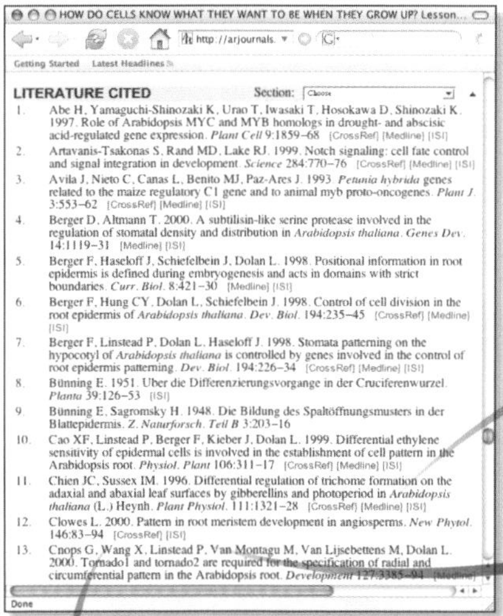

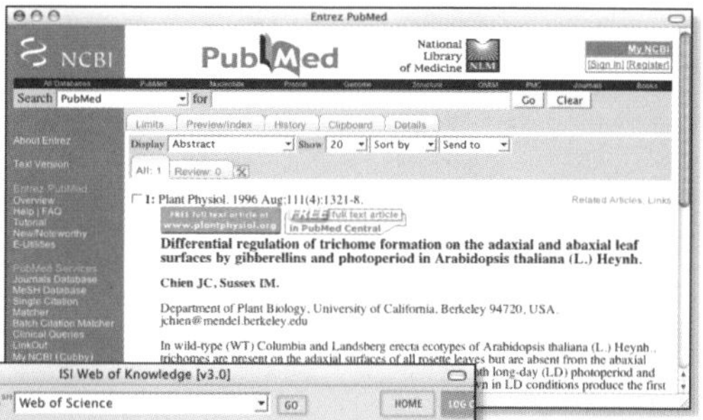

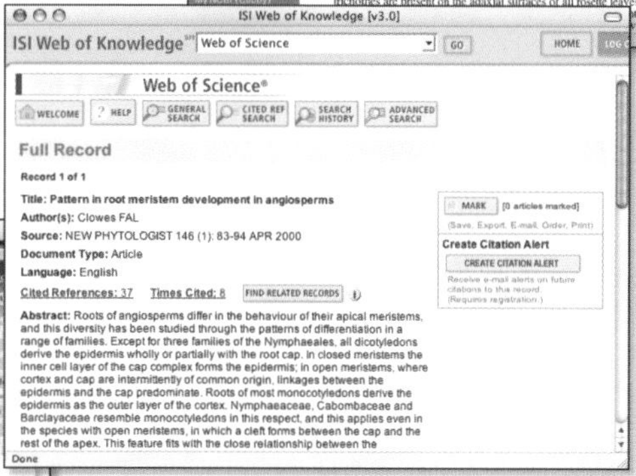

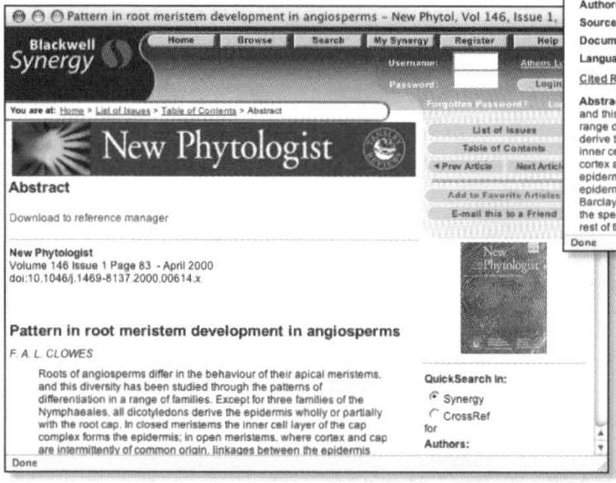

Annual Review of
Plant Biology

Production Editor: Ellen Terry
Managing Editor: Veronica Dakota Padilla
Bibliographic Quality Control: Mary A. Glass
Electronic Content Coordinator: Suzanne K. Moses
Illustration Editor: Eliza K. Jewett

Annual Review of Plant Biology

Volume 58, 2007

Sabeeha Merchant, *Editor*
University of California, Los Angeles

Winslow R. Briggs, *Associate Editor*
Carnegie Institution of Washington, Stanford, California

Vicki L. Chandler, *Associate Editor*
University of Arizona

www.annualreviews.org • science@annualreviews.org • 650-493-4400

Annual Reviews
4139 El Camino Way • P.O. Box 10139 • Palo Alto, California 94303-0139

Annual Reviews
Palo Alto, California, USA

International Standard Serial Number: 1543-5008
International Standard Book Number: 978-0-8243-0658-8
Library of Congress Catalog Card Number: 50-13143

∞ The paper used in this publication meets the minimum requirements of American National Standards for Information Sciences—Permanence of Paper for Printed Library Materials, ANSI Z39.48-1992.

TYPESET BY APTARA, INC.
PRINTED AND BOUND BY FRIESENS CORPORATION, ALTONA, MANITOBA, CANADA

Annual Review of
Plant Biology

Volume 58, 2007

Contents

Indexes

Errata

An online log of corrections to *Annual Review of Plant Biology* chapters (if any, 1997 to the present) may be found at http://plant.annualreviews.org/

Related Articles

From the ***Annual Review of Biochemistry***, Volume 75 (2006)

Energy Converting NADH:Quinone Oxidoreductase (Complex I)
Ulrich Brandt

Energy Transduction: Proton Transfer Through the Respiratory Complexes
Jonathan P. Hosler, Shelagh Ferguson-Miller, and Denise A. Mills

Chromatin Modifications by Methylation and Ubiquitination: Implications in the Regulation of Gene Expression
Ali Shilatifard

Structure and Mechanism of the Hsp90 Molecular Chaperone Machinery
Laurence H. Pearl and Chrisostomos Prodromou

Cold-Adapted Enzymes
Khawar Sohail Siddiqui and Ricardo Cavicchioli

The Biochemistry of Sirtuins
Anthony A. Sauve, Cynthia Wolberger, Vern L. Schramm, and Jef D. Boeke

Introduction to the Membrane Protein Reviews: The Interplay of Structure, Dynamics, and Environment in Membrane Protein Function
Jonathan N. Sachs and Donald M. Engelman

Relations Between Structure and Function of the Mitochondrial ADP/ATP Carrier
H. Nury, C. Dahout-Gonzalez, V. Trézéguet, G.J.M. Lauquin, G. Brandolin, and E. Pebay-Peyroula

Transmembrane Traffic in the Cytochrome b_6f Complex
William A. Cramer, Huamin Zhang, Jiusheng Yan, Genji Kurisu, and Janet L. Smith

From the ***Annual Review of Biophysics and Biomolecular Structure***, Volume 35 (2006)

Spinach on the Ceiling: A Theoretical Chemist's Return to Biology
Martin Karplus

Evolutionary Relationships and Structural Mechanisms of AAA+ Proteins
Jan P. Erzberger and James M. Berger

From the ***Annual Review of Cell and Developmental Biology***, Volume 22 (2006)

From the ***Annual Review of Genetics***, Volume 40 (2006)

D von Wettstein

From Analysis of Mutants to Genetic Engineering

Diter von Wettstein

Department of Crop and Soil Sciences, School of Molecular Biosciences and Center for Integrated Biotechnology, Washington State University, Pullman, Washington 99164-6420; email: diter@wsu.edu

Annu. Rev. Plant Biol. 2007. 58:1–19

The *Annual Review of Plant Biology* is online at plant.annualreviews.org

This article's doi: 10.1146/annurev.arplant.58.032806.104003

First published online as a Review in Advance on October 26, 2006

1543-5008/07/0602-0001$20.00

Key Words

endosperm proteins, (1,3;1,4)-β-glucanase, transgenic barley, proanthocyanidins, (2,3-*trans*) catechin, (2,3-*cis*) epicatechin

Abstract

This chapter describes the research of developing transgenic barley for synthesis of recombinant proteins with practical significance and of metabolic engineering of proanthocyanidin-free barley. The results were obtained by graduate students, postdoctoral researchers, and visiting scientists at the Carlsberg Laboratory from 1972–1996 and during the past ten years at Washington State University. It is written in appreciation of their enthusiasm, skill, and perseverance.

Contents

PREFACE

In the Introductory Review for Progress in Botany 2006, I reviewed two areas of my research interests: fascinations with chloroplasts and chromosome pairing (82). Here I discuss my engagement in genetic engineering of barley for improved feed and in metabolic engineering of proanthocyanidin-free barley. Due to space limitations, my activities in breeding of brewer's yeast and disease resistance will be dealt with on another occasion.

In 1972 I was asked to join the Carlsberg Laboratory as head of the Department of Physiology. The funds for operating the Carlsberg Laboratory for basic research are provided by the Carlsberg Foundation, which receives its income from the ownership of the Carlsberg and Tuborg Breweries. The two breweries fused in 1970 to permit international expansion and thereby provide improved income for the Foundation. Paul Brandt Rehberg, Christian Crone, and Kristof Glamann of the Carlsberg Foundation, Mogens Westergaard and Martin Ottesen of the Chemistry Department, and A.W. Nielsen, Chief Executive Officer of the breweries, as well as their Director of Technology Eigil Bjerl Nielsen, were eager to upgrade the facilities, staff, and research activities. As a result of these negotiations the staff of the Carlsberg Laboratory's two departments increased from 10 to 54 and a new building was erected to house portions of the Carlsberg Laboratory and a newly created research and development department for the breweries, to be known as the Carlsberg Research Laboratory. In exchange for these improvements I was also requested to head Carlsberg Plant Breeding to develop improved barley varieties and to take initiatives of breeding improved strains of brewer's yeast, *Saccharomyces carlsbergensis*.

I was lucky to be joined by Jørgen Larsen, an outstanding plant breeder. Our strategy was to breed spring and winter barley varieties with good malt quality by standard recombination and selection techniques, in order to be up-to-date with regard to yield and quality and to have competitive lines which could be used to introduce novel biochemical characteristics by mutation and genetic engineering. This gave rise at the end of the 1980s to barley varieties such as Canut, Caruso, and Alondra. For several years Canut was the highest-yielding barley variety in Denmark and widely grown. But in looking for novel, useful characteristics we turned our attention to molecular improvement of protein quality and removal of proanthocyanidins.

SYNTHESIS AND MOBILIZATION OF ENDOSPERM PROTEINS IN BARLEY

Barley grain contains about 10–15% protein and the composition of the protein is of major importance for the malting and feeding quality of the grain. A major part of our program therefore endeavored to elucidate the molecular biological basis for storage protein synthesis and deposition in the vacuoles of the developing endosperm. Together with the extensive projects carried out at the John Innes Institute, the Rothamsted Institute, the Max-Planck-Institut für Züchtungsforschung, Purdue University, the University of California, and the Institute of Biological Chemistry, Washington State University, these investigations have led to the situation that the synthesis of storage proteins in wheat, barley, corn, and rice belong to one of the best molecularly analyzed areas in the plant sciences (49). The results provided the insight that cereal endosperms provide outstanding opportunities for the production of recombinant proteins (25, 57).

Fundamental ultrastructural information on the development of protein bodies in the barley endosperm of wild-type and the high lysine mutants *lys* 3a (Risø 1508) and *hor* 2ca (Riso 56) was provided by electron microscopy (9, 13, 50, 78, 79). The former mutant produces practically no B- and C- hordeins while the latter lacks B-hordeins (due to a deletion of the *hor* 2 locus). From these analyses I proposed that the hordein polypeptides are synthesized on the polysomes of the endoplasmic reticulum, cotranslationally transferred into the lumen of the endoplasmic reticulum and from there possibly via the Golgi apparatus into the vacuoles, where they are compacted into the protein bodies together with some other proteins (follow the Supplemental Material link from the Annual Reviews home page at **http://www.annualreviews.org** to see **Supplemental Figure 1**). This was verified by Anders Brandt, John Ingversen, Verena Cameron-Mills, and Susan Madrid using isolated endoplasmatic reticulum (5, 6, 11–13) and isolated protein bodies and by localization studies with monoclonal antibodies developed by Gunilla Høyer-Hansen (58, 65, 66, 77) (see **Supplemental Figures 2** and **3**). Cloning and sequencing of structural genes for B-, C-, and γ-hordein polypeptides have corroborated that the *Hor 1*, *Hor* 2, and *Hor* F loci each consist of a family of closely linked structural genes for these proteins, with each gene having its own promoter (7, 10, 18, 24, 62, 63, 70). Some of these genes contain stop codons, but transient transformation experiments revealed that expression of such a gene in the endosperm can take place by amber codon suppression (19).

B-, C-, D-, γ-hordeins: prolamin storage proteins of the barley grain

PCR: polymerase chain reaction

There are two results that deserve special mentioning. The first is that the nature of the high lysine mutant *lys* 3a was clarified. Steady-state levels of transcripts encoding B-hordein, C-hordein, and protein Z are almost absent in the developing endosperm of the mutant, whereas the transcript levels for D-hordein, glyceraldehyde-3-phosphate dehydrogenase, and histone are normal. Mikael Blom Sørensen showed by genomic sequencing and ligation-mediated PCR that 10 CpGs in the promoters of the B-hordein genes and 4 CpGs of the adjacent coding region of the gene are hypomethylated in the endosperm but fully methylated at the cytosins in the leaf. In the developing endosperm of the mutant, demethylation of the B-hordein promoter does not occur (71). Genomic sequencing of the D-hordein gene promoter using bisulphite-treated DNA revealed a CpG island and confirmed that the promoter is unmethylated in the leaf as well as in the endosperm both in the wild type and in the mutant. Comparison of transient expression in wild-type and lys3a endosperms demonstrated that the activities of an unmethylated D- or C-hordein promoter were equivalent, and in vitro methylation of the promoters prior to expression severely inhibited their activities. Thus, two categories of promoters for endosperm-specific gene expression can be distinguished: One is silenced by methylation in other tissues and has to be demethylated before transactivation of transcription can take place, and the other is solely dependent on removal of repressors or induction of transcription factors specific for the endosperm.

The second result is that it was demonstrated early on (21) that storage protein synthesis during grain development is regulated at the transcriptional level in dependence of the supply of nitrogen. In fact, upregulation of transcript levels could be effectuated within hours if ammonium nitrate was supplied through the peduncle, and equally rapid reduced when the supply was stopped. In a detailed analysis of a C-hordein promoter using particle bombardment of the endosperm with the glucuronidase reporter enzyme, Müller & Knudsen (48) showed that the GCN4 motif ATGA(C/G)TCAT is the dominating *cis*-acting element in this response. However, synergistic interaction with the neighboring endosperm motif TGTAAGT within the bifactorial prolamin element is an absolute requirement for a strong, positive regulation by an optimal nitrogen regime. Low nitrogen levels convert the GCN4 box into a negative motif (which is just the opposite response to that effectuated by the GCN4 element in yeast). The endosperm box on its own exerts silencing activity independent of nitrogen nutrition. These findings should be used to construct storage protein genes with constitutive expression—a desirable goal—in order to reduce the necessity of the large amount of nitrogen fertilizer now required to sustain high protein yields. Characterization of the promoters of the storage proteins in barley and of the targeting pathway of the storage protein precursors to the storage vacuoles in barley (47) has provided the basis for developing the production of large amounts of recombinant proteins in the transgenic barley grain (25).

GENETIC ENGINEERING OF BARLEY FOR IMPROVED FEED AND MALT

When Jeff Fincher from La Trobe University came on sabbatical it was decided to isolate a cDNA clone of his favorite enzyme: barley (1,3;1,4)-β-glucanase (20). Use of barley with the current malting technology for producing beer and other food commodities is possible because the endogenous barley α-amylases, which are responsible for starch degradation in the mashtun, survive the last stage in malting, the kilning process. This implies drying the germinated grain at temperatures of up to 80°C. During germination, (1,3;1,4)-β-glucanase is secreted from the aleurone tissue into the endosperm to degrade the endosperm cell walls, up to 80% of which comprise (1,3;1,4)-β-D-glucans. Secreted α-amylases cannot reach the starch grains unless the β-glucan walls have been depolymerized (e.g., 32). In contrast to α-amylases, β-glucanase does not survive the kilning process. Incomplete depolymerization of the β-glucan walls during germination in the malting process causes their solubilization at increasing temperatures up to 76°C in the preparation of wort for fermentations. The wort then has an unacceptable high viscosity for full-scale filtration or centrifugation. For this reason, most barley varieties cannot be used for malting. The problem is remedied if a heat-stable (1,3;1,4)-β-glucanase is used in the mashtun (4, 51), but it would be more economical if malting barley contained the enzyme. Therefore, we bred transgenic barley plants, which express a protein-engineered thermostable (1,3;1,4)-β-glucanase during germination/malting in the aleurone cells and secrete the enzyme into the endosperm (33).

Attempts to obtain thermotolerant barley (1,3;1,4)-β-glucanase by site-directed mutagenesis of a cloned cDNA were not very successful, but *Bacillus* species produce (1,3;1,4)-β-glucanases with the same substrate specificity as that of the barley enzyme. Among hybrids between the genes for *B. amyloliquefaciens* β-glucanase (which is active at low pH and thus suitable for mashing processes) and for *B. macerans* β-glucanase (which displays slightly better heat stability at high pH), genes were synthesized and selected, coding for enzymes with better heat stability and a wider pH optimum than both parental enzymes (4, 51). This was the first demonstration of mutagenic hybrid vigor (heterosis), a phenomenon of high significance in the natural evolution of enzymes and enzyme-breeding strategies. It is now used with advanced technology in directed evolution of enzymes by iterated DNA shuffling (e.g., 15). The hybrid enzyme H(A16-M), which contains 16 N-terminal amino acids from the *B. amyloliquefaciens* enzyme and 198 amino acids from the *B. macerans* enzyme, displays an increase in the enzymatic half-life from 6 min to 2 h at 70°C. Mashing experiments have shown that this improvement is sufficient to eliminate the viscosity caused by β-glucans in the mashtun (51). The three-dimensional structure of the hybrid and one parent enzyme was determined by Udo Heinemann and his students (22, 35, 36). Besides characterization of the catalytic site (Glu-105), definition of the substrate binding site, and the importance of the bound Ca^{2+}, the structure revealed that heat stability is determined by the amount of hydrogen bonding between the aligned C- and N-terminal domains of the polypeptide chains, which form two seven-stranded antiparallel β-pleated sheets (a jelly-roll β-barrel structure) (**Figure 1**). A further increase in thermotolerance to a half-life of 4 h at 70°C was engineered by removing residue Tyr-13 from the H(AI2-M) hybrid (61). Kinetic analyses showed that the reaction velocity and substrate affinity of the hybrid enzyme did not differ significantly from those of the *B. macerans* parental enzyme.

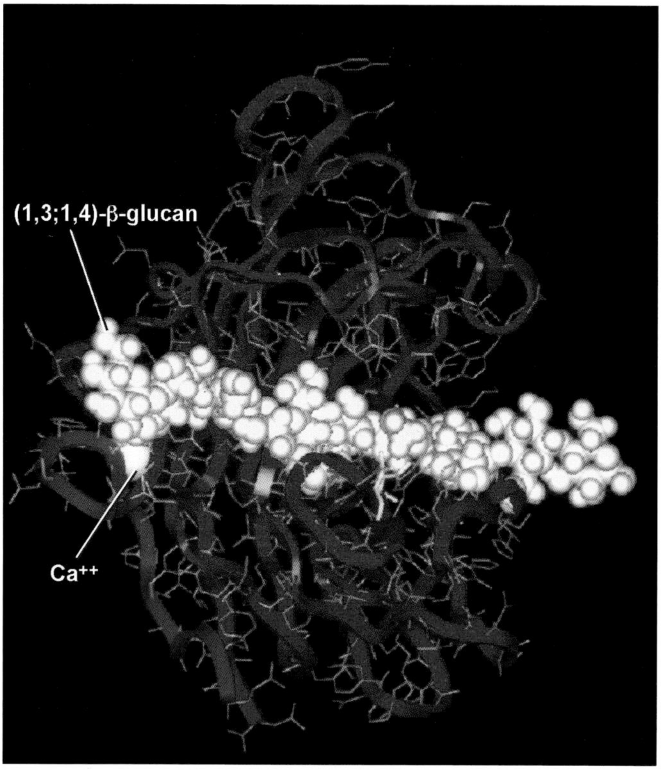

Figure 1

Structure of a heat-stable (1,3;1,4)-β-glucanase. A substrate stretch of (1,3;1,4)-β-D-glucan has been modeled into the active-site cleft. The active-site glutamate is in ball and stick. Activity requires Ca^{2++}. Courtesy of Morten Meldgaard.

A series of hybrid genes was transformed into yeast and using suitable signal peptides production strains that secrete more than 100 mg of enzyme per liter of medium were created (46, 52). The hybrid β-glucanases were glycosylated at Asn-31 and Asn-185 (**Figure 2**), which increased the enzymic half-life to 7 h at 70°C. Efficient expression in aleurone protoplasts and aleurone cells of transgenic barley required construction of a GC-rich version of the H(A12-M) ΔY13 gene with a G+C content of 63.4% (34, 60). Expression was obtained either with the aleurone-specific promoter of the barley (1,3;1,4)-β-glucanase isoenzyme II (91) or the high pI α-amylase promoter both carrying gibberellic acid response elements. The transgene further required the code for the signal sequence of either of these enzymes to ensure cotranslational transfer of the synthesized enzyme precursor into the lumen of the endoplasmic reticulum for entrance into the default pathway for secretion of the enzyme into the endosperm during germination. The transgenic barley plants have retained the genes in stable form over many generations and faithfully express the novel enzyme (27).

The enzyme depolymerizes the mixed linked β-glucans at 1,4 linkages adjacent to 1,3 linkages but leaves stretches of oligosaccharides with only 1,4 linkages undegraded (see **Supplemental Figure 4**). These have to be depolymerized to glucose residues by a

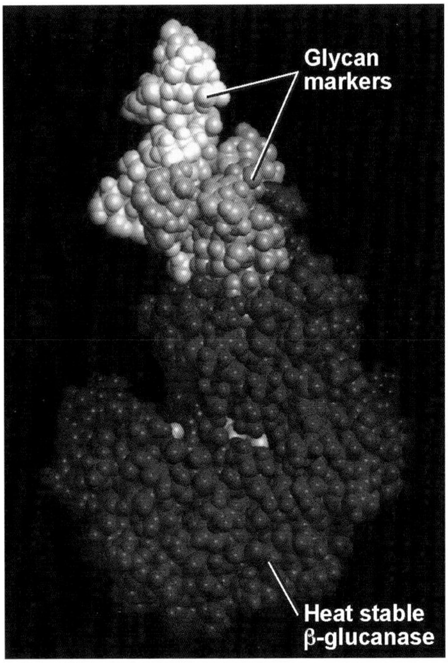

Figure 2

The heat-stable β-glucanase was expressed in the barley endosperm with a signal peptide for transfer into the endoplasmic reticulum and became modified with two glycans. The glycan markers revealed that the enzyme is transported by default into the storage vacuoles. On its way the enzyme depolymerized the newly synthesized (1,3;1,4)-β-D-glucan destined for the cell walls, thereby revealing its site of synthesis in the endoplasmic reticulum.

(1,4)-β-glucanase (cellulase). A gene that encodes a multienzyme that combines both (1,3;1,4) and (1,4) catalytic activity in the same protein was constructed (53). It has been successfully expressed in the yeast *Pichia pastoris* and the multienzyme depolymerizes mixed linked β-glucans into glucose. This is an example for construction of enzymes with altered substrate preference for depolymerization of plant-derived polymers to expand the use of abundant and renewable biomass resources.

Pichia pastoris: a methylotrophic yeast used to export recombinantly synthesized proteins into the medium

At present, breweries and malting companies will not use transgenic barley that is popularly, and nonsensically, called GM or GMO barley (standing for genetically modified or genetically modified organism barley). They are waiting for consumers to be enlightened that transgenic barley is not different and equally well tested as safe food as barley bred by mutation, selection, and hybridization for the past 10,000 years.

CONVERTING BARLEY FROM A LOW- TO A HIGH-ENERGY FEED

Our recent research with Jintai Huang, Henny Horvath-O'Geen, Gamini Kannangara, and enthusiastic students has revealed that barley is a highly suitable crop for producing recombinant proteins, not only in the malting process but also in the endosperm of the developing grain (25, 26, 81). Under suitable conditions as much as 1 g of recombinant protein can be obtained per kilogram of grain. Barley is a self-pollinating small-grain cereal that does not hybridize with any wild species in the surrounding natural habitats of its area of cultivation. Cross-pollination among barley cultivars, especially with two-row spring barley, is low and can be controlled adequately by spatial separation. Patterns and sites of T-DNA integrations into the barley genome from single- and double-cassette vectors are determined for the identification of cultivars with value-added properties as well as for the selection of marker-free transgenic lines (72). T-DNA/Plant DNA junctions are captured as single-stranded DNA with a biotinylated primer annealing to the vector adjacent to the border and an adaptor is ligated to a restriction site overhang in the flanking barley DNA. The captured junction is converted into a double strand and sequenced. Primers of 15–30 nucleotides designed from the genomic DNA at the insertion site can amplify fragments by polymerase chain reactions (PCRs) that identify and thus trace unequivocally any transformant. Adjacent transgene insertions with single-cassette vectors are always in tandem direct repeat configuration. Twelve of the 46 integrations characterized by BLAST searches were within different regions of the

BARE-1 retrotransposon element that is occurring with a frequency of 2×10^5 copies in the barley genome. Transcription and translation of transgenes inserted in retrotransposon elements is of normal efficiency and has the advantage that no essential gene is inactivated through the insertion. Human serum albumin, lysozyme, lactoferrin, α-antitrypsin, antithrombin III, and other proteins and transcription factors have been expressed in the barley endosperm.

Today, 85% of barley harvest is used to feed animals, and the malting industry uses 15% for beverage production. Barley acreage has decreased worldwide by 14–20% since 1989, a decrease that continues. A major reason is that low-priced, transgenic corn with higher nutritional quality is a strong competitor of barley in the feed industry. But barley is needed for crop rotation and it is therefore desirable to keep barley competitive by breeding for improved nutritional and other value-adding characteristics.

Figure 3

Nontransgenic barley diet with addition of only 0.2 g transgenic grain containing the (1,3;1,4)-β-glucanase per kilogram diet provided equal weight gain of broiler chickens as feed containing corn.

The low nutritional value of barley for poultry is due to the absence of an intestinal enzyme for efficient depolymerization of (1,3;1,4)-β-D-glucan, the major polysaccharide of the endosperm cell walls. This leads to high viscosity in the intestine, limited nutrient uptake, decreased growth rate, and unhygenic sticky droppings adhering to chickens and floors of the production cages. Consequently, the 8 billion broiler chickens produced annually in the United States are primarily raised on corn-soybean diets. If barley is used as nonruminant animal feed the diet is supplied with enzymes from *Trichoderma*. For practical purposes the enzyme solutions are produced in fermenters and transferred to mixers with starch carrier material in the form of barley or wheat flour. The coating on a carrier material is required to stabilize the enzyme. After mixing, the material is dried in warm air and the pellets milled, homogenized, and packaged for shipping. We have performed 4 broiler chicken trials with 240 Hubbard High Yield broilers and 200 Cornish Cross broiler chickens to investigate whether transgenic grain expressing the protein-engineered thermostable (1,3;1,4)-β-glucanase from *Bacillus* during malting (33, 34) or in the endosperm during grain maturation (25) as feed additive can increase the nutritive value of barley-based diets to that of maize (86, 88). In both cases equal weight gain, feed consumption, and feed efficiency over a 21-day period could be achieved (**Figure 3**). The frequency of chicks with sticky droppings could be reduced to the same level as observed with the maize diet. Analyses of the different parts of the gut and excreta revealed this to be due to the β-glucan-degrading activity of the enzyme in the duodenum, ileum, caeca, and excreta. With a barley-soybean diet containing 620 g nontransgenic barley/kilogram diet, it was sufficient to add 0.2 g (0.02%) transgenic grain/kilogram diet to achieve the high nutritive value while the commercial Avizyme 1100® is added at a concentration of 1 g/kg diet. The addition of the transgenic grain in the ground diet compares to the amount of trace minerals added to standard diets.

The transgene has been bred into modern barley varieties. With these lines a yield has been achieved that compares favorably with that of Baronesse, the cultivar that is grown on 70% of the barley acreage in the state of Washington. The lines produce constant amounts of recombinant enzyme per kilogram of grain. Surprisingly, the endosperm cell walls of the transgenic lines are highly deficient in (1,3;1,4)-β-D-glucans and are therefore not significantly stainable with calcofluor (see **Supplemental Figure 5**). Using the glycan modifications of the (1,3;1,4)-β-glucanase (23, 45) as a marker, the following picture emerged (D. von Wettstein, H. Horvath-O'Geen & G. Kannangara, unpublished): Isolated protein bodies of the transgenic plants contain the glycosylated form of the thermostable (1,3;1,4)-β-glucanase (see **Supplemental Figure 6**). Thus, during cotranslational transfer of the nascent polypeptide chain into the lumen of the endoplasmic reticulum the preformed glycans are transferred from the dolichol carrier in the membrane to the recombinant (1,3;1,4)-β-glucanase. After removal of the presequence the recombinant β-glucanase is targeted by vesicle transport into the storage vacuoles, where it is sequestered in the protein bodies and protected from the programmed cell death of the endosperm during the final stages of grain maturation. Analysis of early to late endosperm developmental stages reveals a continuous degradation of the newly synthesized (1,3;1,4)-β-D-glucan by the transgenic enzyme. This observation requires the β-glucanase to meet the newly synthesized cell wall polymer. Wilson and coworkers (90) analyzed the appearance of the different cell wall polysaccharides during cellularization of the barley endosperm with monoclonal antibodies specific for (1,3;1,4)-β-D-glucan and other cell wall polysaccharides. (1,3;1,4)-β-D-glucan was deposited five days after pollination and arabinoxylan somewhat later. Although arabinoxylans were detectable in the Golgi apparatus and associated vesicles as well as in cell walls, immunolabeling of (1,3;1,4)-β-D-glucan was only obtained in the cell walls. Electron micrographs of developing prismatic cells in the barley endosperm at the stage when (1,3;1,4)-β-D-glucan is added to the cell walls show a successive widening of the space between the plasmalemma and the wall (14). The endoplasmic reticulum is in contact with the vacuoles accumulating storage protein as well as with the plasmalemma at the thickening cell walls. Thus, it is feasible that synthesis of (1,3;1,4)-β-D-glucan as well as transfer of the recombinant glycosylated (1,3;1,4)-β-glucanase into the storage vacuoles proceeds via the endoplasmic reticulum, which explains the hydrolysis of the newly synthesized (1,3;1,4)-β-D-glucan by the β-glucanase enzyme in the transgenic plants. The cellulose synthase-like *CslF2* gene of rice encodes an enzyme that is required for the synthesis of (1,3;1,4)-β-D-glucan in transgenic *Arabidopsis thaliana* (8). This gene can now be exploited to further study the synthesis of (1,3;1,4)-β-D-glucan in barley.

An efficient way of changing the composition (fiber content) of the cereal endosperm cell walls has been demonstrated. It is also possible to produce recombinant xylanase in barley and wheat and eliminate the antinutritive arabinoxylans (57, 37), i.e., producing cell walls containing only cellulose.

The use of transgenic grain containing recombinant (1,3;1,4)-β-glucanase as feed additive will boost the production of nontransgenic barley in areas where grain maize cannot be grown and thus has to be imported. Feeding 40 million broiler chicks with barley instead of imported maize will require 280,000 tons of nontransgenic barley but only 56 tons of transgenic barley containing 0.14 mg/kg thermotolerant (1,3;1,4)-β-glucanase, which could be produced on 25 acres of farmland. The barley feed with added transgenic grain or malt containing thermotolerant (1,3–1,4)-β-glucanase provides an environmentally friendly alternative to enzyme additives, as it uses photosynthetic energy for production of the enzyme in the grain and thus avoids use of nonrenewable energy for fermentations. The deposition of the enzyme in the protein bodies of the grain in the field makes coating procedures for stabilization of enzyme activity superfluous.

THE DEVELOPMENT OF PROANTHOCYANIDIN-FREE MALTING BARLEY

Proanthocyanidins (condensed tannins) from the testa tissue (seed coat) of the barley grain (**Figure 4**) are carried from the malt into the wort prepared in the mashtun and are also found after fermentation of the wort in the beer. There they cause precipitation of proteins and haze formation especially after refrigeration of the beer, even if it previously had been filtered to be brilliantly clear (**Figure 5**). Breweries that wish to produce brilliant clear beer with a long shelf life have to carry out stabilizing treatments. Chill proofing occurs either by removing proanthocyanidins with polyvinylpyrollidone, or by using enzymes to degrade the proteins, so that they can no longer bind to the proanthocyanidins. In the Schönen process proanthocyanidins and proteins are removed by precipitation with added condensed tannins extracted from plant material. I thought that barley varieties lacking proanthocyanidins should be obtainable. Proanthocyanidins of the barley testa consist of dimers and trimers of (2,3-*trans*) catechin and (2,3-*trans*) gallocatechin (31, 55, 56) and are synthesized by the flavonoid pathway (28–30) (presented in **Supplemental Figure 7**). This pathway branches at an intermediate [identified as 2,3-*trans*-3,4-*cis*-leucocyanidin by Klaus Kristiansen (41)] into one branch giving rise to anthocyanins and another leading to the formation of (2,3-*trans*) catechin and (2,3-*trans*) gallocatechin, which condense in a yet unknown way to the identified dimers, trimers, and higher polymers. Proanthocyanidins containing (2,3-*cis*) epicatechin were not found in the barley testa (31). If the pathway is blocked prior to the branch point the plants will be free of anthocyanins and proanthocyanidins; on the other hand, mutations blocking the branch specific to the synthesis of anthocyanins will contain normal amounts of proanthocyanidins in the seed coat. Accordingly, Barbro Jende-Strid set out to collect induced mutations in barley that are free from anthocyanins and that by chemical analysis were also identified to be free of

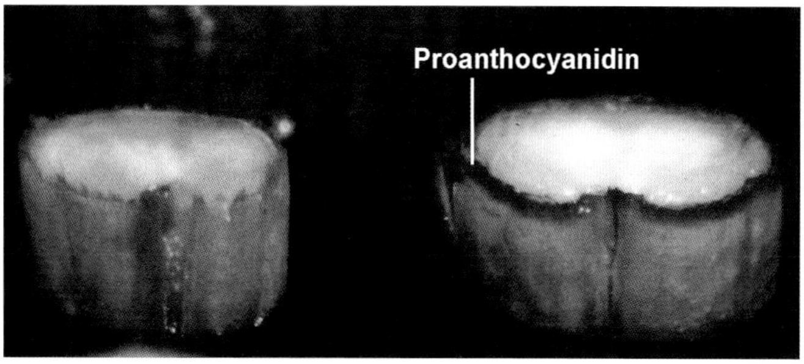

Figure 4

Proanthocyanidin (condensed tannin) in the testa of barley stained with vanillin-HCl. No staining is obtained in malting barley cultivar lacking proanthocyanidin.

Figure 5

Beer brewed from barley with proanthocyanidin causes protein precipitation in the form of chill haze. Proanthocyanidin-free barley produces brilliant clear beer. Today breweries produce brilliant clear beer by chemical treatments to remove the proanthocyanidins prior to bottling.

proanthocyanidins. In 1974 the first such mutant (*ant13-13*) was identified; it was propagated on the Carlsberg farm, malted and beer brewed from the malt. Excellent haze stability was found in all the beers and in beer made from all subsequently isolated and tested proanthocyanidin-free barley mutants and varieties (84, 85, 87). Mutant *ant13-13* had a 25% lower yield than standard malting barley varieties at the time. Both recombination breeding and search for mutations without yield depressions have been pursued (80).

Some 560 mutants that are blocked prior to the above-mentioned branch point [i.e., mutated in the genes *Ant 13*, *Ant 17*, *Ant 18*, *Ant 21*, *Ant 22*, and *Ant 30* (follow the Supplemental Material link from the Annual Reviews home page at **http://www.annualreviews.org** to see **Supplemental Table 1**) in more than 80 spring and winter barley varieties and breeding lines from Europe, the United States, and Japan] have been selected by screening some 18.5 million M_2 plants (frequency ~0.003%) after mutagen treatment with sodium azide, the powerful mutagen developed by R.A. Nilan and coworkers. But perhaps shutting down the flavonoid pathway in all tissues of the barley plant was detrimental. Crop science student Henriette Kristensen and Steen Aastrup (39) were encouraged to explore a new technique that permits the direct nondestructive chemical screening for lack of proanthocyanidin and thereby identify mutants in the proanthocyanidin-specific pathway that is only expressed in the seed coat. The M_3 kernels were embedded in plastic blocks that are abraded to open a window into the testa and endosperm layers and then the vanillin-HCl test is performed (**Figure 4**). Kernels that do not give red-stained seed coats are free of proanthocyanidins. They are taken out of the plastic blocks and germinated. In this way 107 mutants were isolated that were either blocked in the reduction of 2,3-*trans*-3,4-*cis*-leucocyanidin or 2,3-*trans*-3,4-*cis*-leucodelphinidin to (2,3-*trans*) catechin or (2,3-*trans*) gallocatechin (genes *ant 25*, *ant 28*) or in the condensation reaction to form the proanthocyanidin dimers and trimers (*ant 26*, *ant* 27, *ant 29*). These mutants develop anthocyanins in the vegetative parts of the plant in amounts that correspond to the genotype used for the mutagen treatment.

Several of these mutants were high yielding and gave rise to two varieties on the European recommended list of barley. They were Caminant [*ant* 28-484 (Grit) × Blenheim] and Clearity [*ant*27-488(Zenit) × Sewa × Fergie] and were used to produce malt in 5000-ton scale. The malts met all specifications of the Carlsberg Breweries and all-malt as well as adjunct beer has been produced in full scale (5000 hectoliter) at three different breweries with appropriate reference brews again meeting all specifications. The beers showed the expected excellent haze stability. It should be mentioned that haze stability of beer made from proanthocyanidin-free malt is much better than that achieved by the presently employed chemical treatments. This enables the use of appropriately stored hops, which is rich in bitter compounds relative to proanthocyanidins (87). The flavor and flavor stability of the beer has to be tested organoleptically by trained taste panels. This was done in two ways. Beer made from proanthocyanidin-free malt was supplemented with highly purified proanthocyanidins individually and in combination. The taste panel could not distinguish between beer with or without proanthocyanidins. This was the rigorous proof that proanthocyanidins do not belong to flavor compounds. Beers made from malt with and without proanthocyanidins were also assessed and no difference in flavor and flavor stability was recorded.

New higher-yielding proanthocyanidin-free cultivars with improved malting quality have since been bred. Thus, the variety process developed in Denmark is being tested in European Brewery Convention trials and by several malting companies. Since moving my research to Washington, I have been breeding proanthocyanidin-free barley lines with new mutants in the malting barley variety Harrington. This has led to the release and marketing of the variety Radiant [*ant29-667*(Harrington) × Baronesse], with a yield matching that of Baronesse, the variety presently grown on 70% of the barley acreage in Washington state (83). Although excellent as feed barley, it is also promising in areas suitable for producing malting barley. Radiant has produced crops judged positive in malting tests, and large-scale malt tests and brewing trials are in progress. The brewing industry's interest is increasing as attention is focused on the undesirable use of chemicals for chill proofing of beer. This project has provided a model of how metabolic engineering can produce value-added cereal crops, but its commercial introduction in the brewing industry is pending.

PROANTHOCYANIDIN BIOCHEMISTRY IN BARLEY AND *DESMODIUM*

In parallel, biochemical analyses and cloning of structural genes for enzymes in the pathway of barley were pursued. Thus Klaus Kristiansen (40) determined that the maximal rate of catechin, procyanidin B3, and proanthocyanidin C2 synthesis in wild-type barley occurred 8 to 16 days after flowering. Dihydroquercetin was radioactively labeled by feeding (1-^{14}C)- and (2-^{14}C)-acetate to flower buds of a petunia mutant accumulating this flavonoid. When fed to pericarp-testa tissue of wild type labeled (2,3-*trans*) catechin, procyanidin B3 and C2 [i.e., dimers and trimers consisting of (2,3-*trans*) catechin] were synthesized, thus establishing dihydroquercetin as a precursor of these compounds. In addition, labeled 2,3-*trans*-3,4-*cis*-leucocyanidin was synthesized, indicating that this compound is an intermediate.

Such experiments permitted assignment of functions to some of the *Ant* genes (as shown in **Supplemental Table 1**) and the determination of the molecular nature of the mutations that were induced by sodium azide, the most efficient mutagen in barley. cDNA clones were obtained for dihydroflavonol-4-reductase (*Ant* 18) and flavanone-3-hydroxylase (*Ant* 17) (42, 44). These and a cDNA for chalcone synthase cloned by Wolfgang Rohde have been used as probes to study transcription of the corresponding genes in pericarp-testa tissue from the *ant* 13 mutants of barley. No, or very low levels of, transcripts are found for the three structural genes and identify *Ant* 13 as a regulatory gene, probably a transcription factor required for all steps in proanthocyanidin synthesis. As Nilan and coworkers showed, sodium azide induces few if any chromosome aberrations. The molecular basis for four independent sodium azide–induced *ant* 18 mutants was examined by sequencing the genes encoding dihydroflavonol 4-reductase in these mutants (54). Three of these mutants lacking dihydroflavonol 4-reductase activity were due to missense mutations in the respective coding regions. The fourth mutant (*ant* 18–161) did not produce a mature message, as the result of transitions of the dinucleotide GT to AC at the 5′ splice site of intron 3 of the gene. Monocots invariably require the sequence GT in these positions and anthocyanin synthesis could be restored with transient transformation by bombarding the leaves with plasmids containing the wild-type dinucleotide sequence at the splice site (89).

Procyanidin B3: procyanidin dimer consisting of two units of (2,3-trans) catechin

Of importance to the plant breeder is the observation that sodium azide preferentially generates A:T→G:C transitions and that 86% of the nucleotide substitutions are transitions and 14% transversions. Among the 12,704 nucleotides sequenced in the four mutants of the *Ant* 18 gene, 21 base substitutions were found. Although most of the base changes in the coding region were silent, it illustrates the effect of the high concentration of the mutagen used to obtain a maximum yield of mutants. In this case it yielded between four and seven base changes in the gene of interest and probably additional ones in other genes. Comparable results were obtained with mutants in the *ant-30* gene encoding chalcone isomerase (17). Site-directed mutations and stable transformation are now better tools for cereal breeders than randomly induced mutations.

In 1993, Gregory Tanner and Klaus Kristiansen (74) developed an improved synthesis method for 2,3-*trans*-3,4-*cis*-[4-^{3}H]leucocyanidin and demonstrated its NADPH-dependent reduction to (2,3-*trans*) catechin with extracts from barley and the forage legume sainfoin (*Onobrychis viciifolia*). In 2003, Gregory Tanner, Philip Larkin, Anthony Ashton, and coworkers (73) purified leucocyanidin reductase from the leaves of *Desmodium uncinatum* (Papilionaceae) and cloned its cDNA. The cDNA was expressed in *Escherichia coli*, tobacco, and white clover. The 43-kDa enzyme synthesized (2,3-*trans*) catechin, (2,3-*trans*) gallocatechin, and (2,3-*trans*) afzelechin, i.e., flavan-3-ol building blocks of the proanthocyanidins found in the barley testa tissue. Using the sequence of this gene, it is now possible to clone the ortholog of barley and use it for further experiments.

Accumulation of proanthocyanidin in the barley testa layers was visualized by vanillin-HCl staining (69). Proanthocyanidins started to accumulate in the wild type in multiple vesicles with a diameter of 0.4 μ at 12 days after anthesis, filled the whole cell by 20 days, and formed crystalloid sheets by 35 days. Testa layers from *ant* 13, *ant* 17, *ant* 18, *ant* 22, *ant* 25, and *ant* 28 mutants did not stain with vanillin-HCl, whereas *ant* 19, *ant* 26, *ant* 27, and *ant* 29 mutants had vesicles with a pale reddish color, indicating accumulation of catechin or small amounts of proanthocyanidins. The mutation in the *Ant* 19 gene has decreased leucocyanidin reductase activity, producing only a small amount of (2,3-*trans*) catechin and proanthocyanidins, and may be in the structural gene for leucoanthocyanidin reductase. Among the mutants that provided high-yielding proanthocyanidin-free cultivars, mutants in the *Ant* 25, *Ant* 27, and *Ant* 28 genes revealed decreased dihydroflavonol- and leucocyanidin reductase activity. Four *ant 26* mutants synthesized (2,3-*trans*) catechin in almost wild-type amounts (38, 75). These genes are candidates for genes encoding the condensing enzymes that remain to be discovered.

PROANTHOCYANIDIN BIOCHEMISTRY IN *ARABIDOPSIS*

Proanthocyanidins in *Arabidopsis* are synthesized in the endothelial cell layer of the seed coat and have been studied with *transparent testa* (*tt*) and *tannin deficient seed* (*tds*) mutants (1, 2). With the aid of such mutants the genes encoding the steps up to the 2,3-*trans*-3,4-*cis*-leucocyanidin have been cloned and characterized (16) (**Supplemental Table 1**). However, *Arabidopsis* does not contain a gene that encodes a leucocyanidin reductase and produces proanthocyanidins based on (2,3-*cis*) epicatechin (**Supplemental Figure 7**). Leucoanthocyanidin dioxygenase (LDOX) catalyzes the reaction from 2,3-*trans*-3,4-*cis*-leucocyanidin to cyanidin, the precursor of anthocyanins. It is encoded by the gene *LDOX* (59) and mutated in the *tds4-1* mutant, which affects both anthocyanin and proanthocyanidin synthesis (2). *Arabidopsis* contains a gene that upon mutation leads to accumulation of anthocyanins instead of proanthocyanidins, giving the tissue a deep red color called *BANYULS* after the color of a popular wine variety. De-Yu Xie, Shashi Sharma, and Richard Dixon (92) isolated the cDNAs of this gene and the corresponding gene of *Medicago truncatula*, expressed them in *E. coli*, and assayed the recombinant soluble protein with anthocyanidins (cyanidin, pelargonidin, or delphinidin) in the presence of NADPH. Cyanidin yielded (2,3-*cis*) epicatechin, pelargonidin produced (2,3-*cis*) epiaafzelechin, and delphinidin correspondingly produced (2,3-*cis*) epigallocatechin (92). This showed that BANYULS is an anthocyanidin reductase (ANR). Ectopic expression of the *BAN* gene in tobacco flowers leads to a reduction in the anthocyanin color and significant proanthocyanidin production, as demonstrated by staining with dimethylaminocinnamaldehyde and by high-pressure liquid chromatography (HPLC) (93).

Analysis of the transparent testa mutants in *Arabidopsis* has identified at least nine transcription factors regulating different structural genes for enzymes in the biosynthesis of anthocyanins and proanthocyanidins as well as in the development of the endothelial cell layers of the seed. Their actions have been reviewed in the Tansley review by Dixon et al. (16), and by Marles et al. (43). They are listed in **Supplemental Table 1**.

PROANTHOCYANIDINS PROVIDE BENEFITS TO FORAGE LEGUMES

Proanthocyanidins present in the leaves of *Lotus corniculatus* (birdsfoot trefoil) or *Onobrychis viciifolia* (sainfoin) bind in the rumen of cows and sheep to the leaf proteins and thereby retard a rapid protein digestion by rumen organisms. This retardation prevents pasture bloat of the animals and reduces protein deamination that provides a protein bypass into the duodenum and thus increased absorption of amino acids (cf. 68, 76). It is therefore explored how genetic engineering can elicit expression of proanthocyanidins e.g., in alfalfa leaves (*Medicago sativa*), where they are absent but present in the seed coat. Expression of anthocyanin and proanthocyanidin was obtained by Heather Ray and coworkers (64) in leaves of alfalfa by expressing with the CaMV35S promoter the *Lc myc*-like transcription factor that induces anthocyanin expression in maize. The production of proanthocyanidins at high light intensity was the result of an induction of leucoanthocyanidin reductase, leading to proanthocyanidin in amounts 100–250 $\mu g \cdot g^{-1}$ freshweight.

In another approach the PAP1 MYB transcription factor of *Arabidopis* (3) was employed to produce proanthocyanidins in leaves of tobacco (94). Tobacco transformants expressing this transcription factor produced massive accumulation of anthocyanin in the trichomes and epidermal cells of leaves and flowers, and also in stems and roots. The *ANR* gene of *Medicago truncatula* was introduced into these tobacco plants by crossing with *ANR* tobacco transformants. The double transformants produced proanthocyanidins in corollas and leaves derived from (2,3-*cis*) epicatechin and (2,3-*cis*) epigallocatechin in amounts that would be sufficient to prevent bloat. *M. truncatula* transformants overexpressing *ANR* contain threefold more proanthocyanidins than wild-type plants in their leaf spots. In a third approach metabolic engineering of expression of proanthocyanidins in leaves of *Arabidopsis* was attempted by ectopically expressing several transcription factors (67). The *TT2* gene, which activates *DFR*, *BAN*, *LDOX*, *TT8*, and *TT12*, was transformed into the *pap1-D* activation-tagged mutant, which overexpresses the anthocyanin pathway (3). In homozygous transformants transcribing *TT2* and *PAP1*, ectopic expression of *BAN*, *AHA10*, and *TT12* was obtained and proanthocyanidins were synthesized, but only in subsets of cells in which the *BAN* (i.e., *ANR* gene promoter) is normally active. Combined expression of the transcription factors PAP1, TT2, and Lc from maize yielded *Arabidopsis* plants containing anthocyanin, epicatechin, and proanthocyanidin dimers and trimers, but were mostly lethal.

OUTLOOK

In a previous analysis of the leucoanthocyanidin reductase activity at different developmental stages and in different tissues of several legume species, the cell specificity and turnover of the proanthocyanidins was highlighted by histochemical analyses (68). It was apparent that the enzyme activity did not necessarily correlate with the proanthocyanidin content. In *Lotus japonicus*, the papillae of the petals are filled with proanthocyanidins (250 $\mu g \cdot mg^{-1}$ fresh mass), but almost no leucoanthocyanidin reductase was measureable. These proanthocyanidins are likely synthesized by the anthocyanidin reductase pathway (no analyses of the chemical composition of these proanthocyanidins have been made). Using the sequenced anthocyanidin reductase genes and the leucoanthocyanidin reductase genes as probes, it is now possible to determine which synthetic pathway is employed in a given tissue and developmental stage. It can further be determined if a polymer containing a mixture of (2,3-*cis*) epicatechin and (2,3-*trans*) catechin-derived monomers in a given cell requires transcription and translation of both a gene for anthocyanidin reductase and a gene for leucoanthocyanidin reductase. Mutants deficient in condensation of the monomers will lead to the enzymes carrying out these reactions.

SUMMARY POINTS

1. An environmentally friendly barley feed additive for nonruminant animals consisting of grain expressing a protein-engineered (1,3;1,4)-β-glucanase is needed only at concentrations comparable to the amount of trace minerals in the diet.
2. An efficient way to change the composition (fiber content) of cereal endosperm cell walls has been discovered.
3. Proanthocyanidin-free, high-yielding barley mutant cultivars make chemical stabilization of beer against chill haze and for a long shelf life superfluous.
4. The employed mutants contributed to the elucidation of the biosynthetic pathways of proanthocyanidins.
5. These findings are complemented with a critical discussion of two new groundbreaking discoveries at *CSIRO* Plant Industry and the Samuel Roberts Noble Foundation: Leucoanthocyanidin reductase synthesizes (2,3*trans*)-catechin in *Desmodium*, whereas in *Arabidopsis* anthocyanidin reductase synthesizes (2,3*cis*)-epicatechin. Accordingly, *Arabidopsis* assembles proanthocyanidin polymers consisting entirely of (2,3*cis*)-epicatechin and *Desmodium leaves* and barley seed coat synthesize polymers consisting entirely of (2,3*trans*)-catechin units.

LITERATURE CITED

1. Abrahams S, Lee E, Walker AR, Tanner GJ, Larkin P, Ashton AR. 2003. The *Arabidopsis TDS4* gene encodes leucoanthocyanidin dioxygenase (LDOX) and is essential for proanthocyanidin synthesis and vacuole development. *Plant J.* 35:624–36
2. Abrahams S, Tanner GJ, Larkin PJ, Ashton AR. 2002. Identification and biochemical characterization of mutants in the proanthocyanidin pathway in *Arabidopsis*. *Plant Physiol.* 130:561–76
3. Borevitz JO, Xia Y, Blount J, Dixon RA, Lamb C. 2000. Activation tagging identifies a conserved MYB regulator of phenylpropanoid biosynthesis. *Plant Cell* 12:2383–94
4. Borriss R, Olsen O, Thomsen KK, von Wettstein D. 1989. Hybrid bacillus endo- (1–3,1–4)-β-glucanases: construction of recombinant genes and molecular properties of the gene products. *Carlsberg Res. Commun.* 54:41–54
5. Brandt A, Ingversen J. 1976. In vitro synthesis of barley endosperm proteins on wild type and mutant templates. *Carlsberg Res. Commun.* 41:311–20
6. Brandt A, Ingversen J. 1978. Isolation and translation of hordein messenger RNA from wild type and mutant endosperm in barley. *Carlsberg Res. Commun.* 43:451–69

7. Brandt A, Montembault A, Cameron-Mills V, Rasmussen SK. 1985. Primary structure of a B1 hordein gene from barley. *Carlsberg Res. Commun.* 50:333–45
8. Burton RA, Wilson SM, Hrmova M, Harvey AJ, Shirley NJ, et al. 2006. Cellulose synthase-like *CsIF* genes mediate the synthesis of cell wall (1,3;1,4)-β-D-glucans. *Science* 311:1940–42
9. Cameron-Mills V. 1980. The structure and composition of protein bodies purified from barley endosperm by silica sol density gradients. *Carlsberg Res. Commun.* 45:557–76
10. Cameron-Mills V, Brandt A. 1988. A γ-hordein gene. *Plant Mol. Biol.* 11:449–61
11. Cameron-Mills V, Ingversen J. 1978. In vitro synthesis and transport of barley endosperm proteins: Reconstitution of functional rough microsomes from polyribosomes and stripped microsomes. *Carlsberg Res. Commun.* 43:471–89
12. Cameron-Mills V, Ingversen J, Brandt A. 1978. Transfer of in vitro synthesized barley endosperm proteins into the lumen of the endoplasmic reticulum. *Carlsberg Res. Commun.* 43:91–102
13. Cameron-Mills V, Madrid S. 1989. The signal peptide cleavage site of a B1 hordein determined by radiosequencing of the in vitro synthesized and processed polypeptide. *Carlsberg Res. Commun.* 54:181–92
14. Cameron-Mills V, von Wettstein D. 1980. Protein body formation in the developing barley endosperm. *Carlsberg Res. Commun.* 45:577–95
15. Castle LA, Siehl DL, Gorton R, Patten PA, Chen YH, et al. 2004. Discovery and directed evolution of a glyphosphate tolerance gene. *Science* 304:1151–54
16. Dixon RA, Xie DY, Sharma SB. 2005. Proanthocyanidins—a final frontier in flavonoid research? *N. Phytol.* 165:9–28
17. Druka A, Kudrna D, Rostoks N, Brueggeman R, von Wettstein D, Kleinhofs A. 2003. Chalcone isomerase gene from rice (*Oryza sativa*) and barley (*Hordeum vulgare*): physical, genetic and mutation mapping. *Gene* 302:171–78
18. Entwistle J. 1988. Primary structure of a C-hordein gene from barley. *Carlsberg Res. Commun.* 53:247–58
19. Entwistle J, Knudsen S, Müller M, Cameron-Mills V. 1991. Amber codon suppression: the in vivo and in vitro analysis of two C-hordein genes from barley. *Plant Mol. Biol.* 17:1217–31
20. Fincher GB, Lock PA, Morgan MM, Lingelbach K, Wettenhall REH, et al. 1986. Primary structure of the (1→3,1→4)-β-D-glucan 4-glucanohydrolase from barley aleurone. *Proc. Natl. Acad. Sci. USA* 83:2081–85
21. Giese H, Hopp E. 1984. Influence of nitrogen nutrition on the amount of hordein, protein Z and β-amylase messenger RNA in developing endosperms of barley. *Carlsberg Res. Commun.* 49:365–83
22. Hahn M, Olsen O, Politz O, Borriss R, Heinemann U. 1995. Crystal structure and site-directed mutagenesis of *Bacillus macerans* endo-1,3–1,4-β-glucanase. *J. Biol. Chem.* 270:3081–88
23. Harthill JE, Thomsen KK. 1995. Analysis of glycan structures of barley (1–3,1–4)-β-D-glucan 4-glucanohydrolase isoenzyme EII. *Plant Physiol. Biochem.* 33:9–18
24. Hopp HE, Rasmussen SK, Brandt A. 1983. Organization and transcription of B1 hordein genes in high lysine mutants of barley. *Carlsberg Res. Commun.* 48:201–16
25. Horvath H, Huang J, Wong OT, Kohl E, Okita T, et al. 2000. The production of recombinant proteins in transgenic barley grains. *Proc. Natl. Acad. Sci. USA* 97:1914–19
26. Horvath H, Huang J, Wong OT, von Wettstein D. 2002. Experiences with genetic transformation of barley and characteristics of transgenic plants. In *Barley Science*, ed. GA Slafer, JL Molina-Cano, R Savin, JL Araus, J Romagosa, pp. 143–76. New York: Harworth

27. Horvath H, Jensen LG, Wong OT, Kohl E, Ullrich SE, et al. 2001. Stability of transgene expression, field performance and recombination breeding of transformed barley lines. *Theor. Appl. Genet.* 102:1–11
28. Jende-Strid B. 1988. Analysis of proanthocyanidins and phenolic acids in barley, malt, hops and beer. In *Modern Methods of Plant Analysis*. New Ser. Vol. 7: *Beer Analysis*, ed. HF Linskens, JF Jackson, pp. 110–27. Berlin/Heidelberg: Springer-Verlag
29. Jende-Strid B. 1991. Gene-enzyme relations in the pathway of flavonoid biosynthesis in barley. *Theor. Appl. Genet.* 81:668–74
30. Jende-Strid B. 1993. Genetic control of flavonoid biosynthesis in barley. *Hereditas* 119:187–204
31. Jende-Strid B, Møller BL. 1981. Analysis of proanthocyanidins in wild-type and mutant barley (Hordeum vulgare L.). *Carlsberg Res. Commun.* 46:53–64
32. Jensen LG. 1994. Developmental patterns of enzymes and proteins during mobilization of endosperm stores in germinating barley grains. *Hereditas* 121:53–72
33. Jensen LG, Olsen O, Kops O, Wolf N, Thomsen KK, von Wettstein D. 1996. Transgenic barley expressing a protein-engineered, thermostable (1,3–1,4)-β-glucanase during germination. *Proc. Natl. Acad. Sci. USA* 93:3487–91
34. Jensen LG, Politz O, Olsen O, Thomsen KK, von Wettstein D. 1998. Inheritance of a codon optimized transgene expressing heat stable (1,3–1,4)-β-glucanase in scutellum and aleurone of germinating barley. *Hereditas* 129:215–25
35. Keitel T, Meldgaard M, Heinemann U. 1994. Cation binding to a *Bacillus* (1,3–1,4)-β-glucanase: Geometry, affinity and effect on protein stability. *Eur. J. Biochem.* 222:203–14
36. Keitel T, Simon O, Borriss R, Heinemann U. 1993. Molecular and active-site structure of a *Bacillus* 1–3,1–4-β-glucanase. *Proc. Natl. Acad. Sci. USA* 90:5287–91
37. Kohl EA. 2003. *Development of transgenic barley expressing (1,4)-β-xylanase*. M.Sc. thesis. Washington State Univ., Pullman. 74 pp.
38. Kristensen H. 1987. Selektion af proanthocyanidin-frie bygmutantkerner. Hoved opgave I planteforædling. *Inst. Landbr. Plantekult*. Copenhagen: R. Vet. Agric. Univ.
39. Kristensen H, Aastrup S. 1986. A non-destructive screening method for proanthocyanidin-free barley mutants. *Carlsberg Res. Commun.* 51:509–13
40. Kristiansen KN. 1984. Biosynthesis of proanthocyanidins in barley: Genetic control of the conversion of dihydroquercetin to catechin and procyanidins. *Carlsberg Res. Commun.* 49:503–24
41. Kristiansen KN. 1986. Conversion of (+)-dihydroquercetin to (+)-2,3-trans-3,4-*cis*- leucocyanidin and (+)-catechin with an enzyme extract from maturing grains of barley. *Carlsberg Res. Commun.* 51:51–60
42. Kristiansen KN, Rohde W. 1991. Structure of the *Hordeum vulgare* gene encoding dihydroflavonol-4-reductase and molecular analysis of *ant* 18 mutants blocked in flavonoid synthesis. *Mol. Gen. Genet.* 230:49–59
43. Marles MA, Ray H, Gruber MY. 2003. New perspectives on proanthocyanidin biochemistry and molecular regulation. *Phytochemistry* 64:367–83
44. Meldgaard M. 1992. Expression of chalcone synthase, dihydroflavonol reductase, and flavanone-3-hydroxylase in mutants of barley deficient in anthocyanin and proanthocyanidin biosynthesis. *Theor. Appl. Genet.* 83:695–706
45. Meldgaard M, Harthill J, Petersen B, Olsen O. 1995. Glycan modification of a thermostable recombinant (1–3,1–4)-β-glucanase secreted from *Saccharomyces cerevisiae* is determined by strain and culture condition. *Glycoconjug. J.* 12:380–90

46. Meldgaard M, Svendsen I. 1994. Different effects of *N*-glycosylation on the thermostability of highly homologous bacterial (1,3–1,4)-β-glucanases secreted from yeast. *Microbiology* 140:159–66
47. Møgelsvang S, Simpson D. 1998. Protein folding and transport from the endoplasmic reticulum to the Golgi apparatus in plants. *J. Plant Physiol.* 153:1–15
48. Müller M, Knudsen S. 1993. The nitrogen response of a barley C-hordein promoter is controlled by positive and negative regulation of the GCN4 and endosperm box. *Plant J.* 4:343–55
49. Müller M, Muth JR, Gallusci P, Knudsen S, Maddaloni M, et al. 1995. Regulation of storage protein synthesis in cereal seeds: Developmental and nutritional aspects. *J. Plant Physiol.* 145:606–13
50. Munck L, von Wettstein D. 1974. Effects of genes that change the amino acid composition of barley endosperm. In "Workshop on Genetic Improvement of Seed Proteins" at *Natl. Acad. Sci. USA* 1974:71–82
51. Olsen O, Borriss R, Simon O, Thomsen KK. 1991. Hybrid *Bacillus* (1–3,1–4)-β-glucanases: Engineering thermostable enzymes by construction of hybrid genes. *Mol. Gen. Genet.* 225:177–85
52. Olsen O, Thomsen KK. 1991. Improvement of bacterial β-glucanase thermostability by glycosylation. *J. Gen. Microbiol.* 137:579–85
53. Olsen O, Thomsen KK, Weber J, Duus JØ, Svendsen I, et al. 1996. Transplanting two unique β-glucanase catalytic activities into one multienzyme, which forms glucose. *Bio/Technology* 14:71–76
54. Olsen O, Wang X, von Wettstein D. 1993. Sodium azide mutagenesis: Preferential generation of A · T → G · C transitions in the barley *Ant18* gene. *Proc. Natl. Acad. Sci. USA* 90:8043–47
55. Outtrup H. 1981. Structure of prodelphinidins in barley. *Proc. Eur. Brew. Conv. Congr. Copenhagen* 1981:323–33
56. Outtrup H, Schaumburg K. 1981. Structure elucidation of some proanthocyanidins in barley by ^{1}H 270 MHz NMR spectroscopy. *Carlsberg Res. Commun.* 46:43–52
57. Patel M, Johnson JS, Brettell RIS, Jacobsen J, Xue G-P. 2000. Transgenic barley expressing a fungal xylanase gene in the endosperm of the developing grains. *Mol. Breed.* 6:113–24
58. Pelger S, Høyer-Hansen G. 1989. The reaction of monoclonal antibodies with hordeins from five different *Hordeum* species. *Hereditas* 111:273–79
59. Pelletier MK, Murrell JR, Shirley BW. 1997. Characterisation of flavonol synthase and leucoanthocyanidin dioxygenase genes in *Arabidopsis*. *Plant Physiol.* 113:1437–45
60. Phillipson BA. 1993. Expression of a hybrid (1–3,1–4)-β-glucanase in barley protoplasts. *Plant Sci.* 91:195–206
61. Politz O, Simon O, Olsen O, Borriss R. 1993. Determinants for the enhanced thermostability of hybrid (1–3,1–4)-β-glucanases. *Eur. J. Biochem.* 216:829–34
62. Rasmussen SK, Brandt A. 1986. Nucleotide sequences of cDNA clones for C-hordein polypeptides. *Carlsberg Res. Commun.* 51:371–79
63. Rasmussen SK, Hopp HE, Brandt A. 1983. Nucleotide sequences of cDNA clones for B1 hordein polypeptides. *Carlsberg Res. Commun.* 48:187–99
64. Ray H, Yu M, Auser P, Blahut-Beatty L, McKersie B, et al. 2003. Expression of anthocyanins and proanthocyanidins after transformation of alfalfa with maize *Lc*. *Plant Physiol.* 132:1448–63
65. Rechinger KB, Bougri OV, Cameron-Mills V. 1993. Evolutionary relationship of the members of the sulphur-rich hordein family revealed by common antigenic determinants. *Theor. Appl. Genet.* 85:829–40

66. Rechinger KB, Simpson DJ, Svendsen I, Cameron-Mills V. 1993. A role for $\gamma 3$ hordein in the transport and targeting of prolamin polypeptides to the vacuole of developing barley endosperm. *Plant J.* 4:841–53
67. Sharma SB, Dixon RA. 2005. Metabolic engineering of proanthocyanidins by ectopic expression of transcription factors in *Arabidopsis thaliana*. *Plant J.* 44:62–75
68. Skadhauge B, Gruber MY, Thomsen KK, von Wettstein D. 1997. Leucocyanidin reductase activity and accumulation of proanthocyanidins in developing legume tissues. *Am. J. Bot.* 84:494–503
69. Skadhauge B, Thomsen KK, von Wettstein D. 1997. The role of the barley testa layer and its flavonoid content in resistance to *Fusarium* infections. *Hereditas* 126:147–60
70. Sørensen MB. 1989. Mapping of the *Hor2* locus in barley by pulsed field gel electrophoresis. *Carlsberg Res. Commun.* 54:109–20
71. Sørensen MB. 1992. Methylation of B-hordein genes in barley endosperm is inversely correlated with gene activity and affected by the regulatory gene *Lys3*. *Proc. Natl. Acad. Sci. USA* 89:4119–23
72. Stahl R, Horvath H, Van Fleet J, Voetz M, von Wettstein D, Wolf N. 2002. T-DNA integration into the barley genome from single and double cassette vectors. *Proc. Natl. Acad. Sci. USA* 99:2146–51
73. Tanner GJ, Francki KT, Abrahams S, Watson JM, Larkin PJ, Ashton AR. 2003. Proanthocyanidin biosynthesis in plants. Purification of legume leucoanthocyanidin reductase and molecular cloning of its cDNA. *J. Biol. Chem.* 278:31647–56
74. Tanner GJ, Kristiansen KN. 1993. Synthesis of ^{3}H-3,4-*cis*-leucocyanidin and enzymatic reduction to catechin. *Anal. Biochem.* 209:274–77
75. Tanner GJ, Kristiansen KN, Jende-Strid B. 1992. Biosynthesis of proanthocyanidins (condensed tannins) in barley. *Bull. Liaison Groupe Polyphen.* 16:170–73
76. Tanner GJ, Moore AE, Larkin PJ. 1994. Proanthocyanidins inhibit hydrolysis of leaf proteins by rumen microflora in vitro. *Br. J. Nutr.* 71:947–58
77. Ullrich SE, Rasmussen U, Høyer-Hansen G, Brandt A. 1986. Monoclonal antibodies to hordein polypeptides. *Carlsberg Res. Commun.* 51:381–99
78. von Wettstein D. 1979. Biochemical and molecular genetics in the improvement of malting barley and brewers yeast. *Proc. 17th Congr. Eur. Brewery Convent., West Berl.*, pp. 587–629
79. von Wettstein D. 1983. Genetic engineering in the adaptation of plants to evolving human needs. *Experientia* 39:687–713
80. von Wettstein D. 1995. Breeding of value added barley by mutation and protein engineering. *Induced Mutat. Mol. Tech. Crop Improv., Proc. FAO/IAEA Symp.*, pp. 67–76. Vienna: IAEA-SM-340/15
81. von Wettstein D. 2004. Transgenic barley. In Proc. 9th Int. Barley Genet. Symp., Brno, Czech Republ. *Czech J. Genet. Plant Breed.* 40:79
82. von Wettstein D. 2006. Fascination with chloroplasts and chromosome pairing. *Progr. Bot.* 67:1–28
83. von Wettstein D, Cochran JS, Ullrich SE, Kannangara CG, Jitkov VA, et al. 2004. Registration of 'Radiant' Barley. *Crop Sci.* 44:1859–60
84. von Wettstein D, Jende-Strid B, Ahrenst-Larsen B, Erdal K. 1980. Proanthocyanidin-free barley prevents the formation of beer haze. *MBAA Techn. Q.* 17:16–23
85. von Wettstein D, Jende-Strid B, Ahrenst-Larsen B, Sørensen JA. 1977. Biochemical mutant in barley renders chemical stabilization of beer superfluous. *Carlsberg Res. Commun.* 42:341–51

86. von Wettstein D, Mikhaylenko G, Froseth JA, Kannangara CG. 2000. Improved barley broiler feed with transgenic malt containing heat-stable (1,3–1,4)-β-glucanase. *Proc. Natl. Acad. Sci. USA* 97:13512–17
87. von Wettstein D, Nilan RA, Ahrenst-Larsen B, Erdal K, Ingversen J, et al. 1985. Proanthocyanidin-free barley for brewing: Progress in breeding for high yield and research tool in polyphenol chemistry. *MBAA Techn. Q.* 22:41–52
88. von Wettstein D, Warner J, Kannangara CG. 2003. Supplements of transgenic malt or grain containing (1,3–1,4)-β-glucanase to barley based broiler diets lift their nutritive value to that of corn. *Br. J. Poultry Sci.* 44:438–49
89. Wang X, Olsen O, Knudsen S. 1993. Expression of the dihydroflavonol reductase gene in an anthocyanidin-free barley mutant. *Hereditas* 119:67–75
90. Wilson SM, Burton RA, Doblin MS, Stone BA, Newbigin EJ, et al. 2006. Temporal and spatial appearance of wall polysaccharides during cellularization of barley (*Hordeum vulgare*) endosperm. *Planta* 224:655–67
91. Wolf N. 1992. Structure of the genes encoding *Hordeum vulgare* (1→3,1→4)-β-glucanase isoenzymes I and II and functional analysis of their promoters in barley aleurone protoplasts. *Mol. Gen. Genet.* 234:33–42
92. Xie DY, Sharma SB, Dixon RA. 2004. Anthocyanidin reductases from *Medicago truncatula* and *Arabidopsis thaliana*. *Arch. Biochem. Biophys.* 422:91–102
93. Xie DY, Sharma SB, Palva NL, Ferreira DF, Dixon RA. 2003. Role of anthocyanidin reductase encoded by *BANYULS* in plant flavonoid biosynthesis. *Science* 299:396–99
94. Xie DY, Sharma SB, Wright E, Wang ZY, Dixon RA. 2006. Metabolic engineering of proanthocyanidins through coexpression of anthocyanidin reductase and the PAP1 MYB transcription factor. *Plant J.* 45:895–907

NOTE ADDED IN PROOF

This is scientific paper 1001–06 from the College of Agricultural, Human, and Natural Resource Science Research Center, Washington State University.

Phototropin Blue-Light Receptors

John M. Christie

Plant Science Group, Division of Biochemistry and Molecular Biology, Institute of Biomedical and Life Sciences, University of Glasgow, Glasgow G12 8QQ, Scotland, United Kingdom; email: J.Christie@bio.gla.ac.uk

Annu. Rev. Plant Biol. 2007. 58:21–45

The *Annual Review of Plant Biology* is online at plant.annualreviews.org

This article's doi:
10.1146/annurev.arplant.58.032806.103951

First published online as a Review in Advance on October 26, 2006

1543-5008/07/0602-0021$20.00

Key Words

phototropism, LOV domain, kinase, autophosphorylation

Abstract

Phototropins are blue-light receptors controlling a range of responses that serve to optimize the photosynthetic efficiency of plants. These include phototropism, light-induced stomatal opening, and chloroplast movements in response to changes in light intensity. Since the isolation of the *Arabidopsis PHOT1* gene in 1997, phototropins have been identified in ferns and mosses where their physiological functions appear to be conserved. *Arabidopsis* contains two phototropins, phot1 and phot2, that exhibit overlapping functions in addition to having unique physiological roles. Phototropins are light-activated serine/threonine protein kinases. Light sensing by the phototropins is mediated by a repeated motif at the N-terminal region of the protein known as the LOV domain. Photoexcitation of the LOV domain results in receptor autophosphorylation and an initiation of phototropin signaling. Here we summarize the photochemical and biochemical events underlying phototropin activation in addition to the current knowledge of the molecular mechanisms associated with photoreceptor signaling.

Contents

INTRODUCTION

Environmental cues have an extensive regulatory influence on the growth and development of plants. Among the most important environmental factors is light. Light is not only an energy source for photosynthesis but also a stimulus that regulates numerous developmental processes, from seed germination to the onset of flowering. Collectively, these light-dependent responses are known as photomorphogenesis.

Several classes of photoreceptors that absorb light in two spectral regions act to control photomorphogenesis. Phytochromes (phy) predominantly absorb red and far-red light (600–800 nm) and mediate a wide range of photomorphogenic responses (19, 44). Two distinct classes of photoreceptors mediate the effects of UV-A/blue light (320–500 nm): the cryptochromes and the phototropins (7). Cryptochromes (cry), like the phytochromes, play a major role in plant photomorphogenesis (8, 100). Phototropins (phot), on the other hand, are involved in regulating light-dependent processes that serve to optimize the photosynthetic efficiency of plants and promote growth (14, 18). More recently, a third class of putative blue-light receptors related to the phototropins was identified in plants. This novel family of photoreceptors includes proteins that mediate targeted proteolysis of components associated with circadian clock function and flowering (7).

It has been just under a decade since the first phototropin gene was identified in plants, yet our knowledge of these and related blue-light receptors has increased dramatically during this time. In this article, we summarize the major advances in phototropin research with respect to their physiological roles, their mode of action, and the mechanisms underlying phototropin receptor signaling. For additional information, readers are directed to a number of recently published reviews (7, 18, 87, 107) and book chapters on phototropins (16, 20, 25, 146, 147). Historical overviews describing the discovery of plant blue-light receptors, including the phototropins (12, 157), will also provide a valuable resource for the interested reader.

PHOTOTROPISM AND THE DISCOVERY OF PHOTOTROPIN

The effects of blue light on plant development have been studied for almost two centuries (12, 65). Phototropism, for example, is specifically induced by UV-A/blue light and has provided researchers with an excellent experimental system to study blue-light perception and signaling in plants (65, 66, 157). Generally, shoots show positive phototropism, i.e., movement toward the light, whereas roots exhibit negative phototropic movement. Photobiological studies using shoots of dark-grown seedlings from a variety of plant species have uncovered a surprising degree of complexity in the bending response to blue light. Phototropism can be divided into two phases depending on the fluence and time requirements (66). First-positive curvature is generally described as the bending of shoots toward unilateral blue light delivered in brief pulses at very low fluences. These curvature responses obey the Bunsen-Roscoe reciprocity law in that they are the same for a given fluence over a wide range of time-fluence rate combinations (17). Second-positive curvature occurs with prolonged irradiation in a time-dependent manner. Although fluence-response measurements have provided important photobiological information about the photosensors mediating phototropism (65), a greater understanding of the photodetection mechanisms involved has come from biochemical and molecular genetic approaches.

Insights into the biochemical properties of the photoreceptors responsible for phototropism were obtained prior to the isolation of the first phototropin gene back in 1997 (60). Briggs and colleagues were the first to identify a plasma membrane-associated protein from the growing regions of dark-grown pea epicotyls that became phosphorylated upon blue-light irradiation (47). Extensive photochemical and biochemical characterization of the light-induced phosphorylation reaction (15) and its correlation with phototropism (51, 52, 119, 136) indicated that the phosphoprotein in question was a candidate photoreceptor for phototropism that undergoes autophosphorylation in response to blue-light treatment.

The molecular identity of the aforementioned phosphoprotein was later uncovered through the use of the plant genetic model *Arabidopsis thaliana*. The nonphototropic hypocotyl (*nph*) mutants of *Arabidopsis* show impaired hypocotyl phototropism to low fluence rates of unilateral blue light (101). A particular class of *nph* mutant, *nph1*, was found to lack the activity of the plasma membrane-associated protein that becomes heavily phosphorylated upon irradiation with blue light. The encoded protein, originally designated NPH1, was therefore hypothesized to represent a phototropic receptor that undergoes autophosphorylation in response to blue light (101). Subsequent biochemical experiments confirmed this hypothesis and the NPH1 protein was renamed phototropin 1 (phot1) after its functional role in phototropism (13, 22).

PHYSIOLOGICAL ROLES OF THE PHOTOTROPINS

Biological Functions in Higher Plants

Phototropins are ubiquitous in higher plants and have been identified in several plant species (35, 78, 88, 121). Genetic analysis using *Arabidopsis* has been instrumental in identifying the molecular nature of the phototropins and establishing their roles as blue-light receptors. *Arabidopsis* contains two phototropins designated phot1 and phot2 (14, 18). Analysis of phot-deficient mutants has shown that phot1 and phot2 exhibit partially overlapping roles in regulating phototropism. Both phot1 and phot2 act to regulate hypocotyl phototropism in *Arabidopsis* in response to high intensities of unilateral blue light (123). By contrast, hypocotyl phototropism under low light conditions is solely mediated by phot1 (101, 123, 124).

Besides phototropism, phot1 and phot2 mediate other light responses in *Arabidopsis* that serve to regulate and fine-tune the photosynthetic status of plants. Blue light induces the opening of stomatal pores in the leaf and stem epidermis, a response that allows plants to regulate CO_2 uptake for photosynthesis and water loss through transpiration. This response is controlled redundantly by phot1 and phot2, which unlike the situation for phototropism contribute equally to stomatal opening by acting across the same light intensity range (88). Phototropins also control the movement of chloroplasts in response to different light intensities (156). Under low light conditions, phot1 and phot2 induce chloroplast accumulation movement to the upper cell surface to promote light capture for photosynthesis (123). Phot1 is more sensitive than phot2 in activating chloroplast accumulation movement, as phot2 activity requires a higher light threshold (77, 123). In high light conditions, chloroplasts move away from the site of irradiation (156) to prevent photodamage of the photosynthetic apparatus in excess light (79). The avoidance movement of chloroplasts is mediated only by phot2 (74, 76), demonstrating that a given phototropin can play a unique role. Likewise, the rapid inhibition of hypocotyl elongation of dark-grown seedlings by blue light is controlled exclusively by phot1 (43). This process represents the earliest light response initiating the transition from growth dependent on seed reserves to photoautotrophy. Phototropins also promote cotyledon expansion (116) and leaf expansion in *Arabidopsis* (125). In addition, a possible role for phototropins in controlling light-stimulated leaf movement in kidney bean was recently reported (70).

Together, the abovementioned responses serve to enhance photosynthetic performance and promote plant growth under weak light conditions (150). Gene expression analysis indicates that phototropins have only a minor role in transcriptional regulation in light-grown *Arabidopsis* seedlings (116). Nevertheless, phot1 is required for the destabilization of specific nuclear and chloroplast transcripts in response to high-intensity blue light (41). A summary of the known phototropin-mediated responses found in higher plants is illustrated in **Figure 1*a***.

Biological Functions in Lower Plants

Owing to their simplified cell architecture, blue-light responses have been studied extensively in ferns, mosses, and green algae (146). The fern *Adiantum capillus-veneris* has two phototropins (75, 115). Phot2, like its higher plant counterpart, is solely responsible for mediating chloroplast avoidance movement in *Adiantum* (75). *Adiantum* also contains a novel photoreceptor neochrome, formerly known as phy3 (115, 145). *Adiantum* neochrome (neo) is a chimeric protein containing a phytochrome photosensory domain fused to the N terminus of an entire phototropin receptor (115, 145). Neo is required for phototropism and chloroplast relocation in *Adiantum* (82), both of which are regulated by red and blue light in this organism. Fern species such as *Adiantum* appear to lack stomatal responses to blue light despite having functional phototropins (32).

Four phototropins have been identified in the moss *Physcomitrella patens* that mediate chloroplast movements in this organism (80). The function of phototropins in higher and lower plants is therefore likely to be conserved. The filamentous green alga *Mougeotia scalaris* contains two phototropins in addition to two neochromes (145). A comparison of algal and fern *NEO* genes suggests that they have arisen independently, providing an intriguing example of convergent evolution (145). Only one phototropin exists in the biflagellate unicellular green alga *Chlamydomonas reinhardtii* (62, 81) where it is necessary for completion of the sexual life cycle (61). Unlike higher plant phototropins, light activation of *Chlamydomonas* phot causes major changes in the transcription of specific gene targets (61). However, whereas the function of *Chlamydomonas* phot is distinct from its higher plant counterparts, the gene encoding *Chlamydomonas* phot can restore phototropin-mediated responses in the *phot1 phot2* double mutant of *Arabidopsis* (118), indicating that the mechanism of action of higher and lower plant phototropins is highly conserved.

PHOTOTROPIN STRUCTURE AND ACTIVITY

Protein Structure and Receptor Autophosphorylation

Protein structures of plant phototropins can be separated into two segments: a photosensory domain at the N terminus and a serine/threonine kinase domain at the C terminus (**Figure 1*b***). Phototropins belong to the AGC family of kinases (cAMP-dependent protein kinase, cGMP-dependent protein kinase G and phospholipid-dependent protein kinase C) and are members of the AGC-VIIIb subfamily (11). The N-terminal photosensory domain of the phototropins contains two very similar domains of ~110 amino acids designated LOV1 and LOV2. LOV domains are members of the large and diverse superfamily of PAS (Per, ARNT, Sim) domains associated with cofactor binding and mediating protein interactions (152). LOV domains, however, are more closely related to a subset of proteins within the PAS domain superfamily that are regulated by external signals such as light, oxygen, or voltage, hence the acronym LOV (60). LOV domains bind the cofactor flavin mononucleotide (FMN) and function as blue-light sensors for the protein (22, 126). *Adiantum* neo is unique in that it contains a phytochrome chromophore-binding domain in addition to a phototropin protein sequence (**Figure 1*b***), consistent with its function as a dual red-/blue-light photoreceptor (82).

Photoexcitation of the LOV domains by blue light leads to phototropin receptor autophosphorylation (21, 123). Phototropin autophosphorylation can be monitored in vivo (15) and in vitro (21, 23, 123). Whether autophosphorylation occurs intra- or intermolecularly is not known although biochemical evidence suggests that phot1 autophosphorylation involves intermolecular communication between separate phototropin molecules (121). Autophosphorylation, at least for phot1, occurs on multiple serine residues (120, 127, 130, 137). Indeed, phot1 from several plant species exhibits reduced electrophoretic mobility after blue-light irradiation, consistent with autophosphorylation on multiple sites (94, 101, 138). Interestingly, autophosphorylation of oat phot1 is accompanied by a loss in immunoreactivity with an antibody raised against the N-terminal region of *Arabidopsis* phot1 (127). Irradiation with UV-C (280 nm) induces an electrophoretic mobility shift for oat phot1 without any change in immunoreactivity (94), implying that distinct serine residues may be phosphorylated in response to different light qualities.

cGMP: cyclic guanosine monophosphate

cAMP: cyclic adenosine monophosphate

LOV: light, oxygen, or voltage

PAS: domain acronym derived from a class of sensory proteins named after Drosophila Period (Per), vertebrate Aryl Hydrocarbon Receptor Nuclear Translocator (ARNT), and Drosophila Single-minded (Sim)

FMN: flavin mononucleotide

LOV-Domain Structure and Function

Purification of milligram quantities of bacterially expressed LOV domains has greatly facilitated the spectral and structural analysis of these blue-light-sensing motifs. Indeed, enormous progress has been made in elucidating the primary processes associated with LOV-domain photochemistry and readers are directed to several recent excellent reviews for a more extensive description of the reaction mechanisms involved (28, 97, 107).

Owing to their FMN-binding capacity, LOV domains expressed and purified from *Escherichia coli* are yellow in color and emit a strong green fluorescence when irradiated with UV/blue light. Upon illumination, LOV domains undergo a photocycle that can be monitored by absorbance or fluorescence spectroscopy (81, 126). In darkness, LOV domains bind FMN noncovalently forming a spectral species, designated LOV_{447}, which absorbs maximally near 447 nm (22, 126, 148). Irradiation of the domain induces a unique mode of photochemistry that involves the formation of a covalent adduct between the C(4a) carbon of the flavin chromophore and a conserved cysteine residue within the LOV domain (**Figure 2*a***). Mutation of the cysteine to either alanine or serine results in a loss of photochemical reactivity (126). Light-driven FMN-cysteinyl adduct formation occurs in the order of microseconds producing

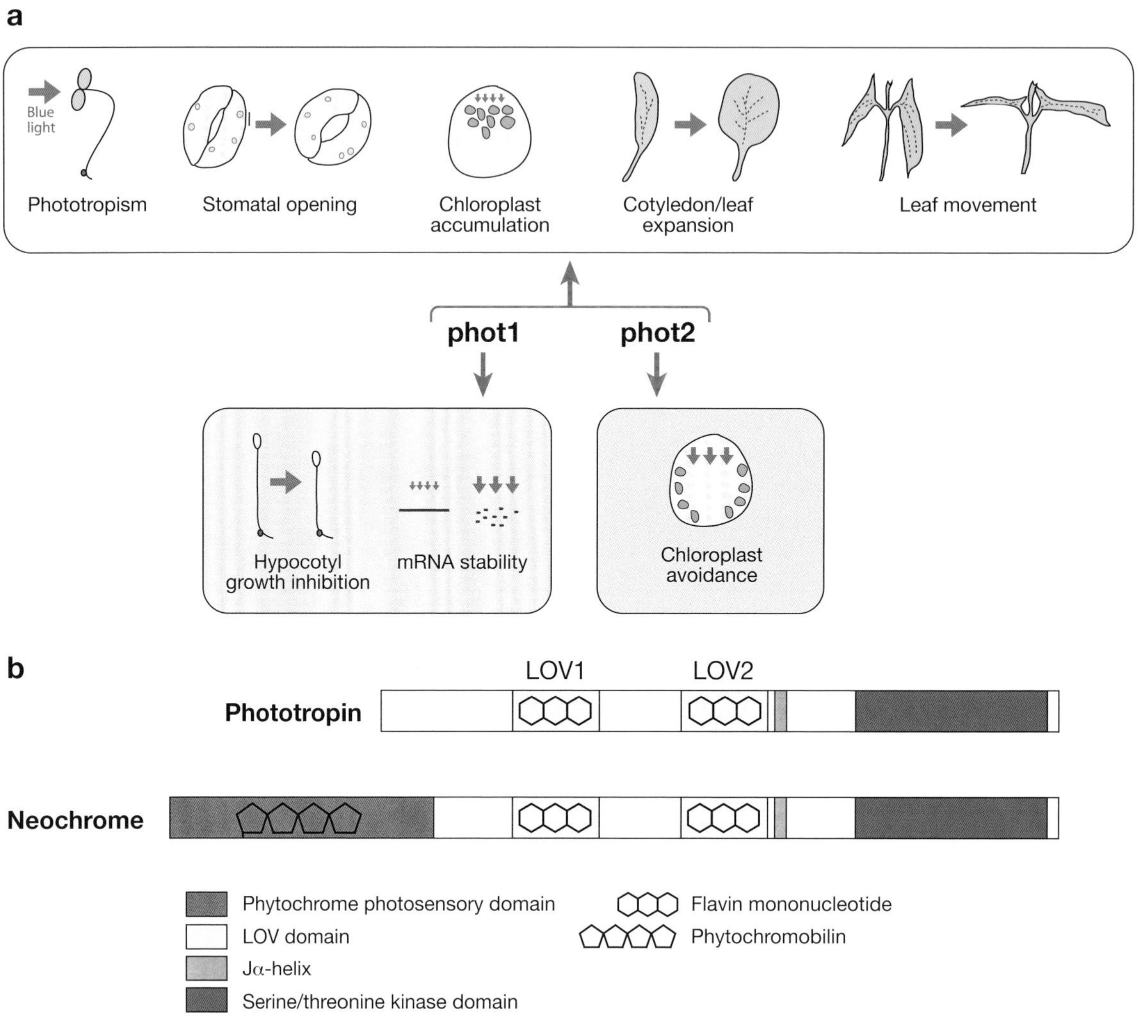

Figure 1

Phototropin structure and function. (*a*) Diagram illustrating the range of phototropin-induced responses in higher plants. Phot1 and phot2 are activated by blue light and overlap in function to mediate several responses. These are enclosed in the yellow rectangle and include phototropism, stomatal opening, chloroplast accumulation movement, and cotyledon and leaf expansion. Phototropins have also been implicated in controlling blue-light-induced leaf movements. Chloroplast avoidance movement is only mediated by phot2. Likewise, phot1 alone plays a role in mediating the rapid inhibition of hypocotyl growth and promoting the destabilization of specific transcripts under high light intensities. (*b*) Protein structures of phototropin and neochrome photoreceptors. Domain structures of these proteins along with their respective chromophores are indicated.

a spectral species (LOV_{390}) that absorbs maximally at 390 nm (81, 126, 148). For phototropin LOV domains, formation of LOV_{390} is fully reversible in darkness, returning the LOV domain back to its initial ground state (LOV_{447}) within the order of tens to hundreds of seconds (81, 126). LOV domains therefore cycle between active (LOV_{390}) and inactive (LOV_{447}) states depending on the light conditions (**Figure 2*b***). Illumination with near

UV-light can revert LOV_{390} to its initial dark state (84). However, the biological significance of this UV-mediated reversibility with respect to receptor function is not known.

Intermediates of the LOV-domain photocycle have been defined spectrally. Initial absorption of blue light by the FMN chromophore results in the formation of an excited singlet state, which subsequently decays into a flavin triplet state (LOV_{660}) that absorbs maximally in the red region of the spectrum (83, 96, 148). The triplet state in turn decays to form the FMN-cysteinyl adduct. Although there is still some debate as to the reaction mechanism for LOV_{390} formation from LOV_{660} (107), it is generally accepted that LOV_{390} represents the active signaling state that leads to photoreceptor activation. Crystal structures of LOV1 and LOV2 from *Chlamydomonas* phot and *Adiantum* neo, respectively, have been solved and show a close resemblance in overall structure to other PAS domains (26, 39). Importantly, structures for dark and illuminated states of LOV1 and LOV2 have been obtained and confirm the reaction mechanism of light-driven adduct formation (26, 27, 39). The structures of LOV1 and LOV2 are almost identical and comprise five antiparallel β–strands interconnected by two α-helices (**Figure 3**). The FMN chromophore is held tightly within a central cavity by hydrogen bonding and van der Waals forces via 11 conserved amino acids (26, 39). The constraints imposed by the protein environment surrounding the flavin chromophore account for the vibronic fine structure observed in the absorption spectrum of the LOV domain, which is not observed for free flavins in solution (126, 148). The ribityl phosphate side chain of FMN is not essential for LOV-domain photochemistry, since flavin exchange analysis has shown that riboflavin can function efficiently as chromophore (33).

Functional Roles of LOV1 and LOV2

LOV1 and LOV2 exhibit different quantum efficiencies and photochemical reaction

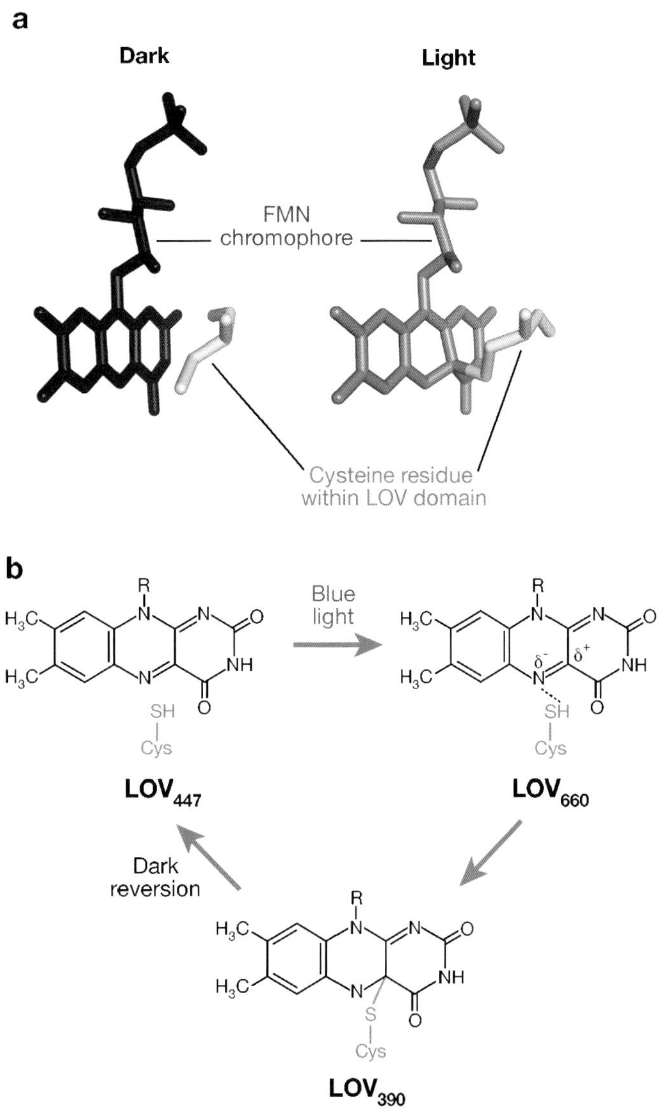

Figure 2

LOV-domain structure and photochemistry. (*a*) Close-ups of the vicinity of the FMN chromophore within the LOV domain under dark and light conditions. Light drives the formation of a covalent adduct between the FMN chromophore and a conserved cysteine residue within the LOV domain. Coordinates were obtained from Protein Data Bank ID codes 1G28 and 1JNU. (*b*) Schematic representation of LOV-domain photochemistry. In darkness, the FMN chromophore is noncovalently bound within the LOV domain, forming a species that absorbs maximally at 447 nm (LOV_{447}). Light drives the production of a highly reactive triplet-state flavin (LOV_{660}) that leads to formation of a covalent bond between the C(4a) carbon of the FMN chromophore and a conserved cysteine residue within the LOV domain (LOV_{390}). The photoreaction process is fully reversible in darkness.

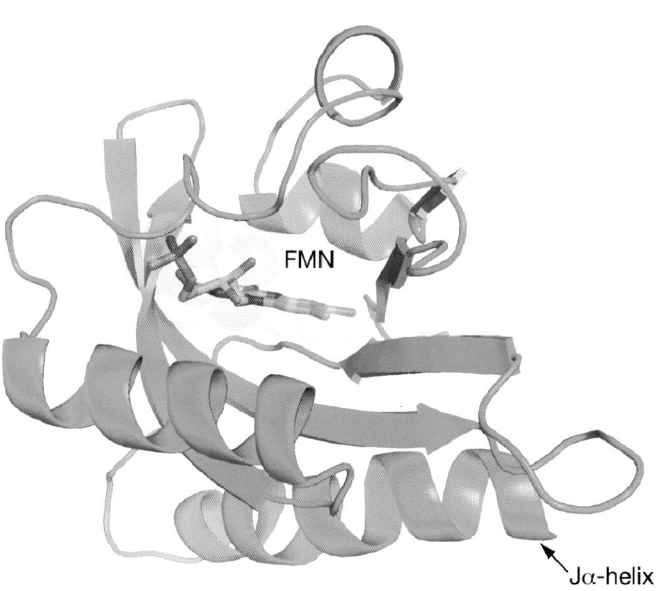

Figure 3

Structural model of the LOV2 domain from oat phot1 in the dark state as determined by nuclear magnetic resonance analysis and homology modeling (57). Positions of the FMN chromophore and the Jα-helix are indicated.

Quantum efficiency: the ratio of photoreceptor molecules that carry out photochemistry to the number of photons absorbed

SAXS: small-angle X-ray scattering

FTIR: Fourier transform infrared spectroscopy

kinetics (81, 126), implying that these domains have different light-sensing roles in regulating phototropin activity. LOV2 has a higher quantum efficiency for light-induced cysteinyl adduct formation than that of LOV1 for both phot1 and phot2 (81). Mutagenesis studies have shown that LOV2 photoreactivity is required for phot1 autophosphorylation and to elicit phot1-mediated hypocotyl phototropism in *Arabidopsis* (23). LOV1 photoreactivity, however, is not sufficient to elicit phot1 autophosphorylation and phot1-induced phototropic curvature. Thus, at least for phototropism, LOV2 is essential for phot1 function in *Arabidopsis*. Similarly, the LOV2 domain of phot2 is sufficient to mediate chloroplast avoidance movement in *Adiantum* (75).

Whereas LOV2 plays an important role in regulating phototropin activity, the exact role of LOV1 remains unknown. Size exclusion chromatography (128) in addition to small-angle X-ray scattering (SAXS) analysis (110) of purified LOV1 suggests that this domain may play a role in receptor dimerization. The function of LOV1 has also been proposed to prolong the lifetime of phototropin receptor activation (75). This hypothesis certainly warrants further investigation given that the LOV1 domain of phot2 has been shown to modulate the activity of a bacterially expressed phot2 kinase (108).

Bacterially expressed fusion proteins containing both LOV domains exhibit photochemical properties that more closely resemble those of the full-length photoreceptor proteins (81) than those of individual LOV1 and LOV2 domains. Although tandem LOV-fusion proteins for phot1 and phot2 exhibit similar relative quantum efficiencies, the dark recovery for phot2 is about 10-times faster than that for phot1 (23, 75, 81). The faster dark recovery might account for why phot2 requires higher light intensities for activity in vivo (77, 123, 150). However, at least for phototropism, the functional activity of phot2 observed under high light conditions results from differential gene expression as *PHOT2* transcripts are induced by light in dark-grown seedlings (74, 76) through the photoactivation of phyA (154). Conversely, long-term exposure of dark-grown seedlings to light results in a decrease in *PHOT1* transcript levels (78, 125), which also depends on phytochrome photoactivation (35).

MODE OF RECEPTOR PHOTOACTIVATION

LOV2-Induced Structural Changes

Light-induced phototropin kinase activity requires LOV2 photoreactivity (23, 108). Therefore, LOV2 can be viewed as a molecular light switch that controls the activity of the C-terminal kinase domain. The photoexcited crystal structure of *Adiantum* neo LOV2 shows only minor, light-induced protein changes within the vicinity of the FMN chromophore (27). Yet, Fourier transform infrared spectroscopy (FTIR) and circular dichroism (CD) studies (24) demonstrate that photoactivation of purified LOV2 in solution is accompanied by changes in the LOV-domain apoprotein (71, 149). In particular, the βE sheet region of *Adiantum* neo LOV2 exhibits a significant conformational change upon cysteinyl adduct formation (114). The βE sheet region contains a conserved glutamine residue (Gln^{1029}) that when mutated to leucine results in a loss of these light-driven protein changes (72, 114). Gln^{1029} forms hydrogen bonds with the FMN chromophore and undergoes side chain rotation upon cysteinyl adduct formation (26, 27). This residue therefore appears to be important for signal transmission from inside the chromophore-binding pocket to protein changes at the LOV2 surface. Only minimal light-induced protein changes are reported for LOV1 despite the presence of the conserved glutamine residue (3, 72, 104).

Nuclear magnetic resonance (NMR) studies involving extended LOV2 fragments derived from oat phot1 have identified a conserved α–helix (designated Jα) that associates with the surface of LOV2 in the dark state (57). The Jα-helix is located at the C terminus of LOV2 and is amphipathic in nature, consisting of polar and apolar sides, the latter of which docks onto the β–sheet strands of the LOV2-core (**Figure 3**). The interaction between Jα and LOV2 is disrupted upon cysteinyl adduct formation. The Jα-helix becomes disordered and more susceptible to proteolysis (57). Independent approaches provide further support for light-induced helical movements associated with extended LOV2 fragments (24, 34, 72). Moreover, artificial disruption of the LOV2-Jα interaction through site-directed mutagenesis results in phot1 autophosphorylation in the absence of light (56), indicating that unfolding of the Jα-helix results in activation of the C-terminal kinase domain.

Kinase Regulation

An important question to address now is whether LOV2 acts as a dark-state repressor of phototropin kinase activity or a light-state activator. In vitro studies involving protein fragments derived from *Arabidopsis* phot2 are in accordance with LOV2 functioning as a dark-state inhibitor of phototropin kinase activity (107, 108). The bacterially expressed kinase domain of *Arabidopsis* phot2 can phosphorylate the artificial substrate casein in vitro. Substrate phosphorylation by phot2 kinase occurs constitutively but becomes light dependent upon the addition of purified LOV2. A similar PAS-kinase interaction mechanism has been proposed for regulating the activities of the bacterial oxygen sensor, FixL (49), and the eukaryotic protein kinase, PAS kinase (2, 122), a regulator of mammalian glycogen synthesis (158). A schematic representation of phototropin receptor activation is illustrated in **Figure 4*a***. Curiously, the transinteraction observed between purified LOV2 and the kinase domain of *Arabidopsis* phot2 occurs in the absence of the Jα-helix (108). Whether the dispensability of the Jα-helix reflects a difference between the mechanisms of substrate phosphorylation and receptor autophosphorylation remains to be clarified. In addition, the LOV1 domain of *Arabidopsis* phot2 has been shown to mediate a small degree of light-activated receptor autophosphorylation (23). Although this is not apparent for phot1 (23), it does raise the question as to how LOV1 can mediate autophosphorylation for phot2. A further challenge will be to uncover the mode of action of *Adiantum* neo and whether red-light-induced photoactivation of its N-terminal phytochrome chromophore-binding domain leads to receptor autophosphorylation.

The kinase domain of plant phototropins is closely related to the catalytic subunit of cAMP-dependent protein kinase A (PKA). Indeed, PKA has been shown to exhibit the same phosphorylation-site specificity as the kinase domain of oat phot1 for autophosphorylation (127). In its inactive state, PKA comprises two regulatory and two catalytic subunits (153). Each regulatory subunit inhibits kinase activity, which is achieved through the presence of a pseudosubstrate sequence within the

NMR: nuclear magnetic resonance

PKA: protein kinase A

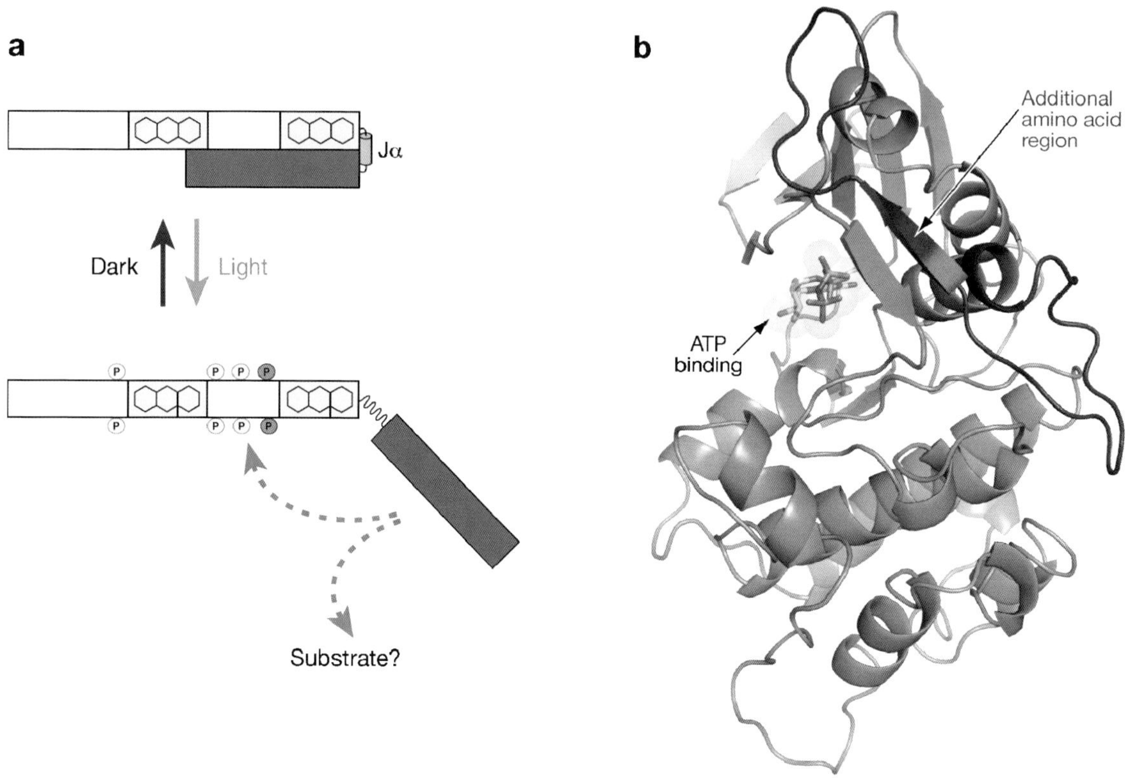

Figure 4

Phototropin kinase regulation. (*a*) Schematic overview of phototropin receptor activation by light. In the dark or ground state, the phototropin receptor is unphosphorylated and inactive. Absorption of light by the predominant light sensor LOV2 results in a disordering of the Jα-helix and activation of the C-terminal kinase domain, which consequently leads to autophosphorylation of the photoreceptor and possibly phosphorylation of an as-yet-unidentified protein substrate(s). Relative positions of known phosphorylation sites are indicated and color-coded based on their hierarchical pattern of occurrence (127): pale blue, low/intermediate fluence phosphorylation sites; dark blue, high fluence phosphorylation sites. (*b*) Structure of the kinase domain of *Arabidopsis* phot1 as determined by homology modeling based on Protein Data Bank ID codes 1O6KA, 1O6LA, and 1ZRZA. The position of the ATP-binding site is indicated. The additional amino acid region present in the phot1 kinase domain is shown in red.

regulatory subunit that specifically occupies the catalytic site of PKA (86). Binding of cAMP to the regulatory subunits leads to a release of the enzymatically active catalytic subunits (153). It is possible that LOV2 may regulate the kinase activity of plant phototropins through a similar mechanism. However, no pseudosubstrate sequence associated with the LOV2 apoprotein has been identified.

The catalytic subunit of PKA adopts a two-lobed structure connected by a deep cleft that functions as the catalytic site (159). Comparative modeling indicates that the kinase domain of *Arabidopsis* phot1 adopts a similar structure to the catalytic subunit of PKA (**Figure 4*b***). One noticeable difference in structure observed for the kinase domains of plant phototropins is the presence of an additional amino acid region within subdomain VIII. This region is rich in polar residues, but its functional significance is unknown. Interestingly, subdomain VIII plays a role in substrate recognition (53). Furthermore, subdomain VIII of many protein kinases, including

PKA, contains a conserved threonine residue that when phosphorylated results in maximal kinase activity (1). This residue is conserved in *Arabidopsis* phot1 and phot2 as a serine (107), raising the possibility that this mode of activation may also occur for plant phototropins. A peptide region C-terminal to the kinase domain in *Adiantum* phot2 is essential for mediating chloroplast avoidance movement (75). The functional significance of this region is currently not known but is likely to serve an important function given that it is conserved among higher and lower plant phototropins (75).

Sites and Function of Receptor Autophosphorylation

Phosphorylation of the N-terminal photosensory region of oat phot1 by PKA exhibits the same phosphorylation-site specificity as that observed for phot1 autophosphorylation in protein extracts from dark-grown oat coleoptiles (127). Mapping of these phosphorylation sites indicates that there are eight serine residues within the N-terminal region of oat phot1 that become phosphorylated (127). Two of these sites (Ser^{27}, Ser^{30}) are located before LOV1, near the N terminus of the protein (**Figure 4*a***). The remaining sites (Ser^{274}, Ser^{300}, Ser^{317}, Ser^{325}, Ser^{332}, and Ser^{349}) are located in the peptide region between LOV1 and LOV2. Light-dependent autophosphorylation of oat phot1 in vivo occurs in a fluence-dependent manner (127). Sites close to LOV1 (Ser^{27}, Ser^{30}, Ser^{274} and Ser^{300}) are phosphorylated at low fluences of blue light. The remaining four sites, on the other hand, are phosphorylated in response to intermediate or high fluences of blue light. Photoactivated phot1 in vivo has been shown to return to its nonphosphorylated state upon incubation in darkness (52, 89, 129, 136). The mechanism underlying this recovery process is not known, but likely involves the activity of an as-yet-unidentified protein phosphatase.

Despite these findings, the biochemical consequences of receptor autophosphorylation are still poorly understood. Autophosphorylation requires more light than is needed for a number of phototropin-mediated responses (15), indicating that it is not the primary signaling event. However, the hierarchical pattern of autophosphorylation observed for oat phot1 (127) is proposed to result in different biochemical consequences. The low/intermediate fluence phosphorylation sites may play a role in receptor signaling, whereas the high fluence phosphorylation sites may be involved in receptor desensitization (127). *Chlamydomonas* phot lacks the N-terminal extension preceding LOV1 found in higher plant phototropins but is still able to restore phot1- and phot2-mediated responses when introduced into the *phot1 phot2* double mutant of *Arabidopsis* (118). Hence, the two-phosphorylation events upstream from LOV1 in higher plant phototropins are not essential for phototropin function. One consequence of autophosphorylation may be to promote receptor dissociation from the plasma membrane. *Arabidopsis* phot1 and phot2 are hydrophilic proteins but are localized to the plasma membrane (95, 125). The nature of membrane association remains unknown but it is thought to involve posttranslational modification or binding of a membrane protein anchor. Studies involving GFP fusions show that a fraction of phot1 is rapidly internalized (within minutes) from the plasma membrane in *Arabidopsis* upon blue-light irradiation (125). Similarly, a fraction of phot2 moves to the Golgi apparatus upon blue-light irradiation (95). Although the functional significance of this partial redistribution is currently unknown, the kinase domain of phot2 appears to be essential for Golgi localization (95).

GFP: green fluorescent protein

So far, there is no direct evidence to indicate that phototropins initiate signaling through the activation of a phosphorylation cascade. A truncated version of phot2 comprising only the LOV2 domain and the C-terminal kinase domain is able to complement chloroplast avoidance movement in a *phot2* mutant of *Adiantum* (75). Given that the sites of phototropin autophosphorylation are located before LOV2 (127), it is tempting to speculate that phot2 uses some means other than autophosphorylation to bring about this response. However, until now the only substrate identified for phot2 kinase activity is the artificial substrate casein (108).

PHOTOTROPIN RECEPTOR SIGNALING

Phototropism

Phototropic curvature is mediated by an increase in growth on the shaded side of the stem resulting from an accumulation of the plant growth hormone auxin (65). At present, little is known about how phototropin activation by blue light leads to an accumulation of auxin on the shaded side of the stem. Unilateral irradiation has been shown to induce a gradient of phot1 autophosphorylation across oat coleoptiles, with the highest level of phosphorylation occurring on the irradiated side (131, 132). A phosphorylation gradient model has therefore been proposed to account for the complex fluence response curve for phototropism (131), but the question still remains as to how a gradient in phototropin phosphorylation across the stem can bring about a lateral gradient in auxin.

Insights into the signaling mechanisms involved in phototropism have again come from *Arabidopsis* mutants. The nonphototropic hypocotyl mutant *nph3* (101, 102) led to the identification of a phot1-interacting protein that is essential for lateral auxin redistribution (50) and phototropism (50, 109). NPH3 is a novel protein containing several protein interaction motifs and has been shown to interact with phot1 in vitro (109) and in vivo (98). Like the phototropins, NPH3 is hydrophilic in nature but is associated with the plasma membrane (109). Although its biochemical function is unknown, NPH3 most likely serves as a scaffold to assemble components of a phototropin receptor complex. NPH3 is a member of a large plant-specific gene family consisting of 31 members in *Arabidopsis* (18) and at least 24 in rice (87). A protein closely related to NPH3, designated Root Phototropism 2 (RPT2), has also been shown to bind phot1 and is required for both phototropism and stomatal opening by blue light (69, 123, 124). Although the interaction between phot1 and RPT2 is unaffected by light (69), blue-light irradiation is reported to alter the phosphorylation status of NPH3. NPH3 appears to be phosphorylated in the dark and becomes dephosphorylated upon exposure to blue light (109). More recently, studies demonstrated that Phytochrome Kinase Substrate (PKS) proteins are required for hypocotyl phototropism in *Arabidopsis* (98). PKS1 binds both phot1 and NPH3 in vivo (98). Because phytochromes influence phototropic curvature in *Arabidopsis* (66), PKS proteins may provide a link between these two photoreceptor families.

Genetic analysis has also shown that auxin responsiveness is necessary for phototropism. Auxin-regulated transcription factors such as Nonphototropic Hypocotyl 4 (NPH4) and Massugu 2 (MSG2) are required for normal phototropism and gravitropism (55, 143, 151), highlighting the need for auxin-regulated gene expression. Activities of NPH4 and MSG2 are likely to be regulated by the recently identified auxin receptor Transport Inhibitor Response 1 (TIR1), a subunit of the Skp1-Cullin-F-box $(SCF)^{TIR1}$ complex, which targets proteins for degradation in the presence of auxin (30, 85). The identities of auxin-regulated gene targets involved in phototropism were recently uncovered using a transcriptomic approach (38). Gene targets of NPH4 action whose expression levels increase on the elongating side of phototropically stimulated *Brassica oleracea* hypocotyls include two members of the α–expansin family, *EXPA1* and *EXPA8*. Because members of the α–expansin family mediate cell wall extension, EXPA1 and EXPA8 may have important roles in the establishment of phototropic curvatures.

Further genetic studies have demonstrated that auxin transport is required for phototropism. Mutants impaired in the localization of the putative auxin efflux carrier Pinformed 1 (PIN1) exhibit altered hypocotyl phototropism (10, 113). PIN3, a second member of the *Arabidopsis* PIN family, also appears to play a role in establishing the lateral auxin gradient required for phototropism (45). Given that phot1 photoactivation results in a change in PIN1 localization in *Arabidopsis* hypocotyls cells (10), a regulation of auxin transporter localization may represent a major point of control in the development of phototropic curvatures. Furthermore, phot1 is more strongly localized to the plasma membrane adjacent to the apical and basal walls rather than the sidewalls in the cells associated with auxin transport (125). This would place phot1 in an ideal location to influence the activity or distribution of auxin influx and efflux carriers. However, such a mechanism is likely to be complex since PIN proteins appear to act in conjunction with members of a second transporter family of p-glycoproteins to bring about active auxin transport in *Arabidopsis* (48).

Stomatal Opening

Stomatal guard cells of the *phot1 phot2* double mutant fail to extrude protons in response to blue-light treatment (88). Proton extrusion is essential for stomatal opening and involves activation of the plasma membrane H^+-ATPase (31). Activation of the guard cell H^+-ATPase involves phosphorylation of the H^+-ATPase and 14-3-3 binding (90, 92). Similarly, phot1 from *Vicia faba* (broad bean) guard cells binds a 14–3-3 protein upon autophosphorylation (89). Specifically, 14–3-3 binding to *Vicia* phot1 requires phosphorylation of Ser^{358} situated between LOV1 and LOV2, which is equivalent to Ser^{325} of oat phot1 that is phosphorylated in response to intermediate fluences of blue light (127). Thus, one consequence of phototropin autophosphorylation is to mediate 14–3-3 binding. 14–3-3 proteins belong to a highly conserved protein family that typically bind to phosphorylated target proteins and regulate signaling in eukaryotic cells (40). One might speculate that the binding of 14–3-3 proteins may facilitate a direct interaction between phototropins and the guard cell H^+-ATPase. Yet, the fungal toxin fusicoccin has been shown to induce phosphorylation of the H^+-ATPase and subsequent 14–3-3 binding in the absence of phot1 and phot2, implying that some other protein kinase is responsible for phosphorylation of the H^+-ATPase (91, 155). Indeed, the signaling pathway between the phototropins and the H^+-ATPase was recently reported to involve the activity of type 1 protein phosphatase in broad bean guard cells (150a). Further work is now needed to clarify the functional significance of 14–3-3 binding, especially since 14–3-3 binding to the phototropins has also been observed in etiolated seedlings (89).

A novel phot1-interacting protein from *Vicia* guard cells was isolated using a yeast two-hybrid approach (37). This protein, designated VfPIP1 (*Vicia faba* Phot1a-interacting Protein 1), shows homology to dyneins, proteins associated with microtubule function in animal cells. VfPIP is localized to cortical microtubules in *Vicia* guard cells (37), and may function in organizing the guard cell cytoskeleton to promote stomatal opening.

Chloroplast Movement

Arabidopsis mutants impaired in blue-light-induced chloroplast movements have also provided insights into the signaling events acting downstream of phototropin receptor activation. The isolation of mutants lacking chloroplast avoidance movement has led to the identification of CHUP1 (79, 117). CHUP1 (Chloroplast Unusual Positioning 1) is a novel F-actin-binding protein (117), consistent with the known requirement for cytoskeletal changes during chloroplast movement (156). CHUP1 confers the ability to target GFP into the chloroplast envelope (117), suggesting that CHUP1 may function at the periphery of the chloroplast outer membrane. In addition to lacking chloroplast

BAPTA: 1,2-bis(2-aminophenoxy)-ethane-N,N,N,M -tetraacetic acid

avoidance movement, *chup1* mutants exhibit aberrant chloroplast positioning in which chloroplasts are constantly gathered at the bottom of palisade cells, in contrast to wild type (117). Thus, CHUP1 most likely represents an essential component of the machinery required for chloroplast positioning and movement. Similarly, the plastid movement–impaired mutant *pmi1* exhibits severely attenuated chloroplast movements under low and high light intensities (29), indicating that PMI1 is necessary for both chloroplast accumulation and avoidance movements. By contrast, lesions in *PMI2* and *PMI5* lead only to impaired chloroplast avoidance movement (105). PMI2 and PMI5, like PMI1, represent novel plant-specific proteins whose modes of action have yet to be elucidated.

A genetic screen designed to isolate mutants altered in chloroplast accumulation movement has identified the protein JAC1 (144). JAC1 (J-domain Protein Required for Chloroplast Accumulation Response 1) is a cytosolic protein specifically required for chloroplast accumulation movement. Although its biochemical function is unknown, the C terminus of JAC1 exhibits homology to auxilin, a protein that plays a role in clathrin-mediated endocytosis in animal cells. The functional significance of this domain in regulating chloroplast accumulation movement now awaits further characterization of the JAC1 protein.

Calcium and the Rapid Inhibition of Hypocotyl Growth

Phototropin activation leads to an increase in cytosolic Ca^{2+} concentrations (9). Pharmacological experiments indicate that changes in cytosolic Ca^{2+} are required for the phot1-mediated inhibition of hypocotyl growth (42). The calcium-specific chelator BAPTA inhibits the phot1-mediated increase in intracellular Ca^{2+} levels observed in *Arabidopsis* seedlings (42). An equivalent BAPTA treatment prevents the rapid inhibition of hypocotyl elongation by blue light but has no effect on phototropism (42). However, phot2 as well as phot1 is reported to stimulate increases in cytosolic Ca^{2+} in *Arabidopsis* leaves (54, 142). Both phot1 and phot2 can mediate an influx of Ca^{2+} from the apoplast, whereas only phot2 can induce a release of Ca^{2+} from intracellular stores (5, 54). It is therefore likely that Ca^{2+} serves as signal messenger in processes other than hypocotyl growth inhibition. Indeed, changes in intracellular calcium levels are important for light-induced stomatal opening (31) and chloroplast movements (156). Likewise, electrophysiological studies indicate that phototropic bending involves changes in ion fluxes, including calcium (4). Intriguingly, as found for chloroplast accumulation movement (123) and the promotion of growth under low light conditions (150), phot2 is less sensitive than phot1 in mediating blue-light-induced calcium fluxes (54), again indicating that phot1 and phot2 exhibit different photosensitivities in vivo. Readers are directed to the following excellent review (54a) for further details regarding phototropin-induced Ca^{2+} changes.

OTHER LOV SENSOR PROTEINS

Light-sensitive LOV domains have also been identified in plant proteins other than the phototropins. These proteins constitute a new class of putative blue-light receptors known as the ZTL/ADO family (7) and comprise three members: Zeitlupe (ZTL, also referred to as Adagio, ADO); Flavin-binding, Kelch Repeat, F-box 1 (FKF1); and LOV Kelch, Protein 2 (LKP2). A detailed account of these proteins is beyond the scope of this article and the reader is directed to the following recent publication for further details (133). Gain- and loss-of-function analyses have shown that these proteins are required for circadian clock function and photoperiodic-dependent flowering in *Arabidopsis* (73, 93, 112, 134, 141).

ZTL, FKF1, and LKP2 share three characteristic domains: a phototropin-like LOV domain at the N terminus followed by an

F-box motif and six kelch repeats at the C terminus (7). The F-box motif is typically found in E3 ubiquitin ligases, which target proteins for degradation via the ubiquitin-proteosome system (140). In fact, evidence now indicates that members of the ZTL/ADO family mediate their effects by targeting key regulatory components for proteolysis (67, 106). Kelch-domain repeats form a β-propeller structure (99) thought to be involved in mediating protein-protein interactions. The LOV domains of ZTL, FKF1, and LKP2 contain the 11 conserved residues necessary for flavin binding, including the cysteine required for photochemical reactivity (28). Indeed, bacterially expressed LOV domains derived from each of these proteins exhibit photochemical properties analogous to those of the phototropin LOV domains (68, 111). Yet, unlike the phototropin LOV domains, the LOV domains of ZTL, FKF1, and LKP2 fail to revert back to their dark state (68, 111). Whether this inability to recover to the dark state is related to their physiological functions remains unknown. Nonetheless, the demonstration of light-driven cysteinyl adduct formation provides strong evidence that these proteins function as blue-light receptors.

Of these three proteins, only FKF1 has been demonstrated to function as a photoreceptor (68). Expression of FKF1 peaks late in the day and on long days is activated by light to mediate the degradation of CDF1 (Cycling Dof Factor 1), a transcriptional repressor of CONSTANS (CO) (67), a key factor required for the photoperiodic control of flowering. FKF1 thus plays a role in detecting long days and activating the photoperiodic flowering pathway in *Arabidopsis*. To date, no specific photoreceptor role has been ascribed to either ZTL or LKP2.

A search of the *Arabidopsis* genome has uncovered a unique LOV-domain-containing protein that is unrelated to the phototropins or the ZTL/ADO family. This protein, referred to as PAS/LOV (28), contains a conventional PAS domain followed by a phototropin-like LOV domain. To date, the function of this protein remains unknown. It will now be important to establish whether PAS/LOV binds a flavin cofactor and represents an as-yet-uncharacterized blue-light receptor in *Arabidopsis*.

LOV-domain-containing proteins are not only restricted to plants. White Collar-1 (WC-1) is a fungal blue-light receptor that contains a single LOV domain and has been shown to function as a photoreceptor for phototropism and other light responses in *Neurospora crassa* and other fungi (6, 46, 58, 63, 64). A second *Neurospora* protein VIVID (VVD) consists mostly of a LOV domain and plays an important role in mediating photoadaptive responses (59, 135, 139) and facilitating circadian clock entrainment (36). In addition, a large number of otherwise very different bacterial proteins contain LOV domains (28, 103). Those bacterial proteins investigated to date show typical LOV-domain photochemistry (28, 103). No biological function has yet been linked to these putative blue-light receptors.

FUTURE PROSPECTS

In the past decade, the progress made in understanding the photochemical and biochemical properties of the phototropins in addition to their physiological roles has been enormous. Identification of the LOV domain as a blue-light-sensing motif and deciphering its structure and photochemical reactivity represent a major advance. Furthermore, the presence of LOV-domain-containing proteins throughout various kingdoms of life clearly demonstrates that this functional light sensor is not only restricted to plants but has been conserved throughout evolution. Yet, we are still far from understanding the complete mechanistic picture with regard to receptor photoactivation. For instance, what is the function of LOV1? And what is the exact role(s) of receptor autophosphorylation? Further biochemical and structure-function analyses are now needed to address such questions. To date, biophysical and structural studies have been directed to regions of the phototropin molecule lacking the C-terminal kinase domain. Extending these studies to LOV2-kinase fragments and ultimately the full-length receptor protein will provide a greater understanding of the photosensory transduction pathway underlying phototropin activation. A major challenge for the future will be to unravel the processes associated with phototropin signaling and how these relate to components that have already been identified, including increases in cytosolic Ca^{2+}, 14–3-3 proteins and members of the NPH3/RPT2 family. The identification of the LOV-sensing motif in proteins other than the phototropins greatly expands the possible directions for future research. Evidently, much work remains to be done and the coming decade will undoubtedly yield exciting advances in our knowledge of phototropin receptor function and signaling.

SUMMARY POINTS

1. Phototropins act to control a number of plant processes that serve to optimize photosynthetic performance, including phototropism after which they were named.
2. Phototropins are plasma membrane-associated serine/threonine photoreceptor kinases that undergo autophosphorylation in response to blue-light excitation.
3. Light sensing by the phototropins is mediated by a conserved motif called the LOV domain that binds the light-absorbing cofactor FMN.
4. Phototropins possess two FMN-binding LOV domains that exhibit distinct functional roles in regulating photoreceptor activation.
5. Induction of phototropin kinase activity by LOV2 occurs through light-driven structural changes involving a conserved α-helix designated Jα.
6. Phototropin autophosphorylation occurs on multiple serine residues in a fluence-dependent manner, the consequences of which may play a role in receptor signaling, receptor desensitization or receptor relocalization.
7. Genetic and biochemical analyses have uncovered novel plant-specific components associated with phototropin receptor signaling, in addition to 14–3-3 proteins and intracellular calcium.
8. LOV-domain-containing proteins besides the phototropins have been identified in plants, fungi, and bacteria and represent further blue-light receptors.

ACKNOWLEDGMENTS

J.M.C. is grateful to the Royal Society for the award of University Research Fellowship and the BBSRC (Biotechnology and Biological Sciences Research Council) and the Gatsby Charitable Foundation for supporting research in his laboratory. J.M.C. is also very grateful to Winslow Briggs for his careful review of the manuscript and providing helpful comments, to Kevin Gardner for providing **Figure 3**, and to Brian Smith for his help in generating figures.

LITERATURE CITED

1. Adams JA, McGlone ML, Gibson R, Taylor SS. 1995. Phosphorylation modulates catalytic function and regulation in the cAMP-dependent protein kinase. *Biochemistry* 34:2447–54

2. Amezcua CA, Harper SM, Rutter J, Gardner KH. 2002. Structure and interactions of PAS kinase N-terminal PAS domain: model for intramolecular kinase regulation. *Structure* 10:1349–61
3. Ataka K, Hegemann P, Heberle J. 2003. Vibrational spectroscopy of an algal Phot-LOV1 domain probes the molecular changes associated with blue-light reception. *Biophys. J.* 84:466–74
4. Babourina O, Godfrey L, Voltchanskii K. 2004. Changes in ion fluxes during phototropic bending of etiolated oat coleoptiles. *Ann. Botany* 94:187–94
5. Babourina O, Newman I, Shabala S. 2002. Blue light-induced kinetics of H+ and Ca2+ fluxes in etiolated wild-type and phototropin-mutant Arabidopsis seedlings. *Proc. Natl. Acad. Sci. USA* 99:2433–38
6. Ballario P, Vittorioso P, Magrelli A, Talora C, Cabibbo A, Macino G. 1996. White collar-1, a central regulator of blue light responses in Neurospora, is a zinc finger protein. *EMBO J.* 15:1650–57
7. Banerjee R, Batschauer A. 2005. Plant blue-light receptors. *Planta* 220:498–502
8. Batschauer A. 2005. Plant cryptochromes: their genes, biochemistry, and physiological roles. In *Handbook of Photosensory Receptors*, ed. WR Briggs, JL Spudich, pp. 2112–46. Weinheim: Wiley-VCH
9. Baum G, Long JC, Jenkins GI, Trewavas AJ. 1999. Stimulation of the blue light phototropic receptor NPH1 causes a transient increase in cytosolic Ca2+. *Proc. Natl. Acad. Sci. USA* 96:13554–59
10. Blakeslee JJ, Bandyopadhyay A, Peer WA, Makam SN, Murphy AS. 2004. Relocalization of the PIN1 auxin efflux facilitator plays a role in phototropic responses. *Plant Physiol.* 134:28–31
11. Bogre L, Okresz L, Henriques R, Anthony RG. 2003. Growth signaling pathways in Arabidopsis and the AGC protein kinases. *Trends Plant Sci.* 8:424–31
12. Briggs WR. 2006. Blue-UV-A receptors: Historical overview. In *Photomorphogenesis in Plants*, ed. E Shäfer, F Nagy, pp. 171–97. Dordrecht: Springer
13. Briggs WR, Beck CF, Cashmore AR, Christie JM, Hughes J, et al. 2001. The phototropin family of photoreceptors. *Plant Cell* 13:993–97
14. Briggs WR, Christie JM. 2002. Phototropins 1 and 2: versatile plant blue-light receptors. *Trends Plant Sci.* 7:204–10
15. **Briggs WR, Christie JM, Salomon M. 2001. Phototropins: a new family of flavin-binding blue light receptors in plants. *Antiox. Redox Signal.* 3:775–88**
16. Briggs WR, Christie JM, Swartz TE. 2006. Phototropins. In *Photomorphogenesis in Plants*, ed. E Shäfer, F Nagy, pp. 223–52. Dordrecht: Springer
17. Bunsen R, Roscoe H. 1862. Photochemische Unterzuchungen. *Ann. Phys. Chem.* 117:529–62
18. Celaya RB, Liscum E. 2005. Phototropins and associated signaling: providing the power of movement in higher plants. *Photochem. Photobiol.* 81:73–80
19. Chen M, Chory J, Fankhauser C. 2004. Light signal transduction in higher plants. *Annu. Rev. Genet.* 38:87–117
20. Christie JM, Briggs WR. 2005. Blue light sensing and signaling by the phototropins. In *Handbook of Photosensory Receptors*, ed. WR Briggs, JL Spudich, pp. 277–304. Weinheim: Wiley-VCH
21. **Christie JM, Reymond P, Powell GK, Bernasconi P, Raibekas AA, et al. 1998. Arabidopsis NPH1: a flavoprotein with the properties of a photoreceptor for phototropism. *Science* 282:1698–701**

15. Comprehensive review providing detailed coverage of the early biochemical and photochemical studies of phototropin blue-light receptors.

21. First evidence showing that recombinant phot1 binds FMN, undergoes blue light-dependent autophosphorylation, and displays spectral properties characteristic of a phototropic receptor.

22. Christie JM, Salomon M, Nozue K, Wada M, Briggs WR. 1999. LOV (light, oxygen, or voltage) domains of the blue-light photoreceptor phototropin (nph1): Binding sites for the chromophore flavin mononucleotide. *Proc. Natl. Acad. Sci. USA* 96:8779–83

23. Christie JM, Swartz TE, Bogomolni RA, Briggs WR. 2002. Phototropin LOV domains exhibit distinct roles in regulating photoreceptor function. *Plant J.* 32:205–19

24. Corchnoy SB, Swartz TE, Lewis JW, Szundi I, Briggs WR, Bogomolni RA. 2003. Intramolecular proton transfers and structural changes during the photocycle of the LOV2 domain of phototropin 1. *J. Biol. Chem.* 278:724–31

25. Crosson S. 2005. LOV domain structure, dynamics, and diversity. In *Handbook of Photosensory Receptors*, ed. WR Briggs, JL Spudich, pp. 323–36. Weinheim: Wiley-VCH

26. Crosson S, Moffat K. 2001. Structure of a flavin-binding plant photoreceptor domain: Insights into light-mediated signal transduction. *Proc. Natl. Acad. Sci. USA* 98:2995–3000

27. Crosson S, Moffat K. 2002. Photoexcited structure of a plant photoreceptor domain reveals a light-driven molecular switch. *Plant Cell* 14:1067–75

28. Crosson S, Rajagopal S, Moffat K. 2003. The LOV domain family: Photoresponsive signaling modules coupled to diverse output domains. *Biochemistry* 42:2–10

29. DeBlasio SL, Luesse DL, Hangarter RP. 2005. A plant-specific protein essential for blue-light-induced chloroplast movements. *Plant Physiol.* 139:101–14

30. Dharmasiri N, Dharmasiri S, Estelle M. 2005. The F-box protein TIR1 is an auxin receptor. *Nature* 435:441–45

31. Dietrich P, Sanders D, Hedrich R. 2001. The role of ion channels in light-dependent stomatal opening. *J. Exp. Bot.* 52:1959–67

32. Doi M, Wada M, Shimazaki K. 2006. The Fern Adiantum capillus-veneris lacks stomatal responses to blue light. *Plant Cell Physiol.* 47:748–55

33. Durr H, Salomon M, Rudiger W. 2005. Chromophore exchange in the LOV2 domain of the plant photoreceptor phototropin1 from oat. *Biochemistry* 44:3050–55

34. Eitoku T, Nakasone Y, Matsuoka D, Tokutomi S, Terazima M. 2005. Conformational dynamics of phototropin 2 LOV2 domain with the linker upon photoexcitation. *J. Am. Chem. Soc.* 127:13238–44

35. Elliott RC, Platten JD, Watson JC, Reid JB. 2004. Phytochrome regulation of pea phototropin. *J. Plant Physiol.* 161:265–70

36. Elvin M, Loros JJ, Dunlap JC, Heintzen C. 2005. The PAS/LOV protein VIVID supports a rapidly dampened daytime oscillator that facilitates entrainment of the Neurospora circadian clock. *Genes Dev.* 19:2593–605

37. Emi T, Kinoshita T, Sakamoto K, Mineyuki Y, Shimazaki K. 2005. Isolation of a protein interacting with Vfphot1a in guard cells of Vicia faba. *Plant Physiol.* 138:1615–26

38. Esmon CA, Tinsley AG, Ljung K, Sandberg G, Hearne LB, Liscum E. 2006. A gradient of auxin and auxin-dependent transcription precedes tropic growth responses. *Proc. Natl. Acad. Sci. USA* 103:236–41

39. Fedorov R, Schlichting I, Hartmann E, Domratcheva T, Fuhrmann M, Hegemann P. 2003. Crystal structures and molecular mechanism of a light-induced signaling switch: the Phot-LOV1 domain from *Chlamydomonas reinhardtii*. *Biophys. J.* 84:2474–82

40. Ferl RJ. 2004. 14-3-3 proteins: regulation of signal-induced events. *Physiol. Plant* 120:173–78

41. Folta KM, Kaufman LS. 2003. Phototropin 1 is required for high-fluence blue-light-mediated mRNA destabilization. *Plant Mol. Biol.* 51:609–18

23. Detailed study demonstrating that LOV2 has a major role in regulating phototropin activity, invoking the question as to the function of LOV1.

26. Breakthrough study presenting the crystal structure of the LOV2-core complete with FMN chromophore.

38. Meticulous study identifying auxin-dependent gene expression changes that occur across the hypocotyl and precede tropic growth.

42. Folta KM, Lieg EJ, Durham T, Spalding EP. 2003. Primary inhibition of hypocotyl growth and phototropism depend differently on phototropin-mediated increases in cytoplasmic calcium induced by blue light. *Plant Physiol.* 133:1464–70
43. Folta KM, Spalding EP. 2001. Unexpected roles for cryptochrome 2 and phototropin revealed by high-resolution analysis of blue light-mediated hypocotyl growth inhibition. *Plant J.* 26:471–78
44. Franklin KA, Larner VS, Whitelam GC. 2005. The signal transducing photoreceptors of plants. *Int. J. Dev. Biol.* 49:653–64
45. Friml J, Wisniewska J, Benkova E, Mendgen K, Palme K. 2002. Lateral relocation of auxin efflux regulator PIN3 mediates tropism in Arabidopsis. *Nature* 415:806–9
46. Froehlich AC, Liu Y, Loros JJ, Dunlap JC. 2002. White Collar-1, a circadian blue light photoreceptor, binding to the frequency promoter. *Science* 297:815–19
47. Gallagher S, Short TW, Ray PM, Pratt LH, Briggs WR. 1988. Light-mediated changes in two proteins found associated with plasma membrane fractions from pea stem sections. *Proc. Natl. Acad. Sci. USA* 85:8003–7
48. Geisler M, Murphy AS. 2006. The ABC of auxin transport: the role of p-glycoproteins in plant development. *FEBS Lett.* 580:1094–102
49. Gong W, Hao B, Mansy SS, Gonzalez G, Gilles-Gonzalez MA, Chan MK. 1998. Structure of a biological oxygen sensor: a new mechanism for heme-driven signal transduction. *Proc. Natl. Acad. Sci. USA* 95:15177–82
50. Haga K, Takano M, Neumann R, Iino M. 2005. The Rice COLEOPTILE PHOTOTROPISM1 gene encoding an ortholog of Arabidopsis NPH3 is required for phototropism of coleoptiles and lateral translocation of auxin. *Plant Cell* 17:103–15
51. Hager A. 1996. Properties of a blue-light-absorbing photoreceptor kinase localized in the plasma membrane of the coleoptile tip region. *Planta* 198:294–99
52. Hager A, Brich M. 1993. Blue-light-induced phosphorylation of plasma-membrane protein from phototropically sensitive tips of maize coleoptiles. *Planta* 189:567–76
53. Hanks SK, Hunter T. 1995. Protein kinases 6. The eukaryotic protein kinase superfamily: kinase (catalytic) domain structure and classification. *FASEB J.* 9:576–96
54. Harada A, Sakai T, Okada K. 2003. Phot1 and phot2 mediate blue light-induced transient increases in cytosolic Ca2+ differently in Arabidopsis leaves. *Proc. Natl. Acad. Sci. USA* 100:8583–88
54a. Harada A, Shimazaki KI. 2006. Phototropins and blue light-dependent calcium signaling in higher plants. *Photochem. Photobiol.* In press
55. Harper RM, Stowe-Evans EL, Luesse DR, Muto H, Tatematsu K, et al. 2000. The NPH4 locus encodes the auxin response factor ARF7, a conditional regulator of differential growth in aerial Arabidopsis tissue. *Plant Cell* 12:757–70
56. Harper SM, Christie JM, Gardner KH. 2004. Disruption of the LOV-J alpha helix interaction activates phototropin kinase activity. *Biochemistry* 43:16184–92
57. Harper SM, Neil LC, Gardner KH. 2003. Structural basis of a phototropin light switch. *Science* 301:1541–44
58. He Q, Cheng P, Yang Y, Wang L, Gardner KH, Liu Y. 2002. White collar-1, a DNA binding transcription factor and a light sensor. *Science* 297:840–43
59. Heintzen C, Loros JJ, Dunlap JC. 2001. The PAS protein VIVID defines a clock-associated feedback loop that represses light input, modulates gating, and regulates clock resetting. *Cell* 104:453–64
60. Huala E, Oeller PW, Liscum E, Han IS, Larsen E, Briggs WR. 1997. Arabidopsis NPH1: a protein kinase with a putative redox-sensing domain. *Science* 278:2120–23

57. Pivotal paper presenting novel insights into the protein structural changes that follow LOV2 photoexcitation.

61. Huang KY, Beck CF. 2003. Photoropin is the blue-light receptor that controls multiple steps in the sexual life cycle of the green alga *Chlamydomonas reinhardtii*. *Proc. Natl. Acad. Sci. USA* 100:6269–74
62. Huang KY, Merkle T, Beck CF. 2002. Isolation and characterization of a Chlamydomonas gene that encodes a putative blue-light photoreceptor of the phototropin family. *Physiologia Plantarum* 115:613–22
63. Idnurm A, Heitman J. 2005. Light controls growth and development via a conserved pathway in the fungal kingdom. *PLoS Biol.* 3:e95
64. Idnurm A, Rodriguez-Romero J, Corrochano LM, Sanz C, Iturriaga EA, et al. 2006. The Phycomyces madA gene encodes a blue-light photoreceptor for phototropism and other light responses. *Proc. Natl. Acad. Sci. USA* 103:4546–51
65. Iino M. 2001. Phototropism in higher plants. In *Photomovement*, ed. D-P Häder, M Lebert, pp. 659–811. Amsterdam: Elsevier Sci.
66. Iino M. 2006. Toward understanding the ecological functions of tropisms: interactions among and effects of light on tropisms. *Curr. Opin. Plant Biol.* 9:89–93
67. Imaizumi T, Schultz TF, Harmon FG, Ho LA, Kay SA. 2005. FKF1 F-box protein mediates cyclic degradation of a repressor of CONSTANS in Arabidopsis. *Science* 309:293–97
68. Imaizumi T, Tran HG, Swartz TE, Briggs WR, Kay SA. 2003. FKF1 is essential for photoperiodic-specific light signaling in Arabidopsis. *Nature* 426:302–6
69. Inada S, Ohgishi M, Mayama T, Okada K, Sakai T. 2004. RPT2 is a signal transducer involved in phototropic response and stomatal opening by association with phototropin 1 in Arabidopsis thaliana. *Plant Cell* 16:887–96
70. Inoue S, Kinoshita T, Shimazaki K. 2005. Possible involvement of phototropins in leaf movement of kidney bean in response to blue light. *Plant Physiol.* 138:1994–2004
71. Iwata T, Nozaki D, Tokutomi S, Kagawa T, Wada M, Kandori H. 2003. Light-induced structural changes in the LOV2 domain of Adiantum phytochrome3 studied by low-temperature FTIR and UV-visible spectroscopy. *Biochemistry* 42:8183–91
72. Iwata T, Nozaki D, Tokutomi S, Kandori H. 2005. Comparative investigation of the LOV1 and LOV2 domains in Adiantum phytochrome3. *Biochemistry* 44:7427–34
73. Jarillo JA, Capel J, Tang RH, Yang HQ, Alonso JM, et al. 2001. An Arabidopsis circadian clock component interacts with both CRY1 and phyB. *Nature* 410:487–90
74. Jarillo JA, Gabrys H, Capel J, Alonso JM, Ecker JR, Cashmore AR. 2001. Phototropin-related NPL1 controls chloroplast relocation induced by blue light. *Nature* 410:952–54
75. Kagawa T, Kasahara M, Abe T, Yoshida S, Wada M. 2004. Function analysis of phototropin2 using fern mutants deficient in blue light-induced chloroplast avoidance movement. *Plant Cell Physiol.* 45:416–26
76. **Kagawa T, Sakai T, Suetsugu N, Oikawa K, Ishiguro S, et al. 2001. Arabidopsis NPL1: a phototropin homolog controlling the chloroplast high-light avoidance response. *Science* 291:2138–41**

76. Key paper employing an elegant genetic approach to isolate a second phototropin gene from *Arabidopsis* and demonstrating its involvement in chloroplast avoidance movement.

77. Kagawa T, Wada M. 2000. Blue light-induced chloroplast relocation in *Arabidopsis thaliana* as analyzed by microbeam irradiation. *Plant Cell Physiol.* 41:84–93
78. Kanegae H, Tahir M, Savazzini F, Yamamoto K, Yano M, et al. 2000. Rice NPH1 homologues, OsNPH1a and OsNPH1b, are differently photoregulated. *Plant Cell Physiol.* 41:415–23
79. Kasahara M, Kagawa T, Oikawa K, Suetsugu N, Miyao M, Wada M. 2002. Chloroplast avoidance movement reduces photodamage in plants. *Nature* 420:829–32
80. Kasahara M, Kagawa T, Sato Y, Kiyosue T, Wada M. 2004. Phototropins mediate blue and red light-induced chloroplast movements in *Physcomitrella patens*. *Plant Physiol.* 135:1388–97

81. Kasahara M, Swartz TE, Olney MA, Onodera A, Mochizuki N, et al. 2002. Photochemical properties of the flavin mononucleotide-binding domains of the phototropins from Arabidopsis, rice, and *Chlamydomonas reinhardtii*. *Plant Physiol.* 129:762–73
82. Kawai H, Kanegae T, Christensen S, Kiyosue T, Sato Y, et al. 2003. Responses of ferns to red light are mediated by an unconventional photoreceptor. *Nature* 421:287–90
83. Kennis JTM, Crosson S, Gauden M, van Stokkum IHM, Moffat K, van Grondelle R. 2003. Primary reactions of the LOV2 domain of phototropin, a plant blue-light photoreceptor. *Biochemistry* 42:3385–92
84. Kennis JTM, van Stokkum NHM, Crosson S, Gauden M, Moffat K, van Grondelle R. 2004. The LOV2 domain of phototropin: a reversible photochromic switch. *J. Am. Chem. Soc.* 126:4512–13
85. Kepinski S, Leyser O. 2005. The Arabidopsis F-box protein TIR1 is an auxin receptor. *Nature* 435:446–51
86. Kim C, Xuong NH, Taylor SS. 2005. Crystal structure of a complex between the catalytic and regulatory (RIalpha) subunits of PKA. *Science* 307:690–96
87. Kimura M, Kagawa T. 2006. Phototropin and light-signaling in phototropism. *Curr. Opin. Plant Biol.* 9:1–6
88. Kinoshita T, Doi M, Suetsugu N, Kagawa T, Wada M, Shimazaki K. 2001. Phot1 and phot2 mediate blue light regulation of stomatal opening. *Nature* 414:656–60
89. Kinoshita T, Emi T, Tominaga M, Sakamoto K, Shigenaga A, et al. 2003. Blue-light- and phosphorylation-dependent binding of a 14-3-3 protein to phototropins in stomatal guard cells of broad bean. *Plant Physiol.* 133:1453–63
90. Kinoshita T, Shimazaki K. 1999. Blue light activates the plasma membrane H(+)-ATPase by phosphorylation of the C-terminus in stomatal guard cells. *EMBO J.* 18:5548–58
91. Kinoshita T, Shimazaki K. 2001. Analysis of the phosphorylation level in guard-cell plasma membrane H+-ATPase in response to fusicoccin. *Plant Cell Physiol.* 42:424–32
92. Kinoshita T, Shimazaki K. 2002. Biochemical evidence for the requirement of 14-3-3 protein binding in activation of the guard-cell plasma membrane H+-ATPase by blue light. *Plant Cell Physiol.* 43:1359–65
93. Kiyosue T, Wada M. 2000. LKP1 (LOV kelch protein 1): a factor involved in the regulation of flowering time in arabidopsis. *Plant J.* 23:807–15
94. Knieb E, Salomon M, Rudiger W. 2005. Autophosphorylation, electrophoretic mobility and immunoreaction of oat phototropin 1 under UV and blue light. *Photochem. Photobiol.* 81:177–82
95. Kong SG, Suzuki T, Tamura K, Mochizuki N, Hara-Nishimura I, Nagatani A. 2006. Blue light-induced association of phototropin 2 with the Golgi apparatus. *Plant J.* 45:994–1005
96. Kottke T, Heberle J, Hehn D, Dick B, Hegemann P. 2003. Phot-LOV1: photocycle of a blue-light receptor domain from the green alga *Chlamydomonas reinhardtii*. *Biophys. J.* 84:1192–201
97. Kottke T, Hegemann P, Dick B, Heberle J. 2006. The photochemistry of the light-, oxygen-, and voltage-sensitive domains in the algal blue light receptor phot. *Biopolymers* 82:373–78
98. Lariguet P, Schepens I, Hodgson D, Pedmale UV, Trevisan M, et al. 2006. PHYTOCHROME KINASE SUBSTRATE 1 is a phototropin 1 binding protein required for phototropism. *Proc. Natl. Acad. Sci. USA* 103:10134–39
99. Li X, Zhang D, Hannink M, Beamer LJ. 2004. Crystal structure of the Kelch domain of human Keap1. *J. Biol. Chem.* 279:54750–58

100. Lin C, Shalitin D. 2003. Cryptochrome structure and signal transduction. *Annu. Rev. Plant Biol.* 54:469–96
101. Liscum E, Briggs WR. 1995. Mutations in the NPH1 locus of Arabidopsis disrupt the perception of phototropic stimuli. *Plant Cell* 7:473–85
102. Liscum E, Briggs WR. 1996. Mutations of Arabidopsis in potential transduction and response components of the phototropic signaling pathway. *Plant Physiol.* 112:291–96
103. Losi A. 2004. The bacterial counterparts of plant phototropins. *Photochem. Photobiol. Sci.* 3:566–74
104. Losi A, Quest B, Gartner W. 2003. Listening to the blue: the time-resolved thermodynamics of the bacterial blue-light receptor YtvA and its isolated LOV domain. *Photochem. Photobiol. Sci.* 2:759–66
105. Luesse DR, Deblasio SL, Hangarter RP. 2006. Plastid Movement Impaired 2, a new gene involved in normal blue-light-induced chloroplast movements in *Arabidopsis thaliana*. *Plant Physiol.* 141:1328–37
106. Mas P, Kim WY, Somers DE, Kay SA. 2003. Targeted degradation of TOC1 by ZTL modulates circadian function in *Arabidopsis thaliana*. *Nature* 426:567–70
107. Matsuoka D, Iwata T, Zikihara K, Kandori H, Tokutomi S. 2006. Primary processes during the light-signal transduction of phototropin. *Photochem. Photobiol.* In press
108. Matsuoka D, Tokutomi S. 2005. Blue light-regulated molecular switch of Ser/Thr kinase in phototropin. *Proc. Natl. Acad. Sci. USA* 102:13337–42

108. Conclusive biochemical study demonstrating that light activation of bacterially expressed phot2 kinase is regulated by LOV2 and not LOV1.

109. Motchoulski A, Liscum E. 1999. Arabidopsis NPH3: a NPH1 photoreceptor-interacting protein essential for phototropism. *Science* 286:961–64
110. Nakasako M, Iwata T, Matsuoka D, Tokutomi S. 2004. Light-induced structural changes of LOV domain-containing polypeptides from Arabidopsis phototropin 1 and 2 studied by small-angle X-ray scattering. *Biochemistry* 43:14881–90
111. Nakasako M, Matsuoka D, Zikihara K, Tokutomi S. 2005. Quaternary structure of LOV-domain containing polypeptide of Arabidopsis FKF1 protein. *FEBS Lett.* 579:1067–71
112. Nelson DC, Lasswell J, Rogg LE, Cohen MA, Bartel B. 2000. FKF1, a clock-controlled gene that regulates the transition to flowering in Arabidopsis. *Cell* 101:331–40
113. Noh B, Bandyopadhyay A, Peer WA, Spalding EP, Murphy AS. 2003. Enhanced gravi- and phototropism in plant mdr mutants mislocalizing the auxin efflux protein PIN1. *Nature* 423:999–1002
114. Nozaki D, Iwata T, Ishikawa T, Todo T, Tokutomi S, Kandori H. 2004. Role of Gln1029 in the photoactivation processes of the LOV2 domain in Adiantum phytochrome3. *Biochemistry* 43:8373–79
115. Nozue K, Kanegae T, Imaizumi T, Fukuda S, Okamoto H, et al. 1998. A phytochrome from the fern Adiantum with features of the putative photoreceptor NPH1. *Proc. Natl. Acad. Sci. USA* 95:15826–30
116. Ohgishi M, Saji K, Okada K, Sakai T. 2004. Functional analysis of each blue light receptor, cry1, cry2, phot1, and phot2, by using combinatorial multiple mutants in Arabidopsis. *Proc. Natl. Acad. Sci. USA* 101:2223–28
117. Oikawa K, Kasahara M, Kiyosue T, Kagawa T, Suetsugu N, et al. 2003. Chloroplast unusual positioning1 is essential for proper chloroplast positioning. *Plant Cell* 15:2805–15
118. Onodera A, Kong SG, Doi M, Shimazaki K, Christie J, et al. 2005. Phototropin from *Chlamydomonas reinhardtii* is functional in *Arabidopsis thaliana*. *Plant Cell Physiol.* 46:367–74
119. Palmer JM, Short TW, Briggs WR. 1993. Correlation of blue light-induced phosphorylation to phototropism in *Zea mays* L. *Plant Physiol.* 102:1219–25

120. Palmer JM, Short TW, Gallagher S, Briggs WR. 1993. Blue light-induced phosphorylation of a plasma membrane-associated protein in *Zea mays* L. *Plant Physiol.* 102:1211–18
121. Reymond P, Short TW, Briggs WR. 1992. Blue light activates a specific protein kinase in higher plants. *Plant Physiol.* 100:655–61
122. Rutter J, Michnoff CH, Harper SM, Gardner KH, McKnight SL. 2001. PAS kinase: an evolutionarily conserved PAS domain-regulated serine/threonine kinase. *Proc. Natl. Acad. Sci. USA* 98:8991–96
123. Sakai T, Kagawa T, Kasahara M, Swartz TE, Christie JM, et al. 2001. Arabidopsis nph1 and npl1: blue light receptors that mediate both phototropism and chloroplast relocation. *Proc. Natl. Acad. Sci. USA* 98:6969–74
124. Sakai T, Wada T, Ishiguro S, Okada K. 2000. RPT2. A signal transducer of the phototropic response in Arabidopsis. *Plant Cell* 12:225–36
125. Sakamoto K, Briggs WR. 2002. Cellular and subcellular localization of phototropin 1. *Plant Cell* 14:1723–35
126. Salomon M, Christie JM, Knieb E, Lempert U, Briggs WR. 2000. Photochemical and mutational analysis of the FMN-binding domains of the plant blue light receptor, phototropin. *Biochemistry* 39:9401–10
127. Salomon M, Knieb E, von Zeppelin T, Rudiger W. 2003. Mapping of low- and high-fluence autophosphorylation sites in phototropin 1. *Biochemistry* 42:4217–25
128. Salomon M, Lempert U, Rudiger W. 2004. Dimerization of the plant photoreceptor phototropin is probably mediated by the LOV1 domain. *FEBS Lett.* 572:8–10
129. Salomon M, Zacherl M, Luff L, Rudiger W. 1997. Exposure of oat seedlings to blue light results in amplified phosphorylation of the putative photoreceptor for phototropism and in higher sensitivity of the plants to phototropic stimulation. *Plant Physiol.* 115:493–500
130. Salomon M, Zacherl M, Rudiger W. 1996. Changes in blue-light-dependent protein phosphorylation during the early development of etiolated oat seedlings. *Planta* 199:336–42
131. Salomon M, Zacherl M, Rudiger W. 1997. Asymmetric, blue light-dependent phosphorylation of a 116-kilodalton plasma membrane protein can be correlated with the first- and second-positive phototropic curvature of oat coleoptiles. *Plant Physiol.* 115:485–91
132. Salomon M, Zacherl M, Rudiger W. 1997. Phototropism and protein phosphorylation in higher plants: Unilateral blue light irradiation generates a directional gradient of protein phosphorylation across the oat coleoptile. *Botanica Acta* 110:214–16
133. Schultz TF. 2005. The ZEITLUPE family of putative photoreceptors. In *Handbook of Photosensory Photoreceptors*, ed. WR Briggs, JL Spudich, pp. 337–47. Weinheim: Wiley-VCH
134. Schultz TF, Kiyosue T, Yanovsky M, Wada M, Kay SA. 2001. A role for LKP2 in the circadian clock of Arabidopsis. *Plant Cell* 13:2659–70
135. Schwerdtfeger C, Linden H. 2003. VIVID is a flavoprotein and serves as a fungal blue light photoreceptor for photoadaptation. *EMBO J.* 22:4846–55
136. Short TW, Briggs WR. 1990. Characterization of a rapid, blue light-mediated change in detectable phosphorylation of a plasma membrane protein from Etiolated Pea (*Pisum sativum* L.) seedlings. *Plant Physiol.* 92:179–85
137. Short TW, Porst M, Palmer J, Fernbach E, Briggs WR. 1994. Blue light induces phosphorylation at seryl residues on a Pea (*Pisum sativum* L.) plasma membrane protein. *Plant Physiol.* 104:1317–24
138. Short TW, Reymond P, Briggs WR. 1993. A Pea plasma membrane protein exhibiting blue light-induced phosphorylation retains photosensitivity following triton solubilization. *Plant Physiol.* 101:647–55

126. First study to show that bacterially expressed LOV domains undergo a unique mode of photochemical reactivity.

127. Insightful study mapping the sites of phototropin autophosphorylation and confirming the earlier proposal by Briggs that phosphorylation occurs in a fluence-dependent manner.

Nutrient Sensing and Signaling: NPKS

Daniel P. Schachtman and Ryoung Shin

Donald Danforth Plant Science Center, St. Louis, Missouri 63132;
email: dschachtman@danforthcenter.org

Annu. Rev. Plant Biol. 2007. 58:47–69

The *Annual Review of Plant Biology* is online at plant.annualreviews.org

This article's doi:
10.1146/annurev.arplant.58.032806.103750

First published online as a Review in Advance on October 26, 2006

1543-5008/07/0602-0047$20.00

Key Words

phosphorus, nitrogen, potassium, sulfur, nutrient deprivation

Abstract

Plants often grow in soils that contain very low concentrations of the macronutrients nitrogen, phosphorus, potassium, and sulfur. To adapt and grow in nutrient-deprived environments plants must sense changes in external and internal mineral nutrient concentrations and adjust growth to match resource availability. The sensing and signal transduction networks that control plant responses to nutrient deprivation are not well characterized for nitrogen, potassium, and sulfur deprivation. One branch of the signal transduction cascade related to phosphorus-deprivation response has been defined through the identification of a transcription factor that is regulated by sumoylation. Two different microRNAs play roles in regulating gene expression under phosphorus and sulfur deprivation. Reactive oxygen species increase rapidly after mineral nutrient deprivation and may be one upstream mediator of nutrient signaling. A number of molecular analyses suggest that both short-term and longer-term responses will be important in understanding the progression of signaling events when the external, then internal, supplies of nutrients become depleted.

Contents

INTRODUCTION

Plants are a diverse group of organisms in part because of the many different environments in which they grow and to which they are adapted. Environmental change occurs due to location, seasonal variation, and daily, weekly, monthly, and yearly changes in climate. Plant diversity in genetic architecture, gene expression, and physiological adaptations is very large (31). Plants are anchored in one place for most of their life cycle and therefore must be able to adapt to a range of abiotic and biotic stresses. Variation in soil nutrient composition is found in agriculture systems and in natural environments and ranges from an extreme lack of nutrients in soils (e.g., 43), to soils containing optimal amounts of fertilizer, and to soils with excess nutrients.

To provide a framework for understanding plant adaptation to the range of nutrients found in soils, one may consider two different modes of plant growth and metabolism. One mode is when plants grow under nutrient-sufficient conditions where mineral resources are not limiting. Growth proceeds at optimal rates and it may be relatively easy for plants to acquire nutrients and water from soils where sunlight is abundant. When nutrients become limiting, growth is reduced and plants alter certain aspects of their acquisition, utilization, and morphology to maximize and acquire scarce resources. Plant species respond differently to nutrient limitations based on the environment in which they evolved (40). Responses to nutrient limitations also depend on the nutrient, such as whether it is mobile or immobile in the soil (48). Crop plants have been bred over the past 30–40 years under high nutrient conditions and have high intrinsic growth rates and yield. Such characteristics may not be the most suitable for lower input environments. Plant response to the deprivation of essential mineral nutrients is fairly well described, but an understanding of how plants sense and signal changes (at the cellular or whole-plant level) in the availability of nutrients is lacking.

As fertilizer becomes more expensive and as farmers reduce fertilizer usage because of the negative environmental impacts, it will be important to gain a better understanding of how crop plants can be designed to grow more efficiently in environments with lower nutrient inputs (39). In addition, access to fertilizers in developing countries is limited (116), so it could be helpful to create higher-yielding crop plants that are adapted to a lower input environment.

In this review we mainly discuss the events that lead to plant responses to nutrient-deficient conditions. First, we discuss general features of deficiency to the macronutrients. Second, we cover what is known about sensing and signaling phosphorus, nitrogen, potassium, and sulfur deficiency. This field of plant research is still in its infancy and has mainly focused on the downstream responses to nutrient deprivation. As such there is still scope for many fundamental discoveries that will be important to basic and more applied aspects of plant biology.

TIMING

One of the many open questions in nutrient signaling is the progression of events and the importance of early versus later signals. Plants respond to nutrients by altering their physiology and morphology (48). Responses vary over time and also with the nutrient deficiency encountered. Compared with pathogen response where important signaling events occur very rapidly after infection (67), important signaling events are likely to occur both very soon (less than six hours) and much later (days or weeks) after nutrient deprivation.

When plants are deprived of nutrients, the root is usually the primary site of perception. Early signals in response to nutrient deprivation may occur when external concentrations of nutrients are reduced and the later signals may be initiated when internal stores of nutrients are reduced below a critical level. At the earliest time points after deprivation, changes in membrane potential may be a dominant effect. During potassium deprivation root cells become hyperpolarized (128), whereas with phosphate- (22), sulfate- and nitrate-deprivation cells become depolarized. It is not yet clear how plant cells decode these changes in membrane potential. Under natural conditions nutrients become depleted slowly in soils due to the activity of roots over time or a lack of soil moisture. Soils may also lack a certain mineral nutrient, roots may encounter a patch of soil that is enriched in nutrients, and other variations in soil fertility may occur. In the laboratory the imposition of nutrient stress can be rapid. This occurs when plants are transferred to conditions lacking or very low in nutrients. As soils become depleted of nutrients or under laboratory conditions, plants will not immediately become deficient because of their ability to remobilize potassium from the vacuole or phosphorus, nitrogen, or sulfur from molecules containing those minerals or the vacuole.

Some clues to how plants respond to nutrient deprivation over time come from several gene expression studies that have used short and longer time courses. Some of the earliest responses to nutrient deprivation come from a study on phosphorus, potassium, and iron starvation of roots (150). After one hour of deprivation the expression of a number of genes changed (150). In that study, genes encoding mitogen activated protein (MAP) kinases, a transcription factor, and a 14-3-3 protein were upregulated, but their role in signaling or adaptation to low nutrients has not yet been determined. In another study, *Arabidopsis thaliana* plants were deprived of phosphorus for a short time (4 h) and a longer time (28 and 100 h) and changes in gene expression were studied using an 8K GeneChip™ (42). The conclusions from that study were that the short-term transcriptional responses in leaves appeared to be more related to general stresses than specifically to nutrient deprivation. Long-term responses were considered to be more specific for phosphorus deprivation. In that study, shoot phosphorus concentrations were significantly lower in the deprived plants between 24 and 72 h after deprivation (42). In published microarrays characterizing sulfur deprivation, the earliest time point studied is 24 h after deprivation, but it is known that sulfate transporters are induced after 6 h of deprivation (156). Based on the sulfate transporter gene induction data, it is likely that important early signaling factors have yet to be identified.

Plant roots in particular will rapidly sense changes in nutrient composition, which will lead to changes in gene expression in 60 min or less. Changes in the tissue concentrations of nutrients do not appear until later. Therefore, a decrease in the tissue concentration of a nutrient acts as a marker for the later events that occur in response to nutrient deprivation. Little is known about the consequences of short- or long-term sensing or signaling in response to nutrient deprivation, so definitive conclusions cannot be drawn regarding the relative importance of short-term and long-term responses.

CROSS TALK AND SPECIFICITY OF RESPONSE

The interaction between nutrients for uptake and the imbalances caused by the deficiency of one mineral are well-described phenomena (84). For example, sulfur and nitrogen are tightly interrelated on the level of plant metabolism (44). In some plant species a decrease in sulfate disrupts nitrogen metabolism, resulting in high levels of nitrate in leaves (109). Ammonium is known to interact with potassium for uptake (118), and that interaction may be genetically determined (11).

ROS: reactive oxygen species

Evidence for interactions are also found in numerous molecular studies. In one study on potassium-deprivation, nitrate transporters were downregulated (4). In another similar study, ammonium transporters were upregulated by potassium deprivation (82). Although the interaction between similar-sized cations ammonium and potassium is expected, the link between nitrate and potassium is less obvious. However, during long-term deprivation of potassium changes in nitrogen assimilation may cause increased glutamine and decreased glutamate concentrations (3). Short-term (6 h) potassium deprivation in corn leads to the rapid upregulation of an *NRT2* homolog in corn (D. Schachtman & K. Huppert, unpublished), suggesting that factors other than changes in nitrogen metabolism may act on nitrate-sensing mechanisms.

Cross talk between potassium and phosphorus deprivation has been demonstrated in tomato. Genes encoding a MAP kinase, transcription factors, and nutrient transporters were all induced by potassium and phosphorus deprivation (150). Analysis of phosphorus-deprived plants also showed that the upregulation of sulfur transporters and iron transporters occurred over the long term (93). The increased sulfur uptake may be required for sulfolipid synthesis, which to a certain extent replaces phospholipids. A decrease in iron transporter expression was linked to the increase in iron content in plants over time (93). Although one study on shoot transcriptional responses to nitrogen, phosphorus, and potassium noted very few overlapping changes in genes expression (42), a 2.5-fold cutoff was used to select up- and downregulated genes; therefore, important overlapping changes may have been overlooked. Another study showed that the expression of the gene encoding the response regulator AAR6 was upregulated by nitrogen, phosphorus, or potassium deprivation (17). Thus, shared nutrient signaling transduction pathways can be expected. Evidence also suggests that some genes are more specifically regulated in response to the deprivation of a particular nutrient (42, 125).

Although most studies on reactive oxygen species (ROS) have focused on leaves, some work has shown that nonphotosynthetic tissues such as roots or tubers also undergo an oxidative burst due to pathogen challenge (141) or deprivation of nutrients (125). The production of ROS in roots is a common feature in response to nitrogen, phosphorus, and potassium deprivation (125, 126). In this review we present data showing that ROS is also produced when plants are deprived of sulfur (**Figure 1**). Although this response occurs due to the deprivation of several macronutrients, it appears that there are some differences in localization as well as differences in the molecules that produce the ROS (125). One specific NADPH oxidase, AtrbohC, appears to be the main source of ROS produced in response to potassium deprivation (125). When this gene is inactive, ROS is still

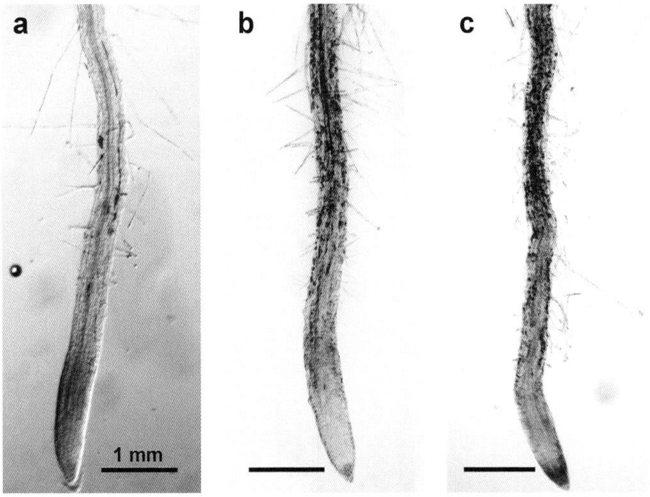

Figure 1

Reactive oxygen species increases in *Arabidopsis* roots after 24 h of nutrient deficiency. Red fluorescence overlays of brightfield root images are shown for (*a*) nutrient-sufficient, (*b*) sulfur-deficient, and (*c*) potassium-deficient *Arabidopsis* wild-type, Col-0 roots after incubation with 50 μM DFFDA.

produced under conditions of nitrogen and phosphorus deprivation.

The amount of ROS visualized after 30 h of sulfur deprivation in roots was greater than in sulfate-sufficient roots (**Figure 1**) and the localization was similar to potassium- or nitrogen-deprived roots (125, 126). The involvement of ROS in sulfur signaling may be more complex than that of potassium deprivation because the ascorbate-glutathione cycle that is downstream of sulfate assimilation is involved in the removal of H_2O_2. Changes in the reduced glutathione (GSH) pool depend on the supply of sulfate and correlate inversely with the activity of key components of the sulfate assimilation pathway (68). GSH pools become depleted under sulfur starvation; therefore, the reduction of dehyroascorbate to ascorbate is limited by decreased GSH availability (57). In contrast, GSH is not decreased under nitrogen and phosphorus starvation and ascorbate content is increased (57). Direct evidence showing that ROS is a signaling component in sulfur-starved plants is not yet available. However, in *Bacillus subtilis* several genes involved in sulfur assimilation and synthesis of sulfur-containing amino acids were induced by adding paraquat or by exposure to O^{2-} (98). The regulation and interaction between ROS and GSH ascorbate impact the synthesis of plant hormones such as salicylic acid, gibberellins, abscisic acid, and ethylene (18, 107), which may signal plant response to nutrient deficiency. Although ROS plays a role in nutrient signaling in roots (125, 126), the differences in GSH and ascorbate production may provide some specificity to each nutrient deficiency.

The consequences of ROS production in roots are not well understood. Calcium may be part of this signal transduction cascade acting both upstream (141) and downstream (97) of the ROS production. Although calcium is not required for ROS production by NADPH oxidases in plants, the activity of these proteins is modulated by calcium (113), presumably through binding to putative calcium-binding sites (59). One consequence of ROS production may be changes in gene expression (125, 142). ROS production may also directly alter the function of proteins through modification of thiols as well as lead to alterations in the redox state of the cell, which would lead to downstream changes in proteins that sense the redox state of the cell. In the longer term, ROS has important consequences that lead to aerenchyma formation in nutrient-deprived roots (60), which may be an important adaptation that lowers the cost of maintaining roots (23).

GSH: reduced glutathione

PLD: phospholipase D

PHOSPHORUS

Phosphorus deprivation is probably the most widely studied macronutrient deficiency. Studies on plant response to phosphorus deprivation are ecologically and agriculturally important because in many parts of the world soils are low in available phosphorus due to numerous factors including immobility, low solubility, or lack of the mineral (119). Several recent reviews have focused on aspects of phosphate sensing and transcriptional changes in response to phosphate deprivation (1, 28, 138). Another area of major interest is the change in root development in response to phosphorus deprivation (79), which is a mechanism for increasing phosphorus acquisition. Upstream regulators of changes in root morphology or architecture involve the hormones auxin and ethylene (75, 80, 81, 101, 152). One important factor involved in multiple signal transduction pathways in plants is phospholipase D (PLD) (148). Studies of mutants revealed that PLD is involved in regulating root development under phosphate deprivation (74).

Several studies using microarrays describe the transcriptional responses that occur when plants are deprived of phosphate (42, 93, 147, 150). These microarray experiments confirm what was previously known about plant response to phosphate deficiency and extend the number of known phosphate responsive genes. When plants are deprived of phosphate, scavenging systems are activated to recover phosphate from lipids and nucleic acids. As a result, lipid composition shifts from phospholipids to more galacto- and sulfolipids, ribonucleases recover phosphate from nucleic acids, acid phosphatase activity increases, and phosphate transporter genes are induced (138). These are well-known downstream responses to phosphorus deprivation. Less is known about phosphate sensing and signaling.

Pi: inorganic phosphate

Promoter analysis of genes that are responsive to phosphate deficiency (93) showed the presence of phosphate starvation response 1 (PHR1) binding sites. PHR1 is a Myb transcription factor involved in certain aspects of response to phosphorus deprivation (112) (**Figure 2**). PHR1 contains a single MYB domain and a predicted coiled-coil domain that may be involved in dimerization that binds to an imperfect palindromic sequence (P1BS element: GNATATNC) in certain promoters. PHR1 was identified through the mutagenesis of lines containing a promoter (*AtIPS1*) driving the expression of β-glucuronidase (GUS) that is upregulated by phosphate deprivation. The expression of *PHR1* itself is not very responsive to phosphate deprivation and therefore it is thought to be downstream in the phosphate signal transduction pathway. PHR1 also plays a role in phosphorus homeostasis under nutrient-sufficient conditions (**Figure 2**). This conclusion is based on the finding that *phr1-1* plants have lower phosphorus content than wild types under nutrient-sufficient conditions (112). The promoter element to which PHR1 binds, an 8-nucleotide imperfect palindromic sequence, is significantly enriched in many genes expressed under inorganic phosphate (Pi)-deprived conditions (9, 112). Confirmation of the importance of this promoter region for gene induction due to phosphate starvation was provided by promoter deletion studies of an inducible phosphate transporter from barley (122). Although many phosphate starvation-induced genes contain the GNATATNC motif (28), the expression of some of these genes is attenuated and not

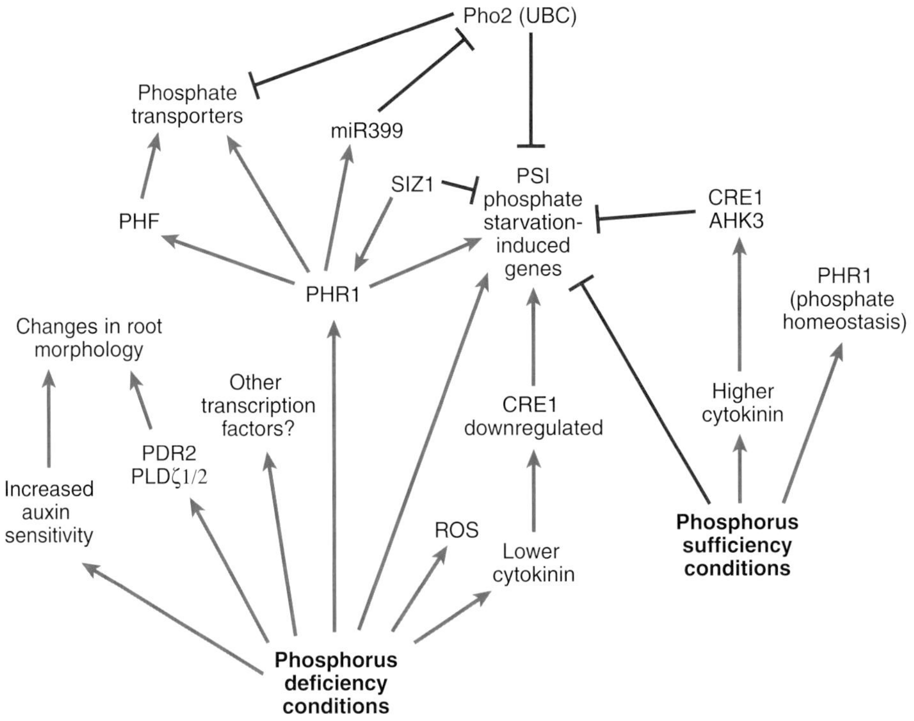

Figure 2
Components involved in plant responses and signaling under phosphorus-deficient and -sufficient conditions as described in the text.

completely abolished in the *phr1* knockout. Therefore, it is likely that other transcription factors may act together with PHR1 in a combinatorial fashion (136), or that other redundant Myb transcription factors are also involved (112).

PHR1 belongs to a larger gene family that includes 11 homologs in *Arabidopsis*. A second member of this family is rapidly induced by phosphorus deprivation (140). Recent studies using microarray technology and qPCR identified additional genes that are regulated by PHR1 (9). These genes include transcription factors, known phosphate starvation-inducible genes, and genes encoding proteins involved in lipid metabolism. Although these genes may not be directly regulated by PHR1, they do provide an important entry point for further characterization of one branch of the signal transduction pathway in response to phosphorus deprivation (**Figure 2**).

One intermediate link in the signal transduction cascade from *PHR1* to the function of high-affinity phosphate transporters is the phosphate transporter traffic facilitator (*PHF1*) (38). This gene encodes a protein involved in the exiting of phosphate transporters from the endoplasmic reticulum to eventually function in the plasma membrane (38). When plants are deprived of phosphate the genes encoding several different phosphate transport proteins are upregulated. PHF1 is involved in the localization of these transporters to the plasma membrane (38). This facilitator protein may be specific for phosphate transporters because it is required for localization of one phosphate transporter, but not an aquaporin (38). The gene encoding the facilitator protein contains the PHR binding motif, and the upregulation of *PHF1* is attenuated, but not abolished, in the *phr1* mutant, again suggesting that other transcription factors are involved in the expression of these phosphate responsive genes, which may provide complexity in their regulation.

Although *PHR1* is not regulated at the transcriptional level, the function of this factor is regulated by sumoylation. SIZ1 functions as a small ubiquitin-like modifier (SUMO) E3 ligase (94) that contributes to the transfer of a peptide to a substrate. The conjugation of a SUMO peptide influences protein function by a number of different mechanisms (100). Under phosphate-starved conditions, SIZ1 plays a transient role in the activation of PHR1 (**Figure 2**), which alters the expression of some genes that are activated by this transcription factor (94). SIZ1 also acts as a repressor of phosphate-deficiency responses because the *siz1* knockout has increased lateral root and root hair growth, increased root-to-shoot ratios, and greater anthocyanin accumulation in response to phosphate deficiency. The hypothesis that phosphate starvation-inducible genes are repressed was previously proposed (99). *siz1* knockout plants exhibit a defect in the signal transduction pathway of phosphate sensing because these responses are observed even though internal phosphate concentrations are maintained at wild-type levels.

SUMO: small ubiquitin related modifier

miR: microRNA

Another posttranscriptional mechanism for regulating plant response to phosphate deprivation is a microRNA (miR) miR399 that targets a putative ubiquitin-conjugating enzyme (UBC) whose expression is reduced under deprived conditions (33) (**Figure 2**). Overexpression of miR399 leads to increased phosphate accumulation under nutrient-sufficient conditions. The increased accumulation appears to be due to a defect in the remobilization of phosphate in leaves (16) similar to the phenotype of the phosphate over accumulator mutant *pho2* (20). Studies show that *pho2* (20) contains a mutation in the gene encoding the UBC targeted by miR399, which suggests that the ubiquitination of specific targets is necessary for the repression of certain phosphorus-deficiency responses (6, 9).

To understand and elucidate phosphate signaling pathways in plants, it is important to consider both local and systemic responses to low phosphate. Mineral deficiencies usually first impact the roots and then transmit signals either to leaves [as with chemical signaling under drought (151)] or the shoots themselves become nutrient deficient. The *Arabidopsis pdr2* mutant provides one example of local phosphate-sensing responses in plants (139). The *pdr2* mutant alters response to low phosphate in a root-specific manner and does not alter shoot growth. This mutant grows normally under phosphorus-sufficient conditions but is unable to maintain root meristem activity under Pi-deprived conditions even though internal concentrations of phosphate were similar to the wild type. The *pdr2* mutant also supports the concept that plants have different programs for growth in sufficient and deficient conditions.

The allocation of phosphate between roots and shoots is an important systemic response to phosphate limitation. In split-root experiments where the shoot and half the roots maintain sufficient concentrations of Pi, the deprived portion of the roots do not exhibit phosphate starvation responses (8, 12). The sufficient phosphate levels in shoots can suppress some of the deprivation responses in roots. One gene involved in the root-to-shoot allocation of phosphate is At4 (85), which is induced by phosphate deprivation and expressed primarily in roots. The function of the gene product is unclear. It may produce small peptides or exert control through the RNA itself (124).

In phosphorus deprivation, it appears that the inorganic ion itself is sensed by plants. Evidence for this comes from experiments such as those on *pdr2* and also from experiments with phosphite. Phosphite is an analog of inorganic phosphate (13, 14) that suppresses the phosphate starvation responses such as increased anthocyanin production and the activation of phosphatases, ribonucleases, and root hair growth. Suppression of the starvation responses occurs in the presence of phosphite even though plant growth is greatly reduced; therefore, it appears that phosphite-grown plants do not sense the phosphate deprivation.

Some systemic and local responses to phosphate deprivation may be controlled by the phytohormone cytokinin. Cytokinin levels decrease when plants are starved of Pi (49) and the reduced flow of cytokinin to leaves likely plays an important role in the resource allocation shift in favor of more root growth. However, it is not known how the low Pi signal is translated into the downregulation of cytokinin synthesis. At the local level in the roots certain genes that are normally induced by Pi deprivation are repressed by exogenous application of cytokinin (85), whereas other responses such as the stimulation of root hair growth are still enhanced even in the presence of this hormone. This highlights the multiple signals and signaling pathways that control plant responses to phosphate deprivation. Using the promoter of a gene induced by phosphate deprivation (*AtIPS1*), changes in repression due to cytokinin were identified using a mutational screening strategy (30). The mutants identified were those that showed AtIPS1::GUS expression in the presence of kinetin. A mutation in *CRE1* was responsible for determining the degree of cytokinin sensitivity. CRE1 is a cytokinin receptor (51) that is involved in repression of phosphate starvation-induced genes (**Figure 2**), providing further evidence for the hypothesis (99) that starvation responses are repressed under nutrient-sufficient conditions. A second cytokinin receptor was also identified as being involved in Pi sensing, demonstrating the redundant roles that these receptors may play in the signal transduction pathway (29).

NITROGEN

Nitrogen and carbon metabolism are linked and therefore cross talk between the signal transduction pathways that regulate nitrogen assimilation and carbon metabolism is expected. Systems analysis provides new insights that should help to further elucidate these complex networks (106, 129). To date, much work on nitrogen sensing has focused on the metabolic and morphological responses to the addition of nitrate, in particular the induction of nitrogen metabolism (121, 129, 146, 147,

149) and long-distance signaling and transporter responses (26). The root responses to nitrate are well documented, and even though the transporter NRT2.1 was reported to be involved in nitrate sensing or transduction (77) in roots, a more recent report suggests an alternative role for NRT2.1 (111). Plant responses and signal transduction pathways for nitrogen limitation are poorly understood because experiments that focus on the switch from nitrogen-sufficient to -deficient conditions have received less attention. However, the importance of nitrogen limitation cannot be overstated as nitrogen-use efficiency is a critical future goal for agriculture because of the need for enhanced stewardship of the environment and for more efficient use of an increasingly expensive input (34).

PII is potentially a key protein involved in the sensing of internal nitrogen supply. PII is a homolog of the well-characterized *Escherichia coli* protein involved in the nitrogen regulatory system (50, 96). The protein PII in *E. coli* is encoded by the *GlnB* gene, which interacts with other proteins to regulate glutamine synthase (GS). In plants, much of carbon and nitrogen metabolism occurs in the chloroplast where PII is localized (50). The plant PII homolog interacts with N-acetyl glutamate kinase, which is a key enzyme specifically involved in arginine biosynthesis (15, 24) and in nitrogen metabolism in general.

Nitrate may be a key molecule that is sensed by plants (as with phosphate) and is involved in controlling the ratio of roots to shoots (120). A highly significant correlation was found between leaf nitrate and shoot:root ratios (120). Another important factor in root-to-shoot allocation that is influenced by nitrogen is the hormone cytokinin (134). Cytokinin levels increase when plants are supplied with nitrogen and decrease when deprived. Cytokinins play a potentially important role in root-to-shoot communication as they are synthesized in roots and are well correlated with observed changes in biomass allocation between roots and shoots (114). Possible response regulators downstream of a cytokinin receptor have been identified, supporting a role for cytokinin in nitrogen responses (115, 135).

GS: glutamine synthase

Under nutrient-deprived conditions, phosphorylation also plays a key role in fine-tuning plant responses. In wheat, a gene encoding an *SNF1* kinase is upregulated by nitrogen, phosphorus, and sulfur deficiency (117). This kinase is also induced by cytokinin, even though cytokinin levels would be expected to decline under nitrogen deficiency. Another phosphorylation event and potential signaling cascade regulated by nitrogen deprivation is the phosphorylation and conversion of the nitrate transporter CHL1 to a high-affinity mode (78). The kinase that phosphorylates CHL1 has not been identified but will be an important component for understanding the signal transduction cascade in response to nitrogen deprivation. Phosphylation and dephosphorylation are also known to activate and inactivate nitrate reductase (56). More work is needed to identify the networks of phosphorylated proteins and their downstream targets involved in the nitrogen signal transduction cascade.

In addition to sensing nitrate, plant cells may also sense carbon status, which leads to the regulation of key nitrate transporters *NRT2.1* and *NRT1* (69). A key metabolite involved in nitrogen metabolism is glutamate. The presence of multiple glutamate receptors in plant genomes (66) may indicate the potential importance of this metabolite in initiating signal transduction cascades. However, the role of glutamate receptors or glutamate is only beginning to be elucidated. Antisense lines of *AtGLR1.1* displayed conditional phenotypes in response to changes in carbon to nitrogen ratios. A model was presented that suggests that this receptor regulates abscisic acid (ABA) biosynthesis during seed germination, which is stimulated by nitrate and inhibited by sucrose (58). Glutamate also leads to changes in cation fluxes (21) as well as changes in root growth (25, 144). At this time these putative receptors cannot be placed in signal transduction cascades. Glutamine is another key metabolite linked to nitrogen metabolism that plays roles in metabolic regulation and possibly signal transduction (130).

Some indications for possible signal transduction networks can be found in studies on global transcriptional changes after two days of nitrogen deprivation and MADS-box transcription factors after 2.5 days of nitrogen deprivation (35, 121). Global transcriptional changes to nitrogen deficiency were mainly described in terms of changes that occurred after reintroduction of nitrate (121), and therefore discussion of this rich data set on the transcriptional changes in response to nitrogen deprivation is beyond the scope of this review. However, extensive changes were found after two days of nitrogen deprivation, including a large number of transcription factors and kinases (see the supplemental table in Reference 121). A previously identified MADS-box transcription factor, *ANR1*, which was identified as being induced by nitrate, is actually upregulated by nitrogen deprivation and not rapidly repressed by nitrate (after 3 h of nitrate) (35). Seven other MADS-box transcription factors are also upregulated by nitrogen deprivation, but to a lesser extent (35).

An Myb transcription factor that is a member of the *PHR* family of transcription factors in plants (112) was found to be upregulated by nitrogen deprivation (140). These Myb factors likely play important roles in the signaling of nitrogen deprivation. Other Mybs in the R2R3 family have also been shown to be responsive to nitrogen deprivation (95), but their targets or physiological roles are unknown. Gene expression data provide potential components of signal transduction pathways, but additional work will be required to link these components into networks.

POTASSIUM

Potassium is a macronutrient required in large quantities by plants and is the most abundant cation in plant cells. One difference between potassium and the other macronutrients discussed in this review is that potassium is not metabolized or incorporated into other macromolecules. It does, however, play a role in the activation of enzymes. This should simplify how plants respond to low potassium because the scavenging responses are limited to remobilization from intracellular compartments such as the vacuole or from older tissues. As with nitrogen and sulfur response to nutrient deprivation, very little is known about signaling pathways and the sensors of potassium deprivation in plants (3). In bacteria, specific sensors for potassium have been identified (131). In *E. coli*, phosphorylation plays a central role in the sensing of potassium deficiency. KdpD is a sensor kinase that undergoes autophosphorylation and transfers a phosphoryl group to a response regulator KdpE (145). This response regulator controls the expression of an operon coding for a high-affinity potassium uptake system in *E. coli*. The sensor kinase transduces changes in turgor caused by low potassium (55).

Few known potassium transporters are upregulated by potassium deprivation (5, 82), but after 6 h of deprivation *Arabidopsis* plants exhibit a shift to high-affinity potassium uptake (126). This shows that plants rapidly sense the changes in external potassium. At least two transporters are upregulated by potassium deprivation, including a high-affinity potassium uptake transporter (HAK5) (2, 36, 126) and KEA5 (126), which may be involved in remobilization of potassium from the vacuole. In *Arabidopsis*, the high-affinity potassium transporter that is induced by potassium deficiency (2, 36) is also regulated by ROS. Therefore, upstream signal transduction components for plant response to potassium deprivation include ROS (126).

A weak transcriptional response to potassium deprivation, even after 96 h, was observed in a microarray study (36) and this may suggest that posttranslational mechanisms contribute to the signal transduction networks involved in the response to potassium deprivation. After two weeks of growth on potassium-free medium, *Arabidopsis* plants exhibited a stronger transcriptional response with about 600 genes potentially upregulated and downregulated (4). After two weeks of deprivation plants developed visual symptoms due to reduced tissue concentrations of potassium. Upon changes in tissue potassium concentrations, it is likely that the activity of certain enzymes requiring potassium may be reduced. Pyruvate kinase is one such enzyme that has been suggested to potentially play a key role as an internal potassium sensor (3).

The importance of posttranscriptional regulation was recently demonstrated through the isolation of the gene encoding an SnRK3 kinase (*CIPK23*) that interacts with two calcineurin B-like proteins (CBL1 and CBL9) to regulate a potassium channel by phosphorylation and increase potassium uptake under both controlled and deprived conditions (73, 155). The upstream factors that trigger the phosphorylation of this potassium channel are unknown. However, this new finding implicates calcium in the signal transduction pathway in response to potassium deprivation because the CBLs contain an EF hand for calcium binding (73, 155). Further support for the role of calcium comes from experiments on potassium-deprived roots that showed a number of genes encoding "calcium-related" proteins, including a calcium-dependent protein kinase, a calcium ATPase, and a calmodulin binding protein, were up- and downregulated 2 and 6 h after potassium deprivation (R. Shin and D.P. Schachtman, unpublished). In yeast, many nutrient-regulated kinases have been identified (153). It is likely that more phosphorylation-related events will be found to be involved as part of the signal transduction pathway in response to nutrient deprivation.

ROS are a component of the signal transduction pathway in roots in response to low potassium (125, 126). ROS are necessary for root hair growth (27) and are involved in root elongation (76) and gene expression under potassium-deprived conditions (125). Using an NADPH oxidase mutant, certain genes involved in response to potassium deprivation were shown to be dependent on ROS production whereas other potassium responsive genes were independent of ROS (126). The upstream factors involved in the initiation of ROS production are unknown, but calcium is one candidate.

JA: jasmonic acid

The hormonal responses to potassium deprivation include ethylene, jasmonic acid (JA), and auxin. After 6–30 h of potassium deprivation, the expression of genes encoding ethylene biosynthetic enzymes and ethylene production in potassium-deprived roots increased (125). This hormonal response is presumably downstream from unknown factors in the signal transduction network. The consequences of increased levels of ethylene are unknown. Although auxin may play a role in controlling the expression of potassium channels (108), it is not yet clear whether auxin levels change in potassium-deprived *Arabidopsis*. However, the DR5-GUS reporter gene expression, a marker for auxin localization, changes in potassium-deprived roots (143). Changes in auxin localization, concentrations, or sensitivity could also lead to the reduced lateral root growth observed over a longer time course of potassium deprivation (4, 126). Long-term potassium starvation resulted in the conspicuous upregulation of genes linked to JA and defense. This occurred after plants showed visual symptoms of potassium deprivation, which may link this late response to other general stress signaling pathways (4).

SULFUR

Sulfur metabolism has been extensively studied from many angles (61, 71). This section addresses how sulfur metabolism is regulated under sulfur deficiency and what is known about the upstream signal components of sulfate assimilation. Sulfate is taken up and then assimilated to cysteine and reduced in the

IAA: indole-3-acetic acid

chloroplast (72). There are five functional subgroups of sulfate transporters in *Arabidopsis* (127). As with other mineral nutrient transporters, the expression of these high-affinity sulfate transporters is upregulated by sulfate deficiency and they facilitate uptake of sulfate from soil when sulfate is less available (86, 90, 132, 133). Recently, an ethylene insensitive-like (EIL) transcription factor, *SLIM1*, was isolated and shown to be involved in the regulation of a high-affinity sulfate transporter in response to sulfate limitation (87).

One regulatory component of sulfate assimilation is a miRNA. MicroRNAs play a regulatory role during development and in response to environmental stress (83) and were recently implicated in response to nutrient deprivation. The miR395 regulates specific targets under sulfur-deprived conditions. The targets of miR395 are ATP sulfurylase (*APS*) [*APS1* (at3g22890), *APS3* (at4g144680), and *APS4* (at5g43780)], which is the first step in sulfate assimilation. The abundance of *APS* transcripts decreases when the miR395 increases under sulfur-deprived conditions (53). The regulation of the gene encoding ATP sulfurylase may also be upregulated by sulfur deprivation (45, 110). Studies show the upregulation and downregulation of ATP sulfurylase by sulfur deprivation. One factor that may explain the differing results with ATP sulfurylase expression may be that studies with miR395 were performed on plants that had been starved of sulfur for two weeks as compared with shorter time courses in other studies. This further highlights how important the consideration of timing is in the responses to nutrient deprivation.

When plants are starved for sulfur, they activate mechanisms for increasing acquisition from soil. However, when plants cannot acquire enough sulfate, the decreased sulfate uptake leads to reduced assimilation activity (45, 46, 132, 133) and affects many different metabolic processes. Eventually, the limited supplies of sulfur in plants result in decreased plant tissue sulfur content (10, 37, 65, 70, 92, 102, 109). Decreases in sulfur content result in the inhibition of sulfate assimilation; decreased glutathione and cysteine; increased amounts of serine, O-acetylserine, and tryptophan (102); reduced amounts of chlorophyll, RNA, and total protein; increased photorespiration; decreased lipids; and nitrogen imbalance. Overall, these changes lead to a reduced rate of metabolism and growth (103).

Despite the fact that the steps and regulation of sulfate metabolism have been well characterized, the upstream factors that trigger changes in sulfate assimilation and the regulatory components involved in signaling sulfur deficiency have only been partially elucidated. Recent microarray experiments provide some clues as to the possible signaling component; however, data analysis has mainly focused on genes involved in sulfate assimilation (45–47, 54, 86, 102) and the earliest time points considered start 24 h after deprivation. Therefore early signals have not been characterized.

The plant hormones cytokinin, auxin, and JA are signaling components in response to sulfur deficiency. The expression of *APR1* (APS reductase 1) is upregulated by sulfur deficiency (123) and also by exogenous cytokinin (104). Exogenous cytokinin downregulates the expression of the high-affinity transporter SULTR1;2 (90), which is upregulated by sulfur deprivation. Cytokinin acts through the cytokinin response receptor (CRE1) to regulate sulfate uptake and transporter expression. In the *cre1-1* mutant, application of cytokinin only partly reduces sulfate uptake, suggesting redundancy as noted for the case of phosphate deprivation (91). Auxin is also a signaling component under sulfate limitation (102). The expression of auxin-inducible genes (*IAA18*, At1g51950, tryptophan synthase beta chain, At5g38530, putative auxin-regulated protein, At2g33830) is upregulated by sulfur starvation (47, 102). The expression of *NIT3* nitrilase, which can convert indole-3-acetonitrile to indole-3-acetic acid (IAA), is strongly increased by sulfur starvation (65, 89). The increased auxin production may result in an increase in lateral root density in *Arabidopsis* under sulfate-limited conditions (79). JA is also a possible signaling component in leaves. Genes involved in JA biosynthesis are upregulated under sulfur deficiency (45, 54). These genes include 12-oxophytodienoate reductase 1 and lipoxygenase (45, 86, 102). JA may regulate the expression of genes involved in sulfate assimilation and GSH synthesis (54, 154). Furthermore, MeJA is involved in regulating the activity of sulfur assimilation enzymes such as serine acetyltransferase (SAT) and APR (54). Although JA is a regulator of sulfur metabolism, its levels in plants are not well characterized under deficient conditions.

Other known components for signaling under sulfur deprivation include kinases and transcription factors. SAC3 kinase, which is a yeast sucrose nonfermenting-type kinase, was isolated from *Chlamydomonas* and regulates response to sulfur deprivation (19). SAC3 is required for depletion of chloroplast RNAs under sulfate-deficient conditions. In the *sac3* mutant, the chloroplast RNA abundance is less reduced than wild type and accumulation of ARS1 RNA is greater than in wild type under sulfate-deficient conditions. In wild type, the reduction in chloroplast transcripts under sulfur-limiting conditions depends on phosphorylation (52). SAC3 kinase regulates photosynthesis under sulfate-limited conditions and sulfate assimilation through the reduction of chloroplast transcripts. R2R3 Myb transcription factors (Myb16, 56, 69, 75, 90, 93 and 94) have been shown to be upregulated (102) by sulfate limitation in plants. However, the role that these transcription factors play in signaling under sulfate deficiency has not been established.

In contrast to plants, many genes have been identified that are involved in signaling sulfur deficiency in bacteria and yeast. The expression of *Cys3*, a bZIP transcription factor in *Neurospora crassa* (32, 105), is regulated by sulfur deficiency and controls the expression of some enzymes involved in sulfur metabolism. Mutations in *Scon1* and *Scon2*, which are components of E3 ubiquitin ligase and an F-box protein, lead to constitutive expression of *Cys3* and enzymes involved in sulfur metabolism (32, 105). Under sulfur-sufficient conditions Scon1 prevents Cys3 function through the multimerization of Scon1-Scon2 and binding to Cys3 (62–64, 105). MET30 (137) is a transcriptional repressor that has similar functions to the Scon1-Scon2 complex. The expression of sulfate assimilation enzymes in yeast depends on *MET4* and *MET28*, which encode bZIP transcription factors that are homologs of *Cys3* (64). For *E. coli*, the expression of genes that are involved in synthesis of flagella, chemotaxis, and methionine synthesis were all downregulated by sulfur limitation. This may be controlled by RpoS, which is a sigma factor required under sulfur deficiency (41). In plants, a chloroplast sigma factor that controls chloroplast transcriptional activity may be deactivated by the SAC3 kinase (52). Sulfur-containing essential amino acids such as Cys and Met are synthesized through the sulfur assimilation pathway, and changes in the concentrations of these amino acids eventually modulate cell cycle processes and cell viability.

Recently, the 16-bp sulfur-responsive element (SURE) and the 5-bp core sequence were identified from the *Arabidopsis* sulfate transporter, SULTR1;1 promoter. The SURE is essential for induction under sulfur-deprived conditions of genes, including *SULTR2*;1, *SULTR4;2*, *APR3*, and *NADPH* oxidoreductase (88). Soybean embryo factors (SEFs) 3 and 4 bind to the seed-specific β-conglycinin promoter, which is a sulfur responsive promoter that does not contain SURE elements. However, the function of SEFs in sulfate signaling has not been determined (7). Identification of transcription factor binding sites and the factors that bind to the SURE elements will help in assembling the signal transduction cascade in response to sulfate deficiency and perhaps provide links to other signal networks.

SUMMARY POINTS

1. Some of the signal transduction pathways in response to nutrient deprivation are beginning to be elucidated and transcriptional responses from microarray studies as well as other approaches are providing insight into possible components of these networks.
2. Interactions between nutrients are commonly observed and suggest that significant cross talk will be identified in signal transduction networks for responses to the deprivation of different nutrients.
3. More work is needed to determine the role of early and late signals in signal transduction networks.
4. ROS production in roots is observed in response to the deprivation of several macronutrients and may be an important component in signaling nutrient deprivation.
5. The PHR1 transcription factor is a central regulator of plant responses to phosphate deprivation.

FUTURE ISSUES

1. The relative importance of short- and longer-term responses to nutrient deprivation needs to be determined.
2. Cross talk between responses to the deprivation of NPKS and other mineral nutrients is likely, but the extent and importance of these overlapping pathways is not yet known.
3. Nutrient sensors have not yet been identified in plants.
4. Will plant scientists be able to use this basic information on sensing and signaling of nutrient deficiencies to create plants that grow better in low-nutrient environments while maintaining relatively high yields?

ACKNOWLEDGMENTS

We thank Joe Jez (Danforth Center) for comments on the manuscript. Danforth start-up funds and Monsanto Company have supported our work on plant response to nutrient deprivation. We apologize to the authors of many publications which we were unable to cite due to space limitations.

LITERATURE CITED

1. Abel S, Ticconi CA, Delatorre CA. 2002. Phosphate sensing in higher plants. *Physiol. Plant* 115:1–8
2. Ahn SJ, Shin R, Schachtman DP. 2004. Expression of KT/KUP genes in *Arabidopsis* and the role of root hairs in K^+ uptake. *Plant Physiol.* 134:1135–45
3. Amtmann A, Hammond J, Armenguad P, White PJ. 2006. Nutrient sensing and signaling in plants: potassium and phosphorus. *Adv. Bot. Res.* 43:209–57

4. Armengaud P, Breitling R, Amtmann A. 2004. The potassium-dependent transcriptome of *Arabidopsis* reveals a prominent role of jasmonic acid in nutrient signaling. *Plant Physiol.* 136:2556–76
5. Ashley MK, Grant M, Grabov A. 2006. Plant responses to potassium deficiencies: a role for potassium transport proteins. *J. Exp. Bot.* 57:425–36
6. Aung K, Lin SI, Wu CC, Huang YT, Su CL, Chiou TJ. 2006. *pho2*, a phosphate overaccumulator, is caused by a nonsense mutation in a microRNA399 target gene. *Plant Physiol.* 141:1000–11
7. Awazuhara M, Kima H, Goto DB, Matsui A, Hayashi H, et al. 2002. A 235-bp region from a nutritionally regulated soybean seed-specific gene promoter can confer its sulfur and nitrogen response to a constitutive promoter in aerial tissues of *Arabidopsis thaliana*. *Plant Sci.* 163:75–82
8. Baldwin JC, Karthikeyan AS, Raghothama KG. 2001. LEPS2, a phosphorus starvation-induced novel acid phosphatase from tomato. *Plant Physiol.* 125:728–37
9. Bari R, Datt Pant B, Stitt M, Scheible WR. 2006. PHO2, microRNA399, and PHR1 define a phosphate-signaling pathway in plants. *Plant Physiol.* 141:988–99
10. Blake-Kalff MMA, Harrison KR, Hawkesford MJ, Zhao FJ, McGrath SP. 1998. Distribution of sulfur within oilseed rape leaves in response to sulfur deficiency during vegetative growth. *Plant Physiol.* 118:1337–44
11. Bloom AJ, Finazzo J. 1986. The influence of ammonium and chloride on potassium and nitrate absorption by barley roots depend on time of exposure and cultivar. *Plant Physiol.* 81:67–69
12. Burleigh SH, Harrison MJ. 1999. The down-regulation of Mt4-like genes by phosphate fertilization occurs systematically and involves phosphate translocation to the shoots. *Plant Physiol.* 119:241–48
13. Carswell C, Grant BR, Theodorou ME, Harris J, Niere JO, Plaxton WC. 1996. The fungicide phosphonate disrupts the phosphate-starvation response in *Brassica nigra* seedlings. *Plant Physiol.* 110:105–10
14. Carswell MC, Grant BR, Plaxton WC. 1997. Disruption of the phosphate-starvation response of oilseed rape suspension cells by the fungicide phosphonate. *Planta* 203:67–74
15. Chen YM, Ferrar TS, Lohmeir-Vogel E, Morrice N, Mizuno Y, et al. 2006. The PII signal transduction protein of *Arabidopsis thaliana* forms an arginine-regulated complex with plastid N-acetyl glutamate kinase. *J. Biol. Chem.* 281:5726–33
16. Chiou TJ, Aung K, Lin SI, Wu CC, Chiang SF, Su CL. 2006. Regulation of phosphate homeostasis by microRNA in *Arabidopsis*. *Plant Cell* 18:412–21
17. Coello P, Polacco JC. 1999. ARR6, a response regulator from *Arabidopsis*, is differentially regulated by plant nutritional status. *Plant Sci.* 143:211–20
18. Conklin PL, Barth C. 2004. Ascorbic acid, a familiar small molecule intertwined in the response of plants to ozone, pathogens, and the onset of senescence. *Plant Cell Environ.* 27:959–70
19. Davies JP, Yildiz FH, Grossman AR. 1999. Sac3, an snf1-like serine/threonine kinase that positively and negatively regulates the responses of *Chlamydomonas* to sulfur limitation. *Plant Cell* 11:1179–90
20. Delhaize E, Randall PJ. 1995. Characterization of a phosphate-accumulator mutant of *Arabidopsis thaliana*. *Plant Physiol.* 107:207–13
21. Demidchik V, Essah PA, Tester M. 2004. Glutamate activates cation currents in the plasma membrane of *Arabidopsis* root cells. *Planta* 219:167–75
22. Dunlop J, Gardiner S. 1993. Phosphate uptake, proton extrusion and membrane electropotentials of phosphorus-deficient *Trifolium repens* L. *J. Exp. Bot.* 44:1801–8

23. Fan M, Zhu J, Richards C, Brown KM, Lynch JP. 2003. Physiological roles for aerenchyma in phosphorus-stressed roots. *Funct. Plant Biol.* 30:493–506
24. Ferrario-Mery S, Besin E, Pichon O, Meyer C, Hodges M. 2006. The regulatory PII protein controls arginine biosynthesis in *Arabidopsis*. *FEBS Lett.* 580:2015–20
25. Filleur S, Walch-Liu P, Gan Y, Forde BG. 2005. Nitrate and glutamate sensing by plant roots. *Biochem. Soc. Trans.* 33:283–86
26. Forde BG. 2002. Local and long-range signaling pathways regulating plant responses to nitrate. *Annu. Rev. Plant Biol.* 53:203–24
27. Foreman J, Demidchik V, Bothwell JH, Mylona P, Miedema H, et al. 2003. Reactive oxygen species produced by NADPH oxidase regulate plant cell growth. *Nature* 422:442–46
28. Franco-Zorrilla JM, Gonzalez E, Bustos R, Linhares F, Leyva A, Paz-Ares J. 2004. The transcriptional control of plant responses to phosphate limitation. *J. Exp. Bot.* 55:285–93
29. Franco-Zorrilla JM, Martin AC, Leyva A, Paz-Ares J. 2005. Interaction between phosphate-starvation, sugar, and cytokinin signaling in Arabidopsis and the roles of cytokinin receptors CRE1/AHK4 and AHK3. *Plant Physiol.* 138:847–57
30. Franco-Zorrilla JM, Martin AC, Solano R, Rubio V, Leyva A, Paz-Ares J. 2002. Mutations at CRE1 impair cytokinin-induced repression of phosphate starvation responses in *Arabidopsis*. *Plant J.* 32:353–60

30. Provides molecular evidence that cytokinins and two component receptors play important roles in plant response to phosphorus deprivation.

31. Fu H, Dooner HK. 2002. Intraspecific violation of genetic colinearity and its implications in maize. *Proc. Natl. Acad. Sci. USA* 99:9573–78
32. Fu Y, Paietta J, Mannix D, Marzluf G. 1989. Cys-3, the positive-acting sulfur regulatory gene of *Neurospora crassa*, encodes a protein with a putative leucine zipper DNA-binding element. *Mol. Cell. Biol.* 9:1120–27
33. Fujii H, Chiou TJ, Lin SI, Aung K, Zhu JK. 2005. A miRNA involved in phosphate-starvation response in *Arabidopsis*. *Curr. Biol.* 15:2038–43
34. Gallais A, Hirel B. 2004. An approach to the genetics of nitrogen use efficiency in maize. *J. Exp. Bot.* 55:295–306
35. Gan Y, Filleur S, Rahman A, Gotensparre S, Forde BG. 2005. Nutritional regulation of ANR1 and other root-expressed MADS-box genes in *Arabidopsis thaliana*. *Planta* 222:730–42
36. Gierth M, Maser P, Schroeder JI. 2005. The potassium transporter AtHAK5 functions in K^+ deprivation-induced high-affinity K^+ uptake and AKT1 K^+ channel contribution to K^+ uptake kinetics in *Arabidopsis* roots. *Plant Physiol.* 137:1105–14
37. Gilbert SM, Clarkson DT, Cambridge M, Lambers H, Hawkesford MJ. 1997. Sulfate-deprivation has an early effect on the content of ribulose 1,5-bisphosphate carboxylase/oxygenase and photosynthesis in young leaves of wheat. *Plant Physiol.* 115:1231–39
38. Gonzalez E, Solano R, Rubio V, Leyva A, Paz-Ares J. 2005. PHOSPHATE TRANSPORTER TRAFFIC FACILITATOR1 is a plant-specific SEC12-related protein that enables the endoplasmic reticulum exit of a high-affinity phosphate transporter in *Arabidopsis*. *Plant Cell* 17:3500–12
39. Good AG, Shrawat AK, Muench DG. 2004. Can less yield more? Is reducing nutrient input into the environment compatible with maintaining crop production? *Trends Plant Sci.* 9:597–605
40. Grime J. 1977. Evidence for the existence of three primary stategies in plants and its relevance to ecological and evolutionary theory. *Am. Nat.* 111:1169–94
41. Gyaneshwar P, Paliy O, McAuliffe J, Jones A, Jordan M, Kustu S. 2005. Lesson from *Escherichia coli* genes similarly regulated in response to nitrogen and sulfur limitation. *Proc. Natl. Acad. Sci. USA* 102:3453–58

42. Hammond JP, Bennett MJ, Bowen HC, Broadley MR, Eastwood DC, et al. 2003. Changes in gene expression in *Arabidopsis* shoots during phosphate starvation and the potential for developing smart plants. *Plant Physiol.* 132:578–96
43. Handreck K. 1997. Phosphorus requirements of Australian native plants. *Aust. J. Soil Res.* 35:241–89
44. Hesse H, Nikiforova V, Gakiere B, Hoefgen R. 2004. Molecular analysis and control of cysteine biosynthesis: integration of nitrogen and sulphur metabolism. *J. Exp. Bot.* 55:1283–92
45. Hirai MY, Fujiwara T, Awazuhara M, Kimura T, Noji M, Saito K. 2003. Global expression profiling of sulfur-starved *Arabidopsis* by DNA microarray reveals the role of O-acetyl-l-serine as a general regulator of gene expression in response to sulfur nutrition. *Plant J.* 33:651–63
46. Hirai MY, Saito K. 2004. Post-genomics approaches for the elucidation of plant adaptive mechanisms to sulphur deficiency. *J. Exp. Bot.* 55:1871–79
47. Hirai MY, Yano M, Goodenowe DB, Kanaya S, Kimura T, et al. 2004. Integration of transcriptomics and metabolomics for understanding of global responses to nutritional stresses in *Arabidopsis thaliana*. *Proc. Natl. Acad. Sci. USA* 101:10205–10
48. **Hodge A. 2004. The plastic plant: root responses to heterogeneous supplies of nutrients. *New Phytol.* 162:9–24**
49. Horgan J, Wareing P. 1980. Cytokinins and the growth responses of seedlings of *Beula pendula* Roth. and *Acer pseudoplatanus* L. to nitrogen and phosphorus deficiency. *J. Exp. Bot.* 31:525–32
50. Hsieh MH, Lam HM, van de Loo FJ, Coruzzi G. 1998. A PII-like protein in *Arabidopsis*: putative role in nitrogen sensing. *Proc. Natl. Acad. Sci. USA* 95:13965–70
51. Hwang I, Sheen J. 2001. Two-component circuitry in *Arabidopsis* cytokinin signal transduction. *Nature* 413:383–89
52. Irihimovitch V, Stern DB. 2006. The sulfur acclimation SAC3 kinase is required for chloroplast transcriptional repression under sulfur limitation in *Chlamydomonas reinhardtii*. *Proc. Natl. Acad. Sci. USA* 103:7911–16
53. Jones-Rhoades MW, Bartel DP. 2004. Computational identification of plant microRNAs and their targets, including a stress-induced miRNA. *Mol. Cell* 14:787–99
54. Jost R, Altschmied L, Bloem E, Bogs J, Gershenzon J, et al. 2005. Expression profiling of metabolic genes in response to methyl jasmonate reveals regulation of genes of primary and secondary sulfur-related pathways in *Arabidopsis thaliana*. *Photosynthesis Res.* 86:491–508
55. Jung K, Veen M, Altendorf K. 2000. K^+ and ionic strength directly influence the autophosphorylation activity of the putative turgor sensor KdpD of *Escherichia coli*. *J.Biol. Chem.* 275:40142–47
56. Kaiser WM, Huber SC. 2001. Post-translational regulation of nitrate reductase: mechanism, physiological relevance and environmental triggers. *J. Exp. Bot.* 52:1981–89
57. Kandlbinder A, Finkemeier I, Wormuth D, Hanitzsch M, Dietz KJ. 2004. The antioxidant status of photosynthesizing leaves under nutrient deficiency: redox regulation, gene expression and antioxidant activity in *Arabidopsis thaliana*. *Physiol. Plant.* 120:63–73
58. Kang JM, Turano FJ. 2003. The putative glutamate receptor 1.1 (AtGLR11) functions as a regulator of carbon and nitrogen metabolism in *Arabidopsis thaliana*. *Proc. Natl. Acad. Sci. USA* 100:6872–77
59. Keller T, Damude HG, Werner D, Doerner P, Dixon RA, Lamb C. 1998. A plant homolog of the neutrophil NADPH oxidase gp91phox subunit gene encodes a plasma membrane protein with Ca^{2+} binding motifs. *Plant Cell* 10:255–66

48. An up-to-date review that provides an overview of how roots respond to the patchy nutrient supplies found in soils.

60. Konings H, Verschuren G. 1980. Formation of aerenchyma in roots of *Zea mays* in aerated solutions, and its relation to nutrient supply. *Physiol. Plant.* 49:265–70

61. Kopriva S. 2006. Regulation of sulfate assimilation in *Arabidopsis* and beyond. *Annals. Bot.* 97:479–95

62. Kumar A, Paietta J. 1995. The sulfur controller-2 negative regulatory gene of *Neurospora crassa* encodes a protein with β-transduction repeats. *Proc. Natl. Acad. Sci. USA* 92:3343–47

63. Kumar A, Paietta JV. 1998. An additional role for the F-box motif: gene regulation within the *Neurospora crassa* sulfur control network. *Proc. Natl. Acad. Sci. USA* 95:2417–22

64. Kuras L, Cherest H, Surdin-Kerjan Y, Thomas D. 1996. A heteromeric complex containing the centromere binding factor1 and two basic leucine zipper factors, Met4 and Met28, mediates the transcription activation of yeast sulfur metabolism. *EMBO J.* 15:2519–29

65. Kutz A, Müller A, Hennig P, Kaiser WM, Piotrowski M, Weiler EW. 2002. A role for nitrilase 3 in the regulation of root morphology in sulphur-starving *Arabidopsis thaliana*. *Plant J.* 30:95–106

66. Lacombe B, Becker D, Hedrich R, DeSalle R, Hollmann M, et al. 2001. The identity of plant glutamate receptors. *Science* 292:1486–87

67. Lamb C, Dixon RA. 1997. The oxidative burst in plant disease resistance. *Annu. Rev. Plant Physiol. Plant Mol. Biol.* 48:251–75

68. Lappartient A, Touraine B. 1997. Glutathione-mediated regulation of ATP sulfurylase activity, SO_4^{2-} uptake and oxidative stress response in intact canola roots. *Plant Physiol.* 114:177–83

69. Lejay L, Tillard P, Lepetit M, Olive F, Filleur S, et al. 1999. Molecular and functional regulation of two NO3- uptake systems by N- and C-status of *Arabidopsis* plants. *Plant J.* 18:509–19

70. Lencioni L, Ranieri A, Fergola S, Soldatini G. 1997. Photosynthesis and metabolic changes in leaves of rapeseed grown under long-term sulfate deprivation. *J. Plant Nutr.* 20:405–15

71. Leustek T, Martin MN, Bick JA, Davies JP. 2000. Pathways and regulation of sulfur metabolism revealed through molecular and genetic studies. *Annu. Rev. Plant Physiol. Plant Mol. Biol.* 51:141–65

72. Leustek T, Saito K. 1999. Sulfate transport and assimilation in plants. *Plant Physiol.* 120:637–44

73. Li L, Kim BG, Cheong YH, Pandey GK, Luan S. 2006. A Ca^{2+} signaling pathway regulates a K^+ channel for low-K response in *Arabidopsis*. *Proc. Natl. Acad. Sci. USA* 103:12625–30

74. Li M, Qin C, Welti R, Wang X. 2006. Double knockouts of phospholipases Dz1 and Dz2 in *Arabidopsis* affect root elongation during phosphate-limited growth but do not affect root hair patterning. *Plant Physiol.* 140:761–70

75. Linkohr BI, Williamson LC, Fitter AH, Leyser HM. 2002. Nitrate and phosphate availability and distribution have different effects on root system architecture of *Arabidopsis*. *Plant J.* 29:751–60

76. Liszkay A, van der Zalm E, Schopfer P. 2004. Production of reactive oxygen intermediates ($O_2^{\cdot -}$, H_2O_2, and $\cdot OH$) by maize roots and their role in wall loosening and elongation growth. *Plant Physiol.* 136:3114–23

77. Little DY, Rao H, Oliva S, Daniel-Vedele F, Krapp A, Malamy JE. 2005. The putative high-affinity nitrate transporter NRT2.1 represses lateral root initiation in response to nutritional cues. *Proc. Natl. Acad. Sci. USA* 102:13693–98

61. A comprehensive review on sulfur assimilation and the known and possible signals/molecules involved in regulation.

78. Liu KH, Tsay YF. 2003. Switching between the two action modes of the dual-affinity nitrate transporter CHL1 by phosphorylation. *EMBO J.* 22:1005–13

79. López-Bucio J, Cruz-Ramirez A, Herrera-Estrella L. 2003. The role of nutrient availability in regulating root architecture. *Curr. Opin. Plant Biol.* 6:280–87

80. López-Bucio J, Hernandez-Abreu E, Sanchez-Calderon L, Nieto-Jacobo MF, Simpson J, Herrera-Estrella L. 2002. Phosphate availability alters architecture and causes changes in hormone sensitivity in the *Arabidopsis* root system. *Plant Physiol.* 129:244–56

81. Ma Z, Bielenberg DG, Brown KM, Lynch JP. 2001. Regulation of root hair density by phosphorus availability in *Arabidopsis thaliana*. *Plant Cell Environ.* 24:459–67

82. Maathuis FJ, Filatov V, Herzyk P, Krijger G, Axelsen K, et al. 2003. Transcriptome analysis of root transporters reveals participation of multiple gene families in the response to cation stress. *Plant J.* 35:675–92

83. Mallory AC, Vaucheret H. 2006. Functions of microRNAs and related small RNAs in plants. *Nat. Genet.* 38 Suppl 1:S31–36

84. Marschner H. 1995. *Mineral Nutrition of Higher Plants*. San Diego: Academic. 889p.

85. Martin AC, del Pozo JC, Iglesias J, Rubio V, Solano R, et al. 2000. Influence of cytokinins on the expression of phosphate starvation responsive genes in *Arabidopsis*. *Plant J.* 24:559–67

86. Maruyama-Nakashita A, Inoue E, Watanabe-Takahashi A, Yamaya T, Takahashi H. 2003. Transcriptome profiling of sulfur-responsive genes in *Arabidopsis* reveals global effects of sulfur nutrition on multiple metabolic pathways. *Plant Physiol.* 132:597–605

87. Maruyama-Nakashita A, Nakamura Y, Tohge T, Saito K, Takahashi H. 2006. *SLIM1/EIL3 transcription factor required for plant growth on low sulfur environment*. Presented at Intl. Conf. on *Arabidopsis* Res., 17th, Madison, Wis.

88. Maruyama-Nakashita A, Nakamura Y, Watanabe-Takahashi A, Inoue E, Yamaya T, Takahashi H. 2005. Identification of a novel cis-acting element conferring sulfur deficiency response in *Arabidopsis* roots. *Plant J.* 42:305–14

89. Maruyama-Nakashita A, Nakamura Y, Watanabe-Takahashi A, Yamaya T, Takahashi H. 2004. Induction of SULTR1;1 sulfate transporter in *Arabidopsis* roots involves protein phosphorylation/dephosphorylation circuit for transcriptional regulation. *Plant Cell Physiol.* 45:340–45

90. Maruyama-Nakashita A, Nakamura Y, Yamaya T, Takahashi H. 2004. Regulation of high-affinity sulphate transporters in plants: towards systematic analysis of sulphur signaling and regulation. *J. Exp. Bot.* 55:1843–49

91. Maruyama-Nakashita A, Nakamura Y, Yamaya T, Takahashi H. 2004. A novel regulatory pathway of sulfate uptake in *Arabidopsis* roots: implication of CRE1/WOL/AHK4-mediated cytokinin-dependent regulation. *Plant J.* 38:779–89

92. Migge A, Bork C, Hell R, Becker T. 2000. Negative regulation of nitrate reductase gene expression by glutamine or asparagine accumulating in leaves of sulfur-deprived tobacco. *Planta* 211:587–95

93. Misson J, Raghothama KG, Jain A, Jouhet J, Block MA, et al. 2005. A genome-wide transcriptional analysis using *Arabidopsis thaliana* Affymetrix gene chips determined plant responses to phosphate deprivation. *Proc. Natl. Acad. Sci. USA* 102:11934–39

94. Miura K, Rus A, Sharkhuu A, Yokoi S, Karthikeyan AS, et al. 2005. The *Arabidopsis* SUMO E3 ligase SIZ1 controls phosphate deficiency responses. *Proc. Natl. Acad. Sci. USA* 102:7760–65

93. A comprehensive study on the many responses in *Arabidopsis* due to phosphorus deprivation.

94. Shows that sumoylation regulates some phosphorus deprivation responses.

95. Miyake K, Ito T, Senda M, Ishikawa R, Harada T, et al. 2003. Isolation of a subfamily of genes for R2R3-MYB transcription factors showing up-regulated expression under nitrogen nutrient-limited conditions. *Plant Mol. Biol.* 53:237–45
96. Moorhead GB, Smith CS. 2003. Interpreting the plastid carbon, nitrogen, and energy status. A role for PII? *Plant Physiol.* 133:492–98
97. Mori IC, Schroeder JI. 2004. Reactive oxygen species activation of plant Ca^{2+} channels. A signaling mechanism in polar growth, hormone transduction, stress signaling, and hypothetically mechanotransduction. *Plant Physiol.* 135:702–8
98. Mostertz J, Scharf C, Hecker M, Homuth G. 2004. Transcriptome and proteome analysis of *Bacillus subtilis* gene expression in response to superoxide and peroxide stress. *Microbiol.* 150:497–512
99. Mukatira UT, Liu C, Varadarajan DK, Raghothama KG. 2001. Negative regulation of phosphate starvation-induced genes. *Plant Physiol.* 127:1854–62
100. Muller S, Hoege C, Pyrowolakis G, Jentsch S. 2001. SUMO, Ubiquitin's mysterious cousin. *Nat. Rev. Mol. Cell Biol.* 2:202–10
101. Nacry P, Canivenc G, Muller B, Azmi A, Van Onckelen H, et al. 2005. A role for auxin redistribution in the responses of the root system architecture to phosphate starvation in *Arabidopsis*. *Plant Physiol.* 138:2061–74
102. Nikiforova V, Freitag J, Kempa S, Adamik M, Hesse H, Hoefgen R. 2003. Transcriptome analysis of sulfur depletion in *Arabidopsis thaliana*: interacting of biosynthetic pathways provides response specificity. *Plant J.* 33:633–50
103. Nikiforova VJ, Kopka J, Tolstikov V, Fiehn O, Hopkins L, et al. 2005. Systems rebalancing of metabolism in response to sulfur deprivation, as revealed by metabolome analysis of *Arabidopsis* plants. *Plant Physiol.* 138:304–18
104. Ohkama N, Takei K, Sakakibara H, Hayashi H, Yoneyama T, Fujiwara T. 2002. Regulation of sulfur-responsive gene expression by exogenously applied cytokinins in *Arabidopsis thaliana*. *Plant Cell Physiol.* 43:1493–501
105. Paietta J. 1992. Production of the CYS3 regulator, a bZIP DNA-binding protein, is sufficient to induce sulfur gene expression in *Neurospora crassa*. *Mol. Cell. Biol.* 12:1568–77
106. Palenchar PM, Kouranov A, Lejay LV, Coruzzi GM. 2004. Genome-wide patterns of carbon and nitrogen regulation of gene expression validate the combined carbon and nitrogen (CN)-signaling hypothesis in plants. *Genome Biol.* 5:R91
107. Pastori GM, Kiddle G, Antoniw J, Bernard S, Veljovic-Jovanovic S, et al. 2003. Leaf vitamin C contents modulate plant defense transcripts and regulate genes that control development through hormone signaling. *Plant Cell* 15:939–51
108. Philippar K, Fuchs I, Luthen H, Hoth S, Bauer CS, et al. 1999. Auxin-induced K^+ channel expression represents an essential step in coleoptile growth and gravitropism. *Proc. Natl. Acad. Sci. USA* 96:12186–91
109. Prosser I, Purves J, Saker L, Clarkson D. 2001. Rapid disruption of nitrogen metabolism and nitrate transport in spinach plants deprived of sulphate. *J. Exp. Bot.* 52:113–21
110. Ravina CG, Chang CI, Tsakraklides GP, McDermott JP, Vega JM, et al. 2002. The *sac* mutants of *Chlamydomonas reinhardtii* reveal transcriptional and posttranscriptional control of cysteine biosynthesis. *Plant Physiol.* 130:2076–84
111. Remans T, Nacry P, Pervent M, Girin T, Tillard P, et al. 2006. A central role for the nitrate transporter NRT2.1 in the integrated morphological and physiological responses of the root system to nitrogen limitation in *Arabidopsis*. *Plant Physiol.* 140:909–21
112. Rubio V, Linhares F, Solano R, Martin AC, Iglesias J, et al. 2001. A conserved MYB transcription factor involved in phosphate starvation signaling both in vascular plants and in unicellular algae. *Genes Dev.* 15:2122–33

112. Identified a key transcription factor in plants similar in structure to a previously identified transcription factor in *Chlamydomonas reinhardtii* that is involved in regulation of phosphate starvation–induced genes.

113. Sagi M, Fluhr R. 2001. Superoxide production by plant homologues of the gp91(phox) NADPH oxidase. Modulation of activity by calcium and by tobacco mosaic virus infection. *Plant Physiol* 126:1281–90
114. Sakakibara H. 2006. Cytokinins: activity, biosynthesis, and translocation. *Annu. Rev. Plant Biol.* 57:431–49
115. Sakakibara H, Suzuki M, Takei K, Deji A, Taniguchi M, Sugiyama T. 1998. A response-regulator homologue possibly involved in nitrogen signal transduction mediated by cytokinin in maize. *Plant J.* 14:337–44
116. Sanchez P. 2002. Soil fertility and hunger in Africa. *Science* 295:2019–20
117. Sano H, Youssefian S. 1994. Light and nutritional regulation of transcripts encoding a wheat protein kinase homolog is mediated by cytokinins. *Proc. Natl. Acad. Sci. USA* 91:2582–86
118. Santa-Maria GE, Danna CH, Czibener C. 2000. High-affinity potassium transport in barley roots. Ammonium-sensitive and -insensitive pathways. *Plant Physiol.* 123:297–306
119. Schachtman DP, Reid RJ, Ayling SM. 1998. Phosphorus uptake by plants: from soil to cell. *Plant Physiol.* 116:447–53
120. Scheible WR, Gonzalez-Fontes A, Lauerer M, Muller-Rober B, Caboche M, Stitt M. 1997. Nitrate acts as a signal to induce organic acid metabolism and repress starch metabolism in tobacco. *Plant Cell* 9:783–98
121. Scheible WR, Morcuende R, Czechowski T, Fritz C, Osuna D, et al. 2004. Genome-wide reprogramming of primary and secondary metabolism, protein synthesis, cellular growth processes, and the regulatory infrastructure of *Arabidopsis* in response to nitrogen. *Plant Physiol.* 136:2483–99

121. A rich data set containing information on transcriptional response to nitrogen deprivation and to the introduction of nitrogen; some metabolic data are also provided.

122. Schunmann PH, Richardson AE, Vickers CE, Delhaize E. 2004. Promoter analysis of the barley Pht1;1 phosphate transporter gene identifies regions controlling root expression and responsiveness to phosphate deprivation. *Plant Physiol.* 136:4205–14
123. Setya A, Murillo M, Leustek T. 1996. Sulfate reduction in higher plants: molecular evidence for a novel 5-adenylylsulfate reductase. *Proc. Natl. Acad. Sci. USA* 93:13383–88
124. Shin H, Shin HS, Chen R, Harrison MJ. 2006. Loss of At4 function impacts phosphate distribution between the roots and the shoots during phosphate starvation. *Plant J.* 45:712–26
125. Shin R, Berg RH, Schachtman DP. 2005. Reactive oxygen species and root hairs in *Arabidopsis* root response to nitrogen, phosphorus and potassium deficiency. *Plant Cell Physiol.* 46:1350–57
126. Shin R, Schachtman DP. 2004. Hydrogen peroxide mediates plant root response to nutrient deprivation. *Proc. Natl. Acad. Sci. USA* 101:8827–32

126. Showed that ROS may be a component in a signal cascade in roots of plants that had been deprived of potassium and that a single NADPH oxidase is important in generating ROS in response to potassium deprivation.

127. Smith F, Ealing P, Hawkesford M, Clarkson D. 1995. Plant members of a family of sulfate transporters reveal functional subtypes. *Proc. Natl. Acad. Sci. USA* 92:9373–77
128. Spalding EP, Hirsch RE, Lewis DR, Qi Z, Sussman MR, Lewis BD. 1999. Potassium uptake supporting plant growth in the absence of AKT1 channel activity. *J. Gen. Phsyiol.* 113:909–18
129. Stitt M, Muller C, Matt P, Gibon Y, Carillo P, et al. 2002. Steps towards an integrated view of nitrogen metabolism. *J. Exp. Bot.* 53:959–70
130. Sugiharto B, Suzuki I, Burnell JN, Sugiyama T. 1992. Glutamine induces the N-dependent accumulation of mRNAs encoding phosphoenolpyruvate carboxylase and carbonic anhydrase in detached maize leaf tissue. *Plant Physiol.* 100:2066–70
131. Sugiura A, Hirokawa K, Nakashima K, Mizuno T. 1994. Signal-sensing mechanisms of the putative osmosensor KdpD in *Escherichia coli*. *Mol. Microbiol.* 14:929–38

132. Takahashi H, Watanabe-Takahashi A, Smith FW, Blake-Kalff M, Hawkesford MJ, Saito K. 2000. The roles of three functional sulphate transporters involved in uptake and translocation of sulphate in *Arabidopsis thaliana*. *Plant J.* 23:171–82
133. Takahashi H, Yamazaki M, Sasakura N, Watanabe A, Leustek T, et al. 1997. Regulation of sulfur assimilation in higher plants: A sulfate transporter induced in sulfate-starved roots plays a central role in *Arabidopsis thaliana*. *Proc. Natl. Acad. Sci. USA* 94:11102–7
134. Takei K, Takahashi T, Sugiyama T, Yamaya T, Sakakibara H. 2002. Multiple routes communicating nitrogen availability from roots to shoots: a signal transduction pathway mediated by cytokinin. *J. Exp. Bot.* 53:971–77
135. Taniguchi M, Kiba T, Sakakibara H, Ueguchi C, Mizuno T, Sugiyama T. 1998. Expression of *Arabidopsis* response regulator homologs is induced by cytokinins and nitrate. *FEBS Lett.* 429:259–62
136. Thanos D, Maniatis T. 1995. Virus induction of human IFN beta gene expression requires the assembly of an enhanceosome. *Cell* 83:1091–100
137. Thomas D, Kuras L, Barbey R, Cherest H, Blaiseau P, Surdin-Kerjan Y. 1995. Met30p, a yeast transcriptional inhibitor that responds to S-adenosylmethionine, is an essential protein with WD40 repeats. *Mol. Cell. Biol.* 15:6526–34
138. Ticconi CA, Abel S. 2004. Short on phosphate: plant surveillance and countermeasures. *Trends Plant Sci.* 9:548–55
139. Ticconi CA, Delatorre CA, Lahner B, Salt DE, Abel S. 2004. *Arabidopsis pdr2* reveals a phosphate-sensitive checkpoint in root development. *Plant J.* 37:801–14

139. The culmination of a novel approach to the isolation of phosphate sensing and signaling mutants identified a locus that is involved in sensing phosphate by roots and regulating root meristem activity.

140. Todd CD, Zeng P, Huete AM, Hoyos ME, Polacco JC. 2004. Transcripts of MYB-like genes respond to phosphorus and nitrogen deprivation in *Arabidopsis*. *Planta* 219:1003–9
141. Torres MA, Dangl JL. 2005. Functions of the respiratory burst oxidase in biotic interactions, abiotic stress and development. *Curr. Opin. Plant Biol.* 8:397–403
142. Vandenabeele S, Van Der Kelen K, Dat J, Gadjev I, Boonefaes T, et al. 2003. A comprehensive analysis of hydrogen peroxide-induced gene expression in tobacco. *Proc. Natl. Acad. Sci. USA* 100:16113–18
143. Vicente-Agullo F, Rigas S, Desbrosses G, Dolan L, Hatzopoulos P, Grabov A. 2004. Potassium carrier TRH1 is required for auxin transport in *Arabidopsis* roots. *Plant J.* 40:523–35
144. Walch-Liu P, Liu LH, Remans T, Tester M, Forde BG. 2006. Evidence that L-glutamate can act as an exogenous signal to modulate root growth and branching in *Arabidopsis thaliana*. *Plant Cell Physiol.* 47:1045–57
145. Walderhaug MO, Polarek JW, Voelkner P, Daniel JM, Hesse JE, et al. 1992. KdpD and KdpE, proteins that control expression of the kdpABC operon, are members of the two-component sensor-effector class of regulators. *J. Bacteriol.* 174:2152–59
146. Wang R, Guegler K, LaBrie ST, Crawford NM. 2000. Genomic analysis of a nutrient response in *Arabidopsis* reveals diverse expression patterns and novel metabolic and potential regulatory genes induced by nitrate. *Plant Cell* 12:1491–509
147. Wang R, Okamoto M, Xing X, Crawford NM. 2003. Microarray analysis of the nitrate response in *Arabidopsis* roots and shoots reveals over 1,000 rapidly responding genes and new linkages to glucose, trehalose-6-phosphate, iron, and sulfate metabolism. *Plant Physiol.* 132:556–67
148. Wang X. 2005. Regulatory functions of phospholipase D and phosphatidic acid in plant growth, development, and stress responses. *Plant Physiol.* 139:566–73
149. Wang YH, Garvin DF, Kochian LV. 2001. Nitrate-induced genes in tomato roots. Array analysis reveals novel genes that may play a role in nitrogen nutrition. *Plant Physiol.* 127:345–59

150. Wang YH, Garvin DF, Kochian LV. 2002. Rapid induction of regulatory and transporter genes in response to phosphorus, potassium and iron deficiencies in tomato roots. Evidence for cross talk and root/rhizosphere-mediated signals. *Plant Physiol.* 130:1361–70
151. Wilkinson S, Davies WJ. 2002. ABA-based chemical signaling: the co-ordination of responses to stress in plants. *Plant Cell Environ.* 25:195–210
152. Williamson LC, Ribrioux SP, Fitter AH, Leyser HM. 2001. Phosphate availability regulates root system architecture in *Arabidopsis*. *Plant Physiol.* 126:875–82
153. Wilson WA, Roach PJ. 2002. Nutrient-regulated protein kinases in budding yeast. *Cell* 111:155–58
154. Xiang C, Oliver D. 1998. Glutatione metabolic genes cordinately respond to heavy metals and jasmoic acid in *Arabidopsis*. *Plant Cell* 10:1539–50
155. Xu J, Li HD, Chen LQ, Wang Y, Liu LL, et al. 2006. A protein kinase, interacting with two calcineurin B-like proteins, regulates K^+ transporter AKT1 in *Arabidopsis*. *Cell* 125:1347–60
156. Yoshimoto N, Takahashi H, Smith FW, Yamaya T, Saito K. 2002. Two distinct high-affinity sulfate transporters with different inducibilities mediate uptake of sulfate in *Arabidopsis* roots. *Plant J.* 29:465–73

Hydrogenases and Hydrogen Photoproduction in Oxygenic Photosynthetic Organisms*

Maria L. Ghirardi,[1] Matthew C. Posewitz,[2] Pin-Ching Maness,[1] Alexandra Dubini,[1] Jianping Yu,[1] and Michael Seibert[1]

[1]National Renewable Energy Laboratory, Golden, Colorado 80401; email: maria_ghirardi@nrel.gov, pinching_maness@nrel.gov, alexandra_dubini@nrel.gov, jianping_yu@nrel.gov, mike_seibert@nrel.gov

[2]Colorado School of Mines, Environmental Science and Engineering Division, Golden, Colorado 80401; email: matthew_posewitz@nrel.gov

Annu. Rev. Plant Biol. 2007. 58:71–91

The *Annual Review of Plant Biology* is online at plant.annualreviews.org

This article's doi: 10.1146/annurev.arplant.58.032806.103848

First published online as a Review in Advance on December 6, 2006

1543-5008/07/0602-0071$20.00

Key Words

green algae, cyanobacteria, oxygen inhibition, photosynthesis

Abstract

The photobiological production of H_2 gas, using water as the only electron donor, is a property of two types of photosynthetic microorganisms: green algae and cyanobacteria. In these organisms, photosynthetic water splitting is functionally linked to H_2 production by the activity of hydrogenase enzymes. Interestingly, each of these organisms contains only one of two major types of hydrogenases, [FeFe] or [NiFe] enzymes, which are phylogenetically distinct but perform the same catalytic reaction, suggesting convergent evolution. This idea is supported by the observation that each of the two classes of hydrogenases has a different metallo-cluster, is encoded by entirely different sets of genes (apparently under the control of different promoter elements), and exhibits different maturation pathways. The genetics, biosynthesis, structure, function, and O_2 sensitivity of these enzymes have been the focus of extensive research in recent years. Some of this effort is clearly driven by the potential for using these enzymes in future biological or biohybrid systems to produce renewable fuel or in fuel cell applications.

Contents

INTRODUCTION

Scientific innovation, resulting in a growing world economy, has improved the lives of people around the world; however, the consequences have been congestion, pollution, growing environmental stress, and the recent recognition of global climate change. With increased demand for energy (oil was about $75 per barrel at this writing), pressures on finite sources of conventional fuel reserves will continue to rise. Alternative "green" energy technologies, including wind, photovoltaics, biofuels, solar thermal, geothermal, as well as hydropower, wood burning, and waste-to-energy technologies are being developed around the world to address the pressure on conventional fuels. Recent promotion of biofuels, including cellulosic ethanol, biodiesel, green diesel, and hydrogen, is good news for plant biologists.

Currently, most of the 9–10 million tons of H_2 (classified as an energy carrier rather than a primary energy source), representing about 1% of the U.S. yearly energy budget, comes from reforming natural gas. Photobiological means to produce H_2 include the use of photosynthesis and the natural water-splitting process, in which the microbes absorb energy from the sun and evolve H_2 directly from water without an intermediate biomass stage. The advantage of this strategy is a tenfold improvement in land area utilization at the maximum potential 10% efficiency of light conversion into H_2. At this efficiency, photobiological systems could produce enough energy to totally displace gasoline use in the United States when cultivated in sealed reactors in an area of less than 5000 square miles.

This review discusses fundamental aspects of [FeFe]- and [NiFe]-hydrogenases found, respectively, in green algae and cyanobacteria, as well as some of the challenges that must be resolved before practical photobiological H_2-production systems can be developed. Specifically, we emphasize current ideas about (*a*) the structural and catalytic properties of both kinds of hydrogenases, including a discussion of the unique aspects of their metalloclusters, (*b*) the diversity of hydrogenases and differences in their gene structures, (*c*) mechanisms for assembly and maturation of their unique metallo catalytic clusters, and (*d*) the O_2 sensitivity of hydrogenases and approaches to surmount this challenge.

GREEN ALGAL [FeFe]-HYDROGENASES

Enzyme Structure and Catalytic Mechanism

Green algal hydrogenases belong to the class of [FeFe]-hydrogenases, which are also found

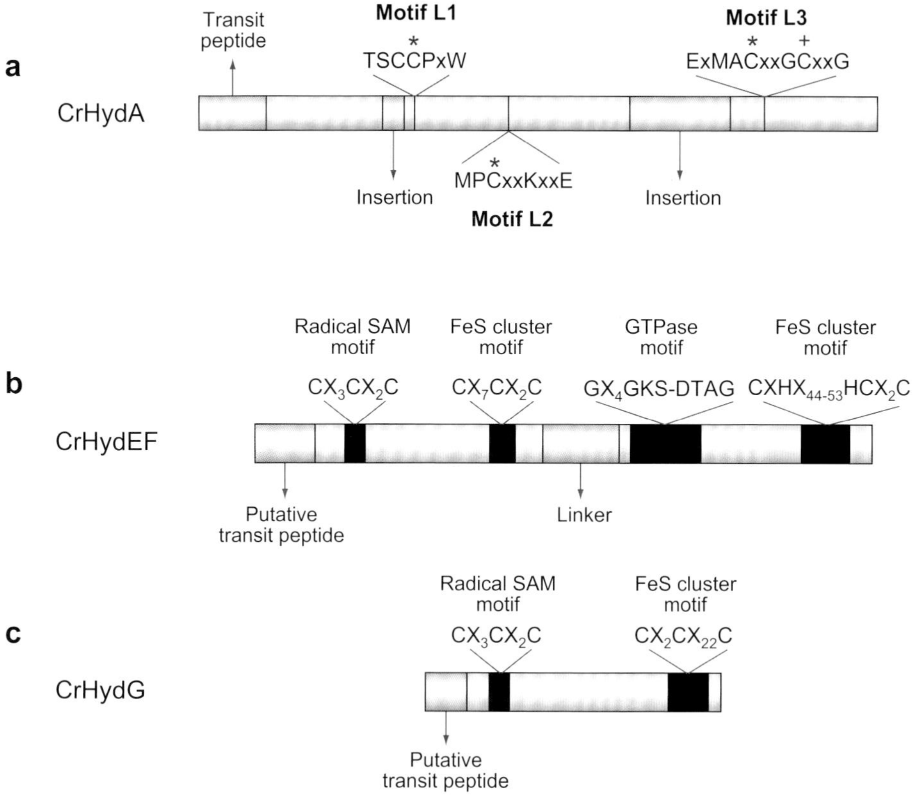

Figure 1

Schematic diagram of the structures of, respectively, *Chlamydomonas reinhardtii* (Cr) HydA (*a*), HydEF (*b*) and HydG (*c*) proteins. The blue areas correspond to transit peptides, insertions, and linker peptides not found in the homologous bacterial proteins. The characteristic Radical SAM and metallo-cluster-binding motif sequences present in each protein are shown in more detail. The location of the three terminal cysteine residues that ligate the [4Fe-4S] center in HydA are indicated by *; the location of the cysteine residue that links the [4Fe-4S] and the 2Fe centers in the H-cluster of HydA is indicated by the + sign.

in strict anaerobes, fungi, and protists (10, 105). The sequence similarity between different [FeFe]-hydrogenase proteins is very high, around 50% (31). This is due to conservation of the residues involved in binding or in providing the appropriate structure and environment for the metallo catalytic cluster. Indeed, as **Figure 1*a*** shows, three main motifs, L1, L2, and L3, have been identified as being present in most [FeFe]-hydrogenases (105). Nevertheless, Mishra et al. (64) recently reported the cloning of an open reading frame (ORF) encoding a putative hydrogenase from *Enterobacter cloacae* that lacks characteristic motif 1 (and its conserved cysteine residue), and has less-conserved motifs 2 and 3 (which, however, still contain characteristic conserved cysteine residues). When expressed in *Escherichia coli*, the gene product from this ORF exhibited H_2-production activity. Although most [FeFe]-hydrogenases are monomeric, enzymes with two, three, and even four subunits have been described in the literature (105).

In contrast to [NiFe]-hydrogenases (see below), [FeFe]-hydrogenases contain only Fe and S in their catalytic site, and are typically involved in H_2 production rather than H_2 oxidation. The metallo catalytic cluster of [FeFe]-hydrogenases, the H-cluster, is

Open reading frame (ORF): the protein-encoding portion of a gene

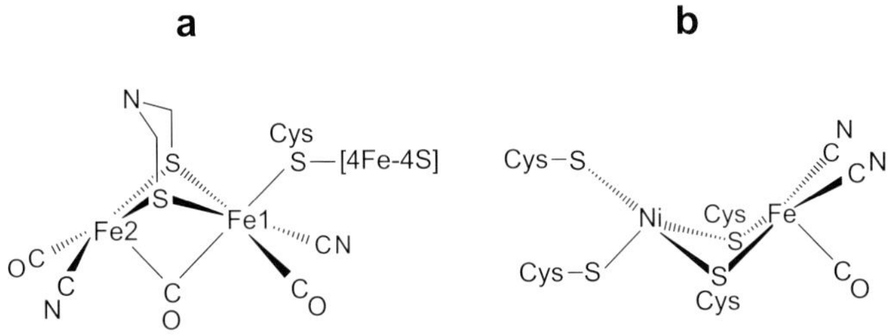

Figure 2

Chemical structure of the (*a*) [FeFe]-hydrogenase H-cluster and the (*b*) [NiFe]-hydrogenase metallo-cluster.

Chlorophycophyta: a division of green algae that includes the H_2-producing taxa Chlorellales (*Scenedesmus*, *Chlorella*, and others) and Volvocales (*Chlamydomonas*, *Pandorina*, *Volvox*, and others)

unique. As seen in **Figure 2*a***, it consists of a [4Fe-4S] cubane linked through a protein cysteine residue (present in motif L3, see **Figure 1*a***) to a 2Fe subcluster. The iron atoms of the [4Fe-4S] center are coordinated to the protein structure by three additional conserved cysteine residues, which are found in motifs L1–L3. Except for the bridging cysteine, the iron atoms of the 2Fe center are coordinated to the nonprotein carbon monoxide (CO) and cyanide (CN) ligands. One of the CO ligands probably bridges both iron atoms when the center is in the oxidized state (100). The presence of these ligands is not common in biological enzymes (33) and suggests that an unusual maturation pathway must occur to assemble the metallo-cluster (see below). Additionally, an organic bridge between the two S moieties has been tentatively identified as di(thiomethyl)amine through both X-ray crystallography and theoretical studies (30, 67).

In algal hydrogenases, electrons are delivered directly to the H-cluster by reduced ferredoxin molecules. All other hydrogenases, however, contain a putative electron relay composed of additional FeS centers (either [2Fe2S] or [4Fe-4S]), the F-clusters, that may be involved in electron transport from an external donor to the H-cluster (15). It has been proposed that the electrons doubly reduce the distal Fe (Fe2 in the 2Fe center of the H-cluster, see **Figure 2*a***), which is stabilized in the reduced form by the presence of the CO and CN ligands (33). According to many research groups, the subsequent generation of H_2 gas would involve the double reduction of a proton bound to the distal Fe2 (68) and recombination between the resulting hydride anion and a second proton donated by either the di(thiomethyl)amine bridge (26) or by a nearby lysine or cysteine residue (68, 78). This question is being addressed through extensive spectroscopic and structural measurements that test predictions of various hypotheses. Finally, although the reducing side of the H-cluster is buried in the interior of the protein cluster, it is accessible to oxygen and carbon monoxide, two of its known inhibitors (see below).

Gene Structure and Diversity

At present, [FeFe]-hydrogenase genes have been identified and sequenced from many green algal species. The first algal hydrogenase gene sequence reported in the literature was that of *Scenedesmus obliquus* (31). Since then, gene sequences for the *Chlamydomonas reinhardtii* HydA1 (43) and HydA2 (32), *Chlorella fusca* (or *Scenedesmus vacuolatus* [111]), and *Chlamydomonas moewusii* (112) proteins have also been published. Moreover, biochemical evidence points to the presence of hydrogenase activity in *Chlorococcum littorale* (90, 101), *Platimonas subcordiformis* (39), and a variety of Chlorophycophyta (12). In all reported cases, the hydrogenase gene encodes for a monomeric protein of about 48 kDa, which represent the smallest class of [FeFe]-hydrogenases identified so far (42).

All the eukaryotic hydrogenase genes are nucleus-encoded, although their gene products are localized either in the chloroplasts (green algae) or in the hydrogenosomes (trichomonads, anaerobic ciliates, and chytrid fungi), except for a few protozoan parasites, including *Entamoeba* and *Spironucleus* (45), where the hydrogenase is localized in the cytosol. The organellar location of most [FeFe]-hydrogenases requires a transit peptide sequence at the N-terminal end of the protein to target it to the specific organelle. Indeed,

Table 1 Structural characteristics of algal [FeFe]-hydrogenases from *Chlamydomonas reinhardtii* (Cr HydA1 and Cr HydA2), *Scenedesmus obliquus* (So HydA1), and *Chlorella fusca*, previously known as *Scenedesmus vacuolatus* (Cf HydA). The respective Genbank accession numbers are AF289201, AY055756, AJ271546, and AJ298227

Characteristics	Cr HydA1	Cr HydA2	So HydA1	Cf HydA
Exons	8	10	6	6
Introns	7	9	5	5
Coding regions (base pairs)	1494	1515	1344	1310
5′ UTR* (base pairs)	158	139	154	
3″ UTR (base pairs)	747	873	1100	
Protein (amino acid residues)	497	505	448	436
Transit peptide (amino acid residues)	50	62	35	21
Insert (base pairs)	45	54	16	16
Location of TATA box	187 bp upstream from 5′ UTR	24 bp upstream from 5′ UTR	25 bp upstream from 5′ UTR	25 bp upstream from 5′ UTR

*UTR: untranslatable terminal region of a DNA gene sequence

all algal hydrogenases have transit peptides of various length, as indicated in **Table 1**. There are also small differences in the length of each gene product, as well as in the number and localization of exons and introns. Characteristically, algal hydrogenases have two extra peptide insertions that are not observed in bacterial [FeFe]-hydrogenases (**Figure 1*a***). The insertions correspond to extra loops found at the N- and C-terminal domains of the protein that may serve as signal transducing components.

Gene Transcription

In *C. reinhardtii*, transcription of the *HYDA1* and *HYDA2* genes, as well as the two [FeFe]-hydrogenase maturation genes, *HYDEF* and *HYDG* (see below), are induced by anaerobiosis, although the precise regulatory mechanisms are unknown. Interestingly, a number of metabolic mutants have been isolated that exhibit attenuated hydrogenase gene transcription under anaerobic conditions. This suggests that additional factors other than just the lack of O_2 have a role in controlling *HYDA1/2* gene transcription in the alga. In two *C. reinhardtii* starchless mutants, *sta6* (5, 113) and *sta7* (83), *HYDA1/2* gene transcription is significantly reduced relative to wild-type cultures under dark, anaerobic conditions (83). These results illustrate the importance of starch metabolism in sustaining *HYDA1/2* transcription under anaerobic conditions. Recent experiments have demonstrated that *sta7* has a more oxidized plastoquinone pool during dark, anaerobic adaptation (M.C. Posewitz, unpublished), and it is possible that the reduction state of cellular redox constituents such as plastoquinones are responsible for initiating signal transduction events controlling the expression of the hydrogenase genes during anaerobiosis. The degradation of starch, which occurs under anerobic conditions, can influence intracellular levels of NAD(P)H and/or the oxidation state of the plastoquinone pool, both of which have been demonstrated to regulate gene transcription either directly or through signal transduction (29, 80, 86). However, further experimentation will be required to more clearly investigate whether this type of regulation applies to hydrogenase gene expression.

The precise DNA regulatory elements controlling hydrogenase gene expression are also currently unknown. Experiments coupling various lengths of the *C. reinhardtii HYDA1* promoter region to a promoterless arylsulfatase (*ARS*) reporter gene have been reported (95). The region between –128 and

Radical SAM: family of proteins that catalyze the formation of radicals through reductive cleavage of *S*-adenosylmethionine

–21 relative to the *HYDA1* transcription start site is required for anaerobic gene expression, but a more detailed examination will be needed to fully define the DNA elements regulating the expression of the genes encoding the HydA1/2 hydrogenases.

Enzyme Maturation

The maturation proteins involved in [FeFe]-hydrogenase assembly were initially discovered in *C. reinhardtii*, where two novel radical *S*-adenosylmethionine (Radical-SAM or AdoMet radical) proteins, HydEF and HydG, are required for [FeFe]-hydrogenase activity (82). The *HYDEF* gene is disrupted in a *C. reinhardtii* strain that was isolated by screening insertional mutants for colonies that were unable to produce H_2 (81, 82). Hydrogenase activity was restored in this mutant by complementation with a wild-type copy of the *HYDEF* gene, demonstrating that *HYDEF* disruption is responsible for the observed phenotype.

Additional evidence that HydEF and HydG are required for [FeFe]-hydrogenase maturation was demonstrated by the heterologous expression of active *C. reinhardtii* HydA1 in *E. coli*, a bacterium that lacks a native [FeFe]-hydrogenase (82). This work revealed that the expression of active [FeFe]-hydrogenase occured only when the *HYDEF* and *HYDG* genes were coexpressed with *HYDA1*.

The HydEF protein contains two unique domains, which are homologous to two distinct prokaryotic proteins, HydE and HydF, that are found exclusively in organisms containing [FeFe]-hydrogenases. In several bacterial genomes, *hydE*, *hydF*, and *hydG* are found in putative operons with the [FeFe]-hydrogenase structural genes (82).

HydE and HydG belong to the emerging Radical-SAM superfamily of proteins, and both contain the signature C-X_3-C-X_2-C motif that is characteristic of these proteins, as indicated in **Figure 1*b*,*c***. In both proteins, a [4Fe-4S] cluster is coordinated by each of the three cysteines of this motif, and the methionine carboxylate and amine of SAM bind the [4Fe-4S] cluster at the open iron coordination site in a bidentate fashion (18, 110). Biophysical characterization of purified and reconstituted *T. maritima* HydE and HydG indicates that both proteins most likely contain two distinct [4Fe-4S] clusters and have the expected SAM cleavage activity (85). A C-terminal C-X_7-C-X_2-C motif that could potentially bind FeS centers is found in several HydE homologs; however, not all HydE proteins have this second motif, and the two forms of HydE are differentiated as A and B forms. The HydG proteins also contain a cysteine-containing C-terminal motif, C-X_2-C-X_{22}-C, that could bind FeS centers as well. Each cysteine in the Radical-SAM motifs of both HydE and HydG has been mutated to serine in the *Clostridium acetobutylicum* (Ca) proteins, and, in combination with three cysteins in the C-terminal of HydG, are critical for [FeFe]-hydrogenase assembly in vivo and in *E. coli* heterologous expression system (51).

HydF contains a putative GTPase domain (**Figure 1*b***), and all HydF homologs have an N-terminal triphosphate nucleotide-binding domain. The C-terminal domain has conserved cysteine and histidine amino acids arranged in a C-X-H-$X_{(44-53)}$-HC-X_2-C motif, suggesting [FeS]-cluster coordination (13, 51, 82). HydF from *T. maritima* was purified, reconstituted, demonstrated to have GTP-hydrolysis activity, and found to possess a [4Fe4S] cluster (13). The GTPase domain and [FeS]-cluster binding in HydF are also required for [FeFe]-hydrogenase assembly as demonstrated by site-directed mutagenesis (51).

The precise mechanism of [FeFe]-hydrogenase assembly is unknown. However, the Radical-SAM superfamily is a highly versatile group of enzymes, and several characteristics exhibited by these proteins are consistent with the requirements necessary for assembling the [FeFe]-hydrogenase catalytic site. First, three different Radical-SAM enzymes are known to incorporate sulfur into novel substrates (19, 20, 23, 48, 102), and it is likely that either HydE and/or HydG is required to synthesize the di-(thiomethyl)amine bridging ligand. Second, a Radical-SAM protein catalyzes the formation of the NifB cofactor, which contains iron and sulfur that ultimately becomes incorporated into the catalytic site of the nitrogenase enzyme (2, 25). HydE or HydG could have similar roles in the biosynthesis of the [FeFe]-hydrogenase active site. Lastly, members of the Radical-SAM superfamily facilitate a number of difficult synthetic reactions, often under anaerobic conditions. In addition to the bridging dithiolate ligand, the hydrogenase 2Fe catalytic center also has CN and CO ligands. It is conceivable that either HydE and/or HydG are responsible for the biosynthesis of these ligands from metabolic precursors as Peters et al. (79) recently proposed. Insight into the functional role of HydF in H-cluster biosynthesis and [FeFe]-hydrogenase maturation may also be garnered from examples of similar proteins known to participate in the maturation of other metalloenzymes. These include CO-dehydrogenase (CODH) (49), urease (65), and [NiFe]-hydrogenase (9), each of which require either GTP or ATP hydrolysis by system-specific maturation proteins.

Oxygen Inhibition

As indicated above, the unique metallo-cluster of [FeFe]-hydrogenases exhibits irreversible sensitivity to O_2 and reversible sensitivity to CO inhibition. In most [FeFe]-hydrogenases, both compounds inhibit the enzyme, possibly by binding to the open coordination site on the distal Fe and preventing protons from binding competitively to the same site (54). The reversibility of O_2 inhibition is a property that has been characterized for most of the [NiFe]-hydrogenases (see below). However, in the presence of a two-electron oxidant, iron, and ethylenediaminetetraacetic acid (EDTA), the [FeFe]-hydrogenase from *Desulfovibrio vulgaris* Hildenborough (DvH) (109) can be exposed to O_2 (104), and then reactivated following anaerobic reduction with sodium dithionite.

The chemical nature of the oxygen species bound to the H-cluster after the exposure of [FeFe]-hydrogenase to O_2 is not known, but density function theory (DFT) calculations on inactivated states of the H-cluster in dithiolate models have proposed a Fe^{II}-Fe^{II} oxidation state for the 2Fe center, with a possible OH group terminally bound to the distal iron (56).

The level of O_2 inhibition varies among [FeFe]-hydrogenases, with I_{50} values ranging from less than a few seconds for *C. reinhardtii* enzymes, to several minutes for the clostridial enzymes (51). Algal hydrogenases, which lack the additional accessory cluster domain found in bacterial enzymes, are typically more sensitive to O_2 inhibition than are the enzymes isolated from bacteria. However, it is clear that this elevated sensitivity is not solely because of the lack of the N-terminal accessory cluster domain (51).

ISSUES AND CHALLENGES RELATED TO THE DEVELOPMENT OF EFFICIENT HYDROGEN-PHOTOPRODUCING ALGAE

The O_2 sensitivity of hydrogenase transcription, maturation, and catalytic activity are all major issues delaying the development of an applied algal photobiological H_2-producing system. Additionally, (*a*) limitations on the rate of electron transport and the predominance of cyclic electron transfer under H_2-producing conditions, (*b*) competition for reductants from ferredoxin among different pathways, and (*c*) low sunlight conversion efficiency of H_2 photoproduction represent major challenges that need to be solved (36a). These challenges must be surmounted before a cost-effective process can be developed. Here we address some of the ongoing molecular engineering research that focuses on the O_2-sensitivity issue. Another approach, based

Molecular dynamics: computational technique that allows the calculation and visualization of the diffusion of small molecules through a known protein structure

on the partial and reversible inhibition of photosynthetic O_2 evolution has provided a tool to help develop processes for sustainable algal H_2 photoproduction (63). This approach, based on sulfur deprivation of algal cultures, has been reviewed extensively (38, 42, 62) and is not discussed here.

The O_2 sensitivity of [FeFe]-hydrogenase gene transcription might be circumvented by expressing the hydrogenase genes under the control of a promoter that is inducible by conditions other than anaerobiosis, as has been achieved in heterologous systems (51). However, the reactions catalyzed by the hydrogenase's maturation proteins may also require anaerobiosis (82). This important challenge can be overcome by increasing the half-life of the hydrogenase activity. An engineered enzyme that has a half-life of about half a day in the presence of photosynthetically generated O_2, only needs to be resynthesized during the night, when the cultures become anaerobic via respiration of metabolic substrate. In practice, it may be possible to engineer an [FeFe]-hydrogenase so that its catalytic site is not accessible to O_2 (37). The following observations support this approach: (*a*) the O_2 tolerance of the H_2-sensing [NiFe] hydrogenase from *Ralstonia eutropha* can be decreased by site-directed mutagenesis that widens its gas channel (14), and (*b*) the O_2 tolerance of the two *Anabaena variabilis* ferredoxins, FdxH1 and FdxH2, can be decreased or increased respectively by mutation of a single amino acid residue (94). The amino acid residue in question is part of the structure of a channel between the surface of the enzyme and its catalytic site.

Engineering the [FeFe]-hydrogenase gas channel for increased O_2 tolerance requires characterization of the diffusion pathways available for H_2 and O_2 gases to and from the catalytic site. Reliable structures for [FeFe]-hydrogenases are necessary to develop this information. Although the algal [FeFe]-hydrogenases have not been crystallized, the X-ray structure of the homologous [FeFe]-hydrogenases from *Clostridium pasteurianum* (78) and *Desulfovibrio desulfuricans* (69) has been solved. The existence of a putative H_2-gas pathway was initially proposed (69) based on the available structures. Molecular dynamics simulations (21) and solvent accessibility maps (22) have demonstrated that, due to its small diameter, H_2-gas molecules can diffuse freely through most of the protein structure, whereas the movement of larger O_2 molecules is restricted to two major pathways. One of the latter corresponds to the H_2 channel proposed by previous research (69). The second channel, however, was discovered on the basis of the dynamic motion of the protein (21, 22). These results suggest the feasibility of preferentially closing the O_2 pathways without significantly interfering with the diffusion of H_2 gas (37). The hypothesis was recently confirmed by the implementation of a mutation, identified through computational simulation, to partially close one of the O_2 gas-diffusion pathways. The resulting mutant showed increased tolerance to O_2 inactivation (36b). Unfortunately, other mutants similarly identified displayed significantly decreased catalytic activity.

CYANOBACTERIAL [NiFe]-HYDROGENASES

Enzyme Structure and Catalytic Mechanism

Cyanobacteria possess two functionally different types of [NiFe]-hydrogenases, an uptake and a bidirectional enzyme. The former is responsible for taking up and recycling the H_2 produced as a side reaction from nitrogenase; it is found mostly in nitrogen-fixing cyanobacteria (46) and in some unicellular non-nitrogen-fixing microorganisms (77). The amino acid sequences of uptake hydrogenases are highly conserved in cyanobacteria, ranging from 93–99.7% similarity (96). The catalytic large subunit (HupL) has putative Ni- and Fe-binding sites at its N- and C-terminals (**Figure 3*a***), suggesting the existence of a classic NiFe active site structure

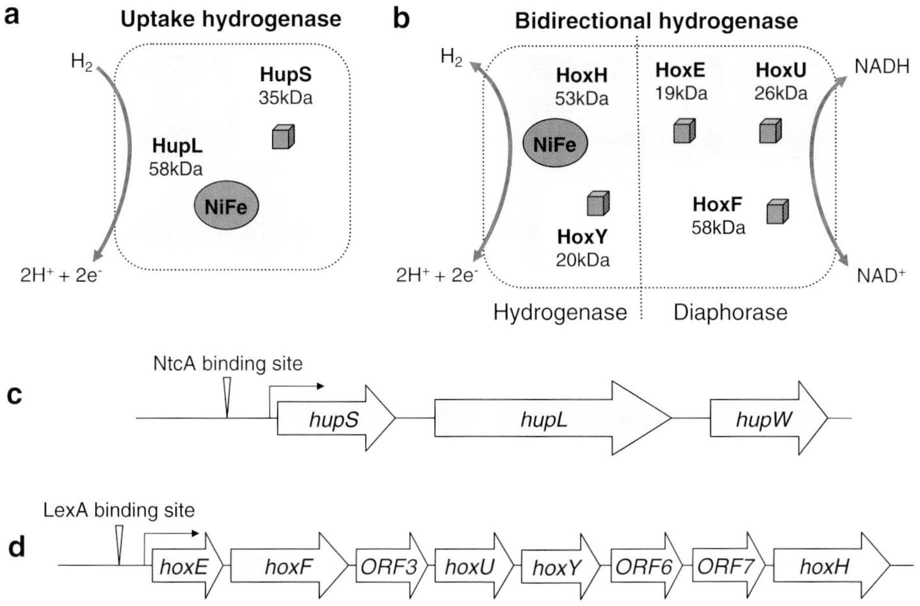

Figure 3

[NiFe]-hydrogenases and their gene structures in cyanobacteria. The molecular masses indicated for the uptake hydrogenase (*a*) are from *Anabaena* strain PCC 7120, and those for the bidirectional enzyme (*b*) are from *Synechocystis* sp. PCC 6803. The gene structure for the uptake hydrogenase (*c*) is based on *Gloeothece* sp. ATCC 27152, whereas the one for the bidirectional hydrogenase (*d*) is from *Synechocystis* sp. PCC 6803.

(**Figure 2*b***) (105). The small subunit (HupS), on the other hand, contains eight cysteine residues that are probably involved in [FeS] cluster formation (**Figure 3*a***) (75). Although a two-subunit enzyme seems to be the current model, no active cyanobacterial uptake hydrogenase has been purified yet, and therefore the exact subunit composition remains uncertain.

Cyanobacterial bidirectional hydrogenases are heteropentameric (**Figure 3*b***), consisting of a hydrogenase structural complex (HoxYH) and a diaphorase component (HoxEFU). The catalytic large subunit (HoxH) harbors conserved cysteines involved in binding nickel and iron to the active site (96). The small subunit (HoxY) also contains cysteine residues believed to be involved in coordinating a putative [4Fe-4S] cluster. The diaphorase moiety is a flavo-protein, encoded by the *hoxEFU* genes. It is composed of a catalytic part (HoxFU) that interacts with the NAD^+/NADH couple, and a third subunit (HoxE) (11). Both the large subunit (HoxF) and the small subunit (HoxU) have typical binding motifs for [2Fe-2S] and [4Fe-4S] clusters, whereas HoxE may be involved as a bridging subunit in membrane attachment and in electron transport, since its gene sequence also indicates the presence of a [2Fe-2S] binding motif (3, 34, 89).

Uptake and bidirectional enzymes have different functions in the cell. The former oxidizes H_2 whereas the latter can both oxidize and produce H_2. The catalytic process is well documented in bacterial [NiFe] uptake hydrogenases (108). Hydrogen is transported via a hydrophobic channel to the active site, where it is oxidized by a [NiFe] cluster. The binuclear metal center is ligated to four conserved cysteine residues (RxCGxCxxxH and DPCxxCxxH) in the N-terminal and C-terminal regions of the large subunit. Two of the cysteines bridge the Fe and the Ni atoms. The Fe is also coordinated by two cyanide

Diaphorase: any enzyme capable of catalyzing oxidation of NADH or NADPH in the presence of an electron acceptor other than oxygen

ATCC: American Type Culture Collection

PCC: Pasteur Culture Collection of Cyanobacteria

IAM: Institute of Applied Microbiology, Culture Collection, The University of Tokyo

CALU: St. Petersburg Culture Collection

ligands and one carbon monoxide ligand (**Figure 2*b***). The Ni atom is probably the site where the H_2 binds as a terminal ligand because the gas channel ends there. However, other studies suggest that H_2 could bind to the Fe atom instead (70).

Hydrogen cleavage is a heterolytic reaction, which forms a hydride and a proton. Then two electrons are extracted from the hydride and a second proton is formed to complete the reaction. Finally, the electrons are transferred to a redox partner and the protons to the protein environment (6). The electron pathway is formed by two [4Fe-4S] clusters, one proximal to the active site and the other exposed to the protein surface. Together, they allow fast tunneling of the electrons between the active site and the distal cluster (6). Less is known about proton transport pathways, but a cysteine ligand and a glutamate residue (28) seem to be involved in the process (27).

Bidirectional enzymes produce and recycle H_2 depending on the physiological requirements of the cell. Their catalytic mechanism remains unclear, although it is often assumed to be similar to uptake enzyme, but to operate in reverse sequence for H_2 production.

Gene Structure and Diversity

Uptake hydrogenases. The first *hupL* sequence was reported in 1995 (16) in *Anabaena* PCC 7120 (also named *Nostoc* PCC 7120). Subsequently, *hupS* and *hupL* genes were sequenced and characterized in *Nostoc* PCC 73102 and *Anabaena variabilis* (reviewed in 96). Since then, *hup* sequences for *Gloeothece* ATCC 27152 (71), *Lyngbya majuscule* CCAP 1446/4 (53), *Lyngbya aestuarii* CCY 9616 (GenBank #DQ375444), *Nostoc* PCC 7422 (GenBank #AB237640), *Gloeothece* PCC 6909 (GenBank #AY260103), *Anabaena siamensis* TISTR8012 (GenBank #AY152844), *Trichodesmium erythraeum* IMS101, and *Crocosphaera watsonii* WH8501 (**http://genome.jgi-psf.org/mic_home.html**) have become available.

As is the case for most [NiFe]-hydrogenases, the structural genes are clustered in one operon, where *hupS* is upstream of *hupL* (**Figure 3*c***). In some heterocystous strains, such as *Anabaena* PCC 7120, the excision of a DNA element by site-specific recombination occurs within *hupL* during the differentiation of a vegetative cell into a heterocyst, allowing the exclusive expression of HupL in the latter cell type (16).

Bidirectional hydrogenase. The first set of structural genes (*hox*) was reported in 1995 from *Anabaena variabilis* ATCC 29413 (88). Since then, *hox* genes have been sequenced in the *Synechocystis* PCC 6803, *Synechococcus* PCC 6301, *Anabaena* PCC 7120 (**http://www.kazusa.or.jp/cyano/anabaena**), and *Anabaena* IAM M58 (GenBank #AB057405), as reviewed in Reference 96, and in *Gloeocapsa alpicola* CALU 743 (93), several strains of *Arthrospira* and *Spirulina* (114), and *Prochlorothrix hollandica* (GenBank #U88400). Not all cyanobacteria have *hox* genes. In a recent survey from a broad range of sources, there was no evidence for their presence in several filamentous strains (97). In the cyanobacterial strains that possess them, there is no data supporting the existence of multiple bidirectional hydrogenases in a single strain.

The physical organizations of the structural genes encoding the bidirectional hydrogenases are similar in different strains. The *hoxEFUYH* genes are usually clustered, with additional ORFs interspersed between some of the structural genes (**Figure 3*d***). These ORFs are unlikely to be components of the enzyme since their products are not known to be involved in hydrogen metabolism, and since their number, order, and identity vary in the different strains (89).

Gene Transcription

The transcription of the cyanobacterial uptake hydrogenases coincides with heterocyst development, as seen when *Anabaena* PCC

7120, *Nostoc muscorum*, and *A. variabilis* ATCC 29413 are transferred from non-N_2-fixing to N_2-fixing conditions (4, 11, 16). Analysis of promoter elements reveals putative binding sequences for the fumarate nitrate reductase regulator (FNRR) in *A. variabilis* and for the global cyanobacterial nitrogen regulator NtcA in the *Nostoc* strain PCC 73102 (**Figure 3*c***); the former is involved in expressing the *E. coli hyp* operon and the latter is required for cyanobacterial heterocyst differentiation and nitrogen fixation (96).

Culturing conditions, such as the constant removal of O_2 and low light intensity promote H_2 production in growing cultures of *Anabaena* sp. 7120, *A. cylindrica*, and *A. variabilis* (47, 92). Surprisingly, the bidirectional hydrogenase polypeptide is present in both the anaerobically induced and noninduced (aerobic culture) cells of *A. variabilis*. When cultured under limited nitrogen (1–2 mM sodium nitrate), enhanced rates of H_2 production in the dark were observed almost immediately in both *Synechocystis* sp. PCC 6803 and *Gloeocapsa alipicola* CALU 743 (91, 93). Even though the hydrogenase activity in *G. alipicola* increased over time in darkness, constant levels of *hoxY* and *hoxH* transcripts were determined at different times. Collectively, these observations suggest that the transcription of the bidirectional hydrogenase gene is constitutive (97) and that the protein is likely present in aerobic cultures to account for the immediate appearance of activity when anaerobic conditions are restored.

The genetic elements controlling the transcription of the *hox* genes are still unclear. The mechanism will likely vary, based on differences in transcript sizes and gene loci in different cyanobacteria. Schmitz et al. (87) detected three separate promoter activities in *Synechococcus* driving the synthesis of three discrete hydrogenase-related transcripts: *hoxEF*, *hoxUYH*, and *hoxWhypAB*. Furthermore, both the P_{hoxE} and P_{hoxU} promoters are under the control of a circadian clock, the first demonstration of circadian control of gene expression for any hydrogenase. In contrast, transcription of the *hoxEFUYH* polycistronic unit in *Synechocystis* sp. PCC 6803 is driven by a single promoter situated upstream of *hoxE* (**Figure 3*d***) (41, 72). Gel shift assays, followed by protein analysis, showed that LexA acts as a transcription factor by binding to two essential regions: −97 bp to +10 bp and −569 bp to −690 bp relative to the initiation codon ATG start of *hoxE*. The LexA-deleted mutant exhibited a decrease in bidirectional hydrogenase activity, reinforcing LexA's involvement in *hox* gene expression. More research is underway to further elucidate the mode of action conferred by LexA in *Synechocystis* and to determine its role in modulating bidirectional hydrogenase gene expression in other cyanobacteria.

Fumarate nitrate reductase regulator (FNRR): an oxygen sensor in *E. coli*. It is a DNA-binding protein that up- or downregulates various modes of metabolism in response to the oxygen levels in the environment

Enzyme Maturation

Biosynthesis of active [NiFe]-hydrogenases is a complex process that has been well studied in *E. coli* [see review by Böck et al. (9)]. The maturation of those enzymes requires several accessory proteins that are encoded by the *hypABCDE-F* genes (57). Homologous gene products from other bacteria are well characterized and homologous Hyp proteins may fulfill similar functions in different organisms (17), including cyanobacteria (see below). A schematic diagram of the biosynthesis and maturation pathway of hydrogenase-3 in *E. coli* is shown in **Figure 4**.

According to the current model, the iron ligand is initially formed on the HypC/HypD dimer. Simultaneously, a thiocyanate ligand is generated from carbamoyl phosphate on the HypE protein (76, 84) in a reaction driven by ATP hydrolysis and catalyzed by the interacting HypF subunit. The CN ligands are subsequently donated to the Fe-HydC/HydD dimer. The partially assembled metallo center is then transferred from HypC to the precursor form of the catalytic subunit, pre-HycE (58), which is present in an open conformation that allows for metal insertion (8). The

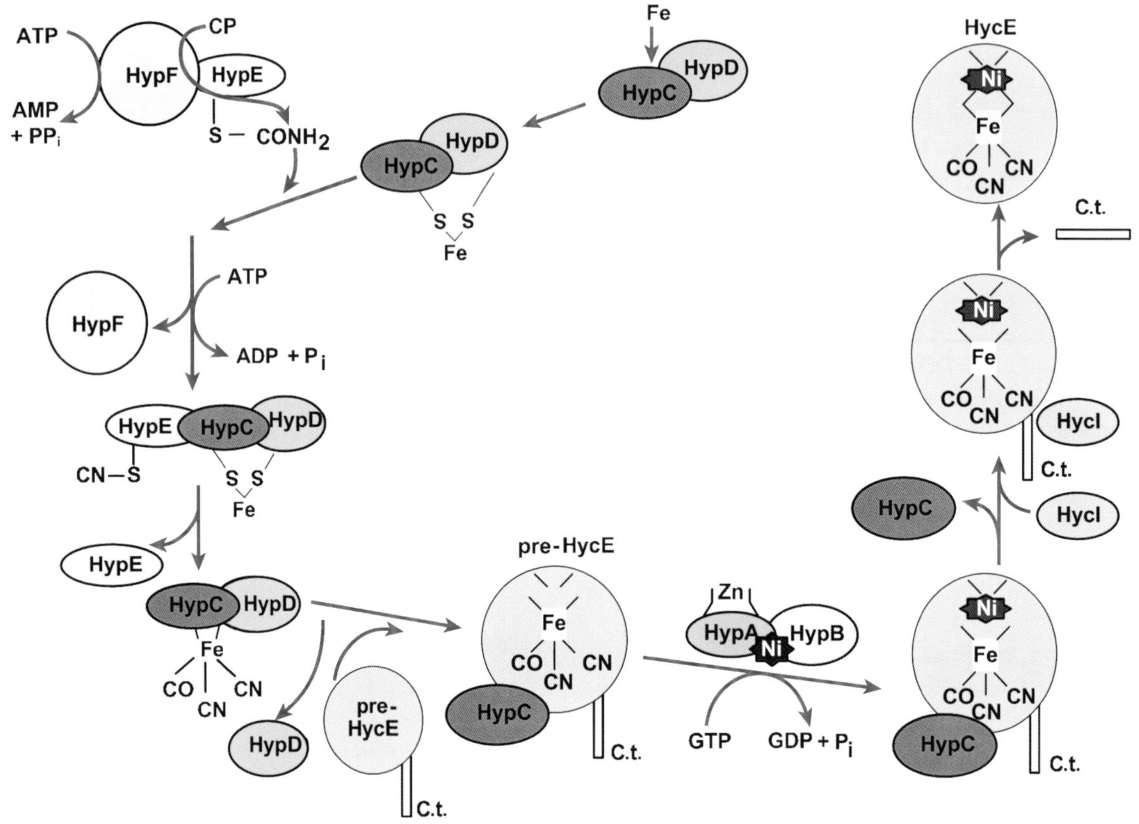

Figure 4

Network of hydrogenase maturation chaperones (adapted from Reference 8). CP is carbomoyl phosphate, C.t. is the C terminus of the precursor to the large subunit pre-HycE.

next step is inserting nickel via the HypB-HypA complex (59), involving GTP hydrolysis by HypB (60). When both metals have been coordinated to the large subunit precursor form, the C terminus is then accessible for cleavage by an endopeptidase. The cleavage reaction is a subunit-specific process and is catalyzed by HycI (106).

In the *Synechococcus* strain PCC 6301, the HypA, B, C, D, E, F, and W proteins (which are homologs of the *E. coli* HypABCDE-F and HycI) are believed to be involved in the maturation of the large subunit, HoxH (98). HypC possesses an N-terminal cysteine domain that could interact with the precursor of HoxH (73), and HypD has five conserved cysteines, which could be involved in metal binding (40). The HypE protein has a guanine-rich motif that could possibly be involved in purine-derivative binding, as well as other motifs indicating that it could have a phosphoribosylformylglycinamidine cycloligase function (35). Its potential partner, HypF, is also a good candidate for CO/CN ligand synthesis and has a conserved motif, potentially corresponding to an acylphosphatase (73). The HypA and HypB protein couple shows putative Ni-binding sites as well as a GTPase domain for the latter (74, 99). A recent report (44b) demonstrates that HypA and HypB are required for Ni insertion into the large subunit of the *Synechocystis* sp. 6803 hydrogenase, while its two homologs, HypA2 and HypB2, probably have a role as chaperones in the maturation of a different set of metalloproteins in this organism. Finally, HoxW in *Alcaligenes eutrophus* (98) and Slr2876 in *Synechocystis* sp. 6803 (44b) are putative proteases for HoxH, and could have a role similar to that of HycI in *E. coli*.

Oxygen Inhibition

As mentioned previously, [NiFe]-hydrogenases are generally more tolerant to O_2 than [FeFe]-hydrogenases, and their inactivation by oxygen and carbon monoxide is reversible (50, 61, 103, 107). Upon exposure to O_2, an oxo- or a hydroxo-group is formed, bridging the Ni and the Fe atoms and rendering the [NiFe]-hydrogenase resistant to further O_2 inactivation (108). Once returned to an anaerobic environment, the addition of a reducing agent such as sodium dithionite or H_2 re-reduces the oxo- or hydroxo-group, and hydrogenase activity is restored (33, 44a). This (*a*) accounts for the reversibility of the [NiFe]-hydrogenase with respect to O_2 inactivation, (*b*) explains the lack of H_2 evolution in cyanobacterial cultures during photosynthesis, and (*c*) provides a rationale for the restoration of hydrogenase activity when cultures are assayed under dark, anaerobic conditions.

Using sensitive mass spectrometry for real-time measurements, light-driven H_2 production was first detected in *Oscillatoria chalybea* and *Synechocystis* sp. PCC 6803 (1, 24). However, the reaction was short-lived (lasting less than 30 s in the light) and was followed immediately by H_2 uptake. More sustained H_2 production was reported in the *Synechocystis ndh*B M55 mutant, defective in type I NAD(P)H dehydrogenase complex, where continuous H_2 production lasted for about 5 min. Cournac et al. (24) attributed this phenomenon to a combination of lower O_2 evolution and negligible H_2-uptake activities in the M55 mutant. These results again underscore the need to address hydrogenase O_2 sensitivity in order to sustain H_2 photoproduction in cyanobacteria.

ISSUES AND CHALLENGES RELATED TO THE DEVELOPMENT OF EFFICIENT HYDROGEN-PHOTOPRODUCING CYANOBACTERIA

Several technical barriers and issues associated with cyanobacterial H_2 production must be solved prior to any biotechnological application. These include (*a*) the identification of more active strains, (*b*) the O_2 sensitivity of the hydrogenase reaction, (*c*) the partitioning of photosynthetic reductant toward H_2 production, (*d*) H_2 consumption by the bidirectional hydrogenase possibly via complex I–mediated respiration, and (*e*) H_2 oxidation mediated by the *hup*-uptake hydrogenase. This review addresses only the O_2-sensitivity issue as the other issues were discussed extensively in a recent review (96).

X-ray crystallography and molecular dynamics modeling has identified a hydrophobic gas channel connecting the active site of [NiFe]-hydrogenase to its surface in *Desulfovibrio gigas* and *D. fructosovorans* (66). Presumably, this putative H_2-diffusion gas channel also provides the path for O_2 to access the active site. Restriction of this gas channel might block O_2 access, thus improving O_2 tolerance at the molecular level, as discussed above for [FeFe]-hydrogenases. Crystal structures of *hox*-type hydrogenases from cyanobacteria are needed to guide this research.

Interestingly, the unusual O_2 tolerance of the soluble hydrogenase in *R. eutropha* was correlated with the presence of an extra cyanide ligand coordinated to the nickel in the active site (7). Engineering an extra ligand to occupy the uncoordinated site of the NiFe metal cluster in other hydrogenases might improve O_2 tolerance, provided that the enzyme remains active. Alternatively, known O_2-tolerant bacterial [NiFe]-hydrogenases [e.g., the hydrogenase from *Rubrivivax gelatinosus* (61)] could be cloned and expressed in cyanobacteria for H_2 photoproduction. However, heterologous expression of foreign [NiFe]-hydrogenase structural genes in cyanobacteria might require the cotransfer of donor-specific assembly/maturation genes to yield a functional hydrogenase. This concept was demonstrated by the heterologous expression of the *R. eutropha* hydrogenase in *Pseudomonas stutzeri* (55). Fortunately, several strains of cyanobacteria are genetically tractable and many tools tailored to cyanobacteria have been developed to meet genetic engineering needs (52). The identification of additional O_2-tolerant hydrogenases and the development of more advanced genetic tools will inevitably accelerate progress.

SUMMARY POINTS

1. Hydrogen photoproduction from water requires the coordinated action of oxygenic photosynthetic electron transport and the catalytic activity of a hydrogenase enzyme. Both are found in green algae and cyanobacteria.
2. Green algae contain [FeFe]-hydrogenases, which are H_2-producing, high-catalytic-turnover enzymes; cyanobacteria contain [NiFe]-hydrogenases, which can be involved either in H_2 uptake or H_2 production but have lower turnover rates.
3. The metallo-clusters of the two types of hydrogenases are unique in that they have CO and CN ligands; both are sensitive to O_2 and CO inactivation, although [NiFe]-hydrogenases can be reversibly inactivated by both inhibitors, whereas [FeFe]-hydrogenases are usually irreversibly inactivated by O_2.
4. The sensitivity of [FeFe]-hydrogenases to O_2 is reflected at the gene transcription and gene maturation levels as well; gene transcription of uptake [NiFe]-hydrogenases is usually coupled to heterocyst development, whereas that of the bidrectional enzyme is constitutive and perhaps under the control of a circadian clock.

FUTURE ISSUES

1. Detailed studies of the catalytic mechanism of different hydrogenases are required, including the analysis of synthetic metallo compounds with properties similar to those of the enzyme cofactors. The latter may show potential for a more immediate impact in the application of novel catalysts for artificial hydrogen-producing systems or fuel cells.
2. Additional crystal structures are also required to complement our current fundamental understanding of [FeFe] and [NiFe] hydrogenases. The structure and function, proton, electron, and gas transfer pathways need to be examined. These studies will guide further protein engineering research aimed at improving overall enzyme performance.
3. Very little is known about the regulation of [FeFe]-hydrogenase gene transcription and maturation, and we foresee a major effort in resolving these issues.
4. The major challenge for developing applied photobiological systems for H_2 production is still the O_2 sensitivity of the hydrogenases. Many different research directions are being pursued to solve this problem, and we expect to see a concentrated effort over the next couple of years in bioprospecting, molecular engineering, and selection technologies to generate desired enzymes.

ACKNOWLEDGMENTS

The authors would like to thank the U.S. Department of Energy's Office of Science (M.L.G. and M.S.), the Hydrogen, Fuel Cell and Infrastructure Technologies Program (M.L.G., M.S., P.C.M.), and the U.S. Air Force Office of Scientific Research (M.P.) for support. We also thank Dr. Paul King, NREL, for fruitful comments and suggestions.

LITERATURE CITED

1. Abdel-Basset R, Bader KP. 1998. Physiological analyses of the hydrogen gas exchange in cyanobacteria. *Photochem. Photobiol. B Biol.* 43:146–51
2. Allen RM, Chatterjee R, Ludden PW, Shah VK. 1995. Incorporation of iron and sulfur from NifB cofactor into the iron-molybdenum cofactor of dinitrogenase. *J. Biol. Chem.* 270:26890–96
3. Appel J, Schulz R. 1996. Sequence analysis of an operon of a NAD(P)–reducing nickel hydrogenase from the cyanobacterium *Synechocystis* sp. PCC 6803 gives additional evidence for direct coupling of the enzyme to NAD(P)H-dehydrogenase (complex I). *Biochim. Biophys. Acta* 1298:141–47
4. Axelsson R, Oxelfelt F, Lindblad P. 1999. Transcriptional regulation of *Nostoc* uptake hydrogenase. *FEMS Microbiol. Lett.* 170:77–81
5. Ball S. 1998. Regulation of starch biosynthesis. In *The Molecular Biology of Chloroplasts and Mitochondria in Chlamydomonas*, ed. JD Rochaix, M Goldschmidt-Clermont, S Merchant, 7:549–67. Dordrecht: Kluwer Acad.
6. Berlier Y, Lespinat PA, Dimon B. 1990. A gas chromatographic-mass spectrometric technique for studying simultaneous hydrogen-deuteron exchange and para-orthohydrogen conversion in hydrogenases of *Desulfovibrio vulgaris* Hildenborough. *Anal. Biochem.* 188:427–31
7. Bleijlevens B, Buhrke T, van der Linden E, Friedrich B, Albracht SPJ. 2004. The auxiliary protein HypX provides oxygen tolerance to the soluble [NiFe]-hydrogenase of *Ralstonia eutropha* H16 by way of a cyanide ligand to nickel. *J. Biol. Chem.* 279:46686–91
8. Blokesch M, Böck A. 2002. Maturation of the [NiFe]-hydrogenases in *Escherichia coli*: the HypC cycle. *J. Mol. Biol.* 324:287–96
9. Böck A, King PW, Blokesch M, Posewitz MC. 2006. Maturation of hydrogenases. *Adv. Microb. Physiol.* 51:1–72
10. Boichenko VA, Greenbaum E, Seibert M. 2004. Hydrogen production by photosynthetic microorganisms. In *Photoconversion of Solar Energy: Molecular to Global Photosynthesis*, ed. MD Archer, J Barber, pp. 397–452. London: Imperial College Press
11. Boison G, Bothe H, Schmitz O. 2000. Transcriptional analysis of hydrogenase genes in the cyanobacteria *Anacystis nidulans* and *Anabaena variabilis* monitored by RT-PCR. *Curr. Microbiol.* 40:315–21
12. Brand JJ, Wright J, Lien S. 1989. Hydrogen production by eukaryotic algae. *Biotechnol. Bioeng.* 33:1482–88
13. Brazzolotto X, Rubach JK, Gaillard J, Gambarelli S, Atta M, Fontecave M. 2006. The [Fe-Fe]-hydrogenase maturation protein HydF from *Thermotoga maritima* is a GTPase with an iron-sulfur cluster. *J. Biol. Chem.* 281:769–74
14. Buhrke T, Lenz O, Krauss N, Freidrich B. 2005. Oxygen tolerance of the H_2-sensing [NiFe] hydrogenase from *Ralstonia eutropha* H16 is based on limited access of oxygen to the active site. *J. Biol. Chem.* 280:23791–96
15. Cammack R. 1999. Hydrogenase sophistication. *Nature* 397:214–15

16. Carrasco CD, Buettner JA, Golden JW. 1995. Programmed DNA rearrangement of a cyanobacterial *hupL* gene in heterocysts. *Proc. Natl. Acad. Sci. USA* 92:791–95
17. Casalot L, Rousset M. 2001. Maturation of the [NiFe] hydrogenases. *Trends Microbiol.* 9:228–37
18. Chen D, Walsby C, Hoffman BM, Frey PA. 2003. Coordination and mechanism of reversible cleavage of *S*-adenosylmethionine by the [4Fe-4S] center in lysine 2,3-aminomutase. *J. Am. Chem. Soc.* 125:11788–89
19. Cicchillo RM, Booker SJ. 2005. Mechanistic investigations of lipoic acid biosynthesis in *Escherichia coli*: Both sulfur atoms in lipoic acid are contributed by the same lipoyl synthase polypeptide. *J. Am. Chem. Soc.* 127:2860–61
20. Cicchillo RM, Lee KH, Baleanu-Gogonea C, Nesbitt NM, Krebs C, Booker SJ. 2004. *Escherichia coli* lipoyl synthase binds two distinct [4Fe-4S] clusters per polypeptide. *Biochemistry* 43:11770–81
21. Cohen J, Kim K, King P, Seibert M, Schulten K. 2005. Finding gas diffusion pathways in proteins: O_2 and H_2 gas transport in CpI [FeFe]-hydrogenase and the role of packing defects. *Structure* 13:1–9
22. Cohen J, Kim K, Posewitz M, Ghirardi ML, Schulten K, et al. 2005. Molecular dynamics and experimental investigation of H_2 and O_2 diffusion in [Fe]-hydrogenase. *Biochem. Soc. Transact.* 33:80–82
23. Cosper MM, Jameson GN, Hernandez HL, Krebs C, Huynh BH, Johnson MK. 2004. Characterization of the cofactor composition of *Escherichia coli* biotin synthase. *Biochemistry* 43:2007–21
24. Cournac L, Guedeney G, Peltier G, Vignais PM. 2004. Sustained photoevolution of molecular hydrogen in a mutant of *Synechocystis* sp. strain PCC 6803 deficient in the type I NADPH-dehydrogenase. *J. Bacteriol.* 186:1737–46
25. Curatti L, Ludden PW, Rubio LM. 2006. NifB-dependent in vitro synthesis of the iron-molybdenum cofactor of nitrogenase. *Proc. Natl. Acad. Sci. USA* 103:5297–301
26. Darensbourg MY, Lyon EJ, Xhao X, Georgakaki IP. 2003. The organometallic active site of [Fe] hydrogenase: models and entatic states. *Proc. Natl. Acad. Sci. USA* 100:3683–88
27. De Lacey AL, Santamaria E, Hatchikian EC, Fernandez VM. 2000. Kinetic characterization of *Desulfovibrio gigas* hydrogenase upon selective chemical modification of amino acid groups as a tool for structure-function relationships. *Biochim. Biophys. Acta* 1481:371–80
28. Dementin S, Burlat B, De Lacey AL, Pardo A, Adryanczyk-Perrier G, et al. 2004. A glutamate is the essential proton transfer gate during the catalytic site of the [NiFe] hydrogenase. *J. Biol. Chem.* 279:10508–13
29. Escoubas JM, Lomas M, LaRoche J, Falkowski PG. 1995. Light intensity regulation of *cab* gene transcription is signaled by the redox state of the plastoquinone pool. *Proc. Natl. Acad. Sci. USA* 92:10237–41
30. Fan HJ, Hall MB. 2001. A capable bridging ligand for Fe-only hydrogenase: density functional calculations of a low-energy route for heterolytic cleavage and formation of dihydrogen. *J. Am. Chem. Soc.* 123:3828–29
31. Florin L, Tsokoglou A, Happe T. 2001 A novel type of [Fe]-hydrogenase in the green alga *Scenedesmus obliquus* is linked to the photosynthetic electron transport chain. *J. Biol. Chem.* 276:6125–32
32. Forestier M, King P, Zhang L, Posewitz M, Schwarzer S, et al. 2003. Expression of two [Fe]-hydrogenases in *Chlamydomonas reinhardtii* under anaerobic conditions. *Eur. J. Biochem.* 270:2750–58
33. Frey M. 2002. Hydrogenases: hydrogen-activating enzymes. *Chem. Bio. Chem.* 3:153–60

34. Friedrich T, Steinmüller K, Weiss H. 1995. The proton-pumping respiratory complex I of bacteria and mitochondria and its homologue in chloroplasts. *FEBS Lett.* 367:107–11
35. Garg RP, Menon AL, Jacobs K, Robson RM, Robson RL. 1994. The *hypE* gene completes the gene cluster for H_2-oxidation in *Azotobacter vinelandii*. *J. Mol. Biol.* 236:390–96
36a. Ghirardi ML. 2006. Hydrogen production by photosynthetic green algae. *Indian J. Biochem. Biophys.* 43:201–10
36b. Ghirardi ML, Cohen J, King P, Schulten K, Kim K, Seibert M. 2006. [FeFe]-hydrogenases and photobiological hydrogen production. *Solar Hydrog. Nanotechnol.* 6340:253–58
37. Ghirardi ML, King P, Kosourov S, Forestier M, Zhang L, Seibert M. 2005. Development of algal systems for hydrogen photoproduction – addressing the hydrogenase oxygen-sensitivity problem. In *Artificial Photosynthesis*, ed. C Collings. Weinheim, Germany: Wiley-VCH Verlag
38. Ghirardi ML, King PW, Posewitz MC, Maness PC, Fedorov A, et al. 2005. Approaches to developing biological H_2-photoproducing organisms and processes. *Biochem. Soc. Trans.* 33:70–72
39. Guan Y, Deng M, Yu X, Zhang W. 2004. Two-stage photobiological production of hydrogen by marine green alga *Platymonas subcordiformis*. *Biochem. Engin. J.* 19:69–73
40. Gubili J, Borthakur D. 1998. Organization of the *hupDEAB* genes within the hydrogenase gene cluster of *Anabaena sp.* strain PCC 7120. *J. Appl. Phycol.* 10:163–67
41. Gutekunst K, Phunpruch S, Schwarz C, Schuchardt S, Schulz-Friedrich R, Appel J. 2005. LexA regulates the bidirectional hydrogenase in the cyanobacterium *Synechocystis* sp. PCC 6803 as a transcription activator. *Mol. Microbiol.* 58:810–23
42. Happe T, Hemschemeier A, Winkler M, Kaminski A. 2002. Hydrogenases in green algae: Do they save the algae's life and solve our energy problems? *Trends Plant Scie.* 7:246–50
43. Happe T, Kaminski A. 2002. Differential regulation of the [Fe]-hydrogenase during anaerobic adaptation in the green alga *Chlamydomonas reinhardtii*. *Eur. J. Biochem.* 269:1022–32
44a. Higuchi Y, Ogata H, Miki K, Yasuoka N, Yagi T. 1999. Removal of the bridging ligand atom at the Ni-Fe active site of [NiFe] hydrogenase upon reduction with H_2, as revealed by X-ray structure analysis at 1.4 Å resolution. *Structure* 7:549–56
44b. Hoffman D, Gutekunst K, Klissenbauer M, Schulz-Friedrich S, Appel J. 2006. Mutagenesis of hydrogenase accessory genes of *Synechocystis* sp. PCC 6803. *FEBS J.* 273:4516–27
45. Horner DS, Heil B, Happe T, Embley TM. 2002. Iron hydrogenases – ancient enzymes in modern eukaryotes. *Trends Biochem. Sci.* 27:148–53
46. Houchins JP. 1984. The physiology and biochemistry of hydrogen metabolism in cyanobacteria. *Biochim. Biophys. Acta* 768:227–55
47. Houchins JP, Burris RH. 1981. Physiological reactions of the reversible hydrogenase from *Anabaena* 7120. *Plant Physiol.* 68:717–21
48. Jarrett J. 2005. Biotin synthase: enzyme or reactant? *Chem. Biol.* 12:409–10
49. Jeon WB, Cheng J, Ludden PW. 2001. Purification and characterization of membrane-associated CooC protein and its functional role in the insertion of nickel into carbon monoxide dehydrogenase from *Rhodospirillum rubrum*. *J. Biol. Chem.* 276:38602–9
50. Kemner J, Zeikus JG. 1994. Purification and characterization of the membrane-bound hydrogenase from *Methanosarcina barkeri* MS. *Arch. Microbiol.* 161:47–54
51. King PW, Posewitz MC, Ghirardi ML, Seibert M. 2006. Functional studies of [FeFe] hydrogenase maturation in an *Escherichia coli* biosynthetic system. *J. Bacteriol.* 188:2163–72

52. Koksharova OA, Wolk CP. 2002. Genetic tools for cyanobacteria. *Appl. Microbiol. Biotechnol.* 58:123–37
53. Leitão E, Oxefelt F, Oliveira P, Moradas-Ferreira P, Tamagnini P. 2005. Analysis of the *hupSL* operon of the nonheterocystous cyanobacterium *Lyngbya majuscule* CCAP 1446/4: regulation of transcription and expression under a light-dark regimen. *Appl. Envir. Microbiol.* 71:4567–76
54. Lemon BJ, Peters JW. 1999. Binding of exogenously added carbon monoxide at the active site of the iron-only hydrogenase (CpI) from *Clostridium pasteurianum*. *Biochemistry* 38:12969–73
55. Lenz O, Gleiche A, Strack A, Friedrich B. 2005. Requirement for heterologous production of a complex metalloenzyme: the membrane-bound [NiFe] hydrogenase. *J. Bacteriol.* 187:6590–95
56. Liu ZP, Hu P. 2002. A density functional theory study on the active center of Fe-only hydrogenase: characterization and electronic structure of the redox states. *J. Am. Chem. Soc.* 124:5175–82
57. Lutz S, Jacobi A, Schlensog V, Böhm R, Sawers G, Böck A. 1991. Molecular characterization of an operon (*hyp*) necessary for the activity of the three hydrogenases isoenzymes in *Escherichia coli*. *Mol. Microbiol.* 5:123–35
58. Magalon A, Böck A. 2000. Dissection of the maturation reactions of the (NiFe) hydrogenase 3 from *Escherichia coli* taking place after nickel incorporation. *FEBS Lett.* 473:254–58
59. Maier T, Jacobi A, Sauter M, Böck A. 1993. The product of the *hypB* gene, which is required for nickel incorporation into hydrogenases, is a novel guanine nucleotide-binding protein. *J. Bacteriol.* 175:630–35
60. Maier T, Lottspeich F, Böck A. 1995. GTP hydrolysis by HypB is essential for nickel insertion into hydrogenases of *Escherichia coli*. *Eur. J. Biochem.* 230:133–38
61. Maness PC, Smolinski S, Dillon AC, Heben MJ, Weaver PF. 2002. Characterization of the oxygen tolerance of a hydrogenase linked to a carbon monoxide oxidation pathway in *Rubrivivax gelatinosus*. *Appl. Environ. Microbiol.* 68:2633–36
62. Melis A, Seibert M, Happe T. 2004. Genomics of green algal hydrogen research. *Photosynth. Res.* 82:277–88
63. Melis A, Zhang L, Forestier M, Ghirardi ML, Seibert M. 2000. Sustained photobiological hydrogen gas production upon reversible inactivation of oxygen evolution in the green alga *Chlamydomonas reinhardtii*. *Plant Physiol.* 122:127–35
64. Mishra J, Kumar N, Ghosh AK, Das D. 2002. Isolation and molecular characterization of hydrogenase gene from a high rate of hydrogen-producing bacterial strain *Enterobacter cloacae* IIT-BT. *Int. J. Hydrogen Energy* 27:1475–79
65. Moncrief MB, Hausinger RP. 1997. Characterization of UreG, identification of a UreD-UreF-UreG complex, and evidence suggesting that a nucleotide-binding site in UreG is required for *in vivo* metallocenter assembly of *Klebsiella aerogenes* urease. *J. Bacteriol.* 179:4081–86
66. Montet Y, Amara P, Volbeda A, Vernede X, Hatchikian EC, et al. 1997. Gas access to the active site of Ni-Fe hydrogenases probed by X-ray crystallography and molecular dynamics. *Nat. Struct. Biol.* 7:523–26
67. Nicolet Y, Cavazza C, Fontecilla-Camps JC. 2002. Fe-only hydrogenases: structure, function and evolution. *J. Inorg. Biochem.* 91:1
68. Nicolet Y, de Lacey AL, Vernede X, Fernandez VM, Hatchikian EC, Fontecilla-Camps JC. 2001. Crystallographic and FTIR spectroscopic evidence of changes in Fe coordination upon reduction of the active site of the Fe-only hydrogenase from *Desulfovibrio desulfuricans*. *J. Am. Chem. Soc.* 123:1596–601

69. Nicolet Y, Piras P, Legrand CE, Hatchikian CD, Fontecilla-Camps JC. 1999. *Desulfovibrio desulfuricans* iron hydrogenase: The structure shows unusual coordination to an active site Fe binuclear center. *Struct. Fold. Des.* 7:13–23
70. Niu S, Hall MB. 2001. Modeling the active sites in metalloenzymes 5. The heterolytic bond cleavage of H_2 in the [NiFe] hydrogenase of *Desulfovibrio gigas* by a nucleophilic addition mechanism. *Inorg. Chem.* 40:6201–03
71. Oliveira P, Leitão E, Tamagnini P, Moradas-Ferreira P, Oxelfelt F. 2004. Characterization and transcriptional analysis of *hupSLW* in *Gloeothece* sp. ATCC 27152: an uptake hydrogenase from a unicellular cyanobacterium. *Microbiology* 150:3647–55
72. Oliveira P, Lindblad P. 2005. LexA, a transcription regulator binding in the promoter region of the bidirectional hydrogenase in the cyanobacterium *Synechocystis* sp. PCC 6803. *FEMS Microbiol. Lett.* 251:59–66
73. Olson JW, Maier RJ. 1997. The sequences of *hypF*, *hypC* and *hypD* complete the *hyp* gene cluster required for hydrogenase activity in *Bradyrhizobium japonicum*. *Gene* 199:93–99
74. Olson JW, Maier RJ. 2000. Dual roles of *Bradyrhizobium japonicum* nickelin protein in nickel storage and GTP-dependent Ni mobilization. *J. Bacteriol.* 182:1702–5
75. Oxelfelt F, Tamagnini P, Lindblad P. 1998. Hydrogen uptake in *Nostoc* sp. strain PCC 73102. Cloning and characterization of a *hupSL* homologue. *Arch. Microbiol.* 169:267–74
76. Paschos A, Glass RS, Böck A. 2001. Carbamoylphosphate requirement for synthesis of the active center of [NiFe]-hydrogenases. *FEBS Lett.* 488:9–12
77. Peschek GA. 1979. Evidence for two functionally distinct hydrogenases in *Anacystis nidulans*. *Arch. Microbiol.* 123:81–92
78. Peters JW, Lanzilotta WN, Lemon BJ, Seefeldt LC. 1998. X-ray structure of the Fe-only hydrogenase (CpI) from *Clostridium pasteurianum* to 1.8 Å resolution. *Science* 282:1853–58
79. Peters JW, Szilagyi RK, Naumov A, Douglas T. 2006. A radical solution for the biosynthesis of the H-cluster of hydrogenase. *FEBS Lett.* 580:363–67
80. Pfannschmidt T, Schutze K, Brost M, Oelmuller R. 2001. A novel mechanism of nuclear photosynthesis gene regulation by redox signals from the chloroplast during photosystem stoichiometry adjustment. *J. Biol. Chem.* 276:36125–30
81. Posewitz MC, King PW, Smolinski SL, Smith RD, Ginley AR, et al. 2005. Identification of genes required for hydrogenase activity in *Chlamydomonas reinhardtii*. *Biochem. Soc. Trans.* 33:102–4
82. Posewitz MC, King PW, Smolinski SL, Zhang L, Seibert M, Ghirardi ML. 2004. Discovery of two novel radical *S*-adenosylmethionine proteins required for the assembly of an active [Fe] hydrogenase. *J. Biol. Chem.* 279:25711–20
83. Posewitz MC, Smolinski SL, Kanakagiri S, Melis A, Seibert M, Ghirardi ML. 2004. Hydrogen photoproduction is attenuated by disruption of an isoamylase gene in *Chlamydomonas reinhardtii*. *Plant Cell* 16:2151–63
84. Reissmann S, Hochleitner E, Wang H, Paschos A, Lottspeich F, et al. 2003. Taming of a poison: biosynthesis of the NiFe-hydrogenase cyanide ligands. *Science* 299:1067–70
85. Rubach JK, Brazzolotto X, Gaillard J, Fontecave M. 2005. Biochemical characterization of the HydE and HydG iron-only hydrogenase maturation enzymes from *Thermatoga maritima*. *FEBS Lett.* 579:5055–60
86. Rutter J, Reick M, Wu LC, McKnight SL. 2001. Regulation of clock and NPAS2 DNA binding by the redox state of NAD cofactors. *Science* 293:510–14
87. Schmitz O, Boison G, Bothe H. 2001. Quantitative analysis of expression of two circadian clock-controlled gene clusters coding for the bidirectional hydrogenase in the cyanobacterium *Synechococcus* sp. PCC7942. *Mol. Microbiol.* 41:1409–17

88. Schmitz O, Boison G, Hilscher R, Hundeshage B, Zimmer W, et al. 1995. Molecular biological analysis of a bidirectional hydrogenase from cyanobacteria. *Eur. J. Biochem.* 233:266–76
89. Schmitz O, Boison G, Salzmann H, Bothe H, Schütz K, et al. 2002. HoxE: A subunit specific for the pentameric bidirectional hydrogenase complex (HoxEFUYH) of cyanobacteria. *Biochim. Biophys. Acta* 1554:66–74
90. Schnackenberg J, Ikemoto H, Miyachi S. 1995. Relationship between oxygen-evolution and hydrogen-evolution in a *Chlorococcum* strain with high CO_2-tolerance. *J. Photochem. Photobiol.* 28:171–74
91. Schutz K, Happe T, Troshina O, Lindblad P, Leitao E, et al. 2004. Cyanobacterial H_2 production – a comparative analysis. *Planta* 218:350–59
92. Serebriakova L, Zorin NA, Lindblad P. 1994. Reversible hydrogenase in *Anabaena variabilis* ATCC 29413. *Arch. Microbiol.* 161:140–44
93. Sheremetieva ME, Troshina OY, Serebryakova LT, Lindblad P. 2002. Identification of *hox* genes and analysis of their transcription in the unicellular cyanobacterium *Gloeocapsa alpicola* CALU 743 growing under nitrate-limiting conditions. *FEMS Microbiol. Lett.* 214:229–33
94. Singh BB, Curdt I, Shomburg D, Bisen PS, Böhme H. 2001. Valine 77 of heterocystous ferredoxin FdxH2 in *Anabaena variabilis* strain ATCC 29413 is critical for its oxygen sensitivity. *Mol. Cell. Biochem.* 217:137–42
95. Stirnberg M, Happe T. 2004. Identification of a *cis*-acting element controlling anaerobic expression of the HYDA gene from *Chlamydomonas reinhardtii*. In *Biohydrogen III*, ed. J Miyake, Y Igarashi, M Rogner, pp. 117–27. New York: Elsevier
96. Tamagnini P, Axelsson R, Lindberg P, Oxelfelt F, Wünschiers R, Lindblad P. 2002. Hydrogenases and hydrogen metabolism of cyanobacteria. *Microbiol. Mol. Biol. Rev.* 65:1–20
97. Tamagnini P, Costa J, Almeida L, Oliveira, Salema R, Lindblad P. 2000. Diversity of cyanobacterial hydrogenases, a molecular approach. *Curr. Microbiol.* 40:356–61
98. Thiemermann S, Dernedde J, Bernhard M, Schroeder W, Massanz C, Friedrich B. 1996. Carboxyl-terminal processing of the cytoplasmic NAD-reducing hydrogenase of *Alcaligenes eutrophus* requires the *hoxW* gene product. *J. Bacteriol.* 178:2368–74
99. Tibelius KH, Du L, Tito D, Stejskal F. 1993. The *Azotobacter chroococcum* hydrogenase gene cluster: sequences and genetic analysis of four accessory genes, *hupA*, *hupB*, *hupY and hupC*. *Gene* 127:53–61
100. Tye JW, Hall MB, Darensbourg MY. 2005. Better than platinum? Fuel cells energized by enzymes. *Proc. Nat. Acad. Sci. USA* 102:16911–12
101. Ueno Y, Kurano N, Miyachi S. 1999. Purification and characterization of hydrogenase from the marine green alga, *Chlorococcum littorale*. *FEBS Lett.* 443:144–48
102. Ugulava NB, Gibney BR, Jarrett JT. 2001. Biotin synthase contains two distinct iron-sulfur cluster binding sites: chemical and spectroelectrochemical analysis of iron-sulfur cluster interconversions. *Biochemistry* 40:8343–51
103. Van der Linden E, Faber BW, Bleijlevens B, Burgdorf T, Berhard M, et al. 2004. Selective release and function of one of the two FMN groups in the cytoplasmic NAD+-reducing hydrogenase from *Ralstonia eutropha*. *Eur. J. Biochem.* 271:801–8
104. Van Dijk C, Van Berkel-Arts A, Veeger C. 1983. The effect of reoxidation on the reduced hydrogenase of *Desulfovibrio vulgaris* strain Hildenborough and its oxygen stability. *FEBS Lett.* 156:340–44
105. Vignais PM, Billoud B, Meyer J. 2001. Classification and phylogeny of hydrogenases. *FEMS Microbiol. Rev.* 25:455–501

106. Vignais PM, Colbeau A. 2004. Molecular biology of microbial hydrogenases. *Curr. Iss. Mol. Biol.* 6:159–88
107. Volbeda A, Charon MH, Piras C, Hatchikian EC, Frey M, Fontecilla-Camps JC. 1995. Crystal structure of the nickel-iron hydrogenase from *Desulfovibrio gigas*. *Nature* 373:580–87
108. Volbeda A, Montet Y, Vernède X, Hatchikian EC, Fontecilla-Camps JC. 2002. High resolution crystallographic analysis of *Desulfovibrio fructosovorans* [NiFe] hydrogenase. *Int. J. Hydrog. Ener.* 27:1449–61
109. Voordouw G, Hagen WR, Krüse-Wolters KM, van Berkel-Arts A, Veeger C. 1987. Purification and characterization of *Desulfovibrio vulgaris* (Hildenborough) hydrogenase expressed in *Escherichi coli*. *Eur. J. Biochem.* 162:31–36
110. Walsby C, Ortillo D, Yang J, Nnyepi M, Broderick W, et al. 2005. Spectroscopic approaches to elucidating novel iron-sulfur chemistry in the "Radical-SAM" protein superfamily. *Inorg. Chem.* 44:727–41
111. Winkler M, Hemschemeier A, Gotor C, Melis A, Happe T. 2002. [Fe]-hydrogenases in green algae: photo-fermentation and hydrogen evolution under sulfur deprivation. *Int. J. Hydrog. Ener.* 27:1431–39
112. Winkler M, Mauerer C, Hemschemeier A, Happe T. 2004. The isolation of green algal strains with outstanding H_2-productivity. In *Biohydrogen III*, ed. J Miyake, Y Igarashi, M Roegner, pp. 103–16. Oxford: Elsevier Sci.
113. Zabawinski C, van Den Koornhuyse N, D'Hulst C, Schlichting R, Giersch C, et al. 2001. Starchless mutants of *Chlamydomonas reinhardtii* lack the small subunit of a heterotetrameric ADP-glucose pyrophosphorylase. *J. Bacteriol.* 183:1069–77
114. Zhang X, Zhang X, Shiraiwa Y, Mao Y, Sui Z, Liu J. 2005. Cloning and characterization of *hoxH* genes from *Arthrospira* and *Spirulina* and application in phylogenetic study. *Mar. Biotechnol.* 7:287–96

Hidden Branches: Developments in Root System Architecture

Karen S. Osmont,* Richard Sibout,*
and Christian S. Hardtke

Department of Plant Molecular Biology, University of Lausanne, CH-1015 Lausanne, Switzerland; email: karen.osmont@unil.ch, richard.sibout@unil.ch, christian.hardtke@unil.ch

Annu. Rev. Plant Biol. 2007. 58:93–113

The *Annual Review of Plant Biology* is online at plant.annualreviews.org

This article's doi:
10.1146/annurev.arplant.58.032806.104006

First published online as a Review in Advance on December 19, 2006

1543-5008/07/0602-0093$20.00

*These authors contributed equally to this manuscript.

Key Words

root branching, root meristem, monocotyledons, dicotyledons, phytohormones, nutrients

Abstract

The root system is fundamentally important for plant growth and survival because of its role in water and nutrient uptake. Therefore, plants rely on modulation of root system architecture (RSA) to respond to a changing soil environment. Although RSA is a highly plastic trait and varies both between and among species, the basic root system morphology and its plasticity are controlled by inherent genetic factors. These mediate the modification of RSA, mostly at the level of root branching, in response to a suite of biotic and abiotic factors. Recent progress in the understanding of the molecular basis of these responses suggests that they largely feed through hormone homeostasis and signaling pathways. Novel factors implicated in the regulation of RSA in response to the myriad endogenous and exogenous signals are also increasingly isolated through alternative approaches such as quantitative trait locus analysis.

Contents

INTRODUCTION

Root system architecture (RSA): the three-dimensional structure of the root system, including the primary root, branch roots, and root hairs

LR: lateral root

Adventitious roots: roots that emerge from shoot structures

The root system primarily functions to acquire essential macro- and micronutrients and water from the soil, and to provide anchorage. It also participates in secondary functions, including, for example, photoassimilate storage, phytohormone synthesis, or clonal propagation. During evolution, root organization has gradually progressed from very simple, such as rhizomes, to highly hierarchical, including specialized tissues (12). This elaboration was likely driven by the fact that the root system is instrumental in facing major constraints to plant growth and reproductive success: the availability of nutrients and water, and the subterranean competition of other plants for these resources. Thus, depending on soil composition, differences in root system morphology or architecture (RSA) may affect competitive ability for soil resources (37). RSA reflects the spatial configuration of roots of different age and order, implying that the overall structure has functional significance (76). RSA can be modulated in several ways: through promotion or inhibition of primary root growth, through growth of lateral roots (LRs), through the formation of adventitious roots, and through an increase in root hairs.

The primary root is formed during embryogenesis, whereas LRs are derived postembryonically from existing roots. LRs originate from the pericycle tissue layer (sometimes also from the endodermis), generally either adjacent to xylem pole cells (in dicotyledons) or phloem pole cells (in monocotyledons) (14). Anticlinal divisions of pericycle cells initiate LRs, whose development mimics the organogenesis of the primary root in terms of tissue composition and organization (30). Notably, this process can be reiterated in the subsequent LRs of higher order. In many species, the root system also enlarges through adventitious roots, which can be broadly defined as roots originating from shoot structures, usually stems. Finally, any type of root can increase its absorptive surface by growing extensions of epidermal cells, the root hairs.

Root hair development and its molecular control have been well described elsewhere (102). Other recent reviews focus on the transcription factor networks functioning in the root (9, 44, 81). In this review, we concentrate on recent advances in the understanding of RSA modulation in angiosperms, with a special focus on root branching. We refer to root branching as the formation of LRs or other postembryonically formed roots. We emphasize recent advances regarding the role of nutrients and phytohormones in root branching, mainly achieved by genetic approaches in the model dicotyledon *Arabidopsis thaliana*, as well as alternative approaches to investigate the control of root system growth.

The Morphological Diversity of Root Systems

A wide variation in RSA is observed across species, suggesting that it is determined by inherent genetic factors. Although there is no simple classification scheme for RSA (37), it is well accepted that two main root system morphologies are recurrent in angiosperms. The first one is typically found in dicotyledons, including model species (*Arabidopsis*, tomato, pea), and is termed allorhizic (**Figure 1*a***). It

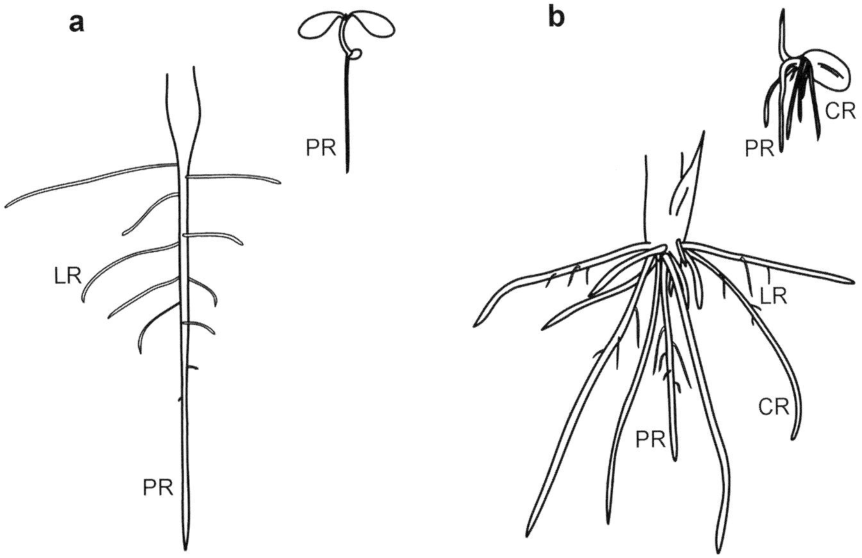

Figure 1

(*a*) Schematic representation of a typical allorhizic root system architecture as found in most dicotyledons, exemplified by *Arabidopsis*, in 5-day-old and 12-day-old seedlings. Root hairs are not represented. PR, primary root; LR, lateral root. (*b*) Schematic representation of a typical secondary homorhizic root system architecture as found in most monocotyledons, exemplified by rice, in 7-day-old and 14-day-old seedlings. Root hairs are not represented. PR, primary root; LR, lateral root; CR, crown root.

ABA: abscisic acid

is usually comprised of at least two root types: primary (tap) roots and LRs (33). Allorhizic root systems are dominated by the primary root, which produces LRs that can form higher-order LRs. The major growth axis of allorhizic root systems parallels primary root growth. Adventitious roots are rare in allorhizic systems, but occasionally emerge from the hypocotyl or stems, in particular upon wounding. In contrast to the allorhizic root system of dicotyledons, the root system of monocotyledons (**Figure 1*b***) is characterized by the development of many adventitious roots in parallel to the primary root (33). In fact, the majority of the root system in maize and rice is formed from postembryonic shoot-borne roots. Depending on the tissue from which they emerge, they also possess specific names, such as crown or brace roots (53). Furthermore, some monocotyledons, such as maize, form additional embryogenic roots. These emerge from the scutellar node and are called seminal roots. Notably, primary roots of monocotyledons are often also referred to as seminal roots. All the root types can branch by forming LRs (35), giving the root system a bushy appearance. This typical monocotyledonous root system is called secondary homorhizic. Although the primary root is important throughout the life cycle in allorhizic systems, in secondary homorhizic systems it only seems to be important during very early stages of seedling development (35). However, maize seminal roots have a higher water uptake capacity than other root types and are thought to be important throughout the plant life cycle (80).

Most insight into the environmental and genetic control of RSA has been obtained by analyses of model species. Among them, *Arabidopsis* displays a typical allorhizic root system, whereas maize and rice form highly similar homorhizic root systems. However, maize possesses seminal roots, which are missing in rice. In both species, the primary root arrests growth shortly after emergence (34).

THE DEVELOPMENTAL PLASTICITY OF ROOT BRANCHING: RESPONSE TO EXOGENOUS FACTORS

RSA is influenced by numerous biotic and abiotic factors that make up the heterogeneous composition of the soil environment. Thus, RSA is a highly plastic trait, meaning that genotypically identical plants can heavily differ in RSA, depending on their macro- and microenvironment. Although this plasticity is well documented, the underlying molecular mechanisms are poorly understood. However, an increasing number of mutant studies in *Arabidopsis* start to address this issue. For instance, one recent study examined LR formation as a function of water availability. Indeed, lateral root primordia emergence is repressed when water becomes limiting, and this response requires abscisic acid (ABA) and the *LATERAL ROOT DEVELOPMENT2* gene (23). Apart from water, the other most important abiotic factor for root system growth is nutrient availability (32, 72). Among the different nutrients, the abundance of the two growth-limiting macronutrients, nitrogen (N) and phosphorus (P), exerts the greatest effect on RSA.

The Molecular Basis of Root System Response to Nitrogen

The main sources of N in the soil are ammonium, nitrite, and nitrate. Two distinct N-uptake mechanisms have been described (42). The low-affinity uptake system functions when N is plentiful, whereas the high-affinity transport system (HATS) functions when N is limiting. In *Arabidopsis*, high external N concentrations reduce primary as well as lateral root elongation. In contrast, LR elongation is induced under N-limiting conditions (67). LR density is relatively constant across a range of N concentration. However, although the above is true for homogeneous nutrient concentrations in media, the situation is different in the more heterogeneous soil environment, where nutrients are often unequally distributed. Indeed, localized zones of high N concentration promote rather than suppress LR growth (32). Therefore, LR growth is systemically inhibited in response to globally high levels of N, but locally induced in response to N-rich patches. An important component in the latter response is the nitrate-inducible *Arabidopsis* MADS box transcription factor NITRATE-REGULATED1 (ANR1) (118). In plants with decreased *ANR1* activity, LRs do not proliferate in response to nitrate-rich zones. However, systemic inhibition of LR growth is still maintained, suggesting that *ANR1* exclusively mediates localized responses to N.

Conceivably, N sensing and thereby root branching are also affected by N transport. In particular, the *Arabidopsis* NRT2 family of transporter-like proteins is required for HATS (42). One of its members, *NRT2.1*, was recently implicated in LR initiation (68, 96). Lateral root initiation is strongly repressed in *Arabidopsis* by a high external carbon to nitrogen source ratio, and this response is abolished in the *lin1* mutant (79), which is defective in NRT2.1 (68).

Another natural N source that strongly influences RSA is glutamate, which results from the breakdown of organic matter. L-glutamate-treated seedlings have a shorter, more branched root system, and this response is strain-dependent (113). Finally, nitric oxide is required for LR primordia formation in tomato (22).

The Molecular Basis of Root System Response to Phosphate

After N, P is the second limiting macronutrient for plant growth. In *Arabidopsis*, increasing levels of inorganic phosphate (P_i) stimulate primary root elongation, but decrease LR density and suppress LR elongation (67). However, under low to moderate levels of P_i, LR growth is favored over primary root growth (115). In severely P_i-starved plants, primary root growth is strongly inhibited and LR number can increase up to five times as compared to optimal P_i concentrations (73). The determinate primary root growth on low P_i media results from cessation of meristematic activity and loss of auxin responsiveness in the root meristem (72). Indeed, auxin has been implicated in the P_i-starvation response (2, 72, 73, 88), although this subject remains controversial (44, 67, 115).

Several mutant studies have shed light on P_i-sensing pathways. *Arabidopsis pho2* mutants overaccumulate P_i, demonstrating that the internal P_i concentration is important for RSA (115). *PHO2* encodes an unusual E2 conjugase. Mutants in *PHOSPHATE RESPONSE REGULATOR 1 (PHR1)* display altered shoot/root growth ratio in P_i-starved conditions and altered low P_i-induced gene expression (98). *PHR1* encodes a MYB-like transcription factor, which binds promoter sequences of low P_i-induced genes. Recent evidence suggests that *PHO2* and *PHR1*, together with the microRNA *miR-399*, define a P_i signaling pathway (6). *miR-399* is induced under P_i starvation, and its overexpression represses *PHO2* expression and results in high P_i concentrations in leaves (19, 40). Conversely, *miR399* expression is repressed by high P_i, stabilizing *PHO2* transcripts. However, *PHO2* expression is also P_i regulated in a *miR399*-independent manner. Both *miR399* and *PHO2* are thought to act downstream of *PHR1* (6). The identification of matching *PHO2* and *miR399* orthologs in other higher plants suggests that this P_i signaling pathway is evolutionarily conserved.

Other Nutrients and Exogenous Abiotic Factors Modulating Root Branching

Apart from N and P, other nutrients also influence root development. For instance, iron limitation suppresses LR growth but promotes LR formation in *Arabidopsis* (82). Also, limiting sulfate results in a highly branched root system because of increased LR formation (72). Finally, more general abiotic factors also impinge on RSA. For example, light is a major positive regulator of root branching (20, 105).

IAA: indole acetic acid

Interestingly, light also limits adventitious root formation in *Arabidopsis* (105). However, much less is known about the influence of the other factors (discussed above) on adventitious rooting in model systems. This might result from the focus of these studies on early stages in *Arabidopsis* development, when adventitious roots are rarely formed outside the context of wounding.

ENDOGENOUS REGULATORS OF ROOT SYSTEM ARCHITECTURE: THE ROLE OF PHYTOHORMONES

In addition to exogenous factors, several endogenous molecules are pivotal determinants of RSA. For instance, vitamin C antagonizes the effects of N and sugar in LR growth (91). Also, root growth is perturbed in the *rsr4* (*reduced sugar response 4*) mutant, which requires vitamin B6 supplementation for normal development (112). However, the most important endogenous modulators of root system development are the different phytohormones.

The bulk of genetic and physiological evidence suggests that nutrient status and availability and other abiotic factors trigger changes in RSA by modulating hormone homeostasis and/or signaling (44, 77). This notion is supported by the observation that responsiveness to certain abiotic factors is lost if a certain hormone response pathway is no longer fully functional. For instance, mutants in the auxin transport mediator *AXR4* (28) are unable to respond to local N-rich zones (119). Also, the systemic inhibition of LR formation in high N conditions is abolished in the ABA-insensitive mutant *abi4* (104). Thus, auxin might mediate localized responses to N, whereas ABA mediates systemic responses. Further, suboptimal sulfate promotes root system growth by modulating auxin homeostasis through transcriptional induction of NITRILASE3 (NIT3), an enzyme that converts indole-3-acetonitrile into the major active endogenous auxin indole-3-acetic acid (IAA) (62). Finally, recent advances also suggest that auxin signaling mediates LR as well as adventitious root formation in response to light stimulus (20, 105, 107). Below we highlight the significance of individual hormones in modulating RSA.

Auxin

The minor active endogenous auxin indole-3-butyric acid is an efficient promoter of adventitious root formation and commonly used as such in horticulture. However, the more abundant IAA appears to be most important for RSA in planta. Generally, IAA promotes LR development, for instance when applied exogenously (63, 64). Few exceptions to this rule have been observed [for example, exogenous IAA application does not promote LR formation in fern (55)]. Importantly, auxin is necessary for all stages of lateral root development, i.e., initiation, emergence, and growth (15, 16). In *Arabidopsis*, acropetal auxin transport from young aerial tissues into the primary root triggers lateral root formation by promoting cell division of pericycle cells adjacent to xylem vessels (16). Basipetal auxin transport in the root tip also appears to influence LR emergence, despite the fact that LRs emerge relatively far from root tips (16). The precise site of LR initiation is probably more a consequence of auxin transport than of de novo auxin synthesis, although data to determine when and why a lateral root is initiated in a certain position along a primary root are lacking (3).

Genetic studies have confirmed the predominant role of auxin in LR formation. For instance, mutants or transgenic lines with elevated auxin biosynthesis display significantly increased root branching (10, 60). Consistently, mutants with low auxin content or impaired auxin signaling have a very short, poorly branched root system (17), also confirming the requirement of auxin for primary root growth. However, overall auxin content is not always positively correlated with LR number. For instance, some mutants with fewer LRs have elevated rather than decreased auxin levels, but are impaired in auxin signaling (57). Nevertheless, as expected, mutants impaired in root polar auxin transport are affected in LR development (51), and the same is true for auxin signaling pathway mutants. For instance, the gain-of-function *solitary root* (*slr*) mutant, which carries a stabilizing mutation in the IAA14 negative regulator of auxin signaling, displays deficient early LR primordium formation because cell division is not maintained in the pericycle (41). Similarly, the gain-of-function mutant *massugu2* (*msg2*), defective in IAA19, displays a significantly decreased LR number. IAA19 inhibits the activity of the auxin response factor (ARF) NON-PHOTOTROPIC HYPOCOTYL 4 (NPH4)/ARF7, a positive regulator of LR formation (110). In summary, both auxin transport and signaling strongly influence RSA, but, because of the inherent interdependency of these two processes (65), their individual impact is difficult to evaluate.

Cytokinins

In contrast to auxin, exogenous cytokinin (CK) application suppresses rather than promotes LR formation, and transgenic *Arabidopsis* plants with decreased cytokinin levels display increased root branching and also enhanced primary root growth (114). Thus, the general auxin-cytokinin antagonism also appears to hold true for root branching (20). The notion that cytokinin negatively regulates root growth has also been verified by studies of cytokinin perception and signaling mutants. For instance, double mutants for the redundant *Arabidopsis* cytokinin receptors AHK2 and AHK3 display a faster-growing primary root and greatly increased root branching (97).

Ethylene

Ethylene is thought to mediate some auxin-induced responses because several auxin signaling mutants of *Arabidopsis* are also ethylene-insensitive, and because expression of ACC synthase, a rate-limiting enzyme for ethylene biosynthesis, is strongly auxin-inducible (109). Indeed, moderate concentrations of ethylene inhibit root growth, likely by impinging on the maintenance of the quiescent center and root cap both in maize and *Arabidopsis* (93). Root-specific mutants overcoming this inhibition have been isolated in *Arabidopsis* (108). Interestingly, these mutants display decreased auxin levels in the root and can suppress the enhanced branching phenotype of auxin overproducers, suggesting that auxin not only influences ethylene homeostasis, but also vice versa. However, the suppression effect could also be due to reduced auxin transport from the shoot into the root. A role for ethylene in root growth was also confirmed in rice, where it is necessary for cell proliferation in adventitious roots (74). Ethylene might also have a role during LR emergence by promoting the breakdown of cortical cells (63).

CK: cytokinin

GA: gibberellic acid

Gibberellins

Gibberellic acid (GA) biosynthesis has been detected in root tips of different plants and GA signaling is indeed required for primary root growth (39, 58). However, a major role for GA in root branching has never been clearly demonstrated, although GA acts synergistically together with ethylene to promote both initiation and growth of adventitious roots in flooded rice plants (106).

Abscisic Acid

Recently, ABA was also implicated in root system development because exogenous ABA application inhibits primary and lateral root development in *Arabidopsis* (7, 25). ABA action

BRASSINOSTEROIDS AND ROOT GROWTH

The role of brassinosteroids (BRs) in root development has been largely overlooked in the literature, likely because most BR biosynthesis and signaling mutants have a pleiotropic dwarf phenotype, making it hard to judge whether or not an associated root phenotype is a primary defect. Several classic BR mutants show an overall reduction in root growth as compared to wild type. For example, the *bri1* mutant, resulting from a lesion in the BR receptor, shows an approximate 50% reduction in primary root growth. Also, the *cpd* mutant, which is defective in a rate-limiting BR biosynthesis enzyme, displays an approximate 40% reduction in primary root growth as well as decreased LR number and growth (87; K.S. Osmont, unpublished observations). The recent finding that loss of function of the novel plant gene *BRX* results in a root-specific BR deficiency substantiates a crucial role for BR in root system development. *brx* loss of function results in strongly reduced primary root growth, but also in increased formation of (slower-growing) lateral roots. Interestingly, such compensation of slower primary root growth by increased root branching is a general feature in *Arabidopsis* accessions.

BR: brassinosteroids

in LR development is thought to be auxin-independent (25), but is modulated by nutrient availability (104). Interestingly, in rice ABA stimulates rather than suppresses LR formation from primary roots (18). However, whether this is a common feature of monocotyledons remains to be determined. Interaction between IAA and ABA has been also proposed, because some mutants with pleiotropic defects in ABA responses in roots are also impaired in auxin-mediated LR development (11).

Brassinosteroids

Brassinosteroids (BRs) are abundant in shoots, but have also been detected in substantial amounts in roots of maize and tomato (117). Indeed, gene expression studies suggest that BRs are synthesized in roots, albeit at lower levels than in shoots (59). Like auxin, BRs promote primary root growth at low concentrations, but are inhibitory at higher concentrations (87). It also has been proposed that BRs affect LR development by regulating auxin transport (5). However, evidence is now mounting that there is significant interaction between the BR and auxin hormone signaling pathways (85, 90). Indeed, many auxin signaling genes involved in root growth and development are induced by both auxin and BRs, and this induction requires BR biosynthesis (59, 89). Also, microarray analyses indicate that many BR-responsive genes are also auxin-responsive (43, 86, 90).

In general, the role of BRs in root development has been largely overlooked because of the pleiotropic dwarf phenotypes of many BR biosynthesis and signaling mutants. However, a recent study of an *Arabidopsis* accession with a short and highly branched root system revealed a root-specific BR deficiency, which is due to loss of function of *BREVIS RADIX* (*BRX*). This deficiency is responsible for the RSA phenotype and connects BR biosynthesis and auxin signaling in *Arabidopsis* roots (85). Finally, rice mutants affected in BR biosynthesis or sensitivity also show significant root morphology phenotypes (83).

MODULATION OF ROOT SYSTEM ARCHITECTURE BY BIOTIC FACTORS

RSA is not only influenced by abiotic factors, but also by a range of biotic factors. Notably, in the wild, plant roots are in contact with saprophytic and pathogenic microorganisms, which can drastically change RSA. For example, after petunia is infected by specific *Ralstonia solanacearum* strains, LR elongation is inhibited, but new lateral roots with abnormal morphology are induced (120). Most frequently, however, changes in RSA are triggered by symbiotic interactions. Unfortunately, the best-characterized plant model system, *Arabidopsis thaliana*, does not form root symbioses and thus cannot be used to address questions of plant-symbiont interactions and their impact on RSA. Nevertheless, an increasing amount of data has emerged from the analyses of alternative systems.

Root System Architecture Changes Triggered by Interaction with Mycorrhizae

Unlike *Arabidopsis*, more than 80% of higher plants associate with mycorrhizal fungi, which elicit profound changes in the root morphology of host plants (47). In particular, ectomycorrhizae suppress root elongation and induce dichotomous branching of short LRs, culminating in the formation of coralloid structures resulting from higher-order dichotomous branching. All of these anatomical structures are variable depending on the plant and fungal species. Once the fungus is established, root branching is suppressed, which makes the plant more dependent on the nutrients provided by the fungus (47, 94). Whether this modification of RSA is a direct consequence of symbiosis or an indirect effect of improved nutrient status of the plant is not clear. However, it appears that symbionts can trigger RSA changes by promoting LR initiation very early in the interaction (46). Moreover, the maize mutant *lrt1* normally lacks LRs, but displays extensive LR development following inoculation with the mycorrhizae *Glomus mosseae* (92). Notably, many microorganisms that interact with plants can produce plant hormone analogs. Thus, symbiotic association might employ hormone signaling pathways to regulate RSA. Indeed, numerous studies have tried to determine the importance of modifications in hormone homeostasis in symbiosis establishment and the associated changes in root branching (e.g., 24, 71). However, it remains to be determined which pathways are definitely recruited.

Nodulation

The second most important symbiosis of plant roots is their association with N-fixing bacteria in legumes, a process termed nodulation. Nodules and LRs share some common features. For instance, both organs form adjacent to xylem poles, develop meristems, and break cell layers to emerge. In support of this idea, the *lateral root organ-defective* mutant of *Medicago truncatula* initiates both nodule and LR formation, but does not complete either process. Moreover, nodule formation shares common molecular processes with LR development (24, 50).

Dichotomous: branching of a meristem into direct, nearly equivalent branches

ROOT MERISTEM FORMATION

Biotic and abiotic factors can feed into RSA at different levels. For instance, although some factors mainly promote LR emergence, others influence LR initiation. Moreover, such control could occur at multiple entry points because a multitude of factors required for normal root branching has been identified, mainly by molecular genetic approaches in *Arabidopsis* (26).

Cell Cycle Control of Lateral Root Initiation

The initiation of an LR starts with the reentry of a pericycle cell into the cell cycle and subsequent anti- and periclinal cell divisions, which eventually lead to the differentiation of a LR meristem. Thus, it is not surprising that cell cycle control is important for LR growth and development (77). It has been hypothesized that pericycle cells are blocked at the G2 stage of the cell cycle in *Arabidopsis* (15). However, most pericycle cells appear to remain in G1, and only xylem-adjacent pericycle cells, which include the LR founder cells, progress into G2 (8). Nevertheless, the G1-specific cyclin D4 and the G1 to S transition inhibitor Kip-related protein 2 (KRP2) have both been implicated in LR initiation (27), suggesting that both G1- and G2-specific cell cycle blocks must be overcome to form LR primordia. Interestingly, *KRP2* expression is auxin-repressible and overexpression of KRP2 leads to reduced LR initiation (48), substantiating a role of auxin in LR initiation through control of cell cycle progression.

Thus, despite their obvious importance, it appears that cell cycle genes are not the primary determinants of LR initiation.

Making and Maintaining Root Meristems

Beyond cell cycle activation, root meristem formation requires the activity of a number of factors that organize the differentiation and reiterative production of the root tissues from the apical meristems. In molecular terms, this process has been best described for the primary root of *Arabidopsis*. Whether the same factors function in elaborating and maintaining LR meristems largely remains to be determined. Nevertheless, it is reasonable to assume that lateral and primary root meristem development are driven by principally equivalent factors (78), a notion corroborated by recent analyses of mutants and transgenic lines that are impaired in root meristem maintenance.

The embryonic primary root apical meristem (RAM) comprises a pool of stem cells that are required for continuous root growth. In the *Arabidopsis* RAM, a small group of slowly dividing, so-called quiescent center cells maintains the stem cell identity of the immediately surrounding cells (101, 111). The two *PLETHORA* (*PLT*) genes encode AP2-type transcription factors that act redundantly to pattern this root stem cell niche. Strikingly, their ectopic expression is sufficient for initiating root meristems (1). Loss of function of both *PLT* genes results in failure to maintain the quiescent center, leading to stem cell differentiation and eventually breakdown of the RAM. Importantly, *plt1 plt2* double mutants also display defective LR meristems, indicating that these genes are likely required for the maintenance of all types of *Arabidopsis* roots. Likewise, overexpression of CLAVATA 3 (CLV3)-like peptides results in the consumption of both primary and LR meristems (13, 36, 52), suggesting that a CLV-like pathway might function in all root types as well.

A novel factor required for maintaining the rice RAM is *GLUTAMATE RECEPTOR-LIKE 3.1* (*GLR3.1*) (66). In contrast to rice *glr3.1* mutants, Arabidopis *glr* mutants have no discernable root phenotype, suggesting that *GLR3.1* may have been co-opted for a specialized function in the rice root. However, although *GLR3.1* is a single-copy gene in rice, there are 20 *GLR* genes in *Arabidopsis*. Thus, the absence of phenotypes in the *Arabidopsis* mutants might simply reflect genetic redundancy. This example also demonstrates that all critical factors for RSA cannot be easily isolated in a single system. Genome architecture must be considered, making the parallel exploitation of different model systems desirable.

New Protagonists in Lateral Root Formation

Both forward- and reverse-genetic approaches have proven fruitful in the identification of numerous factors that influence root branching (see 26). However, for most factors identified by gain-of-function approaches, it remains to be seen whether they actually have a genuine role in LR formation (26). Among the loss-of-function mutants, many display pleiotropic phenotypes. Thus, only a few genes have been shown to be involved specifically in LR initiation and/or growth (26). For example, the *Arabidopsis aberrant lateral root formation 4* (*alf4*) mutant forms a normal primary root, but fails to form LRs. Notably, this defect cannot be rescued by auxin treatment (17). Thus, *ALF4*, which encodes a protein of unknown function, is hypothesized to promote competency to respond to LR developmental cues in the pericycle (29).

By contrast, the fully redundant *ARABIDILLO-1* and *2* genes (which encode Armadillo/beta-catenin-type proteins) are involved in promoting LR development at a later stage (21). *arabidillo 1/2* double mutants display reduced LR formation, and, conversely, *ARABIDILLO-1* overexpressing lines show an increase in LRs. Overexpression of truncated ARABIDILLO-1 protein fragments containing an F-box motif act in a dominant negative manner to reduce LR formation to the levels of the double mutant. Collectively, these data define ARABIDILLO proteins as a new class of root promoting factors and suggest that they may target an inhibitor of LR development for degradation by the ubiquitin-proteasome pathway (21). A novel F-box protein involved in this process is CEGENDUO, which negatively regulates auxin-mediated LR production (31).

Novel insight into the regulation of root branching has also been obtained from analysis of plants other than *Arabidopsis*. Of particular interest is the *diageotropica* (*dgt*) mutant of tomato. *DGT* encodes a type-A cyclophilin and is required for LR primordia formation. Notably, *dgt* uncouples LR primordium patterning from cell proliferation of xylem-adjacent pericycle cells (57), classifying it as one of the earliest known factors in LR initiation after activation of the cell cycle machinery.

Regulators of Adventitious Root Formation

As compared to LR formation, little is known about the molecular basis of adventitious root formation. In part, this might be due to the rare occurrence of adventitious roots in the early stages of normal *Arabidopsis* development. Although it is conceivable that the same principal mechanisms as in LR formation are at work, some significant variation could be expected, mainly because adventitious roots generally form from differentiated stem tissues rather than from within other roots. Nevertheless, auxin signaling again appears to be limiting for adventitious root formation in *Arabidopsis* (105), a notion that is supported by a recent study of a rice mutant, *crown rootless1* (*crl1*).

CRL1 is necessary for crown root initiation (56). However, *crl1* defects can be overcome by auxin treatment, suggesting that *CRL1* acts upstream of the auxin stimulus. Nevertheless, canonical auxin signaling is essential for the auxin-dependent expression of *CRL1*, suggesting that *CRL1* might be involved in a feed-forward loop required for adventitious root formation. Another rice mutant, *adventitious rootless1* (*arl1*), is defective in crown root formation as well (69). However, unlike *crl1*, *arl1* cannot be rescued by auxin treatment, suggesting that it acts further downstream in the process. Notably, both *CRL1* and *ARL1* encode LATERAL ORGAN BOUNDARIES (LOB) domain family transcription factors. Interestingly, so far LOB domain transcription factors have not been reported to modulate RSA in *Arabidopsis*, suggesting either specificity for adventitious rooting, or simply an increased redundancy among LOB-encoding genes in *Arabidopsis*.

ALTERNATIVE APPROACHES TO STUDY ROOT BRANCHING

Genetic redundancy generally limits the applicability of mutant analysis for the molecular genetic dissection of root branching in *Arabidopsis*. Thus, alternative approaches are a welcome complement to the described physiological and standard genetic techniques.

Microarray and Proteomics Analyses of Root Branching

The common availability and affordability of microarray experiments in *Arabidopsis* have also left their mark in the analysis of root development. For instance, transcript profiling of early LR initiation in *Arabidopsis* has been performed (49). In these experiments, a major problem in global gene expression analysis of LR formation had to be overcome: LR primordia cannot be easily isolated in sufficient quantity, and the proportion of primary root cells involved in LR initiation is very small. Thus, locally significant expression changes are easily masked if the whole root tissue is used for analysis. These problems were circumvented by blocking LR formation through pharmacological inhibition of

QTL: quantative trait locus

auxin transport and, after release of the block, massive parallel induction of LR primordia by auxin application. The respective transcription profile established a timeline of differential gene activities. Not surprisingly, many of them were involved in cell cycle progression. Cell cycle genes are also misexpressed in a cell type-specific transcriptome profiling experiment of primary roots of wild-type maize as compared to primary roots of the *rootless with undetectable meristems 1* (*rum1*) mutant, which lack lateral roots (116).

Compared to microarray analyses, systematic proteomic analyses of root branching are very rare. One of the few available studies again investigated the proteome of the primary roots of *rum1* mutants, revealing significant expression-level differences for proteins involved in primary metabolism, signal transduction, and secondary metabolism (53, 70). A similar study has been performed for adventitious roots by comparing wild-type maize and the *rtcs* mutant, which is completely devoid of shoot-borne roots as well as seminal roots (100). Although these proteomic approaches are surely interesting for studying the dynamics of metabolic changes during root initiation, they might be less suitable for detecting regulatory proteins found in low abundance, such as transcription factors.

Isolation of Modifiers of Root System Architecture by Quantitative Trait Locus Analyses

Although RSA is a highly plastic trait, practically all species elaborate a genetically programmed, basic layout of their root system morphology. Nevertheless, significant variation in RSA is also observed on the intraspecific level, as illustrated by the diversity of RSA between different wild isolates (accessions) of *Arabidopsis thaliana* grown in identical conditions (**Figure 2**). So far, this variability has been largely neglected. However, several recent studies report attempts to isolate genes that are responsible for this variability, generally by using quantitative trait locus (QTL) analysis.

QTL analysis of naturally occurring intraspecific allelic variation enables identification of genes that are difficult to find by mutant analysis (for instance, because they only weakly contribute to the studied phenotype). The fully sequenced genome of *Arabidopsis* and established high-throughput genotyping techniques make QTL analysis highly feasible in this model plant species (4). Most importantly, accessions collected from the wild should not contain major deleterious alleles. Therefore, QTL analysis of RSA should be biased toward the identification of modifiers

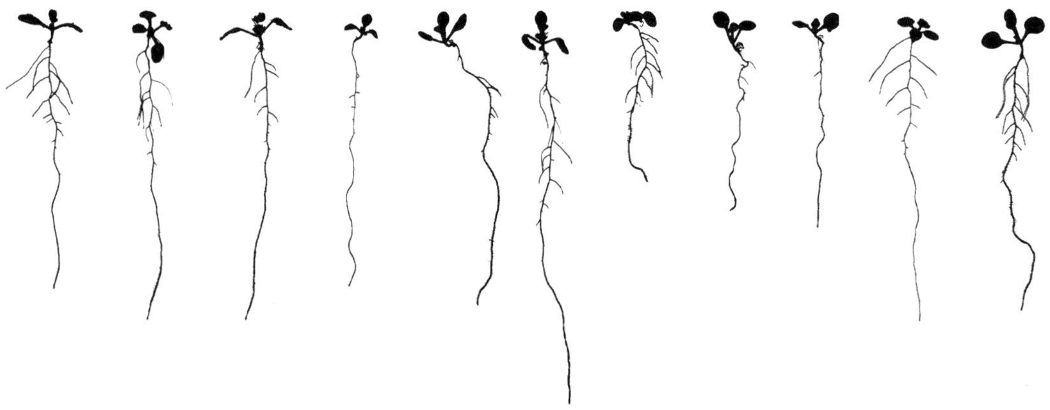

Figure 2

Root morphologies of different *Arabidopsis thaliana* wild accessions. Root system architecture of different accessions at young stage grown in identical tissue culture conditions.

of RSA rather than genes with a major role in root formation and/or maintenance. These modifiers can be expected to influence either the temporal or spatial control of root initiation or root growth rate.

Another interesting aspect of QTL analysis of root development is the fact that genetically complex interactions can be detected. For instance, two of 13 recently detected quantitative trait loci for primary root length, LR number and LR density, show epistatic interaction (75). Notably, many of these loci specifically affect LR formation. In another study, it was noted that the root system of the accession Landsberg *erecta* (Ler) is larger and less affected by osmotic stress conditions than the system of the accession Columbia (Col) (38). Among several QTLs involved in this difference, two loci named *ELICITORS OF DROUGHT GROWTH* (*EDG*) were mapped. Surprisingly, however, the *EDG1* allele of Ler promotes root system growth as expected, but the Ler *EDG2* allele represses root growth, and vice versa for Col. However, epistasis between the *EDG1* and *2* loci was not detected, suggesting that additional modifiers (to be found among the other detected QTLs?) mask the effect of the Col *EDG2* allele in this background.

Importantly, many of the QTL described so far quantitatively affect RSA to a degree that allows their detection in other genetic backgrounds (for, instance in near isogenic lines), which to some degree mimic mutants in the respective loci. This enables their identification at the molecular level by standard map-based cloning procedures. For example, a QTL that significantly contributes to the difference in primary root growth between the Ler and Cape Verde Islands accessions encodes an invertase enzyme (103). Also, the already mentioned *BRX* gene was identified through a major QTL responsible for most of the primary root growth difference between the Umkirch-1 and Slavice-0 accessions (84).

Traditionally, QTL analyses have been intensively performed in crop plants (99). The increasing availability of genomic sequences has now also enabled the identification of genes underlying QTL in these systems, as successfully reported for QTL of grain development in rice (e.g., 61). Thus, the cloning of QTL affecting root branching in rice (e.g., 54) should also become feasible soon.

CONCLUSIONS

Recent advances through forward- and reverse-genetic approaches confirm the major role of hormones, especially auxin, in LR initiation and development. An emerging theme from these studies is the idea that most environmental responses of RSA are mediated by modulating hormone homeostasis or signaling pathways. Novel insights in this process also surface from alternative approaches to understand RSA, notably QTL analysis of natural genetic variation. Also, other systems and models are being increasingly exploited to understand fully the variation and conservation of mechanisms underlying root growth and RSA.

However, despite the recent efforts and progress in understanding the molecular cues involved in LR development, root branching, and RSA modulation, many unanswered questions remain. For example, many factors involved in initiating, patterning, and maintaining root meristems have been isolated in *Arabidopsis*. Most of these have been defined for their requirement in primary root development. Although recent evidence suggests that some of the pivotal factors have an equivalent role in LR development, it remains to be seen whether this is a general rule. For example, the *MONOPTEROS* (*MP*) gene is absolutely required for primary root formation in embryogenesis. However, *mp* loss-of-function mutants can form perfectly normal adventitious roots (95). Does this mean that homologous redundant factors take over *MP*'s role in different contexts (45), or is it an indication for an altogether alternative pathway of adventitious root formation in *Arabidopsis*? Many issues that are central to RSA have so far not been elaborated on at the molecular genetic level. For instance, LRs emerge from the parent root at a specific angle, often growing horizontally. What are the molecular determinants of this feature and how do lateral roots adjust themselves with such accuracy with respect to the gravity vector? Also, most roots do not grow indeterminately. Is their longevity genetically programmed from the start or, rather, under environmental control? Whatever the answers, surely plant researchers mining the rhizosphere will not run out of interesting questions for a long time.

SUMMARY POINTS

1. Although RSA is highly diverse, angiosperm root systems can be broadly classified into allorhizic, as in most dicotyledons, or secondary homorhizic, as in most monocotyledons.
2. Despite the fact that RSA is a highly plastic trait and thus is difficult to study, several molecular factors controlling root system morphology have been identified.
3. Although it is well established that RSA and root growth are strongly affected by the nutrient status of the soil, the molecular mechanisms mediating root system responses to nutrients are just beginning to be understood.
4. Almost all known plant hormones affect RSA, including auxin, CK, ethylene, BR, ABA, and GA. However, recent evidence suggests that significant interactions occur between these hormone pathways to regulate root system growth. Moreover, it appears that hormones mediate root developmental responses to both biotic and abiotic stresses downstream of the respective stimuli.
5. The auxin signaling pathway appears to be particularly important in mediating environmental responses of RSA.
6. Physiological data and recent molecular evidence suggest that symbiont establishment shares common processes with LR development and strongly modulates RSA.
7. Many loss- and gain-of-function mutants affect LR development, but few only affect LR development and/or root branching in a specific manner.
8. Several alternative approaches have proven useful in defining novel factors and global responses in LR and adventitious root development, namely microarray, proteomic, and, notably, QTL analyses.

FUTURE ISSUES

1. One outstanding question is whether one distinct pathway controls the initiation and maintenance of root meristems in all root types, i.e., are the same molecular pathways at play in primary roots as well as lateral and adventitious roots? Are differences in the pathways initiating distinct root types due to context-specific use of homologous factors, or do they reflect the existence of altogether alternative pathways for root formation?
2. What is the level of conservation of molecular mechanisms involved in root formation and RSA modulation across distantly related species?

3. What are the mechanistic modes of action of some important *Arabidopsis* proteins of unknown function involved in root branching, such as ALF4 or BRX? Is their function conserved in other plant species? Conversely, several novel genes important for RSA have been discovered in model systems other than *Arabidopsis*. What are the specific roles of genes like tomato *DGT1* or rice LOB domain transcription factors in RSA, and do they function in the same way in *Arabidopsis*?
4. One challenge facing those working in the rhizosphere is the inherent plasticity of the root system. This challenge is mostly tackled by investigating root systems in standardized tissue culture conditions. Is RSA modulation observed in tissue culture really relevant for root systems in their native soil environment? Applying novel imaging techniques for root systems grown in soil (e.g., X-ray imaging, computer tomography) might help to address this question.
5. Analysis of natural genetic variation can be expected to become an increasingly useful tool to determine the molecular genetic basis of RSA modulation.
6. Variation of RSA between and within species is commonly observed. Is this variation functionally significant in an ecological context? For instance, do intraspecific variations in RSA confer competitive advantages in particular conditions?
7. Model systems such as rice, maize, and poplar are becoming more tractable systems for studying root development because of increasing availability of genetic and genomic tools. To understand fully the conserved underlying mechanisms of root growth, simple vascular plants (for example, ferns) should also be investigated in detail at the molecular genetic level.
8. During the plant life cycle, the geotropism of roots changes. For instance, LRs emerge from the parent root at a specific angle, but become geotropic after being diageotropic or ageotropic for a long period of time. What molecular factors determine their orientation and changes thereof with respect to the gravity vector?

ACKNOWLEDGMENTS

We apologize to those authors whose work could not be discussed owing to space restrictions. Work in our lab is supported by the Swiss National Science Foundation and the Canton de Vaud.

LITERATURE CITED

1. Aida M, Beis D, Heidstra R, Willemsen V, Blilou I, et al. 2004. The *PLETHORA* genes mediate patterning of the *Arabidopsis* root stem cell niche. *Cell* 119:109–20
2. Al-Ghazi Y, Muller B, Pinloche S, Tranbarger TJ, Nacry P, et al. 2003. Temporal responses of *Arabidopsis* root architecture to phosphate starvation: evidence for the involvement of auxin signaling. *Plant Cell Environ.* 26:1053–66
3. Aloni R, Aloni E, Langhans M, Ullrich CI. 2006. Role of cytokinin and auxin in shaping root architecture: regulating vascular differentiation, lateral root initiation, root apical dominance and root gravitropism. *Ann. Bot.* 97:883–93
4. Alonso-Blanco C, Koornneef M. 2000. Naturally occurring variation in *Arabidopsis*: an underexploited resource for plant genetics. *Trends Plant Sci.* 5:22–29

5. Bao F, Shen J, Brady SR, Muday GK, Asami T, Yang Z. 2004. Brassinosteroids interact with auxin to promote lateral root development in *Arabidopsis*. *Plant Physiol.* 134:1624–31
6. Bari R, Pant BD, Stitt M, Scheible WR. 2006. PHO2, microRNA399, and PHR1 define a phosphate-signaling pathway in plants. *Plant Physiol.* 141:988–99
7. Beaudoin N, Serizet C, Gosti F, Giraudat J. 2000. Interactions between abscisic acid and ethylene signaling cascades. *Plant Cell* 12:1103–15
8. Beeckman T, Burssens S, Inze D. 2001. The peri-cell-cycle in *Arabidopsis*. *J. Exp. Bot.* 52:403–11
9. Benfey PN. 2005. Developmental networks. *Plant Physiol.* 138:548–49
10. Boerjan W, Cervera MT, Delarue M, Beeckman T, Dewitte W, et al. 1995. *Superroot*, a recessive mutation in *Arabidopsis*, confers auxin overproduction. *Plant Cell* 7:1405–19
11. Brady SM, Sarkar SF, Bonetta D, McCourt P. 2003. The *ABSCISIC ACID INSENSITIVE 3 (ABI3)* gene is modulated by farnesylation and is involved in auxin signaling and lateral root development in *Arabidopsis*. *Plant J.* 34:67–75
12. Brundrett MC. 2002. Coevolution of roots and mycorrhizas of land plants. *New Phytol.* 154:275–304
13. Casamitjana-Martinez E, Hofhuis HF, Xu J, Liu CM, Heidstra R, Scheres B. 2003. Root-specific *CLE19* overexpression and the *sol1/2* suppressors implicate a *CLV*-like pathway in the control of *Arabidopsis* root meristem maintenance. *Curr. Biol.* 13:1435–41
14. Casero PJ, Casimiro I, Lloret PG. 1995. Lateral root initiation by asymmetrical transverse divisions of pericycle cells in 4 plant-species – *Raphanus sativus, Helianthus annuus, Zea mays, and Daucus carota*. *Protoplasma* 188:49–58
15. Casimiro I, Beeckman T, Graham N, Bhalerao R, Zhang HM, et al. 2003. Dissecting *Arabidopsis* lateral root development. *Trends Plant Sci.* 8:165–71
16. Casimiro I, Marchant A, Bhalerao RP, Beeckman T, Dhooge S, et al. 2001. Auxin transport promotes *Arabidopsis* lateral root initiation. *Plant Cell* 13:843–52
17. Celenza JL J, Grisafi PL, Fink GR. 1995. A pathway for lateral root formation in *Arabidopsis thaliana*. *Genes Dev.* 9:2131–42
18. Chen CW, Yang YW, Lur HS, Tsai YG, Chang MC. 2006. A novel function of abscisic acid in the regulation of rice (*Oryza sativa L.*) root growth and development. *Plant Cell Physiol.* 47:1–13
19. Chiou TJ, Aung K, Lin SI, Wu CC, Chiang SF, Su CL. 2006. Regulation of phosphate homeostasis by MicroRNA in *Arabidopsis*. *Plant Cell* 18:412–21
20. Cluis CP, Mouchel CF, Hardtke CS. 2004. The *Arabidopsis* transcription factor HY5 integrates light and hormone signaling pathways. *Plant J.* 38:332–47
21. Coates JC, Laplaze L, Haseloff J. 2006. Armadillo-related proteins promote lateral root development in *Arabidopsis*. *Proc. Natl. Acad. Sci. USA* 103:1621–26
22. Correa-Aragunde N, Graziano M, Lamattina L. 2004. Nitric oxide plays a central role in determining lateral root development in tomato. *Planta* 218:900–5
23. Deak KI, Malamy J. 2005. Osmotic regulation of root system architecture. *Plant J.* 43:17–28
24. de Billy F, Grosjean C, May S, Bennett M, Cullimore JV. 2001. Expression studies on *AUX1*-like genes in *Medicago truncatula* suggest that auxin is required at two steps in early nodule development. *Mol. Plant Microbe. Interact.* 14:267–77
25. De Smet I, Signora L, Beeckman T, Inze D, Foyer CH, Zhang HM. 2003. An abscisic acid-sensitive checkpoint in lateral root development of *Arabidopsis*. *Plant J.* 33:543–55
26. De Smet I, Vanneste S, Inze D, Beeckman T. 2006. Lateral root initiation or the birth of a new meristem. *Plant Mol. Biol.* 60:871–87

27. De Veylder L, de Almeida Engler J, Burssens S, Manevski A, Lescure B, et al. 1999. A new D-type cyclin of *Arabidopsis thaliana* expressed during lateral root primordia formation. *Planta* 208:453–62
28. Dharmasiri S, Swarup R, Mockaitis K, Dharmasiri N, Singh SK, et al. 2006. AXR4 is required for localization of the auxin influx facilitator AUX1. *Science* 312:1218–20
29. DiDonato RJ, Arbuckle E, Buker S, Sheets J, Tobar J, et al. 2004. *Arabidopsis ALF4* encodes a nuclear-localized protein required for lateral root formation. *Plant J.* 37:340–53
30. Dolan L, Janmaat K, Willemsen V, Linstead P, Poethig S, et al. 1993. Cellular organisation of the *Arabidopsis thaliana* root. *Development* 119:71–84
31. Dong L, Wang L, Zhang Y, Deng X, Xue Y. 2006. An auxin-inducible F-box protein CEGENDUO negatively regulates auxin-mediated lateral root formation in *Arabidopsis*. *Plant Mol. Biol.* 60:599–615
32. Drew MC. 1975. Comparison of effects of a localized supply of phosphate, nitrate, ammonium and potassium on growth of seminal root system, and shoot, in barley. *New Phytol.* 75:479–90
33. Esau K. 1965. *Plant Anatomy*. New York: Wiley
34. Feix G, Hochholdinger F, Park WJ. 2002. Maize root system and genetic analysis of its formation. In *Plant Roots: The Hidden Half*, ed. Y Waisel, A Eshel, U Kafkafi, pp. 239–48. New York: Marcel Dekker
35. Feldman LJ. 1994. The maize root. In *The Maize Handbook*, ed. M Freeling, V Walbot, pp. 29–37. New York/Berlin: Springer-Verlag
36. Fiers M, Hause G, Boutilier K, Casamitjana-Martinez E, Weijers D, et al. 2004. Misexpression of the *CLV3/ESR*-like gene *CLE19* in *Arabidopsis* leads to a consumption of root meristem. *Gene* 327:37–49
37. Fitter AH. 1987. An architectural approach to the comparative ecology of plant-poot systems. *New Phytol.* 106:61–77
38. Fitz Gerald JN, Lehti-Shiu MD, Ingram PA, Deak KI, Biesiada T, Malamy JE. 2006. Identification of quantitative trait loci that regulate *Arabidopsis* root system size and plasticity. *Genetics* 172:485–98
39. Fu X, Harberd NP. 2003. Auxin promotes *Arabidopsis* root growth by modulating gibberellin response. *Nature* 421:740–43
40. Fujii H, Chiou TJ, Lin SI, Aung K, Zhu JK. 2005. A miRNA involved in phosphate-starvation response in *Arabidopsis*. *Curr. Biol.* 15:2038–43
41. Fukaki H, Tameda S, Masuda H, Tasaka M. 2002. Lateral root formation is blocked by a gain-of-function mutation in the *SOLITARY-ROOT/IAA14* gene of *Arabidopsis*. *Plant J.* 29:153–68
42. Glass ADM, Britto DT, Kaiser BN, Kinghorn JR, Kronzucker HJ, et al. 2002. The regulation of nitrate and ammonium transport systems in plants. *J. Exp. Bot.* 53:855–64
43. Goda H, Shimada Y, Asami T, Fujioka S, Yoshida S. 2002. Microarray analysis of brassinosteroid-regulated genes in *Arabidopsis*. *Plant Physiol.* 130:1319–34
44. Hardtke CS. 2006. Root development–branching into novel spheres. *Curr. Opin. Plant Biol.* 9:66–71
45. Hardtke CS, Ckurshumova W, Vidaurre DP, Singh SA, Stamatiou G, et al. 2004. Overlapping and non-redundant functions of the *Arabidopsis* auxin response factors MONOPTEROS and NONPHOTOTROPIC HYPOCOTYL 4. *Development* 131:1089–1100
46. Harrison MJ. 2005. Signaling in the arbuscular mycorrhizal symbiosis. *Annu. Rev. Microbiol.* 59:19–42

47. Hetrick BAD. 1991. Mycorrhizas and root architecture. *Experientia* 47:355–62
48. Himanen K, Boucheron E, Vanneste S, de Almeida Engler J, Inze D, Beeckman T. 2002. Auxin-mediated cell cycle activation during early lateral root initiation. *Plant Cell* 14:2339–51
49. Himanen K, Vuylsteke M, Vanneste S, Vercruysse S, Boucheron E, et al. 2004. Transcript profiling of early lateral root initiation. *Proc. Natl. Acad. Sci. USA* 101:5146–51
50. Hirsch AM, Lum MR, Downie JA. 2001. What makes the rhizobia-legume symbiosis so special? *Plant Physiol.* 127:1484–92
51. Hobbie L, Estelle M. 1995. The *axr4* auxin-resistant mutants of *Arabidopsis thaliana* define a gene important for root gravitropism and lateral root initiation. *Plant J.* 7:211–20
52. Hobe M, Muller R, Grunewald M, Brand U, Simon R. 2003. Loss of CLE40, a protein functionally equivalent to the stem cell restricting signal CLV3, enhances root waving in *Arabidopsis*. *Dev. Genes Evol.* 213:371–81
53. Hochholdinger F, Park WJ, Sauer M, Woll K. 2004. From weeds to crops: genetic analysis of root development in cereals. *Trends Plant Sci.* 9:42–48
54. Horii H, Nemoto K, Miyamoto N, Harada J. 2005. Quantitative trait loci for adventitious and lateral roots in rice. *Plant Breed.* 125:198–200
55. Hou GC, Hill JP, Blancaflor EB. 2004. Developmental anatomy and auxin response of lateral root formation in *Ceratopteris richardii*. *J. Exp. Bot.* 55:685–93
56. Inukai Y, Sakamoto T, Ueguchi-Tanaka M, Shibata Y, Gomi K, et al. 2005. *Crown rootless1*, which is essential for crown root formation in rice, is a target of an AUXIN RESPONSE FACTOR in auxin signaling. *Plant Cell* 17:1387–96
57. Ivanchenko MG, Coffeen WC, Lomax TL, Dubrovsky JG. 2006. Mutations in the *Diageotropica (Dgt)* gene uncouple patterned cell division during lateral root initiation from proliferative cell division in the pericycle. *Plant J.* 46:436–47
58. Kaneko M, Itoh H, Inukai Y, Sakamoto T, Ueguchi-Tanaka M, et al. 2003. Where do gibberellin biosynthesis and gibberellin signaling occur in rice plants? *Plant J.* 35:104–15
59. Kim HB, Kwon M, Ryu H, Fujioka S, Takatsuto S, et al. 2006. The regulation of *DWARF4* expression is likely a critical mechanism in maintaining the homeostasis of bioactive brassinosteroids in *Arabidopsis thaliana*. *Plant Physiol.* 140:548–57
60. King JJ, Stimart DP, Fisher RH, Bleecker AB. 1995. A mutation altering auxin homeostasis and plant morphology in *Arabidopsis*. *Plant Cell* 7:2023–37
61. Konishi S, Izawa T, Lin SY, Ebana K, Fukuta Y, et al. 2006. An SNP caused loss of seed shattering during rice domestication. *Science* 312:1392–96
62. Kutz A, Muller A, Hennig P, Kaiser WM, Piotrowski M, Weiler EW. 2002. A role for nitrilase 3 in the regulation of root morphology in sulphur-starving *Arabidopsis thaliana*. *Plant J.* 30:95–106
63. Laskowski M, Biller S, Stanley K, Kajstura T, Prusty R. 2006. Expression profiling of auxin-treated *Arabidopsis* roots: toward a molecular analysis of lateral root emergence. *Plant Cell Physiol.* 47:788–92
64. Laskowski MJ, Williams ME, Nusbaum HC, Sussex IM. 1995. Formation of lateral root meristems is a two-stage process. *Development* 121:3303–10
65. Leyser O. 2006. Dynamic integration of auxin transport and signaling. *Curr. Biol.* 16:R424–33
66. Li J, Zhu SH, Song XW, Shen Y, Chen HM, et al. 2006. A rice glutamate receptor-like gene is critical for the division and survival of individual cells in the root apical meristem. *Plant Cell* 18:340–49

67. Linkohr BI, Williamson LC, Fitter AH, Leyser HM. 2002. Nitrate and phosphate availability and distribution have different effects on root system architecture of *Arabidopsis*. *Plant J.* 29:751–60
68. Little DY, Rao H, Oliva S, Daniel-Vedele F, Krapp A, Malamy JE. 2005. The putative high-affinity nitrate transporter NRT2.1 represses lateral root initiation in response to nutritional cues. *Proc. Natl. Acad. Sci. USA* 102:13693–98
69. Liu HJ, Wang SF, Yu XB, Yu J, He XW, et al. 2005. ARL1, a LOB-domain protein required for adventitious root formation in rice. *Plant J.* 43:47–56
70. Liu Y, Lamkemeyer T, Jakob A, Mi GH, Zhang FS, et al. 2006. Comparative proteome analyses of maize (*Zea mays L.*) primary roots prior to lateral root initiation reveal differential protein expression in the lateral root initiation mutant *rum1*. *Proteomics* 6:4300–8
71. Lohar DP, Schaff JE, Laskey JG, Kieber JJ, Bilyeu KD, Bird DM. 2004. Cytokinins play opposite roles in lateral root formation, and nematode and Rhizobial symbioses. *Plant J.* 38:203–14
72. Lopez-Bucio J, Cruz-Ramirez A, Herrera-Estrella L. 2003. The role of nutrient availability in regulating root architecture. *Curr. Opin. Plant Biol.* 6:280–87
73. Lopez-Bucio J, Hernandez-Abreu E, Sanchez-Calderon L, Nieto-Jacobo MF, Simpson J, Herrera-Estrella L. 2002. Phosphate availability alters architecture and causes changes in hormone sensitivity in the *Arabidopsis* root system. *Plant Physiol.* 129:244–56
74. Lorbiecke R, Sauter M. 1999. Adventitious root growth and cell-cycle induction in deepwater rice. *Plant Physiol.* 119:21–29
75. Loudet O, Gaudon V, Trubuil A, Daniel-Vedele F. 2005. Quantitative trait loci controlling root growth and architecture in *Arabidopsis thaliana* confirmed by heterogeneous inbred family. *Theor. Appl. Genet.* 110:742–53
76. Lynch J. 1995. Root architecture and plant productivity. *Plant Physiol.* 109:7–13
77. Malamy JE. 2005. Intrinsic and environmental response pathways that regulate root system architecture. *Plant Cell Environ.* 28:67–77
78. Malamy JE, Benfey PN. 1997. Organization and cell differentiation in lateral roots of *Arabidopsis thaliana*. *Development* 124:33–44
79. Malamy JE, Ryan KS. 2001. Environmental regulation of lateral root initiation in *Arabidopsis*. *Plant Physiol.* 127:899–909
80. McCully ME. 1999. Roots in soil: unearthing the complexities of roots and their rhizospheres. *Annu. Rev. Plant Physiol. Plant Mol. Biol.* 50:695–718
81. Montiel G, Gantet P, Jay-Allemand C, Breton C. 2004. Transcription factor networks. Pathways to the knowledge of root development. *Plant Physiol.* 136:3478–85
82. Moog PR, van der Kooij TA, Bruggemann W, Schiefelbein JW, Kuiper PJ. 1995. Responses to iron deficiency in *Arabidopsis thaliana*: the Turbo iron reductase does not depend on the formation of root hairs and transfer cells. *Planta* 195:505–13
83. Mori M, Nomura T, Ooka H, Ishizaka M, Yokota T, et al. 2002. Isolation and characterization of a rice dwarf mutant with a defect in brassinosteroid biosynthesis. *Plant Physiol.* 130:1152–61
84. Mouchel CF, Briggs GC, Hardtke CS. 2004. Natural genetic variation in *Arabidopsis* identifies *BREVIS RADIX*, a novel regulator of cell proliferation and elongation in the root. *Genes Dev.* 18:700–14
85. Mouchel CF, Osmont KS, Hardtke CS. 2006. *BRX*-mediated feedback between brassinosteroid levels and auxin signaling in root growth. *Nature.* 443:458–61
86. Mussig C, Fischer S, Altmann T. 2002. Brassinosteroid-regulated gene expression. *Plant Physiol.* 129:1241–51

87. Mussig C, Shin GH, Altmann T. 2003. Brassinosteroids promote root growth in *Arabidopsis*. *Plant Physiol.* 133:1261–71
88. Nacry P, Canivenc G, Muller B, Azmi A, Van Onckelen H, et al. 2005. A role for auxin redistribution in the responses of the root system architecture to phosphate starvation in *Arabidopsis*. *Plant Physiol.* 138:2061–74
89. Nakamura A, Higuchi K, Goda H, Fujiwara MT, Sawa S, et al. 2003. Brassinolide induces *IAA5*, *IAA19*, and *DR5*, a synthetic auxin response element in *Arabidopsis*, implying a cross talk point of brassinosteroid and auxin signaling. *Plant Physiol.* 133:1843–53
90. Nemhauser JL, Mockler TC, Chory J. 2004. Interdependency of brassinosteroid and auxin signaling in *Arabidopsis*. *PLoS Biol.* 2:E258
91. Olmos E, Kiddle G, Pellny T, Kumar S, Foyer C. 2006. Modulation of plant morphology, root architecture, and cell structure by low vitamin C in *Arabidopsis thaliana*. *J. Exp. Bot.* 57:1645–55
92. Paszkowski U, Kroken S, Roux C, Briggs SP. 2002. Rice phosphate transporters include an evolutionarily divergent gene specifically activated in arbuscular mycorrhizal symbiosis. *Proc. Natl. Acad. Sci. USA* 99:13324–29
93. Ponce G, Barlow PW, Feldman LJ, Cassab GI. 2005. Auxin and ethylene interactions control mitotic activity of the quiescent center, root cap size, and pattern of cap cell differentiation in maize. *Plant Cell Environ.* 28:719–32
94. Price NS, Roncadori RW, Hussey RS. 1989. Cotton root-growth as influenced by phosphorus-nutrition and vesicular arbuscular mycorrhizas. *New Phytol.* 111:61–66
95. Przemeck GK, Mattsson J, Hardtke CS, Sung ZR, Berleth T. 1996. Studies on the role of the *Arabidopsis* gene *MONOPTEROS* in vascular development and plant cell axialization. *Planta* 200:229–37
96. Remans T, Nacry P, Pervent M, Girin T, Tillard P, et al. 2006. A central role for the nitrate transporter NRT2.1 in the integrated morphological and physiological responses of the root system to nitrogen limitation in *Arabidopsis*. *Plant Physiol.* 140:909–21
97. Riefler M, Novak O, Strnad M, Schmulling T. 2006. *Arabidopsis* cytokinin receptor mutants reveal functions in shoot growth, leaf senescence, seed size, germination, root development, and cytokinin metabolism. *Plant Cell* 18:40–54
98. Rubio V, Linhares F, Solano R, Martin AC, Iglesias J, et al. 2001. A conserved MYB transcription factor involved in phosphate starvation signaling both in vascular plants and in unicellular algae. *Genes Dev.* 15:2122–33
99. Salvi S, Tuberosa R. 2005. To clone or not to clone plant QTLs: present and future challenges. *Trends Plant Sci.* 10:297–304
100. Sauer M, Jakob A, Nordheim A, Hochholdinger F. 2006. Proteomic analysis of shoot-borne root initiation in maize (*Zea mays* L.). *Proteomics* 6:2530–41
101. Scheres B, Wolkenfelt H, Willemsen V, Terlouw M, Lawson E, et al. 1994. Embryonic origin of the *Arabidopsis* primary root and root meristem initials. *Development* 120:2475–87
102. Schiefelbein J. 2003. Cell-fate specification in the epidermis: a common patterning mechanism in the root and shoot. *Curr. Opin. Plant Biol.* 6:74–78
103. Sergeeva LI, Keurentjes JJ, Bentsink L, Vonk J, van der Plas LH, et al. 2006. Vacuolar invertase regulates elongation of *Arabidopsis thaliana* roots as revealed by QTL and mutant analysis. *Proc. Natl. Acad. Sci. USA* 103:2994–99
104. Signora L, De Smet I, Foyer CH, Zhang HM. 2001. ABA plays a central role in mediating the regulatory effects of nitrate on root branching in *Arabidopsis*. *Plant J.* 28:655–62
105. Sorin C, Bussell JD, Camus I, Ljung K, Kowalczyk M, et al. 2005. Auxin and light control of adventitious rooting in *Arabidopsis* require ARGONAUTE1. *Plant Cell* 17:1343–59

106. Steffens B, Wang JX, Sauter M. 2006. Interactions between ethylene, gibberellin and abscisic acid regulate emergence and growth rate of adventitious roots in deepwater rice. *Planta* 223:604–12
107. Steindler C, Matteucci A, Sessa G, Weimar T, Ohgishi M, et al. 1999. Shade avoidance responses are mediated by the ATHB-2 HD-zip protein, a negative regulator of gene expression. *Development* 126:4235–45
108. Stepanova AN, Hoyt JM, Hamilton AA, Alonso JM. 2005. A link between ethylene and auxin uncovered by the characterization of two root-specific ethylene-insensitive mutants in *Arabidopsis*. *Plant Cell* 17:2230–42
109. Swarup R, Parry G, Graham N, Allen T, Bennett M. 2002. Auxin cross-talk: integration of signaling pathways to control plant development. *Plant Mol. Biol.* 49:411–26
110. Tatematsu K, Kumagai S, Muto H, Sato A, Watahiki MK, et al. 2004. *MASSUGU2* encodes Aux/IAA19, an auxin-regulated protein that functions together with the transcriptional activator NPH4/ARF7 to regulate differential growth responses of hypocotyl and formation of lateral roots in *Arabidopsis thaliana*. *Plant Cell* 16:379–93
111. van den Berg C, Willemsen V, Hendriks G, Weisbeek P, Scheres B. 1997. Short-range control of cell differentiation in the *Arabidopsis* root meristem. *Nature* 390:287–89
112. Wagner S, Bernhardt A, Leuendorf JE, Drewke C, Lytovchenko A, et al. 2006. Analysis of the *Arabidopsis rsr4–1/pdx1–3* mutant reveals the critical function of the PDX1 protein family in metabolism, development, and vitamin B6 biosynthesis. *Plant Cell* 18:1722–35
113. Walch-Liu P, Ivanov II, Filleur S, Gan Y, Remans T, Forde BG. 2006. Nitrogen regulation of root branching. *Ann. Bot.* 97:875–81
114. Werner T, Motyka V, Laucou V, Smets R, Van Onckelen H, Schmulling T. 2003. Cytokinin-deficient transgenic *Arabidopsis* plants show multiple developmental alterations indicating opposite functions of cytokinins in the regulation of shoot and root meristem activity. *Plant Cell* 15:2532–50
115. Williamson LC, Ribrioux SP, Fitter AH, Leyser HM. 2001. Phosphate availability regulates root system architecture in *Arabidopsis*. *Plant Physiol.* 126:875–82
116. Woll K, Borsuk LA, Stransky H, Nettleton D, Schnable PS, Hochholdinger F. 2005. Isolation, characterization, and pericycle-specific transcriptome analyses of the novel maize lateral and seminal root initiation mutant *rum1*. *Plant Physiol.* 139:1255–67
117. Yokota T, Sato T, Takeuchi Y, Nomura T, Uno K, et al. 2001. Roots and shoots of tomato produce 6-deoxo-28-norcathasterone, 6-deoxo-28-nortyphasterol and 6-deoxo-28-norcastasterone, possible precursors of 28-norcastasterone. *Phytochemistry* 58:233–38
118. Zhang H, Forde BG. 1998. An *Arabidopsis* MADS box gene that controls nutrient-induced changes in root architecture. *Science* 279:407–9
119. Zhang H, Jennings A, Barlow PW, Forde BG. 1999. Dual pathways for regulation of root branching by nitrate. *Proc. Natl. Acad. Sci. USA* 96:6529–34
120. Zolobowska L, Van Gijsegem F. 2006. Induction of lateral root structure formation on petunia roots: A novel effect of GMI1000 *Ralstonia solanacearum* infection impaired in *Hrp* mutants. *Mol. Plant-Microbe Interact.* 19:597–606

Leaf Senescence

Pyung Ok Lim,[1] Hyo Jung Kim,[2]
and Hong Gil Nam[2]

[1] Department of Science Education, Cheju National University, Jeju, Jeju, 690-756, Korea

[2] Division of Molecular Life Sciences and National Core Research Center for Systems Bio-Dynamics, POSTECH, Pohang, Kyungbuk, 790-784, Korea; email: nam@postech.ac.kr

Annu. Rev. Plant Biol. 2007. 58:115–36

The *Annual Review of Plant Biology* is online at plant.annualreviews.org

This article's doi:
10.1146/annurev.arplant.57.032905.105316

First published online as a Review in Advance on December 19, 2006

1543-5008/07/0602-0115$20.00

Key Words

longevity, developmental aging, programmed cell death, nutrient remobilization, environmental factors

Abstract

Leaf senescence constitutes the final stage of leaf development and is critical for plants' fitness as nutrient relocation from leaves to reproducing seeds is achieved through this process. Leaf senescence involves a coordinated action at the cellular, tissue, organ, and organism levels under the control of a highly regulated genetic program. Major breakthroughs in the molecular understanding of leaf senescence were achieved through characterization of various senescence mutants and senescence-associated genes, which revealed the nature of regulatory factors and a highly complex molecular regulatory network underlying leaf senescence. The genetically identified regulatory factors include transcription regulators, receptors and signaling components for hormones and stress responses, and regulators of metabolism. Key issues still need to be elucidated, including cellular-level analysis of senescence-associated cell death, the mechanism of coordination among cellular-, organ-, and organism-level senescence, the integration mechanism of various senescence-affecting signals, and the nature and control of leaf age.

Contents

INTRODUCTION

Senescence is the age-dependent deterioration process at the cellular, tissue, organ, or organismal level, leading to death or the end of the life span (48). Leaf senescence is an organ-level senescence but is often intimately associated with cellular or organismal death. Annual plants undergo leaf senescence along with the organismal-level senescence when they reach the end of their temporal niche, as we observe at the grain-filling and maturation stage of the crop fields of soybean, corn, or rice. For trees and other perennial plants, leaf senescence is illustrated by the splendid autumn scenery of color changes in leaves.

Leaf senescence is not a passive and unregulated degeneration process. During senescence, leaf cells undergo rather orderly changes in cell structure, metabolism, and gene expression. The earliest and most significant change in cell structure is the breakdown of the chloroplast, the organelle that contains up to 70% of the leaf protein. Metabolically, carbon assimilation is replaced by catabolism of chlorophyll and macromolecules such as proteins, membrane lipids, and RNA. Increased catabolic activity is responsible for converting the cellular materials accumulated during the growth phase of leaf into exportable nutrients that are supplied to developing seeds or to other growing organs. Thus, although leaf senescence is a deleterious process for the sake of the leaf organ, it can be seen as an altruistic process: It critically contributes to the fitness of whole plants by ensuring optimal production of offspring and better survival of plants in their given temporal and spatial niches. Leaf senescence is thus an evolutionarily selected developmental process and comprises an important phase in the plant life cycle (7, 40, 46, 48). In agricultural aspects, however, leaf senescence may limit yield in crop plants by limiting the growth phase and may also cause postharvest spoilage such as leaf yellowing and nutrient loss in vegetable crops. Thus, studying leaf senescence will not only enhance our understanding of a fundamental biological process, but also may provide means to control leaf senescence to improve agricultural traits of crop plants.

Leaf senescence is basically governed by the developmental age. However, leaf senescence is also influenced by various internal and environmental signals that are integrated into the age information: Leaf senescence is an integrated response of leaf cells to age information and other internal and environmental signals. This integrated senescence response provides plants with optimal fitness by incorporating the environmental and endogenous status of plants in a given ecological setting by fine-tuning the initiation timing, progression rate, and nature of leaf senescence. The environmental factors that influence leaf senescence include abiotic and biotic factors. The abiotic factors include drought, nutrient limitation, extreme temperature, and oxidative stress by UV-B irradiation and ozone, etc. The biotic factors include pathogen infection and shading by other plants. Leaf senescence can occur prematurely under these unfavorable environmental conditions (39).

In naturally senescing leaves, senescence occurs in a coordinated manner at the whole-leaf level, usually starting from the tips or the margins of a leaf toward the base of a leaf. However, when the uneven environmental stress is targeted locally on a leaf, the stressed leaf region undergoes earlier senescence than do the other parts. Thus, leaf cells show some degree of locality in a senescence program.

Leaf senescence can occur without an obvious correlation with senescence of other organs in some plants, such as many tree species, although it is often developmentally coordinated with senescence of other organs or whole plants, especially monocarpic plants. In some monocarpic plants, the reproductive development often governs senescence of leaves. This so-called correlative control is dramatically observed in pea and soybean, where removal of the reproductive organ can actually reverse the fate of senescing leaves to juvenile leaves. However, in some plants such as *Arabidopsis*, leaf senescence does not appear to be under correlative control, but the leaf senescence at the whole-plant level is somewhat correlated with the life span of the whole plant.

Arabidopsis thaliana is a favorite model for the molecular genetic study of leaf senescence (5, 7, 39). As a monocarpic plant, it has a short life cycle. Its leaves undergo readily distinguishable developmental stages and show a well-defined and reproducible senescence program (**Figure 1**), which makes genetic analysis of leaf senescence feasible. Extensive genomic resources available for *Arabidopsis* allow rapid identification and functional analysis of senescence regulatory genes.

In this review, we discuss recent progress toward molecular and genetic understanding of leaf senescence and longevity that has been achieved mostly from *Arabidopsis*. It is cautioned that *Arabidopsis* leaves have a senescence character different from that of some other monocarpic plants in that the leaf longevity in *Arabidopsis* is not controlled by the developing reproductive structures. Thus, the findings in *Arabidopsis* might not reveal some of the mechanisms involved in leaf senescence of other plants. Thus, wherever appropriate we also discuss the discoveries achieved from other plants.

Monocarpic plant: a plant that reproduces once and then dies at the end of its reproductive phase

Leaf longevity: reflects the period of a whole life span of a leaf from its emergence as a leaf primordium to death

LEAF SENESCENCE-ASSOCIATED CELL DEATH AS A PROGRAMMED CELL DEATH

Leaf senescence involves cell death that is controlled by age under the influence of other endogenous and environmental factors. Programmed cell death (PCD) is a self-destructing cellular process triggered by external or internal factors and mediated through an active genetic program. Cell death in leaf senescence is controlled by many active genetic programs (10). The cell death occurring in leaf senescence is thus a type of PCD. Leaf organs are composed of various cell types. Cell death in leaf senescence starts from mesophyll cells and then proceeds to other cell types. It also appears that cell death does not occur coherently but starts with local patches of early-dying cells and then propagates into the whole-leaf area.

PCD plays crucial roles in various developmental and defense responses in plants. Typical examples of PCD in plants are observed in the formation of tracheary elements, germination-related degeneration of aleuron

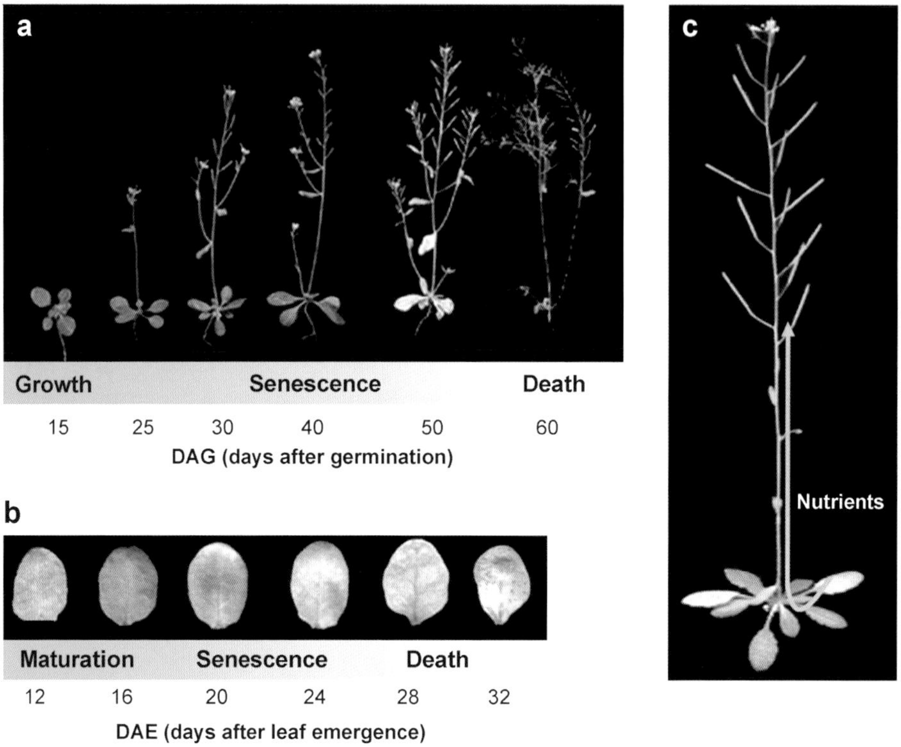

Figure 1

Characteristics of whole-plant senescence and leaf senescence in *Arabidopsis*. (*a*) Stages in the life cycle of whole plants. Plants are pictured at 15, 25, 30, 40, 50, and 60 days after germination. (*b*) An age-dependent senescence phenotype in the third rosette leaf. Leaves are pictured at 12, 16, 20, 24, 28, and 32 days after emergence. (*c*) As leaves senesce, nutrients such as nitrogen, phosphorus, and metals are relocated to other parts of the plants such as developing seeds and leaves.

layer cells, and pathogen-induced hypersensitive response (HR) (34, 70). PCD in leaf senescence has some features distinctive from other PCDs (71). First, leaf senescence involves an organ-level cell death that eventually encompasses the entire leaf, whereas other PCDs involve rather localized cell death or occur in limited tissues and cell types. Second, cell death rate during leaf senescence is slower than that in the other PCDs. Third, in terms of the biological function, PCD in leaf senescence is mostly for remobilization of nutrients from the leaf to other organs including developing seeds. The leaf organ is the major photosynthetic organ. Thus, optimal utilization of nutrients accumulated during the photosynthetic period is critical for plants' fitness and is critically affected by fine control of senescence process. In this regard, the slow degeneration of cells during leaf senescence is in part to ensure effective remobilization of nutrients that are generated by macromolecular hydrolysis during senescence. Many molecular events during leaf senescence can be readily understood from the viewpoint of this altruistic remobilization activity.

Structural and Biochemical Changes in Leaf Senescence-Associated Cell Death

Leaf cells at the senescence stage show some distinctive structural and biochemical changes. A notable feature of cellular structural change during leaf senescence is the order of disintegration of intracellular organelles (48, 68). The earliest structural changes occur in the chloroplast, i.e., changes in the grana structure and content and formation of lipid droplet called plastoglobuli. In contrast, the nucleus and mitochondria that are essential for gene expression and energy production, respectively, remain intact until the last stages of senescence. This reflects that the leaf cells need to remain functional for progression of senescence until a late stage of senescence, possibly for effective mobilization of the cellular materials. In the last stage of leaf senescence, typical symptoms of PCD such as controlled vacuolar collapse, chromatin condensation, and DNA laddering are detected in naturally senescing leaves from a variety of plants including *Arabidopsis*, tobacco, and five trees (10, 62, 78). These observations imply that leaf senescence involves cellular events that ultimately lead to PCD. Eventually, visible disintegration of the plasma and vacuolar membranes appears. The loss of integrity of the plasma membrane then leads to disruption of cellular homeostasis, ending the life of a cell in senescing leaves.

The cellular biochemical changes in senescing leaves are first accompanied by reduced anabolism (4, 5, 76). The overall cellular content of polysomes and ribosomes decreases fairly early, reflecting a decrease in protein synthesis. This occurs concomitantly with reduced synthesis of rRNAs and tRNAs. Further cellular biochemical changes are most easily understood from the viewpoint of nutrient salvage, e.g., hydrolysis of macromolecules and subsequent remobilization, which requires operating a complex array of metabolic pathways. Chloroplast degeneration is accompanied by chlorophyll degradation and the progressive loss of proteins in the chloroplast, such as ribulose biphosphate carboxylase (Rubisco) and chlorophyll *a/b* binding protein (CAB). Hydrolysis of proteins to free amino acids depends on the actions of several endo- and exopeptidases (6, 28, 52). Senescence-associated cystein proteases, which are accumulated in the vacuole, also play a role in protein degradation. Lipid-degrading enzymes, such as phospholipase D, phosphatidic acid phosphatase, lytic acyl hydrolase, and lipoxygenase appear to be involved in hydrolysis and metabolism of the membrane lipid in senescing leaves (66, 67). Most of the fatty acids are either oxidized to provide energy for the senescence process or converted to α-ketoglutarate via the glyoxylate cycle. The α-ketoglutarate can be converted into phloem-mobile sugars through gluconeogenesis or used to mobilize amino acids released during leaf protein degradation (28, 66). A massive decrease in nucleic acids occurs during leaf senescence (65). Total RNA levels are rapidly reduced along with progression of senescence. The initial decrease in the RNA levels is distinctively observed for the chloroplast rRNAs and cytoplasmic rRNAs. The amount of various rRNA species is likely regulated coordinately, although this aspect has not been analyzed. The decrease of the amount of rRNAs is followed by that of the cytoplasmic mRNA and tRNA. The decrease in the RNA levels is accompanied by increased activity of several RNases.

Molecular Comparison of Leaf Senescence-Associated Cell Death with Other Programmed Cell Deaths

One obvious question regarding leaf senescence-associated cell death is how the cell death pathways during leaf senescence are distinct from those of other types of PCDs at the molecular level. Cell death in pathogen-induced HR is best-characterized among plant PCDs. Pathogenesis-related (PR) proteins are associated with PCD in HR. A few earlier works showed that many PR genes are induced during leaf senescence in several plant species (55, 56). A comparative study of leaf senescence and HR showed that *HIN1*, an HR cell death marker, is also expressed at late stages of leaf senescence (63). Furthermore, defense-related genes

Photochemical efficiency: deduced from the characteristics of chlorophyll fluorescence of PSII. The ratio of maximum variable fluorescence (Fv) to maximum yield of fluorescence (Fm), which corresponds to the potential quantum yield of the photochemical reactions of PSII, is used as the measure of the photochemical efficiency of PSII

including the *Arabidopsis ELI3* gene showed a senescence-associated induction as well (56). The *LSC54* gene encoding a metallothionine is also highly induced during both senescence and pathogen-related cell death (9). These observations indicate that, at the molecular level, some common steps or crosstalks exist between senescence-associated and pathogen-induced cell death.

In contrast, a few molecular markers that are specific for each of the senescence-associated and HR-associated PCDs were also identified. For example, *HSR203J* is upregulated during HR but not during leaf senescence (54). Similarly, the *Arabidopsis SAG12* gene expression is associated with leaf senescence but is not detected in the HR PCD in tobacco. Thus, these two genes may be a specific part of signaling steps for HR PCD and senescence-associated cell death, respectively, indicating that there are specific branches of molecular pathways leading to these two types of PCD.

A comparison of changes in global gene expression patterns during natural leaf senescence with those during starvation-induced death of suspension culture cells has shown similarities as well as considerable differences between these two PCDs (8). Of the 827 senescence-enhanced genes, 326 showed at least threefold upregulation in the starvation-induced PCD of suspension culture. In contrast, the rest of the senescence-upregulated genes were not significantly upregulated in starvation-induced PCD of suspension culture. The result implies that distinctive pathways for the two PCD processes are present.

MOLECULAR AND GENETIC APPROACHES FOR ANALYZING LEAF SENESCENCE

As with any other biological phenomena, it was critical to develop an accurate and proper assay for leaf senescence. Two main points must be seriously considered in analyzing leaf senescence. First, leaf senescence should be measured on a single leaf base along with its age information. Measuring senescence parameters with a mixture of several leaves at a given age of a plant is not a valid analysis for leaf senescence because the individual leaves of a plant have different ages. Second, the senescence symptom should be measured with various senescence parameters and ideally with markers that cover various aspects of senescence physiology. Senescence results from a sum of various physiological changes and it is often possible that a single parameter may not reflect senescence but only the change of a specific physiology related to the measuring parameter.

Assay of Leaf Senescence

To quantitatively measure the leaf senescence symptom, a range of physiological and molecular parameters can be utilized. Well-established senescence markers include chlorophyll content, photochemical efficiency, senescence-associated enzyme activities, change of protein levels, membrane ion leakage, and gene expression, etc. Leaf yellowing is a convenient visible indicator of leaf senescence and reflects mainly chloroplast senescence of mesophyll cells, which is the first step in senescence-associated PCD. The survivorship curve assay based on visual examination of leaf yellowing (the time when the half of a leaf turns yellow) provides a reliable measure, although the assay is somewhat subjective. Measuring chlorophyll loss and photochemical efficiency is another convenient assay for chloroplast senescence (49, 74). The activation of catabolic or hydrolytic activities, such as RNase or peroxidase activity occurs during leaf senescence (1, 65). Thus, measuring these enzyme activities is also a reliable and quantitative way to assay leaf senescence. Senescence involves disruption of plasma membrane integrity as the final step of cell death, which can be conveniently quantified by monitoring membrane ion leakage (74). This provides one of most reliable assays for senescence-associated cell death, although it measures the later step of senescence.

Leaf senescence is accompanied by decreased expression of genes related to photosynthesis (e.g., *CAB2*) and protein synthesis (e.g., *RPS*, *RBC*) and by increased expression of senescence-associated genes (*SAG*s). The expression pattern of these genes during leaf development can be monitored by RNA gel blot analysis or reverse transcription-polymerase chain reaction (RT-PCR) (74). Microarray analysis would provide a quantifiable and global picture of the senescence process at the gene expression level with a clue of which downstream pathways of leaf senescence are affected by a specific mutation or by a specific environmental condition (8, 11). Microarray data may be further utilized for systems-level analysis of leaf senescence.

Genetic Analysis of Leaf Senescence

Two main approaches were utilized for understanding regulation of leaf senescence: the genetic and molecular approaches. The genetic approach involves isolation and characterization of mutants that show altered senescence phenotypes. *Arabidopsis* is a suitable model plant in this regard (25). Considering the complex nature of leaf senescence, it is expected that regulation of senescence involves many regulatory elements composed of positive and negative elements to finely tune the initiation and progression of senescence. The positive elements must exist for senescence to proceed. The negative elements are also important to prevent senescence from occurring prematurely. Many of these regulatory elements may contribute subtly in the senescence phenotype due to redundant functions in senescence. In addition, senescence is inevitably affected by the previous developmental stages including leaf formation and growth. Thus, a screening scheme suitably focused on leaf senescence symptoms was developed and successfully employed for isolating senescence mutants. So far, most of the genetic screening was focused on identifying delayed senescence mutants from T-DNA or a chemical mutant pool, which allowed identification of various important positive elements of senescence (50, 74, 75, 79). Early-senescence mutants screened from T-DNA or chemical mutant pools would enable identification of negative factors involved in the leaf senescence process (80). However, this approach should be taken with the caution that mutations with apparent early-senescence symptoms may not be directly associated with control of senescence because mutations in many homeostatic or housekeeping genes could also give apparent early-senescence symptoms.

Molecular Approaches to Understanding Leaf Senescence

The alternative approach was to identify and characterize genes that show enhanced or reduced expression during leaf senescence. Recent technological advances have allowed investigation of the *SAG*s at the genome-wide scale (3, 8, 17, 19, 41, 72). For example, a DNA microarray with 13,490 aspen expressed sequence tags (ESTs) was used to analyze the transcriptom of aspen leaves during autumn senescence (3). In *Arabidopsis*, Affymetrix GeneChip arrays representing 24,000 genes were utilized for analyzing changes in global expression pattern during leaf senescence (8, 72). This analysis has identified more than 800 *SAG*s, illustrating the dramatic alteration in cellular physiology that underlies the developmental transition to the senescence stage. Unlike the genome-wide microarray analysis, microarray analysis of 402 potential transcription factors was carried out at different developmental stages and under various biotic and abiotic stresses, providing a clue to the transcriptional regulatory network during leaf senescence (11). Similar approaches for other crop plants should allow comparison of molecular pictures of leaf senescence in different species. The collection of T-DNA insertion lines available in *Arabidopsis* was effectively utilized for functional analysis of individual *SAG*s.

In particular, functional characterization of potential regulators such as signal transduction-related proteins and transcription factors was the primary target for this analysis. Analysis of these mutant lines has provided and will continue to provide important information for understanding regulatory pathways of leaf senescence.

MOLECULAR GENETIC REGULATION OF LEAF SENESCENCE

Leaf senescence is an integral part of plant development and constitutes the final stage of development. The timing of leaf senescence is thus controlled by developmental age. However, the senescence process including senescence rate and molecular nature is intimately influenced by various environmental and internal factors. The environmental cues that affect leaf senescence include stresses such as high or low temperature, drought, ozone, nutrient deficiency, pathogen infection, and shading, etc. The internal factors include various phytohormones and reproductive development as well as developmental age (**Figure 2**). It is obvious that multiple pathways responding to various internal and external factors should exist and are interconnected to form a complex network of regulatory pathways for senescence (24). It is also obvious that, although the apparent symptoms of leaf senescence appear similar during senescence,

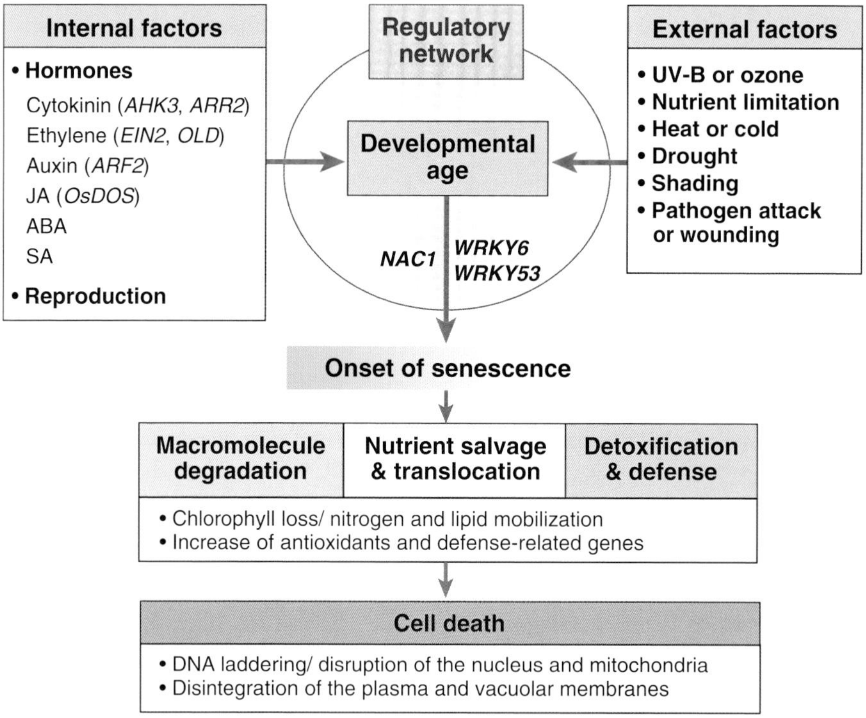

Figure 2

A model for regulatory pathways in leaf senescence. Leaf senescence is considered a complex process in which the effects of various internal and external signals are integrated into the developmental age-dependent senescence pathways. Multiple pathways that respond to various factors are possibly interconnected to form regulatory networks. These regulatory pathways activate distinct sets of senescence-associated genes, which are responsible for executing the degeneration process and ultimately lead to cell death.

the molecular nature of the senescence state influenced by these factors will be distinctive (53).

Leaf senescence should be a finely regulated process, considering its potential role in plants' fitness and the various factors involved in senescence control. Below we discuss the progress regarding molecular and genetic understanding of leaf senescence.

Onset of Leaf Senescence

A few of the central and unanswered questions regarding leaf senescence are how the leaf senescence is initiated, what the nature of the threshold that triggers leaf senescence is, how the developmental age is recognized to initiate the senescence program, and what the nature of the developmental age is. There are some indications that lead to answers to these questions. In plants, sugar status modulates and coordinates internal regulators and environmental cues that govern growth and development. Several lines of evidence suggest that a high concentration of sugars lowers photosynthetic activity and induces leaf senescence (12, 31, 44, 55). Senescence would be triggered when the level of sugars is above an acceptable window. In that sense, sugar metabolic rate would affect leaf longevity and might be the mechanism that regulates the developmental aging process, as shown in a range of organisms from yeast to mammals (15, 36).

An interesting finding was obtained from studies of the *oresara 4-1* (*ore4-1*) mutation, which causes a delay in leaf senescence during age-dependent senescence, but not in hormone- or dark-induced senescence (75). The *ore4-1* mutant has a partial lesion in chloroplast functions, including photosynthesis, which results from reduced expression of the plastid ribosomal protein small subunit 17 (*PRPS17*) gene. It was suggested that the delayed leaf senescence phenotype observed in the *ore4-1* mutant is likely due to a reduced metabolic rate because the chloroplasts, the major energy source for plant growth via photosynthesis, are only partially functional in the mutant. Reduced metabolic rate could lead to less oxidative stress, which might be a crucial factor in senescence.

Leaf senescence should be intimately related to the previous developmental stages of leaf, such as leaf initiation, growth, and maturation. Thus, it is possible that genes controlling these processes, including meristematic activity, could influence age-dependent senescence. In this respect, we observed that the leaves of the *blade on petiole 1-1* (*bop1-1*) mutant that showed enhanced meristematic activity in leaves exhibited a prolonged life span (21). Exact mechanisms by which this gene regulates leaf senescence need to be investigated.

Aging: an addition of timing to a cell, organ, or a whole plant that occurs throughout development. In this sense, aging would be a major determinant of senescence but not senescence itself

Environmental Factors and Leaf Senescence

Senescence is an integrated response of plants to endogenous developmental and external environmental signals. Thus, some of the genes involved in the response to environmental changes are expected to regulate leaf senescence. A comparison of gene expression patterns between stress responses and leaf senescence indicated that considerable crosstalk exists between these processes. For example, among the 43 transcription factor genes that are induced during senescence, 28 genes are also induced by various stresses. Our current understanding of the relationship between environmental responses and leaf senescence mostly comes from the study of senescence response to the phytohormones such as abscisic acid (ABA), jasmonic acid (JA), ethylene, and salicylic acid (SA) that are extensively involved in response to various abiotic and biotic stresses. These stresses affect synthesis and/or signaling pathways of the hormones to eventually trigger expression of stress-responsive genes, which in turn appears to affect leaf senescence. The involvement of these hormonal pathways is discussed below. However, we emphasize the need to directly examine the relationship between the stress

Abscission: the shedding of leaves, flowers, or fruits, usually at a weak area termed the abscission zone

Arabidopsis and tomato plants that constitutively overproduce ethylene do not exhibit earlier-onset leaf senescence, suggesting that ethylene alone is not sufficient to initiate leaf senescence. This is consistent with the postulation that age-dependent factors are required for ethylene-regulated leaf senescence. Furthermore, potential regulators involved in integrating ethylene signaling into age-dependent pathways have been reported. The *onset of leaf death 1* (*old1*) mutant of *Arabidopsis* displays a phenotype with earlier-onset senescence in an age-dependent manner (33). The early-senescence phenotype was further accelerated by exposure to ethylene, showing that the *old1* mutation resulted in alternation of both of the age- and ethylene signaling-dependent leaf senescence. However, in the *old1etr1* double mutant where ethylene perception was blocked by the mutation in the *ETR1* gene, age-dependent earlier-onset leaf senescence still occurred but was not further accelerated by ethylene treatment. These observations suggested that OLD1 negatively regulates integration of ethylene signaling into leaf senescence. Recent studies with several *old* mutants that exhibited an altered senescence response to ethylene treatment further supported the notion that the effect of ethylene on leaf senescence depends on age-related changes through these *OLD* genes (32).

ABA is a key plant hormone mediating plant responses to environmental stresses. It also functions in plant development such as seed germination and plant growth. Furthermore, it has been well known that exogenous application of ABA promotes leaf abscission and senescence (81). However, the role of ABA in leaf senescence has not been clearly defined aside from some circumstantial evidence. The ABA level increases in senescing leaves and exogenously applied ABA induces expression of several *SAG*s (73), which is consistent with the effect on leaf senescence. Environmental stresses such as drought, high salt condition, and low temperature positively affect leaf senescence, and under these stress conditions ABA content increases in leaves. Concurrently with the increased ABA level in senescing leaves (18), the genes encoding the key enzyme in ABA biosynthesis, 9-*cis*-epoxycarotenoid dioxygenase (*NECD*), and two aldehyde oxidase genes *AAO1* and *AAO3* show increased expression (8, 72). The ABA-inducible receptor-like kinase gene of *Arabidopsis*, *RPK1*, was found to be gradually up-regulated during leaf senescence (J.C. Koo & H.G. Nam, unpublished data). Inducible expression of RPK1 hastened the onset of leaf senescence, supporting a role for ABA in leaf senescence. A recent report argued that ABA induces accumulation of H_2O_2 in senescing rice leaf, which in turn accelerates leaf senescence (30). Another possibility is that senescence accelerated by exogenous ABA treatment might cause increased H_2O_2 generation, since it is well known that there is an increase of reactive oxygen species during leaf senescence. ABA also induces expression of antioxidant genes and enhances the activities of antioxidative enzymes such as superoxide dismutase (SOD), ascorbate peroxidase (APOD), and catalase (CAT) (29). These activities may play at least a partial role in protecting the cellular functions required for progression and completion of senescence. It appears that ABA controls activities of both the cellular protection activities and senescence activities. The balance between these two activities seems to be important in controlling progression of leaf senescence and may be adjusted by other senescence-affecting factors such as age. A crucial link between ABA and leaf senescence has yet to be discovered via genetic analysis.

Methyl jasmonate (MeJA) and its precursor JA promote senescence in detached oat (*Avena sativa*) leaves (69). Exogenously applied MeJA to detached *Arabidopsis* leaves leads to a rapid loss of chlorophyll content and photochemical efficiency of photosystem II (PSII) and increased expression of *SAG*s such as *SEN4*, *SEN5*, and *γVPE*. A more convincing support for the role of JA in leaf senescence comes from the observation that JA-dependent senescence is defective in the

JA-insensitive mutant *coronatine insensitive 1* (*coi1*), implying that the JA signaling pathway is required for JA to promote leaf senescence (23). Functional studies on a nuclear-localized CCCH-type zinc finger protein, OsDOS (*Oryza sativa* Delay of the Onset of Senescence), also supports involvement of MeJA in leaf senescence (37). The expression of the *OsDOS* gene was downregulated during leaf senescence. Notably, RNAi knockdown of *OsDOS* accelerated age-dependent leaf senescence, whereas its overexpression resulted in a marked delay of leaf senescence, showing that OsDOS acts as a negative regulator for leaf senescence. A genome-wide expression analysis revealed that many of the JA signaling-dependent genes in particular were upregulated in the RNAi transgenic lines but downregulated in the overexpressing transgenic lines. This implies that OsDOS acts as a negative regulator of leaf senescence in integrating the JA signaling pathway into age-dependent senescence.

SA is the hormone involved in pathogen response and pathogen-mediated cell death. A recent intriguing discovery in leaf senescence was the role of SA in age-dependent leaf senescence. The concentration of endogenous SA is four times higher in senescing leaves of *Arabidopsis*. The higher SA level in senescing leaves appears to be involved in upregulation of several *SAG*s during leaf senescence (45): expression of a number of *SAG*s such as *PR1a*, *chitinase*, and *SAG12* is considerably reduced or undetectable in *Arabidopsis* plants defective in the SA signaling or biosynthetic pathway (*npr1* and *pad4* mutants, and *NahG* transgenic plants). A surprising discovery was derived from transcriptome analysis: The change of transcriptome mediated by the SA pathway is highly similar to that mediated by age-dependent senescence. The fact that the SA pathway is specifically involved in age-dependent leaf senescence is further supported by the finding that age-dependent but not dark-induced leaf senescence is delayed in *NahG* overexpressing transgenic plants that produce dramatically reduced SA levels.

There is evidence indicating that SA might be involved in senescence-associated cell death. Leaves from *Arabidopsis pad4* mutants that are defective in the SA signaling pathway do not appear to undergo cell death as efficiently as the wild type (45). In this mutant, leaves often remain yellow during the senescence stage with a much-delayed cell death (57, 58). This result shows a clear involvement of the SA pathway in senescence-associated cell death. A hypersenescence mutant *hys1*, which showed an early-senescence phenotype, was found to be allelic to *cpr5*, which was isolated based on its constitutive expression of defense responses and spontaneous cell death. The enhanced levels of SA and defense-related gene expressions might cause precocious senescence, supporting a role of SA pathways in the senescence-associated cell death process.

NahG: a gene-encoding bacterial salicylate hydroxylase that destroys SA by converting it to catechol

The role of auxin in leaf senescence has been elusive, particularly due to its involvement in various aspects of plant development. However, evidence suggests that auxin is also involved in the senescence process (61). The auxin level increases during leaf senescence. Consequently, IAA biosynthetic genes encoding tryptophan synthase (*TSA1*), IAAld oxidase (*AO1*), and nitrilases (*NIT1-3*) are upregulated during age-dependent leaf senescence (72). Exogenous application of auxin represses transcription of some *SAG*s (27, 47). Together, this implies that the auxin level increases during leaf senescence due to increased expression of auxin biosynthetic genes, which leads to delayed leaf senescence, leaving auxin as a negatively acting factor of leaf senescence. It has also been suggested that changes in auxin gradients rather than the endogenous auxin level itself could be important in modulating the senescence process (2). Expression of more than half of the genes related to auxin transport is reduced during senescence (72). This may cause aberrant distribution of auxin following leaf senescence.

Studies on the genetic mutation altered in auxin signaling support the involvement of auxin in controlling leaf senescence (14, 50).

AUXIN RESPONSE FACTOR 2 (ARF2) is one of the transcription repressors in the auxin signaling pathway. Microarray analysis shows that expression of the *ARF2* gene is induced in senescing leaves. Disruption of *ARF2* by T-DNA insertion causes delay in leaf senescence. The phenotype canonically puts ARF2 as a positive regulator of leaf senescence. We also isolated another allele of *arf2* from an ethylmethane sulfonate (EMS)-mutagenized pool, which showed delayed leaf senescence along with an increased sensitivity to the exogenous auxin in hypocotyls growth inhibition. Together, these imply that the reduced ARF2 function in the mutant can cause reduced repression of auxin signaling with increased auxin sensitivity, leading to delayed senescence. However, it has yet to be seen whether the effect of auxin pathways is directly involved in leaf senescence or whether it indirectly influences leaf senescence because auxin and the *arf2* mutants also cause a pleiotropic effect in plant development.

Other Regulatory Genes of Senescence

Besides the regulatory genes mentioned above, several other regulatory genes of leaf senescence have been identified through genetic screening of senescence mutants and through functional identification of some *SAG*s.

Ubiquitin-dependent proteolysis is likely involved in regulation of leaf senescence. The *ORE9* gene encodes an F-box protein, a component of the SCF complex, which acts as an E3 ligase in ubiquitin-dependent proteolysis. Leaf senescence was delayed in the *ore9* mutant (74). It was also known that proteolysis by the N-end rule pathway has a function in senescence progression. The *delayed-leaf-senescence 1* (*dls1*) mutant, which is defective in arginyl tRNA:protein transferase (R-transferase), showed delayed development of leaf senescence symptoms (79). R-transferase is a component of the N-end rule proteolytic pathway.

A few regulatory genes identified in our laboratory by genetic screening include ORE7, ORE1, and SOR12. ORE7 is an AT-hook transcription factor that may be involved in controlling chromatin architecture. ORE1 is NO APICAL MERISTEM (NAM), ATAF1, and CUP-SHAPED COTYLEDONS2 (CUC2) (NAC) family transcription factor. SOR12 suppresses the delayed senescence phenotype of ORE12/AHK3. It also appears that miRNA is involved in controlling leaf senescence. Functional characterization of these genes and further genetic isolation of the senescence regulatory elements should be a critical asset in understanding leaf senescence.

A large number of *SAG*s have been identified in various plants through microarray analysis. Some of them have been found to encode potential regulatory factors that are components of signal perception and transductions, such as transcription factors and receptor-like kinases. Characterization of these potential regulatory genes led to discovery of a few important senescence regulatory genes and provided some insight into the regulatory mechanism of leaf senescence.

Genes for 96 transcription factors were identified in *Arabidopsis* to be upregulated at least threefold in senescing leaves. These belong to 20 different transcription factor families, the largest groups being NAC, WRKY, C2H2-type zinc finger, AP2/EREBP, and MYB proteins. Among the WRKY transcription factors, AtWRKY53 and WRKY6 have been further characterized in relation to leaf senescence. *WRKY53* is upregulated at a very early stage of leaf senescence but decreases again at later stages, implying that WRKY53 might play a regulatory role in the early events of leaf senescence (26). Putative target genes of WRKY53 include various *SAG*s, PR genes, stress-related genes, and transcription factors including other WRKY factors. A knockout line of the *WRKY53* gene showed delayed leaf senescence, whereas inducible overexpression caused precocious senescence, showing that it functions as a positive element in leaf senescence (43). Identification of direct target genes of WRKY53 should further reveal the WRKY53-mediated senescence regulatory pathways. Another WRKY transcription factor gene, *WRKY6*, shows high-level upregulation during leaf senescence as well as during pathogen infection (60). WRKY6 regulates a set of genes through the W-box sequences in their promoter. Many WRKY6-regulated genes are associated with senescence and pathogen response, including the senescence-induced receptor-like kinase gene (*SIRK*). Although WRKY6 appears to have a functional role both in pathogen defense as well as senescence, the *SIRK* gene appeared to be expressed only during senescence but not during pathogen infection. The *wrky6* knockout mutation alters expression of *SAG*s but does not have any apparent effect on leaf senescence. The altered expression of *SAG*s in the knockout mutation may not be enough to be manifested into the apparent change of leaf senescence. It is also likely that functional redundancy exists among the WRKY transcription factors, considering the large number of members in the family.

NAC proteins are one of the largest families of plant-specific transcription factors with more than 100 members in *Arabidopsis*. NAC family genes play a role in embryo and shoot meristem development, lateral root formation, auxin signaling, and defense response. A total of 20 genes encoding the NAC transcription factor, representing almost one fifth of the NAC family members, showed enhanced expression during natural senescence and in dark-induced senescence (20). The T-DNA knockout mutant of *AtNAP*, a gene encoding an NAC family transcription factor, showed significantly delayed leaf senescence. Thus, AtNAP functions as a positive element in leaf senescence. *AtNAP* orthologs exist in kidney bean and rice and are also upregulated during leaf senescence. We also isolated a delayed leaf senescence mutant, which is due to a nonsense mutation in one of the NAC transcription factors (J.H. Kim & H.G. Nam, unpublished data). It is likely that several other senescence-upregulated NAC transcription factors play a regulatory role in leaf senescence. Transcriptional autoregulation and inter-regulation, as well as homodimerization and heterodimerization, among the NAC family members are important mechanisms in regulating NAC transcription factor-mediated developmental processes. Similar mechanisms are expected to be involved in the NAC transcription factor-mediated regulatory network of leaf senescence. The potential functions of most leaf senescence-associated transcription factors remain to be elucidated. Functional characterization of these genes including the signaling pathways they are involved in and the target genes they regulate will be invaluable in understanding the complex molecular pathways regulating leaf senescence.

Another example of *SAG*s with which in vivo function was assayed is the autophagy genes. Autophagy is an intracellular process for vacuolar bulk degradation of cytoplasmic components and is required for nutrient cycling. Mutants carrying a T-DNA insertion within the *Arabidopsis* autophagy genes, *AtAPG7*, *AtAPG9*, and *AtAPG18a*, exhibited premature leaf senescence (13, 22, 77). In these mutants, nutrients may be less efficiently utilized during execution of senescence, or some of the components needed for progression of senescence may not be efficiently provided.

A few of the genes involved in lipid metabolism have a role in leaf senescence. Reduced expression of the *Arabidopsis* acyl hydrolase gene by antisense RNA interference in transgenic plants delayed the onset of leaf senescence, whereas chemically induced overexpression of the gene caused precocious senescence (20). In addition, transgenic plants with reduced expression of a senescence-induced lipase also showed delayed leaf senescence (67). It is likely that the delayed senescence in these transgenic lines with reduced lipase expression is due to prolonged maintenance of membrane integrity, indicating the importance of membrane integrity during senescence.

Autophagy: a regulated recycling process whereby cytosol and organelles are encapsulated in vesicles, which are then engulfed and digested by lytic vacuoles/lysosomes

CONCLUSIONS AND FUTURE CHALLENGES

With the aid of microarray, we now know that more than 800 genes are distinctively upregulated during senescence, which illustrates the dramatic alteration in cellular physiology that underlies leaf senescence. With the knowledge of the nature of these *SAGs*, we can now figure out the molecular landscape of leaf senescence. We need to see the dynamic changes of the transcriptome along more detailed windows of leaf senescence stages, which will enable identification of more *SAGs*. This will also allow us to understand the dynamic changes of the physiology undergoing senescence. The microarray data could be further examined using various bioinformatic tools for classifying the *SAGs*, for establishing a hierarchical relationship among the *SAGs*, and for systems-level analysis of molecular events underlying senescence. The senescence pathways affected by various signals are being revealed. Further microarray analyses of leaf senescence in various senescence mutants and under various senescence-affecting conditions should reveal a detailed molecular level of the senescence pathways. One important challenge will be to investigate how the *SAGs* are coordinately regulated during leaf senescence. DNA microarray and chromatin immunoprecipitation approaches could be helpful to answering this question.

Although isolating a few key regulatory genes of leaf senescence greatly aided the understanding of leaf senescence, there are many more regulatory elements of leaf senescence. Finding the senescence regulatory genes has been and will continue to be one of the main challenges. Some regulatory elements could be found fairy easily by functionally characterizing the potential regulatory *SAGs*. However, it should be noted that there are many other senescence regulatory genes that do not belong to *SAGs*. The in vivo function of the *SAGs* can now be easily assayed using the genomic tools available in *Arabidopsis*, including the large collection of T-DNA insertion lines or the Targeting-Induced Local Lesions in Genomes (TILLING) approach.

It is encouraging that senescence is now well assayed in the realm of genetics. The genetic screening of senescence mutants was fruitful in understanding the genetic regulatory mode and in isolating senescence regulatory genes. However, considering the complex nature of senescence, current genetic screening is far from saturated. It will be important to use various mutant pools to identify novel senescence regulatory elements. Chemically mutagenized pools in particular are valuable because they might provide novel alleles that cannot be obtained by T-DNA mutagenesis, as illustrated by the discovery of ORE12/AHK3. Global gene expression analysis of the senescence mutants could provide important clues for dissecting the regulatory pathways. Identifying suppressors to known mutants will be also useful to dissect genetic mechanisms governing senescence processes. More genetic mutants can certainly be isolated by designing a more elegant screening scheme, for example, for mutations defective in integrating environmental effects into a senescence program.

Leaf senescence occurs at the last step of leaf development. Accordingly, some genes that function in senescence could also be involved in other biological processes. The mutations in this type of gene may show difficulties in assaying their function in senescence because their effects on senescence can be masked by other early-mutant phenotypes. Through senescence-specific gene silencing or senescence-specific induction, using a senescence-specific promoter or chemically inducible promoters will partially circumvent the problem.

Although most of the molecular analyses on leaf senescence were based on mRNA expression, it should be noted that mRNA expression is only one aspect of functional gene regulation. Other regulatory mechanisms such as protein-level expression, protein stability, or localization of regulatory proteins involved in senescence process should certainly be involved. An integrated informational analysis involving proteomic and metabolomic analyses during leaf senescence will eventually be needed to better understand leaf senescence.

One serious pitfall in the current assay for leaf senescence is that senescence symptoms are measured at the organ level. However, within a senescing leaf, individual cells are usually at different stages of developmental age or senescence. It is also unlikely that all the cells within an individual leaf undergo coherent cell death. To better understand the senescence process, it will be necessary to develop assays that can monitor senescence symptoms and senescence-associated cell death symptoms at the individual cellular level.

Leaf senescence is certainly an evolutionarily acquired process and thus the plants evolved in different ecological settings with different evolutionary tracks will show differences in the pattern and regulation of senescence. It would be interesting to do a comparative study utilizing the information obtained from *Arabidopsis*. This may even be pursued in various ecotypes of *Arabidopsis*.

Another important challenge in the area of leaf senescence is determining its biotechnological applications. Although manipulation of leaf senescence can greatly improve crop yield and other characteristics such as increased shelf life, the knowledge and materials obtained so far have been poorly utilized for this purpose. Considering the potential future food shortage and the use of plants as a source of bioenergy, improving crop productivity should be a top priority.

SUMMARY POINTS

1. Leaf senescence is a finely regulated and complex process that incorporates multiple developmental and environmental signals. Microarray analysis revealed the complex molecular network of the senescence pathways.
2. Leaf senescence involves age-dependent PCD. Senescence-associated cell death and other PCDs show common as well as distinctive signaling pathways.
3. The genetic scheme was established in screening leaf senescence mutants. Genetic analysis revealed that leaf senescence is controlled by various negative and positive genetic elements.
4. More than 800 genes were identified as *SAGs*, reflecting the dramatic alteration in cellular physiology that underlies the leaf senescence. Potential regulatory elements among the *SAGs* were characterized for their function in leaf senescence. The transcription factors WRKY53 and AtNAP act as positive regulators of leaf senescence.
5. Metabolic rate appears to be one mechanism involved in age-dependent leaf senescence.
6. A key molecule for cytokinin-mediated leaf longevity was identified as AHK3, one of the cytokinin receptors. ARR2 phosphorylation mediated by AHK3 is essential for controlling leaf longevity.
7. Signaling pathways of various phytohormones including ABA, SA, JA, and ethylene are intimately linked to leaf senescence. Interestingly, SA-mediated senescence pathway is highly similar to that of natural leaf senescence, as revealed by microarray analysis.

8. Ubiquitin-dependent proteolysis is likely involved in controlling leaf senescence. Other senescence-regulatory factors include acyl hydrolase, invertase, and autophagy, which suggest involvement of membrane integrity, apoplasitc sugar levels, and nutrient recycling, respectively.

ACKNOWLEDGMENTS

We apologize to all our colleagues whose work could not be properly reviewed here because of space limitation. The work by H.G.N. was supported in part by MOST (KOSEF) through the National Core Research Center for Systems Bio-Dynamics (R15-2004-033-05002-0) and in part by the Crop Functional Genomics Research Program (CG1312). The work by P.O.L. was supported by the Korea Research Foundation Grant funded by the Korea Government (MOEHRD, Basic Research Promotion Fund, KRF-2005–261-C00075).

LITERATURE CITED

1. Abeles FB, Dunn LJ, Morgens P, Callahans A, Dinterman RE, Schmidit J. 1988. Induction of 33-kD and 60-kD peroxidases during ethylene-induced senescence of cucumber cotyledons. *Plant Physiol.* 87:609–15
2. Addicott FT, Lynch RS, Carns HR. 1955. Auxin gradient theory of abscission regulation. *Science* 121:644–45
3. Andersson A, Keskitalo J, Sjodin A, Bhalerao R, Sterky F, et al. 2004. A transcriptional timetable of autumn senescence. *Genome Biol.* 5:R24
4. Bate NJ, Rothstein SJ, Thompson JE. 1990. Expression of nuclear and chloroplast photosynthesis-specific genes during leaf senescence. *J. Exp. Bot.* 239:801–11
5. Bleecker AB, Patterson SE. 1997. Last exit: senescence, abscission, and meristem arrest in *Arabidopsis*. *Plant Cell* 9:1169–79
6. Brouquisse R, Evrard A, Rolin D, Raymond P, Roby C. 2001. Regulation of protein degradation and protease expression by mannose in maize root tips. Pi sequestration by mannose may hinder the study of its signaling properties. *Plant Physiol.* 125:1485–98
7. Buchanan-Wollaston V, Earl S, Harrison E, Mathas E, Navabpour S, et al. 2003. The molecular analysis of leaf senescence: a genomics approach. *Plant Biotechnol. J.* 1:3–22
8. **Buchanan-Wollaston V, Page T, Harrison E, Breeze E, Lim PO, et al. 2005. Comparative transcriptome analysis reveals significant differences in gene expression and signaling pathways between developmental and dark/starvation-induced senescence in *Arabidopsis*. *Plant J.* 42:567–85**

8. A key report describing the importance of the SA pathway in natural leaf senescence.

9. Butt A, Mousley C, Morris K, Beynon J, Can C, et al. 1998. Differential expression of a senescence-enhanced metallothionein gene in *Arabidopsis* in response to isolates of *Peronospora parasitica* and *Pseudomonas syringae*. *Plant J.* 16:209–21
10. Cao J, Jiang F, Sodmergen, Cui K. 2003. Time-course of programmed cell death during leaf senescence in *Eucommia ulmoides*. *J. Plant Res.* 162:7–12
11. Chen W, Provart NJ, Glazebrook J, Katagiri F, Chang HS, et al. 2002. Expression profile matrix of *Arabidopsis* transcription factor genes suggests their putative functions in response to environmental stresses. *Plant Cell* 14:559–74

12. Dai N, Schaffer A, Petreikov M, Shahak Y, Giller Y, et al. 1999. Overexpression of *Arabidopsis* hexokinase in tomato plants inhibits growth, reduces photosynthesis, and induces rapid senescence. *Plant Cell* 11:1253–66
13. Doelling JH, Walker JM, Friedman EM, Thompson AR, Vierstra RD. 2002. The APG8/12-activating enzyme APG7 is required for proper nutrient recycling and senescence in *Arabidopsis thaliana*. *J. Biol. Chem.* 277:33105–14
14. **Ellis CM, Nagpal P, Young JC, Hagen G, Guilfoyle TJ, Reed JW. 2005. AUXIN RESPONSE FACTOR1 and AUXIN RESPONSE FACTOR2 regulate senescence and floral organ abscission in *Arabidopsis thaliana*. *Development* 132:4563–74**
15. Ewbank JJ, Barnes TM, Lakowski B, Lussier M, Bussey H, Hekimi S. 1997. Structural and functional conservation of the *Caenorhabditis elegans* timing gene *clk-1*. *Science* 275:980–83
16. Gan S, Amasino RM. 1995. Inhibition of leaf senescence by autoregulated production of cytokinin. *Science* 270:1986–88
17. Gepstein S, Sabehi G, Carp MJ, Hajouj T, Nesher MF, et al. 2003. Large-scale identification of leaf senescence-associated genes. *Plant J.* 36:629–42
18. Gepstein S, Thimann KV. 1980. Changes in the abscisic acid content of oat leaves during senescence. *Proc. Natl. Acad. Sci. USA* 77:2050–53
19. Guo Y, Cai Z, Gan S. 2004. Transcriptome of *Arabidopsis* leaf senescence. *Plant Cell Environ.* 27:521–49
20. **Guo Y, Gan S. 2006. AtNAP, a NAC family transcription factor, has an important role in leaf senescence. *Plant J.* 46:601–12**
21. Ha CM, Kim GT, Kim BC, Jun JH, Soh MS, et al. 2003. The *BLADE-ON-PETIOLE 1* gene controls leaf pattern formation through the modulation of meristematic activity in *Arabidopsis*. *Development* 130:161–72
22. Hanaoka H, Noda T, Shirano Y, Kato T, Hayashi H, et al. 2002. Leaf senescence and starvation-induced chlorosis are accelerated by the disruption of an *Arabidopsis* autophagy gene. *Plant Physiol.* 129:1181–93
23. He Y, Fukushige H, Hildebrand DF, Gan S. 2002. Evidence supporting a role of jasmonic acid in *Arabidopsis* leaf senescence. *Plant Physiol.* 128:876–84
24. **He Y, Tang W, Swain JD, Green AL, Jack TP, Gan S. 2001. Networking senescence-regulating pathways by using *Arabidopsis* enhancer trap lines. *Plant Physiol.* 126:707–16**
25. Hensel L, Grbic V, Baumgarten DA, Bleecker AB. 1993. Developmental and age-related processes that influence the longevity and senescence of photosynthetic tissues in *Arabidopsis*. *Plant Cell* 5:553–64
26. Hinderhofer K, Zentgraf U. 2001. Identification of a transcription factor specifically expressed at the onset of leaf senescence. *Planta* 213:469–73
27. Hong SB, Sexton R, Tucker ML. 2000. Analysis of gene promoters for two tomato polygalacturonases expressed in abscission zones and the stigma. *Plant Physiol.* 123:869–81
28. Hörtensteiner S, Feller U. 2002. Nitrogen metabolism and remobilization during senescence. *J. Exp. Bot.* 53:927–37
29. Hung KT, Kao CH. 2003. Nitric oxide counteracts the senescence of rice leaves induced by abscisic acid. *J. Plant Physiol.* 160:871–79
30. Hung KT, Kao CH. 2004. Hydrogen peroxide is necessary for abscisic acid-induced senescence of rice leaves. *J. Plant Physiol.* 161:1347–57
31. Jang JC, Leon P, Zhou L, Sheen J. 1997. Hexokinase as a sugar sensor in higher plants. *Plant Cell* 9:5–19
32. Jing HC, Schippers JH, Hille J, Dijkwel PP. 2005. Ethylene-induced leaf senescence depends on age-related changes and OLD genes in *Arabidopsis*. *J. Exp. Bot.* 56:2915–23

14. A key report regarding the involvement of auxin in controlling leaf senescence.

20. Describes functional analysis of *AtNAP*, a *Arabidopsis* gene encoding a NAC family transcription factor, one of the senescence-enhanced transcription factors.

24. Describes a large-scale screen of enhancer trap lines to identify 147 lines in which the expression of the reporter gene is upregulated in senescing leaves. Analysis of these lines revealed the existence of a network of senescence-promoting pathways.

33. First report identifying a repressor for integrating ethylene action into leaf senescence.

35. A key report describing the possible molecular mechanisms of cytokinin-mediated leaf longevity control.

37. A report describing functional analysis of a novel nuclear protein, OsDOS, which plays a role as a novel negative regulator in integrating the JA pathway to the developmental age factor.

33. Jing HC, Sturre MJG, Hille J, Dijkwel PP. 2002. *Arabidopsis* onset of leaf death mutants identify a regulatory pathway controlling leaf senescence. *Plant J.* 32:51–63
34. Jones AM, Dangl JL. 1996. Logjam at the Styx: programmed cell death in plants. *Trends Plant Sci.* 1:114–19
35. Kim HJ, Ryu H, Hong SH, Woo HR, Lim PO, et al. 2006. Cytokinin-mediated control of leaf longevity by AHK3 through phosphorylation of ARR2 in *Arabidopsis*. *Proc. Natl. Acad. Sci. USA* 103:814–19
36. Kimura KD, Tissenbaum HA, Liu Y, Ruvkun G. 1997. *daf-2*, an insulin receptor-like gene that regulates longevity and diapause in *Caenorhabditis elegans*. *Science* 277:942–46
37. Kong Z, Li M, Yang W, Xu W, Xue Y. 2006. A novel nuclear-localized CCCH-type zinc finger protein, OsDOS, is involved in delaying leaf senescence in rice (*Oryza sativa* L.). *Plant Physol.* 141:1376–88
38. Lara MEB, Garcia MCG, Fatima T, Ehne R, Lee TK, et al. 2004. Extracellular invertase is an essential component of cytokinin-mediated delay of senescence. *Plant Cell* 16:1276–87
39. Lim PO, Woo HR, Nam HG. 2003. Molecular genetics of leaf senescence in *Arabidopsis*. *Trends Plant Sci.* 8:272–78
40. Lim PO, Nam HG. 2005. The molecular and genetic control of leaf senescence and longevity in *Arabidopsis*. *Curr. Top. Dev. Biol.* 67:49–83
41. Lin JF, Wu SH. 2004. Molecular events in senescing *Arabidopsis* leaves. *Plant J.* 39:612–28
42. McCabe MS, Garratt LC, Schepers F, Jordi WJ, Stoopen GM, et al. 2001. Effects of P_{SAG12}-*IPT* gene expression on development and senescence in transgenic lettuce. *Plant Physiol.* 127:505–16
43. Miao Y, Laun T, Zimmermann P, Zentgraf U. 2004. Targets of the WRKY53 transcription factor and its role during leaf senescence in *Arabidopsis*. *Plant Mol. Biol.* 55:853–67
44. Moore B, Zhou L, Rolland F, Hall Q, Cheng WH, et al. 2003. Role of the *Arabidopsis* glucose sensor HXK1 in nutrient, light, and hormonal signaling. *Science* 300:332–36
45. Morris K, Mackerness SA, Page T, John CF, Murphy AM, et al. 2000. Salicylic acid has a role in regulating gene expression during senescence. *Plant J.* 23:677–85
46. Nam HG. 1997. Molecular genetic analysis of leaf senescence. *Curr. Opin. Biotech.* 8:200–7
47. Noh YS, Amasino R. 1999. Identification of a promoter region responsible for the senescence-specific expression of *SAG12*. *Plant Mol. Biol.* 41:181–94
48. Noodén LD. 1988. The phenomena of senescence and aging. In *Senescence and Aging in Plants*, ed. LD Noodén, AC Leopold, pp. 1–50. San Diego: Academic
49. Oh SA, Park JH, Lee GI, Paek KH, Park SK, Nam HG. 1997. Identification of three genetic loci controlling leaf senescence in *Arabidopsis thaliana*. *Plant J.* 12:527–35
50. Okushima Y, Mitina I, Quach HL, Theologis A. 2005. AUXIN RESPONSE FACTOR 2 (ARF2): a pleiotropic developmental regulator. *Plant J.* 43:29–46
51. Ori N, Juarez MT, Jackson D, Yamaguchi J, Banowetz GM, Hake S. 1999. Leaf senescence is delayed in tobacco plants expressing the maize homeobox gene knotted1 under the control of a senescence-activated promoter. *Plant Cell* 11:1073–80
52. Otegui MS, Noh YS, Martinez DE, Vila Petroff MG, Staehelin LA, et al. 2005. Senescence-associated vacuoles with intense proteolytic activity develop in leaves of *Arabidopsis* and soybean. *Plant J.* 41:831–44
53. Park JH, Oh SA, Kim YH, Woo HR, Nam HG. 1998. Differential expression of senescence-associated mRNAs during leaf senescence induced by different senescence-inducing factors in *Arabidopsis*. *Plant Mol. Biol.* 37:445–54
54. Pontier D, Gan S, Amasino RM, Roby D, Lam E. 1999. Markers for hypersensitive response and senescence show distinct patterns of expression. *Plant Mol. Biol.* 39:1243–55

55. Quirino BF, Noh YS, Himelblau E, Amasino RM. 2000. Molecular aspects of leaf senescence. *Trends Plant Sci.* 5:278–82
56. Quirino BF, Normanly J, Amasino RM. 1999. Diverse range of gene activity during *Arabidopsis thaliana* leaf senescence includes pathogen-independent induction of defense-related genes. *Plant Mol. Biol.* 40:267–78
57. Rao MV, Davis KR. 2001. The physiology of ozone-induced cell death. *Planta* 213:682–90
58. Rao MV, Lee HI, Davis KR. 2002. Ozone-induced ethylene production is dependent on salicylic acid, and both salicylic acid and ethylene act in concert to regulate ozone-induced cell death. *Plant J.* 32:447–56
59. Richmond AE, Lang A. 1957. Effect of kinetin on protein content and survival of detached *Xanthium* leaves. *Science* 125:650–51
60. Robatzek S, Somssich IE. 2002. Targets of AtWRKY6 regulation during plant senescence and pathogen defense. *Genes Dev.* 16:1139–49

60. First paper describing the target genes of senescence- and defense-associated transcription factor AtWRKY6. A SIRK gene was identified as a potential target of AtWRKY6.

61. Sexton R, Roberts JA. 1982. Cell biology of abscission. *Annu. Rev. Plant Physiol.* 33:133–62
62. Simeonova E, Sikira S, Charzynska M, Mostowska A. 2000. Aspects of programmed cell death during leaf senescence of mono- and dicotyledonous plants. *Protoplasma* 214:93–101
63. Takahashi Y, Berberich T, Yamashita K, Uehara Y, Miyazaki A, Kusano T. 2004. Identification of tobacco HIN1 and two closely related genes as spermine-responsive genes and their differential expression during the Tobacco mosaic virus-induced hypersensitive response and during leaf- and flower-senescence. *Plant Mol. Biol.* 54:613–22
64. Tang D, Christiansen KM, Innes RW. 2005. Regulation of plant disease resistance, stress responses, cell death, and ethylene signaling in *Arabidopsis* by the EDR1 protein kinase. *Plant Physiol.* 138:1018–26
65. Taylor CB, Bariola PA, del Cardayre SB, Raines RT, Green PJ. 1993. RNS2: a senescence-associated RNase of *Arabidopsis* that diverged from the S-RNase before speciation. *Proc. Natl. Acad. Sci. USA* 90:5118–22
66. Thompson JE, Froese CD, Madey E, Smith MD, Hong Y. 1998. Lipid metabolism during plant senescence. *Prog. Lipid Res.* 37:119–41
67. Thompson J, Taylor C, Wang TW. 2000. Altered membrane lipase expression delays leaf senescence. *Biochem. Soc. Trans.* 28:775–77
68. Thomson WW, Plat-Aloia KA. 1987. Ultrastructure and senescence in plants. In *Plant Senescence: Its Biochemistry and Physiology*, ed. WW Thomson, EA Nothnagei, RC Huffaker, pp. 20–30. Rockville: Am. Soc. Plant Physiol.
69. Ueda J, Kato J. 1980. Identification of a senescence-promoting substance from wormwood (*Artemisia absinthum* L.). *Plant Physiol.* 66:246–49
70. van Doorn WG. 2005. Plant programmed cell death and the point of no return. *Trends Plant Sci.* 10:478–83
71. van Doorn WG, Woltering EJ. 2004. Senescence and programmed cell death: substance or semantics? *J. Exp. Bot.* 55:2147–53
72. van der Graaff E, Schwacke R, Schneider A, Desimone M, Flugge UI, Kunze R. 2006. Transcription analysis of *Arabidopsis* membrane transporters and hormone pathways during developmental and induced leaf senescence. *Plant Physiol.* 141:776–92

72. A report describing the expression profiles of biosynthesis, metabolism, signaling, and response genes of plant hormones during leaf senescence.

73. Weaver LM, Gan S, Quirino B, Amasino RM. 1998. A comparison of the expression patterns of several senescence-associated genes in response to stress and hormone treatment. *Plant Mol. Biol.* 37:455–69
74. Woo HR, Chung KM, Park JH, Oh SA, Ahn T, et al. 2001. ORE9, an F-box protein that regulates leaf senescence in *Arabidopsis*. *Plant Cell* 13:1779–90

75. First paper describing the involvement of metabolic rate in controlling age-dependent leaf senescence.

75. **Woo HR, Goh CH, Park JH, Teyssendier B, Kim JH, et al. 2002. Extended leaf longevity in the *ore4-1* mutant of *Arabidopsis* with a reduced expression of a plastid ribosomal protein gene. *Plant J.* 31:331–40**
76. Woolhouse HW. 1984. The biochemistry and regulation of senescence in chloroplasts. *Can. J. Bot.* 62:2934–42
77. Xiong Y, Contento AL, Bassham DC. 2005. AtATG18a is required for the formation of autophagosomes during nutrient stress and senescence in *Arabidopsis thaliana*. *Plant J.* 42:535–46
78. Yen CH, Yang CH. 1998. Evidence for programmed cell death during leaf senescence in plants. *Plant Cell Physiol.* 39:922–27
79. Yoshida S, Ito M, Callis J, Nishida I, Watanabe A. 2002. A delayed leaf senescence mutant is defective in arginyl-tRNA:protein arginyltransferase, a component of the N-end rule pathway in *Arabidopsis*. *Plant J.* 32:129–37
80. Yoshida S, Ito M, Nishida I, Watanabe A. 2002. Identification of a novel gene HYS1/CPR5 that has a repressive role in the induction of leaf senescence and pathogen-defense responses in *Arabidopsis thaliana*. *Plant J.* 29:427–37
81. Zeevaart JAD, Creelman RA. 1988. Metabolism and physiology of abscisic acid. *Annu. Rev. Plant Physiol. Plant Mol. Biol.* 39:439–73

The Biology of Arabinogalactan Proteins

Georg J. Seifert and Keith Roberts

Department of Cell and Developmental Biology, John Innes Centre, Norwich Research Park, Colney, Norwich, NR4 7UH, United Kingdom; email: georg.seifert@boku.ac.at, keith.roberts@bbsrc.ac.uk

Annu. Rev. Plant Biol. 2007. 58:137–61

The *Annual Review of Plant Biology* is online at plant.annualreviews.org

This article's doi: 10.1146/annurev.arplant.58.032806.103801

First published online as a Review in Advance on January 2, 2007

1543-5008/07/0602-0137$20.00

Key Words

proteoglycan, xylogenesis, pollen tube guidance, embryogenesis, lipid raft

Abstract

Arabinogalactan proteins is an umbrella term applied to a highly diverse class of cell surface glycoproteins, many of which contain glycosylphosphatidylinositol lipid anchors. The structures of protein and glycan moieties of arabinogalactan proteins are overwhelmingly diverse while the "hydroxproline contiguity hypothesis" predicts arabinogalactan modification of members of many families of extracellular proteins. Descriptive studies using monoclonal antibodies reacting with carbohydrate epitopes on arabinogalactan proteins and experimental work using β-Yariv reagent implicate arabinogalactan proteins in many biological processes of cell proliferation and survival, pattern formation and growth, and in plant microbe interaction. Advanced structural understanding of arabinogalactan proteins and an emerging molecular genetic definition of biological roles of individual arabinogalactan protein species, in conjunction with potentially analogous extracellular matrix components of animals, stimulate hypotheses about their mode of action. Arabinogalactan proteins might be soluble signals, or might act as modulators and coreceptors of apoplastic morphogens; their amphiphilic molecular nature makes them prime candidates of mediators between the cell wall, the plasma membrane, and the cytoplasm.

Contents

INTRODUCTION

Arabinogalactan proteins (AGPs) are plant glycoproteins[1] that consist of a core-protein backbone O-glycosylated by one or more complex carbohydrates consisting of galactan and arabinose as main components. This article does not attempt a census of "all *Arabidopsis thaliana* genes encoding ...," because, based on the presently available information, it is impossible to accurately include uncharacterized genes in such a list or even to confidently exclude a large number of genes. Originally, AGPs were seen as a subgroup of hydroxyproline-rich glycoproteins; however, genomic information suggests overlap between AGPs and other groups of hydroxyproline-rich glycoproteins, and recent molecular genetic and proteomic studies suggest that many more proteins might be regarded as AGPs than previously assumed. Most, if not all, introductions to recent AGP papers contain the notion that "AGPs are implicated in many processes of growth and development." In addition to carbohydrate chemistry and molecular genetics, these roles have been studied employing two specific experimental tools, a group of synthetic phenylglycoside dyes, here called β-Yariv reagents, that specifically bind to AGPs, and a set of monoclonal antibodies recognizing AGP glycan epitopes (see sidebar Tools to Study AGPs). Here we summarize the key evidence implicating AGPs in various biological roles and try to outline several hypotheses for how AGPs might exert such roles on a molecular level. A marvelous review covering the AGP literature up to 1996 was written by Nothnagel (76). To study AGP structure, function, localization, and their economic and medical benefits, we also recommend excellent reviews covering these topics in great depth (19, 37, 64, 93, 103).

ARABINOGALACTAN PROTEIN STRUCTURE

AGPs consist of a core protein of highly varying length and domain complexity and one or more arabinogalactan (AG) side chains, and often contain a glycosylphosphatidylinositol (GPI) lipid anchor. The relative ratio of

[1] We use the generic and more inclusive term glycoprotein as opposed to proteoglycan to avoid inconsistencies with definitions of animal proteoglycans.

glycan to protein is sometimes >9, but can vary strongly for the same AGP core protein isolated from the same tissue (108).

Core Protein

Every known AGP is processed via the secretory pathway and contains an N-terminal cleavable signal sequence. A hydrophobic C-terminal domain on many AGP precursor peptides generally indicates posttranslational modification by a GPI lipid anchor (see below). Some mature AGP backbone peptides are only 10 to 13 residues long and are termed AG peptides (94). These and some longer AGP sequences consist of a single central domain rich in Pro, Ala, Ser, and Thr (94) flanked by an N-terminal signal peptide and a C-terminal GPI-modification signal and were termed classical in reference to the two full-length cDNAs encoding AGPs that were first cloned (17, 31). Most groups of AGPs deviate from this simple domain structure. Some AGP species contain a short lysine-rich domain between the proline-rich domain and the hydrophobic C terminus (95, 114). There is a large group of fasciclin-like AGPs (FLA) that contain one or two fasciclin domains thought to be involved in protein-protein interactions that are variably GPI modified (48). Early nodulin (ENOD) sequences that are related to the copper-containing phytocyanins were repeatedly identified as β-Yariv reagent-reactive glycopeptides (30, 48, 65). The occurrence of (AP) repeats suggests that they might be AG modified. A further gene family repeatedly found to contain AGPs are the nonspecific lipid transfer proteins (ns-LTPs) that contain the *Zinnia elegans* xylogen, its *A. thaliana* homologs (73) and a β-Yariv reagent-reactive glycopeptide from rice (65).

The identification of different types of hydroxyproline (Hyp) repeats in various plant glycoproteins and subsequent analysis of artificial peptides, led to the Hyp-contiguity hypothesis, which predicts arabinosylation of contiguous Hyp residues and arabinogalactosylation of clustered noncontiguous Hyp residues, giving rise to extensin-like or AGP-like glycomodules, respectively (104, 105). Although many known AGPs contain (XP) repeats (where X stands for A, S, or T), and artificial (SP) repeats are efficiently AG modified (104, 119), other known AGPs lack clustered noncontiguous (XP) repeats. For example, the AG-modified SOS5/FLA4 contains two [AT]PPP motifs (100) and AtAGP30 contains three PPAKAPIKLP repeats (128). Similarly, the sequence motif for efficient arabinogalactosylation of sporamin, including the elongation

Arabinogalactan protein (AGP): plant glycoprotein O-glycosylated by complex carbohydrates that consist of galactan and arabinose as main components

Development (used in conjunction with growth): qualitative changes such as pattern formation and differentiation

TOOLS TO STUDY AGPs

β-Yariv reagents

Brown-red-colored compounds with the generic structure 1,3,5-tri-(pglycosyloxyphenylazo)-2,4,5-trihydroxybenzene (139), called Yariv reagents, are important tools for AGP research. Although (β-glucosyl) and (β-galactosyl) Yariv reagents bind to AGPs, the α-forms or (β-mannosyl) Yariv do not bind and are often used as negative controls. There is no consensus on whether the carbohydrate or the protein part of AGPs interact with β-Yariv reagent, but it is thought that AGPs do not bind to individual β-glycosides like lectins but rather bind to the stacked, self-associated form of the compound (76). β-Yariv reagents are extremely useful tools for AGP purification and are frequently used to probe the role of AGPs in biological processes. Such experiments are especially informative when backed up by independent genetic approaches. The affinity of AGPs to β-Yariv reagents led to the assumption that similar interactions with β-linked polysaccharides in the cell wall matrix might take place in vivo.

Monoclonal antibodies

AGP-specific mAbs have been instrumental in revealing the developmental dynamics of AGP glycan structure and represent a diagnostic tool for AGPs. Frequently used anti-AGP mAbs are JIM8, JIM13, JIM14, LM2, and CCRC-M7.

of the glycan side chain, is [not basic]-[not T]-[AVSG]-Pro-[AVST]-[GAVPSTC]-[APS] (101). According to the Hyp-contiguity hypothesis, 40% of *A. thaliana* proteins predicted to be GPI modified are also AG modified (5). If this prediction is correct, members of Extensin-related proteins, *COBRA*-like proteins, glycerophosphodiesterase-like proteins, SKU5-like proteins, and receptor-like proteins might be AG modified. Generally speaking, any peptide sequence containing a secretion signal and AG glycomodules can be considered AGP-like, pending experimental proof.

Growth (used in conjunction with development): quantitative increase in size, number, or mass

AG: arabinogalactan

GPI: glycosylphosphatidylinositol

FLA: fasciclin-like AGP

ns-LTP: nonspecific lipid transfer protein

Glycomodules: short stretches of peptide sequence containing contiguous or noncontiguous Hyp residues that are arabinosylated (extensin glycomodule) and AG modified (AG -glycomodule), respectively

To allow O-glycosylation, the backbone peptides are posttranslationally modified by prolyl hydroxylation, which requires the action of proly 4-hydroxylases (P4H), oxidoreductases that depend on Fe^{2+}, 2-oxoglutarate, O_2, and ascorbate. The *A. thaliana* genes *P4H-1* and *P4H-2* encode active P4H, but with different substrate specificities (45, 123). P4H-1 efficiently hydroxylates various prolyl peptides including $(SPPPV)_3$, $(PPG)_{10}$, and $(APG)_5$. PH4-2 hydroxylates $(SPPPV)_3$, but not $(PPG)_{10}$ and $(APG)_5$, peptides. The *Arabidopsis* genome contains at least six, and possibly up to twelve, putative P4H genes, and prolyl hydroxylation pattern might be influenced by the activity of specific P4H isoforms. The efficiency of prolyl hydroxylation also depends on the peptide sequence, as synthetic $(SP)_n$ and $(AP)_n$ glycomodules are completely hydroxylated; whereas, $(TP)_n$ and $(VP)_n$ glycomodules are only partially hydroxylated (119).

Glycan

Most AGPs are O-glycosylated at one or more hydroxyproline residues by AG type II groups. These consist of (1→3)-β-galactan and (1→6)-β-linked galactan chains connected to each other by (1→3, 1→6)-linked branch points, O-3 and O-6 positions substituted with terminal arabinosyl residues (13). There is large variability in the side-chain modifications that can also contain D-glucuronic acid, L-fucose, L-rhamnose, D-xylose, and other sugars (reviewed in 19, 76). A three-dimensional (3D) model of a Hyp-arabinogalactan polysaccharide was elaborated from nuclear magnetic resonance (NMR) analysis of synthetic AG glycomodules, suggesting multiple glycosidic branch points (120) resulting in a bulky glycan that provides contact surface for interaction with other matrix molecules (**Figure 1**). Due to its similarity with other AGP glycans, this structure might be representative of many AG glycomodules (120). Divergent types of AG side chains on AGPs are present in the *Artemisia vulgaris* pollen allergen, which is O-glycosylated at hydroxyproline residues with a (1→6)-β-galactan core with 5-; 2,5-; 3,5-; and 2,3,5-linked arabinan side chains (59) and in the moss *Physcomitrella patens* AGP where the (1→3)-β-galactan backbone appears to contain linear (1→5)-α-arabinan side chains (57). The large molecular weight heterogeneity of individual AGP species on denaturing gels indicates a high variation in glycosylation in vivo (73, 128). Other AGP species appeared more uniformly glycosylated (115).

Glycosylphosphatidylinositol Lipid Anchor

GPI anchoring of many AGP species is generally accepted (5, 6, 77, 92, 93, 117, 140). In membrane AGP from pear, the C terminus of the core protein is coupled via a phosphoethanolamine linker to a D-Man-(1→2)-α-D-Man-(1→6)-α-D-Man-(1→4)-α-D-GlcNac-oligosaccharide that is linked to an inositylphosphoceramide lipid residue, which is largely consistent with the animal, protozoan, and yeast GPI anchor structure. Approximately half of the AGPs contained a galactosyl substitution on O-4 of the six-linked mannosyl residue, a novel structure compared with GPI anchors from other kingdoms. Another anomaly was the presence of a ceramide lipid instead of a glycerolipid (77).

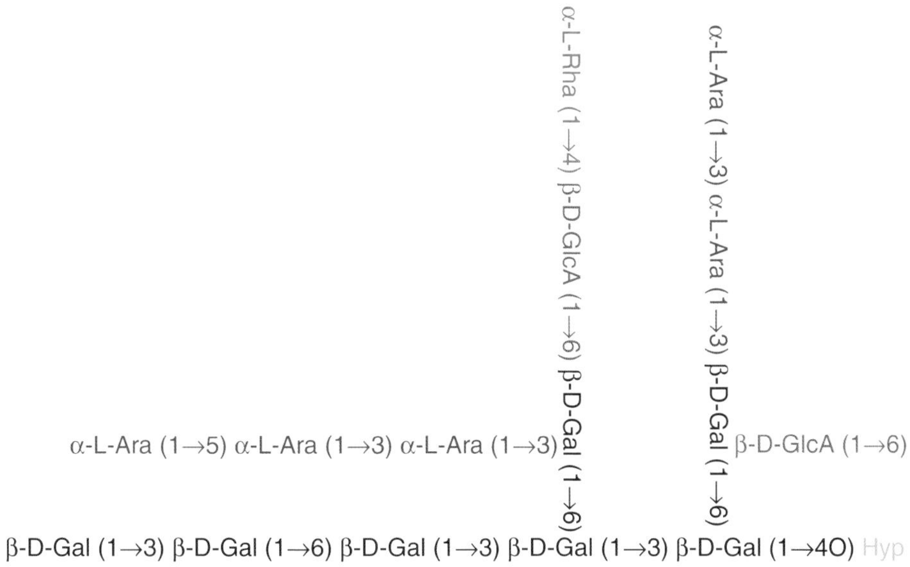

Figure 1

The first complete structure of an arabinogalactan glycan, derived from synthetic GFP:$(AP)_{51}$, expressed in tobacco BY2 cells, reveals surprising complexity (120).

ROLES OF ARABINOGALACTAN PROTEINS IN BIOLOGICAL PROCESSES

Cell Division and Programmed Cell Death

In one of its earliest pharmacological applications, adding β-Yariv reagent to a 40-year-old rose cell suspension line inhibited cell division without apparent loss of cell viability (56, 98), suggesting a role for AGPs in cell division. In contast, newly established rose cell suspensions as well as suspension cultures of *Nicotiana edwardsonii*, *A. thaliana*, and tobacco BY2 cells (16, 35, 43, 56, 98) are rapidly killed by treatment with β-Yariv reagent. The two different rose suspension lines differ in the relation between cell wall–bound and secreted AGPs, which might be important for the different response to β-Yariv reagent. Even if AGPs are required for cell division, it is unlikely they require GPI anchoring, as seedling lethal *peanut* mutants, deficient in GPI anchor precursor biosynthesis, proliferate as callus cultures (40). The killing of cells by β-Yariv reagent involves programmed cell death (PCD), evidenced by the morphology of cell corpses, abundant DNA breaks, and nucleosomal DNA laddering (16, 35). β-Yariv reagent-triggered PCD in *A. thaliana* suspension cells is preceded by massive transcriptional alterations resembling wounding responses (43). Carrot cell cultures die when grown at low cell density, but cell survival and proliferation is rescued with conditioned culture medium. The responsible survival factors are so far unknown, but related studies of the role of AGPs in somatic embryogenesis also suggest a role in cell survival (66, 67, 129). The mechanism of PCD triggered by AGP damage is not known. Although the molecular machinery that acts in the final stage of PCD in plants is becoming better understood, very little is known about the nature and perception of signals involved in regulating cell survival.

Programmed Cell Death (PCD): a genetically defined physiological process leading to the destruction, and sometimes removal, of a cell

Pattern formation: the generation of asymmetry and periodicity from symmetry and uniformity, respectively

Pattern Formation

Biological pattern formation (the generation of asymmetries and periodicities from symmetrical and uniform structures, respectively) often involves a spatially confined source of positional information and the establishment of long-range inhibitory and short-range stimulatory interactions between the system components (69). The role of GPI anchors in development is highlighted by the Arabidopsis *peanut* mutants, which are defective in GPI precursor biosynthesis. These plants lack root cap cells, develop ectopic cotyledons, and display increased shoot apical meristem size and a disorganized root quiescent center (40). Significantly, the transition from the radially symmetric, spherical embryos to the bilaterally symmetric, torpedo-stage embryo is dramatically delayed in *peanut* embryos. Many GPI-anchored proteins exist and could be responsible for some aspects of the *peanut* phenotype, but 40% of GPI-anchored proteins contain AG glycomodules (5, 6), indicating that AGPs may fulfill roles in intercellular signaling that are essential for pattern formation. Descriptive and functional studies conducted in different systems are providing increasing support for this hypothesis.

Embryo pattern. Application of β-Yariv reagent to plants has been used to test for a possible involvement of AGPs in pattern formation. The flowering ornamental *Streptocarpus proxilus* is distinct from other streptocarpus species in the suppression of one cotyledon and the absence of shoot apical meristem (SAM) and leaves, resulting in a vegetative body derived from the growth of only one of its cotyledons. Treatment with β-Yariv reagent and with inhibitors of hydroxyproline biosynthesis results in relatively equal development of both cotyledons and emergence of SAM-like and leaf-like plumules and phyllomorphs, respectively, thereby resembling presumably more primitive related species. This suggests that AGPs might be involved in inhibitory signaling, leading to the suppression of various organs in *S. proxilus* and that β-Yariv reagent blocks this inhibition (86, 87). An involvement of AGPs in early embryonic patterning is suggested by the observation that certain AGP carbohydrate epitopes are differentially expressed during embryogenesis. An AGP epitope recognized by the monoclonal antibody JIM8 is expressed in the two-cell embryo in oilseed rape, but subsequently JIM8 expression is lost from the embryo proper and continues to be expressed in the suspensor (**Figure 2**), whereas the AGP epitope recognized by MAC207 is distributed in an essentially opposite fashion (80, 81). AGP epitopes not only mark different cell types in the zygotic embryo, but during indirect somatic embryogenesis from carrot hypocotyls the AGP epitope recognized by the JIM4 antibody is first expressed at a very low level in proembryogenic masses. Subsequently, globular and early heart stage somatic embryos express the JIM4-reactive epitope in a strictly polar fashion in the shoot apical domain (110). Furthermore, embryogenic and nonembryogenic carrot cells can be separated using magnetic beads coupled to the JIM8 antibody. In this system, asymmetric cell divisions result in a JIM8 expressing ($JIM8^+$) and in a JIM8 nonexpressing ($JIM8^-$) cell (**Figure 3**). If $JIM8^+$ cells are present in the suspension, $JIM8^-$ cells undergo somatic embryogenesis. However, if $JIM8^+$ cells are removed,

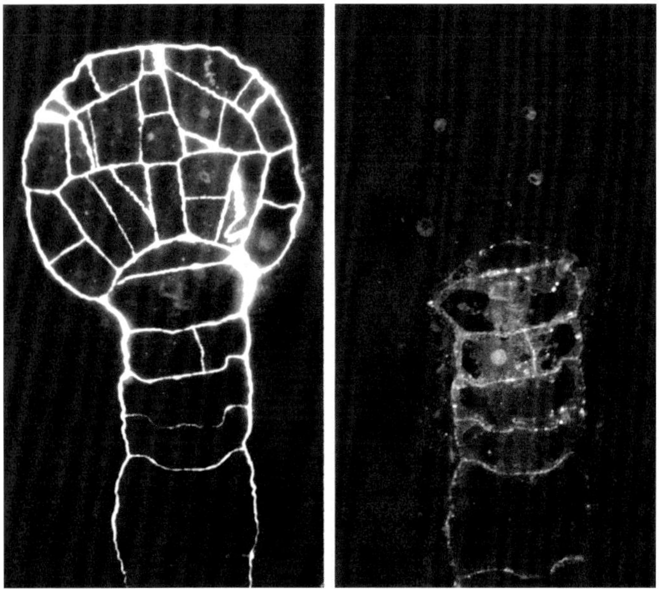

Figure 2

The carbohydrate epitope recognized by the JIM8 monoclonal antibody (mAb) (*right*) is absent from the embryo proper but is expressed in the suspensor, counterstained with aniline blue (*left*) (80).

embryogenesis is lost. JIM8$^+$ cells, per se, do not form somatic embryos (67, 124), but adding growth medium conditioned by JIM8$^+$ cells fully restores embryogenic capacity to JIM8$^-$ cells, demonstrating the involvement of a soluble signal (67). The nature of this factor is still unknown, but it might be an oligosaccharide released from AGPs with JIM8-reactive epitopes (67) since β-Yariv reagent inhibits somatic embryogenesis in carrot suspension cells and in early globular embryos (122). Biomass formation is unaffected by β-Yariv reagent, but abnormal roots are produced, the root cap facing the proembryogenic clusters and the shoots replaced by globular unorganized tissue. This indicates a requirement for AGPs in shoot apical patterning and in the orientation of root patterning (122). Similarly, direct somatic embryogenesis from cichorium roots involves anticlinal cell divisions giving rise to embryogenic cells, but β-Yariv reagent induces periclinal divisions instead, and suppresses further somatic embryo development (15). Excised tobacco zygotes divide asymmetrically to form two-cell pro-embryos consisting of a small apical and a large basal cell. β-Yariv reagent suppresses asymmetric divisions without altering division frequency, indicating a requirement for AGPs in the pattern of the first zygotic division but not in cell division per se (85). AGPs have also been reported to improve the embryogenic potential of microspore cultures (60) or cell cultures (32, 52).

Due to the crudeness of the AGP preparations used in these studies, it is unclear if specific protein backbones or AG-glycosylation patterns are essential or if AGP-derived oligosaccharides affected the process. Attempts to clarify this situation have shown that AGP subfractions, immunoaffinity purified with monoclonal antibodies ZUM15 and ZUM18, inhibit and stimulate somatic embryogenesis, respectively (53). Identical effects are obtained with immunoaffinity-purified AGPs from two different species and because anti-AGP mAbs typically react with the glycan, these data suggest that crucial biological information is contained in the complex carbohydrate of AGPs. Pre-incubating AGP preparations with an endochitinase at first decrease, and after repurification of the AGPs, increase its stimulatory effect on somatic embryogenesis in carrot cells (129), confirming, together with other data in this study, that the AGP glycan is essential for biological activity and that even subtle structural alterations of the glycan can profoundly alter its effect. Moreover, these data support the notion that AGPs and AGP-derived signals can have antagonistic effects (53, 129). The abundance of endochitinase cleavage sites on carrot seed AGP as well as the expression of endochitinase are developmentally correlated (126, 130). The fine-tuned regulation of AGP glycosylation as well as the expression and activity of AGP-specific glycosyl hydrolases might therefore provide mechanisms to control the biological activity of AGPs.

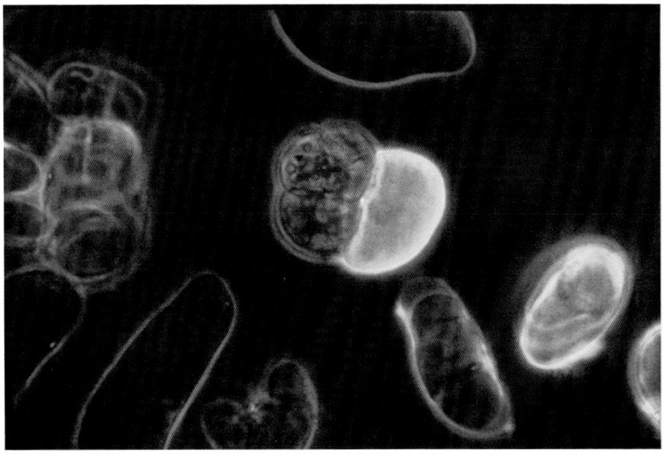

Figure 3

Carrot suspension cells can be divided in JIM8$^+$ cells not undergoing somatic embryogenesis (*green*) and JIM8$^-$ cells that form somatic embryos (*brown*). Removal of JIM8$^+$ cells from the suspension blocks somatic embryogenesis (67).

mAb: monoclonal antibody

Postembryonic pattern. Involvement of AGPs in the formation or maintenance of postembryonic pattern might be inferred from the cell type–specific expression of particular AGP epitopes in roots (14, 28, 29, 51) and flowers (80, 82) and from the inability

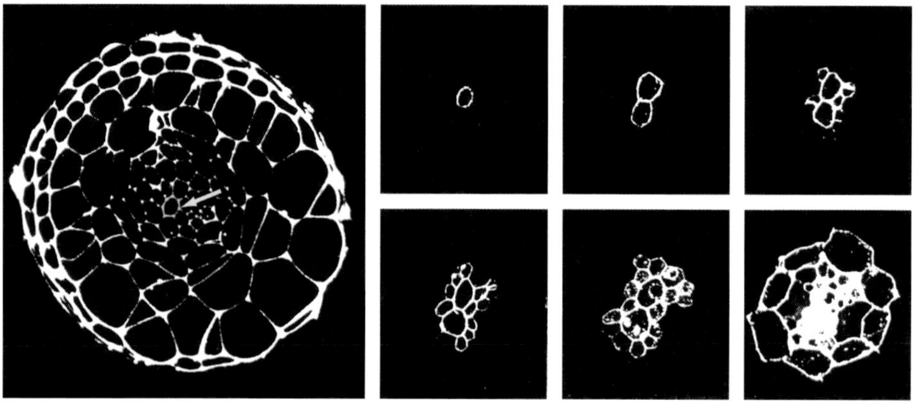

Figure 4

The arabinogalactan protein–reactive JIM13 antibody labels a single xylem initial in the *A. thaliana* root meristem (*arrow to green cell, left*). During further cell division, elongation, and differentiation, the epitope is expressed by proto and metaxylem and in xylem parenchyma and the endodermis (*series panel, right*).

XYP: *xylogen protein*

RNAi: RNA interference

of the *atagp30* mutant to form roots from callus cultures (128). Ectopic expression of *AtAGP30* permits normal root development but completely disrupts shoot development (128). Because certain AGP epitopes spatially and temporarily correlate with xylem differentiation (91, 109) and mark the initial cell that gives rise to proto and metaxylem (28) (**Figure 4**), an involvement of AGPs in vascular pattern formation was proposed. This possibility is strongly supported by the discovery of an AGP from *Z. elegans* termed xylogen (71–73). Isolated *Z. elegans* mesophyll cells can transdifferentiate into tracheary elements in the presence of growth regulators at sufficiently high cell density. Xylogen, which triggers transdifferentiation at low cell density (71, 72), is a hybrid molecule with homology to ns-LTPs and contains AG-glycomodules and a GPI-anchor modification motif (73). Xylogen is an AGP, as indicated by its large molecular weight range of 50 to 100 kDa that exceeds its predicted peptide mass of 16 kDa, and its reactivity to β-Yariv reagent and JIM13 antibody. Transgenic (nonxylogenic) tobacco BY2 cells secrete functional xylogen, indicating that the type of glycosylation required for xylogenic activity does not depend on the cell type or species. Xylogen expression precedes differentiation of tracheary elements in vitro, and in *Z. elegans*, xylogen mRNA is expressed in cambium and immature xylem. Xylogen protein is localized in a highly polar fashion at the apical side of immature xylem cells, a property that may be related to the GPI anchor and that also recalls the asymmetric localization of the JIM8-reactive epitope in carrot cells that marks embryogenic and nonembryogenic daughter cells (67). *A. thaliana* contains several genes encoding xylogen-like peptides and the double-mutant combination of T-DNA insertions in two xylogen homologs *AtXYP1* and *-2* results in defective vascular development. Double mutants contained discontinuous leaf veins and tracheary elements that lacked proper interconnection. Although not an essential component for vascular tissue differentiation, xylogen might act as a mediator of inductive cell-cell communication in vascular development (73).

The requirement for *AtAGP18* in female gametogenesis is demonstrated using RNA interference (RNAi) (1). In the silenced lines, ovules develop normally up to the formation of a functional megaspore, but a large proportion fails to undergo the subsequent mitotic divisions (1). Reporter construct analysis of *AGP18*–RNAi lines shows that the functional megaspore remains viable but neither undergoes further mitotic divisions nor expresses a marker specific for the mitotic division stage of female gametophyte development. The absence of ovules aborted at an intermediate mitotic stage implicates *AGP18* in the initiation of the first mitotic division of the functional megaspore. The high ovule abortion frequency in *AGP18*-silenced plants and its independence from zygosity and transgene copy number suggest that AGP18 does not play its crucial role in the (haploid) functional megaspore, which is expected to be transgene free in 50% of a heterozygous single-copy plant, but possibly in the surrounding (diploid) tissue. This supports a role of AGP18 as a diffusible signal in the interaction between sporophyte and gametophyte to positively regulate the meiotic division of the functional megaspore. This notion is also fully consistent with the differentiation of AGP epitopes between sporophytic and gametophytic tissues in pea and brassica (80, 82), which reflects the pattern observed in *A. thaliana* where MAC207 labeling is specifically absent and JIM8 labeling specifically present in the gametophtic tissue (L. Bruun & K. Roberts, unpublished observations).

Pollen tube guidance. The stigma contains AGPs as a major component that were proposed to act as a pollen adhesive (20). Upon pollination the pollen tube penetrates the stigma, which also contains a high amount of AGPs, and grows through the transmitting tissue of the style following a highly directional growth pattern. Lily pollen adheres to a semiartificial nitrocellulose substratum coated with style exudates containing pectin and AGP, but generic preparations of pectin and AGPs do not mediate pollen adhesion, indicating the requirement of a specific factor (30). Directional pollen growth might be guided by chemotropic factors and AGPs might be involved in pollen tube guidance (18, 137). Transmitting tissue-specific protein (TTS) from *Nicotiana tabacum* is an AGP that supports pollen tube growth in vitro and pollen chemotropism in vivo (18). However, depending on the extraction protocol, the highly homologous galactose-rich style glycoprotein (NaTTS, also know as GaRSGP) from *N. alata* (97% identical to TTS) promotes (138) and not only inhibits pollen tube growth but makes pollen tube tips burst (107, 108). NaTTS extracted with high salt appears more highly glycosylated than the low salt–extracted species (22, 138), which is consistent with the idea that the degree and pattern of AGP glycosylation species modulate the biological activity. It is currently unclear whether the differences between the two proteins are of technical or biological origin (108). However, it is possible that pollen tube guidance and pollen incompatibility reactions (see below) both involve the interaction of stylar AGPs with the pollen.

Pollen incompatibility. Numerous plant species prevent self-pollination and most prevent interspecific pollination. A key player in this process in nicotiana and other plant groups is S-RNase, a glycoprotein abundant in the style transmitting tract. S-RNase is transported from the transmitting tissue into a membrane-bound compartment of both compatible and incompatible pollen, and from there is released into the cytoplasm only of incompatible pollen, where it might exert its cytotoxic effect (42). S-RNase binds to several stylar proteins (22), including the AGP extensin hybrid glycoproteins TTS (18, 138), 120-kDa protein (120k) (61), and PELPIII (8, 26). It is unclear whether binding to S-RNase occurs on an individual basis or if it might involve a higher-order complex (22). RNAi of 120k suppresses self-specific pollen rejection in *N. alata*, supporting a role of this AGP in self-recognition. 120k is specifically required for self-pollen rejection, but not for interspecies pollen rejection, and it is imported into compatible and incompatible pollen where it remains close to the vacuolar membrane. However, 120k is not required for the uptake of S-RNase into pollen and it might be involved in a yet unknown step at a later stage, also suggested by its failure to degrade the S-RNase-containing compartment in 120k-RNAi pollen (42). 120k shows size heterogeneity between various nicotiana species the self-incompatible species generally containing larger 120k glycoproteins than the self-compatible ones. There are only subtle sequence polymorphisms between different nicotiana 120k species, suggesting variability in O-glycosylation as the reason for size heterogeneity.

Uge4: *UDP-glucose 4-epimerase*

Growth

Plant cells enlarge in two fundamentally different ways. Diffuse cell expansion involves deposition of new wall material and controlled cell wall extension in a relatively large area of the cell wall, whereas in tip growth new cell wall material is deposited in a small focused area and gradually reinforced. After cell expansion stops, certain cell types reinforce their cell walls by secondary depositions of cellulose, hemicellulose, and lignin. The potential role of AGPs in this processes and in organ abscission as well as interactions between growth regulator signaling and AGPs are discussed here.

Diffuse expansion. When seedlings, explanted hypocotyls, or cultured cells are grown in the presence of β-Yariv reagent, cell elongation is strongly inhibited (24, 27, 63, 79, 132, 135). The suppression of root elongation by β-Yariv reagent is reversible, but it is not known if this also applies to individual cells (135). The *uge4* mutant, defective in AGP galactosylation (2, 27, 75, 97) (**Figure 5**), and the fucosylation-defective *mur1* mutant (127) show root epidermal bulging and reduced root elongation. A short root phenotype is also induced by a lectin that probably recognizes t-α-L-Fuc-(1→2)-α-L-Ara-(1→ groups on AGP, suggesting that the short root phenotype of *mur1* might be caused by under-fucosylation of AGPs (127). A point mutation in the protein interaction domain of *SOS5/FLA4* leads to a conditional root expansion defect (100) (**Figure 5**). The over-expression of a classical AGP from cucumber leads to increased stem elongation in tobacco (79). So far, none of these observations has been ascribed to a basic physiological process. Possible modes of action involve a function of AGPs in cell wall polymer biosynthesis, cell wall remodeling, and growth regulator or stress signaling (see below).

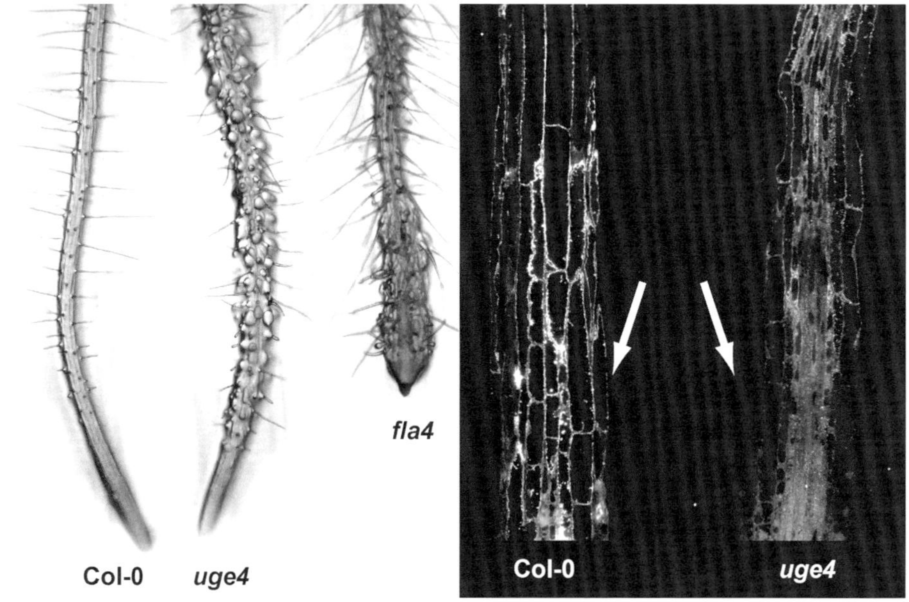

Figure 5

Two *A. thaliana* mutants that are defective in arabinogalactan proteins due to undergalactosylation, in *uge4* (97), or due to a point mutation in *FLA4* (100), display dramatic cell expansion defects. AG type II galactosylation, revealed by the CCRC-M7 antibody (111), is specifically affected in the root epidermis of *uge4* (*arrows, right*).

Tip growth. In higher plants most cells expand by diffuse expansion; whereas, root hairs and pollen tubes expand by tip growth. On one hand AGP genes are highly over-represented in the root hair transcriptome (49), and on the other hand a mutation in the *COBRA LIKE 9* (*COBL9*) gene is required for normal root hair growth. Although not experimentally confirmed to be AG modified, *COBL9* contains a $(PA)_4$ repeat and a GPI anchor motif, both typical AGP features. AGPs are deposited into the tip of growing pollen tubes (47, 62). Transgenically expressed synthetic AG peptides are selectively secreted at the pollen growth tip, indicating that GPI anchoring or other specific peptide motifs are not required for polarized AGP secretion (25). Tip-growing pollen tubes are inhibited by genetic interference with GPI-anchor biosynthesis (54) and also by β-Yariv reagent treatment, leading to ectopic callose deposition and pectin alteration (47, 62, 70, 89). When exposure to β-Yariv reagent does not exceed 1 h, 60% of pollen tubes resume tip growth. However, the inhibition of tip growth is not truly reversible because the original tip remains cemented by the β-Yariv reagent-AGP precipitate and a new growth cone is established at the flanks of the pollen tube. Any attempt to derive a normal AGP function from these results dramatically illustrates the problems with evidence obtained from β-Yariv reagent treatment but lacking genetic confirmation. The β-Yariv reagent-AGP cement at pollen tips might simply form a rigidified barrier impeding growth, but this does not necessarily mean that AGPs are involved in the normal process.

The relationship between AGPs and tip growth also has been evaluated in the moss model *Physcomitrella patens*, which consists of protonemal and caulonemal cells growing by apical tip growth and serial division of the apical cells (57). *P. patens* contains β-Yariv reagent binding glycoproteins and expresses at least two different genes encoding classical AGPs, three AG-peptides and two fasciclin-like AGPs. Tip-growth inhibition of protonema by β-Yariv reagent is rapid and completely reversible; after withdrawal of β-Yariv reagent, and after a certain recovery period, cell extension resumes at its normal position and at a normal rate. *PpAGP1*, one of the *P. patens* genes encoding a classical AGP, is required for full protonemal cell length, which is consistent with a role of AGPs in tip growth.

Secondary wall deposition. The presence of AGP carbohydrate epitopes (28, 109) and AGP backbone peptides in secondary wall thickenings (34, 36), and the tight physical association of AGP-like glycoproteins with flax fiber cellulose (41), suggest that GPI-anchored AGPs might be secreted to the cell surface in parallel with cellulose synthase, and while binding to cellulose they might be released from their GPI anchor and incorporated into cell wall thickening. PtaAGP6 is specifically localized in the differentiating xylem of *Pinus taeda* (141) and the expression of *AtFLA11* and *AtFLA12* is closely linked to secondary cell wall biosynthesis in *A. thaliana* (11, 46, 83), consistent with its expression in sclerenchyma and the expression of its close homolog *ZeFLA11* in reticulate xylem of *Z. elegans* (23). The role of *AtFLA11* for xylogenesis is not certain because *atfla11* mutants were reported to display either a subtle irregular xylem phenotype or no phenotype by different groups (11, 46, 83). However, *AtFLA11* and *AtFLA12* might fulfill redundant roles.

Abscission. A possible role of AGPs in organ abscission is highlighted by the identification and characterization of the *INFLORESCENCE DEFICIENT IN ABSCISSION (IDA)* gene (12, 112). Loss of *IDA* function results in delayed abscission of floral organs and overexpression of *IDA* produces the opposite effect: Premature abscission correlates with strong production of β-Yariv reagent-reactive AGP in the abscission zone. The *ida* mutant lacks the wild-type level of abscission-zone AGPs, indicating that *IDA* is a positive regulator of organ abscission and abscission-zone secreted AGPs. Indirectly, this might implicate AGPs in organ abscission. In particular, the mRNA level of the AG peptide AtAGP24 correlates with *IDA* expression and AGP level. Although not a major abscission-zone AGP itself, IDA fulfills some criteria expected from an AGP. It is predicted to be a secreted protein, it contains a short Pro-rich discontinuous region, and its apparent molecular weight is much higher than predicted (112). However, its apparent molecular weight is well defined and a more thorough structural characterization of IDA will clarify the nature of its posttranslational modifications.

Interactions with growth regulators. AGPs might interact with plant-growth regulator signaling and thereby influence development and growth. Treatment of barley aleurone protoplasts with β-Yariv reagent suppresses the induction of α-amylase enzyme and promoter activity by gibberelic acid without affecting the ABA-regulated dehydrin promoter or cell viability. This was interpreted as AGPs possibly acting in the transduction of the gibberellic acid signal regulating starch degradation (116). This is an interesting observation in an economically important system, but the reservations described above concerning the exclusive use of β-Yariv reagent apply here as well. Overexpression of tomato *LeAGP1* leads to a reduction of apical dominance, suppression of flower maturation and fruit production, and delayed senescence, a phenomenology that suggests an altered balance of auxin versus cytokinin signaling (113). The opposing effects of auxin and cytokinin on *LeAGP1* expression support this interpretation. Strikingly, a GPI-anchor-deficient version of *LeAGP1* does not cause an abnormal phenotype. This could mean that the GPI anchor is essential for LeAGP1 biosynthesis, stability, localization, or function. Interestingly, overexpression of a different AGP in tobacco causes an opposite effect on growth (79). Loss of function of *AtAGP30* results in reduced sensitivity to ABA, including the expression of ABA-responsive genes in seedling roots and ABA-induced seed dormancy (128). AGPs could have functions that are antagonistic to those of growth regulators. AGPs could either act downstream of growth regulators to stimulate cell division, or might modulate the sensitivity of receptors, or might serve as a coreceptor to present the growth regulator to a low-affinity receptor in a more concentrated form (113, 128). It would be interesting to further investigate the potential involvement of AGPs in growth regulator signaling using epistasis analysis with well-established signaling mutants in *A. thaliana*.

Plant Microbe Interactions

A number of observations implicate AGPs in plant microbe interactions. Several early nodulins are likely to be AG modified (30, 33, 48, 90), and AGP-like genes are expressed during symbiotic associations with arbuscular mycorrhizal fungi (125). AGP-like cell wall proteins induced during nodulation might act in rhizobial adhesion and infection thread morphogenesis. Reduced expression of *AtAGP17/RAT1* due to a T-DNA insertion in the promoter region and treatment of roots with β-Yariv reagent trigger resistance of roots to transformation with *Agrobacterium tumefaciens (rat)* (38, 74). *AtAGP17* might either be required for agrobacterium binding or might be involved in a signal transduction mechanism that enables agrobacterium to infect the wild-type plant (38). However, in the *atagp17/rat1* mutant the expression of marker genes for salicylic acid dependent systemic acquired resistance are elevated in the absence and presence of agrobacterium, suggesting a nonredundant function of *AGP17* in the uninfected plant. Related to this work is the observation that β-Yariv reagent suppresses the attachment of rhizobium to *A. thaliana* roots (131). Because *A. thaliana* is not colonized by rhizobium, β-Yariv reagent might have an unspecific effect on microbial adhesion in roots. A significant sequence of papers has shown that alterations to the cell wall induce biotic stress resistance.

Constitutive powdery mildew resistance is triggered in mutants in cellulose synthase, callose synthase, and a GPI-anchored pectate lyase, suggesting a flow of information from the cell wall to intracellular signaling networks (133). It was hypothesized that a novel "cell wall performance and integrity control" system senses cell wall alterations during pathogenesis, wounding, and mechanical stress and also balances cell wall biosynthesis and remodeling (96, 106). Colonization of roots by the endophytic fungus *Piriformospora indica*, which triggers broad-spectrum disease resistance (134) in a similar fashion, induces accumulation of a GPI-anchored pectate lyase and several receptor-like kinases (RLKs) in a detergent-resistant plasma membrane microdomain (99). It is intriguing to speculate that GPI-anchored AGPs might also reside in such microdomains as a close association between RLKs and AGPs might provide a link between extracellular matrix polysaccharides and cell signaling.

MODE OF ACTION OF AGPs: WHAT IS THE OUTLOOK?

As outlined above, most of the evidence involving AGPs in various biological roles is based on genetic manipulation of individual AGP backbone peptides or the global alteration of AGP glycosylation in mutants or transgenic lines, or by the binding of AGPs by β-Yariv reagent. Although important stepping stones to find modes of action, these observations are currently difficult to interpret. In a few cases, biological roles can be inferred from an effect of AGPs added to a biological assay, and so far only two biologically active AGPs, TTS (18) and xylogen (73), have been purified to homogeneity and have had their corresponding cDNAs cloned. The molecular mechanism of action is still unclear. However, the lack of a widely accepted mechanistic paradigm is not so much a case of plant research lagging behind, but it simply reflects the gargantuan technical difficulties posed by highly complex glycoproteins for which there exist no stereotypical protocols for purification, quantification, structural elucidation, activity, or intermolecular interactions. Sugars cannot be cloned! A comparison of AGP biology and analogous molecules in animals is nevertheless a useful stimulant for generating hypotheses for their modes of action.

RLK: receptor-like kinase

Morphogen: a diffusible substance that forms a concentration gradient and provides stimulatory or inhibitory information required for pattern formation

Arabinogalactan Proteins as (Sources of) Soluble and Lipid Signals

The simplest way to envisage a signaling role for an AGP is direct binding to a receptor that activates a signal transduction cascade, and indeed some observations are compatible with this idea. Because purified or recombinant xylogen is biologically active, its soluble form contains sufficient information for the transdifferentiation assay. Its polarized cellular localization in vivo in conjunction with its rapid release from the plasma membrane might set up a morphogen gradient that sends distinct positional information to the two clonal neighbors of a differentiating xylem cell. Soluble AGPs might act as cell survival signals that prevent cells grown at low density from dying (66), or they might be released from $JIM8^+$ cells to induce embryogenesis in $JIM8^-$ cells (67). However, the complex structure of AGPs makes them a potential source of small signaling molecules. Cleavage of the GPI anchor by GPI-specific phospholipases might release lipid signals (93), and endoglycanases could release biologically active oligosaccharides from AGP. The observation that endochitinase treatment of AGPs initially suppressed its positive effect on somatic embryogenesis, and that the stimulatory effect was restored and even improved only after repurification (129), suggests that not only is the relatively bulky AGP a stimulator, but it might also release a small oligosaccharin-like inhibitor. The bulky nature and potential affinity to β-glycans in the cell wall matrix could impede AGP mobility in the apoplast, which might restrict the radius of action to one cell, whereas AGP-derived oligosaccharins might travel for longer distances.

Differential action radii of activators and inhibitors are often encountered in biological patterning (69).

Extracellular Matrix Glycoproteins as Modulators of Morphogen Gradients

In animal development it is thought that cell surface–bound and –secreted proteoglycans convey positional information. Heparansulfateproteoglycan (HSPG) is either linked to GPI anchors (called glypicans) or exists in the form of transmembrane proteins (called syndecans). HSPGs play a role in animal pattern formation, as shown by a number of mutant loci involved in the interconversion and transport of nucleotide sugar precursors and glycan biosynthesis (44). Specific HSPG molecules are required for the full function of the *Hedgehog* developmental pathway. The peptide Hedgehog acts as a morphogen in fruitflies, and its secretion, movement, and perception are modulated by the glypicans Dally and Dally-like protein (136). Although the precise mechanisms of Dally action in Hedgehog signaling is not clear, it was speculated that they could facilitate ligand presentation to receptive cells, concentrate *Hedgehog* in microdomians, or particitpate in a large receptor complex (136). CLAVATA3 is a secreted peptide acting as a morphogen upstream of the RLK CLAVATA1 in stem cell homeostasis in the shoot apical meristem. CLV3 diffusion through the apoplast is restricted by CLV1, but it is presently unknown if apoplastic factors mediate this restriction (58). The properties of AGPs would make them suited to fulfill this role and it will be interesting to test if CLV3 diffusion is altered in an AGP-deficient mutant background.

Arabinogalactan Proteins as (Co)-Receptors

The demonstration that AGPs are dynamic plasma membrane–localized glycoproteins that bind to β-Yariv reagent suggested that they might connect the plasma membrane to cell wall matrix carbohydrates (81). The plasma membrane localization of several AGPs via their GPI anchor has since been demonstrated. However, whether AGPs interact with cell wall polymers and transmembrane proteins remains speculative. Immunofluorescence labeling of tobacco protoplasts reveals a polygonal network of AG epitopes whose vertices colocalized with wall-associated kinase (WAK) (39). It is presently unclear if AGP and WAKs interact and if their colocalization with WAKs is a result of immunolabeling of unfixed protoplasts with AGP-reactive antibodies, a procedure that can also cause aggregation of AGPs (81). Some AGP core proteins share a domain of six conserved cysteine residues, termed the (proline-rich, AGP, conserved cysteine) PAC domain (3), with the small glycoprotein LAT52 that has been reported to interact with the receptor-like kinase(RLK), LePRK2 (121). It has been suggested that the PAC domain might mediate binding between some AGPs and RLKs (128). This hypothesis may be supported by the observation that treating protoplasts and growing pollen tubes with β-Yariv reagent triggers a rapid increase in intracellular [Ca^{2+}], which could involve the activation of Ca^{2+} channels by protein phosphorylation (84, 89). The modulation of Ca^{2+} fluctuations by β-Yariv reagent might lead to PCD, modulate growth regulator responses, and trigger pathogen resistance. However, out of 620 RLK sequences in the *A. thaliana* genome (102), approximately 60 contain their own potential AG glycomodules (G.J. Seifert, unpublished observations) and might therefore display AGP-like properties.

Arabinogalactan Proteins as Hydrophilic Templates of Lipid Domains

The remarkable observation that many AGPs, whether classical, fasciclin-like, plastocyanin-like, or nsLTP-like, contain a GPI anchor and also that many GPI-anchored proteins contain AG-glycomodules (5), merits further discussion. GPI-containing proteins are often associated with lipid microdomains or rafts. The biological significance of rafts is under intense scrutiny, and separate fluid phases containing artificial raft-associated proteins (including GPI-anchored GFP) in a mobile, freely diffusing phase, and nonraft proteins in a diffusion-restricted phase, exist in apical domains of mammalian epithelial cells (68). In plant cells numerous plasma membrane proteins, including AGPs and other GPI-achored peptides, are associated with a detergent-resistant membrane subfraction, interpreted as evidence of the existence of lipid microdomains in plants (7). Oriented cell expansion depends on the directionally restricted mobility of cellulose synthase complexes, which might be influenced by plasma membrane fluidity. Oriented bidirectional movement of cellulose synthase depends on well-oriented cortical microtubules (78) but persists on a seemingly predefined trajectory after the collapse of microtubules by dynamic instability. After prolonged pharmacological microtubule disruption, cellulose synthase movement follows a microtubule-independent pattern similar to the untreated controls, suggesting that a second extrinsic organizational mechanism might be involved (78). GPI-anchored protein-containing plasma membrane microdomains might interact with cortical microtubules and form relatively stable tracks of high membrane fluidity that facilitate cellulose synthase mobility even in the absence of underlying microtubules. GPI-anchored COBRA, required for normal organization of cellulose deposition, is localized in a microtubule-dependent transverse orientation in a fashion that is consistent with this model (88). GPI-anchored AGPs might not only partition into lipid microdomains, but might also interact with pre-existing cell wall polymers. This could be the basis of a reciprocal interaction between the cell wall and the cortical microtubule cytoskeleton that was recently postulated (4), and could explain the loss of microtubule organization in the *uge4* root epidermis (2). A reciprocal interaction between the cortical cytoskeleton and AGPs is supported by the demonstration that β-Yariv treatment rapidly disrupts the normal organization of the microtubule and actin cytoskeleton and that drug interference with microtubules and actin has the same effect on GFP-tagged LeAGP1, in tobacco BY2 cells (89a).

Arabinogalactan Proteins as Modulators of Cell Wall Mechanics

The binding of AGPs to β-Yariv reagent has led to the suggestion that analogous lectin-like interactions of AGP with β-linked glycans might occur at the cell surface and in the cell wall. Potential interaction partners are pectic (1→4)-β-galacturonan and hemicellulosic (1→4)-β-glucans. Tight association of type II AG with pectin has sometimes been observed (reviewed in 76), but the significance of this association remains to be elucidated. One possible mechanical role of AGP in the cell wall might be rigidification by oxidative crosslinking. This process might take place after the end of cell expansion, after wounding (9), and during the formation of the infection thread in legumes, and could involve the action of peroxidases (50). It is unclear if peroxide-mediated crosslinking is a property typical for AGPs or if it might be a property of AGP-like extensins, as suggested by Brewin (10). His intriguing hypothesis proposed that the infection thread-localized AGP-like root nodule extensin might undergo a regulated fluid to solid transition by removing AG side chains through the action of endogalactanase, which would expose a more conventional extensin for peroxide-mediated crosslinking (10). Another interesting observation was the in vitro stimulation of xyloglucan endotransglycosylase (XET) by an AGP-like preparation (118). XET is thought to remodel the cellulose xyloglucan network thereby facilitating cell wall deposition and cell extension (21). Although cell wall extensibility is not directly affected by β-Yariv reagent (24), it is conceivable that acidic AGP-like molecules modulate XET action and are thus involved in cell expansion. A function of AGPs as a "pectin plasticizer" has recently been postulated based on the relative distribution of different AGP fractions in salt-adapted suspension cells (55). It was proposed that a relatively small proportion of AGPs incorporated into a pectin gel might increase its porosity. Although the experimental foundation for this prediction remains to be established, modification of pectin pore sizes could regulate the mobility of apoplastic enzymes and signals.

CONCLUSIONS AND KEY PROBLEMS

The main conclusion that emerges from the integration of the cloning of AGP core-protein genes, the Hyp-contiguity hypothesis, and the availability of entire genome sequences is that AG modification of extracellular proteins is (*a*) widespread among numerous gene families and (*b*) cannot be reliably predicted in any individual case. Another important notion is that the AG-glycan moiety of AGPs is crucial for their function, but virtually nothing is known about AG biosynthesis and remodeling.

Key future challenges are the elucidation of the enzymatic machinery that synthesizes the AG carbohydrate structure and the molecular nature and biological role of endogenous AG-specific carbohydrate hydrolases. The molecular genetic analysis of these enzymes will clarify the biological significance of tissue- and cell-type-specific expression of AGP carbohydrate epitopes. This will tie in with efforts to test whether the precise structure of AGP glycans is genetically determined by the core protein or under developmental control. Observations of transgenic plants expressing synthetic AG glycomodules suggest the latter (32a). On the cellular level it will be interesting to investigate whether GPI-anchored AGP species establish lipid microdomains that are informed by, as well as generate the structure of, the inner face of the cell wall and the cortical cytoskeleton. A further challenge is to identify molecular interaction partners of AG-modified glycoproteins. Ultimately, the study of these widespread and complex lipo-glycoprotein hybrid molecules will be crucial to understanding the mysterious relationship between the cell wall, the plasma membrane, and the cytoplasm.

SUMMARY POINTS

1. AG modification is widespread among many families of GPI-anchored and -secreted proteins.
2. The *A. thaliana* AGP genes *xylogen*, *AGP18*, and *FLA4* play nonredundant roles in vascular and gemetophyte development and cell expansion.
3. The structural, localization, and genetic roles of AGPs point to molecular functions as diffusible and anchored signals, and as modulators of the mobility of extracellular ligands and plasma membrane proteins.

FUTURE ISSUES

1. AG biosynthesis will be elucidated at the molecular level.
2. Mechanisms of structural and developmental control of AG heterogeneity will be identified.

3. Localization and dynamics of AGPs in lipid microdomains will be investigated in vivo.
4. Molecular interaction partners of AGPs will be found.

ACKNOWLEDGMENTS

GJS received financial support from the Biotechnology and Biological Science Research Council Grant 208/D10332.

LITERATURE CITED

1. Acosta-Garcia G, Vielle-Calzada JP. 2004. A classical arabinogalactan protein is essential for the initiation of female gametogenesis in *Arabidopsis*. *Plant Cell* 16:2614–28
2. Andeme-Onzighi C, Sivaguru M, Judy-March J, Baskin TI, Driouich A. 2002. The *reb1-1* mutation of *Arabidopsis* alters the morphology of trichoblasts, the expression of arabinogalactan-proteins and the organization of cortical microtubules. *Planta* 215:949–58
3. Baldwin TC, van Hengel AJ, Roberts K. 2000. The C-terminal PAC domain of a secreted arabinogalactan-protein from carrot defines a family of basic proline-rich proteins. See Ref. 76a, pp. 43–49
4. Baskin TI. 2001. On the alignment of cellulose microfibrils by cortical microtubules: a review and a model. *Protoplasma* 215:150–71
5. Borner GH, Lilley KS, Stevens TJ, Dupree P. 2003. Identification of glycosylphosphatidylinositol-anchored proteins in arabidopsis. A proteomic and genomic analysis. *Plant Physiol.* 132:568–77
6. Borner GH, Sherrier DJ, Stevens TJ, Arkin IT, Dupree P. 2002. Prediction of glycosylphosphatidylinositol-anchored proteins in arabidopsis. A genomic analysis. *Plant Physiol.* 129:486–99
7. Borner GH, Sherrier DJ, Weimar T, Michaelson LV, Hawkins ND, et al. 2005. Analysis of detergent-resistant membranes in arabidopsis. Evidence for plasma membrane lipid rafts. *Plant Physiol.* 137:104–16
8. Bosch M, Knudsen JS, Derksen J, Mariani C. 2001. Class III pistil-specific extensin-like proteins from tobacco have characteristics of arabinogalactan proteins. *Plant Physiol.* 125:2180–88
9. Bradley DJ, Kjellbom P, Lamb CJ. 1992. Elicitor- and wound-induced oxidative cross-linking of a proline-rich plant cell wall protein: a novel, rapid defense response. *Cell* 70:21–30
10. Brewin NJ. 2004. Plant cell wall remodelling in the rhizobium-legume symbiosis. *Crit. Rev. Plant Sci.* 23:293–316
11. Brown DM, Zeef LA, Ellis J, Goodacre R, Turner SR. 2005. Identification of novel genes in arabidopsis involved in secondary cell wall formation using expression profiling and reverse genetics. *Plant Cell* 17:2281–95
12. Butenko MA, Patterson SE, Grini PE, Stenvik GE, Amundsen SS, et al. 2003. Inflorescence deficient in abscission controls floral organ abscission in arabidopsis and identifies a novel family of putative ligands in plants. *Plant Cell* 15:2296–307

13. Carpita NC, Gibeaut DM. 1993. Structural models of primary cell walls in flowering plants: consistency of molecular structure with the physical properties of the walls during growth. *Plant J.* 3:1–30
14. Casero PJ, Casimiro I, Knox JP. 1998. Occurrence of cell surface arabinogalactan-protein and extensin epitopes in relation to pericycle and vascular tissue development in the root apex of four species. *Planta* 204:252–59
15. Chapman A, Blervacq AS, Vasseur J, Hilbert JL. 2000. Arabinogalactan-proteins in Cichorium somatic embryogenesis: effect of β-glucosyl Yariv reagent and epitope localisation during embryo development. *Planta* 211:305–14
16. Chaves I, Regalado AP, Chen M, Ricardo CP, Showalter AM. 2002. Programmed cell death induced by (β-D-galactosyl)(3) Yariv reagent in *Nicotiana tabacum* BY-2 suspension-cultured cells. *Physiol. Plant* 116:548–53
17. **Chen CG, Pu ZY, Moritz RL, Simpson RJ, Bacic A, et al. 1994. Molecular cloning of a gene encoding an arabinogalactan-protein from pear (*Pyrus communis*) cell suspension culture. *Proc. Natl. Acad. Sci. USA* 91:10305–9**
18. Cheung AY, Wang H, Wu HM. 1995. A floral transmitting tissue-specific glycoprotein attracts pollen tubes and stimulates their growth. *Cell* 82:383–93
19. **Clarke AE, Anderson RL, Stone BA. 1979. Form and function of arabinogalactans and arabinogalactan-proteins. *Phytochemistry* 18:521–40**
20. Clarke AE, Gleeson P, Harrison S, Knox RB. 1979. Pollen-stigma interactions: Identification and characterization of surface components with recognition potential. *Proc. Natl. Acad. Sci. USA* 76:3358–62
21. Cosgrove DJ. 1999. Enzymes and other agents that enhance cell wall extensibility. *Annu. Rev. Plant Physiol. Plant Mol. Biol.* 50:391–417
22. Cruz-Garcia F, Hancock CN, Kim D, McClure B. 2005. Stylar glycoproteins bind to S-RNase in vitro. *Plant J.* 42:295–304
23. Dahiya P, Findlay K, Roberts K, McCann MC. 2006. A fasciclin-domain containing gene, ZeFLA11, is expressed exclusively in xylem elements that have reticulate wall thickenings in the stem vascular system of *Zinnia elegans* cv Envy. *Planta* 223:1281–91
24. Darley CP, Forrester AM, McQueen-Mason SJ. 2001. The molecular basis of plant cell wall extension. *Plant Mol. Biol.* 47:179–95
25. de Graaf BH, Cheung AY, Andreyeva T, Levasseur K, Kieliszewski M, Wu HM. 2005. Rab11 GTPase-regulated membrane trafficking is crucial for tip-focused pollen tube growth in tobacco. *Plant Cell* 17:2564–79
26. de Graaf BH, Knuiman BA, Derksen J, Mariani C. 2003. Characterization and localization of the transmitting tissue-specific PELPIII proteins of *Nicotiana tabacum*. *J. Exp. Bot.* 54:55–63
27. Ding L, Zhu JK. 1997. A role for arabinogalactan-proteins in root epidermal cell expansion. *Planta* 203:289–94
28. Dolan L, Linstead P, Roberts K. 1995. An AGP epitope distinguishes a central metaxylem initial from other vascular initials in the arabidopsis root. *Protoplasma* 189:149–55
29. Dolan L, Roberts K. 1995. Secondary thickening in roots of *Arabidopsis thaliana* - anatomy and cell-surface changes. *New Phytol.* 131:121–28
30. dos Santos ALW, Wietholter N, El Gueddari NE, Moerschbacher BM. 2006. Protein expression during seed development in *Araucaria angustifolia*: transient accumulation of class IV chitinases and arabinogalactan proteins. *Physiol. Plant.* 127:138–48
31. Du H, Simpson RJ, Moritz RL, Clarke AE, Bacic A. 1994. Isolation of the protein backbone of an arabinogalactan-protein from the styles of *Nicotiana alata* and characterization of a corresponding cDNA. *Plant Cell* 6:1643–53

17 & 31. Early report on full-length AGP cDNA.

19. Comprehensive overview on sugar composition of AGP from various sources.

32. Egertsdotter U, von Arnold S. 1995. Importance of Arabinogalactan-proteins for the development of somatic embryos of Norway spruce (*Picea abies*). *Physiol. Plant* 93:334–45
32a. Estévez JM, Kieliszewski MJ, Khitrov N, Somerville C. 2006. Characterization of synthetic hydroxyproline-rich proteoglycans with arabinogalactan protein and extensin motifs in *Arabidopsis*. *Plant Physiol.* 142: 458–70
33. Fruhling M, Hohnjec N, Schroder G, Kuster H, Puhler A, Perlick AM. 2000. Genomic organization and expression properties of the *VfENOD5* gene from broad bean (*Vicia faba* L.). *Plant Sci.* 155:169–78
34. Gao M, Kieliszewski MJ, Lamport DT, Showalter AM. 1999. Isolation, characterization and immunolocalization of a novel, modular tomato arabinogalactan-protein corresponding to the *LeAGP-1* gene. *Plant J.* 18:43–55
35. Gao M, Showalter AM. 1999. Yariv reagent treatment induces programmed cell death in arabidopsis cell cultures and implicates arabinogalactan protein involvement. *Plant J.* 19:321–31
36. Gao M, Showalter AM. 2000. Immunolocalization of LeAGP-1, a modular arabinogalactan-protein, reveals its developmentally regulated expression in tomato. *Planta* 210:865–74
37. Gaspar Y, Johnson KL, McKenna JA, Bacic A, Schultz CJ. 2001. The complex structures of arabinogalactan-proteins and the journey towards understanding function. *Plant Mol. Biol.* 47:161–76
38. Gaspar YM, Nam J, Schultz CJ, Lee LY, Gilson PR, et al. 2004. Characterization of the arabidopsis lysine-rich arabinogalactan-protein *AtAGP17* mutant (*rat1*) that results in a decreased efficiency of agrobacterium transformation. *Plant Physiol.* 135:2162–71
39. Gens JS, Fujiki M, Pickard BG. 2000. Arabinogalactan protein and wall-associated kinase in a plasmalemmal reticulum with specialized vertices. *Protoplasma* 212:115–34
40. Gillmor CS, Lukowitz W, Brininstool G, Sedbrook JC, Hamann T, et al. 2005. Glycosylphosphatidylinositol-anchored proteins are required for cell wall synthesis and morphogenesis in arabidopsis. *Plant Cell* 17:1128–40
41. Girault R, His I, Andeme-Onzighi C, Driouich A, Morvan C. 2000. Identification and partial characterization of proteins and proteoglycans encrusting the secondary cell walls of flax fibres. *Planta* 211:256–64
42. Goldraij A, Kondo K, Lee CB, Hancock CN, Sivaguru M, et al. 2006. Compartmentalization of S-RNase and HT-B degradation in self-incompatible nicotiana. *Nature* 439:805–10
43. Guan Y, Nothnagel EA. 2004. Binding of arabinogalactan proteins by Yariv phenylglycoside triggers wound-like responses in arabidopsis cell cultures. *Plant Physiol.* 135:1346–66
44. Hacker U, Nybakken K, Perrimon N. 2005. Heparan sulphate proteoglycans: the sweet side of development. *Nat. Rev. Mol. Cell Biol.* 6:530–41
45. Hieta R, Myllyharju J. 2002. Cloning and characterization of a low molecular weight prolyl 4-hydroxylase from *Arabidopsis thaliana*. Effective hydroxylation of proline-rich, collagen-like, and hypoxia-inducible transcription factor α-like peptides. *J. Biol. Chem.* 277:23965–71
46. Ito S, Suzuki Y, Miyamoto K, Ueda J, Yamaguchi I. 2005. AtFLA11, a fasciclin-like arabinogalactan-protein, specifically localized in sclerenchyma cells. *Biosci. Biotechnol. Biochem.* 69:1963–69
47. Jauh GY, Lord EM. 1996. Localization of pectins and arabinogalactan-proteins in lily (*Lilium longiflorum* L) pollen tube and style, and their possible roles in pollination. *Planta* 199:251–61

48. Johnson KL, Jones BJ, Bacic A, Schultz CJ. 2003. The fasciclin-like arabinogalactan proteins of arabidopsis. A multigene family of putative cell adhesion molecules. *Plant Physiol.* 133:1911–25
49. Jones MA, Raymond MJ, Smirnoff N. 2006. Analysis of the root-hair morphogenesis transcriptome reveals the molecular identity of six genes with roles in root-hair development in arabidopsis. *Plant J.* 45:83–100
50. Kjellbom P, Snogerup L, Stohr C, Reuzeau C, McCabe PF, Pennell RI. 1997. Oxidative cross-linking of plasma membrane arabinogalactan proteins. *Plant J.* 12:1189–96
51. Knox JP, Linstead PJ, Peart J, Cooper C, Roberts K. 1991. Developmentally regulated epitopes of cell-surface arabinogalactan proteins and their relation to root-tissue pattern-formation. *Plant J.* 1:317–26
52. Kreuger M, van Holst GJ. 1993. Arabinogalactan-proteins are essential in somatic embryogenesis of *Daucus carota* L. *Planta* 189:243–48
53. Kreuger M, van Holst GJ. 1995. Arabinogalactan-protein epitopes in somatic embryogenesis of Daucus-carota L. *Planta* 197:135–41
54. Lalanne E, Honys D, Johnson A, Borner GH, Lilley KS, et al. 2004. SETH1 and SETH2, two components of the glycosylphosphatidylinositol anchor biosynthetic pathway, are required for pollen germination and tube growth in arabidopsis. *Plant Cell* 16:229–40
55. Lamport DTA, Kieliszewski MJ. 2005. Stress upregulates periplasmic arabinogalactan-proteins. *Plant Biosyst.* 139:60–64
56. Langan KJ, Nothnagel EA. 1997. Cell surface arabinogalactan-proteins and their relation to cell proliferation and viability. *Protoplasma* 196:87–98
57. Lee KJD, Sakata Y, Mau SL, Pettolino F, Bacic A, et al. 2005. Arabinogalactan proteins are required for apical cell extension in the moss *Physcomitrella patens*. *Plant Cell* 17:3051–65
58. Lenhard M, Laux T. 2003. Stem cell homeostasis in the arabidopsis shoot meristem is regulated by intercellular movement of *CLAVATA3* and its sequestration by *CLAVATA1*. *Development* 130:3163–73
59. Leonard R, Petersen BO, Himly M, Kaar W, Wopfner N, et al. 2005. Two novel types of O-glycans on the mugwort pollen allergen Art v 1 and their role in antibody binding. *J. Biol. Chem.* 280:7932–40
60. Letarte J, Simion E, Miner M, Kasha KJ. 2006. Arabinogalactans and arabinogalactan-proteins induce embryogenesis in wheat (*Triticum aestivum* L.) microspore culture. *Plant Cell Rep.* 24:691–98
61. Lind JL, Bacic A, Clarke AE, Anderson MA. 1994. A style-specific hydroxyproline-rich glycoprotein with properties of both extensins and arabinogalactan proteins. *Plant J.* 6:491–502
62. Lord EM, Holdaway-Clarke TL, Roy S, Jauh GY, Hepler PK. 2000. Arabinogalactan-proteins in pollen tube growth. See Ref. 76a, pp. 153–67
63. Lu H, Chen M, Showalter AM. 2001. Developmental expression and perturbation of arabinogalactan-proteins during seed germination and seedling growth in tomato. *Physiol. Plant.* 112:442–50
64. Majewska-Sawka A, Nothnagel EA. 2000. The multiple roles of arabinogalactan proteins in plant development. *Plant Physiol.* 122:3–10
65. Mashiguchi K, Yamaguchi I, Suzuki Y. 2004. Isolation and identification of glycosylphosphatidylinositol-anchored arabinogalactan proteins and novel β-glucosyl Yariv-reactive proteins from seeds of rice (*Oryza sativa*). *Plant Cell Physiol.* 45:1817–29
66. McCabe PF, Levine A, Meijer PJ, Tapon NA, Pennell RI. 1997. A programmed cell death pathway activated in carrot cells cultured at low cell density. *Plant J.* 12:267–80

67. McCabe PF, Valentine TA, Forsberg LS, Pennell RI. 1997. Soluble signals from cells identified at the cell wall establish a developmental pathway in carrot. *Plant Cell* 9:2225–41

68. Meder D, Moreno MJ, Verkade P, Vaz WL, Simons K. 2006. Phase coexistence and connectivity in the apical membrane of polarized epithelial cells. *Proc. Natl. Acad. Sci. USA* 103:329–34

69. Meinhardt H, Gierer A. 2000. Pattern formation by local self-activation and lateral inhibition. *BioEssays* 22:753–60

70. Mollet JC, Kim S, Jauh GY, Lord EM. 2002. Arabinogalactan proteins, pollen tube growth, and the reversible effects of Yariv phenylglycoside. *Protoplasma* 219:89–98

71. Motose H, Fukuda H, Sugiyama M. 2001. Involvement of local intercellular communication in the differentiation of zinnia mesophyll cells into tracheary elements. *Planta* 213:121–31

72. Motose H, Sugiyama M, Fukuda H. 2001. An arabinogalactan protein(s) is a key component of a fraction that mediates local intercellular communication involved in tracheary element differentiation of zinnia mesophyll cells. *Plant Cell Physiol.* 42:129–37

73. Motose H, Sugiyama M, Fukuda H. 2004. A proteoglycan mediates inductive interaction during plant vascular development. *Nature* 429:873–78

74. Nam J, Mysore KS, Zheng C, Knue MK, Matthysse AG, Gelvin SB. 1999. Identification of T-DNA tagged arabidopsis mutants that are resistant to transformation by agrobacterium. *Mol. Gen. Genet.* 261:429–38

75. Nguema-Ona E, Andeme-Onzighi C, Aboughe-Angone S, Bardor M, Ishii T, et al. 2006. The *reb1-1* mutation of arabidopsis: Effect on the structure and localization of galactose-containing cell wall polysaccharides. *Plant Physiol.* 140:1406–17

76. Nothnagel EA. 1997. Proteoglycans and related components in plant cells. *Int. Rev. Cytol.* 174:195–91

76a. Nothnagel EA, Bacic A, Clarke AE, eds. 2000. *Cell and Developmental Biology of Arabinoglactan-Proteins*. New York: Kluwer Acad./Plenum

77. Oxley D, Bacic A. 1999. Structure of the glycosylphosphatidylinositol anchor of an arabinogalactan protein from *Pyrus communis* suspension-cultured cells. *Proc. Natl. Acad. Sci. USA* 96:14246–51

78. Paredez AR, Somerville CR, Ehrhardt DW. 2006. Visualization of cellulose synthase demonstrates functional association with microtubules. *Science* 312:1491–95

79. Park MH, Suzuki Y, Chono M, Knox JP, Yamaguchi I. 2003. CsAGP1, a gibberellin-responsive gene from cucumber hypocotyls, encodes a classical arabinogalactan protein and is involved in stem elongation. *Plant Physiol.* 131:1450–59

80. Pennell RI, Janniche L, Kjellbom P, Scofield GN, Peart JM, Roberts K. 1991. Developmental regulation of a plasma membrane arabinogalactan protein epitope in oilseed rape flowers. *Plant Cell* 3:1317–26

81. Pennell RI, Knox JP, Scofield GN, Selvendran RR, Roberts K. 1989. A family of abundant plasma membrane-associated glycoproteins related to the arabinogalactan proteins is unique to flowering plants. *J. Cell Biol.* 108:1967–77

82. Pennell RI, Roberts K. 1990. Sexual development in the pea is presaged by altered expression of arabinogalactan-protein. *Nature* 344:547–49

83. Persson S, Wei H, Milne J, Page GP, Somerville CR. 2005. Identification of genes required for cellulose synthesis by regression analysis of public microarray data sets. *Proc. Natl. Acad. Sci. USA* 102:8633–38

67. Highly creative use of AGP-reactive monoclonal antibodies.

73. The most convincing evidence that AGPs act as morphogens.

76. Landmark review covering the literature up to the cloning of the first AGP cDNA.

84. Pickard BG, Fujiki M. 2005. Ca^{2+} pulsation in BY-2 cells and evidence for control of mechanosensory Ca^{2+}-selective channels by the plasmalemmal reticulum. *Funct. Plant Biol.* 32:863–79
85. Qin Y, Zhao J. 2006. Localization of arabinogalactan proteins in egg cells, zygotes, and two-celled proembryos and effects of β-D-glucosyl Yariv reagent on egg cell fertilization and zygote division in *Nicotiana tabacum* L. *J. Exp. Bot.* 57:2061–74
86. Rauh RA, Basile DV. 2000. Induction of phylogenetic phenocopies in streptocarpus (gesneriaceae) by three antagonists of hydroxyproline-protein synthesis. See Ref. 76a, pp. 191–203
87. Rauh RA, Basile DV. 2003. Phenovariation induced in *Streptocarpus prolixus* (gesneriaceae) by β-glucosyl Yariv reagent. *Can. J. Bot.* 81:338–44
88. Roudier F, Fernandez AG, Fujita M, Himmelspach R, Borner GH, et al. 2005. COBRA, an *Arabidopsis* extracellular glycosyl-phosphatidyl inositol-anchored protein, specifically controls highly anisotropic expansion through its involvement in cellulose microfibril orientation. *Plant Cell* 17:1749–63
89. Roy SJ, Holdaway-Clarke TL, Hackett GR, Kunkel JG, Lord EM, Hepler PK. 1999. Uncoupling secretion and tip growth in lily pollen tubes: evidence for the role of calcium in exocytosis. *Plant J.* 19:379–86
89a. Sardar HS, Yang J, Showalter A. 2006. Molecular interactions of arabinogalactan-proteins (AGPs) with cortical microtubules and F-actin in bright yellow-2 (BY-2) tobacco cultured cells. *Plant Physiol.* 142:1469–79
90. Scheres B, van Engelen F, van der Knaap E, van de Wiel C, van Kammen A, Bisseling T. 1990. Sequential induction of nodulin gene expression in the developing pea nodule. *Plant Cell* 2:687–700
91. Schindler T, Bergfeld R, Schopfer P. 1995. Arabinogalactan proteins in maize coleoptiles: developmental relationship to cell death during xylem differentiation but not to extension growth. *Plant J.* 7:25–36
92. Schultz CJ, Ferguson KL, Lahnstein J, Bacic A. 2004. Post-translational modifications of arabinogalactan-peptides of *Arabidopsis thaliana*. Endoplasmic reticulum and glycosylphosphatidylinositol-anchor signal cleavage sites and hydroxylation of proline. *J. Biol. Chem.* 279:45503–11
93. Schultz CJ, Gilson P, Oxley D, Youl JJ, Bacic A. 1998. GPI-anchors an arabinogalactan-proteins: implications for signaling in plants. *Trends Plant Sci.* 3:426–31
94. Schultz CJ, Johnson KL, Currie G, Bacic A. 2000. The classical arabinogalactan protein gene family of arabidopsis. *Plant Cell* 12:1751–68
95. Schultz CJ, Rumsewicz MP, Johnson KL, Jones BJ, Gaspar YM, Bacic A. 2002. Using genomic resources to guide research directions. The arabinogalactan protein gene family as a test case. *Plant Physiol.* 129:1448–63
96. Schulze-Lefert P. 2004. Knocking on heaven's wall: pathogenesis of and resistance to biotrophic fungi at the cell wall. *Curr. Opin. Plant Biol.* 7:377–83
97. Seifert GJ, Barber C, Wells B, Dolan L, Roberts K. 2002. Galactose biosynthesis in arabidopsis: genetic evidence for substrate channeling from UDP-D-galactose into cell wall polymers. *Curr. Biol.* 12:1840–45
98. Serpe MD, Nothnagel EA. 1994. Effects of Yariv phenylglycosides on rosa cell-suspensions-evidence for the involvement of arabinogalactan-proteins in cell-proliferation. *Planta* 193:542–50
99. Shahollari B, Varma A, Oelmuller R. 2005. Expression of a receptor kinase in arabidopsis roots is stimulated by the basidiomycete *Piriformospora indica* and the protein accumulates in Triton X-100 insoluble plasma membrane microdomains. *J. Plant Physiol.* 162:945–58

100. Shi H, Kim Y, Guo Y, Stevenson B, Zhu JK. 2003. The arabidopsis *SOS5* locus encodes a putative cell surface adhesion protein and is required for normal cell expansion. *Plant Cell* 15:19–32
101. Shimizu M, Igasaki T, Yamada M, Yuasa K, Hasegawa J, et al. 2005. Experimental determination of proline hydroxylation and hydroxyproline arabinogalactosylation motifs in secretory proteins. *Plant J.* 42:877–89
102. Shiu SH, Bleecker AB. 2001. Receptor-like kinases from arabidopsis form a monophyletic gene family related to animal receptor kinases. *Proc. Natl. Acad. Sci. USA* 98:10763–68
103. Showalter AM. 1993. Structure and function of plant-cell wall proteins. *Plant Cell* 5:9–23
104. Shpak E, Barbar E, Leykam JF, Kieliszewski MJ. 2001. Contiguous hydroxyproline residues direct hydroxyproline arabinosylation in *Nicotiana tabacum*. *J. Biol. Chem.* 276:11272–78
105. Shpak E, Leykam JF, Kieliszewski MJ. 1999. Synthetic genes for glycoprotein design and the elucidation of hydroxyproline-O-glycosylation codes. *Proc. Natl. Acad. Sci. USA* 96:14736–41
106. Somerville C, Bauer S, Brininstool G, Facette M, Hamann T, et al. 2004. Toward a systems approach to understanding plant cell walls. *Science* 306:2206–11
107. Sommer-Knudsen J, Clarke AE, Bacic A. 1996. A galactose-rich, cell-wall glycoprotein from styles of *Nicotiana alata*. *Plant J.* 9:71–83
108. Sommer-Knudsen J, Lush WM, Bacic A, Clarke AE. 1998. Re-evaluation of the role of a transmitting tract-specific glycoprotein on pollen tube growth. *Plant J.* 13:529–35
109. Stacey NJ, Roberts K, Carpita NC, Wells B, McCann MC. 1995. Dynamic changes in cell surface molecules are very early events in the differentiation of mesophyll cells from *Zinnia elegans* into tracheary elements. *Plant J.* 8:891–906
110. Stacey NJ, Roberts K, Knox JP. 1990. Patterns of expression of the JIM4 arabinogalactan-protein epitope in cell-cultures and during somatic embryogenesis in *Daucus carota* L. *Planta* 180:285–92
111. Steffan W, Kovac P, Albersheim P, Darvill AG, Hahn MG. 1995. Characterization of a monoclonal antibody that recognizes an arabinosylated (1→6)-β-D-galactan epitope in plant complex carbohydrates. *Carbohydr. Res.* 275:295–307
112. Stenvik GE, Butenko MA, Urbanowicz BR, Rose JK, Aalen RB. 2006. Overexpression of *INFLORESCENCE DEFICIENT IN ABSCISSION* activates cell separation in vestigial abscission zones in arabidopsis. *Plant Cell* 18:1467–76
113. Sun W, Kieliszewski MJ, Showalter AM. 2004. Overexpression of tomato LeAGP-1 arabinogalactan-protein promotes lateral branching and hampers reproductive development. *Plant J.* 40:870–81
114. Sun W, Xu J, Yang J, Kieliszewski MJ, Showalter AM. 2005. The lysine-rich arabinogalactan-protein subfamily in Arabidopsis: gene expression, glycoprotein purification and biochemical characterization. *Plant Cell Physiol.* 46:975–84
115. Sun W, Zhao ZD, Hare MC, Kieliszewski MJ, Showalter AM. 2004. Tomato LeAGP-1 is a plasma membrane-bound, glycosylphosphatidylinositol-anchored arabinogalactan-protein. *Physiol. Plant.* 120:319–27
116. Suzuki Y, Kitagawa M, Knox JP, Yamaguchi I. 2002. A role for arabinogalactan proteins in gibberellin-induced α-amylase production in barley aleurone cells. *Plant J.* 29:733–41
117. Svetek J, Yadav MP, Nothnagel EA. 1999. Presence of a glycosylphosphatidylinositol lipid anchor on rose arabinogalactan proteins. *J. Biol. Chem.* 274:14724–33
118. Takeda T, Fry SC. 2004. Control of xyloglucan endotransglucosylase activity by salts and anionic polymers. *Planta* 219:722–32

119. Tan L, Leykam JF, Kieliszewski MJ. 2003. Glycosylation motifs that direct arabinogalactan addition to arabinogalactan-proteins. *Plant Physiol.* 132:1362–69
120. Tan L, Qiu F, Lamport DT, Kieliszewski MJ. 2004. Structure of a hydroxyproline (Hyp)-arabinogalactan polysaccharide from repetitive Ala-Hyp expressed in transgenic *Nicotiana tabacum*. *J. Biol. Chem.* 279:13156–65
121. Tang W, Ezcurra I, Muschietti J, McCormick S. 2002. A cysteine-rich extracellular protein, LAT52, interacts with the extracellular domain of the pollen receptor kinase LePRK2. *Plant Cell* 14:2277–87
122. Thompson HJM, Knox JP. 1998. Stage-specific responses of embryogenic carrot cell suspension cultures to arabinogalactan protein-binding β-glucosyl Yariv reagent. *Planta* 205:32–38
123. Tiainen P, Myllyharju J, Koivunen P. 2005. Characterization of a second *Arabidopsis thaliana* prolyl 4-hydroxylase with distinct substrate specificity. *J. Biol. Chem.* 280:1142–48
124. Toonen MAJ, Schmidt EDL, Hendriks T, Verhoeven HA, van Kammen A, deVries SC. 1996. Expression of the JIM8 cell wall epitope in carrot somatic embryogenesis. *Planta* 200:167–73
125. van Buuren ML, Maldonado-Mendoza IE, Trieu AT, Blaylock LA, Harrison MJ. 1999. Novel genes induced during an arbuscular mycorrhizal (AM) symbiosis formed between *Medicago truncatula* and *Glomus versiforme*. *Mol. Plant Microbe Interact.* 12:171–81
126. van Hengel AJ, Guzzo F, Van Kammen A, de Vries HT. 1998. Expression pattern of the carrot EP3 endochitinase genes in suspension culture and developing seeds. *Plant Physiol.* 117:43–53
127. van Hengel AJ, Roberts K. 2002. Fucosylated arabinogalactan-proteins are required for full root cell elongation in arabidopsis. *Plant J.* 32:105–13
128. van Hengel AJ, Roberts K. 2003. AtAGP30, an arabinogalactan-protein in the cell walls of the primary root, plays a role in root regeneration and seed germination. *Plant J.* 36:256–70
129. van Hengel AJ, Tadesse Z, Immerzeel P, Schols H, van Kammen A, de Vries SC. 2001. N-acetylglucosamine and glucosamine-containing arabinogalactan proteins control somatic embryogenesis. *Plant Physiol.* 125:1880–90
130. van Hengel AJ, van Kammen A, de Vries SC. 2002. A relationship between seed development, Arabinogalactan-proteins (AGPs) and the AGP mediated promotion of somatic embryogenesis. *Physiol. Plant.* 114:637–44
131. Vicre M, Santaella C, Blanchet S, Gateau A, Driouich A. 2005. Root border-like cells of *Arabidopsis*. Microscopical characterization and role in the interaction with rhizobacteria. *Plant Physiol.* 138:998–1008
132. Vissenberg K, Feijo JA, Weisenseel MH, Verbelen JP. 2001. Ion fluxes, auxin and the induction of elongation growth in *Nicotiana tabacum* cells. *J. Exp. Bot.* 52:2161–67
133. Vorwerk S, Somerville S, Somerville C. 2004. The role of plant cell wall polysaccharide composition in disease resistance. *Trends Plant Sci.* 9:203–9
134. Waller F, Achatz B, Baltruschat H, Fodor J, Becker K, et al. 2005. The endophytic fungus *Piriformospora indica* reprograms barley to salt-stress tolerance, disease resistance, and higher yield. *Proc. Natl. Acad. Sci. USA* 102:13386–91
135. Willats WG, Knox JP. 1996. A role for arabinogalactan-proteins in plant cell expansion: evidence from studies on the interaction of β-glucosyl Yariv reagent with seedlings of *Arabidopsis thaliana*. *Plant J.* 9:919–25
136. Wilson CW, Chuang PT. 2006. New "hogs" in Hedgehog transport and signal reception. *Cell* 125:435–38

120. An impressively detailed structure of an AGP-carbohydrate.

129. Highlights the potential role of glycoside hydrolases acting on AGP carbohydrate, but molecular confirmation is needed.

137. Wu HM, Wang H, Cheung AY. 1995. A pollen tube growth stimulatory glycoprotein is deglycosylated by pollen tubes and displays a glycosylation gradient in the flower. *Cell* 82:395–403

138. Wu HM, Wong E, Ogdahl J, Cheung AY. 2000. A pollen tube growth-promoting arabinogalactan protein from nicotiana alata is similar to the tobacco TTS protein. *Plant J.* 22:165–76

139. Yariv J, Rapport MM, Graf L. 1962. The interaction of glycosides and saccharides with antibody to the corresponding phenylazo glycosides. *Biochem. J.* 85:383–88

140. Youl JJ, Bacic A, Oxley D. 1998. Arabinogalactan-proteins from *Nicotiana alata* and *Pyrus communis* contain glycosylphosphatidylinositol membrane anchors. *Proc. Natl. Acad. Sci. USA* 95:7921–26

140. The discovery of GPI-anchors on AGPs revealed a crucial structural feature.

141. Zhang Y, Brown G, Whetten R, Loopstra CA, Neale D, et al. 2003. An arabinogalactan protein associated with secondary cell wall formation in differentiating xylem of loblolly pine. *Plant Mol. Biol.* 52:91–102

RELATED REVIEWS

Turner S, Gallois P, Brown D. 2007. Tracheary element differentiation. *Annu. Rev. Plant Biol.* 58:407–33

Stomatal Development

Dominique C. Bergmann[1] and Fred D. Sack[2]

[1]Biological Sciences, Stanford University, Stanford, California 94305; email: dbergmann@stanford.edu

[2]Botany, University of British Columbia, Vancouver, Canada BC V6T 1Z4; email: fsack@interchange.ubc.ca

Annu. Rev. Plant Biol. 2007. 58:163–81

The *Annual Review of Plant Biology* is online at plant.annualreviews.org

This article's doi: 10.1146/annurev.arplant.58.032806.104023

First published online as a Review in Advance on January 2, 2007

1543-5008/07/0602-0163$20.00

Key Words

patterning, division, differentiation, receptor, guard cell

Abstract

Stomata are cellular epidermal valves in plants central to gas exchange and biosphere productivity. The pathways controlling their formation are best understood for *Arabidopsis thaliana* where stomata are produced through a series of divisions in a dispersed stem cell compartment. The stomatal pathway is an accessible system for analyzing core developmental processes including position-dependent patterning via intercellular signaling and the regulation of the balance between proliferation and cell specification. This review synthesizes what is known about the mechanisms and genes underlying stomatal development. We contrast the functions of genes that act earlier in the pathway, including receptors, kinases, and proteases, with those that act later in the cell lineage. In addition, we discuss the relationships between environmental signals, stomatal development genes, and the capacity for controlling shoot gas exchange.

Contents

INTRODUCTION

Stomata are epidermal valves that are essential for plant survival because they control the entry of carbon dioxide assimilated in photosynthesis and optimize water use efficiency (58, 63). Collectively, these valves enhance plant performance and influence global water and carbon cycles. Each stoma originates through a series of divisions that generate the stomatal spacing pattern, control stomatal frequency, and produce the two guard cells of the valve (49). The importance of these divisions is underscored by the finding that all genes shown to act during stomatal development function by regulating division, with different sets of genes controlling different transitions in the progression toward terminal cell fates. This review focuses on *Arabidopsis thaliana*, the source of most currently known "stomatal" genes.

Meristemoid: intermediate precursor cell that undergoes one or more amplifying asymmetric divisions that regenerate the meristemoid and increase the number of larger daughter cells capable of founding the next generation in the lineage; converts into a GMC

Major research efforts have elucidated the signal transduction pathways that control the movement of mature stomata, especially in response to abiotic stresses (66) at the whole-plant, canopy, and global levels. For developmental biology, this technically accessible epidermal system is valuable for understanding how plant cells are specified and patterned and how these events are integrated with specific division modes. For example, relatively little is known about how division polarities are generated and oriented in plants compared with animals and yeast, but the stomatal pathway is a promising system for studying these questions.

Stomatal Cell Lineage

Stomata are produced by a dedicated and specialized cell lineage (24, 49, 50, 84). This lineage is prevalent in the developing shoot epidermis (such as in young leaves) and is inactive after epidermal maturation. The lineage starts with an asymmetric division and ends with a symmetric one (**Figure 1**). The former division marks pathway entry and produces a small stomatal precursor cell, the meristemoid. The latter produces the two cells of the stoma. Between these events, the meristemoid converts into a guard mother cell (GMC), the end-stage precursor cell. Meristemoids usually divide asymmetrically several times, whereas GMCs divide just once symmetrically. Thus, a stoma is produced after a series of cell fate changes with each precursor cell undergoing a specific cell division program.

Each asymmetric division produces a meristemoid and a larger sister cell. The latter can divide or become a pavement cell, the generic type of cell in the epidermis. Asymmetric divisions in the lineage can be grouped by context and function into three types. Entry divisions occur in division-competent postprotodermal cells called meristemoid mother cells (MMCs) and

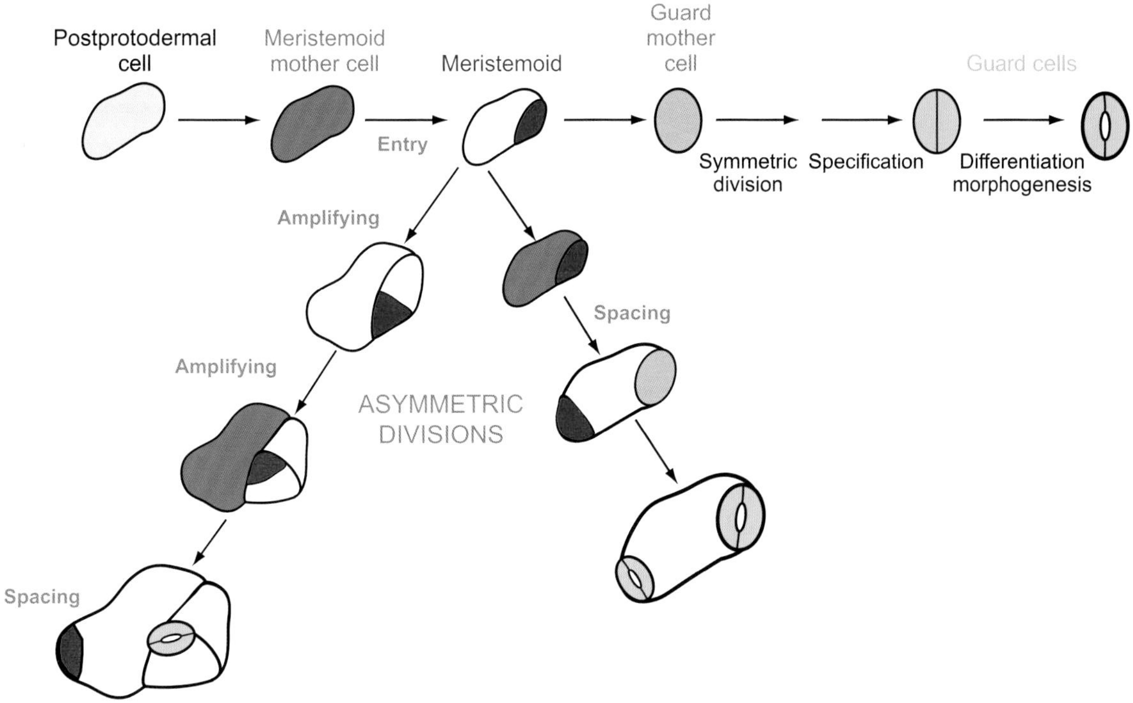

Figure 1

Diagram of key stages and divisions in *Arabidopsis thaliana* stomatal development. Protodermal cells in the epidermis are converted into meristemoid mother cells (MMCs) through an unknown process. MMCs undergo an asymmetric entry division to create a meristemoid. Meristemoids may undergo additional asymmetric amplifying divisions, or convert into a guard mother cell (GMC). The GMC will divide a single time, symmetrically, to form the two guard cells. Later, morphogenesis and pore formation create the mature stoma. The division process is reiterative. Cells next to meristemoids, GMCs, and guard cells can become MMCs and undergo spacing divisions to create new meristemoids. The plane of this division is oriented so that the new meristemoid is placed away from the preexisting stoma or precursor cell.

initiate the stomatal lineage. Amplifying divisions occur in meristemoids and increase the number of larger daughter cells, thereby increasing the epidermal cell total. The meristemoid is regenerated in each amplifying division. Spacing divisions take place in cells next to a stoma or precursor, and establish the one-cell spacing pattern in which stomata do not directly contact each other. Both entry and spacing divisions produce new meristemoids and thus directly increase the total number of stomata formed. Amplifying divisions increase stomatal number indirectly by producing larger daughter cells that undergo spacing divisions. The net effect of all these divisions is an increase in the number of pavement cells as well as stomata. Output estimates suggest that most of the epidermal cells in a leaf are generated by the stomatal lineage (24).

The number of stomata produced depends on the frequency of the different types of asymmetric divisions. If no larger daughter cells divide then only one stoma will be produced per lineage even if there are many amplifying divisions. If larger daughter cells do divide (via spacing divisions) then the initial entry division can lead to many generations of stomata (2, 26). These variations provide a developmental mechanism to generate the observed differences in stomatal number between different parts of a plant (for example, between the adaxial and abaxial leaf

Guard mother cell (GMC): last precursor cell in the lineage; it divides symmetrically, producing the two guard cells of a stoma

Pavement cell: the general or ground type of epidermal cell; often produced by asymmetric divisions in the stomatal pathway

Meristemoid mother cell (MMC): earliest precursor cell that founds the cell lineage by undergoing an entry asymmetric division

Spacing division: an asymmetric division in a neighbor cell that establishes the one-celled stomatal spacing pattern when the new wall is placed away from the stoma or precursor cell

Neighbor cell: an epidermal cell adjacent to a stoma, meristemoid, or GMC

epidermis, or between stems and leaves), between plants in bright and shaded environments, and among different taxa (25, 26, 62).

The attributes and timing of asymmetric and symmetric divisions provide a rough framework for analyzing the function of genes during stomatal development. Earlier-acting genes largely control asymmetric divisions and thus cell proliferation, patterning, and stomatal number. Later-acting genes function at or after the terminal symmetric division in processes such as stopping cell proliferation and promoting stomatal morphogenesis. Genes that act in mature stomata are considered elsewhere in this volume (66).

Patterning

Although stomatal density varies, virtually all stomata are separated by at least one intervening cell (24, 60). This one-celled spacing is probably adaptive in generating solute reservoirs between stomata, in minimizing mechanical interference between adjacent valves, and in optimizing the distribution of gas diffusion shells. The spacing pattern arises when a new asymmetric division takes place next to an existing stoma (24). Here the division is oriented so that the new wall is placed away from the stoma, resulting in the larger daughter cell—not the new meristemoid—contacting the stoma.

Patterning likely involves the transmission of spatial cues from the stoma to the adjacent cell that are used to correctly orient the plane of the spacing division. Patterning is unlikely to require mitosis-allocated factors because stomata are correctly spaced even when the cells involved are clonally unrelated. Intercellular signaling probably occurs through the free space of the cell wall (apoplasm) because mature stomata can signal even though they lack plasmodesmata. In addition to the divisions occurring next to stomata, divisions next to stomatal precursors (meristemoids and GMCs) are also correctly oriented, suggesting that these cells can also broadcast spatial cues (24, 43). The earliest marker of a cell about to undergo a spacing division is the appearance of a preprophase band of microtubules surrounding a nucleus that is located away from a neighboring stoma or precursor cell (43).

Many tissues are patterned by interactions between differentiated cells and their immature neighbors (21). Some plant epidermal cells, like trichomes and root hairs, are probably patterned by lateral inhibition (38). The stomatal spacing mechanism differs from lateral inhibition because the neighbor cell is not prevented from acquiring a particular cell fate (49). Neither is stomatal spacing due to the generation of a border of surrounding nonstomatal cells via a series of stereotyped asymmetric divisions. In fact, the neighbors, far from providing a "boundary," are more likely to produce stomata than other epidermal cells. Instead, spacing appears to result from cell-cell signaling that orients the plane of asymmetric division in cells situated next to a stoma or precursor.

A secondary spacing module generates the "anisocytic" arrangement of cells around the stoma typical of the Brassicaceae (including *Arabidopsis*) (54, 55). Here three successive asymmetric divisions form an inward spiral (23, 65). Although this pattern is common in *Arabidopsis*, it is, in contrast to the one-cell spacing pattern, not invariant. Because these spiral divisions all take place in same-cell lineage, the mechanism could involve the control of wall placement by landmarks deposited during successive mitoses—a process similar to that used in yeast bud site selection.

EARLY-ACTING GENES

Perhaps the most dramatic recent progress in the field of stomatal development has been the identification of genes that control the production and spacing of *Arabidopsis* stomata. These genes, which comprise putative receptors, a processing protease, and a kinase, act primarily by modulating the number and placement of asymmetric divisions in the stomatal cell lineage.

Receptor-Mediated Cell-Cell Signaling and Pattern Formation

Mutations in the *TOO MANY MOUTHS (TMM)* gene lead to alterations in all types of asymmetric divisions in the stomatal lineage. *tmm* mutants form excess stomata in leaves (see **Figure 2**) indicating that the normal role of *TMM* is to repress divisions. *tmm-1* mutants fail to orient the asymmetry of spacing divisions, fail to inhibit asymmetric divisions in cells adjacent to two or more stomata or their precursor cells, and have a reduced number of amplifying divisions, which leads to the premature conversion of meristemoids into GMCs (24). The first two of these *tmm-1* phenotypes appear to arise from a failure of neighbor cells to respond to positional cues.

TMM encodes a putative cell-surface receptor that is expressed in meristemoids, in GMCs, and in sister cells of the asymmetric divisions that create meristemoids (49). The protein also appears in cells likely to undergo entry division. *TMM* expression is absent from fully differentiated guard cells and from pavement cells. This expression pattern is consistent with TMM normally being required for stomatal lineage cells to perceive and respond to signals that control the number and orientation of spacing divisions.

Receptor-mediated signal transduction in *Arabidopsis* follows the logic of animal signaling, but does not employ the receptor tyrosine kinases typical of animals. In *Arabidopsis*, more than 200 receptor-like kinases share a common structure of leucine-rich extracellular domains, a single transmembrane domain, and a cytoplasmic kinase (LRR-RLKs) (67, 76). TMM is a member of a related class of proteins that contains extracellular LRRs and a transmembrane domain, but lacks the kinase domain (LRR-RLPs) (48). Based on studies of the shoot meristem size control genes *CLAVATA1 (CLV1)* and *CLAVATA2 (CLV2)* that encode an LRR-RLK and LRR-RLP, respectively, it was hypothesized that LRR-RLPs require an LRR-RLK to participate in signal transduction (16).

ERECTA (ER) is an LRR-RLK required for diverse processes including growth and development, as well as responses to biotic and abiotic stresses (39, 68). Many processes affected by loss of ER function are united in their requirement for cell proliferation. Expression of a form of ER missing the kinase domain (ER-Δkinase) yields growth phenotypes more severe than in *er* null mutants, suggesting that the truncated ER protein acts as a dominant negative, probably by forming unproductive heterodimers with other LRR-RLKs (69). Likely partners are *ER's* closest homologs *ERECTA-LIKE 1 (ERL1)* and *ERECTA-LIKE 2 (ERL2)*, which can each partially rescue the *er* growth phenotype when driven by the *ER* promoter (68). Under normal growth conditions neither *erl1* or *erl2* single mutants nor the *erl1;erl2* double mutant has any obvious growth phenotype, but *er;er1;er2* triple mutants are dwarfed and sterile (68).

Examination of the *er;erl1;erl2* shoot epidermis reveals striking stomatal overproliferation and spacing defects (70). These phenotypes, like the growth phenotype, reveal functional redundancy. For example, only the triple mutant shows stomatal spacing defects, suggesting that the activity of any ER-family member on its own is sufficient for pattern generation. However, ER, ERL1, and ERL2 have subtly different roles in epidermal development. Meristemoid differentiation, for example, is consistently inhibited by ERL1 and is often promoted by ER. ER also appears to function in repressing entry divisions. The contrasting phenotypes of different double and triple *ER*-family mutant combinations highlight different genetic requirements during various stages of stomatal development.

TMM and ER-family members control asymmetric divisions in stomatal development. *TMM* and the *ER*-family also have overlapping domains of gene expression. ER is rather broadly expressed, but *TMM*, *ERL1*, and *ERL2* all show overlapping expression patterns in aerial organs (48, 70). LRR-RLKs are surmised to act as homodimers

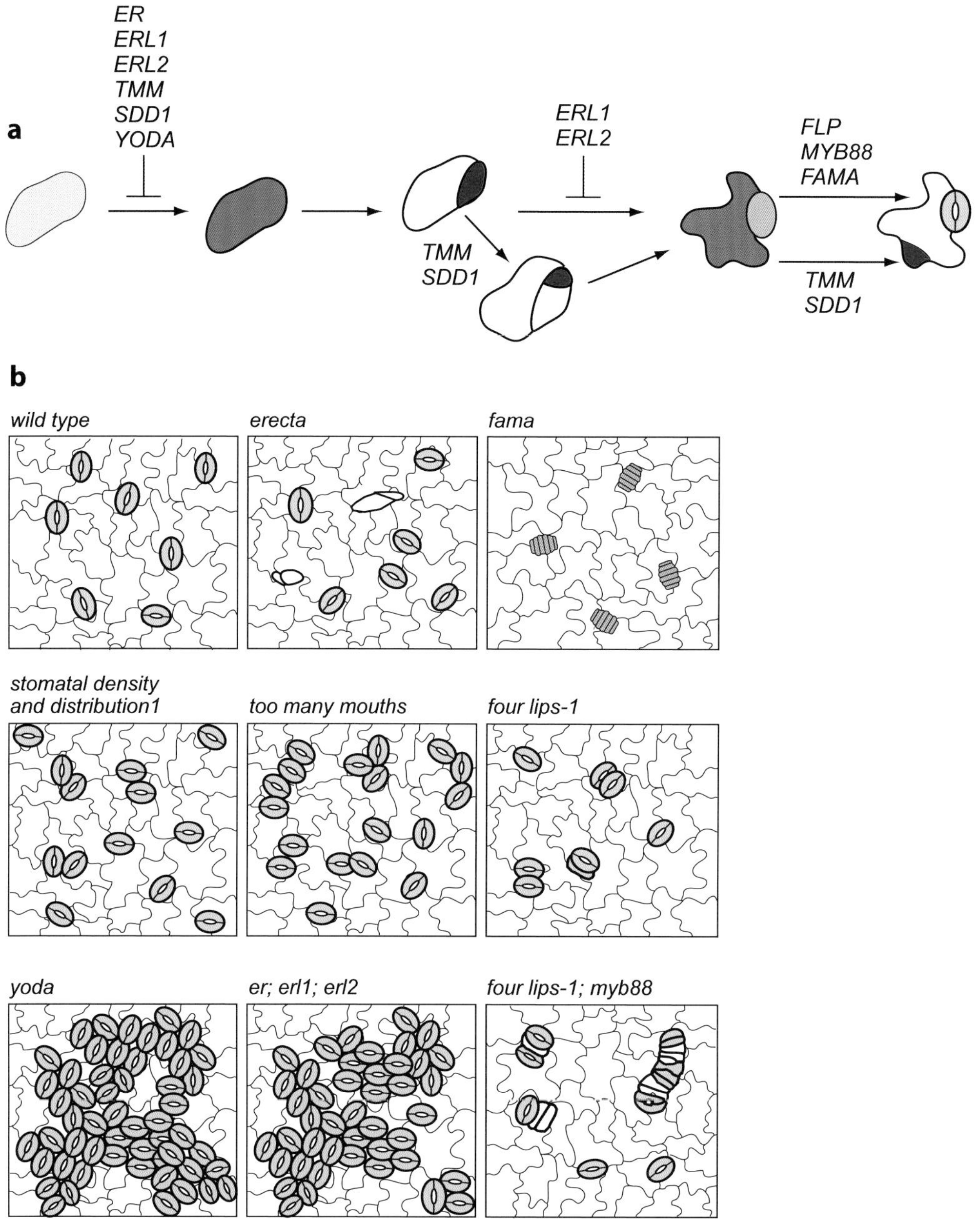

Figure 2

Genetic control of stomatal development. (*a*) Summary of the major stomatal development stages annotated with the presumed point of action of genes. Negative regulation is indicated by T-shaped lines; positive regulation is indicated when just the gene abbreviation is shown. Note that with the current markers, it is not possible to determine whether mutations affect the transition from protodermal cell to meristemoid mother cell (MMC), or the ability of a MMC to divide asymmetrically. (*b*) Diagrams of terminal leaf phenotypes in stomatal mutants indicating the typical number and arrangement of stomata (*green*) or terminal cell type (*pink* for guard mother cells). White cells in *erecta* and *flp;myb88* panels represent cells of indeterminate identity.

or heterodimers with other LRR-RLKs or LRR-RLPs. To work as a complex, TMM and the ER-family proteins must also have overlapping subcellular localizations. TMM-GFP appears to be expressed in both the cell membrane and the endoplasmic reticulum of stomatal lineage cells (48). The localization of the ER-family proteins has yet to be determined.

Genetic data from the CLV signaling system suggest that the association of an RLP with its partner RLK serves to activate the complex (32, 73). However, TMM has contrasting functions in different organs that suggest that if TMM does form a complex with the ER-family, it cannot consistently serve as an activator. In stems, *TMM* has an opposite mutant phenotype to that of the *ER*-family and the published genetic interactions do not distinguish whether TMM is independent of or in the same complex as any member of the ER-family. In siliques, a neomorphic phenotype (failure of stomatal lineage cells to differentiate into guard cells) appears in plants doubly mutant for *TMM* and *ER*. Neomorphism is not expected when two proteins are exclusive partners, but it can arise when two proteins compete for a common partner. In this specific case, ERL1 was proposed as a possible target of TMM inhibition (via the formation of inactive heterodimers) (70). To be consistent with the epistatic relationships among *ER*, *ERL1*, and *TMM*, the "ERL1 target" model requires there to be exquisite control on the levels of each receptor in silique stomatal lineage cells. Quantitative data on TMM and ER-family protein levels and dosage studies using heterozygotes and mild overexpression would allow this interaction model to be tested more rigorously. Other, currently unidentified, partners might also be involved in the tissue-specific regulation of stomatal formation.

Signaling Through a MAP Kinase Cascade

Regardless of whether the *ER*-family and *TMM* participate in shared signaling complexes, loss of these proteins leads to changes in multiple cellular behaviors including altered gene expression and division plane determination. Intracellular signaling is required to transduce signals from the cell periphery to nuclear and cytoplasmic targets. The signaling cascades downstream of plant LRR-RLKs are diverse and not easily predicted by the sequence of the receptors. For example, two well-studied pathways employing LRR-RLKs—brassinosteroid signaling and response to bacterial pathogens—utilize widely conserved, but distinct intracellular signaling cascades (1, 22, 77).

A mitogen-activated protein (MAP) kinase signaling pathway has been implicated in the control of cell division and cell fate during stomatal development (3). Loss-of-function mutations in the MAP kinase kinase kinase (MAPKKK) gene *YODA* profoundly alter stomatal density and spacing. The phenotype of the *yoda* mutant is similar to the *er;erl1;erl2* triple mutant and includes a severe reduction in overall plant height and internode length as well as excess production of guard cells (3, 70). Cells in the epidermis of *yoda* cotyledons exhibit excessive entry divisions, fail to orient spacing divisions and fail to prevent division of neighbor cells that contact two cells of the stomatal lineage. Asymmetry of cell fates is also compromised in *yoda;* often both daughters of a stomatal lineage cell division become stomata without the obvious production of an intervening cell.

YODA is a member of a class of MAPKKKs that possess a long N-terminal extension with negative regulatory activity (44). Expression of N-terminally deleted YODA (*CA-YODA)* results in dose-dependent effects on stomatal development—in the strongest lines, no stomata are produced (3). The block in stomatal development occurs early and results in the production of a leaf epidermis composed entirely of pavement cells and occasional trichomes. *CA-YODA/+* is capable of suppressing the *tmm-1* mutant phenotype, consistent with YODA acting downstream in a common signaling pathway, but it cannot be ruled out that YODA participates in an independent and parallel signaling pathway in young epidermal cells. A physical or biochemical connection between the YODA MAPKK kinase and the stomatal regulators at the membrane will be needed to answer this question, but obtaining this information may be challenging. Even in the best-studied *Arabidopsis* LRR-RLK/MAPK pathways, it is not clear whether the LRR-RLK directly phosphorylates the MAPKKK or whether phosphorylation requires an intermediary protein or protein complex (1).

MAP kinase cascades are organized into a core module of three protein kinases consisting of a MAPKKK, a MAP kinase kinase (MKK), and a MAP kinase (MPK). Transmission of a signal to downstream targets is achieved by sequential phosphorylation and activation of the core MAP kinase components. One or more MKKs and MPKs are predicted to act downstream of *YODA*. Finding the distinct downstream kinases is complicated by the fact that multiple, interdependent, MAP kinase pathways are at work in *Arabidopsis* cells. The *Arabidopsis* genome is predicted to contain 20 MEKK1/STE11 class MAPKKKs, 10 MAPKK genes, and 20 MAPK genes (28). Despite this gene family expansion, which might provide the raw material for specialization, several lines of evidence suggest that the specific outcomes of *Arabidopsis* MAPK signaling do not arise from the use of dedicated kinases for each biological event. For example, redundant kinase pairs MKK4/5 and MPK3/6 are required for biological processes as diverse as immune recognition (1) and responses to hormones and ozone (42). No single MKK or MPK mutant has been reported to have a significant effect on stomatal development suggesting overlapping or compensatory functions for MPKs and MKKs in this biological process. It will be intriguing to see whether, like with the ER-family kinases, mutations in multiple MKK or MPK genes reveal roles in stomatal development.

Receptor-Ligand Interactions

Stomatal receptors may induce cellular responses by activating a MAP kinase cascade, but how are they themselves activated? Animal receptor tyrosine kinases dimerize, and ligand binding induces phosphorylation of the receptors and activation of downstream signaling pathways. Plant LRR-RLKs probably share some of these activities based on genetic and protein interaction studies. Evidence from the brassinosteroid perception pathway suggests that plant LRR-RLKs may not need a ligand to dimerize (59). Therefore, physical interactions among ER-family members or between this trio and *TMM* could be ligand-independent. Yet even if it is possible to activate LRR-RLKs without a ligand, stomatal development and patterning require positional information. TMM localization in the cell membrane of stomatal lineage cells is uniform (49). It is difficult to imagine how TMM (or any receptor) could instruct cell division orientation unless its activation was restricted to only a small part of the cell periphery. A simple way to selectively activate a receptor is to provide a spatially restricted ligand from a neighboring cell.

LRR-RLKs respond to a diverse set of ligands including steroids [brassinosteroid (35)], small secreted proteins [CLV3 (20)], and exogenous peptides [flagellin (42)]. Indirect evidence for a protein-based signal involved in stomatal development comes from analysis of the subtilisin protease *STOMATAL DENSITY AND DISTRIBUTION 1 (SDD1)*. *sdd1* mutants undergo excessive entry divisions and fewer amplification divisions and can fail to orient spacing divisions (2). SDD1 appears to be secreted from the cell and is expressed in meristemoids and GMCs (78). Overexpression of *SDD1* represses stomatal divisions and also causes arrest of meristemoids and GMCs. Although *sdd1* mutants increase the stomatal index and misorient spacing divisions, the overall stomatal phenotypes of *sdd1* are distinct from those in *tmm*, *yoda*, and the *er*-family. However, the *SDD1* overexpression phenotype depends on a functional *TMM* (78), and loss of *SDD1* function can be dominantly suppressed by *CA-YODA* (3). All of these data point to interdependent relationships among known early-acting stomatal genes, but also indicate that placing them in a single linear pathway is likely to be oversimplistic.

Stomatal Pattern by Negative Regulation

All of these earlier-acting genes are broadly required for the number, distribution, and patterning of stomata. However, the cellular mechanisms by which these genes control stomatal development may vary. For example, TMM and SDD1 directly generate pattern by orienting spacing divisions (2, 24), whereas patterning defects may be a secondary consequence of the cell fate transformations in *YODA* mutants (3). The genes discussed in this section encode proteins that are predicted to act in inter- and intracellular signaling. Moreover, they all act essentially as negative regulators of stomatal formation. While this review was in press, several transcription factors were identified that actively promote stomatal development at the entry, proliferation, and differentiation stages (44a, 52a, 57a). The regulatory relationships between these proteins and the negative regulators discussed here have yet to be determined.

Dispersed and Specialized Stem Cell Compartment

The stomatal cell lineage can be considered a specialized stem cell compartment (50). Like multipotent stem cells, the lineage produces only a few mature cell types (guard cells and pavement cells), and the stem cell population can be renewed when a larger daughter cell produced by asymmetric division undergoes an entry division. This compartment is marked by TMM expression, which is found in both daughter cells produced by all asymmetric divisions in the lineage. ERL1 and ERL2 expression overlap with that of TMM, but these genes are also more widely expressed (70).

A unifying way to think about the *tmm*, *sdd1*, *yoda*, *er*, *erl1*, and *erl2* phenotypes is to look at the relative and combinatorial effects of these genes in forming and/or maintaining the stem cell–like compartment. Compartment size increases when cells undergo entry and amplifying divisions and decreases when cells terminally differentiate. How protodermal cells are chosen to enter this compartment is unknown, and cells so chosen cannot be distinguished by size or location. The above genes mostly restrict compartment size, for example by limiting the number of entry divisions (**Figure 2**). But many of these genes also positively regulate compartment size by promoting amplifying divisions. Thus, the number of stomata produced depends on the net integration of genetic and signaling inputs at each developmental node in the pathway.

Although TMM is expressed in this specialized compartment, it is not required for the specification of the cell types in it. Instead, it seems to receive signals that modulate the placement and number of divisions in the compartment in a cell-type and cell-position appropriate manner. This stem cell/TMM-marked compartment exists in a spatially dispersed and temporally transient developmental window in the postprotodermal shoot epidermis (48). The compartment is critical for leaf development because many epidermal cells are produced in the stomatal cell lineage. This function was graphically demonstrated when the *TMM* promoter was used to drive the expression of a cyclin-dependent kinase inhibitor (KRP1), which blocks cell cycle progression, resulting in a severe reduction in asymmetric divisions and in much smaller leaves (80).

The stomatal stem cell compartment differs from apical meristems, which are coherent and perpetually active groups of stem cells that produce the progenitors of all shoot and root cells (79). Bünning recognized the similar but more restricted division potential of the stomatal compartment by calling the distinctive precursor cell a meristemoid instead of a meristem (8). Other dispersed self-renewing populations of cells such as the procambium and the cork cambium (which produce the primary vascular tissues and the bark, respectively) act outside the apical meristems. Study of the stomatal pathway should contribute to understanding how all such dispersed and specialized stem cell compartments help build the plant.

LATE-ACTING GENES

After asymmetric divisions initiate and populate the cell lineage and establish the spacing pattern, the developmental program changes to symmetric division and terminal differentiation. Genes known to act during these later stages help end cell proliferation, execute GMC cytokinesis, and regulate the timing of guard cell specification and differentiation.

Ending Cell Proliferation

A common feature of eukaryotic cell lineages is that precursor cells divide a limited number of times and then differentiate into specialized cell types (41, 75). The symmetric division of the GMC is the last division in the stomatal cell lineage. The resulting daughter cells mature into guard cells that withdraw from the cell cycle either in G_1 or G_0 (46). The relative timing of cell cycle withdrawal and guard cell specification has not been established. However, ending cell cycling before terminal differentiation appears to be tightly regulated and adaptive for valve function because stomata in virtually all plant taxa consist of just two cells.

Three putative transcriptional regulators, FOUR LIPS (FLP), MYB88, and FAMA, restrict cell cycling at the end of the stomatal lineage (36). Loss-of-function mutations in *FLP* induce clusters of laterally aligned cells that are clonal in origin. These clusters result from the reiteration of a GMC program in daughter cells that would usually differentiate directly into guard cells. The excess divisions delay rather than block stomatal specification because clusters contain normal and arrested stomata. *FLP* encodes an R2R3 MYB protein whose expression starts before GMC mitosis and is downregulated as guard cells differentiate. FLP limits GMCs to one symmetric division and promotes a timely transition to terminal differentiation. As a putative transcription factor, FLP could halt proliferation directly by regulating the expression of cell cycle genes, and/or indirectly by promoting the developmental transition to a terminal cell fate.

MYB88 and *FLP* (*MYB124*) are paralogs (36). The genes overlap in function and expression. *myb88* mutants do not have a phenotype alone, but enhance the stomatal cluster phenotype of *flp*, and extra copies of the *MYB88* genomic region complement *flp*. These data hint at a possible gene dosage mechanism where a threshold level of FLP and/or MYB88 controls the number of symmetric divisions at the end of the cell lineage.

Mutations in *FAMA*, which encodes a basic helix-loop-helix (bHLH) protein, cause clusters reminiscent of severe *flp* alleles and of *flp/myb88* double mutants, suggesting that *FAMA* also limits symmetric divisions at the end of the stomatal cell lineage (3). However, unlike cells in *flp* clusters, those in *fama* lack cytological traits characteristic of guard cells, revealing that *FAMA* is also required for proper cell specification and differentiation. FAMA is also sufficent to promote at least partial guard cell identity because FAMA overexpression or misexpression in ectopic domains causes cells to take on a guard cell morphology and express molecular markers of guard cell identity (52a).

The specification of other epidermal cell types, such as trichomes and root hairs, involves the physical interaction of R-like bHLH proteins with R2R3 MYB proteins to form a complex that transcriptionally activates cell fate factors (4, 57). The phenotypes of *FLP*, *MYB88*, and *FAMA* mutants and the genes' overlapping expression patterns in the late stages of stomatal development suggest that they might act in the same transcriptional complex to limit divisions and to promote guard cell specification. However, tests of both genetic and physical interactions between FAMA and FLP/MYB88 indicate that these particular bHLH and MYB proteins work independently (52a). FLP/MYB88 might have a primary role in cell division that feeds back on differentiation, whereas FAMA is likely to be responsible for differentiation and indirectly for cell cycle control (52a). In addition, FAMA is not an R-like bHLH protein, and FLP/MYB88 lacks the amino acid signature implicated in MYB binding to such bHLH proteins (29, 85). Whereas early-acting genes directly establish stomatal patterning via asymmetric division, *FLP*, *MYB88*, and *FAMA* act only indirectly in patterning by preventing extra symmetric divisions in correctly positioned GMCs.

Division and Differentiation: Independent and Coordinated

The final stages of stomatal development comprise GMC cytokinesis, guard cell specification and differentiation, and stomatal morphogenesis. The latter includes the elaboration of pore thickenings in each guard cell followed by the controlled separation of their walls to form the opening of the valve.

These events are normally coordinated in time and space, but many of them can be uncoupled artificially. Because the GMC division is stereotyped, cytokinetic defects can be readily identified when division is mutationally blocked (5, 19, 31, 71, 72, 83). These studies show that guard cell specification, differentiation, and morphogenesis can all continue without proper GMC cytokinesis (31). Specification markers include *KAT1* (potassium channel) gene expression and microtubule arrays that radiate out from the future pore site (22, 43, 51, 53). GMCs with defective division still acquire a guard cell identity (19, 74). In the absence of GMC cytokinesis, wall deposition and secretion can continue, but are redirected as evidenced by the presence of abnormal swellings in the center of the wall facing the atmosphere (6, 71, 83). Single cells can even become kidney shaped in the absence of GMC division, such as when FAMA or a dominant negative version of *CYCLIN-DEPENDENT KINASE B1;1* (*CDKB1;1*) is overexpressed (6, 52a).

Excess divisions, as well as no division, can also permit guard cell morphogenesis. Extra symmetric divisions in *flp* GMCs still allow the formation of normally shaped, although ectopic, stomata (36). Stomatal morphogenesis can also tolerate an abnormal extra division in guard cells as shown by the formation of four-celled stomata in dark-grown cucumber hypocotyls transiently exposed to ethylene and red light (33, 34).

That these processes can be artificially separated underscores the temporal and spatial coordination normally needed for stomatal morphogenesis. Some guard cell differentiation can take place without cell division, but forming a functional valve requires that equal-sized daughter cells be produced and specified at the same time. The beautiful and adaptive mirror-like symmetry of the stoma might arise cell autonomously, or it might use signaling between developing guard cells to orchestrate wall deposition and pore formation.

Cell Cycle Regulators

Stomatal lineage proteins like TMM, FLP, and FAMA presumably regulate division by interacting directly or indirectly with the cell cycle machinery. The identities of most such cell cycle regulators are unknown because loss-of-function phenotypes are often uninformative due to lethality or redundancy and because gene overexpression has mostly not produced a dramatic stomatal phenotype (7, 15, 18).

Three classes of cell cycle regulators have been implicated in the stomatal pathway so far. *CDKB1;1* positively regulates stomatal production in addition to promoting GMC mitosis and cytokinesis as described above (6, 7).

The second class includes *CDT1* and *CDC6*, which regulate the licensing of origins of replication. Both are normally expressed in stomatal precursor cells, and their overexpression increases stomatal density twofold, suggesting that stomatal fate acquisition is normally kept in check by regulating the number of cells permitted to initiate DNA replication (11). The third class includes the RETINOBLASTOMA RELATED (RBR) protein, which represses the activity of the heterodimeric transcription factor complex E2F-DP. The number of asymmetric divisions in the stomatal lineage increases strongly when E2Fa is overexpressed as well as when RBR is inducibly inactivated (13, 14, 56). RBR inactivation by virus-induced gene silencing also results in *tmm*-like stomatal clusters (56), raising the possibility that TMM restricts asymmetric divisions via the RBR, E2F, and DP pathway. RBR was recently shown to play a major role in maintaining stem-cell competence in root apical meristems (81), highlighting some common molecular requirements among the self-renewing cell populations in plants.

Based on their phase-specific expression pattern, other cell cycle genes likely act in stomatal pathway, including *CDKA;1* (CDC2a) and *CYCLINA2;2*, *2;3*, and *B1;1* (formerly cyca1At) (6, 30, 64). All of these are also expressed outside the stomatal pathway; there are no reports of a cell cycle regulator whose expression is restricted to the stomata lineage. Of particular interest will be determining how stomatal developmental regulators work with the cell cycle machinery. One could imagine that cell cycle regulators are regulated transcriptionally by FLP and FAMA, or posttranslationally by phosphorylation of kinases downstream of YODA or the ER-family. More broadly, the stomatal system is valuable for studying cell cycles in multicellular development because division behavior and gene expression can be visualized in living tissues and because the pathway includes a rich sampling of division types and events.

INFLUENCE OF ENVIRONMENT

The experimental focus on the "developmental" genes that act within the epidermis to control cell identity and division behavior will identify many of the signaling pathways and cell autonomous factors required for stomatal development and pattern. However, interactions with underlying tissues and the environment also influence the final density and distribution of stomata. These long-range signals could act by modulating the activity of genes like *TMM*, *YODA*, *SDD1*, and the *ER*-family or they could impinge directly on the cell cycle machinery and other downstream targets. In this section we examine the nature of the environmental response including potential signals and possible connections between regulation at the levels of development and physiology.

Inputs from Old to New

Paleontologists and ecophysiologists have long noticed a correlation between stomatal density and environmental parameters such as the levels of humidity, light, and carbon dioxide (CO_2). A strong inverse correlation between stomatal density and atmospheric [CO_2] (10) was observed in preserved and fossil plant specimens and this correlation was used to retrospectively estimate global [CO_2] (10) over the past 450 million years (82). Changes in stomatal density can also occur over much shorter timescales. *Arabidopsis* plants of the Col ecotype grown at double the normal [CO_2] produce fewer stomata per unit area than siblings grown at ambient [CO_2] (37). This response depends on the activity of the *HIC1* gene, which encodes an enzyme required for the synthesis of the very long chain fatty acids that are components of the cuticle (27). Interestingly, the selective application of high [CO_2] to mature leaves causes newly formed leaves to exhibit a decrease in stomatal density similar to leaves of plants grown continuously at high [CO_2]. However, directly exposing only developing (as opposed to mature) leaves to high [CO_2] has no effect on stomatal density (37). This suggests that the environmental stimulus and stomatal response are spatially distinct and, consequently, plants require long-range signals to transmit environmental information.

Precedent for systemic signaling comes from studies on a wide variety of plant behaviors including flowering, pathogen and herbivore responses, and inhibition of lateral branching. Structurally diverse molecules can serve as the signals in these events, including proteins, peptides, and phytohormones. Many of the classical plant hormones [including ethylene, abscisic acid (ABA), gibberellins, and cytokinins] mediate plant responses to the environment, including regulation of stomatal opening. Some recent evidence points to a role for these hormones in regulating stomatal development—both gibberellins and ethylene promote cell divisions, leading to stomatal formation in hypocotyls (33, 61)—but these studies do not distinguish local vs. long-range effects.

Subsequent studies on long-range signals in maize, *Arabidopsis*, and poplar suggest that environmental perception by old leaves and response by new leaves is universal, but the details of the response can vary among species (12, 17, 47). Several cellular mechanisms could account for changes in stomatal density including altered expansion of pavement cells, changes in the number of entry and amplifying divisions in the stomatal lineage, and the arrest or dedifferentiation of meristemoids or GMCs. Combining environmental treatments with developmental methods such as lineage tracing could reveal which steps and which genes in the stomatal pathway are targets of environmental regulation. Some evidence already points to the expansion of pavement cells and the division of stomatal lineage cells being under independent control (47).

As much as environmental studies would benefit from careful examination of development, so would the understanding of developmental genes benefit from testing their response to environmental change. This has been done to some extent with the *SDD1* gene. *sdd1* plants have increased stomatal density in ambient conditions, a phenotype that could reflect either a developmental defect or an inability to correctly sense environmental signals (similar to *hic1*). When tested for response to light intensity changes, *sdd1* mutants responded similarly to wild type, suggesting that the circuit that controls light responses is still intact in *sdd1* (62). Whether *sdd1* is deficient in response to other environmental parameters, or whether other stomatal genes like *TMM* and *ER* mediate both developmental and environmental responses, remains to be tested.

Stomatal Development and Physiology

Mutations in stomatal development genes can affect the physiology of the entire plant. *sdd1* plants, with their higher stomatal density, can assimilate 30% more carbon than wild-type plants when transferred to high light (62). Conversely, *35S::SDD1* plants with reduced stomatal density fare worse than wild type in similar assays (9). *ERECTA* also has a major effect on transpiration efficiency (45). Transpiration efficiency is the ratio of carbon fixation to water loss and requires coordination between photosynthesis and transpiration and is therefore closely related to stomatal activity.

The independent identification of ER as a factor influencing both stomatal development and transpiration efficiency raises the question of how this single protein might affect these two processes. ER's role in development could be completely independent from its role in transpiration efficiency. Because plant LRR-RLKs sit at the top of signaling cascades with many potential targets and because the RLKS can homo- and heterodimerize (40, 52), ER might act with one set of proteins in stomatal development and a different set in coordinating transpiration and photosynthesis. Alternatively, the effect of *ER* on transpiration efficiency could be an indirect consequence of *ER*'s developmental roles (such as altered stomatal density, plant height, and leaf thickness). Identifying partners and downstream targets of ER through protein interactions or genetic screens may reveal how diverse plant activities are mechanistically coupled to ER function.

SUMMARY POINTS

1. Stomata are produced through a stereotyped series of asymmetric and symmetric cell divisions within a dispersed stem cell compartment. The activity of this compartment is a major source of cells that build the *Arabidopsis* leaf epidermis.
2. Stomata are spaced via intercellular signaling pathways that appear to involve several types of receptors and a MAPK phosphorylation cascade.
3. Transcription factors help end proliferation in the stomatal lineage and promote timely cell differentiation.
4. Stomatal development represents a tractable system for analyzing the cell and molecular biology of division site selection, cytokinesis, and cell cycle progression and withdrawal.
5. Just as they influence the functioning of mature stomata, environmental signals also regulate the development of the stomatal lineage.

FUTURE ISSUES

1. Despite the progress described, the genes needed for stomatal specification and morphogenesis are mostly unknown. Sensitized genetic screens and genome-based analysis to target genes expressed in the stomatal lineage were very recently used to identify genes involved in promoting pathway entry and in "counting" proliferative divisions of meristemoids (44a, 52a, 57a), and these approaches hold promise for identifying the complete network of stomatal regulatory genes.
2. Uncertainties about signal transduction and other regulatory pathways might be resolved by biochemically characterizing the interactions and activities among known genes and by identifying their transcriptional and signaling targets. This, in combination with new gene discovery in *Arabidopsis*, should generate a core molecular and biochemical framework for understanding stomatal development. Identifying these components and relationships will enable testing the extent to which these players and pathways are conserved in the plant kingdom.
3. The relationships between asymmetric divisions and cell fate are still poorly understood for plants. The stomatal lineage is a promising system for revealing the mechanisms of cell polarity, intercellular signaling, and division site selection in plants. This pathway is also favorable for revealing relationships between cell specification and the context-specific regulation of the cell cycle machinery in a green cell lineage.
4. The roles of plant growth regulators and environmental signals in regulating stomatal number and development are largely unexplored.

5. Despite the complexity of events in the stomatal pathway, these events are visually accessible on the leaf surface. In *Arabidopsis*, the fate decision can be reduced to a binary choice between pavement and guard cell. These traits and progress to date bode well for learning about how intrinsic and extrinsic factors combinatorially affect a developmental decision.

ACKNOWLEDGMENTS

This work was supported by NSF grant IBN-0237016 to F.S. and NSF grant IOB-0544895 and DOE grant DE-FG02-06ER15810 to D.C.B. Thanks to Keiko Torii and members of our laboratories for helpful discussions. Cora MacAlister, Jessica Lucas, and EunKyoung Lee provided the original data, sketches, and diagrams on which the figures are based.

LITERATURE CITED

1. Asai T, Tena G, Plotnikova J, Willmann MR, Chiu WL, et al. 2002. MAP kinase signaling cascade in *Arabidopsis* innate immunity. *Nature* 415:977–83
2. Berger D, Altmann T. 2000. A subtilisin-like serine protease involved in the regulation of stomatal density and distribution in *Arabidopsis thaliana*. *Genes Dev.* 14:1119–31
3. **Bergmann DC, Lukowitz W, Somerville CR. 2004. Stomatal development and pattern controlled by a MAPKK kinase. *Science* 304:1494–97**
4. Bernhardt C, Lee MM, Gonzalez A, Zhang F, Lloyd A, Schiefelbein J. 2003. The bHLH genes GLABRA3 (GL3) and ENHANCER OF GLABRA3 (EGL3) specify epidermal cell fate in the *Arabidopsis* root. *Development* 130:6431–39
5. Binarováa P, Cenklováb V, Procházkováa J, Doskoilováa A, Volca J, et al. 2006. γ-tubulin is essential for acentrosomal microtubule nucleation and coordination of late mitotic events in *Arabidopsis*. *Plant Cell* 18:1199–212
6. **Boudolf V, Barroco R, Engler JD, Verkest A, Beeckman T, et al. 2004. B1-type cyclin-dependent kinases are essential for the formation of stomatal complexes in *Arabidopsis thaliana*. *Plant Cell* 16:945–55**
7. Boudolf V, Vlieghe K, Beemster GT, Magyar Z, Acosta JAT, et al. 2004. The plant-specific cyclin-dependent kinase CDKB1;1 and transcription factor E2Fa-DPa control the balance of mitotically dividing and endoreduplicating cells in *Arabidopsis*. *Plant Cell* 16:2683–92
8. Bünning E. 1965. Die Enstehung von Mustern in der Entwicklung von Pflanzen. In *Encyclopedia of Plant Physiology*, ed. W Ruhland, pp. 383–408. Berlin: Springer-Verlag
9. Büssis D, von Groll U, Fisahn J, Altmann T. 2006. Stomatal aperture can compensate altered stomatal density in *Arabidopsis thaliana* at growth light conditions. *Func. Plant Biol.* 33:1037–43
10. Case AL, Curtis PS, Snow AA. 1998. Heritable variation in stomatal responses to elevated CO_2 in wild radish, *Raphanus raphanistrum* (Brassicaceae). *Am. J. Bot.* 85:253–58
11. Castellano MD, Boniotti MB, Caro E, Schnittger A, Gutierrez C. 2004. DNA replication licensing affects cell proliferation or endoreplication in a cell type-specific manner. *Plant Cell* 16:2380–93
12. Coupe SA, Palmer BG, Lake JA, Overy SA, Oxborough K, et al. 2006. Systemic signaling of environmental cues in *Arabidopsis* leaves. *J. Exp. Bot.* 57:329–41

3. Reports the functions of the MAPKK kinase, YODA, during stomatal development and presents a transcriptional profiling approach to identify new pathway regulators.

6. Describes the first core cell cycle gene shown to have a significant role in cytokinesis during stomatal formation as well as in stomatal production.

13. Desvoyes B, Ramirez-Parra E, Xie Q, Chua NH, Gutierrez C. 2006. Cell type-specific role of the retinoblastoma/E2F pathway during *Arabidopsis* leaf development. *Plant Physiol.* 140:67–80

14. De Veylder L, Beeckman T, Beemster GT, de Almeida Engler J, Ormenese S, et al. 2002. Control of proliferation, endoreduplication and differentiation by the *Arabidopsis* E2Fa-DPa transcription factor. *EMBO J.* 21:1360–68

15. Dewitte W, Riou-Khamlichi C, Scofield S, Healy JM, Jacqmard A, et al. 2003. Altered cell cycle distribution, hyperplasia, and inhibited differentiation in *Arabidopsis* caused by the D-type cyclin CYCD3. *Plant Cell* 15:79–92

16. Dievart A, Clark SE. 2004. LRR-containing receptors regulating plant development and defense. *Development* 131:251–61

17. Driscoll S, Prins A, Olmos E, Kunert K, Foyer C. 2006. Specification of adaxial and abaxial stomata, epidermal structure and photosynthesis to CO_2 enrichment in maize leaves. *J. Exp. Bot.* 57:381–90

18. Ebel C, Mariconti L, Gruissem W. 2004. Plant retinoblastoma homologues control nuclear proliferation in the female gametophyte. *Nature* 429:776–80

19. Falbel TG, Koch LM, Nadeau JA, Segui-Simarro JM, Sack FD, Bednarek SY. 2003. SCD1 is required for cell cytokinesis and polarized cell expansion in *Arabidopsis thaliana*. *Development* 130:4011–24

20. Fletcher JC, Brand U, Running MP, Simon R, Meyerowitz EM. 1999. Signaling of cell fate decisions by *CLAVATA3* in *Arabidopsis* meristems. *Science* 283:1911–14

21. Freeman M, Gurdon JB. 2002. Regulatory principles of developmental signaling. *Annu. Rev. Cell Dev. Biol.* 18:515–39

22. Galatis B, Apostolakos P. 2004. The role of the cytoskeleton in the morphogenesis and function of stomatal complexes. *New Phytol.* 161:613–39

23. Galatis B, Mitrakos K. 1979. On the differential divisions and preprophase microtubule bands involved in the development of stomata of *Vigna sinensis L. J. Cell Sci.* 37:11–37

24. Geisler M, Nadeau J, Sack FD. 2000. Oriented asymmetric divisions that generate the stomatal spacing pattern in *Arabidopsis* are disrupted by the *too many mouths* mutation. *Plant Cell* 12:2075–86

24. Tests models of stomatal pattern formation in *Arabidopsis* wild-type and *tmm-1* mutant development.

25. Geisler M, Yang M, Sack FD. 1998. Divergent regulation of stomatal initiation and patterning in organ and suborgan regions of the *Arabidopsis* mutants *too many mouths* and *four lips*. *Planta* 205:522–30

26. Geisler MJ, Sack FD. 2002. Variable timing of the developmental progression in the stomatal pathway in *Arabidopsis* cotyledons. *New Phytol.* 153:469–76

27. Gray JE, Holroyd GH, van der Lee FM, Bahrami AR, Sijmons PC, et al. 2000. The HIC signaling pathway links CO_2 perception to stomatal development. *Nature* 408:713–16

28. Hamel LP, Nicole MC, Sritubtim S, Morency MJ, Ellis M, et al. 2006. Ancient signals: comparative genomics of plant MAPK and MAPKK gene families. *Trends Plant Sci.* 11:192–98

29. Heim MA, Jakoby M, Werber M, Martin C, Weisshaar B, Bailey PC. 2003. The basic helix-loop-helix transcription factor family in plants: a genome-wide study of protein structure and functional diversity. *Mol. Biol. Evol.* 20:735–47

30. Imai KK, Ohashi Y, Tsuge T, Yoshizumi T, Matsui M, et al. 2006. The A-type cyclin CYCA2;3 is a key regulator of ploidy levels in *Arabidopsis* endoreduplication. *Plant Cell* 18:382–96

31. Jakoby M, Schnittger A. 2004. Cell cycle and differentiation. *Curr. Opin. Plant Biol.* 7:661–69

32. Jeong S, Trotochaud AE, Clark SE. 1999. The *Arabidopsis CLAVATA2* gene encodes a receptor-like protein required for the stability of the *CLAVATA1* receptor-like kinase. *Plant Cell* 11:1925–33
33. Kazama H, Dan H, Imaseki H, Wasteneys GO. 2004. Transient exposure to ethylene stimulates cell division and alters the fate and polarity of hypocotyl epidermal cells. *Plant Physiol.* 134:1614–23
34. Kazama H, Mineyuki Y. 1997. Alteration of division polarity and preprophase band orientation in stomatogenesis by light. *J. Plant Res.* 110:489–93
35. Kinoshita T, Cano-Delgado A, Seto H, Hiranuma S, Fujioka S, et al. 2005. Binding of brassinosteroids to the extracellular domain of plant receptor kinase BRI1. *Nature* 433:167–71
36. Lai LB, Nadeau JA, Lucas J, Lee EK, Nakagawa T, et al. 2005. The *Arabidopsis* R2R3 MYB proteins FOUR LIPS and MYB88 restrict divisions late in the stomatal cell lineage. *Plant Cell* 17:2754–67
37. Lake JA, Quick WP, Beerling DJ, Woodward FI. 2001. Plant development. Signals from mature to new leaves. *Nature* 411:154
38. Larkin JC, Brown ML, Schiefelbein J. 2003. How do cells know what they want to be when they grow up? Lessons from epidermal patterning in *Arabidopsis*. *Annu. Rev. Plant Biol.* 54:403–30
39. Lease KA, Lau NY, Schuster RA, Torii KU, Walker JC. 2001. Receptor serine/threonine protein kinases in signaling: analysis of the ERECTA receptor-like kinase of *Arabidopsis thaliana*. *New Phytol.* 151:133–43
40. Li J, Wen JQ, Lease KA, Doke JT, Tax FE, Walker JC. 2002. BAK1, an *Arabidopsis* LRR receptor-like protein kinase, interacts with BRI1 and modulates brassinosteroid signaling. *Cell* 110:213–22
41. Li L, Vaessin H. 2000. Pan-neural Prospero terminates cell proliferation during *Drosophila* neurogenesis. *Genes Dev.* 14:147–51
42. Liu Y, Zhang S. 2004. Phosphorylation of 1-aminocyclopropane-1-carboxylic acid synthase by MPK6, a stress-responsive mitogen-activated protein kinase, induces ethylene biosynthesis in *Arabidopsis*. *Plant Cell* 16:3386–99
43. Lucas JR, Nadeau JA, Sack FD. 2006. Microtubule arrays and *Arabidopsis* stomatal development. *J. Exp. Bot.* 57:71–79
44. Lukowitz W, Roeder A, Parmenter D, Somerville C. 2004. A MAPKK kinase gene regulates extraembryonic cell fate in *Arabidopsis*. *Cell* 116:109–19
44a. MacAlister CA, Ohashi-Ito K, Bergmann DC. 2007. The bHLH *SPEECHLESS* controls asymmetric cell divisions to establish the stomatal lineage. *Nature* doi:10.1038/nature05491
45. Masle J, Gilmore SR, Farquhar GD. 2005. The ERECTA gene regulates plant transpiration efficiency in *Arabidopsis*. *Nature* 436:866–70
46. Melaragno JE, Mehrotra BM, Coleman AW. 1993. Relationship between endopolyploidy and cell size in epidermal tissue of *Arabidopsis*. *Plant Cell* 5:1661–68
47. Miyazawa S, Livingston N, Turpin D. 2005. Stomatal development in new leaves is related to the stomatal conductance of mature leaves in poplar (*Populus trichocarpa* x *P. deltoides*). *J. Exp. Bot.* 57:373–80
48. Nadeau JA, Sack FD. 2002. Control of stomatal distribution on the *Arabidopsis* leaf surface. *Science* 296:1697–700
49. Nadeau JA, Sack FD. 2002. Stomatal development in *Arabidopsis*. In *The Arabidopsis Book*, ed. CR Somerville, EM Meyerowitz, doi/10.1199/tab.0066. Rockville MD: Am. Soc. Plant Biol. ISSN:1543-8120

36. The first in-depth description of transcription factors required for proper stomatal formation.

37. Using a cuvette system, the authors found that stomatal density is influenced by an environmental signal perceived in mature leaves and transmitted to new leaves.

44a. The authors identify a bHLH related to FAMA that is required for the asymmetric entry divisions that establish the stomatal lineage.

45. Using Δ13C discrimination as a method for estimating transpiration efficiency, the authors identified QTLs between *Arabidopsis* accessions and mapped a major QTL to ERECTA.

48. Identifies TMM as a receptor-like protein expressed in the endomembrane system of stomatal lineage cells.

50. Nadeau JA, Sack FD. 2003. Stomatal development: cross talk puts mouths in place. *Trends Plant Sci.* 8:294–99
51. Nakamura R, McKendree WJ, Hirsch R, Sedbrook J, Gaber R, Sussman M. 1995. Expression of an *Arabidopsis* potassium channel gene in guard cells. *Plant Physiol.* 109:371–74
52. Nam KH, Li J. 2002. BRI1/BAK1, a receptor kinase pair mediating brassinosteroid signaling. *Cell* 110:203–12
52a. Ohashi-Ito K, Bergmann DC. 2006. Arabidopsis FAMA controls the final proliferation/differentiation switch during stomatal development. *Plant Cell* 18:2493–505
53. Palevitz BA. 1981. The structure and development of stomatal cells. In *Stomatal Physiology*, ed. P Jarvis, T Mansfield, pp. 1–23. Cambridge: Cambridge Univ. Press
54. Paliwal GS. 1967. Ontogeny of stomata in some Cruciferae. *Can. J. Bot.* 45:495–500
55. Pant DD, Kidwai PF. 1967. Development of stomata in some Cruciferae. *Ann. Bot.* 31:513–21
56. Park JA, Ahn JW, Kim YK, Kim SJ, Kim JK, et al. 2005. Retinoblastoma protein regulates cell proliferation, differentiation, and endoreduplication in plants. *Plant J.* 42:153–63
57. Payne CT, Zhang F, Lloyd AM. 2000. GL3 encodes a bHLH protein that regulates trichome development in arabidopsis through interaction with GL1 and TTG1. *Genetics* 156:1349–62
57a. Pillitteri LJ, Sloan DB, Bogenschutz NL, Torii KU. 2007. Termination of asymmetric cell division and differentiation of stomata. *Nature* In press
58. Raven J. 2002. Selection pressures on stomatal evolution. *New Phytol.* 153:371–86
59. Russinova E, Borst JW, Kwaaitaal M, Cano-Delgado A, Yin Y, et al. 2004. Heterodimerization and endocytosis of *Arabidopsis* brassinosteroid receptors BRI1 and AtSERK3 (BAK1). *Plant Cell* 16:3216–29
60. Sachs T. 1991. *Pattern Formation in Plant Tissues*. New York: Cambridge Univ. Press
61. Saibo NJ, Vriezen WH, Beemster GT, Van der Straeten D. 2003. Growth and stomata development of *Arabidopsis* hypocotyls are controlled by gibberellins and modulated by ethylene and auxins. *Plant J.* 33:989–1000
62. Schluter U, Muschak M, Berger D, Altmann T. 2003. Photosynthetic performance of an *Arabidopsis* mutant with elevated stomatal density (sdd1–1) under different light regimes. *J. Exp. Bot.* 54:867–74
63. Schroeder JI, Allen GJ, Hugouvieux V, Kwak JM, Waner D. 2001. Guard cell signal transduction. *Annu. Rev. Plant Physiol. Plant Mol. Biol.* 52:627–58
64. Serna L, Fenoll C. 1997. Tracing the ontogeny of stomatal clusters in *Arabidopsis* with molecular markers. *Plant J.* 12:747–55
65. Serna L, Torres-Contreras J, Fenoll C. 2002. Clonal analysis of stomatal development and patterning in *Arabidopsis* leaves. *Dev. Biol.* 241:24–33
66. Shimazaki KI. 2007. Stomatal regulation. *Annu. Rev. Plant Biol.* 58:In press
67. Shiu SH, Bleecker AB. 2001. Receptor-like kinases from *Arabidopsis* form a monophyletic gene family related to animal receptor kinases. *Proc. Natl. Acad. Sci. USA* 98:10763–68
68. Shpak ED, Berthiaume CT, Hill EJ, Torii KU. 2004. Synergistic interaction of three ERECTA-family receptor-like kinases controls *Arabidopsis* organ growth and flower development by promoting cell proliferation. *Development* 131:1491–501
69. Shpak ED, Lakeman MB, Torii KU. 2003. Dominant-negative receptor uncovers redundancy in the *Arabidopsis* ERECTA Leucine-rich repeat receptor-like kinase signaling pathway that regulates organ shape. *Plant Cell* 15:1095–110
70. Shpak ED, McAbee JM, Pillitteri LJ, Torii KU. 2005. Stomatal patterning and differentiation by synergistic interactions of receptor kinases. *Science* 309:290–93

52a. A transcription factor identified through stomatal-lineage transcriptional profiling is shown to be necessary and sufficient to promote the conversion from GMC to guard cells.

57a. The authors identify a bHLH related to FAMA that is required for terminating amplifying divisions and promoting the formation of stomata.

70. Use of various mutant combinations of TMM and the ERECTA-family genes reveals that closely related receptor-like kinases have redundant yet unique functions in stomatal development.

71. Sollner R, Glasser G, Wanner G, Somerville CR, Jurgens G, Assaad FF. 2002. Cytokinesis-defective mutants of *Arabidopsis*. *Plant Physiol.* 129:678–90
72. Soyano T, Nishihama R, Morikiyo K, Ishikawa M, Machida Y. 2003. NQK1/NtMEK1 is a MAPKK that acts in the NPK1 MAPKKK-mediated MAPK cascade and is required for plant cytokinesis. *Genes Dev.* 17:1055–67
73. Takayama S, Isogai A. 2003. Molecular mechanism of self-recognition in Brassica self-incompatibility. *J. Exp. Bot.* 54:149–56
74. Terryn N, Arias MB, Engler G, Tire C, Villarroel R, et al. 1993. rha1, a gene encoding a small GTP binding protein from *Arabidopsis*, is expressed primarily in developing guard cells. *Plant Cell* 5:1761–69
75. Tokumoto YM, Apperly JA, Gao FB, Raff MC. 2002. Posttranscriptional regulation of p18 and p27 Cdk inhibitor proteins and the timing of oligodendrocyte differentiation. *Dev. Biol.* 245:224–34
76. Torii KU. 2004. Leucine-rich repeat receptor kinases in plants: structure, function, and signal transduction pathways. *Int. Rev. Cytol.* 234:1–46
77. Vert G, Nemhauser JL, Geldner N, Hong F, Chory J. 2005. Molecular mechanisms of steroid hormone signaling in plants. *Annu. Rev. Cell Dev. Biol.* 21:177–201
78. von Groll U, Berger D, Altmann T. 2002. The subtilisin-like serine protease SDD1 mediates cell-to-cell signaling during *Arabidopsis* stomatal development. *Plant Cell* 14:1527–39
79. Weigel D, Jurgens G. 2002. Stem cells that make stems. *Nature* 415:751–54
80. Weinl C, Marquardt S, Kuijt SJ, Nowack MK, Jakoby MJ, et al. 2005. Novel functions of plant cyclin-dependent kinase inhibitors, ICK1/KRP1, can act non-cell-autonomously and inhibit entry into mitosis. *Plant Cell* 17:1704–22
81. Wildwater M, Campilho A, Perez-Perez JM, Heidstra R, Blilou I, et al. 2005. The RETINOBLASTOMA-RELATED gene regulates stem cell maintenance in *Arabidopsis* roots. *Cell* 123:1337–49
82. Woodward FI. 1987. Stomatal numbers are sensitive to CO_2 increases from preindustrial levels. *Nature* 327:617–18
83. Yang M, Nadeau JA, Zhao L, Sack FD. 1999. Characterization of a cytokinesis defective (*cyd1*) mutant of *Arabidopsis*. *J. Exp. Bot.* 50:1437–46
84. Zhao L, Sack FD. 1999. Ultrastructure of stomatal development in *Arabidopsis* (Brassicaceae) leaves. *Am. J. Bot.* 86:929–39
85. Zimmermann IM, Heim MA, Weisshaar B, Uhrig JF. 2004. Comprehensive identification of *Arabidopsis thaliana* MYB transcription factors interacting with R/B-like BHLH proteins. *Plant J.* 40:22–34

78. *SDD1* is expressed in stomatal precursor cells, and *SDD1* overexpression reduces stomatal formation (opposite phenotype to *sdd1*).

Gibberellin Receptor and Its Role in Gibberellin Signaling in Plants

Miyako Ueguchi-Tanaka,[1] Masatoshi Nakajima,[2] Ashikari Motoyuki,[1] and Makoto Matsuoka[1]

[1]Bioscience and Biotechnology Center, Nagoya University, Nagoya, Aichi 464-8601, Japan; email: mueguchi@agr.nagoya-u.ac.jp, ashi@agr.nagoya-u.ac.jp, makoto@nuagr1.agr.nagoya-u.ac.jp

[2]Department of Applied Biological Chemistry, University of Tokyo, Tokyo 113-8657, Japan; email: nkjm@pgr1.ch.a.u-tokyo.ac.jp

Annu. Rev. Plant Biol. 2007. 58:183–98

The *Annual Review of Plant Biology* is online at plant.annualreviews.org

This article's doi:
10.1146/annurev.arplant.58.032806.103830

1543-5008/07/0602-0183$20.00

Key Words

GA-insensitive dwarf1 (*GID1*), DELLA proteins, GA receptor

Abstract

Gibberellins (GAs) are a large family of tetracyclic, diterpenoid plant hormones that induce a wide range of plant growth responses. It has been postulated that plants have two types of GA receptors, including soluble and membrane-bound forms. Recently, it was determined that the rice *GIBBERELLIN INSENSITIVE DWARF1* (*GID1*) gene encodes an unknown protein with similarity to the hormone-sensitive lipases that has high affinity only for biologically active GAs. Moreover, GID1 binds to SLR1, a repressor of GA signaling, in a GA-dependent manner in yeast cells. Based on these observations, it has been concluded that GID1 is a soluble receptor mediating GA signaling in rice. More recently, *Arabidopsis thaliana* was found to have three GID1 homologs, AtGID1a, b, and c, all of which bind GA and interact with the five *Arabidopsis* DELLA proteins.

Contents

INTRODUCTION

Gibberellin (GA) is a well-known phytohormone that affects a wide range of plant growth, development, and environmental responses, including seed germination, stem elongation, leaf expansion, pollen maturation, and induction of flowering (reviewed in 7). Recently, the GA receptor GIBBERELLIN INSENSITIVE DWARF1 (GID1) was identified by a combination of biochemical and genetic techniques (42, 69). Because of the identification of the GA receptor, the molecular mechanisms of GA perception and signal transduction are much better understood. In this article, we review the history of attempts to identify receptor candidates for GA, the biochemical and physiological characteristics of the GID1 GA receptor, and its roles in GA signaling.

BIOCHEMICAL SEARCH FOR THE GIBBERELLIN RECEPTOR

Because GAs are relatively hydrophobic molecules, they are believed to be able to transverse plant cell plasma membranes by passive diffusion. In the early stages of the study of GA perception, some researchers expected that GA perception occurs by a mechanism similar to hydrophobic steroid hormones in mammalian cells. In accordance with this concept, Johri & Varner (27) showed that the composition of RNA in isolated nuclei from pea seedlings changed before and after the application of GA. Much later, Sechley & Srivastava (56) reported a similar effect of GA on nuclear transcription rates in cucumber hypocotyls. Further, Witham & Hendry (75) pointed to the possibility of direct interactions between double-stranded DNA and GA based on computer modeling.

In contrast to the idea that GA molecules directly affect gene transcription, there were many attempts to identify proteins with GA-binding activity [GA-binding protein (GBP)]. Stoddart et al. (61) first reported detection of GA-binding activity in crude protein extracts from lettuce hypocotyls. Since this first observation of GBP, a number of GBPs have been proposed as GA receptor candidates (**Table 1**). Once a GBP has been detected and reported, the candidate protein is tested to see if it meets the four criteria of a GA receptor. GA receptors must reversibly bind GA, have GA saturability and high affinity for biologically active GAs, and must have reasonable ligand specificity for biologically active GAs.

Thus far, two GA-responsive systems have been used to detect GBPs: promotion of stem elongation and induction of hydrolytic enzymes in aleurone layers of cereal seed. The stem elongation assay has been used to identify GA-binding activity in the soluble protein fractions of pea epicotyls and cucumber

Table 1 Gibberellin-binding proteins in various plant materials

Materials	Site	MW (kDa)	Amount (pmol/mg)	Method	Ref.
Lettuce hypocotyl	CW	n.d.	n.d.	SDG	(61)
Wheat aleurone	n.d.	n.d.	0.45	UC	(26)
Pea epicotyl	CY	60/500	n.d.	GPC/EQD	(62)
Pea epicotyl	CY	40–70/600	0.9	GPC	(32)
Cucumber hypocotyl	CY	n.d.	0.4	GPC/EQD	(30)
Cucumber hypocotyl	CY	n.d.	0.4	DEAE filter	(29)
Cucumber hypocotyl	CY	n.d.	0.4	DEAE filter	(76)
Maize leaf sheath	CY	40–90/500	n.d.	GPC	(31)
Cucumber hypocotyl	CY	n.d.	0.25	DEAE filter	(77)
Bean epicotyl	N/CH	80–100	330	PEI filter	(66)
Mung bean hypocotyl	CY	150–200	65	Salting-out	(41)
Oat aleurone	MC	60[a]	n.d.	Photoaffinity	(20)
Maize mesocotyl	CY	n.d.	0.62	Salting-out	(53)
Oat aleurone	CY	50[a]	n.d.	Photoaffinity	(72)
Pea epicotyl	CY	n.d.	0.66	DEAE filter	(38)
Pea epicotyl	CY	40–110	0.21	DEAE filter	(37)
Rice leaf	CY	47[a]	n.d.	LB	(34)
Adzuki bean epicotyl	CY	25–40	0.0001	GPC	(43)
Oat aleurone	MC	18/68[a]	n.d.	Photoaffinity	(39)

[a]Estimation under denatured condition (SDS-PAGE).

CH, chloroplast; CW, cell wall; CY, cytosol or soluble fraction; DEAE, diethylaminoethyl; EQD, equilibrium dialysis; GPC, gel-permeation column; LB, ligand blotting; MC, microsome or membrane-bound fraction; n.d., no data or not described clearly; N, nuclear; PEI, polyetyleneimine; SDG, sucrose density-gradient centrifugation; UC, ultracentrifugation.

hypocotyls (29, 30, 32, 76), although none of these GBPs has been further characterized (31, 37, 38, 77). Komatsu et al. (34) identified a soluble GBP from rice leaves by detecting a GA-binding activity for membrane-blotted proteins. The GBP was homologous to RuBisCO activase (ribulose-1,5-biphosphate carboxylase/oxygenase activase) and was phosphorylated in the presence of Ca^{2+}, Mg^{2+}, ATP, and GA. They suggested that a Ca^{2+}-dependent protein kinase (CDPK) might be involved in the signaling pathway from this GBP (57). Nakajima et al. (43) also detected GA-binding activity in the soluble protein fraction from adzuki bean (*Vigna angularis*) seedlings by using a gel-permeation column. Further studies have revealed that this GBP in the partially purified fraction fulfills all four of the GA receptor criteria (47, 48).

Using the GA-dependent induction of aleurone hydrolytic enzymes, Jelsema et al. (26) first reported GBP activity in wheat seed aleurone homogenates. Later, the researchers in the United Kingdom demonstrated that α-amylase can be induced in aleurone protoplasts in a GA-dependent manner. Moreover, such induction occurs even with the application of GA derivatives that cannot pass through the plasma membrane (2, 18). Gilroy & Jones (14) also reported that there was no induction of α-amylase when GA was injected into the cytoplasm of barley aleurone protoplasts. These biochemical experiments strongly suggested that the GA-perception site is outside the plasma membrane and, consequently, that at least one GA receptor is in the plasma membrane of aleurone cells. Based on these observations, Hooley and his colleagues attempted to identify a

GA receptor located on the plasma membrane of oat aleurone cells using a photoaffinity-labeled GA, and succeeded in detecting two GBPs: a 60-kDa protein localized in the microsomal fraction (20), and a 50-kDa protein in the cytosolic fraction (19, 72). Two other GBPs of 68/18 kDa were also detected in the plasma membrane fraction from oat aleurone by the same photoaffinity-labeling method (39). Even though partial amino acid sequences for the 18-kDa GBP have been identified, there is no further information about these GBPs.

There is an alternative biochemical approach for isolating receptor candidates using an immunological method. Hooley et al. (17, 19) prepared antisera raised against a monoclonal antibody for GA (i.e., anti-idiotypic antibodies) that competes with GA molecules for binding to the parental monoclonal antibody. The anti-idiotypic antibodies inhibit GA action in aleurone protoplasts, suggesting that the anti-idiotypic antibodies bind to the GA-interacting domains of GA receptor(s). An oat cDNA library was screened using these antibodies and a ubiquitin gene was isolated as a candidate, though there was no report of its GA-binding activity (52).

DELLA PROTEIN IS A KEY REGULATOR IN GIBBERELLIN SIGNALING

Derepression of the repressed state is currently considered to be the key step of GA action in the GA signaling pathway. In this model, DELLA subfamily proteins of the GRAS superfamily play an important role in the negative control of GA signaling. Members of the GRAS family, which was originally defined by the presence of the conserved domains VHIID and RVER of GAI, RGA, and SCR in *Arabidopsis thaliana* (51), are thought to function as transcription factors, although there is as yet no direct evidence. Members of the DELLA subfamily contain the conserved amino acid motifs DELLA (hence its name) and TVHYNP near their N-terminal portion. DELLA proteins are highly conserved in *Arabidopsis* (GAI, RGA, RGL1, RGL2, RGL3) (9, 35, 49, 59, 73), and in several crop plants, including rice (SLR1) (22), wheat (Rht) (50), barley (SLN1) (5), maize (d8) (50), and grape (VvGAI) (3). The wheat DELLA gene *Rht* is well known for its contribution to increases in crop yield in the "green revolution" wheat breeding program of the middle of the twentieth century. Gain-of-function mutations in this gene family result in dwarfism and reduced GA response, whereas loss of function results in the GA-constitutive response phenotype, even in the presence of GA-biosynthesis inhibitors. For example, loss-of-function mutants of rice *SLR1* show a slender phenotype with an elongated stem and leaf, and reduced root number and length (22, 24). Also, GA-inducible α-amylase is produced in embryo half-seeds in the absence of GA application. These *slr1* phenotypes are typical of plants treated with exogenous GA, even though levels of endogenous GA are lower than in wild-type plants. Moreover, the GA-overdose phenotype of *slr1* is not affected by the GA-biosynthesis inhibitor uniconazol (22). Barley *sln1* loss-of-function mutants also have a similar GA-constitutive response phenotype (5), indicating that DELLA proteins function as negative regulators in GA signaling.

In contrast to the clear GA-overdose phenotype in the loss-of-function mutants of rice and barley, loss of DELLA protein functions in *Arabidopsis* does not induce an obvious GA-overdose phenotype. Among the five DELLA proteins, RGA plays the most prominent role in stem elongation, leaf expansion, and induction of flowering. Its loss-of-function alleles partially suppress most of the phenotype of the GA-deficient mutant *ga1-3*, except for seed germination and floral development (60). The loss-of-function *gai* allele, *gai-t6*, has wild-type features, but has slightly increased resistance to a GA-biosynthesis inhibitor, paclobutrazol, in vegetative growth (49). The ambiguous phenotypes of loss-of-function *Arabidopsis* mutants are due to the functional redundancy of five DELLA proteins. For example, *RGA* and *GAI* encode proteins sharing 82% amino acid identity (59, 60). A double knockout of *RGA* and *GAI* produces a clearer phenotype of the GA-constitutive phenotype, including increased stem elongation and early flowering in the wild-type background, and also suppresses the severely dwarfed phenotype (9, 33) and stunted root growth (11) in the GA-deficient mutant *ga1-3*. The absence of *RGA* and *GAI* was not sufficient to rescue the *ga1-3* mutant from abnormal germination or flower development, suggesting that other DELLA proteins, besides RGA and GAI, have decisive functions in GA-dependent germination and flower development (9, 33). RGL1 and RGL2 have been reported to be involved in seed germination (35, 68, 73). More recently, by preparing triple and quadruple knockout mutants in the *ga1-3* background, Cao et al. (4) reported that RGL2 is the predominant repressor of seed germination, and that GAI, RGA, and RGL1 are functional enhancers of RGL2 and tuners for environmental conditions. RGA and RGL2 also have been reported to act dominantly in floral development, and RGL1 can function as a minor repressor (68, 81).

In contrast to the GA-constitutive phenotype of loss-of-function mutations in DELLA proteins, dominant alleles in the *Arabidopsis gai* (49), wheat *Rht* (50), and maize *D8* loci (50) confer a GA-insensitive phenotype with characteristic dwarfism. These dominant alleles have in-frame deletions in their conserved N-terminal domains, such as DELLA and TVHYNP, resulting in constitutive DELLA protein function. Similarly, transgenic rice plants that produce a SLR1 protein truncated in the DELLA or TVHYNP domain have a dominant dwarf phenotype similar to the spontaneous mutants (24). All of these mutants and transgenic plants show GA-insensitive characters, suggesting that the N-terminal region that includes the DELLA and TVHYNP domains functions in the perception of an upstream GA signal. Further domain analysis of the rice DELLA protein SLR1 has shown that the C-terminal region containing the VHIID, PFYRE, and SAW domains, which are shared with other GRAS family proteins, is involved in the suppressive function of DELLA proteins against GA action. The proteins also contain leucine-heptad repeats (LHR), which may mediate protein-protein interaction, and Ser/Thr residues, which may be involved in the regulation of their repression activity. It has been proposed that the activity or stability of DELLA proteins is regulated by *O*-GlcNAc modification or phosphorylation via the action of SPINDLY (SPY), which is another negative regulator of GA signaling, or kinase with the Ser/Thr residues as the target site (67).

F-BOX-DEPENDENT DEGRADATION OF DELLA PROTEINS IS A KEY EVENT IN GIBBERELLIN SIGNALING

All available evidence indicates that DELLA proteins are subject to GA-dependent proteolysis via the ubiquitin-proteasome pathway. The model of DELLA protein degradation by 26S-proteasome-mediated proteolysis was first suggested by the observation that the level of a barley DELLA protein, SLN1, increases in the presence of 26S proteasome inhibitors (12). This model was later greatly substantiated by the cloning of F-box genes from rice (*OsGID2*) and *Arabidopsis* (*AtSLY1*) (40, 55). Loss-of-function mutation of *OsGID2* or *AtSLY1* results in GA-insensitive phenotypes of the host plant. Positional cloning of these mutated genes revealed that *OsGID2* and *AtSLY1* are orthologous and encode F-box domain-containing proteins. An F-box protein is a component of the SCF complex, which is named for its Skp1, cullin, and F-box protein subunits. The SCF complex catalyses the transfer of ubiqutin from E2 to the target protein (13). Rbx1 is another component of the SCF complex and binds to the C terminus of cullin and to the E2 ubiquitin conjugating enzyme. Adding a polyubiquitin chain to the target protein induces degradation of the target protein by the 26S proteasome, which is a large protein degradation complex. F-box proteins contain an F-box domain, which is usually located at their N terminus and involved in interaction with Skp1, and also generally contain interaction domains with target proteins at the C terminus, such as WD40 repeats and leucine-rich repeats. OsGID2 and AtSLY1 contain F-box domains at the N termini, as do other F-box proteins, but they lack known protein-protein interaction domains at their C termini. However, OsGID2 and AtSLY1 share conserved amino acid sequences not only at their N termini but also within their C-terminal regions, and deletions of the conserved C-terminal regions cause a loss of function (40, 55). Yeast two-hybrid assays and in vivo immunoprecipitation experiments demonstrate that OsGID2 is a component of the SCF complex through interaction with one of the rice Skp1-like proteins, OsSkp15 (15). *Arabidopsis* has an *AtSLY1* homologous gene, *SNE*, which can functionally replace *AtSLY1* in the knock-down plants of the *SLY1* function by its antisense construct, suggesting that *SNE* has at least partial overlapping function with *SLY1* (63).

Several lines of evidence support the notion that the target of SCF^{GID2} and SCF^{SLY1} are the DELLA proteins, SLR1 and RGA, respectively. First, high levels of SLR1 and RGA protein accumulation are observed in *gid2* and *sly1* mutant plants. Second, double mutants carrying rice *gid2-1/slr1-1* and *Arabidopsis sly1-10/rga-24* show the *slr1* and *rga* phenotypes (10, 55). This suggests that the GA-insensitive phenotype of *gid2* or *sly1* depends on the function of SLR1 or RGA. Finally, SLY1 interacts directly with RGA and GAI via their C-terminal GRAS domains in yeast two-hybrid and in vitro pull-down assays (10). The direct interaction between DELLA proteins and SLY1 was confirmed by the observation that the product of the gain-of-function allele of *SLY1*, *gar2/sly1-d*, has higher affinity for RGA and GAI than does the wild-type SLY1 protein. In contrast, OsGID2 does not interact directly with SLR1 in yeast cells (H.Tsuji, unpublished results). Recombinant GID2 protein produced in *Escherichia coli* interacts with SLR1 in rice crude extracts in vitro (23), indicating that additional components are required for GID2-SLR1 interaction in rice cells.

Biochemical studies in yeast and mammals have shown that the interaction of F-box proteins with protein substrates depends on modifications such as phosphorylation (8), glycosylation (80), and hydroxylation (25). There are some supportive observations that the GA-induced degradation of DELLA proteins depends on their phosphorylation. For instance, treatment with protein Tyr kinase inhibitors, such as genistein and Tyrophostin B46, blocked the GA-induced degradation of SLN1 in barley seedlings (12). Furthermore, the level of phosphorylated SLR1 increases in response to GA signaling in rice seedlings, and phosphorylated SLR1 binds to recombinant glutathione S-transferase (GST)-GID2 (15, 55). On the other hand, more recent observations suggest that phosphorylation of DELLA proteins is not directly involved in GA-induced degradation (23). For example, exogenously applied GA induces both phosphorylated and nonphosphorylated forms of SLR1 with similar induction kinetics in *gid2* cells. Both phosphorylated and nonphosphorylated SLR1 proteins are degraded by GA treatment with a similar half-life in rice wild-type cells, and both proteins interact with recombinant GST-GID2. Furthermore, Ser/Thr phosphatase inhibitors effectively block RGL2 degradation in tobacco BY2 cells, but Ser/Thr kinase inhibitors have no visible effect, suggesting that the default state of RGL2 is in a phosphorylated form in BY2 cells (21). Transgenic rice plants containing RNAi or an antisense construct for rice *SPY*, which is another negative regulator of GA signaling, alter the phosphorylation state of SLR1 without changing the SLR1 level (58). This also indicates that the amount of SLR1 is not controlled by its phospholylation state. To make sense of all of these apparently contradicting observations, it could be that the phosphorylation of DELLA proteins does not directly lead to their degradation and could be independent of their interaction with F-box proteins.

GID1 IS A SOLUBLE GIBBERELLIN RECEPTOR AND DIRECTLY INTERACTS WITH THE RICE DELLA PROTEIN, SLR1, IN A GIBBERELLIN-DEPENDENT MANNER

The rice *gid1* recessive mutant shows a typical GA-insensitive phenotype (69). A *gid1-1/slr1-1* double mutant exhibits the slr1 phenotype, indicating that SLR1 is epistatic to GID1. GA treatment does not diminish the amount of SLR1 in *gid1-1* plants like the *gid2* mutant. Although the GA-insensitive phenotype of *gid1* is similar to *gid2* mutants, there are some differences, namely that *gid1* dwarfism is more severe than that of *gid2*, and that the amount of accumulated SLR1 in *gid1* is lower than in *gid2*. The *gid1* phenotype is thus similar to the GA-deficient mutant *cps*, indicating that *GID1* functions upstream from *SLR1* in the GA signaling pathway, but not in SLR1 degradation.

GID1 protein fused with a GST tag (GST-GID1) binds to 16,17-dihydro-GA_4 with a reasonable dissociation constant (K_d) of 1.4×10^{-6} M. The ligand specificity of GST-GID1 for various GAs in vitro is generally consistent with the physiological activity of GAs. That is, biologically active GAs generally have higher binding affinity whereas biologically inactive GAs have lower affinity. However, GA_4-binding affinity to GID1 is about 20 times higher than GA_3, but the physiological activity of GA_4 is lower than that of GA_3. This discrepancy between the GA-GID1-binding affinity in vitro and physiological activity in planta has been attributed to differences in the stability of GA_4 and GA_3 in planta. In this case, GA_4 is rapidly inactivated by a GA-inactivating enzyme, GA 2-oxidase (54). This hypothesis is confirmed by the observation that SLR1 degradation starts at a much lower concentration of GA_4 than GA_3 in rice culture cells that contain no bioactive GA or GA 2-oxidase activity (M. Ueguchi-Tanaka, unpublished results). The GA-perception activity of GID1 in vivo is also confirmed by the GA-hypersensitive phenotype of transgenic rice plants that overproduce GID1.

GID1 encodes an unknown protein with similarity to the hormone-sensitive lipase (HSL) family, including the conserved HSL motifs HGG and GXSXG (45). The importance of this GXSXG motif is confirmed by the severe phenotype of *gid1-1* carrying a single amino acid exchange of the first G for D in the motif. Furthermore, GA binding is completely abolished by the deletion of its shared regions with the HSL family, indicating that the entire conserved region between GID1 and HSL is essential for GA binding (M. Ueguchi-Tanaka, unpublished results). However, GID1 may not have a lipase activity, because GID1 shares only two of the three conserved amino acid residues essential for HSL activity. The third residue, H, is replaced by V, which is essential for forming the catalytic triad in the HSL family. Furthermore, recombinant GID1 does not hydrolyze an artificial substrate for HSL, *p*-nitrophenyl acetate. The cellular localization of GID1 is predicted to be mainly in nuclei and its localization does not change with the endogenous GA level (69).

GID1 interacts with SLR1 in a GA-dependent manner in yeast two-hybrid assays (69). This indicates that the GA-GID1 complex interacts directly with SLR1 and probably transduces the GA signal to SLR1. The GA-binding activity of GID1 is increased about threefold in the presence of SLR1 (M. Ueguchi-Tanaka, unpublished results). This enhanced GA binding is caused by the decreased dissociation rate between GID1 and GA although its association rate is not affected

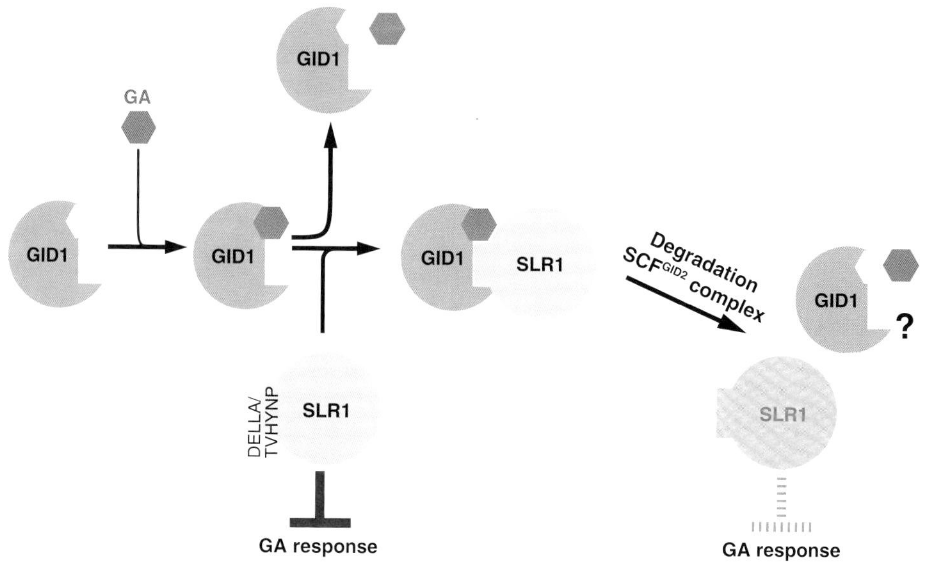

Figure 1

Model of gibberellin signaling in rice. Under low GA concentrations, SLR1 represses the GA responses. Under high GA concentrations, a soluble receptor, GID1, binds to GA; however, the binding is unstable and easily dissociates from the other. The GID1-GA complex specifically interacts with SLR1 at the site of DELLA and TVHYNP domains. The triple complex composed of GID1-GA-SLR1 is stable and does not easily dissociate. The triple complex is in turn targeted by the SCF^{GID2} complex and the SLR1 protein is degraded by the 26S proteosome, which releases the repressive state of GA responses.

by SLR1 (M. Ueguchi-Tanaka, unpublished results). Thus, SLR1 stabilizes the interaction between GID1 and GA. Domain analysis of SLR1 using a yeast two-hybrid assay revealed that the DELLA and TVHYNP domains are essential for its GA-dependent interaction with GID1 (M. Ueguchi-Tanaka, unpublished results). Deleting the Leu-heptad domain, or the conserved regions of the GRAS family such as the VHIID, PFYRE, and SAW domains, does not result in a complete loss of GID1 interaction. These observations indicate that the N-terminal portion of SLR1 is essential and sufficient for the GA-dependent interaction between GID1 and SLR1. This hypothesis has also been confirmed in vitro. Based on these observations, Ueguchi-Tanaka et al. proposed a model of GA perception mediated by GID1 (**Figure 1**). When GID1 binds GA, the GA-GID1 complex can interact with SLR1 probably by some conformational change. The region containing the DELLA/TVHYNP domains of SLR1 and the conserved HSL regions of GID1 are essential for the interaction between GID1 and SLR1. The association and dissociation of GID1 and the GA molecule occur rapidly in the absence of SLR1, but when the GID1-GA complex interacts with SLR1, the GID1-GA complex is greatly stabilized. The stabilized trio-complex consisting of GA, GID1, and SLR1 might be a target of GID2, leading to the degradation of SLR1 by 26S proteasomes through ubiquitination of the SCF^{GID2} complex.

GIBBERELLIN PERCEPTION AND THE GID1-DELLA INTERACTION IN *ARABIDOPSIS*

Three genes, *AtGID1a* (*At3g05120*), *AtGID1b* (*At3g63010*), and *AtGID1c* (*At5g27320*), have been predicted to be GA receptors in *Arabidopsis* based on their structural similarity to the rice GA receptor (42). Like rice GID1, the *AtGID1* protein products have a high affinity only for biologically active GAs, especially for GA_4. Among these four GA receptors, only AtGID1b has a strong pH dependency and about a four-times-higher affinity for GA (K_d for 16,17-dihydro-GA_4 = 4.8×10^{-7} M), setting it apart biochemically from the other receptors. Phylogenic analysis also supports the uniqueness of AtGID1b, and AtGID1a and AtGID1c are classified within the same group as rice GID1, whereas AtGID1b is located in an independent subgroup.

As *Arabidopsis* has three *GID1*s and five DELLA proteins (RGA, GAI, and RGL1, 2, and 3), 15 GID1/DELLA combinations are possible. A yeast two-hybrid analysis confirmed that the GA-dependent interaction between AtGID1s and AtDELLAs occurs in all 15 combinations with different affinities in each combination. This suggests that there are combinatorial biases for AtGID1-DELLA in the complicated GA-signal transduction of *Arabidopsis*, but unambiguous biological differences for each combination have not been observed. However, it is possible that preferential interactions between a specific AtGID1 and AtDELLA occur in each GA-dependent biological event, because all five DELLA proteins are differentially involved in GA-dependent events, as mentioned previously. Precise analysis on the AtGID1/AtDELLA interaction in planta should be examined for each GA-dependent event, because of the possibility of modifying the interaction between AtGID1s and AtDELLAs, including the DELLA protein modification (21, 64), the presence of SLY1, or combining DELLA proteins under completive conditions. The affinity of the AtGID1c-GA interaction was increased about 100-fold by RGA or GAI, suggesting that the GID1-GA complex is stabilized by DELLA proteins (42). The expression of each *AtGID1* in rice *gid1* mutants rescued its GA-insensitive dwarf phenotype, demonstrating that AtGID1s function as GA receptors in rice, and suggesting that a common GA signaling pathway operates in both rice and *Arabidopsis* (42).

IS GID1 THE SOLE GIBBERELLIN RECEPTOR IN PLANTS?

As previously described, several experiments using the induction of α-amylase expression in cereal aleurone cells indicate that binding of GA to a plasma membrane–localized receptor is required for GA signaling. On the other hand, the GA-dependent α-amylase induction hardly occurs in *gid1* aleurone cells (69). In terms of GA perception, cereal aleurone cells are unusual, because these cells cannot produce bioactive GA themselves but can only perceive transported GAs produced in the embryo, while almost all rice cells except aleurone cells also can synthesize bioactive GAs (28). Under such situations, it is possible that cereal aleurone cells may have gained an additional, unique GA-perception system besides the constitutive GID1-DELLA-mediated GA-perception system. Another supporting bit of evidence for the presence of a plasma membrane–localized receptor is the involvement of trimeric G proteins in GA signaling (1, 70). In animal cells, trimeric G proteins function as mediators from hormone receptors, which carry a membrane-spanning structure and are localized within the plasma membrane, to the cytosolic signaling pathway. By analogy, involvement of trimeric G protein in GA signaling would suggest that GA is perceived by a membrane-localized receptor. It is now known that the trimeric G proteins are involved not only in GA signaling but also in various kinds of signal pathways, such as ABA (46), auxin (71), brassinosteroids (6), and pathogen-resistance (36). Thus, the dwarfism of G protein–deficient mutants may not be simply caused by a defect in GA signaling. Actually, rice *d1* Gα mutants show a semidwarf phenotype similar to GA-related mutants, but the global morphology of the *d1* plants is not the same as that of rice GA-deficient mutants (M. Ashikari, unpublished results). The involvement of trimeric G proteins in GA signaling should be reviewed with care to eliminate the effects of other signaling pathways.

Hartweck & Olszewski (16) raised the question of whether GA responses are always mediated by degradation of DELLA proteins based on the observation that the fastest GA response, an increase in calcium concentration, occurs in wheat aleurone cells during the 2–5 min following GA treatment, whereas the fastest documented GA-dependent decrease in a DELLA protein occurs 5–10 min after GA treatment. According to the time course of GA-induced responses of barley and wheat aleurone cells, which was deduced from the results of several independent experiments, the response of SLN1 degradation to GA treatment is a little faster than calcium uptake (65). Either way, it will be necessary to precisely examine the time course of GA signaling and response events under the same experimental conditions. As GA perception by GID1 directly transmits to DELLA proteins and continuously induces the degradation of DELLA proteins to release the suppressive state of GA action, if there is some level of GA action(s) not accounted for by the DELLA protein degradation mechanism, the presence of an alternative GA receptor would be the primary candidate. Because DELLA proteins and GID1 are specifically and preferentially localized in nuclei, GA signaling mediated by the GID1/DELLA system should be directly linked with the regulation of transcription. In this context, GA actions not involved in transcriptional regulation would need to be known to address whether there is GA signaling independent from the GID1/DELLA pathway.

Establishing when the GID1/DELLA-mediated GA-perception system evolved is an alternative way to answer the above question. Some tree ferns contain GA_1 and GA_4 as the dominant bioactive GA forms (78, 79), and the lycophyte *Selaginella moellendorffii*, which is a member of one of the oldest lineages of vascular plants (74), contains some homologous genes to the angiosperm GID1, DELLA, and GID2 proteins (M. Matsuoka, unpublished results), suggesting that moniliforms with true leaves may also use the GID1/DELLA-mediated GA-perception system. However, genes encoding these homologous proteins in a model moss plant, *Physcomitrella patens*, which as a member of the bryophytes diverged from the ancestors of vascular plants early in land plant evolution, around 430 mya (44), are not detected, and it is also unknown whether this plant uses GAs as a growth regulator (M. Matsuoka, unpublished results). This suggests that the growth hormone GA and the GID1/DELLA-perception system may have been established in ancestors of vascular plants soon after divergence from bryophytes at an early stage of land plant evolution, although direct evidence that the *S. moellendorffii* homologous genes actually function in GA signaling should be confirmed.

SUMMARY POINTS

1. The biochemical search for GA receptors has gone on for a long time, and through these experiments it has been postulated that there are two types of GA receptors, including soluble and membrane-bound forms.
2. Degradation of DELLA proteins is the key step of the GA signaling pathway. DELLA protein degradation causes derepression of the repression state of GA action. The DELLA protein is degraded by 26S proteasome-mediated proteolysis, whereas an F-box protein specific for DELLA protein degradation is necessary for SCF targeting to DELLA protein.

3. The single rice GID1 protein, which is similar to the hormone-sensitive lipases, was identified as a soluble GA receptor by its affinity to GAs, specific interactions with bioactive GAs, and the GA-hypersensitive response of GID1 overproducing plants. *Arabidopsis* has three GID1 proteins, which have similar characteristics to the rice protein.
4. Binding of GA with GID1 induces interaction of the DELLA protein to the GA-GID1 complex in yeast cells. Furthermore, the DELLA protein promotes the binding of GA with GID1 by stabilizing the GA-GID1 complex. Interaction between GID1 and DELLA proteins suggests that the GA-perception signal mediated by GID1 is transduced to the DELLA protein directly and probably induces its degradation, mediated by the F-box protein.
5. Several lines of evidence suggest the presence of a plasma membrane–bound GA receptor. Further studies are necessary to determine whether an alternative GA receptor is actually located on the plasma membrane.

FUTURE ISSUES

1. The X-ray and nuclear magnetic resonance (NMR) analyses of GID1 are important to gain the three-dimensional (3D) structure of the GA receptor. The information of the 3D structure of GID1 alone and interacting with GA will hint at the molecular mechanism of formation of the GID1-GA-DELLA trio complex.
2. Is the GID1-GA-DELLA trio complex targeted by an F-box, GID2/SLY1, directly for degradation of DELLA protein? If so, what is the molecular mechanism for direct interaction between DELLA and F-box? If not, what is protein necessary for DELLA protein degradation?
3. It is necessary to clarify the molecular function of DELLA proteins. When DELLA proteins are transcription factors, as expected, what are their target genes?
4. Identification and isolation of alternative GA receptor localized on the plasma membrane are necessary.

ACKNOWLEDGMENTS

This work was supported in part by a Grant-in-Aid for the Center of Excellence, the MAFF Rice Genome Project, IP1003 (M.A. and M.M.), and by the Ministry of Education, Culture, Sports, Science and Technology of Japan (M.M., M.N., and M.U.-T.).

LITERATURE CITED

1. Ashikari M, Wu J, Yano M, Sasaki T, Yoshimura A. 1999. Rice gibberellin-insensitive dwarf mutant gene *Dwarf 1* encodes the α-subunit of GTP-binding protein. *Proc. Natl. Acad. Sci. USA* 96:10284–89

2. Beale MH, Ward JL, Smith SJ, Hooley R. 1992. A new approach to gibberellin perception in aleurone: novel, hydrophylic, membrane-impermeant, GA-sulphonic acid derivatives induce α-amylase formation. *Physiol. Plant.* 85:A136
3. Boss PK, Thomas MR. 2002. Association of dwarfism and floral induction with a grape 'green revolution' mutation. *Nature* 416:847–50
4. Cao D, Hussain A, Cheng H, Peng J. 2005. Loss of function of four DELLA genes leads to light- and gibberellin-independent seed germination in *Arabidopsis*. *Planta* 223:105–13
5. Chandler PM, Marion-Poll A, Ellis M, Bubler F. 2002. Mutants at the *Slender1* locus of barley cv Himalaya. Molecular and physiological characterization. *Plant Physiol.* 129:181–90
6. Chen JG, Pandey S, Huang J, Alonso JM, Ecker JR, et al. 2004. GCR1 can act independently of heterotrimeric G-protein in response to brassinosteroids and gibberellins in *Arabidopsis* seed germination. *Plant Physiol.* 135:907–15
7. Davies PJ, ed. 1995. *Plant Hormones: Physiology, Biochemistry and Molecular Biology*. Dordrecht: Kluwer Acad.
8. Deshaies RJ. 1999. SCF and Cullin/Ring H2-based ubiquitin ligases. *Annu. Rev. Cell Dev. Biol.* 15:435–67
9. Dill A, Sun TP. 2001. Synergistic derepression of gibberellin signaling by removing RGA and GAI function in *Arabidopsis thaliana*. *Genetics* 159:777–85
10. Dill A, Thomas SG, Hu J, Steber CM, Sun TP. 2004. The *Arabidopsis* F-box protein SLEEPY1 targets gibberellin signaling repressors for gibberellin-induced degradation. *Plant Cell* 16:1392–405
11. Fu X, Harberd NP. 2003. Auxin promotes *Arabidopsis* root growth by modulating gibberellin response. *Nature* 421:740–43
12. Fu X, Richards DE, Ait-Ali T, Hynes LW, Ougham H, et al. 2002. Gibberellin-mediated proteasome-dependent degradation of the barley DELLA protein SLN1 repressor. *Plant Cell* 14:3191–200
13. Gagne JM, Downes BP, Shin-Han S, Durski AM, Vierstra RD. 2002. The F-box subunit of the SCF E3 complex is encoded by a diverse superfamily of genes in *Arabidopsis*. *Proc. Natl. Acad. Sci. USA* 99:11519–24
14. Gilroy S, Jones RL. 1994. Perception of gibberellin and abscisic acid at the external face of the plasma membrane of barley (*Hordeum vulgare* L.) aleurone protoplasts. *Plant Physiol.* 104:1185–92
15. Gomi K, Sasaki A, Itoh H, Ueguchi-Tanaka M, Ashikari M, et al. 2004. GID2, an F-box subunit of the SCF E3 complex, specifically interacts with phosphorylated SLR1 protein and regulates the gibberellin-dependent degradation of SLR1 in rice. *Plant J.* 37:626–34
16. Hartweck LM, Olszewski NE. 2006. Rice GIBBERELLIN INSENSITIVE DWARF1 is a gibberellin receptor that illuminates and raises questions about GA signaling. *Plant Cell* 18:278–82
17. Hooley R, Beale MH, Smith SJ. 1990. Gibberellin perception in the *Avena fatua* aleurone. *Symp. Soc. Exp. Biol.* 44:79–86
18. Hooley R, Beale MH, Smith SJ. 1991. Gibberellin perception at the plasma membrane of *Avena fatua* aleurone protoplasts. *Planta* 183:274–80
19. Hooley R, Beale MH, Smith SJ, Walker RP, Rushton PJ, et al. 1992. Gibberellin perception and the *Avena fatua* aleurone: do our molecular keys fit the correct locks? *Biochem. Soc. Trans.* 20:85–89
20. Hooley R, Smith SJ, Beale MH, Walker RP. 1993. In vivo photoaffinity labeling of gibberellin-binding proteins in *Avena fatua* aleurone. *Aust. J. Plant Physiol.* 20:573–84

21. Hussain A, Cao D, Cheng H, Wen Z, Peng J. 2005. Identification of the conserved serine/threonine residues important for gibberellin-sensitivity of *Arabidopsis* RGL2 protein. *Plant J.* 44:88–99
22. Ikeda A, Ueguchi-Tanaka M, Sonoda Y, Kitano H, Koshioka M, et al. 2001. Slender rice, a constitutive gibberellin response mutant is caused by a null mutation of the *SLR1* gene, an ortholog of the height-regulating gene *GAI/RGA/RHT/D8*. *Plant Cell* 13:999–1010
23. Itoh H, Sasaki A, Ueguchi-Tanaka M, Ishiyama K, Kobayashi M, et al. 2005. Dissection of the phoshorylation of rice DELLA protein, SLENDER RICE1. *Plant Cell Physiol.* 46:1392–99
24. Itoh H, Ueguchi-Tanaka M, Sato Y, Ashikari M, Matsuoka M. 2002. The gibberellin signaling pathway is regulated by the appearance and disappearance of SLENDER RICE1 in nuclei. *Plant Cell* 14:57–70
25. Ivan M, Kondo K, Yang H, Kim W, Valiando J, et al. 2001. HIF α targeted for VHL-mediated destruction by proline hydroxylation: implications for O-2 sensing. *Science* 292:464–68
26. Jelsema CL, Ruddat M, Morre DJ, Williamson FA. 1977. Specific binding of gibberellin A_1 to aleurone grain fractions from wheat endosperm. *Plant Cell Physiol.* 18:1009–19
27. Johri MM, Varner JE. 1968. Enhancement of RNA synthesis in isolated pea nuclei by gibberellic acid. *Proc. Natl. Acad. Sci. USA* 59:269–76
28. Kaneko M, Itoh H, Inukai Y, Sakamoto T, Ueguchi-Tanaka M, et al. 2003. Where do gibberellin biosynthesis and gibberellin signaling occur in rice plants? *Plant J.* 35:104–15
29. Keith B, Brown S, Srivastava LM. 1982. In vivo binding of gibberellin A_4 to extracts of cucumber measured by using DEAE-cellulose filters. *Proc. Natl. Acad. Sci. USA* 79:1515–19
30. Keith B, Foster NA, Bonettemaker M, Srivastava LM. 1981. In vitro gibberellin A_4 binding to extracts of cucumber hypocotyls. *Plant Physiol.* 68:344–48
31. Keith B, Rappaport L. 1987. In vitro gibberellin A_1 binding in *Zea mays* L. *Plant Physiol.* 85:934–41
32. Keith B, Srivastava LM. 1980. In vivo binding of gibberellin A_1 in dwarf pea epicotyls. *Plant Physiol.* 66:962–67
33. King KE, Moritz T, Harberd NP. 2001. Gibberellins are not required for normal stem growth in *Arabidopsis thaliana* in the absence of GAI and RGA. *Genetics* 159:767–76
34. Komatsu S, Masuda T, Hirano H. 1996. Rice gibberellin-binding phosphoprotein structurally related to ribulose-1,5-bisphosphate carboxylase/oxygenase activase. *FEBS Lett.* 384:167–71
35. Lee S, Cheng H, King KE, Wang W, He Y, et al. 2002. Gibberellin regulates *Arabidopsis* seed germination via RGL2, a GAI/RGA-like gene whose expression is up-regulated following imbibition. *Genes Dev.* 16:646–58
36. Lieberherr D, Thao NP, Nakashima A, Umemura K, Kawasaki T, Shimamoto K. 2005. A sphingolipid elicitor-inducible mitogen-activated protein kinase is regulated by the small GTPase OsRac1 and heterotrimeric G-protein in rice 1. *Plant Physiol.* 138:1644–52
37. Liu ZH, Ger MJ. 1995. Partial purification of gibberellin-binding proteins from dwarf pea. *Plant Physiol. Biochem.* 33:675–81
38. Liu ZH, Lee BH. 1995. In vitro binding of gibberellin A_4 in epicotyls of dwarf pea. *Bot. Bull. Acad. Sin.* 36:73–79
39. Lovegrove A, Barratt DH, Beale MH, Hooley R. 1998. Gibberellin-photoaffinity labeling of two polypeptides in plant plasma membranes. *Plant J.* 15:311–20
40. McGinnis KM, Thomas SG, Soule JD, Strader LC, Zale JM, et al. 2003. The *Arabidopsis SLEEPY1* gene encodes a putative F-box subunit of an SCF E3 ubiquitin ligase. *Plant Cell* 15:1120–30

41. Nakajima M, Sakai S, Kanazawa K, Kizawa S, Yamaguchi I, et al. 1993. Partial purification of a soluble gibberellin-binding protein from mung bean hypocotyls. *Plant Cell Physiol.* 34:289–96
42. Nakajima M, Shimada A, Takashi Y, Kim YC, Park SH, et al. 2006. Identification and characterization of *Arabidopsis* gibberellin receptors. *Plant J.* 46:880–89
43. Nakajima M, Takita K, Wada H, Mihara K, Hasegawa M, et al. 1997. Partial purification and characterization of a gibberellin-binding protein from seedlings of *Azukia angularis*. *Biochem. Biophys. Res. Commun.* 241:782–86
44. Nishiyama T, Fujita T, Shin-I T, Seki M, Nishide H, et al. 2003. Comparative genomics of *Physcomitrella patens* gametophytic transcriptome and *Arabidopsis thaliana*: Implication for land plant evolution. *Proc. Natl. Acad. Sci. USA* 100:8007–12
45. Osterlund T. 2001. Structure-function relationships of hormone-sensitive lipase. *Eur. J. Biochem.* 268:1899–907
46. Pandey S, Assmann SM. 2004. The *Arabidopsis* putative G protein-coupled receptor GCR1 interacts with the G protein α subunit GPA1 and regulates abscisic acid signaling. *Plant Cell* 16:1616–32
47. Park SH, Nakajima M, Hasegawa M, Yamaguchi I. 2005. Similarities and differences between the characteristics of gibberellin-binding protein and gibberellin 2-oxidases in adzuki bean (*Vigna angularis*) seedlings. *Biosci. Biotechnol. Biochem.* 69:1508–14
48. Park SH, Nakajima M, Sakane M, Xu ZJ, Tomioka K, et al. 2005. Gibberellin 2-oxidases from seedlings of adzuki bean (*Vigna angularis*) show high gibberellin-binding activity in the presence of 2-oxoglutarate and Co^{2+}. *Biosci. Biotechnol. Biochem.* 69:1498–507
49. Peng J, Carol P, Richards DE, King KE, Cowling RJ, et al. 1997. The *Arabidopsis GAI* gene defines a signaling pathway that negatively regulates gibberellin responses. *Genes Dev.* 11:3194–205
50. Peng J, Richards DE, Hartley NM, Murphy GP, Devos KM, et al. 1999. 'Green revolution' genes encode mutant gibberellin response modulators. *Nature* 400:256–61
51. Pysh LD, Wysocka-Diller JW, Camilleri C, Bouchez D, Benfey PN. 1999. The GRAS gene family in *Arabidopsis*: sequence characterization and basic expression analysis of the SCARECROW-LIKE genes. *Plant J.* 18:111–19
52. Reynolds GJ, Hooley R. 1992. cDNA cloning of a tetraubiquitin gene, and expression of ubiquitin-containing transcripts, in aleurone layers of *Avena fatua*. *Plant Mol. Biol.* 20:753–58
53. Sakai S, Nakajima M, Yamaguchi I. 1994. Isolation of gibberellin-binding proteins by affinity chromatography. *Biosci. Biotech. Biochem.* 58:963–64
54. Sakamoto T, Kobayashi M, Itoh H, Tagiri A, Kayano T, et al. 2001. Expression of a gibberellin 2-oxidase gene around the shoot apex is related to phase transition in rice. *Plant Physiol.* 125:1508–16
55. Sasaki A, Itoh H, Gomi K, Ueguchi-Tanaka M, Ishiyama K, et al. 2003. Accumulation of phosphorylated repressor for gibberellin signaling in an F-box mutant. *Science* 299:1896–98
56. Sechley KA, Srivastava LM. 1991. Gibberellin-enhanced transcription by isolated nuclei from cucumber hypocotyls. *Physiol. Plant* 82:543–50
57. Sharma A, Komatsu S. 2002. Involvement of a Ca^{2+}-dependent protein kinase component downstream to the gibberellin-binding phosphoprotein, RuBisCO activase, in rice. *Biochem. Biophys. Res. Commun.* 290:690–95
58. Shimada A, Ueguchi-Tanaka M, Sakamoto T, Fujioka S, Takatsuto S, et al. 2006. The rice *SPINDLY* gene functions as a negative regulator of gibberellin signaling by controlling the suppressive function of the DELLA protein, SLR1, and modulating brassinosteroid synthesis. *Plant J.* 48:390–402

59. Silverstone AL, Ciampaglio CN, Sun T. 1998. The *Arabidopsis RGA* gene encodes a transcriptional regulator repressing the gibberellin signal transduction pathway. *Plant Cell* 10:155–69
60. Silverstone AL, Mak PY, Martínez EC, Sun T. 1997. The new *RGA* locus encodes a negative regulator of gibberellin response in *Arabidopsis thaliana*. *Genetics* 146:1087–99
61. Stoddart JL. 1979. Interaction of [^{3}H] gibberellin A_1 with a subcellular fraction from lettuce (*Lactuca sativa* L.) hypocotyls. I. Kinetics of labeling. *Planta* 146:353–61
62. Stoddart JL, Breidenbach W, Nadeau R, Rappaport L. 1974. Selective binding of [^{3}H] gibberellin A_1 by protein fractions from dwarf pea epicotyls. *Proc. Natl. Acad. Sci. USA* 71:3255–59
63. Strader LC, Ritchie S, Soule JD, McGinnis KM, Steber CM. 2004. Recessive-interfering mutations in the gibberellin signaling gene SLEEPY1 are rescued by overexpression of its homologue, SNEEZY. *Proc. Natl. Acad. Sci. USA* 101:12771–76
64. Swain SM, Tseng TS, Thornton TM, Gopalraj M, Olszewski NE. 2002. SPINDLY is a nuclear-localized repressor of gibberellin signal transduction expressed throughout the plant. *Plant Physiol.* 129:605–15
65. Sun T, Gubler F. 2004. Molecular mechanism of gibberellin signaling in plants. *Annu. Rev. Plant Biol.* 55:197–223
66. Tevzadze NN, Dzhokhadze DI. 1991. Gibberellin-binding proteins from cell nuclei of bean epicotyls and their effect on transcription in nuclei. *Fiziol. Rast.* 38:890–96
67. Thornton TM, Swain SM, Olszewski NE. 1999. Gibberellin signal transduction presents . . . the SPY who O-GlcNAc'd me. *Trends Plant Sci.* 4:424–28
68. Tyler L, Thomas SG, Hu J, Dill A, Alonso JM, et al. 2004. Della proteins and gibberellin-regulated seed germination and floral development in *Arabidopsis*. *Plant Physiol.* 135:1008–19
69. Ueguchi-Tanaka M, Ashikari M, Nakajima M, Itoh H, Katoh E, et al. 2005. *GIBBERELLIN INSENSITIVE DWARF1* encodes a soluble receptor for gibberellin. *Nature* 437:693–98
70. Ueguchi-Tanaka M, Fujisawa Y, Kobayashi M, Ashikari M, Iwasaki Y, et al. 2000. Rice dwarf mutant *d1*, which is defective in the α subunit of the heterotrimeric G protein, affects gibberellin signal transduction. *Proc. Natl. Acad. Sci. USA* 97:11638–43
71. Ullah H, Chen JG, Temple B, Boyes DC, Alonso JM, et al. 2003. The β-subunit of the *Arabidopsis* G protein negatively regulates auxin-induced cell division and affects multiple developmental processes. *Plant Cell* 15:393–409
72. Walker RP, Waterworth WM, Beale MH, Hooley R. 1994. Gibberellin-photoaffinity labeling of wild oat (*Avena fatua* L.) aleurone protoplasts. *Plant Growth Regul.* 15:271–79
73. Wen CK, Chang C. 2002. *Arabidopsis* RGL1 encodes a negative regulator of gibberellin responses. *Plant Cell* 14:87–100
74. Weng JK, Tanurdzic M, Chapple C. 2005. Functional analysis and comparative genomics of expressed sequence tags from the lycophyte *Selaginella moellendorffii*. *BMC Genomics* 6:85
75. Witham FW, Hendry LB. 1992. Computer modeling of gibberellin-DNA binding. *J. Theor. Biol.* 155:55–67
76. Yalpani N, Srivastava LM. 1985. Competition for in vitro [^{3}H] gibberellin A_4 binding in cucumber by gibberellins and their derivatives. *Plant Physiol.* 79:963–67
77. Yalpani N, Suttle JC, Hultstrand JF, Rodaway SJ. 1989. Competition for in vitro [^{3}H] gibberellin A_4 binding in cucumber by substituted phthalimides: Comparison with in vivo gibberellin-like activity. *Plant Physiol.* 91:823–28
78. Yamane H, Fujioka S, Spray CR, Phinney BO, MacMillan J, et al. 1988. Endogenous gibberellins from sporophytes of two tree ferns, *Cibotium glaucum* and *Dicksonia antarctica*. *Plant Physiol.* 86:857–62

79. Yamane H, Yamaguchi I, Kobayashi M, Takahashi M, Sato Y, et al. 1985. Identification of ten gibberellins from sporophytes of the tree fern, *Cyathea australis*. *Plant Physiol.* 78:899–903
80. Yoshida Y, Tokunaga F, Chiba T, Iwai K, Tanaka K, Tai T. 2003. Fbs2 is a new member of the E3 ubiquitin ligase family that recognizes sugar chains. *J. Biol. Chem.* 278:43877–84
81. Yu H, Ito T, Zhao Y, Peng J, Kumar P, Meyerowitz EM. 2004. Floral homeotic genes are targets of gibberellin signaling in flower development. *Proc. Natl. Acad. Sci. USA* 101:7827–32

RELATED RESOURCES

Itoh H, Matsuoka M, Steber CM. 2003. A role for the ubiquitin-26s proteasome pathway in gibberellin signaling. *Trends Plant Sci.* 8:492–97

Olszewski N, Sun T-P, Gubler F. 2002. Gibberellin signaling: biosynthesis, catabolism, and response pathways. *Plant Cell* 14(Suppl.):S61–80

Cyclic Electron Transport Around Photosystem I: Genetic Approaches

Toshiharu Shikanai

Graduate School of Agriculture, Kyushu University, Fukuoka, Japan 812-8581;
email: shikanai@agr.kyushu-u.ac.jp

Annu. Rev. Plant Biol. 2007. 58:199–217

The *Annual Review of Plant Biology* is online at plant.annualreviews.org

This article's doi:
10.1146/annurev.arplant.58.091406.110525

First published online as a Review in Advance on January 3, 2007

1543-5008/07/0602-0199$20.00

Key Words

alternative electron transport, chloroplast, photosynthesis

Abstract

The light reactions in photosynthesis convert light energy into chemical energy in the form of ATP and drive the production of NADPH from $NADP^+$. The reactions involve two types of electron flow in the chloroplast. While linear electron transport generates both ATP and NADPH, photosystem I cyclic electron transport is exclusively involved in ATP synthesis. The physiological significance of photosystem I cyclic electron transport has been underestimated, and our knowledge of the machineries involved remains very limited. However, recent genetic approaches using *Arabidopsis thaliana* have clarified the essential functions of this electron flow in both photoprotection and photosynthesis. Based on several lines of evidence presented here, it is necessary to reconsider the fundamental mechanisms of chloroplast energetics.

Contents

INTRODUCTION

The light reactions in photosynthesis result in electron transport through the thylakoid membrane of the chloroplast. Electrons excised from water in photosystem II (PSII) are ultimately transferred to $NADP^+$, resulting in accumulation of NADPH (**Figure 1**). This process is termed linear electron transport (LET). It is driven by the two photochemical reactions, PSII and photosystem I (PSI), functioning in series. At the same time, electron transport through the cytochrome (cyt) b_6f complex intermediating PSII and PSI generates a proton gradient across the thylakoid membrane (ΔpH), which is utilized in ATP synthesis. In contrast, PSI cyclic electron transport (CET) depends solely on the PSI photochemical reaction (10, 22, 58). Because electrons are recycled from NAD(P)H or ferredoxin (Fd) to plastoquinone (PQ), PSI CET can generate ΔpH without accumulating NADPH (**Figure 2**).

ATP and NADPH, which are generated by light reactions, are utilized in various metabolic pathways not only in chloroplasts but also in other organelles. Their major consumer is CO_2 fixation, which theoretically requires an ATP/NADPH ratio of 1.5. In air, photorespiration inevitably takes place in C_3 plants, resulting in a need for the theoretical ATP/NADPH ratio to rise to 1.66 (74). The fundamental question of how this requirement of ATP/NADPH ratio is fulfilled by light reactions remains unanswered. The historical debate concerns the contribution of PSI CET in photosynthesis, which can generate ATP without accumulation of NADPH. In addition to its crucial role in ATP synthesis, ΔpH is a key factor for inducing the rapid response of photosynthesis to fluctuating light intensity (66). It has long been discussed whether PSI CET is involved in regulating photosynthesis via the lumen acidification (25).

Recently, genetic investigations have clarified the physiological significance of PSI CET (56, 57). Serious re-evaluation of this enigmatic electron transport has just begun. This review covers current views on PSI CET in higher plants, especially as revealed by genetics.

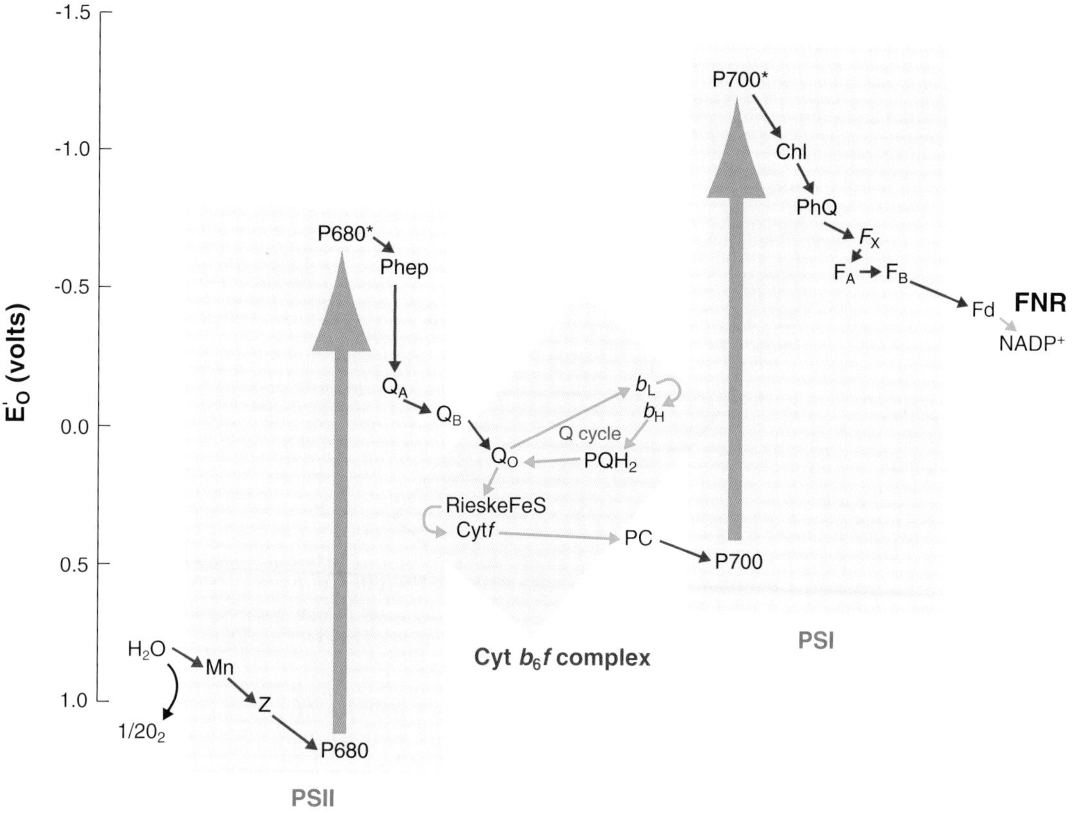

Figure 1

Diagram of linear electron transport (Z scheme). This figure was modified from Reference 50 with permission of the authors.

HISTORICAL VIEW OF PHOTOSYSTEM I CYCLIC ELECTRON TRANSPORT

It is difficult to view the early studies out of their historical context. A review by Arnon (2) helps young scientists to follow the story. In this review, two papers (3, 4) are cited, in which he reports the discovery of photophosphorylation in chloroplasts. Because the phenazine methosulfate (or flavin mononucleotide, vitamin K_5)-mediated, photosynthetic ATP synthesis does not produce oxygen, these papers are generally accepted as the first reports of PSI CET. This reaction was distinguished from photophosphorylation coupled with oxygen evolution and $NADP^+$ reduction (LET) in the paper published later (5), where the term "cyclic phosphorylation" was used.

In 1963 (90), Fd was reported as a cofactor required for cyclic phosphorylation in place of artificial carriers. Another important discovery reported in this paper is that this Fd-dependent cyclic phosphorylation is sensitive to antimycin A (AA). In this study, the concept of cyclic phosphorylation, in which ATP synthesis is driven by cyt-mediated electron transport, was established. Because electron transport is solely driven by PSI, the electron flow is called PSI CET.

In the 1960s, a model of the Z scheme was proposed to explain LET (27), which is now accepted as a dogma. We now understand electron transport at the level of the crystal structures of each component

PSI CET: photosynthetic electron transport that is driven solely by PSI and generates ΔpH, and consequently ATP, without accumulating NADPH

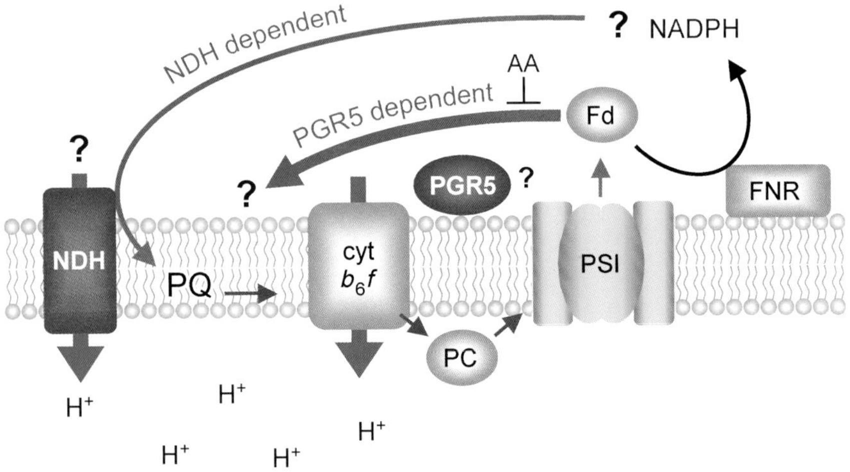

Figure 2

A schematic representation of photosystem I (PSI) cyclic electron transport (CET). In vascular plants, PSI CET consists of two partially redundant pathways, the NDH-dependent and the PGR5-dependent pathway. The PGR5-dependent pathway is inhibited by antimycin A (AA). The electron donor to the NDH complex is not known. The NDH complex may have proton-pumping activity. The route taken by electrons in the PGR5-dependent pathway is not known. The exact localization of PGR5 is unclear. PC indicates plastocyanin.

NDH complex: a protein complex involved in chlororespiration and PSI CET. The cyanobacterial and chloroplast NDH complexes are structurally and functionally divergent from the bacterial and mitochondrial complexes

involved in LET (50). In contrast, our knowledge of PSI CET is still based on information gained in the early 1960s. In their review, Bendall & Manasse (10) used the expression "the Cinderella of chloroplast energetics" to refer to PSI CET. Because the Z scheme can explain both ATP and NADPH synthesis in light reactions, it was not necessary to consider the physiological function of the PSI CET to explain the mechanism of photosynthesis.

DISCOVERY OF THE NDH-DEPENDENT PHOTOSYSTEM I CYCLIC ELECTRON TRANSPORT

In higher plants, there are two partially redundant routes taken by electrons in PSI CET (56). The main route is the classical one discovered by Arnon and coworkers (3, 4). Another pathway depends on the activity of the chloroplast NAD(P)H dehydrogenase (NDH) complex (12, 84).

One surprising feature of the chloroplast genomes is the presence of 11 genes (*ndhA-ndhK*) that encode homologs of the NADH dehydrogenase complex (complex I) subunits in the mitochondria (48). Why would the chloroplast genome encode homologs of the mitochondrial machinery for respiration? The answer was provided by the discovery of the M55 mutant in *Synechocystis* PCC6803, which requires a high CO_2 concentration for growth (69). M55 is defective in *ndhB* encoding a subunit of the cyanobacterial NDH complex and has impaired PSI CET activity (51–53, 69). Although the exact mechanism in which PSI CET contributes to the CO_2 concentration still remains unclear, this discovery raised the possibility that the NDH complex is also involved in PSI CET in chloroplasts. The chloroplast NDH complex is more similar to the cyanobacterial complex than to the mitochondrial complex I in the same species, suggesting an evolutionary and functional similarity between the NDH complexes in cyanobacteria and chloroplasts.

In the 1990s, the technique of chloroplast transformation facilitated the knockout of chloroplast genes in tobacco (89). The 11 *ndh* genes were also early targets of this technique (12, 38, 84). The knockout lines displayed a minor but clear alteration in electron transport in chloroplasts. In wild-type tobacco, chlorophyll fluorescence level transiently increased after turning off actinic light. This fluorescence change is ascribed to PQ reduction and was impaired in the knockout tobacco defective in NDH activity (12, 38, 84). This result indicates that the chloroplast NDH complex is involved in electron transport from the stromal electron pool to PQ. Once PQ is reduced, it is oxidized in the light by PSI via the cyt b_6f complex (PSI CET).

PGR5-DEPENDENT PHOTOSYSTEM I CYCLIC ELECTRON TRANSPORT: REDISCOVERY

Light energy absorption by light-harvesting complex II results in excitation of a chlorophyll molecule. The energy is finally transferred to the PSII reaction center P680. To prevent generation of singlet oxygen under excessive light conditions, chlorophyll is de-excited by a thermal dissipation process (17, 66). Induction of thermal dissipation is triggered by acidification of the thylakoid lumen, and can be monitored as a qE component of nonphotochemical quenching (NPQ). Using a two-dimensional chlorophyll fluorescence imaging system, *npq1* mutants defective in violaxanthin de-epoxidase activity have been isolated in both *Chlamydomonas reinhardtii* (64) and *Arabidopsis thaliana* (65). The genetic approach has confirmed that the xanthophyll cycle, in which violaxanthin is converted to zeaxanthin via antheraxanthin at high light intensity, is essential for qE induction. Much more important progress was made with the discovery of the *Arabidopsis* mutant *npq4* (45). Although its xanthophyll cycle activity is normal, *npq4* lacks qE induction. *NPQ4* encodes a PSII subunit, PsbS, which is not required for PSII photochemistry but is essential for qE induction as a pH sensor (46).

Because PSI CET is involved in ΔpH generation, the contribution of PSI CET in NPQ induction has been discussed (25). If this is true, a mutant defective in PSI CET should be identifiable by its *npq* phenotype. The contribution of the NDH-dependent pathway is too minor to affect NPQ induction (84), and thus the target was a mutant defective in the non-NDH pathway.

PGR5-Dependent Photosystem I Cyclic Electron Transport Revealed by Genetics: Function in Photoprotection

PSI CET mediated by the chloroplast NDH complex is different from the classical version. Although Arnon's pathway is sensitive to AA (90), the NDH complex is resistant to the same concentration of AA (20). The *Arabidopsis* mutant *proton gradient regulation* (*pgr5*) was identified based on its *npq* phenotype using chlorophyll fluorescence imaging (86). In *pgr5*, the ratio of P700$^+$/P700 (oxidized/reduced PSI reaction center) is lowered at high light intensities (57), although it elevates in response to increased light intensity in the wild type. The ratio was restored by infiltrating methylviologen, an artificial electron acceptor from PSI, to *pgr5* leaves. Even under low light conditions, the electron acceptance from PSI was affected in CO_2-free air but not in air including CO_2. Furthermore, LET was not affected in thylakoids isolated from *pgr5* leaves. These results indicate that *pgr5* is defective in alternative electron acceptance from PSI, which is most probably PSI CET. However, NDH activity was unaffected in *pgr5*, suggesting that *pgr5* is defective in the non-NDH pathway of PSI CET.

To show more direct evidence for the conclusion that *pgr5* is defective in PSI CET, Fd-dependent PQ reduction, which has been thought to be the route taken by electrons in Arnon's pathway, was assayed in ruptured chloroplasts (57). The PQ reduction activity was significantly suppressed in *pgr5*, and adding AA to the wild-type chloroplasts

Chlorophyll fluorescence: fluorescence that is emitted from PSII at physiological temperatures and reflects of the status of photosynthetic electron transport

Nonphotochemical quenching (NPQ): a quenching of chlorophyll fluorescence based on regulatory processes of photosynthesis. A qE component reflects the dissipation of absorbed light energy from PSII as heat (thermal dissipation), which is induced by acidification of the thylakoid lumen

P700: a reaction center chlorophyll in PSI. P700$^+$ is an oxidized form and its level can be monitored as absorbance at 810 nm in leaves

PGR5: a small thylakoid protein discovered in an *Arabidopsis* mutant, *pgr5*, that is defective in NPQ induction. PGR5 is essential for PSI CET

mimicked this phenotype. Although our assay system was not identical to that used by Arnon and coworkers, it is highly likely that we observed the same activity. Although there may be an alternative explanation, I believe that it is essential to assess the simplest hypothesis.

Our conclusion depends on the in vitro Fd-dependent PQ reduction assay (57). A possible problem is the slow speed of PQ reduction, which is inconsistent with an extreme phenotype, suggesting a high rate of operation in vivo. Our measurement conditions were not optimal, as the activity of the NDH- and PGR5-dependent PQ reduction was approximately the same in vitro (56). Based on the mutant phenotypes, the in vivo rate of electron transport should be much higher in the PGR5-dependent pathway than in the NDH-dependent pathway. It is probable that the rate of electron donation to PQ is also limited by other factors, such as the reverse reaction of Fd-NADP$^+$ reductase (FNR), which is essential to providing electrons to Fd, in the dark. It is also possible that PQ reduction activity competes with other electron acceptors in vitro, and also that redox conditions are not optimal for PGR5-dependent electron transport.

Arnon's PSI cyclic pathway has been often referred to as an Fd-dependent pathway, compared with the NDH-dependent pathway that probably requires NAD(P)H as an electron donor. However, the NDH-dependent PQ reduction also requires Fd in our in vitro assay system (20, 56). The electron donor to the chloroplast NDH complex is still a matter of debate (83), so it is not a good idea to distinguish two pathways by electron donor. To eliminate any possible confusion, we cautiously refer to the electron flow that we found as PGR5-dependent. Although the NDH complex is a mediator of electron flow, PGR5 may be indirectly involved in electron transport. The putative mediator for the Arnon pathway is often referred to as Fd-PQ reductase (FQR), but it is confusing to use this enzyme name without giving it any specific molecular identity. Strictly speaking, the PGR5-dependent PSI CET corresponds to the electron flow that is impaired in *pgr5*, but it is likely identical to Arnon's pathway. When it is not necessary or impossible to distinguish them rigorously, we use the term "non-NDH pathway" here.

PGR5-dependent PSI CET is essential to inducing qE, suggesting that it is essential for photoprotection. However, *pgr5* is more sensitive to excessive light than *npq4*, which is totally defective in qE induction (45, 57). Another extremely important function of PGR5-dependent PSI CET is protecting PSI from photodamage. Exposing *pgr5* plants to high light intensities causes photodamage to PSI (57). The mechanism of the PSI photoinhibition can be explained by a similar process, identified by chilling sensitive plants (87). At low temperatures, a shortage of electron acceptors (NADP$^+$ or oxidized Fd) from PSI due to low CO_2 fixation activity causes electrons to be trapped on the acceptor side of PSI (F_X and F_A/F_B). The reduced form of these electron carriers interacts with H_2O_2, resulting in the generation of a reactive hydroxyl radical, which damages PSI. A defect in the PSI CET causes stromal over-reduction (19), leading to PSI photoinhibition in *pgr5* even at room temperature.

ATP Synthesis and Photosystem I Cyclic Electron Transport

In addition to qE induction, ΔpH is a driving force for ATP synthesis. The *pgr5* mutant plants can grow as well as the wild type at low light intensities, suggesting the preferential function of PGR5 in photoprotection. However, the yield of PSII photochemistry is reduced at high light intensities in *pgr5* (57), a phenotype which cannot be explained by only a defect in photoprotection. Because NDH-dependent PSI CET is still active, this alternative route may complement the function of PSI CET in *pgr5*. To assess this possibility, we generated double mutants in which both pathways of PSI CET were impaired. Because it was not routinely possible to transform the chloroplast in *Arabidopsis*, we used a modified system of chlorophyll fluorescence imaging to identify *chlororespiratory reduction* (*crr*) mutants, in which NDH activity was specifically impaired due to defects in the nuclear genome (23).

In the double mutants, LET is severely affected, even at low light intensities. Consequently, the growth of the double mutants is severely impaired and the seedlings suffer photodamage. This phenotype can be explained according to the classical idea that PSI CET contributes to ATP synthesis during steady-state photosynthesis. Provided that PSI CET generates ΔpH and its contribution is not negligible, it is inevitably involved in ATP synthesis.

Our assay system characterizing the Fd-dependent PQ reduction activity in ruptured chloroplasts was also applied to *crr2* and the double mutant *crr2 pgr5* (56). *crr2* is specifically defective in the expression of *ndhB* encoding a subunit of the NDH complex (23). The activity was significantly lower in *crr2* than in the wild type, as it was in *pgr5* (56). In the double mutant *crr2 pgr5*, PQ-reducing activity was completely lost, and the remaining activity in *crr2* was totally inhibited by AA. Thus, two routes via PGR5- and NDH-dependent pathways explain almost all of the PSI CET activity in *Arabidopsis*.

Although thermal dissipation is indispensable to plants under fluctuating light conditions (42), it is not necessarily essential for plants to acclimate to constant high light intensity (64). Plants can acclimate to conditions by alternative strategies. However, *pgr5* is sensitive to high light intensity, even under constant light conditions, indicating that this *pgr5* phenotype is not directly caused by a defect in the induction of thermal dissipation (57).

LET activity at PSII is specifically lowered at high light intensity in *pgr5*. The reduced activity of PSI CET is unlikely to affect PSII photochemistry directly, and the balance of ATP/NADPH production should be taken into account. Because PSI CET generates ΔpH, the production ratio of ATP/NADPH is likely to be reduced in *pgr5*, under conditions where ATP limits metabolism, including CO_2 fixation. Ultimately, depletion of $NADP^+$ reduces the efficiency of PSII photochemistry due to a lack of electron acceptors from PSI. We refer to this condition as stromal over-reduction. It can be monitored as a low ratio of $P700^+/P700$ (19). It is likely that ATP and NADPH production are out of balance, even at low light intensities, although this defect cannot be monitored in chlorophyll fluorescence. The flexibility of photosynthetic electron transport relieves the resulting stromal over-reduction at low light intensities. This mechanism may include the water-water cycle, which can also generate ΔpH without accumulating NADPH (6, 47). Because the double mutants lacking both PSI CET exhibited an extreme phenotype, it is clear that the NDH complex is essential for this flexibility in photosynthesis (56).

There has long been a debate about whether the ATP/NADPH ratio required by the Calvin cycle is fulfilled solely by LET (1, 68, 85). The issue concerns the operation of the Q-cycle in the cyt b_6f complex, which doubles the efficiency of ΔpH generation (37), and also how many protons are required to synthesize one molecule of ATP (82). It is also necessary to consider the exchange of the reducing equivalents across chloroplast envelopes (77). Furthermore, electrons generated at PSII are utilized in various metabolic pathways except for in CO_2 fixation, which is incorporated into the calculation. For example, it is estimated that the net rate of N assimilation is around 6–13% of net C assimilation, and both processes occur simultaneously in the same photosynthetic cells (67). At present, the mutant phenotype is more reliable than the calculation, including several hypothetical values of parameters. The stroma is over-reduced in the light in *pgr5*, suggesting that the ATP/NADPH ratio is reduced.

Table 1 Distribution of two photosystem I cyclic pathways in plants

	NDH	PGR5
Arabidopsis thaliana	+	+
Oryza sativa (rice)	+	+
Zea mays (maize)	+	+
Poplus euphratica	+[1]	+
Pinus thunbergii (black pine)	−[2]	+[3]
Physcomitrella patens	+	+
Chlamydomonas reinhardtii	−[2]	+
Cyanidioschyzon merolae	−[2]	+
Synechocystis PCC6803	+	+
Thermosynechococcus elongatus BP-1	+	−

[1]Confirmed by the presence of expressed sequence tags (ESTs) for nucleus-encoded subunits.

[2]The chloroplast genomes do not encode any *ndh* gene.

[3]Estimated from the information of a closely related species, *Pinus taeda* (personal communication by T. A. Long).

Physiological Function of the Chloroplast NDH Complex: Re-evaluation Based on the Phenotype of the Double Mutants

Because there was no apparent phenotype in NDH knockout tobacco under greenhouse conditions, its tolerance to various environmental stresses was extensively surveyed. The knockout tobacco was sensitive to extremely high light intensity (21), low humidity stress (28), drought stress (60), and high and low temperature stresses (93). All these results suggest that the chloroplast NDH complex is involved in alleviating oxidative stresses. In *Arabidopsis*, the NDH complex is essential even at low light intensities under the *pgr5* mutant background (56). The phenotype of the double mutants lacking both activities of PSI CET most markedly demonstrated the physiological function of the chloroplast NDH complex. The NDH complex is a crucial piece of machinery, even for survival under the *pgr5* mutant background, and is also probably essential, even in the wild type in certain stress conditions under which the plants experience stromal over-reduction.

The NDH complex mediates PSI CET, which alleviates oxidative stress in chloroplasts, although its exact mechanisms are unclear. Plants conserve many genes (more than 30 genes, including those involved in chloroplast gene expression) to maintain this complex. The exception is black pine, in which all the *ndh* genes are absent in the chloroplast genome (92) (**Table 1**). Despite the apparent phenotype of *pgr5* even under nonstressed conditions, the mutants defective in NDH activity do not show any phenotype under these conditions. PGR5-dependent PSI CET is essential for maintaining the correct production ratio of ATP/NADPH. In contrast, the NDH complex may act as a safety valve that prevents over-reduction of the stroma. Black pine may have acquired an alternative valve. Interestingly, the NDH complex appears to be absent in *Chlamydomonas* (75), in which high chlororespiration activity has been detected (11). In *Chlamydomonas*, NDH-2, a single subunit NDH, rather than NDH-1, may participate in chlororespiration (63).

In Vivo Characterization of Photosystem I Cyclic Electron Transport

In the 1950s, cyclic photophosphorylation was shown to exist, but it is not yet clear what exactly PSI CET is. The problem has been caused due to the lack of a definitive method of measuring it in vivo. This topic was extensively discussed in a current review (33). From results demonstrating the significant contribution of PSI CET (14, 34, 55, 72), to those raising doubts about the existence of this form of electron transport (32, 44, 80), the conclusions depend on the measuring techniques. These problems are partly due to the fact that two partially redundant pathways, the PGR5- and NDH-dependent pathways, are in operation. For example, the post-illumination rise in chlorophyll fluorescence reflects NDH-dependent activity but not PGR5-dependent activity (56). This suggests that PGR5-dependent PSI CET does not operate in the dark or at very low light intensities, even when the electron donor, reduced Fd, is available.

What has molecular genetics contributed to the elucidation of PSI CET? *PGR5* was identified as a gene essential to non-NDH PSI CET, and does not code any motif that suggests direct involvement in Fd-dependent PQ reduction (57). It may not be easy to clarify the route taken by electrons by a simple application of biochemistry utilizing the PRG5 protein information if PGR5 is not a mediator of the electron flow. However, one advantage of recent progress in genetics is the ability to have the mutants specifically defective in each pathway of PSI CET. The mutants can be used to evaluate the in vivo assay system, especially when clarifying which pathway of PSI cyclic electron flow is monitored by each method (24). It is also essential to reevaluate the mutant phenotypes by different approaches in detail.

Avenson et al. (8) estimated the contribution of the PGR5-dependent PSI CET by monitoring the decay in electrochromic shift that reflects ΔpH. Because the decay kinetics were analyzed upon perturbation of the light conditions with 300-ms dark intervals, it was possible to evaluate the electron flow during steady-state photosynthesis at relatively high light intensity. The ratio of steady-state proton flux into the lumen (v_{H^+}) compared with the rate of LET was 13% smaller in *pgr5* than in the wild type (8). Although it should be noted that the mutant has somehow compensated for this defect, we can roughly estimate the contribution of PGR5-dependent PSI CET in the wild type. The contribution of approximately 10% of ΔpH generation by PGR5-dependent PSI CET is not negligible and is consistent with even the strong phenotype observed in the double mutants (56). Based on a calculation by Noctor & Foyer (68), a 1% increase in the ATP/NADPH production ratio gives a tenfold higher ATP/ADP ratio within 30–40 s than the typical ratio of ATP/ADP = 2.4 used as a starting point. During photosynthesis, chloroplast adenylate and reductant pools turn over rapidly; thus, it is essential to control the generation balance precisely.

MOLECULAR IDENTITY OF PHOTOSYSTEM I CYCLIC ELECTRON TRANSPORT

Machinery Involved in PGR5-Dependent Cyclic Electron Transport

Although Arnon's PSI CET was discovered 50 years ago (3, 4), the molecules involved are still unknown. An *Arabidopsis* mutant, *pgr5*, is most likely related to this PSI CET, as discussed in this review. Once a mutant is isolated, molecular genetics make it possible to identify which gene causes the phenotype. PGR5 is a small protein without any known motifs and conserved in eukaryotic phototrophs including some cyanobacteria (**Table 1**), and the *pgr5* mutation alters an amino acid near the C-terminal end, which totally destabilizes the protein (57). Because PGR5 does not have any motifs for prosthetic group binding, it is unlikely that PGR5 is directly involved in electron transport. It may, however, form a complex with other proteins that are directly involved in the electron transport. Although PGR5 is a soluble protein, it localizes to the thylakoid membranes. The extremely basic nature of PGR5 implies this model, in which PGR5 interacts with the cyt b_6f complex (15) or PSI via their acidic domains. However, PGR5 is stable even in mutant contexts that lack these complexes (57). We recently discovered that overexpressed PGR5 under the control of the cauliflower mosaic virus 35S promoter accumulates stably in the thylakoid membranes and activates PSI CET (Y. Okegawa, T.A. Long, M. Iwano, S. Takayama, Y. Kobayashi, S.F. Covert & T. Shikanai, unpublished). It is also possible that PGR5 is a regulator of photosynthetic electron transport and is indirectly required to operate PSI CET.

In 2003, the crystal structure of the cyt b_6f complex was determined in the thermophilic cyanobacterium *Mastigocladis laminosus* (43) and in *Chlamydomonas* (88). An unexpected discovery was heme *x* (also referred to as heme c_i) occupying a position adjacent to heme b_H. This novel heme is not conserved in the cyt bc_1 complex, which has a close similarity to the cyt b_6f complex in both structure and function. An interesting possibility is that heme *x* is involved in PSI CET, which operates in the chloroplasts but not in the mitochondria (43, 88, 95). Unfortunately, site-directed mutagenesis designed to destabilize this heme generated a *Chlamydomonas* mutant in which the cyt b_6f complex could not be assembled (18). To assess the possibility that Fd-dependent PQ reduction occurs via heme *x*, we assayed this form of electron transport in an *Arabidopsis* mutant, *pgr1* (72). The activity of the cyt b_6f complex was conditionally impaired in *pgr1*, which has an amino acid alteration in the Rieske subunit (30, 59). The activity of the cyt b_6f complex was impaired specifically when the lumen pH was lowered both in isolated thylakoids and in intact leaves. If the *pgr1* defect affects the putative heme *x*-dependent PSI CET, the defect may be monitored in Fd-dependent PQ reduction in vitro or in leaves only when the *pgr1* defect is evident. However, neither analysis showed any evidence to suggest that PGR5-dependent, Fd-dependent PQ reduction activity takes place through the Q-cycle of the cyt b_6f complex (72). This result cannot exclude the possibility that electron transport via heme *x* is unaffected by the *pgr1* mutation in the Rieske subunit, although Q-cycle activity is severely impaired.

The most urgent tasks are for biochemists (*a*) to clarify the function of PGR5 protein, and (*b*) to clarify whether the cyt b_6f complex is involved in the PGR5-dependent PQ reduction. In our working hypothesis, PGR5 is indirectly involved in electron transport as a regulator of electron transport, rather than a subunit of a putative FQR complex. If this were true, we would need more information in order to clarify the machinery of electron transport. Since the discovery of PSI CET, there has been a debate as to whether or not electron transport takes place through the cyt b_6f complex (10). We do not yet have any solid evidence to answer this longstanding question.

There is additional fragmentary information related to non-NDH PSI CET. The most important and reliable information is its sensitivity to AA (31, 56, 57, 90). Although AA binds to the quinone-reducing (Q_i) site of the cyt bc_1 complex (29), the same is not true for the cyt b_6f complex (78). It is possible that AA binds the machinery specifically involved in PGR5-dependent PSI CET. This Fd-dependent PQ reduction may take place via novel cyt b_{559} (54). This spectroscopic information may act as a criterion for evaluating mutants that are possibly defective in PSI CET.

Subunit Composition of the Thylakoid NAD(P)H Dehydrogenase Complex

In *Synechocystis* PCC6803, the NDH complex is involved in multiple functions—respiration, PSI CET, and CO_2 uptake into cells—by modifying subunit compositions (96). The largest NDH-1L complex is essential for photoheterotrophic growth, and thus is involved in respiration and probably also in PSI CET (71). The NDH-1L complex contains all 15 Ndh subunits (NdhA-O) identified so far (96). Although the NDH-1M complex lacks NdhD1 and NdhF1 subunits, it is associated with two versions of the NDH-1S complex that is involved in CO_2 uptake (96, 97). From the analysis of $P700^+$ reduction kinetics after far-red light illumination, NDH-1M was also shown to be involved in PSI CET (96). PSI CET was extensively characterized in M55, in which both NDH-1L and NDH-1M are absent due to a defect in a common subunit, NdhB (51–53). However, it is not clear whether the characterized electron transport was a mixture of two distinct electron flows mediated by NDH-1L and NDH-1M. In higher plants, *ndhD* and *ndhF* are single-copy genes that are similar to *ndhD1/D2* and *ndhF1* in cyanobacteria, respectively. This suggests that the chloroplast NDH complex is involved in PSI CET mediated by NDH-1L in cyanobacteria.

In *Escherichia coli*, the NDH-1 complex consists of 14 subunits (94) that are conserved in the mitochondrial complex. The *E. coli* NDH-1 complex is therefore a useful model that consists of a minimum set of subunits. Three subunits, NuoE, NuoF, and NuoG, whose homologs are not encoded in the chloroplast genome, function in NADH oxidization. Homologs of these subunits were not found either by the complete nucleotide determination of the *Arabidopsis* nuclear genome or in cyanobacterial genomes. This suggests that the NDH complex of phototrophs is equipped with different subunits functioning in electron donor binding. The discovery of these unidentified subunits may help solve the long-debated issue of the electron donor to the chloroplast NDH complex (83). The NDH complex, purified from *Synechocystis* PCC 6803, accepts electrons from NADPH but not from NADH (49, 51). However, the purified NDH complex from higher plants, barley (13), pea (81), and tobacco (79) prefers NADH as an electron donor. In contrast, Fd and NADPH are required for PQ reduction in ruptured chloroplasts isolated from tobacco (20) and *Arabidopsis* (56, 57). This discrepancy is partly due to the instability of the NDH complex during purification.

One strategy for identifying subunits functioning in electron donor binding is based on a proteomics technique. Purification and/or separation of the NDH complex on a blue-native gel were followed by mass spectrometry. These strategies led to the discovery of four new subunits (NdhL–NdhO) in cyanobacteria (9, 70, 76, 96). Among them, NdhM, NdhN, and NdhO were also identified in the chloroplast complex and were encoded in the nuclear genome (79). None of the newly discovered subunits has any motif suggesting binding to an electron donor. Despite exhaustive biochemical investigations, the missing subunits are still unknown. It is probable that the subcomplex containing the binding site to the electron donor is very fragile and easily dissociated from the membrane and connecting subcomplexes. This is consistent with the difficulty of isolating complexes with high activity.

An alternative approach to identifying the subunits is to apply genetics. NDH activity can be monitored as a transient increase in chlorophyll fluorescence after turning off actinic light (12, 38, 84). This change in chlorophyll fluorescence was used to screen *Arabidopsis crr* mutants specifically defective in NDH activity (23). The mutants are related to divergent processes of the expression or stabilization of the NDH complex, including the regulation of chloroplast *ndh* gene expression (23, 39). To select candidate genes possibly encoding unidentified subunits of the chloroplast NDH complex, our criteria are based on (*a*) the genome comparison among organisms that contain or do not contain the NDH complex (**Table 1**), (*b*) the gene that is essential for stabilizing the NDH complex, and (*c*) the NDH complex that is essential for stabilizing the gene product. So far, CRR7 is applicable to all the criteria and is a candidate for a subunit of the most fragile subcomplex including an electron donor binding site (61). However, CRR7 does not contain any motifs and may interact with other subunits to form the electron donor binding subcomplex.

Although CRR6 fits the first two criteria, it is stable in the mutant background of *crr2* defective in the expression of *ndhB* (62). The most straightforward interpretation is that CRR6 is a nonsubunit factor required for expression or stabilization of the NDH complex. However, we cannot eliminate the possibility that CRR6 is a component of the peripheral subcomplex that is essential to stabilize the main complex but is stable without the main complex. Although the NDH-1MS complex is composed of NDH-1M and NDH-1S in *Thermosynechococcus elongatus* BP-1, NDH-1S is stable in the absence of NDH-1M (96, 97). Our genetic analysis indicates that the factor required for stabilizing CRR6 is indirectly essential for the accumulation of the NDH complex.

C_4 Photosynthesis and Photosystem I Cyclic Electron Transport

The C_4 photosynthesis pathway reduces the activity of photorespiration by operating CO_2 concentration mechanisms. In C_4 plants, bundle sheath cells (BSC) and mesophyll cells (MC) form a radial pattern around the vascular system. In NADP-malic enzyme (ME)-type C_4 photosynthesis, CO_2 is fixed by phosphoenolpyruvate (PEP) carboxylase, and the resulting oxaloacetate is reduced to malate using NADPH in MC. Subsequently, malate is transferred to BSC, where CO_2 and NADPH are released by NADP-ME. Finally, pyruvate is recycled to PEP by pyruvate phosphate dikinase, a reaction that requires an additional two ATPs to fix one CO_2 molecule.

In contrast to C_3 plants, the significant contribution of PSI CET was shown in C_4 plants (7, 26). Is C_4 photosynthesis energized by PSI CET? Consistent with the movement of metabolites between two cell types, the gene expression related to PSI CET is upregulated in BSC, like that related to LET in MC (40). Interestingly, the chloroplast *ndh* genes are highly expressed in BSC (16, 41).

Molecular information is now available for two types of PSI CET in higher plants. In NADP-ME-type C_4 plants, the NDH complex accumulated more in BSC than in MC. In contrast, in NAD-ME-type C_4 plants, in which more ATP is required in MC, the NDH complex is overaccumulated in MC. In contrast to the cell-type-specific accumulation of the NDH complex, PGR5 is equally accumulated both in MC and BSC. These results suggest that the NDH complex energizes the C_4 photosynthesis (91). For the final conclusion, it is essential to knock out the gene involved in the electron transport and characterize the phenotype in C_4 plants. Although a lack of technique for knocking out the chloroplast *ndh* genes in C_4 plants has been problematic, the information is now available for the nuclear genes encoding the Ndh subunits (79), and also for the factors specifically required for expressing the chloroplast *ndh* genes (23, 39).

As well as the NDH complex, two Fd isoproteins, Fd I (MC type) and Fd II (BSC type), are cell-types specifically expressed in maize (35). Both types of Fd were introduced into cyanobacteria, *Plectonema boryanum*, in which the endogenous Fd gene was disrupted (36). Although Fd I promotes LET, Fd II activates PSI CET, suggesting that Fd is a determinant of electron flow around PSI. If Fd II is specifically involved in PSI CET, the most straightforward interpretation is that C_4 photosynthesis is energized by PGR5-dependent PSI CET, which requires Fd as an electron donor, in BSC. The conclusion is inconsistent with molecular biological data, suggesting the preferential involvement of the NDH complex in C_4 photosynthesis. It is also possible that Fd II effectively transfers electrons to the NDH complex via selective electron donation to a specific FNR (73). It is essential to determine the electron acceptor from Fd II to clarify the route taken by electrons in PSI CET of C_4 plants.

CONCLUDING REMARKS

PSI CET was discovered more than 50 years ago and is often claimed to be physiologically significant. Although the complete genome sequences have been determined in both *Arabidopsis* and rice, we are still not sure of the machinery involved in PGR5-dependent electron transport. Although the era of system biology is at hand, the fundamental mechanisms that balance the generation ratio of ATP and NADPH are unclear. In plant cells, the chloroplast plays a central role in energy generation and various other metabolisms, as does the mitochondrion. To carry plant science to the next step, it is essential to elucidate the regulation mechanisms of energetics in these organelles. Recent genetic approaches provide the chance to revisit this classic topic. While respecting pioneering work, "old research" still has much to teach us. We need to reconsider the fundamental concepts. Our knowledge of photosynthetic electron transport still cannot explain the plasticity of energetics as required by developmental and environmental cues.

SUMMARY POINTS

1. PSI CET was discovered more than 50 years ago. Its physiological significance has been overlooked and the machinery involved is still unclear.
2. PSI CET consists of partially redundant routes: PGR5-dependent and NDH-dependent pathways in higher plants.
3. In cyanobacteria, the NDH complex is involved in multiple processes by modifying the subunit compositions.
4. In higher plants, the main pathway of PSI CET depends on PGR5, which was discovered in an *Arabidopsis* mutant defective in NPQ. PGR5-dependent PSI CET is essential for both photosynthesis and photoprotection.
5. PSI CET is essential for balancing the generation ratio of ATP and NADPH to prevent the over-reduction of the stroma.
6. The chloroplast NDH complex alleviates the stromal over-reduction and possibly energizes C_4 photosynthesis.

ACKNOWLEDGMENTS

I am grateful to Prof. Kozi Asada for his critical reading of the manuscript. Unpublished results from the author's laboratory are supported by a grant-in-aid for Scientific Research on Priority Areas (16085296) and for Creative Scientific Research (17GS0316) from the Ministry of Education, Culture, Sports, Science and Technology, Japan.

LITERATURE CITED

1. Allen JF. 2002. Photosynthesis of ATP-electrons, proton pumps, rotors, and poise. *Cell* 110:273–76
2. Arnon DI. 1991. Photosynthetic electron transport: Emergence of a concept, 1949–59. *Photosynth. Res.* 29:117–31
3. Arnon DI, Allen MB, Whatley FR. 1954. Photosynthesis by isolated chloroplasts. *Nature* 174:394–96
4. Arnon DI, Allen MB, Whatley FR. 1954. Photosynthesis by isolated chloroplasts. II. Photosynthetic phosphorylation, the conversion of light into phosphate bound energy. *J. Am. Chem. Soc.* 76:6324–29
5. Arnon DI, Allen MB, Whatley FR. 1958. Assimilatory power in photosynthesis. *Science* 127:1026–34
6. Asada K. 2006. Production and scavenging of reactive oxygen species in chloroplasts and their functions. *Plant Physiol.* 141:391–96
7. Asada K, Heber U, Schreiber U. 1992. Pool size of electrons that can be donated to $P700^+$, as determined in intact leaves: donation to $P700^+$ from stromal components via the intersystem chain. *Plant Cell Physiol.* 33:927–32

8. PGR5-dependent PSI CET was estimated to contribute 13% of the total ΔpH generation during steady-state photosynthesis.

8. **Avenson TJ, Cruz JA, Kanazawa A, Kramer DM. 2005. Regulating the proton budget of higher plant photosynthesis. *Proc. Natl. Acad. Sci. USA* 102:9709–13**
9. Battchikova N, Zhang P, Rudd S, Ogawa T, Aro EM. 2004. Identification of NdhL and Ssl1690 (NdhO) in NDH-1L and NDH-1M complexes of *Synechocystis* sp. PCC 6803. *J. Biol. Chem.* 280:2587–95
10. Bendall DS, Manasse RS. 1995. Cyclic photophosphorylation and electron transport. *Biochim. Biophys. Acta* 1229:23–38
11. Bennoun P. 1982. Evidence for a respiratory chain in the chloroplast. *Proc. Natl. Acad. Sci. USA* 79:4352–56
12. Burrows PA, Sazanov LA, Svab Z, Maliga P, Nixon PJ. 1998. Identification of a functional respiratory complex in chloroplasts through analysis of tobacco mutants containing disrupted plastid *ndh* genes. *EMBO J.* 17:868–76
13. Casano LM, Zapata JM, Martín M, Sabater B. 2000. Chlororespiration and poising of cyclic electron transport. Plastoquinone as electron transporter between thylakoid NADH dehydrogenase and peroxidase. *J. Biol. Chem.* 275:942–48
14. Clarke JE, Johnson GN. 2001. In vivo temperature dependence of cyclic and pseudocyclic electron transport in barley. *Planta* 212:808–16
15. Cramer WA, Zhang HM, Yan JS, Kurisu G, Smith JL. 2006. Transmembrane traffic in the cytochrome b_6f complex. *Annu. Rev. Biochem.* 75:769–90
16. Darie CC, de Pascalis L, Mutschler B, Haehnel W. 2005. Studies of the Ndh complex and photosystem II from mesophyll and bundle sheath chloroplasts of the C_4-type plant *Zea mays*. *J. Plant Physiol.* 163:800–8
17. Demmig-Adams B, Adams WW III, Baker DH, Logan BA, Bowling DR, et al. 1996. Using chlorophyll fluorescence to assess the fraction of absorbed light allocated to thermal dissipation of excessive excitation. *Physiol. Plant.* 98:253–64
18. de Vitry C, Desbois A, Redeker V, Zito F, Wollman FA. 2004. Biochemical and spectroscopic characterization of the covalent binding of heme to cytochrome b_6. *Biochemistry* 43:3956–68
19. Endo T, Kawase D, Sato F. 2005. Stromal over-reduction by high-light stress as measured by decreases in P700 oxidation by far-red light and its physiological relevance. *Plant Cell Physiol.* 46:775–81
20. Endo T, Mi H, Shikanai T, Asada K. 1997. Donation of electrons to plastoquinone by NAD(P)H dehydrogenase and by ferredoxin-quinone reductase in spinach chloroplasts. *Plant Cell Physiol.* 38:1272–77
21. Endo T, Shikanai T, Takabayashi A, Asada K, Sato F. 1999. The role of chloroplastic NAD(P)H dehydrogenase in photoprotection. *FEBS Lett.* 457:5–8
22. Fork DC, Herbert SK. 1993. Electron transport and photophosphorylation by photosystem I in vivo in plants and cyanobacteria. *Photosynth. Res.* 36:149–68
23. Hashimoto M, Endo T, Peltier G, Tasaka M, Shikanai T. 2003. A nucleus-encoded factor, CRR2, is essential for the expression of chloroplast *ndhB* in *Arabidopsis*. *Plant J.* 36:541–49
24. Havaux M, Rumeau D, Ducruet JM. 2005. Probing the FQR and NDH activities involved in cyclic electron transport around photosystem I by the 'afterglow' luminescence. *Biochim. Biophys. Acta* 1709:203–13
25. Heber U, Walker D. 1992. Concerning a dual function of coupled cyclic electron transport in leaves. *Plant Physiol.* 100:1621–26
26. Herbert SK, Fork DC, Malkin S. 1990. Photoaccoustic measurements in vivo of energy storage by cyclic electron flow in algae and higher plants. *Plant Physiol.* 94:926–34

27. Hill R, Bendall F. 1960. Function of the cytochrome components in chloroplasts: A working hypothesis. *Nature* 186:136–37
28. Horváth EM, Peter SO, Joët T, Rumeau D, Cournac L, et al. 2000. Targeted inactivation of the plastid *ndhB* gene in tobacco results in an enhanced sensitivity of photosynthesis to moderate stomatal closure. *Plant Physiol.* 123:1337–50
29. Huang LH, Cobessi D, Tung EY, Edward A, Berry EA. 2005. Binding of the respiratory chain inhibitor antimycin to the mitochondrial bc_1 complex: a new crystal structure reveals an altered intramolecular hydrogen-bonding pattern. *J. Mol. Biol.* 351:573–97
30. Jahns P, Graf M, Munekage Y, Shikanai T. 2002. Single point mutation in the Rieske iron-sulfur subunit of cytochrome b_6/f leads to an altered pH dependence of plastoquinol oxidation in *Arabidopsis*. *FEBS Lett.* 519:99–102
31. Jöet T, Cournac L, Horváth EM, Medgyesy P, Peltier G. 2001. Increased sensitivity of photosynthesis to antimycin A induced by inactivation of the chloroplast *ndhB* gene. Evidence for a participation of the NADH-dehydrogenase complex to cyclic electron flow around photosystem I. *Plant Physiol.* 125:1919–29
32. Joët T, Cournac L, Peltier G, Havaux M. 2002. Cyclic electron flow around photosystem I in C_3 plants. In vivo control by the redox state of chloroplasts and involvement of the NADH-dehydrogenase complex. *Plant Physiol.* 128:760–69
33. Johnson GN. 2005. Cyclic electron transport in C_3 plants: fact or artifact? *J. Exp. Bot.* 56:407–16
34. Joliot P, Joliot A. 2005. Quantification of cyclic and linear flows in plants. *Proc. Natl. Acad. Sci. USA* 102:4913–18
35. Kimata Y, Hase T. 1989. Localization of ferredoxin isoproteins in mesophyll and bundle sheath cells in maize leaf. *Plant Physiol.* 89:1193–97
36. **Kimata-Ariga Y, Matsumura T, Kada S, Fujimoto H, Fujita Y, et al. 2000. Differential electron flow around photosystem I by two C_4-photosynthetic-cell-specific ferredoxins. *EMBO J.* 19:5041–50**

36. C_4-type Fd preferentially functions in PSI CET in cyanobacteria. The result may provide a clue to clarifying the route of electrons in PSI CET in C_4 plants.

37. Kobayashi Y, Neimanis S, Heber U. 1995. Coupling ratios $H^+/e = 3$ versus $H^+/e = 2$ in chloroplasts and quantum requirements of net oxygen exchange during the reduction of nitrate, ferricyanide or methylviologen. *Plant Cell Physiol.* 36:1613–20
38. Kofer W, Koop HU, Wanner G, Steinmüller K. 1998. Mutagenesis of the genes encoding subunits A, C, H, I, J and K of the plastid NAD(P)H-plastoquinone-oxidoreductase in tobacco by polyethylene glycol-mediated plastome transformation. *Mol. Gen. Genet.* 258:166–73
39. Kotera E, Tasaka M, Shikanai T. 2005. A pentatricopetide repeat protein is essential for RNA editing in chloroplasts. *Nature* 433:326–30
40. Kubicki A, Funk E, Westhoff P, Steinmüller K. 1994. Differential transcription of plastome-encoded genes in the mesophyll and bundle-sheath chloroplasts of the monocotyledonous NADP-malic enzyme-type C_4 plants maize and Sorghum. *Plant Mol. Biol.* 25:669–79
41. Kubicki A, Steinmüller K, Westhoff P. 1996. Differential expression of plastome-encoded *ndh* genes in mesophyll and bundle-sheath chloroplasts of the C_4 plant *Sorghum bicolor* indicates that the complex I-homologous NAD(P)H-plastoquinone oxidoreductase is involved in cyclic electron transport. *Planta* 199:276–81
42. Külheim C, Ågren J, Jansson S. 2002. Rapid regulation of light harvesting and plant fitness in the field. *Science* 297:91–93
43. Kurisu G, Zhang H, Smith JL, Cramer WA. 2003. Structure of the cytochrome b_6f complex of oxygenic photosynthesis: tuning the cavity. *Science* 302:1009–14

44. Laisk A, Eichelmann H, Oja V, Peterson RB. 2005. Control of cytochrome b_6f at low and high light intensity and cyclic electron transport in leaves. *Biochim. Biophys. Acta* 1708:79–90

45. Li XP, Björkman O, Shih C, Grossman AR, Rosenquist M, et al. 2000. A pigment-binding protein essential for regulation of photosynthetic light harvesting. *Nature* 403:391–95

45. The most successful example of recent genetics in photosynthetic research. A pH sensor for NPQ induction was discovered.

46. Li XP, Müller-Moulé P, Gilmore AM, Niyogi KK. 2002. PsbS-dependent enhancement of feedback de-excitation protects photosystem II from photoinhibition. *Proc. Natl. Acad. Sci. USA* 99:15222–27

47. Makino A, Miyake C, Yokota A. 2002. Physiological functions of the water-water cycle (Mehler reaction) and the cyclic electron flow around PSI in rice leaves. *Plant Cell Physiol.* 43:1017–26

48. Matsubayashi T, Wakasugi T, Shinozaki K, Yamaguchi-Shinozaki K, Zaita N, et al. 1987. Six chloroplast genes (*ndhA-F*) homologous to human mitochondrial genes encoding components of the respiratory chain NADH dehydrogenase are actively expressed: determination of the splice sites in *ndhA* and *ndhB* pre-mRNAs. *Mol. Gen. Genet.* 210:385–93

49. Matsuo M, Endo T, Asada K. 1998. Isolation of a novel NAD(P)H-quinone oxidoreductase from the cyanobacterium *Synechocystis* PCC6803. *Plant Cell Physiol.* 39:751–55

50. Merchant S, Sawaya MR. 2005. The light reactions: a guide to recent acquisitions for the picture gallery. *Plant Cell* 17:648–63

51. Mi H, Endo T, Ogawa T, Asada K. 1995. Thylakoid membrane-bound pyridine nucleotide dehydrogenase complex mediates cyclic electron transport in the cyanobacteria *Synechocystis* PCC 6803. *Plant Cell Physiol.* 36:661–68

52. Mi H, Endo T, Schreiber U, Ogawa T, Asada K. 1992. Electron donation from cyclic and respiratory flows to the photosynthetic intersystem chain is mediated by pyridine nucleotide dehydrogenase in the cyanobacterium *Synochocystis* PCC6803. *Plant Cell Physiol.* 33:1233–37

53. Mi HL, Endo T, Schreiber U, Ogawa T, Asada K. 1994. NAD(P)H-dehydrogenase-dependent cyclic electron flow around photosystem-I in the cyanobacterium *Synochocystis*—a study of dark-starved cells and spheroplasts. *Plant Cell Physiol.* 35:163–73

54. Miyake C, Schreiber U, Asada K. 1995. Ferredoxin-dependent and antimycin A-sensitive reduction of cytochrome *b-559* by far red light in maize thylakoids: participation of a menadiol-reducible cytochrome *b-559* in cyclic electron flow. *Plant Cell Physiol.* 36:743–48

55. Miyake C, Shinzaki Y, Miyata M, Tomizawa K. 2004. Enhancement of cyclic electron flow around PSI at high light and its contribution to the induction of nonphotochemical quenching of Chl fluorescence in intact leaves of tobacco plants. *Plant Cell Physiol.* 45:1426–33

56. Munekage Y, Hashimoto M, Miyake C, Tomizawa K, Endo T, et al. 2004. Cyclic electron flow around photosystem I is essential for photosynthesis. *Nature* 429:579–82

56. A discovery of PGR5 involved in PSI CET. The physiological significance of the electron transport in photoprotection was experimentally shown.

57. Munekage Y, Hojo M, Meurer J, Endo T, Tasaka M, et al. 2002. *PGR5* is involved in cyclic electron flow around photosystem I and is essential for photoprotection in *Arabidopsis*. *Cell* 110:361–71

57. Characterization of the double mutants defective in both pathways of PSI CET in *Arabidopsis*.

58. Munekage Y, Shikanai T. 2005. Cyclic electron transport though photosystem I. *Plant Biotechnol.* 22:361–69

59. Munekage Y, Takeda S, Endo T, Jahns P, Hashimoto T, et al. 2001. Cytochrome b_6f mutation specifically affects thermal dissipation of absorbed light energy in *Arabidopsis*. *Plant J.* 28:351–59

60. Munné-Bosch S, Shikanai T, Asada K. 2005. Enhanced ferredoxin-dependent cyclic electron flow around photosystem I and α-tocopherol quinone accumulation in water-stressed *ndhB*-inactivated tobacco mutants. *Planta* 222:502–11
61. Munshi MK, Kobayashi Y, Shikanai T. 2005. Identification of a novel protein CRR7 required for the stabilization of the chloroplast NAD(P)H dehydrogenase complex in *Arabidopsis*. *Plant J.* 44:1036–44
62. Munshi MK, Kobayashi Y, Shikanai T. 2006. CRR6 is a novel factor required for accumulation of the chloroplast NAD(P)H dehydrogenase complex in *Arabidopsis*. *Plant Physiol.* 141:737–44
63. Mus F, Cournac L, Cardettini V, Caruana A, Peltier G. 2005. Inhibitor studies on non-photochemical plastoquinone reduction and H_2 photoproduction in *Chlamydomonas reinhardtii*. *Biochim. Biophys. Acta* 1708:322–32
64. Niyogi KK, Björkman O, Grossman AR. 1997. The roles of specific xanthophylls in photoprotection. *Proc. Natl. Acad. Sci. USA* 84:14162–67
65. Niyogi KK, Grossman AR, Björkman O. 1998. *Arabidopsis* mutants define a central role for the xanthophyll cycle in the regulation of photosynthetic energy conversion. *Plant Cell* 10:1121–34
66. Niyogi KK, Li XP, Rosenberg V, Jung HS. 2005. Is PsbS the site of nonphotochemical quenching in photosynthesis? *J. Exp. Bot.* 56:375–82
67. Noctor G, Foyer CH. 1998. A re-evaluation of the ATP:NADPH budget during C_3 photosynthesis: a contribution from nitrate assimilation and its associated respiratory activity? *J. Exp. Bot.* 49:1895–908
68. Noctor G, Foyer CH. 2000. Homeostasis of adenylate status during photosynthesis in a fluctuating environment. *J. Exp. Bot.* 51:347–56
69. Ogawa T. 1991. A gene homologous to the subunit-2 gene of NADH dehydrogenase is essential to inorganic carbon transport of *Synechocystis* PCC6803. *Proc. Natl. Acad. Sci. USA* 88:4275–79
70. Ogawa T. 1992. Identification and characterization of the *ictA/ndhL* gene product essential to inorganic carbon transport of *Synechocystis* PCC6803. *Plant Physiol.* 99:1604–8
71. **Ohkawa H, Pakrasi HB, Ogawa T. 2000. Two types of functionally distinct NAD(P)H dehydrogenases in *Synechocystis* sp. strain PCC6803. *J. Biol. Chem.* 275:31630–34**

71. This report clarified that the cyanobacterial NDH complex is involved in multiple processes by altering its subunit composition.

72. Okegawa Y, Tsuyama M, Kobayashi Y, Shikanai T. 2005. The *pgr1* mutation in the Rieske subunit of the cytochrome b_6f complex does not affect PGR5-dependent cyclic electron transport around photosystem I. *J. Biol. Chem.* 280:28332–36
73. Okutani S, Hanke GT, Satomi Y, Takao T, Kurisu G, et al. 2005. Three maize leaf ferredoxin:NADPH oxidoreductases vary in subchloroplast location, expression, and interaction with ferredoxin. *Plant Physiol.* 139:1451–59
74. Osmond CB. 1981. Photorespiration and photoinhibition: some implication for the energetics of photosynthesis. *Biochim. Biophys. Acta* 639:77–98
75. Peltier G, Cournac L. 2002. Chlororespiration. *Annu. Rev. Plant Biol.* 53:523–50
76. Prommeenate P, Lennon AM, Markert C, Hippler M, Nixon PJ. 2004. Subunit composition of NDH-1 complexes of *Synechocystis* sp. PCC 6803: Identification of two new *ndh* gene products with nuclear-encoded homologues in the chloroplast Ndh complex. *J. Biol. Chem.* 279:28165–73
77. Raghavendra AS, Padmasree K. 2003. Beneficial interactions of mitochondrial metabolism with photosynthetic carbon assimilation. *Trends Plant Sci.* 8:546–53

78. Rich PR, Madgwick SA, Brown S, von Jagow G, Brandt U. 1992. MOA-stilbene: A new tool for investigation of the reactions of the chloroplast cytochrome *bf* complex. *Photosynth. Res.* 34:465–77
79. Rumeau D, Bécuwe-Linka N, Beyly A, Louwagie M, Garin J, Peltier G. 2005. New subunits NDH-M, -N, and -O, encoded by nuclear genes, are essential for plastid Ndh complex functioning in higher plants. *Plant Cell* 17:219–32
80. Sacksteder CA, Kanazawa A, Jacoby ME, Kramer DM. 2000. The proton to electron stoichiometry of steady-state photosynthesis in living plants: A proton-pumping Q cycle is continuously engaged. *Proc. Natl. Acad. Sci. USA* 97:14283–88
81. Sazanov LA, Burrows PA, Nixon PJ. 1998. The plastid *ndh* genes code for an NADH-specific dehydrogenase: Isolation of a complex I analogue from pea thylakoid membranes. *Proc. Natl. Acad. Sci. USA* 95:1319–24
82. Seelert H, Poetsch A, Dencher NA, Engel A, Stahlberg H, Müller DJ. 2000. Structural biology. Proton-powered turbine of a plant motor. *Nature* 405:418–19
83. Shikanai T, Endo T. 2000. Physiological function of a respiratory complex, NAD(P)H dehydrogenase in chloroplasts: Dissection by chloroplast reverse genetics. *Plant Biotechnol.* 17:79–86
84. **Shikanai T, Endo T, Hashimoto T, Yamada Y, Asada K, et al. 1998. Directed disruption of the tobacco *ndhB* gene impairs cyclic electron flow around photosystem I. *Proc. Natl. Acad. Sci. USA* 95:9705–9**

84. The plastid transformation technique clarified the involvement of the chloroplast NDH complex in PSI CET.

85. Shikanai T, Munekage Y, Kimura K. 2002. Regulation of proton-to-electron stoichiometry in photosynthetic electron transport: physiological function in photoprotection. *J. Plant Res.* 115:3–10
86. Shikanai T, Munekage Y, Shimizu K, Endo T, Hashimoto T. 1999. Identification and characterization of *Arabidopsis* mutants with reduced quenching of chlorophyll fluorescence. *Plant Cell Physiol.* 40:1134–42
87. Sonoike K, Terashima I. 1994. Mechanism of photosystem I photoinhibition in leaves of *Cucumis sativus* L. *Planta* 194:287–93
88. Stroebel D, Choquet Y, Popot JL, Picot D. 2003. An atypical haem in the cytochrome b_6f complex. *Nature* 426:413–18
89. Svab Z, Maliga P. 1993. High-frequency plastid transformation in tobacco by selection for a chimeric *aadA* gene. *Proc. Natl. Acad. Sci. USA* 90:913–17
90. **Tagawa K, Tsujimoto HY, Arnon DI. 1963. Role of chloroplast ferredoxin in the energy conversion process of photosynthesis. *Proc. Natl. Acad. Sci. USA* 49:567–72**

90. A historical report that establishes the concept of PSI CET.

91. **Takabayashi A, Kishine M, Asada K, Endo T, Sato F. 2006. Differential use of two cyclic electron flows around photosystem I for driving CO_2-concentration mechanism in C_4 photosynthesis. *Proc. Natl. Acad. Sci. USA* 102:16898–903**

91. Tissue-specific localization of NdhH and PGR5 proteins suggests that the NDH complex energizes C_4 photosynthesis.

92. Wakasugi T, Tsudzuki J, Ito S, Nakashima K, Tsudzuki T, Sugiura M. 1994. Loss of all *ndh* genes as determined by sequencing the entire chloroplast genome of the black pine *Pinus thunbergii*. *Proc. Natl. Acad. Sci. USA* 91:9794–98
93. Wang P, Duan W, Takabayashi A, Endo T, Shikanai T, et al. 2006. Chloroplastic NAD(P)H dehydrogenase in tobacco leaves functions in alleviation of oxidative damage caused by temperature stress. *Plant Physiol.* 141:465–74
94. Yagi T, Matsuno-Yagi A. 2003. The proton-translocating NADH-quinone oxidoreductase in the respiratory chain: the secret unlocked. *Biochemistry* 42:2266–74
95. Zhang H, Primak A, Cape J, Bowman MK, Kramer DM, et al. 2004. Characterization of the high-spin heme *x* in the cytochrome b_6f complex of oxygenic photosynthesis. *Biochemistry* 43:16329–36

96. Zhang P, Battchikova N, Jansen T, Appel J, Ogawa T, et al. 2004. Expression and functional roles of the two distinct NDH-1 complexes and the carbon acquisition complex NdhD3/NdhF3/CupA/Sll1735 in *Synechocystis* sp PCC 6803. *Plant Cell* 16:3326–40

97. Zhang P, Battchikova N, Paakkarinen V, Katoh H, Iwai M, et al. 2005. Isolation, subunit composition and interaction of the NDH-1 complexes from *Thermosynechococcus elongatus* BP-1. *Biochem J.* 390:513–20

96. Proteomics analysis confirmed the subunit compositions of the distinct NDH complexes in cyanobacteria.

Light Regulation of Stomatal Movement

Ken-ichiro Shimazaki,[1] Michio Doi,[2] Sarah M. Assmann,[3] and Toshinori Kinoshita[1,4]

[1]Department of Biology, Faculty of Science, [2]Research and Development Center for Higher Education, Kyushu University, Ropponmatsu, Fukuoka, 810-8560, Japan; email: kenrcb@mbox.nc.kyushu-u.ac.jp, doircb@mbox.nc.kyushu-u.ac.jp

[3]Department of Biology, Pennsylvania State University, University Park, Pennsylvania 16802-5301; email: sma3@psu.edu

[4]PRESTO, Japan Science and Technology Agency, 4-1-8 Honcho Kawaguchi, Saitama, Japan; email: toshircb@mbox.nc.kyushu-u.ac.jp

Annu. Rev. Plant Biol. 2007. 58:219–47

The *Annual Review of Plant Biology* is online at plant.annualreviews.org

This article's doi:
10.1146/annurev.arplant.57.032905.105434

First published online as a Review in Advance on January 5, 2007

1543-5008/07/0602-0219$20.00

Key Words

blue light, chloroplasts, guard cells, phototropin, plasma membrane H^+-ATPase, stomata

Abstract

Stomatal pores, each surrounded by a pair of guard cells, regulate CO_2 uptake and water loss from leaves. Stomatal opening is driven by the accumulation of K^+ salts and sugars in guard cells, which is mediated by electrogenic proton pumps in the plasma membrane and/or metabolic activity. Opening responses are achieved by coordination of light signaling, light-energy conversion, membrane ion transport, and metabolic activity in guard cells. In this review, we focus on recent progress in blue- and red-light-dependent stomatal opening. Because the blue-light response of stomata appears to be strongly affected by red light, we discuss underlying mechanisms in the interaction between blue-light signaling and guard cell chloroplasts.

Contents

INTRODUCTION

Stomata regulate gas exchange between plants and atmosphere, optimize photosynthetic CO_2 fixation, and minimize transpirational water loss (8, 17, 144, 189, 193). Stomata move rapidly to adjust plants to the ever-changing environment. The opening of stomata is driven by the accumulation of K^+ salts (37, 42, 64, 191) and/or sugars (124, 174) in guard cells, which results in a decrease in water potential and subsequent water uptake. Turgor elevation from water uptake increases guard cell volume, which widens stomatal apertures because of the asymmetric positioning of microfibrils in the cell wall. The volume increase requires an increase in surface area of the guard cell plasma membrane, and this needed area is provided by the internal membranes of guard cells (162). Guard cells possess a number of small vacuoles in the closed state of stomata (117). Such small vacuoles fuse with each other and generate bigger vacuoles during stomatal opening (43). Prior to these processes, a large number of ions move from the cytosol to the vacuole via channels and pumps in the tonoplast (186). Stomatal closure is caused by the release and/or removal of osmotica from guard cells under drought, darkness, elevated CO_2, or low humidity, resulting in the internalization of the plasma membrane and the generation of small vacuoles.

Stomatal opening is induced by light, including blue and red light, and distinct mechanisms underly stomatal opening in response to these different wavelengths (193). Blue light acts as a signal and red light as both a signal and an energy source. Blue light activates the plasma membrane H^+-ATPase (21, 76), hyperpolarizing the membrane potential with simultaneous apoplast acidification, and drives K^+ uptake through voltage-gated K^+ channels. Red light drives photosynthesis in mesophyll and guard cell chloroplasts and decreases the intercellular CO_2 concentration (*Ci*). Red-light-induced stomatal opening may result from a combination of guard cell response to the reduction in *Ci* and a direct response of the guard cell chloroplasts to red light (132, 151, 183). In the afternoon, sugars also accumulate in guard cells as osmotica and maintain stomatal opening (174). The accumulation of positively charged K^+ ions in guard cells must be compensated by anions, mainly in the form of the organic acid malate^{2-} (189). Malate forms in response to weak blue light under a red-light background, and the formation does not occur without red light (109, 110). Guard cell chloroplasts are responsible for malate formation (151), and the chloroplasts also act as a reservoir for starch and catabolize it as a precursor of malate (183, 189). Guard cells also utilize Cl^- and NO_3^- as counterions for K^+, although such utilization varies with plant species and growth conditions (51).

In this review, we focus on recent progress concerning light-stimulated stomatal opening. On the basis of the highly specialized metabolism in guard cells, which facilitates the rapid transport of ions across membranes, a synergistic effect between blue and red light on stomatal opening has been proposed. For other stomatal responses, particularly stomatal closure responses induced by the phytohormone abscisic acid (ABA) under drought, please refer to other recent reviews on this subject (11, 17, 35, 56, 132, 144).

BLUE-LIGHT RESPONSE OF STOMATA

Properties

Blue-light-specific stomatal opening is found in a number of C_3 and C_4 plants under strong background red light, and in facultative Crassulacean acid metabolism plants functioning in the C_3, but not the CAM, mode (89).

In intact leaves, blue light, on a quantum basis, is several to 20 times more effective than red light in opening stomata (19, 59, 71, 151, 152). The high sensitivity of this blue-light response becomes prominent in the presence of background red light. Simultaneous measurements of stomatal conductance, photosynthetic CO_2 fixation, and intercellular CO_2 concentration (*Ci*) in leaves of *Arabidopsis thaliana* and rice (*Oryza sativa* L. cv, Taichung 65) plants in response to light are presented in **Figure 1**. When a leaf kept in the dark for 1 h was illuminated with strong red light (600 μmol m^{-2} s^{-1}), photosynthetic CO_2 fixation occurred instantly with a sharp drop in *Ci* to 250–270 ppm, followed by a gradual increase in photosynthetic rate for 20 min, until a steady state was achieved. Stomata showed a gradual increase in stomatal conductance with a short lag time, and conductance reached a maximum within 20 min, then showed small fluctuations (**Figure 1*a***). Weak blue light (5 μmol m^{-2} s^{-1}), superimposed on the red light for 10 min, elicited rapid stomatal opening, with a threefold faster rate than that caused by red light (151). The opening responses were observed repeatedly in response to blue light, and the magnitude of the responses showed a slight decrease from morning to afternoon (28). Stomata showed virtually no opening in response to the weak blue-light stimulus in the absence of red light (**Figure 1*b***). These properties are in accord with those shown in previous reports on other plant species (63, 86, 151). When the rice leaf was illuminated with red light, photosynthetic CO_2 fixation increased instantly but remained at a low level for 10 min due to the low stomatal conductance, and then exhibited a large gradual increase (**Figure 1*c***). The gradual increase in photosynthesis resulted from the removal of stomatal limitation, because the increase in CO_2 fixation paralleled that in stomatal conductance of this phase. Weak blue light superimposed on the red light induced a very rapid increase in the stomatal aperture in this plant species (**Figure 1*c***), as has also been shown for other monocots, such as wheat (70, 71) and sugarcane (9).

Blue light acts as a signal, and even a short period of light (pulse, 30–60 s) induces stomatal opening, which is sustained for more than 10 min after the pulse (63). Blue-light-specific stomatal responses can be induced by illuminating the leaves with blue light superimposed on high-intensity red light, in which photosynthesis was not further activated by the blue-light stimulus (110, 148). The magnitude of the response was proportional to the photon flux of blue light and reached saturation. Once a single saturating pulse was given to the leaf, the responsiveness to a second pulse was greatly reduced when the second pulse was immediately applied, and the responsiveness was gradually restored with a half-life time ($T_{1/2}$) of 9 min in *Commelina communis* (63). The blue-light receptor was suggested to exist in two interconvertible forms: one physiologically active and the other inactive (63). The conversion from the inactive form to the active form was a light-induced fast reaction and the active form gradually returned to the inactive form via a thermal reaction. These properties seem to be attributable to the molecular properties of the responsible blue-light receptors, which were discovered later (60, 72, 138).

Electrogenic H^+ Pump Drives K^+ Uptake and Stomatal Opening

Medium acidification by guard cells in *Vicia* epidermal peels was reported, and the

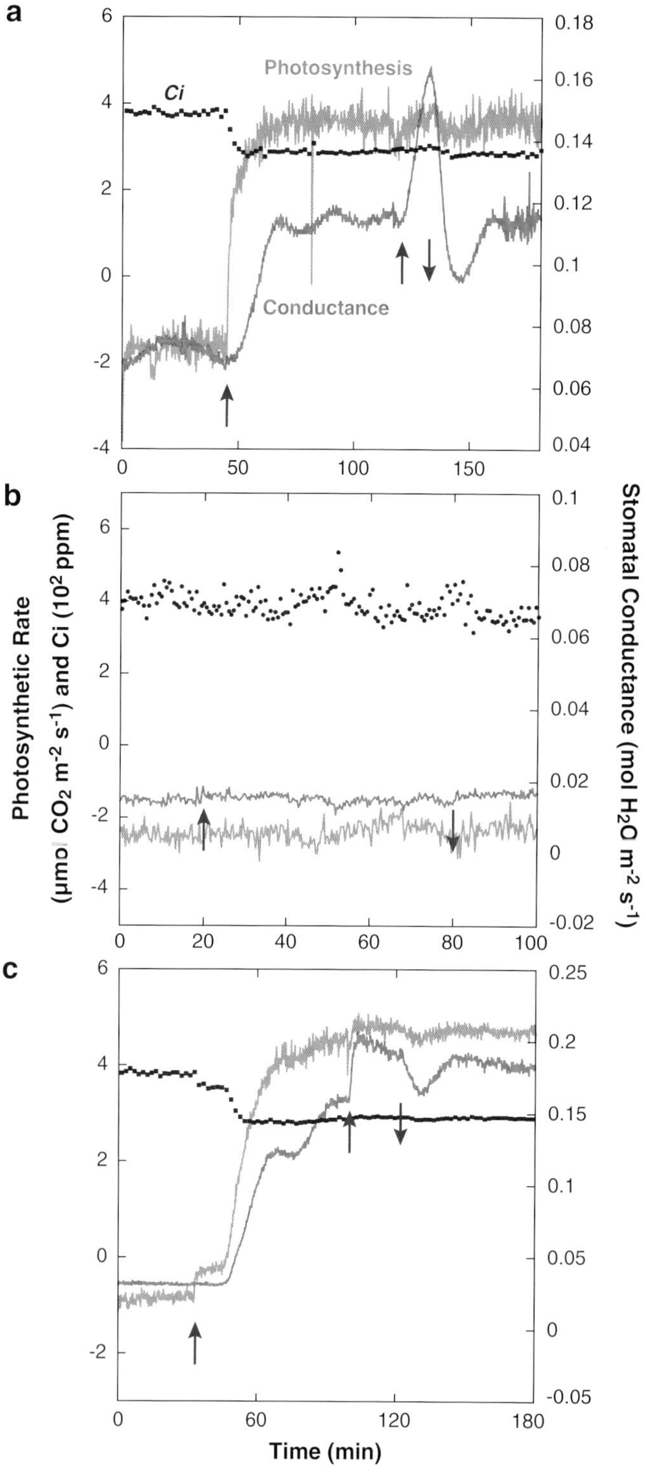

Figure 1

Stomatal conductance, photosynthetic CO_2 fixation, and intercellular CO_2 concentration in response to light in leaves of *Arabidopsis thaliana* (*a*, *b*) and *Oryza sativa* (*c*). Red upward arrows indicate continuous irradiation of the leaf with a strong red light (600 μmol $s^{-1}m^{-2}$). Upward and downward blue arrows indicate onset and termination of irradiation, respectively, with a weak blue light (5 μmol $s^{-1}m^{-2}$). (*a*) When an *Arabidopsis* leaf was illuminated with red light, stomatal conductance gradually increased and reached a steady state within 20 min. Photosynthetic CO_2 fixation showed an initial rapid increase with a subsequent gradual increase. This gradual increase is probably due to removal of stomatal limitation. When weak blue light was superimposed on red light, the stomatal conductance increased rapidly and decreased after turning off the blue light. Photosynthetic CO_2 fixation was only slightly enhanced by blue light. (*b*) When an *Arabidopsis* leaf was illuminated with weak blue light for 60 min in the absence of red light, neither stomatal conductance nor photosynthetic CO_2 fixation was increased. (*c*) When a rice leaf was illuminated with red light, stomatal conductance did not change for 10 min, whereas photosynthetic CO_2 fixation showed a rapid increase to a very low level. Upon the gradual increase in stomatal conductance, photosynthetic CO_2 fixation showed a parallel increase. When the leaf was irradiated with weak blue light superimposed on red light, a very rapid increase in stomatal conductance was observed.

acidification was required for K^+ uptake and stomatal opening (126). The H^+ release from guard cells was stimulated by light and was inhibited by vanadate (46). Zeiger & Hepler (197) found that isolated guard cell protoplasts of onion (*Allium cepa*) increased their volume in response to blue light, which is indicative of stomatal opening. This finding demonstrated that all components responsible for blue-light-specific stomatal opening are localized in guard cells. Using a large number of guard cell protoplasts from *Vicia faba*, it was found that guard cells extruded H^+ in response to a pulse of blue light (157). The properties of H^+ extrusion were very consistent with those of blue-light-induced stomatal opening in intact leaves (63). The H^+ extrusion was mediated by electrogenic pumping, and patch clamp experiments demonstrated that blue light induced a transient membrane hyperpolarization (12). These results indicate that blue light increases the inside-negative electrical potential across the plasma membrane by activating the electrogenic proton pump (**Figure 2**). Because acidification of the external medium by epidermal strips and swelling of guard cell protoplasts by blue light were inhibited by vanadate, an inhibitor of the plasma membrane H^+-ATPase (5, 46), and the blue-light-stimulated electrogenic current required cytosolic ATP (12), the H^+ pump was proposed to be the plasma membrane H^+-ATPase. In this initial report, the blue-light-induced H^+ current across the plasma membrane as measured by whole-cell patch clamp was not of the magnitude required for sufficient K^+ accumulation in guard cells to drive stomatal opening. This may be due to loss of essential cytoplasmic components in the whole-cell configuration, and a much larger current was obtained in a slow whole-cell configuration (8, 143). A sufficiently large current was also measured in whole-cell patch clamp experiments (177) and in intact *Vicia* guard cells (135). A blue-light-sensitive electron transport chain in the guard cell plasma membranes, which releases H^+ to the medium, was hypothesized as an alternative

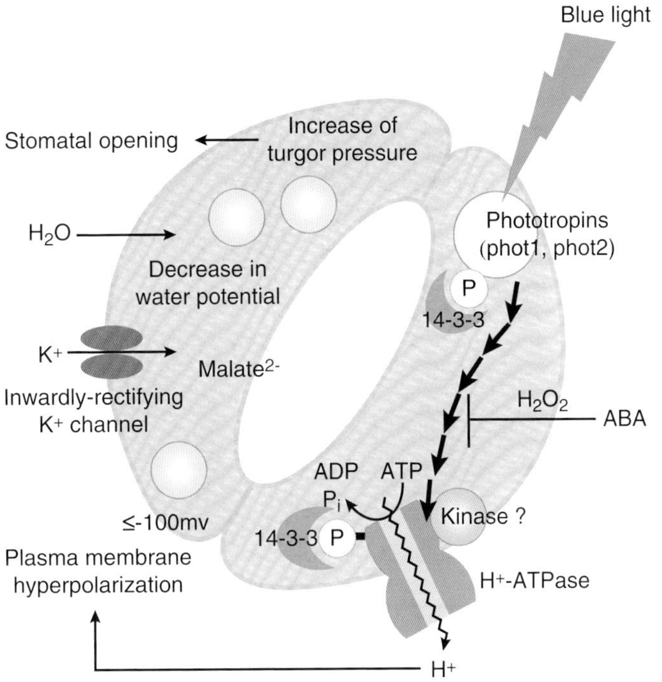

Figure 2

Processes of stomatal opening in response to blue light.

candidate mechanism (45, 125). Redox regulation of the blue-light-activated H^+ pump may be possible (181, 183), although manipulation of cytosolic NADH and NADPH concentrations did not alter the magnitude of blue-light-stimulated pump current (177).

After discovery of the blue-light-induced H^+ pump, voltage-gated K^+ channels, activated by membrane hyperpolarization, were discovered in the guard cell plasma membrane of the same plant species (146). The channels (inward-rectifying K^+ channels, $K^+{}_{in}$) function to take up K^+ when guard cells are hyperpolarized to values more negative than the equilibrium potential for K^+ (**Figure 2**). $K^+{}_{in}$ channels require ATP for the activation and maintenance of their activity (47, 190), and are sensitive to inhibition by Ca^{2+} (34, 145). The first genes for $K^+{}_{in}$ channels to be cloned from *Arabidopsis* were designated as *KAT1* (6) and *AKT1* (149). These genes encoded $K^+{}_{in}$ channels, as determined by expressing the genes in *Xenopus oocytes* and evaluating the resultant currents (104, 141, 184). Subsequent research showed that in addition to *KAT1* and *AKT1*, the *Arabidopsis* K^+ channel genes *AKT2/3*, *AtKC1*, and *KAT2* are also expressed in guard cells (107, 170). *KST1* and *KPT1* are highly expressed in guard cells of potato and poplar plants, respectively (84, 104, 123). Introducing a dominant-negative nonfunctional *KAT1* gene lowered the inward K^+ current by 75% in guard cells and reduced light-induced stomatal opening by 40% (82), suggesting that the inward current largely requires KAT1, although other K^+_{in} channels are also functional (170).

Entity of H^+ Pump and Regulation

The pump was subsequently demonstrated to be a plasma membrane H^+-ATPase that is activated by phosphorylation (76). H^+-ATPase activity in guard cell protoplasts was increased by blue light, and the activity was closely correlated with the rate of H^+ pumping. Levels of phosphorylation of the H^+-ATPase were increased by blue light and the amount of increase was proportional to the rate of H^+ pumping. The phosphorylation occurred exclusively on the C terminus of the H^+-ATPase, with subsequent binding of a 14-3-3 protein to the phosphorylated C terminus, which has been shown to function as an autoinhibitory domain of the H^+-ATPase (119). Phosphorylation of the Ser and Thr residues within this domain promotes H^+-ATPase activity (165), indicating that a serine/threonine protein kinase is involved in this regulation (**Figure 2**).

Binding of a 14-3-3 protein to the H^+-ATPase was absolutely required for its activation; phosphorylation alone was not sufficient to activate the H^+-ATPase (78). Four isoforms of 14-3-3 proteins (Vf14-3-3a, b, c, and d) were detected in *Vicia* guard cells, and mass analysis of the 14-3-3 protein that had bound to the H^+-ATPase in vivo identified Vf14-3-3a as a specific isoform binding to the plasma membrane H^+-ATPase (33). The isoform Vf14-3-3a had a higher binding affinity to the H^+-ATPase than Vf14-3-3b, and Vf14-3-3a might more effectively activate the H^+-ATPase than Vf14-3-3b. The binding site of a 14-3-3 protein was determined to be the penultimate Thr residue in the C terminus by competition experiments with synthetic phosphopeptides (78). The site identified by this method was the same as that determined by genetic analysis of fusicoccin-dependent activation of the plasma membrane H^+-ATPase in spinach leaves (169).

Because the plasma membrane H^+-ATPase shows primary and pivotal functions in many plant tissues by coupling with various carriers and channels, the regulatory mechanisms of these enzymes have been extensively investigated using the fungal toxin fusicoccin (119, 168). However, activation of the H^+-ATPase by this toxin occurrs irreversibly and such mechanisms might not be the same as those occurring in vivo. Guard cells are the ideal systems to manipulate the activity of this enzyme in vivo, and the associated regulatory mechanisms can be investigated under physiological conditions (76).

There are several to 10 isoforms of the plasma membrane H^+-ATPase in plant cells, and 11 functional H^+-ATPases are present in *Arabidopsis* (119). Tissue-specific expression of the H^+-ATPase isoforms has been observed in several organs, including *AHA3* in phloem companion cells (26), *AHA9* in anther cells (58), and *AHA4* in roots (185). Therefore, it is possible that the H^+-ATPase(s), which is specifically activated by blue light, is expressed in guard cells in a cell-specific manner. In *Vicia* guard cells, isoproteins of VHA1 (*Vicia* H^+-ATPase1) and VHA2 were activated by blue light (76). However, these isoproteins were also expressed in roots, leaves, stems, and flowers (55, 106). These observations suggest that guard cells have an additional blue-light signaling system that is cell specific, and the light signal is delivered to the usual H^+-ATPase(s) via this system. Interestingly, all 11 isogenes of AHA were expressed in guard cells, whereas only 4 isogenes were expressed in mesophyll cells of *Arabidopsis* plants

(180). In pulvinar motor cells of *Phaseolus vulgaris*, the plasma membrane H^+-ATPase exists in a highly active and heavily phosphorylated state in the dark, and is inactivated by dephoshorylation via phototropin-mediated reactions (66), suggesting that the blue-light signaling systems are flexible and can be modulated by cell-specific regulatory mechanisms. The H^+-ATPase inactivation caused water loss from the motor cells and resulted in leaf movement.

Blue-Light Receptors

Flavin and zeaxanthin, a component of the xanthophyll cycle (108), are candidate chromophores for the blue-light receptor of guard cells on the basis of action spectra for stomatal opening in intact leaves and malate synthesis in epidermal peels (71, 110, 199). The sensitivity of stomata to blue light increased with their zeaxanthin content (198), and inhibition of zeaxanthin formation suppressed blue-light-stimulated stomatal opening. The zeaxanthin-less mutant of *Arabidopsis*, *npq1*, failed to respond to blue light (40). Strong green light suppressed the blue-light-dependent stomatal opening in epidermal strips of *Arabidopsis* (39), and this effect was also found in isolated epidermal peels from light-treated leaves (173). These results suggest the involvement of zeaxanthin in the blue-light response of stomata. However, further confirmation is needed because the stomatal aperture responses reported for the *npq1* mutant could not be reproduced in leaves, epidermes, or guard cell protoplasts (29, 72, 180).

In 1997, a blue-light receptor for phototropism was identified by the Briggs group (60). The photoreceptor has two Light, Oxygen, Voltage (LOV) domains in the N terminus and a serine/threonine protein kinase in the C terminus, and undergoes autophosphorylation in response to blue light (21, 24). The LOV domains (LOV1 and LOV2) function as binding sites for flavin mononucleotide and absorb blue light (25). A homolog of this protein was found (67) and these proteins were initially called nonphototropic hypocotyl (NPH1) and nonphototropic hypocotyl like (NPL1), and were subsequently renamed phototropins 1 (phot1) and 2 (phot2), respectively (20). Later, phot2 was demonstrated to be the photoreceptor that mediates the chloroplast avoidance response to high-intensity light (68, 69).

The absorption spectrum of the LOV2 domain has peaks at 378 nm in the UV-A region, and 447 nm and 475 nm in the blue region (138). The action spectrum for stomatal opening in wheat leaves closely matches the absorption spectrum of the LOV domains (71). It was thus plausible to hypothesize that phototropins function as the blue-light receptors in guard cells (**Figure 2**). However, in the *Arabidopsis phot1* mutant, transpiration in the leaf still responded to blue light (87). Thus, phot2 might act as a blue-light receptor in guard cells, or phot1 and phot2 might function redundantly in the cells. To resolve this issue, the stomatal response was investigated using a single mutant of *phot2*, and the *phot1 phot2* double mutant of *Arabidopsis*. In the epidermal peels of *Arabidopsis*, stomata in the *phot1 phot2* double mutant did not open in response to blue light, but stomata opened in the *phot1*, *phot2*, and *npq1* single mutants (72). Epidermal strips from the double mutant were unable to extrude H^+ in response to blue light. In this mutant, no change was found in the amount of plasma membrane H^+-ATPase, and the mutant could respond to fusicoccin by opening stomata. Thus, the lack of stomatal opening and H^+ extrusion by blue light is due to an impairment in blue light perception. Stomata in the *phot2* mutant retain more sensitivity to blue light than those of *phot1* in the opening response.

Involvement of phototropins in stomatal response was confirmed by the complementation of the *phot1 phot2* double mutant with the *PHOT1* gene (28): Stomatal conductance in leaves of the *phot1 phot2* double mutant did not increase following blue-light irradiation, and the response was restored in the transformant. *Chlamydomonas reinhardtii* phototropin

(Crphot), which acts as a blue-light receptor in the sexual life cycle of algae (61), also restored stomatal response when the *phot1 phot2* double mutant was transformed with *CrPHOT* (113).

Biochemical Evidence for Phototropins as Blue-Light Receptors

There are at least two *Vicia faba* phototropins, 1a and 1b (Vfphot1a and 1b), with molecular masses of 125 kDa, and both are expressed in guard cells. If phototropins function upstream of the H^+-ATPase, then they should be phosphorylated more quickly than the H^+-ATPase. Simultaneous determination of the phosphorylation levels of phototropins and the H^+-ATPase in *Vicia* guard cells revealed that this was indeed the case (73). The maximum phosphorylation levels of Vfphots and the H^+-ATPase appeared around 1 and 5 min, respectively, after the onset of blue light. When phototropin phosphorylation was inhibited by diphenyleneiodonium chloride (DPI), a flavoprotein inhibitor, or by protein kinase inhibitors (K-252a and staurosporine), the H^+-ATPase phosphorylation was inhibited to the same degree as that of phototropins. Phosphorylation of phototropins and the H^+-ATPase showed similar photon flux dependency. Phototropin 1 is associated with the plasma membrane in guard cells (137), where the H^+-ATPase is also localized, and this is consistent with the idea that phot1 plays a role in blue-light-induced stomatal opening. Thus, biochemical and pharmacological evidence indicate that phototropins act upstream of the plasma membrane H^+-ATPase (**Figure 2**).

The binding of 14-3-3 proteins to target proteins can modify enzyme activity, change intracellular localization, and generate a scaffold for the interaction with other proteins (15, 36, 62, 78, 101). Phototropin binds reversibly to a 14-3-3 protein upon its autophosphorylation in guard cells (73). The 14-3-3-phototropin complex might confer a signaling state to phototropins, which then could transmit the light signal to downstream elements (**Figure 2**). Binding of a 14-3-3 protein to phototropin was also seen in the tissues of both etiolated seedlings and green leaves.

Binding of a 14-3-3 protein to target proteins was phosphorylation-dependent in general. However, the amount of 14-3-3 protein bound did not parallel the levels of phototropin phosphorylation. The maximum level of 14-3-3 binding in guard cells was reached faster than that of phosphorylation, and dissociation of a 14-3-3 protein from phototropin was also faster than the dephosphorylation of phototropin. This is probably because phototropins are phosphorylated on multiple sites (139, 163), and only some of these sites act as the binding sites for 14-3-3 protein; these sites appeared to be phosphorylated and dephosphorylated faster than other sites. Because the binding sites of 14-3-3 protein had phospho-Ser consensus motifs, including R/KXXpSXP, R/XXXpSXP, and R/XXpS/T (1, 105), the binding sites were determined by substituting all candidate Ser residues in phototropins. The binding sites were located in the hinge region between LOV1 and LOV2 in *Vicia* phototropins (73). Two phosphorylation sites were determined to be localized in the N terminus and six in the hinge region between LOV1 and LOV2 in *Avena sativa* phot1 by combining in vitro and in vivo phosphorylation using protein kinase A (139). The phosphorylation sites had different blue-light sensitivities.

Phototropin in etiolated seedlings underwent phosphorylation by blue light and was dephosphorylated gradually under darkness (118), and the long-lived phosphorylated state could act as the desensitized state of the photoreceptor (163). The recovery of sensitivity to blue light in phototropism occurred with a kinetics similar to that for recovery from the phosphorylated state. Essentially the same desensitization process was found in the blue-light response of stomata. The responsiveness of stomata to a second blue-light pulse decreased when guard cells were irradiated immediately after a first pulse, and sensitivity was gradually restored during the dark interval (63, 73, 157).

Blue-Light Signaling in Guard Cells

The signaling mechanism from phototropins to the plasma membrane H^+-ATPase is largely unknown. However, several signaling molecules appear to be present between phototropins and the H^+-ATPase (**Figure 2**).

14-3-3 proteins. The 14-3-3 proteins are involved in blue-light signaling in stomata as regulators of both phototropins and H^+-ATPase. A 14-3-3 protein functions as an activator of the plasma membrane H^+-ATPase via phosphorylation of the H^+-ATPase in response to blue light (33, 78). A 14-3-3 protein also binds phototropin upon its phosphorylation, and the binding appears to precede the activation of the plasma membrane H^+-ATPase in guard cells (73). However, the functional significance of the phototropin-14-3-3 protein complex has not been determined.

RPT2. Root Phototropism 2 (RPT2), a member of a unique, plant-specific family of proteins (102), was suggested to act as a signal transducer from phot1 to the downstream components of the signaling chain in phototropism and stomatal opening, but was not involved in chloroplast movement (65). RPT2 interacted genetically and physically with phot1 but not phot2, and the *phot2 rpt2* double mutant lost the capacity for blue-light-induced stomatal opening in the epidermis (**Figure 3**). Because the determination of stomatal aperture in these experiments was performed only in the epidermis, other separate lines of evidence will be needed to confirm this result.

VfPIP. A protein interacting with *Vicia* phot1a was isolated from guard cells (32). The *Vicia faba* phot1a interacting protein (VfPIP) bound to the N terminus of Vfphot1a but not to Vfphot1b. The *VfPIP* transcript was predominantly expressed in guard cells. VfPIP has sequence homology to dynein light chain and localizes to cortical microtubules. Vfphot1a seemed to exert its function through the microtubules, because treating guard cells with microtubule depolymerizing compounds resulted in partial inhibition of blue-light-stimulated stomatal opening and H^+ extrusion. A recent investigation indicated that the microtubules in guard cells are organized in parallel, straight, and dense arrays by blue light and are not affected by red light (83). Reorganization of the microtubules was not stimulated by activating the H^+-ATPase by fusicoccin. VfPIP may accelerate stomatal opening via phototropin-mediated organization and reorientation of microtubules (**Figure 3**).

Ca^{2+}. Ca^{2+} was suggested to be responsible for blue-light signaling in guard cells by pharmacological tools. Calmodulin antagonists inhibited both blue-light-dependent H^+ pumping and stomatal opening (158). The pumping was inhibited by verapamil, a Ca^{2+} channel blocker, albeit at high concentrations (160). The H^+ pumping was inhibited reversibly by caffeine, which releases Ca^{2+} from intracellular stores, and by inhibitors of endoplasmic reticulum Ca^{2+}-ATPase. Although stimulation of blue-light-induced acidification by external Ca^{2+} was reported in *Arabidopsis* epidermis (134), neither Ca^{2+} channel blockers nor change in external Ca^{2+} concentration affected H^+ pumping in *Vicia* (154). Therefore, the Ca^{2+} that is thought to be required for stomatal opening might originate from intracellular Ca^{2+} stores, most likely from the endoplasmic reticulum. In accord with this possibility, the anion channels that facilitate uptake of Cl^- and malate^{2-} into the vacuole, as occurs during stomatal opening, were activated by CDPK, a Ca^{2+}-activated protein kinase in guard cells (122).

Recent investigations demonstrated that phot1 activation elicited an increase in cytosolic Ca^{2+} in *Arabidopsis* and tobacco seedlings (16). The Ca^{2+} increase exhibited a transient change with a possible lag period of 3–6 s and

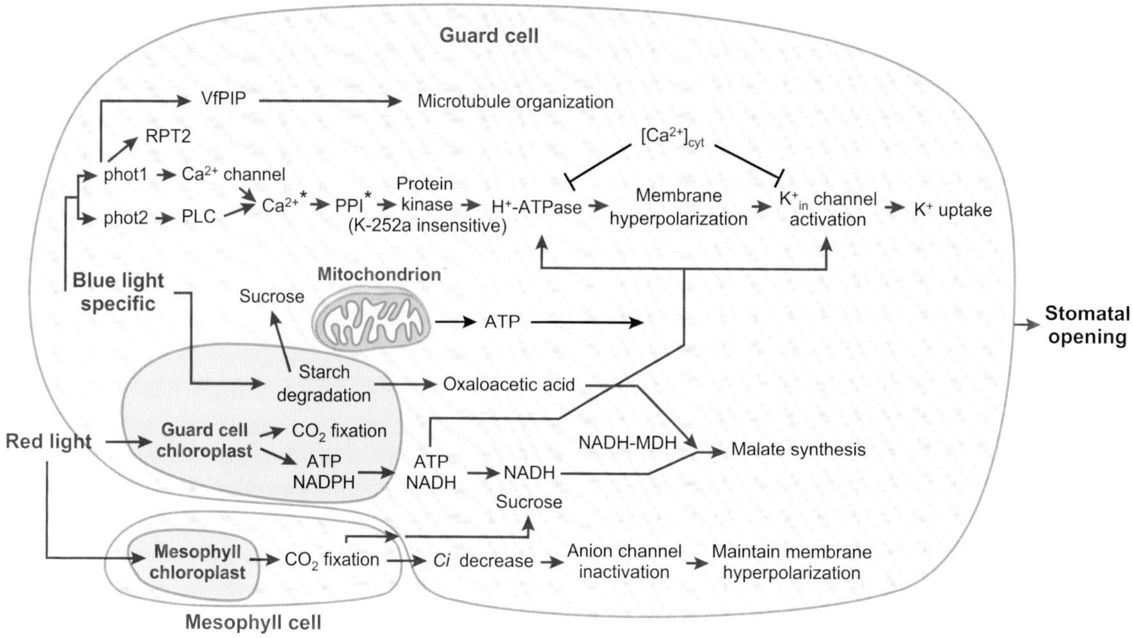

Figure 3

Cooperation of blue-light signaling and metabolism in guard cells for stomatal opening. Blue light is perceived by phot1 and phot2. The light signals are transmitted and activate the plasma membrane H^+-ATPase, thereby inducing K^+ uptake. Blue light also stimulates starch degradation. These responses are specific to blue light under normal conditions. Red light is absorbed by guard cells and mesophyll chloroplasts. Guard cell chloroplasts provide ATP and NADH in the cytosol, and mesophyll chloroplasts fix CO_2, thereby resulting in *Ci* decrease. The *Ci* decrease activates transport mechanisms that result in membrane hyperpolarization. These reactions are not specific to red light. Blue light is also absorbed by chloroplasts and has the same effect as red light, but the role of blue light through the chloroplasts is not depicted in this figure. *We placed Ca^{2+} upstream of protein phosphatase in this signaling pathway, but the order is speculative. K-252a, a broad inhibitor of Ser/Thr protein kinases; NADH-MDH, NADH-malate dehydrogenase; phot1, phototropin 1; phot2, phototropin 2; PLC, phospholipase C; RPT2, ROOT PHOTOTROPISM 2; VfPIP, *Vicia faba* phototropin interacting protein; PP1, type 1 protein phosphatase.

was sustained for 80 s after the pulse of blue light. The responsiveness to blue light was desensitized after the first pulse and gradually restored within 3–4 h. Based on mutant analysis, cry1 and cry2 do not participate in the signal transduction chain triggering this Ca^{2+} increase, whereas both phot1 and phot2 are involved and appear to mediate the cytosolic Ca^{2+} increase via calcium-permeable channels in the plasma membrane of mesophyll cells (16, 54, 167). phot2 may also function via phospholipase C-mediated Ca^{2+} release from intracellular stores (**Figure 3**), as inferred from assays with pharmacological inhibitors of phospholipase C (PLC) (54). Phototropin-mediated Ca^{2+} increase was likely to be required for rapid inhibition of hypocotyl elongation (38), chloroplast movement (178), and phototropism (14, 16), and the rapid growth inhibition was phot1-dependent. However, note that an increase in cytosolic Ca^{2+} by blue light has not yet been reported in guard cells.

It has been well documented that Ca^{2+} is a second messenger for ABA-induced stomatal closure (17, 144). A high concentration of cytosolic Ca^{2+} caused stomatal closure or inhibited stomatal opening (42, 96, 147) through activation of anion channels, and inactivation of the plasma membrane H^+-ATPase and K^+_{in} channels (74, 145). These results appear to conflict with the notion that Ca^{2+} is required for stomatal opening. ABA and a high concentration of external Ca^{2+} induced the oscillation of cytosolic concentration of Ca^{2+} with a definite period, and such oscillation was absolutely required for the maintenance of stomatal closure (3, 95). However, if cytosolic Ca^{2+} increased continuously without oscillation, or the intervals between oscillations were too short or too long (3), the Ca^{2+} response did not cause sustained stomatal closure (3, 4, 140). We note that, in other tissues, the Ca^{2+} increase elicited by blue light via phototropins did not show oscillations (16, 54, 167), and such a nonoscillatory increase might encode information opposite to that for stomatal closure.

Protein kinase. A serine/threonine protein kinase that directly phosphorylates the plasma membrane H^+-ATPase is present in guard cells, and this kinase may be activated by blue light (76) (**Figure 2**). This kinase is unlikely to be phototropin itself, because the kinase was almost completely insensitive to kinase inhibitors that are effective against phototropin kinase activity (73, 77, 158, 167). A protein kinase activity that phosphorylates the H^+-ATPase has been demonstrated in the plasma membrane of spinach leaves in vitro (169) (**Figure 3**).

Protein phosphatase. Okadaic acid and calyculin A inhibit both blue-light-dependent H^+ pumping and stomatal opening (75). This result implies that type 1 or type 2A protein phosphatase(s) mediate the signaling between phototropins and the plasma membrane H^+-ATPase in guard cells. However, it has been unclear which type of protein phosphatase is involved in this signaling, because okadaic acid and calyculin A inhibit both phosphatases. Recent work indicates that type 1 protein phosphatase is responsible for this signaling (172) (**Figure 3**). When the dominant-negative form of type 1 protein phosphatase or inhibitor-2, a proteinaceous-specific inhibitor of type 1 protein phosphatase, was expressed in *Vicia* guard cells via particle bombardment, stomatal opening by blue light was inhibited. Tautomycin, a preferential inhibitor of type 1 protein phosphatase in vivo, suppressed blue-light-induced phosphorylation of the H^+-ATPase but did not affect the autophosphorylation of phototropins.

Energy Source for Blue-Light-Dependent Proton Pumping

Protons are pumped out at the expense of ATP by the plasma membrane H^+-ATPase. It is likely that the energy for blue-light-dependent proton pumping is provided mainly by mitochondria and partly by chloroplasts (**Figure 3**), although the energy source may vary with the prevailing environment (13, 120, 189). Several chloroplasts and numerous mitochondria exist in guard cells, which exhibit high rates of respiration (2, 156, 182). Inhibited respiration reduced ATP levels drastically in guard cells, but the reduction was not severe in mesophyll cells (44, 156, 181). Stomatal opening specific to blue light was inhibited by potassium cyanide (KCN) and anoxia and partially affected by 3-(3,4-dichlorophenyl)-1,1-dimethylurea (DCMU), an inhibitor of photosystem II (PSII) in photosynthetic electron transport (148). Oligomycin, KCN, and low oxygen tension largely reduced blue-light-dependent H^+ pumping, and DCMU partially inhibited these responses (93).

Guard cell chloroplasts synthesize ATP via cyclic and noncyclic photophosphorylation (161, 195) (**Figure 3**) and can be the main energy source when mitochondrial activity is reduced. In the presence of oligomycin, fusicoccin-induced H^+ pumping in guard cells was greatly reduced, whereas the addition of red light enhanced both H^+ pumping and stomatal opening, and this enhancement was eliminated by DCMU (179).

Metabolism Supporting Blue-Light-Dependent Stomatal Opening

Malate. The main events of blue-light-dependent stomatal opening are an activation of the plasma membrane H^+-ATPase and K^+ uptake from the medium. To compensate the positively charged K^+ accumulated in guard cells, malate^{2-} is synthesized, and Cl^- and NO_3^- are taken up from the medium. Most plant species (except onion, which does not accumulate starch in guard cells) accumulate malate^{2-} preferentially over other anions. A close correlation has been obtained between stomatal opening and malate accumulation in guard cells (183, 189). Malate^{2-} can be produced through the degradation of starch in guard cells under blue light (**Figure 3**). The opening of stomata by blue light was severely impaired in *Arabidopsis* phosphoglucomutase mutant plants, which did not accumulate starch in guard cell chloroplasts (88). However, this blue-light response was restored in the presence of high concentrations of Cl^-, possibly because Cl^- replaced malate^{2-} as a necessary counterion for K^+. The mechanisms mediating uptake of anions, including Cl^- and NO_3^-, across the plasma membranes in guard cells are largely unknown, but it is likely that a cotransporter of H^+/Cl^- (NO_3^-) acts as the anion uptake carrier in guard cells.

Recent work indicated that *AtSTP1* (the H^+-monosaccharide symporter gene) expression was quickly and largely upregulated in guard cells in darkness (166). The AtSTP1 transporter may function in the guard cell import of apoplastic glucose derived from mesophyll cells under darkness and contribute to the accumulation of starch. The products of starch breakdown could be exported to the cytosol from guard cell chloroplasts and act as osmotica under the light (114, 116, 129), i.e., in the opposite direction as in mesophyll chloroplasts, where starch is broken down in darkness (164). Exported compounds might be converted to phospho*enol*pyruvate (PEP) in the cytosol, and PEP catalyzed by PEP carboxylase to produce OAA, which is then reduced to malate (189) (**Figure 3**). High activities of PEP carboxylase and NAD^+ malate dehydrogenase, required for these reactions, exist in guard cells.

Malate could be produced through light-activated $NADP^+$ malate dehydrogenase in guard cell chloroplasts (49), but the malate formed by this pathway marginally contributes to stomatal opening, because the activity of $NADP^+$ malate dehydrogenase is much lower than that of NAD^+- malate dehydrogenase in guard cells (142, 189). Because guard cell chloroplasts are likely to overreduce electron acceptors due to their low Rubisco activity and resulting low CO_2 fixation, the light-activated $NADP^+$-MDH may function as a malate valve to prevent damage to guard cells under strong sunlight (41).

Sucrose. Early investigations indicated that starch breakdown produced sugars that acted as principal osmotica for stomatal opening. This starch-sugar hypothesis had been replaced by the theory of guard cell osmoregulation via K^+ and its counterions (37, 42, 189). However, some studies have reported a parallel relationship between stomatal apertures and sucrose content in epidermal peels (128). Recent investigations demonstrate that sucrose accumulates in guard cells and replaces K^+ in the afternoon, thereby serving as a dominant osmoticum to maintain stomatal apertures (176). Sugar accumulates in guard cells under the light; however, the mechanism by which the sucrose accumulation occurs is largely unknown (183). Because guard cells are not connected with neighboring cells via plasmodesmata (188), two other pathways must operate in the sucrose accumulation: either uptake from the apoplast across the plasma membrane and/or production in these symplastically isolated cells. It is unlikely that photosynthetic CO_2 fixation in guard cell chloroplasts provides all of the sucrose, because the chlorophyll content in guard cells is less than 2% of that of mesophyll cells on a cell basis and Rubisco activity is low on a chlorophyll basis (50, 115, 127, 153, 156). It is possible that sugars are produced in guard cells, but fast starch degradation is required for sufficient and continuous provision of sugars (112, 176) (**Figure 3**). Import of sucrose/hexose from the apoplast will likely occur (183). It is very likely that blue light stimulates sucrose/hexose uptake by coupling with a H^+/sucrose or glucose symporter across the plasma membrane of guard cells through acidification of the apoplast by the H^+-ATPase (**Figure 3**). In support of this idea, fusicoccin-induced activation of the H^+-ATPase enhanced sucrose uptake in guard cell protoplasts (128, 130). Sucrose uptake by guard cells was reported in epidermal peels when they were overlayed onto mesophyll tissue (27). Because apoplastic sucrose reaches a high concentration in the daytime (91, 183), the H^+/sucrose and H^+/monosaccharide symporters could function coupled with the H^+-ATPase. Transient upregulation of *AtSTP1* H^+/monosaccharide symporter gene in the midday might contribute to guard cell-specific accumulation of sucrose (166). Expression of the sucrose transporter *AtSUC3* was demonstrated in guard cells of *Arabidopsis* (100).

Role of Blue-Light-Specific Stomatal Opening

The opening of stomata in response to blue light is fast and sensitive (151). Photosynthetic CO_2 fixation starts upon illumination and reaches maximum levels within several minutes. In general, stomatal opening occurs more slowly than photosynthesis, requiring more than 20 min to reach the maximum aperture in many plant species. At dawn, sunlight is rich in blue light, and the consequent rapid stomatal opening would enhance CO_2 uptake for photosynthesis (194).

Blue-light-induced stomatal opening is mediated via phototropins and enhances photosynthetic CO_2 fixation by removing the stomatal limitation to CO_2 entry. Phototropins also mediate phototropic bending, chloroplast accumulation, leaf expansion, and leaf movement (21), and these responses maximize photosynthetic electron transport by improving the efficiency of light capture. Thus, the blue-light response of stomata optimizes photosynthesis in increasing the CO_2 provision for stromal reactions in coordination with these phototropin-mediated thylakoid reactions. Such functions of phototropins were demonstrated by a dramatic growth enhancement of *Arabidopsis* under low and moderate light environments (171).

A fast response of stomata to blue light is suggested to decrease stomatal limitation of CO_2 uptake and facilitate CO_2 fixation under sun flecks in the canopy (79). More importantly, the sustained nature of stomatal opening after stimulation by a short period of (blue) light, such as a sun fleck (63), will enhance light-energy capture by providing CO_2 when the leaf encounters successive sun flecks.

Interaction of Blue-Light Signaling with Abscisic Acid

When plants are under drought stress, stomata close in the daytime to minimize water loss by transpiration. ABA induces anion-channel activation with simultaneous membrane depolarization and a subsequent activation of K^+_{out} channels (11, 144). The activated anion and K^+_{out} channels allow sustained K^+ salt efflux from guard cells, ultimately resulting in stomatal closure. ABA also stimulates stomatal closure by inhibiting H^+ pumping (48, 157) (**Figure 2**). The pump inhibition is important to maintain membrane depolarization and to reduce ATP consumption by the H^+-ATPase. ABA inhibits blue-light-dependent phosphorylation of the H^+-ATPase, and the inhibition may be mediated by H_2O_2 in guard cells (202). ABA produces H_2O_2 in guard cells through NADPH oxidases in the plasma membrane (81, 121, 201, 202), resulting in NO production that stimulates stomatal closure

(22). H_2O_2 also induces an increase through activation of hyperpolarization-activated Ca^{2+}-permeable channels (52, 121). It is possible that the increase in cytosolic Ca^{2+} via H_2O_2 inhibits the H^+-ATPase, although this seems contradictory if Ca^{2+} is required for blue-light signaling, as mentioned above (74, 90). ABA inhibited blue-light-induced apoplastic acidification in the epidermis by the H^+-ATPase, and the inhibition was not found in *abi1* and *abi2* mutants of *Arabidopsis* (134). Phototropins could be the target of H_2O_2, because phototropins have cysteine residues that are essential for their activities (138).

Other Photoreceptors

Phytochrome might be involved in the stomatal movements of several plant species, but the response was quite small (136). Phytochrome was reported as a modulator of the blue-light response of stomata in *Phaseolus vulgaris* seedlings (57). The time required to initiate stomatal opening was shortened by R and lengthened by FR, and the response showed R/FR reversibility. In the orchid *Paphiopedilum*, stomata opened in response to low photon flux density of red light and this opening was reversed by far-red light, suggesting phytochrome involvement (175). Negative results were also reported on phytochrome involvement, e.g., in wheat (71) and *Vicia* (112). Recently, a triple mutation of gene families of phytochrome kinase substrates (*PKS1*, *PKS2*, and *PKS4*) was reported to impair phototropism (85). PKS proteins interacted with phot1 in vivo and in vitro, and PKS proteins were suggested to represent a molecular link between phytochrome and phototropin signaling. However, no data were obtained on the involvement of PKS proteins in the blue-light response of stomata.

The action spectrum of stomatal opening in the UV region was obtained using *Vicia* epidermis (31). The spectrum showed a major peak at 280 nm and a minor peak at 360 nm. The response at 280 nm was three times larger than that at 459 nm. The UV-B-dependent response was antagonized by green light (30). It is suggested that the response to UV is mediated by a blue-light receptor, and that the energy is directly transferred to the signal molecule from the protein-pigment complex upon absorption of UV. However, *Arabidopsis* mutants of both *npq1* and *phot1 phot2* responded to UV-B (287 nm), suggesting the presence of a separate, unidentified UV-B photoreceptor.

Stomata from the *cry1 cry 2* double mutant exhibited a normal response to blue light, which was enhanced by red light in a manner similar to the response in the wild type, and this result excluded the involvement of cryptochromes in these stomatal responses to light (87). By contrast, a recent detailed investigation indicated that stomata of the *cry1 cry2* double mutant showed a reduced blue-light response, whereas those of CRY1-overexpressing plants revealed a hypersensitive response (92), suggesting the involvement of cryptochromes. In agreement with these findings, stomata in epidermal peels from the *phot1 phot2* double mutant opened slightly in response to relatively strong blue light. Note that in these experiments the blue light stimulus used might have activated chlorophyll-dependent stomatal opening in the epidermis of the *phot1 phot2* double mutant, because the background red light intensity employed would not have sufficed to saturate guard cell photosynthesis. In other cryptochrome-mediated responses, COP1 is implicated as a downstream element. Therefore, in an analysis of stomatal responses in either plants overexpressing *COP1* or plants lacking functional *cop1*, the COP1 protein was implicated as a negative regulator of stomatal opening that functions downstream of both cryptochromes and phototropins. However, stomata in CRY1 overexpressing plants and in *cop1* mutant plants are barely closed (widely open) in darkness. Taken together, these results suggest that cry signaling is likely related to the closure system in stomata, and that COP1 activates the closure system, but is not directly related to phototropin signaling. This conclusion is also consistent with other reports showing that neither cry1 nor cry2 participates in stomatal opening in response to blue light (111). Crys thus seem to function in a blue-light-independent manner to inhibit stomatal closure and thereby promote stomatal opening.

RED-LIGHT RESPONSE OF STOMATA

The opening response of stomata to red light requires a high light intensity (see **Figure 1**). The red-light response also requires continuous illumination in most plant species, with *Zea mays* being a notable exception (10). The red-light response may be driven partly by accumulation of K^+ salts (59) and partly by sugar accumulation (112, 174, 176). Potassium accumulation may be driven by the membrane potential, generated by a red-light-activated H^+ pump in the plasma membrane (112, 150). The pump activation might be brought about by the increase in ATP concentration in the cytosol resulting from photophosphorylation in guard cell chloroplasts (161, 179). However, the pump activation by red light reported in initial patch clamp experiments (150) was not reproduced (135, 177), and confirmation of this finding is needed. Sugars might be produced in guard cells from a combination of starch degradation, photosynthesis, and import from the apoplast (114).

A localized beam of red light applied to individual guard cells did not induce stomatal opening, giving rise to the hypothesis that the red-light response is actually a response to the decrease in *Ci* (**Figure 1**) resulting from red-light-driven mesophyll photosynthesis (131). This result is supported by the recent observation that stomata in albino leaf portions of variegated leaves of *Chlorophytum comosum* do not open by red light (133). The reduction of *Ci* brought about by mesophyll photosynthesis is one of the triggers that cause stomatal opening (**Figure 3**), and it is generally accepted that guard cells sense *Ci* rather than *Ca* (ambient CO_2 concentration) (103). CO_2 modulates cytosolic Ca^{2+} transients in guard cells, thereby stimulating Ca^{2+}-induced stomatal closure in a process analogous to that triggered by ABA (144, 187, 192). In addition, guard cells were hyperpolarized in CO_2-free air and depolarized under high CO_2, probably due to the activation of anion channels (18, 53, 131), which results in stomatal closure.

Several lines of evidence indicate that the red-light response is not just an indirect response, mediated by the effects of red light on mesophyll photosynthesis and *Ci*, but also results from a direct response of the guard cells to this light stimulus. First, it is well established that red light can induce stomatal opening in the isolated epidermis. This response depends on guard cell chloroplasts: The response is suppressed by DCMU (112, 148), an inhibitor of PSII, and is not observed in the epidermis of the orchid, *Paphiopedilum*, which has guard cells lacking chlorophyll (193). Experiments in intact leaves also demonstrate that the red-light response is not solely a response to *Ci*: Red light stimulates stomatal opening even when the intracellular concentration of CO_2 is held constant via gas exchange methodology (99).

SYNERGISTIC EFFECT OF BLUE AND RED LIGHT IN STOMATAL OPENING

Stomata in *Arabidopsis* opened rapidly in response to a weak blue light under a background of strong red light, but barely opened without red light (**Figure 1**). Stomatal conductance in leaves illuminated with both red and blue light is typically larger than the sum of the conductances under blue and red light alone, suggesting a synergistic action of stomatal opening in intact leaves. This phenomenon is illustrated in **Figure 1** and has been observed in numerous species (7, 63, 70, 110, 151, 194).

It is likely that both *Ci* and guard cell chloroplasts play roles in the synergistic effect of blue and red light on stomatal opening. That the synergistic effects of blue and red light could occur in part via a stomatal response to the reduction of *Ci* induced by mesophyll photosynthesis is suggested by the observation that manipulation of *Ci* by gas exchange techniques affects the magnitude of the blue-light response. Reducing *Ci* to a low level increases the magnitude of the blue-light response in several plant species (7, 70, 86). In addition, red-light enhancement of stomatal conductance in response to blue light was found in leaves of the orchid *Paphiopedilum*, which has a chlorophyllous mesophyll but achlorophyllous guard cells, whereas *Paphiopedilum* stomata in isolated epidermal peels showed no red-light response (196). Taken together, these results suggest that mesophyll photosynthesis can indirectly stimulate the blue-light response.

That guard cell chloroplasts also play a role in the synergistic effect of red and blue light is indicated by the fact that this synergism is also found in the isolated epidermis for both stomatal opening (148) and malate formation (109, 110). In addition, in wheat, red light enhanced the magnitude of the blue-light-induced conductance response, even in the presence of CO_2-free air (70). Such results indicate that there are also *Ci* independent aspects to this synergism. We suggest that these *Ci*-independent aspects could be accounted for by cooperation of blue-light signaling and guard cell chloroplasts with respect to malate formation. We propose that guard cell chloroplasts are primed to translocate NADH and ATP into the cytosol under red light. These compounds are required for malate synthesis and H^+ pumping, respectively, when blue light is applied (**Figure 4**). Malate is formed by the reduction of oxaloacetate (OAA) by NAD-malate dehydrogenase at the expense of NADH. The large amount of NADH necessary for the reaction can be produced by reverse reactions of NAD^+-glyceraldehyde 3-phosphate dehydrogenase (NAD^+-GAPD) in the cytosol using DHAP/GAP as a reductant, which can be supplied by guard cell chloroplasts via the triose phosphate translocator (TPT) under light (**Figure 4**). DHAP/GAP is oxidized to 1,3-bisphosphate glyceric acid (1,3-BPGA), and the ATP is generated by the reverse reaction via phosphoglycerate kinase (PGAK) from 1,3-BPGA, with simultaneous production of 3-PGA (**Figure 4**).

Carbon skeletons for malate formation are derived from starch in the chloroplasts. Blue light induces starch degradation and ultimately provides Triose-P (DHAP/GAP) in the cytosol, and Triose-P is finally converted to PEP. OAA is then produced through PEP carboxylase at the expense of PEP and HCO_3^- (189). It is interesting to note that, in potato, transcript levels of NAD^+-GAPD, H^+-ATPase, PEP carboxylase, and inward-rectifying K^+ channel (KST1), all of which are

Figure 4

Hypothetical scheme of provision of ATP and NADH from guard cell chloroplasts for blue-light-dependent H^+ pumping in guard cells. Bold black arrows indicate high enzyme activities demonstrated in guard cells. When guard cells are illuminated with blue light, the plasma membrane H^+-ATPase is activated via phototropin signaling. The activated H^+-ATPase would consume ATP and induce K^+ uptake through voltage-gated K^+ channels. The accumulated K^+ would promote malate synthesis to compensate the positive charges at the expense of NADH, resulting in the regeneration of ATP and NADH. The resultant 3-PGA would be transported from the cytosol to the chloroplasts, then phosphorylated and reduced to produce DHAP/GAP. DHAP/GAP could be shuttled out from chloroplasts to the cytosol in guard cells, and the phosphorylation and reduction reactions in guard cell chloroplasts would be facilitated under a strong red light. 1,3-BPGA, 1,3-Bisphosphoglycerate; DHAP, Dihydroxyacetone phosphate; FBP, Fructose1,6-bisphosphate; G-1-P, glucose 1-phosphate; GAP, Glyceraldehyde 3-phosphate; GAPD, Glyceraldehyde-3-phosphate dehydrogenase; MDH, Malate dehydrogenase; OAA, Oxaloacetate; PEP, phospho*enol*pyruvate; PEPC, phospho*enol*pyruvate carboxylase; PGAK, Phosphoglycerate kinase; 3-PGA, 3-phosphoglycerate; Rubisco, Ribulose-1,5-bisphosphate carboxylase/oxygenase; TPI, Triose-phosphate isomerase.

involved in stomatal opening, were downregulated simultaneously in a guard cell–specific manner under drought (80).

Guard cell chloroplasts have suitable properties to meet this hypothesis (**Figure 4**). (*a*) A considerable proportion of reducing equivalents and ATP are available for reactions other than photosynthetic CO_2 fixation (153); for example, only 8% of reducing equivalents were used for CO_2 fixation in *Vicia* guard cells (50). (*b*) The chloroplasts have enzyme activities that favor the accumulation of DHAP;

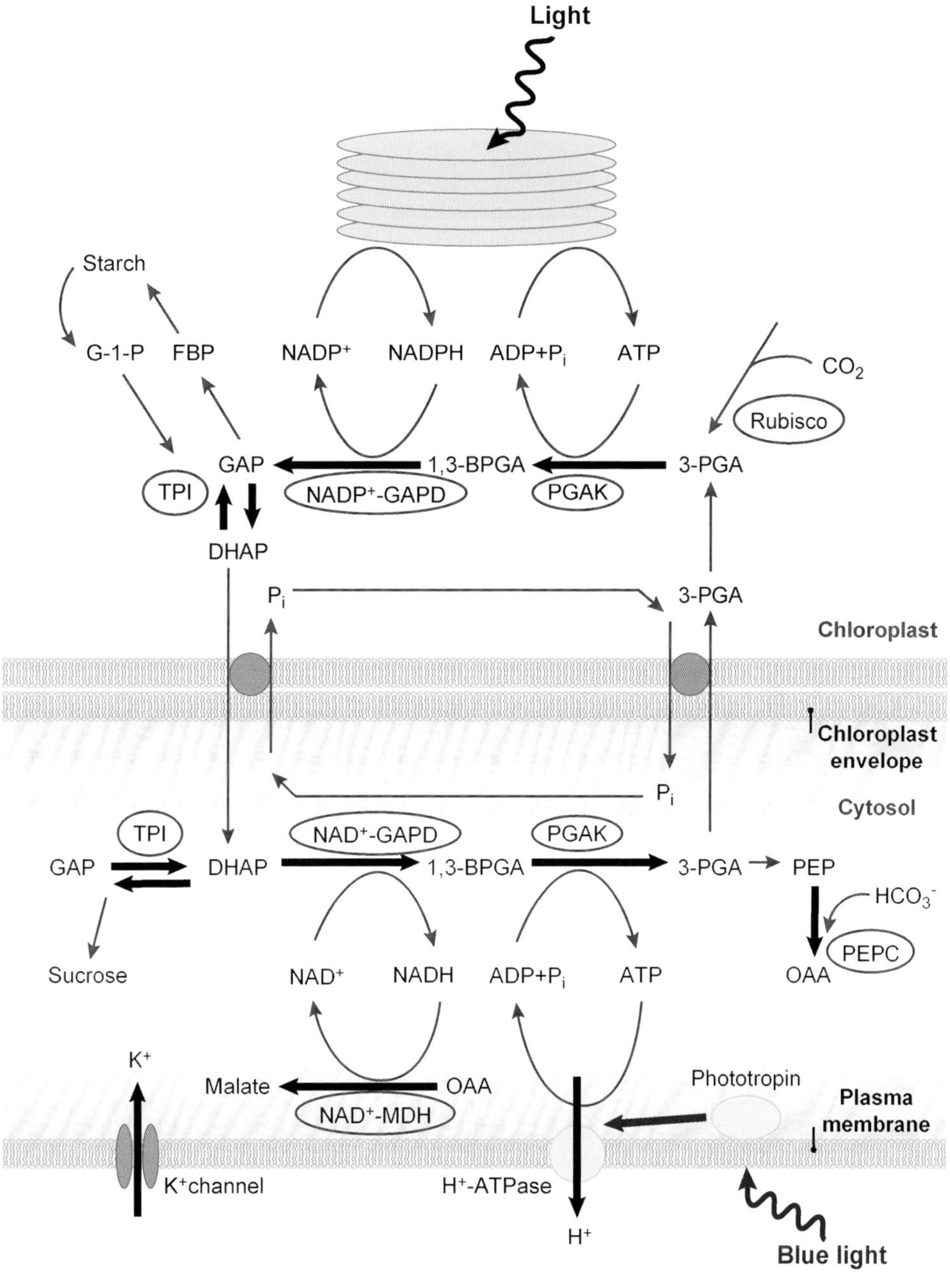

activities of PGAK, $NADP^+$-GAPD, and TPI in guard cell chloroplasts were several to more than 10 times higher than those of mesophyll chloroplasts, but the activity of FBPase in guard cell chloroplasts was roughly half of that in mesophyll chloroplasts. Such accumulation favors the export of DHAP to the cytosol via the phosphate translocator (116, 159). (*c*) Very high activities of NAD^+-GAPD and PGAK were found in the guard cell cytosol (159). These enzymes utilize DHAP and catalyze the production of NADH and ATP. (*d*) Indirect export of ATP, which was generated in guard cell chloroplasts, to the cytosol was suggested by an increase in H^+ pump activity under red light (179). The H^+ pump activation was sensitive to DCMU. The ATP production can indicate simultaneous production of NADH in the cytosol.

Therefore, when guard cells are illuminated with blue light under background red light, the activated plasma membrane H^+-ATPase utilizes ATP for H^+ pumping, with subsequent K^+ uptake. This situation requires malate formation, and results in the production of NAD^+, ADP, 3-PGA, and Pi in the cytosol, and thereby stimulates the export of DHAP/GAP from the chloroplasts (**Figure 4**). A low but significant Rubisco activity in guard cell chloroplasts (23, 127, 153, 200) would generate 3-PGA under red light, and the 3-PGA would act as an intermediate for DHAP production. The enhancement of malate formation is brought about by the activation of the H^+-ATPase in guard cells (**Figure 4**). In accord with this idea, suppression of PEPC activity through inhibition of H^+-ATPase was reported (97).

Because blue light likely stimulates the transfer of metabolites between chloroplasts and cytosol (**Figure 4**), it may also affect the energy state of thylakoid membranes. In accord with this idea, two distinct energy states of thylakoid membranes of guard cells were reported. The chlorophyll fluorescence transient in guard cells evoked by blue light had no prominent M peak (98, 155). Such unique kinetics of the fluorescence was transformed into the standard kinetics of mesophyll cells, with a prominent M peak, when guard cells were excited by green light, which did not activate blue-light signaling (94).

Under CO_2-free conditions, where red-light enhancement of the blue-light response is still observed (70) despite the absence of CO_2 as a substrate for Rubisco, perhaps red light enhances the blue-light stimulation of starch breakdown. Red light promotes starch breakdown under low CO_2 conditions (112), although whether this red-light effect also operates in the presence of blue light has yet to be evaluated.

CONCLUDING REMARKS

Stomatal pores surrounded by a pair of guard cells are important structures and allow uptake of CO_2 for photosynthesis by opening, and reduce the water loss from plants by closing in response to ever-changing environments, thereby extending the land plant growth area. Guard cells are machineries that transduce external and internal signals into the transport of various ions and metabolites across the membranes, and adjust the pore size of stomata. Thus, guard cells are excellent model systems to investigate the mechanisms of perception and transduction of signals, such as light, phytohormones, chemicals, temperature, and humidity, into stomatal movement. In this review, we focused on the light signaling in guard cells and described the properties of the plasma membrane H^+-ATPase and those of blue-light receptor phototropins. The plasma membrane H^+-ATPases are distributed in most plant tissues and drive the secondary transport of numerous kinds of inorganic ions, organic acids, and sugars by coupling with tissue- or organ-specific transporters. Because the plasma membrane H^+-ATPase is activated by blue light in guard cells, the regulatory mechanism of the H^+-ATPase is elucidated under physiological conditions, which has been difficult in other tissues. Phototropins (phot1 phot2), light receptor-type Ser/Thr protein kinases, are identified as blue-light receptors, and both phot1 and phot2 function for stomatal opening. Phototropins also regulate divergent responses, including phototropism, chloroplast movement, leaf movement, and leaf flattening. Because the signaling from phototropins to immediate downstream component(s) remains unknown in any of these responses, identification of the kinase substrate or the signaling component in guard cells will provide valuable information. Type 1 protein phosphatase mediates the signaling between phototropins and the plasma membrane H^+-ATPase in guard cells. Although type 1 protein phosphatase regulates many fundamental processes in animal cells, the role of this phosphatase has not been demonstrated in plant cells. Because the type 1 protein phosphatase works together with a number of regulatory subunits, identification of the subunit functioning in blue-light responses is important and will provide new insight into the role of the phosphatase in plant cells. Finally, stomatal opening is supported by metabolic activity unique to guard cells, and $malate^{2-}$ formation is an important process required for the opening. Reducing equivalents and carbon skeletons needed for the $malate^{2-}$ formation are likely provided from guard cell chloroplasts. Guard cell chloroplasts seem to adapt their role to an organ-specific function of stomatal movement.

SUMMARY POINTS

1. Stomatal opening is induced by light and is mediated by two distinct photosystems: blue-light photosystems and chloroplasts. Stomata open in response to a weak blue light and the opening is enhanced by background red light.
2. Blue light activates the plasma membrane H^+-ATPase via the phosphorylation of the C terminus and increases the inside-negative electrical potential across the plasma membrane in guard cells. The potential drives K^+ uptake through voltage-gated K^+ channels, and the accumulated positive K^+ charges are compensated mainly by $malate^{2-}$ formed in guard cells.
3. Red light induces stomatal opening at high intensity. Red light likely mediates stomatal opening via reduction of the intercellular concentration of CO_2 (*Ci*) by mesophyll photosynthesis, but the role of guard cell chloroplasts in the response could not be excluded.
4. Phototropins (phot1 phot2), light-responsive serine/threonine protein kinases, are identified as blue-light receptors for stomatal opening, and possess flavin mononucleotides as chromophores. Involvement of other photoreceptors in the light-induced opening response of stomata is reported.
5. The signaling between phototropins to the plasma membrane H^+-ATPase is largely unknown. Results indicate that type 1 protein phosphatase mediates this signaling.
6. Abscisic acid inhibits phototropin-mediated phosphorylation of the plasma membrane H^+-ATPase via H_2O_2, thereby suppressing stomatal opening.
7. Enhancement of the blue-light response of stomata by red light is brought about by guard cell chloroplast activity. The chloroplasts likely provide ATP for the H^+-ATPase and NADH for $malate^{2-}$ formation in the cytosol through the translocation of triose phosphate (DHAP/GAP) across the chloroplast envelope.

ACKNOWLEDGMENTS

This work was supported by Grants-in-Aid for Scientific Research (on Priority Areas, no. 17084005 and A, no. 16207003) to K. S., and by grants from the National Science Foundation (MCB02-09694) and the US Department of Agriculture (2006-35100-17254) to S.M.A.

LITERATURE CITED

1. Aitken A. 2002. Functional specificity in 14-3-3 isoform interactions through dimer formation and phosphorylation: Chromosome location of mammalian isoforms and variants. *Plant Mol. Biol.* 50:993–1010
2. Allaway WG, Setterfield G. 1972. Ultrastructural observations on guard cells of *Vicia faba* and *Allium porrum*. *Can. J. Bot.* 50:1405–13
3. Allen GJ, Chu SP, Harrington CL, Schumacher K, Hoffman T, et al. 2001. A defined range of guard cell calcium oscillation parameters encodes stomatal movements. *Nature* 411:1053–57
4. Allen GJ, Chu SP, Schumacher K, Shimazaki CT, Vafeados D, et al. 2000. Alteration of stimulus-specific guard cell calcium oscillations and stomatal closing in *Arabidopsis* det3 mutant. *Science* 289:2338–42
5. Amodeo G, Srivastava A, Zeiger E. 1992. Vanadate inhibits blue light-stimulated swelling of *Vicia* guard cell protoplasts. *Plant Physiol.* 100:1567–70
6. Anderson JA, Huprikar SS, Kochian LV, Lucas WJ, Gaber RF. 1992. Functional expression of a probable *Arabidopsis thaliana* potassium channel in *Saccharomyces cerevisiae*. *Proc. Natl. Acad. Sci. USA* 89:3736–40
7. Assmann SM. 1988. Enhancement of the stomatal response to blue light by red light, reduced intercellular concentrations of CO_2, and low vapor pressure differences. *Plant Physiol.* 87:226–31
8. Assmann SM. 1993. Signal transduction in guard cells. *Annu. Rev. Cell Biol.* 9:345–75
9. Assmann SM, Grantz DA. 1990. Stomatal response to humidity in sugarcane and soybean: effect of vapor pressure difference on the kinetics of the blue light response. *Plant Cell Environ.* 13:163–69
10. Assmann SM, Lee DM, Markus P. 1992. Rapid stomatal response to red light in *Zea mays*. *Photochem. Photobiol.* 56:685–89
11. Assmann SM, Shimazaki K. 1999. The multisensory guard cell: Stomatal responses to blue light and abscisic acid. *Plant Physiol.* 119:809–15
12. Assmann SM, Simoncini L, Schroeder JI. 1985. Blue light activates electrogenic ion pumping in guard cell protoplasts of *Vicia faba* L. *Nature* 318:285–87
13. Assmann SM, Zeiger E. 1987. Guard cell bioenergetics. In *Stomatal Function*, ed. E Zeiger, G Farquhar, I Cowan, pp. 163–94. Stanford, CA: Stanford Univ. Press
14. Babourina O, Newman I, Shabara S. 2002. Blue light-induced kinetics of H^+ and Ca^{2+} fluxes in etiolated wild-type and phototropin-mutant *Arabidopsis* seedlings. *Proc. Natl. Acad. Sci. USA* 99:2433–38
15. Bachmann M, Huber JL, Athwal GS, Wu K, Ferl RJ, Huber SC. 1996. 14-3-3 proteins associate with the regulatory phosphorylation site of spinach leaf nitrate reductase in an isoform-specific manner and reduce dephosphorylation of Ser-543 by endogenous protein phosphatases. *FEBS Lett.* 398:26–30
16. Baum G, Long JC, Jenkins GI, Trewavas AJ. 1999. Stimulation of the blue light phototropic receptor NPH1 causes a transient increase in cytosolic Ca^{2+}. *Proc. Natl. Acad. Sci. USA* 96:13554–59

17. Blatt MR. 2000. Cellular signaling and volume control in stomatal movements in plants. *Annu. Rev. Cell Dev. Biol.* 16:221–41
18. Brearley J, Venis MA, Blatt MR. 1997. The effect of elevated CO_2 concentrations on K^+ and anion channels of *Vicia faba* L. guard cells. *Planta* 203:145–54
19. Briggs WR. 2005. Phototropin overview. In *Light Sensing in Plants*, ed. M Wada, K Shimazaki, M Iino, pp. 139–46. Tokyo: The Bot. Soc. Jpn./Springer-Verlag
20. Briggs WR, Beck CF, Cashmore AR, Christie JM, Hughes J, et al. 2001. The phototropin family of photoreceptors. *Plant Cell* 13:993–97
21. Briggs WR, Christie JM. 2002. Phototropins 1 and 2: versatile plant blue-light receptors. *Trends Plant Sci.* 7:204–10
22. Bright J, Desikan R, Hancock JT, Weir IS, Neill SJ. 2006. ABA-induced NO generation and stomatal closure in *Arabidopsis* are dependent on H_2O_2 synthesis. *Plant J.* 45:113–22
23. Cardon ZG, Berry J. 1992. Effects of O_2 and CO_2 concentration on the steady-state fluorescence yield of single guard cell pairs in intact leaf discs of *Tradescantia albiflora*. *Plant Physiol.* 99:1238–44
24. Christie JM, Reymond P, Powell GK, Bernasconi P, Raibekas AA, et al. 1998. *Arabidopsis* NPH1: A flavoprotein with the properties of a photoreceptor for phototropism. *Science* 282:1698–701
25. Christie JM, Salomon M, Nozue K, Wada M, Briggs WR. 1999. LOV (light, oxygen, or voltage) domains of the blue light photoreceptor phototropin (nph1): Binding sites for the chromophore flavin mononucleotide. *Proc. Natl. Acad. Sci. USA* 96:8779–83
26. DeWitt ND, Sussman MR. 1995. Immunocytological localization of an epitopte-tagged plasma membrane proton pump (H^+-ATPase) in phloem companion cells. *Plant Cell* 7:2053–67
27. Dittrich P, Raschke K. 1977. Uptake and metabolism of carbohydrates by epidermal tissue. *Planta* 134:83–90
28. Doi M, Shigenaga A, Emi T, Kinoshita T, Shimazaki K. 2004. A transgene encoding a blue-light receptor, phot1, restores blue-light responses in the *Arabidopsis phot1 phot2* double mutant. *J. Exp. Bot.* 55:517–23
29. Eckert M, Kaldenhoff R. 2000. Light-induced stomatal movement of selected *Arabidopsis thaliana* mutants. *J. Exp. Bot.* 51:1435–42
30. Eisinger W, Bogomolni RA, Taiz L. 2003. Interaction between a blue-green reversible photoreceptor and separate UV-B receptor in stomatal guard cells. *Am. J. Bot.* 90:1560–66
31. Eisinger W, Swartz TE, Bogomolni RA, Taiz L. 2000. The UV action spectrum for stomatal opening in broad bean. *Plant Physiol.* 122:99–105
32. Emi T, Kinoshita T, Sakamoto K, Mineyuki Y, Shimazaki K. 2005. Isolation of a protein interacting with Vfphot1a in guard cells of *Vicia faba*. *Plant Physiol.* 138:1615–26
33. Emi T, Kinoshita T, Shimazaki K. 2001. Specific binding of vf14-3-3a isoform to the plasma membrane H^+-ATPase in response to blue light and fusicoccin in guard cells of broad bean. *Plant Physiol.* 125:1115–25
34. Fairley-Grenot K, Assmann SM. 1992. Whole cell K^+ current across the plasma membrane of guard cells from a grass: *Zea mays*. *Planta* 186:282–93
35. Fan LM, Zhao Z, Assmann SM. 2004. Guard cells: a dynamic signaling model. *Curr. Opin. Plant Biol.* 7:537–46
36. Ferl RJ. 1996. 14-3-3 proteins and signal transduction. *Annu. Rev. Plant Physiol. Plant Mol. Biol.* 47:49–73
37. Fischer RA. 1968. Stomatal opening: role of potassium uptake by guard cells. *Science* 160:784–85

38. Folta KM, Lieg EJ, Durham T, Spalding EP. 2003. Primary inhibition of hypocotyl growth and phototropism depend differentially on phototropin-mediated increases in cytoplasmic calcium induced by blue light. *Plant Physiol.* 133:1464–70
39. Frechilla S, Talbott LD, Bogomolni RA, Zeiger E. 2000. Reversal of blue light-stimulated stomatal opening by green light. *Plant Cell Physiol.* 41:171–76
40. Frechilla S, Zhu J, Talbott LD, Zeiger E. 1999. Stomata from *npq1*, a zeaxanthin-less *Arabidopsis* mutant, lack a specific response to blue light. *Plant Cell Physiol.* 40:949–54
41. Fridlyand LE, Backhausen JE, Scheibe R. 1998. Flux control of the malate valve in leaf cells. *Arch. Biochem. Biophys.* 349:290–98
42. Fujino M. 1967. Role of ATP and ATPase in stomatal movements. *Sci. Bull. Fac. Educ. Nagasaki Univ.* 18:1–47
43. Gao XQ, Li CG, Wei PC, Zhang XY, Chen J, Wang XC. 2005. The dynamic changes of tonoplasts in guard cells are important for stomatal movement in *Vicia faba*. *Plant Physiol.* 139:1207–16
44. Gautier H, Vavasseur A, Gans P, Lascève G. 1991. Relationship between respiration and photosynthesis in guard cell and mesophyll cell protoplasts of *Commelina communis* L. *Plant Physiol.* 95:636–41
45. Gautier H, Vavasseur A, Lascève G, Boudet AM. 1992. Redox processes in the blue light response of guard cell protoplasts of *Commelina communis L. Plant Physiol.* 98:34–38
46. Gepstein S, Jacobs M, Taiz L. 1982. Inhibition of stomatal opening in *Vicia faba* epidermal tissue by vanadate and abscisic acid. *Plant Sci. Lett.* 28:63–72
47. Goh CH, Dietrich P, Steinmeyer R, Schreiber U, Nam HG, Hedrich R. 2002. Parallel recordings of photosynthetic electron transport and K^+-channel activity in single guard cells. *Plant J.* 32:623–30
48. Goh CH, Kinoshita T, Oku T, Shimazaki K. 1996. Inhibition of blue light dependent H^+ pumping by abscisic acid in Vicia guard-cell protoplasts. *Plant Physiol.* 111:433–40
49. Gotow K, Tanaka K, Kondo N, Kobayashi K, Syono K. 1985. Light activation of NADP-malate dehydrogenase in guard cell protoplasts from *Vicia faba* L. *Plant Physiol.* 79:829–32
50. Gotow K, Taylor S, Zeiger E. 1988. Photosynthetic carbon fixation in guard cell protoplasts of *Vicia faba* L.: Evidence from radiolabel experiments. *Plant Physiol.* 86:700–5
51. Guo FQ, Young J, Crawford NM. 2003. The nitrate transporter AtNRT1.1 (CHL1) functions in stomatal opening and contributes to drought susceptibility in *Arabidopsis Plant Cell* 15:107–17
52. Hamilton DW, Hills A, Köhler B, Blatt MR. 2000. Ca^{2+} channels at the plasma membrane of stomatal guard cells are activated by hyperpolarization and abscisic acid. *Proc. Natl. Acad. Sci. USA* 97:4967–72
53. Hanstein SM, Felle HH. 2002. CO_2-triggered chloride release from guard cells in intact fava bean leaves. Kinetics of the onset of stomatal closure. *Plant Physiol.* 130:940–50
54. Harada A, Sakai T, Okada K. 2003. phot1 and phot2 mediate blue light-induced transient increases in cytosolic Ca^{2+} differently in *Arabidopsis* leaves. *Proc. Natl. Acad. Sci. USA* 100:8583–88
55. Hentzen AF, Smart LB, Wimmers LE, Fang HH, Schroeder JI, Bennett AB. 1996. Two plasma membrane H^+-ATPase genes expressed in guard cells of *Vicia faba* are also expressed throughout the plant. *Plant Cell Physiol.* 37:650–59
56. Hetherington AM. 2001. Guard cell signaling. *Cell* 107:711–14
57. Holmes MG, Klein WH. 1985. Evidence for phytochrome involvement in light-mediated stomatal movements in *Phaseolus vulgaris* L. *Planta* 166:348–53

58. Houlne G, Boutry M. 1994. Identification of an *Arabidopsis thaliana* gene encoding a plasma membrane H^+-ATPase whose expression is restricted to anther tissue. *Plant J.* 5:311–17
59. Hsiao TC, Allway WG, Evans LT. 1973. Action spectra for guard cell Rb^+ uptake and stomatal opening. *Plant Physiol.* 51:82–88
60. Huala E, Oeller PW, Liscum E, Han IS, Larsen E, Briggs WR. 1997. *Arabidopsis* NPH1: a protein kinase with a putative redox-sensing domain. *Science* 278:2120–23
61. Huang K, Beck CF. 2003. Phototropin is the blue-light receptor that controls multiple steps in the sexual life cylce of the green algae *Chlamydomonas reinhardtii*. *Proc. Natl. Acad. Sci. USA* 100:6269–74
62. Huber ST, MacKintosh C, Kaiser WM. 2002. Metabolic enzymes as targets for 14-3-3 proteins. *Plant Mol. Biol.* 50:1053–63
63. Iino M, Ogawa T, Zeiger E. 1985. Kinetic properties of the blue-light response of stomata. *Proc. Natl. Acad. Sci. USA* 82:8019–23
64. Imamura S. 1943. Untersuchungen über den mechanismus der turgorschwankung der spaltöffnungsschliesszellen. *Jpn. J. Bot.* 12:251–346
65. Inada S, Ohgishi M, Mayama T, Okada K, Sakai T. 2004. RPT2 is a signal transducer involved in phototropic response and stomatal opening by association with phototropin 1 in *Arabidopsis thaliana*. *Plant Cell* 16:887–96
66. Inoue S, Kinoshita T, Shimazaki K. 2005. Possible involvement of phototropins in leaf movement of kidney bean in response to blue light. *Plant Physiol.* 138:1994–2004
67. Jarillo JA, Ahmad M, Cashmore AR. 1998. NPL1: A second member of the NPH1 serine/threonine protein kinase family of *Arabidopsis*. *Plant Physiol.* 117:719
68. Jarillo JA, Gabrys H, Capel J, Alonso JM, Ecker JR, Cashmore AR. 2001. Phototropin-related NPL1 controls chloroplast relocation induced by blue light. *Nature* 410:952–54
69. Kagawa T, Sakai T, Suetsugu N, Oikawa K, Ishiguro S, et al. 2001. *Arabidopsis* NPL1: A phototropin homolog controlling the chloroplast high-light avoidance response. *Science* 291:2138–41
70. Karlsson PE. 1986. Blue light regulation of stomata in wheat seedlings. I. Influence of red background illumination and initial conductance level. *Physiol. Plant.* 66:202–6
71. Karlsson PE. 1986. Blue light regulation of stomata in wheat seedlings. II. Action spectrum and search for action dichroism. *Physiol. Plant.* 66:207–10
72. Kinoshita T, Doi M, Suetsugu N, Kagawa T, Wada M, Shimazaki K. 2001. phot1 and phot2 mediate blue light regulation of stomatal opening. *Nature* 414:656–60
73. Kinoshita T, Emi T, Tominaga M, Sakamoto K, Shigenaga A, et al. 2003. Blue-light- and phosphorylation-dependent binding of a 14-3-3 protein to phototropins in stomatal guard cells of broad bean. *Plant Physiol.* 133:1453–63
74. Kinoshita T, Nishimura M, Shimazaki K. 1995. Cytosolic concentration of Ca^{2+} regulates the plasma membrane H^+-ATPase in guard cells of fava bean. *Plant Cell* 7:1333–42
75. Kinoshita T, Shimazaki K. 1997. Involvement of calyculin A- and okadaic acid-sensitive protein phosphatase in the blue light response of stomatal guard cells. *Plant Cell Physiol.* 38:1281–85
76. Kinoshita T, Shimazaki K. 1999. Blue light activates the plasma membrane H^+-ATPase by phosphorylation of the C-terminus in stomatal guard cells. *EMBO J.* 18:5548–58
77. Kinoshita T, Shimazaki K. 2001. Analysis of the phosphorylation level in guard-cell plasma membrane H^+-ATPase in response to fusicoccin. *Plant Cell Physiol.* 42:424–32
78. Kinoshita T, Shimazaki K. 2002. Biochemical evidence for the requirement of 14-3-3 protein binding in activation of the guard-cell plasma membrane H^+-ATPase by blue light. *Plant Cell Physiol.* 43:1359–65

79. Kirschbaum MUF, Pearcy RW. 1988. Gas exchange analysis of the relative importance of stomatal and biochemical factors in photosynthetic induction in *Alocasia macrorrhiza*. *Plant Physiol*. 86:782–85
80. Kopka J, Provart NJ, Müller-Röber B. 1997. Potato guard cells respond to drying soil by a complex change in the expression of genes related to carbon metabolism and turgor regulation. *Plant J*. 11:871–82
81. Kwak JM, Mori IC, Pei ZM, Leonhardt N, Torres MA, et al. 2003. NADPH oxidase *AtrbohD* and *AtrbohF* genes function in ROS-dependent ABA signaling in *Arabidopsis*. *EMBO J*. 22:2623–33
82. Kwak JM, Murata Y, Baizabal-Aguirre VM, Merrill J, Wang M, et al. 2001. Dominant negative guard cell K^+ channel mutants reduce inward-rectifying K^+ currents and light-induced stomatal opening in *Arabidopsis*. *Plant Physiol*. 127:473–85
83. Lahav M, Abu-Abied M, Belausov E, Schwartz A, Sadot E. 2004. Microtubules of guard cells are light sensitive. *Plant Cell Physiol*. 45:573–82
84. Langer K, Levchenco V, Fromm J, Geiger D, Steinmeyer R, et al. 2004. The poplar K^+ channel KPT1 is associated with K^+ uptake during stomatal opening and bud development. *Plant J*. 37:828–38
85. Lariguet P, Schepens I, Hodgson D, Pedmale UV, Trevisan M, et al. 2006. PHYTOCHROME KINASE SUBSTRATE1 is a photoropin1 binding protein required for phototropism. *Proc. Natl. Acad. Sci. USA* 103:10134–39
86. Lascève G, Gautier H, Jappe J, Vavasseur A. 1993. Modulation of the blue light response of stomata of *Commelina communis* by CO_2. *Physiol. Plant*. 88:453–59
87. Lascève G, Leymarie J, Olney MA, Liscum E, Christie JM, et al. 1999. *Arabidopsis* contains at least four independent blue-light-activated signal transduction pathways. *Plant Physiol.* 120:605–14
88. Lascève G, Leymarie J, Vavasseur A. 1997. Alterations in light-induced stomatal opening in a starch-deficient mutant of *Arabidopsis thaliana* L. deficient in chloroplast phosphoglucomutase activity. *Plant Cell Environ*. 20:350–58
89. Lee DM, Assmann SM. 1992. Stomatal responses to light in the facultative Crassulacean acid metabolism species, *Portulacaria afra*. *Physiol. Plant*. 85:35–42
90. Lino B, Baizabal-Aguirre VM, Vara LEG. 1998. The plasma-membrane H^+-ATPase from beet root is inhibited by a calcium-dependent phosphorylation. *Planta* 204:352–59
91. Lu P, Outlaw WHJ, Smith BG, Freed GA. 1997. A new mechanism for the regulation of stomatal aperture size in intact leaves (accumulation of mesophyll-derived sucrose in the guard-cell wall of *Vicia faba*). *Plant Physiol*. 114:109–18
92. Mao J, Zhang YC, Sang Y, Li QH, Yang HQ. 2005. A role for *Arabidopsis* cryptochromes and COP1 in the regulation of stomatal opening. *Proc. Natl. Acad. Sci. USA* 102:12270–75
93. Mawson BT. 1993. Modulation of photosynthesis and respiration in guard and mesophyll cell protoplasts by oxygen concentration. *Plant Cell Environ*. 16:207–14
94. Mawson BT, Zeiger E. 1991. Blue light-modulation of chlorophyll *a* fluorescence transients in guard cell chloroplasts. *Plant Physiol*. 96:753–60
95. McAinsh MR, Hetherington AM. 1998. Encoding specificity in Ca^{2+} signaling systems. *Trends Plant Sci*. 3:32–36
96. McAinsh MR, Webb AAR, Taylor JE, Hetherington AM. 1995. Stimulus-induced oscillation in guard cell cytosolic free calcium. *Plant Cell* 7:1207–19
97. Meinhard M, Schnabl H. 2001. Fusicoccin- and light-induced activation and in vivo phosphorylation of phosphoenolpyruvate carboxylase in *Vicia* guard cell protoplasts. *Plant Sci*. 160:635–46

98. Melis A, Zeiger E. 1982. Chlorophyll *a* fluorescence transients in mesophyll and guard cells. *Plant Physiol.* 69:642–47
99. Messinger SM, Buckley TN, Mott KA. 2006. Evidence for involvement of photosynthetic processes in the stomatal responses to CO_2. *Plant Physiol.* 140:771–78
100. Meyer S, Lauterbach C, Niedermeier M, Barth I, Sjolund RD, Sauer N. 2004. Wounding enhances expression of ATSUC3, a sucrose transporter from *Arabidopsis* sieve elements and sink tissues. *Plant Physiol.* 134:684–93
101. Morrison D. 1994. 14-3-3: Modulators of signaling proteins. *Science* 266:56–57
102. Motchoulski A, Liscum E. 1999. *Arabidopsis* NPH3: A NPH1 photoreceptor-interacting protein essential for phototropism. *Science* 286:961–64
103. Mott KA. 1988. Do stomata respond to CO_2 concentrations other than intercellular? *Plant Physiol.* 86:200–3
104. Müller-Röber B, Ellenberg J, Provart N, Willmitzer L, Busch H, et al. 1995. Cloning and electrophysiological analysis of KST1, an inward rectifying K^+ channel expressed in potato guard cells. *EMBO J.* 14:2409–16
105. Muslin AJ, Tanner JM, Allen PM, Shaw AS. 1996. Interaction of 14-3-3 protein with signaling proteins is mediated by the recognition of phosphoserine. *Cell* 84:889–97
106. Nakajima N, Saji H, Aono M, Kondo N. 1995. Isolation of cDNA for a plasma membrane H^+-ATPase from guard cells of *Vicia faba* L. *Plant Cell Physiol.* 36:919–24
107. Nakamura RL, McKendree WL, Hirsch RE, Sedbrook JC, Gaber RF, Sussman MR. 1995. Expression of an *Arabidopsis* potassium channel gene in guard-cells. *Plant Physiol.* 109:371–74
108. Niyogi KK, Grossman AR, Bjorkman O. 1998. *Arabidopsis* mutants define a central role for the xanthophyll cycle in the regulation of photosynthetic energy conversion. *Plant Cell* 10:1121–34
109. Ogawa T. 1981. Blue light response of stomata with starch-containing (*Vicia faba*) and starch-deficient (*Allium cepa*) guard cells under background illumination with red light. *Plant Sci. Lett.* 22:103–8
110. Ogawa T, Ishikawa H, Shimada K, Shibata K. 1978. Synergistic action of red and blue light and action spectra for malate formation in guard cells of *Vicia faba* L. *Planta* 142:61–65
111. Ohgishi M, Saji K, Okada K, Sakai T. 2004. Functional analysis of each blue light receptor, cry1, cry2, phot1, and phot2, by using combinational multiple mutants in *Arabidopsis*. *Proc. Natl. Acad. Sci. USA* 101:2223–28
112. Olsen RL, Pratt RB, Gump P, Kemper A, Tallman G. 2002. Red light activates a chloroplast-dependent ion uptake mechanism for stomatal opening under reduced CO_2 concentrations in Vicia spp. *N. Phytol.* 153:497–508
113. Onodera A, Kong SG, Doi M, Shimazaki K, Christie J, et al. 2005. Phototropin from *Chlamydomonas reinhardtii* is functional in *Arabidopsis thaliana*. *Plant Cell Physiol.* 46:367–74
114. Outlaw WHJ, Manchester J. 1979. Guard cell starch concentration quantitatively related to the stomatal aperture. *Plant Physiol.* 64:79–82
115. Outlaw WHJ, Manchester J, DiCamelli CA, Randall DD, Rapp B, Veith GM. 1979. Photosynthetic carbon reduction pathway is absent in chloroplasts of *Vicia faba* guard cells. *Proc. Natl. Acad. Sci. USA* 76:6371–75
116. Overlach S, Diekmann W, Raschke K. 1993. Phosphate translocator of isolated guard-cell chloroplasts from *Pisum sativum* L. transports glucose-6-phosphate. *Plant Physiol.* 101:1201–7

117. Palevitz BA, O'Kane DJ, Korbes RE, Raikhel NV. 1981. The vacuole system in stomatal cells of *Allium*. Vacuole movements and changes in morphology in differentiating cells as revealed by epifluorescence, video and electron microscopy. *Protoplasma* 109:23–35
118. Palmer JM, Short TM, Briggs WR. 1993. Correlation of blue light-induced phosphorylation to phototropism in *Zea mays* L. *Plant Physiol.* 102:1219–25
119. Palmgren MG. 2001. Plant plasma membrane H^+-ATPase: power houses for nutrient uptake. *Annu. Rev. Plant Physiol. Plant Mol. Biol.* 52:817–45
120. Parvathi K, Raghavendra AS. 1995. Bioenergetic processes in guard cells related to stomatal function. *Physiol. Plant.* 93:146–54
121. Pei ZM, Murata Y, Benning G, Thomine S, Klüsener B, et al. 2000. Calcium channels activated by hydrogen peroxide mediate abscisic acid signaling in guard cells. *Nature* 406:731–34
122. Pei ZM, Ward JM, Harper JF, Schroeder JI. 1996. A novel chloride channel in *Vicia faba* guard cell vacuoles activated by the serine/threonine kinase, CDPK. *EMBO J.* 15:6564–74
123. Philippar K, Ivashikina N, Ache P, Christian M, Lüthen H, et al. 2004. Auxin activates *KAT1* and *KAT2*, two K^+-channel genes expressed in seedlings of *Arabidopsis thaliana*. *Plant J.* 37:815–27
124. Poffenroth M, Green DB, Tallman G. 1992. Sugar concentrations in guard cells of *Vicia faba* illuminated with red or blue light: analysis by high performance liquid chromatography. *Plant Physiol.* 98:1460–71
125. Raghavendra AS. 1990. Blue light effects on stomata are mediated by the guard cell plasma membrane redox system distinct from the proton translocating ATPase. *Plant Cell Environ.* 13:105–10
126. Raschke K, Humble GD. 1973. No uptake of anions required by opening stomata of *Vicia faba*: guard cells release hydrogen ions. *Planta* 115:47–57
127. Reckmann U, Sheibe R, Raschke K. 1990. Rubisco activity in guard cells compared with the solute requirement for stomatal opening. *Plant Physiol.* 92:246–53
128. Reddy AR, Das VSR. 1986. Stomatal movements and sucrose uptake by guard cell protoplasts of *Commelina benghalensis* L. *Plant Cell Physiol.* 27:1565–70
129. Ritte G, Raschke K. 2003. Metabolite export of isolated guard cell chloroplasts of *Vicia faba*. *N. Phytol.* 159:195–202
130. Ritte G, Rosenfeld J, Rohrig K, Raschke K. 1999. Rates of sugar uptake by guard cell protoplasts of *Pisum sativum* L. related to the solute requirement for stomatal opening. *Plant Physiol.* 121:647–56
131. Roelfsema MRG, Hanstein S, Felle HH, Hedrich R. 2002. CO_2 provides an intermediate link in the red light response of guard cells. *Plant J.* 32:65–75
132. Roelfsema MRG, Hedrich R. 2005. In the light of stomatal opening: New insights into 'the Watergate'. *N. Phytol.* 167:665–91
133. Roelfsema MRG, Konrad KR, Marten H, Psaras GK, Hartung W, Hedrich R. 2006. Guard cells in albino leaf patches do not respond to photosynthetically active radiation, but are sensitive to blue light, CO_2, and abscisic acid. *Plant Cell Environ.* 29:1595–605
134. Roelfsema MRG, Staal M, Prins ABA. 1998. Blue light-induced apoplastic acidification of *Arabidopsis thaliana* guard cells: Inhibition by ABA is mediated through protein phosphatases. *Physiol. Plant.* 103:466–74
135. Roelfsema MRG, Steinmeyer R, Staal M, Hedrich R. 2001. Single guard cell recordings in intact plants: light-induced hyperpolarization of the plasma membrane. *Plant J.* 26:1–13
136. Roth-Bejerano N, Itai C. 1987. Phytochrome involvement in stomatal movement in *Pisum sativum*, *Vicia faba*, and *Pelargonium*, *sp. Physiol. Plant.* 70:85–89

137. Sakamoto K, Briggs WR. 2002. Cellular and subcellular localization of phototropin 1. *Plant Cell* 14:1723–35
138. Salomon M, Christie JM, Knieb E, Lempert U, Briggs WR. 2000. Photochemical and mutational analysis of the FMN-binding domains of the plant blue light receptor, phototropin. *Biochemistry* 39:9401–10
139. Salomon M, Knieb E, von Zeppelin T, Rüdiger W. 2003. Mapping of low- and high-fluence autophosphorylation sites in phototropin 1. *Biochemistry* 42:4217–25
140. Sanders D, Pelloux J, Brownlee C, Harper JF. 2002. Calcium at the crossroads of signaling. *Plant Cell* 14:S401–17
141. Schachtman DP, Schroeder JI, Lucas WJ, Anderson JA, Gaber RF. 1992. Expression of an inward-rectifying potassium channel by the *Arabidopsis* KAT1 cDNA. *Science* 258:1654–58
142. Scheibe R, Reckmann U, Hedrich R, Raschke K. 1990. Malate dehydrogenase in guard cells of *Pisum sativum*. *Plant Physiol.* 93:1358–64
143. Schroeder JI. 1988. K^+ transport properties of K^+ channels in the plasma membrane of *Vicia faba* guard cells. *J. Gen. Physiol.* 92:667–83
144. Schroeder JI, Allen GJ, Hugouvieux V, Kwak JM, Waner D. 2001. Guard cell signal transduction. *Annu. Rev. Plant Physiol. Plant Mol. Biol.* 52:627–58
145. Schroeder JI, Hagiwara S. 1989. Cytosolic calcium regulates ion channels in the plasma membrane of *Vicia faba* guard cells. *Nature* 338:427–30
146. Schroeder JI, Raschke K, Neher E. 1987. Voltage dependence of K^+ channels in guard-cell protoplasts. *Proc. Natl. Acad. Sci. USA* 84:4108–12
147. Schwartz A. 1985. Role of calcium and EGTA on stomatal movements in *Commelina communis*. *Plant Physiol.* 79:1003–5
148. Schwartz A, Zeiger E. 1984. Metabolic energy for stomatal opening: Roles of photophosphorylation and oxidative phosphorylation. *Planta* 161:129–36
149. Sentenac H, Bonneaud N, Minet M, Lacroute F, Salmon JM, et al. 1992. Cloning and expression in yeast of a plant potassium-ion transport-system. *Science* 256:663–65
150. Serrano EE, Zeiger E, Hagiwara S. 1988. Red light stimulates an electrogenic proton pump in *Vicia* guard cell protoplasts. *Proc. Natl. Acad. Sci. USA* 85:436–40
151. Sharkey TD, Ogawa T. 1987. Stomatal responses to light. In *Stomatal Function*, ed. E Zeiger, G Farquhar, I Cowan, pp. 195–208. Stanford, CA: Stanford Univ. Press
152. Sharkey TD, Raschke K. 1981. Separation and measurement of direct and indirect effects of light on stomata. *Plant Physiol.* 68:33–40
153. Shimazaki K. 1989. Ribulosebisphosphate carboxylase activity and photosynthetic O_2 evolution rate in *Vicia* guard cell protoplasts. *Plant Physiol.* 91:459–63
154. Shimazaki K, Goh CH, Kinoshita T. 1999. Involvement of intracellular Ca^{2+} in blue light-dependent proton pumping in guard cell protoplasts from *Vicia faba*. *Physiol. Plant.* 105:554–61
155. Shimazaki K, Gotow K, Kondo N. 1982. Photosynthetic properties of guard cell protoplasts from *Vicia faba* L. *Plant Cell Physiol.* 23:871–79
156. Shimazaki K, Gotow K, Sakaki T, Kondo N. 1983. High respiratory activity of guard cell protoplasts from *Vicia faba* L. *Plant Cell Physiol.* 23:871–79
157. Shimazaki K, Iino M, Zeiger E. 1986. Blue light-dependent proton extrusion by guard-cell protoplasts of *Vicia faba*. *Nature* 319:324–26
158. Shimazaki K, Kinoshita T, Nishimura M. 1992. Involvement of calmodulin and calmodulin-dependent myosin light chain kinase in blue light-dependent H^+ pumping by guard cell protoplasts from *Vicia faba* L. *Plant Physiol.* 99:1416–21
159. Shimazaki K, Terada J, Tanaka K, Kondo N. 1989. Calvin-Benson cycle enzymes in guard-cell protoplasts from *Vicia faba* L. *Plant Physiol.* 90:1057–64

160. Shimazaki K, Tominaga M, Shigenaga A. 1997. Inhibition of the stomatal blue light response by verapamil at high concentration. *Plant Cell Physiol.* 38:747–50
161. Shimazaki K, Zeiger E. 1985. Cyclic and noncyclic photophosphorylation in isolated guard cell chloroplasts from *Vicia faba* L. *Plant Physiol.* 78:211–14
162. Shope JC, DeWald DB, Mott KA. 2003. Changes in surface area of intact guard cells are correlated with membrane internalization. *Plant Physiol.* 133:1314–21
163. Short TW, Briggs WR. 1994. The transduction of blue light signals in higher plants. *Annu. Rev. Plant Physiol. Plant Mol. Biol.* 45:143–71
164. Smith AM, Zeeman SC, Smith SM. 2005. Strach degradation. *Annu. Rev. Plant Biol.* 56:73–98
165. Sondergaard TE, Schulz A, Palmgren MG. 2004. Energization of transport processes in plants. Roles of the plasma membrane H^+-ATPase. *Plant Physiol.* 136:2475–82
166. Stadtler R, Büttner M, Ache P, Hedrich R, Ivashikina N, et al. 2003. Diurnal and light-regulated expression of *AtSTP1* in guard cells of *Arabidopsis*. *Plant Physiol.* 133:528–37
167. Stoelzle S, Kagawa T, Wada M, Hedrich R, Dietrich P. 2003. Blue light activates calcium-permeable channels in *Arabidopsis* mesophyll cells via the phototropin signaling pathway. *Proc. Natl. Acad. Sci. USA* 100:1456–61
168. Sussman MR. 1994. Molecular analysis of proteins in the plant plasma membrane. *Annu. Rev. Plant Physiol. Plant Mol. Biol.* 44:253–61
169. Svennelid F, Olsson A, Piotrowski M, Rosenquist M, Ottman C, et al. 1999. Phosphorylation of Thr-948 at the C terminus of the plasma membrane H^+-ATPase creates a binding site for the regulatory 14-3-3 protein. *Plant Cell* 11:2379–91
170. Szyroki A, Ivashikina N, Dietrich P, Roelfsema MRG, Ache P, et al. 2001. KAT1 is not essential for stomatal opening. *Proc. Natl. Acad. Sci. USA* 98:2917–21
171. Takemiya A, Inoue S, Doi M, Kinoshita T, Shimazaki K. 2005. Phototropins promote plant growth in response to blue light in low light environment. *Plant Cell* 17:1120–27
172. Takemiya A, Kinoshita T, Asanuma M, Shimazaki K. 2006. Protein phosphatase 1 positively regulates stomatal opening in response to blue light in *Vicia faba*. *Proc. Natl. Acad. Sci. USA* 103:13549–54
173. Talbott LD, Hammad JW, Harn LC, Nguyen VH, Patel J, Zeiger E. 2006. Reversal by green light of blue light-stimulated stomatal opening in intact, attached leaves of *Arabidopsis* operates only in the potassium-dependent, morning phase of movement. *Plant Cell Physiol.* 47:332–39
174. Talbott LD, Zeiger E. 1998. The role of sucrose in guard cell osmoregulation. *J. Exp. Bot.* 49:329–37
175. Talbott LD, Zhu JX, Han SW, Zeiger E. 2002. Phytochrome and blue light-mediated stomatal opening in the orchid, *Paphiopedilum*. *Plant Cell Physiol.* 43:639–46
176. Tallman G, Zeiger E. 1988. Light quality and osmoregulation in *Vicia* guard cells. Evidence for involvement of three metabolic pathways. *Plant Physiol.* 88:887–95
177. Taylor AR, Assmann SM. 2001. Apparent absence of a redox requirement for blue light activation of pump current in broad bean guard cells. *Plant Physiol.* 125:329–38
178. Tlalka M, Fricker M. 1999. The role of calcium in blue-light-dependent chloroplast movement in *Lemna trisulca* L. *Plant J.* 20:461–73
179. Tominaga M, Kinoshita T, Shimazaki K. 2001. Guard-cell chloroplasts provide ATP required for H^+ pumping in the plasma membrane and stomatal opening. *Plant Cell Physiol.* 42:795–802
180. Ueno K, Kinoshita T, Inoue S, Emi T, Shimazaki K. 2005. Biochemical characterization of the plasma membrane H^+-ATPase activation in guard-cell protoplasts of *Arabidopsis thaliana* in response to blue light. *Plant Cell Physiol.* 46:955–63

181. Vani T, Raghavendra AS. 1989. Tetrazolium reduction by guard cells in abaxial epidermis of *Vicia faba*. Blue light stimulation of a plasmalemma redox system. *Plant Physiol.* 90:59–62
182. Vani T, Raghavendra AS. 1994. High mitochondrial activity but incomplete engagement of the cyanide-resistant alternative pathway in guard cell protoplasts of pea. *Plant Physiol.* 105:1263–68
183. Vavasseur A, Raghavendra AS. 2005. Guard cell metabolism and CO_2 sensing. *N. Phytol.* 165:665–82
184. Vèry AA, Gaymard F, Bosseux C, Sentenac H, Thibaud JB. 1995. Expression of a cloned plant K^+ channel in Xenopus oocytes: analysis of macroscopic currents. *Plant J.* 7:321–32
185. Vitart V, Baxter I, Doerner P, Harper JF. 2001. Evidence for a role in growth and salt resistance of a plasma membrane H^+-ATPase in the root endodermis. *Plant J.* 27:191–201
186. Ward JM, Pei ZM, Schroeder JI. 1995. Roles of ion channels in initiation of signal transduction in higher plants. *Plant Cell* 7:833–44
187. Webb AAR, McAinsh MR, Mansfield TA, Hetherington AM. 1996. Carbon dioxide induces increases in guard cell cytosolic free calcium. *Plant J.* 9:297–304
188. Wille AC, Lucas WJ. 1984. Ultrastructural and histochemical studies on guard cells. *Planta* 160:129–42
189. Willmer CM, Fricker MD. 1996. *Stomata*. London, UK: Chapman & Hall. 375 pp. 2nd ed.
190. Wu WH, Assmann SM. 1995. Is ATP required for K^+ channel activation in *Vicia* guard cells? *Plant Physiol.* 107:101–9
191. Yamashita T. 1952. Influences of potassium supply upon various properties and movement of guard cell. *Sielboldia Acta Biol.* 1:51–70
192. Young JJ, Mehta S, Israelsson M, Godoski J, Grill E, Schroeder JI. 2006. CO_2 signaling in guard cells: Calcium sensitivity response modulation, a Ca^{2+}-independent phase, and CO_2 insensitivity of the *gca2* mutant. *Proc. Natl. Acad. Sci. USA* 103:7506–11
193. Zeiger E. 1983. The biology of stomatal guard cells. *Annu. Rev. Plant Physiol.* 34:441–75
194. Zeiger E. 1984. Blue light and stomatal function. In *Blue Light Effects in Biological Systems*, ed. H Senger, pp. 484–94. New York/Tokyo: Springer-Verlag
195. Zeiger E, Armond P, Melis A. 1981. Fluorescence properties of guard cell chloroplasts. *Plant Physiol.* 67:17–20
196. Zeiger E, Assmann SM, Meidner H. 1983. The photobiology of *Paphiopedilum* stomata–opening under blue but not red-light. *Photochem. Photobiol.* 38:627–30
197. Zeiger E, Hepler PK. 1977. Light and stomatal function: Blue light stimulates swelling of guard cell protoplasts. *Science* 196:887–89
198. Zeiger E, Talbott LD, Frechilla S, Srivastava A, Zhu J. 2002. The guard cell chloroplast: a perspective for the twenty-first century. *N. Phytol.* 153:415–24
199. Zeiger E, Zhu J. 1998. Role of zeaxanthin in blue light photoreception and the modulation of light-CO_2 interactions in guard cells. *J. Exp. Bot.* 49:433–42
200. Zemel E, Gepstein S. 1985. Immunological evidence for the presence of ribulose bisphosphate carboxylase in guard cell chloroplasts. *Plant Physiol.* 78:586–90
201. Zhang X, Miao YC, An GY, Zhou Y, Shangguan ZP, et al. 2001. K^+ channels inhibited by hydrogen peroxide mediate abscisic acid signaling in guard cells. *Cell Res.* 11:195–202
202. Zhang X, Wang H, Takemiya A, Song CP, Kinoshita T, Shimazaki K. 2004. Inhibition of blue light-dependent H^+ pumping by abscisic acid through hydrogen peroxide-induced dephosphorylation of the plasma membrane H^+-ATPase in guard cell protoplasts. *Plant Physiol.* 136:4150–58

The Plant Heterotrimeric G-Protein Complex

Brenda R. S. Temple[1] and Alan M. Jones[2]

[1]R. L. Juliano Structural Bioinformatics Core Facility, [2]Departments of Biology and Pharmacology, University of North Carolina at Chapel Hill, Chapel Hill, North Carolina, 27599; email: alan_jones@unc.edu

Annu. Rev. Plant Biol. 2007. 58:249–66

The *Annual Review of Plant Biology* is online at plant.annualreviews.org

This article's doi:
10.1146/annurev.arplant.58.032806.103827

First published online as a Review in Advance on January 3, 2007

1543-5008/07/0602-0249$20.00

Key Words

AtRGS1, GPCR, Phosducin, Pirin, PLD, THF1

Abstract

Heterotrimeric G-protein complexes couple extracellular signals via cell surface receptors to downstream enzymes called effectors. Heterotrimeric G-protein complexes, together with their cognate receptors and effectors, operate at the apex of signal transduction. In plants, the number of G-protein complex components is dramatically less than in other multicellular eukaryotes. An understanding of how multiple signals propagate transduction through the G-protein node can be found in the unique structural and kinetic properties of the plant heterotrimeric G-protein complex. This review addresses these unique features and speculates on why the repertoire of G-protein signaling elements is dramatically simpler than that in all other multicellular eukaryotes.

Contents

INTRODUCTION

As defined in metazoans and in yeast, heterotrimeric G proteins serve as physical couplers between cell surface, 7 transmembrane (7TM) G-protein-coupled receptors (GPCRs) to downstream enzymes known as effectors (**Figure 1**). G proteins have three subunits: Gα, Gβ, and Gγ. The Gα subunit binds the guanine nucleotides, GDP and GTP. In its GDP-bound state, the heterotrimer assembles and associates with the receptor through the receptor's cytoplasmic transmembrane loops interfacing with specific regions on the G-protein subunits. Occupancy of the 7TM receptor by its cognate ligand induces exchange of GDP for GTP, leading to G-protein complex dissociation by the realignment of at least three polyamino acid loops, which are designated switches (Switch I, II, III). Thus, the GPCRs should be viewed as enzymes with guanine exchange factor (GEF) activity. The activated Gα directly modulates the activity of cytoplasmic enzymes' designated effectors. Effectors interact with Gα in a guanine-nucleotide-dependent manner and many effectors have GTPase-accelerating protein (GAP) activity, which facilitates the return to the basal heterotrimeric state. The released Gβγ dimer also has interacting effectors; some are the same as the Gα targets for which Gβγ can act upon synergistically or antagonistically to Gα. The combined activity of Gα and Gβγ results in cellular changes that ultimately affect instantaneous/transient and delayed/sustained cell behavior or development.

GPCR: G-protein-coupled receptor

GEF: Guanine exchange factor

GAP: GTPase-accelerating protein

RGS: *R*egulator of *G S*ignaling

An intrinsic GTPase hydrolysis activity of the Gα subunit returns the complex to the receptor-associated, GDP-bound basal state. Aside from the GAP activity that many effectors possess, there are *R*egulator of *G S*ignaling (RGS) proteins that specifically accelerate the intrinsic GTPase activity of Gα to return to the GDP-bound basal state. RGS proteins bind at the same interface that effectors bind to Gα; thus, not only do RGS proteins oppose the action of the receptor GEFs, but they also block signaling through Gα.

Consequently, the basic Gα core contains a number of common functionalities, namely nucleotide binding domains, e.g., switches that establish the basal and activated conformations, and protein interfaces such as for the Gβγ dimer, the receptor, the RGS proteins, the modulators, and the effectors.

Four classes of Gα subunits have been designated in mammals; Gαi, Gαs, Gαq, and Gα12/13 are comprised in humans by 23 well-characterized members (54). There are 5 Gβ and 12 Gγ subunits, ~950 GPCRs, dozens of effectors, and an unknown number of scaffold types that could be found within this

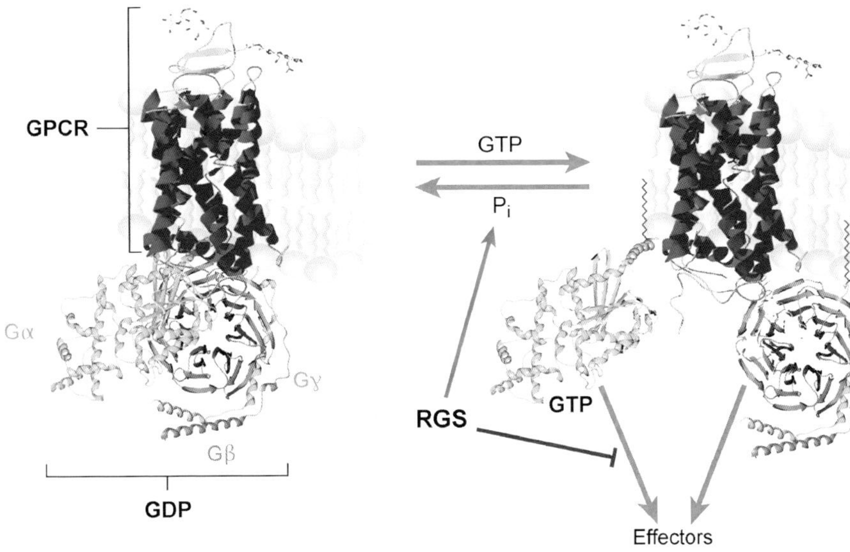

Figure 1

The classical model for G-protein cycling. G-protein-coupled receptors (GPCRs) have a 7 transmembrane (7TM) spanning domain as represented here by the structure for rhodopsin. GPCRs interact with a heterotrimeric G-protein complex, comprised of Gα , Gβ, and Gγ subunits. In the heterotrimer, Gα is bound to GDP. In animals, release of GDP is the rate-limiting step in G-protein cycling. Activation of GPCRs through ligand occupancy favors guanine-nucleotide exchange factor activity of the GPCR and consequently Gα becomes loaded with GTP and dissociates into a free Gα subunit and a free Gβγ dimer. Each of these, in turn, interacts with enzymes (effectors) to alter the amount of secondary messengers produced and/or the movement of ions across the plasma membrane. The Gα subunit has an intrinsic GTPase that can be accelerated by *R*egulator of *G*-protein *S*ignaling (RGS) proteins. RGS proteins can also block interactions between Gα and its effectors.

signaling complex that shares in common the heterotrimeric G protein. Thus, it is easy to imagine how intermolecular protein-protein interactions in this interacting network have constrained the evolution of structure to core functions associated with all the subunits. One must also consider the pressures associated with intermolecular interactions that become specific to unique functions and structures of individual components.

Plants, in contrast to metazoans, have a greatly simplified repertoire of G-protein signaling elements. Thus, it is informative to ask why and how the evolutionary constraints that operated on plant G proteins lead to their ability to regulate cellular processes. This is important not only to the plant biologist, but also to the clinician. Despite the simplicity in the working parts, genetic approaches reveal that plant G proteins have varied and wide functions. Discussion of the various aspects of plant physiology involving G proteins and the phenotypes of G-protein mutants can be found in several excellent reviews (4, 5, 16, 20, 21, 41). This review will serve the plant geneticist who is dissecting the mechanism of G-protein regulation of various physiologies by illuminating the unique structural and kinetic aspects of elements in the newly discovered plant G-protein complex, including potential 7TM receptors, effectors, and modulators. In particular, our intent is to illustrate important surfaces on the complex that may be vital for the physical interactions between components of the complex to be revealed in future studies designed to identify the "working parts" and the mechanism of action of the complex writ large.

HETEROTRIMER STRUCTURE

The Plant G-Protein Heterotrimer

As **Figure 2** shows, the Gα subunit consists of an N-terminal helical tail that binds to the Gβ subunit at a groove, a Ras-like GTPase domain, and an α-helical domain. A guanine nucleotide coordinates with residues lining the region between these two domains. A single Gα gene was reported for 3 monocots and 10 dicots (for original citations see Reference 41). Two Gα genes were reported each for pea and soybean.

The Gβ subunit acts as a scaffold for interacting with other proteins, including Gα, in the signal transduction cascade, but, as first defined in yeast, also acts as a predominant signaling agent. This is also the case in plants. Only a single gene encoding the Gβ subunit has been found in all plants except for tobacco, where three genes for a Gβ subunit have been found. Gβ subunits have been identified from one lower plant, moss, and from 11 higher plants (for most of the original citations see Reference 4).

Two genes proposed to encode Gγ subunits were identified in *Arabidopsis*. *Arabidopsis* Gγ1 (AGG1) (31) and *Arabidopsis* Gγ2 (AGG2) (32) were identified using yeast two-hybrid (Y2H) screens, and interaction with the *Arabidopsis* Gβ1 subunit AGB1 was confirmed using in vitro binding assays. Two sequences annotated as Gγ subunits from pea, PGG1 and PGG2, are found in the sequence databases, but supporting biochemical data verifying these as Gγ subunits have not yet been published.

Figure 2

The plant heterotrimeric G-protein complex. Three views of the heterotrimer (PDB ID 1GP2) are shown. On the left is the structure of a mammalian complex (27, 45), with the three subunits depicted as ribbon structures in separate colors. (*a*) The three main domains for the alpha subunit are indicated as N-terminal helix, Ras domain, and helical domain. Critical substructures of the Gα subunit for GTP binding and hydrolysis as well as for interaction with the Gβγ dimer are highlighted and discussed in the text. Roman numerals mark the position of the three switches. (*b*) Two views of the heterotrimer with conserved, variable, and plant-specific residues are shown as a ribbon/sphere structure for the Gα and Gγ subunits and as a surface model for the Gβ subunit. Plant residues conserved to the invariant or class-specific (see text) values on mammalian G-protein subunits are purple and dark blue spheres on Gα and Gγ, respectively. This illustrates that residues that are critical for G-protein function and structure are conserved between plant and animal G proteins. (*c*) Critical residues in the plant heterotrimeric G-protein complex. Shown are the primary sequences of the α subunit (AtGPA1), β subunit (AtAGB1), and γ subunit (AtAGG1) of the heterotrimeric G-protein complex in *Arabidopsis*. The circled numbers indicate positions of structures and motifs discussed in the text. (1) The myristoylated glycine. (2) A potential S-acetylated cysteine. (3) Residues shaded in gold contact the Gβγ dimer. (4) These lysines are potentially sumoylated. (5) P-loop for nucleotide recognition. (6) The green residues are conserved inserts unique to the plant G-protein subunits. (7) Cholera-toxin-mediated ribosylated cysteine. (8) Critical threonine residue for Gα and the RGS-box interaction. (9) Switch I residues are underlined. (10) A mutation to leucine destroys the GTPase activity without disrupting other functions. (11) Glutamic acid contacts the RGS-box. (12) Switch II residues are underlined. (13) Switch III residues are underlined. (14) The NKxD motif for guanine recognition. (15) In plants, this residue is not a cysteine, as found in the mammalian Gαi class; therefore, pertusis toxin may not mediate its ribosylation. (16) Residues on Gβ (*dark blue*) contact Gγ. (17) Residues on Gβ (*lavender*) contact Gα at the switches. (18) Residues on Gβ (*light blue*) contact Gα at Gα's N-terminal helix. These form the groove shown in **Figure 3**. (19) The WD motif critical for the 7-bladed β-propeller structure is orange. (20) The gray residues form the farnesyl pocket on Gβ. (21) These residues are important for hydrophobic contacts with Gβ; others form the interface. (22) This cysteine is modified with an isoprenyl group. (23) These residues (*yellow*) form hydrogen bonds with residues on Gβ. (24) These red residues form the Gβ and phosducin interface. Many residues in this interface (K_{67}, Y_{69}, W_{109}, M_{111}, Y_{158}, D_{206}, D_{250}, and W_{361}) also form the Gβ and Gα (switch) interface.

The first subunit interaction to be established, albeit in a Y2H configuration, was between *Arabidopsis* Gβ/Gγ1 (31) and was quickly followed by Gβ/Gγ2 (32). Interaction by Y2H between Gα and Gβ in rice (24) and in *Arabidopsis* (10) has been demonstrated. The formation and subcellular localization of full G-protein complexes, including the Gα, Gβ, and Gγ subunits, were probed in rice cells using various experimental techniques. All subunits of the G-protein complex were localized in the plasma membrane fraction using antibodies specific to each subunit (24). Antibodies to rice Gγ2 were raised against the C-terminal, Gγ-like domain, but the protein isolated in the plasma membrane fraction has a molecular weight of 18 kDa, compared to the 10 kDa molecular weight of rice Gγ1, consistent with the full-length cDNA sequence for RGG2. Gel filtration of solubilized membrane proteins showed the presence of large complexes on the order of 400 kDa containing rice Gα, Gβ, Gγ1, and Gγ2 subunits and additional unidentified protein components (24). Complexes of just the rice Gβ/Gγ1 or Gβ/Gγ2 dimers were also evident, and in another experiment, free Gα subunits were evident when the cells were preincubated with

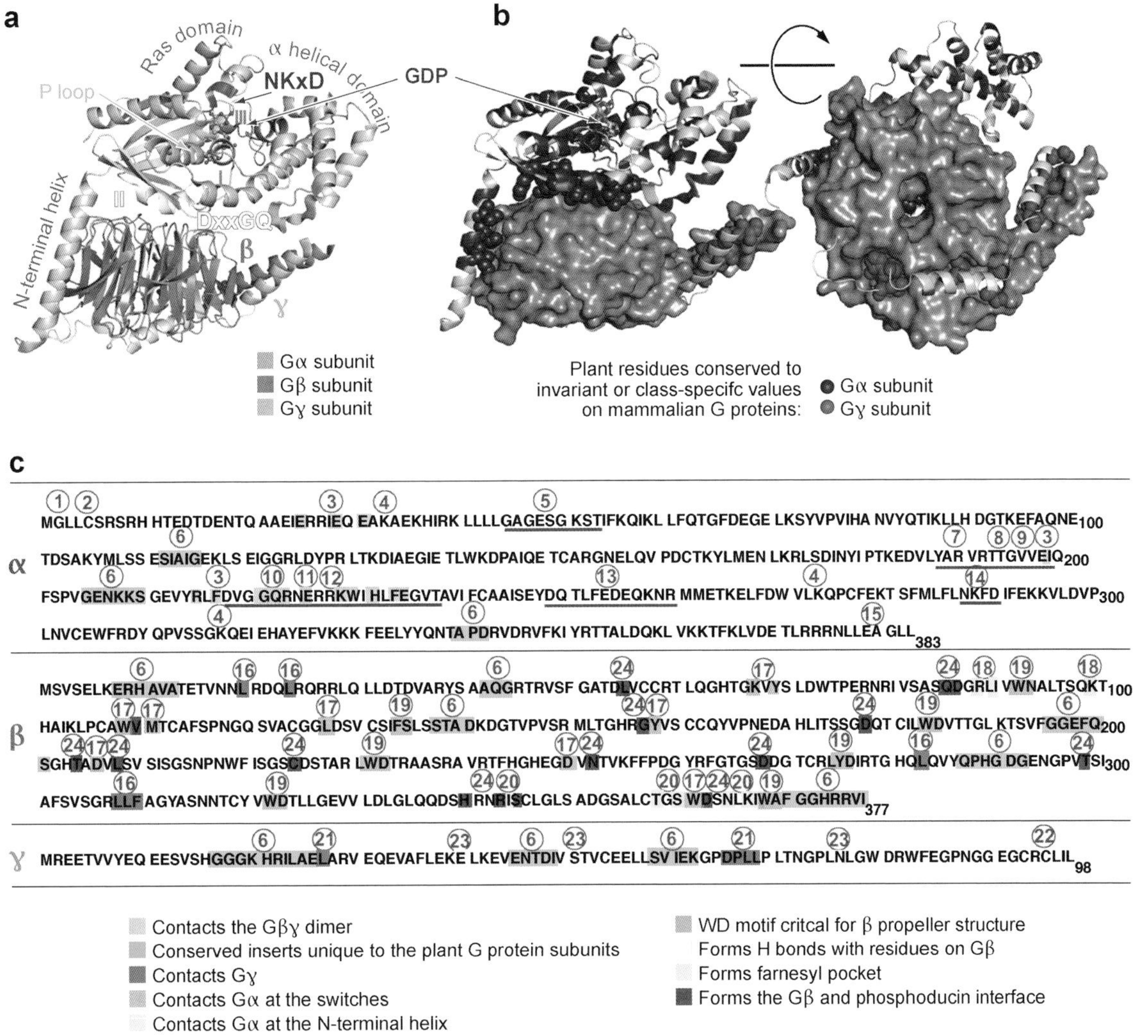

GTPγS. Y2H assays verified an interaction between Gα and Gβ and the loss of interaction between constitutively active Gα and Gβ, whereas coimmunoprecipitation identified only the interactions between the Gβ and Gγ subunits (24). In that study, most Gα did not comigrate with the higher-order complex containing the Gβγ dimer, suggesting that Gα is predominantly activated, unlike in animals. An explanation for this unusual behavior is discussed below.

The presence of two gamma subunit genes raises the intriguing possibility that functional selectivity for Gβγ dimer signaling is provided by the Gγ subunit. Consistent with this is the observation that the two Gγ genes have non-overlapping expression patterns in tissues that when summed together cover the expression pattern of the single Gβ gene in *Arabidopsis* (J.G. Chen & J.R. Botella, unpublished data).

Conservation of Atomic Interactions within the G-Protein Complex

Molecular modeling of the *Arabidopsis* G-protein complex based on an experimentally determined structure of a mammalian G-protein complex revealed known atomic interactions within the mammalian complex that are conserved in the plant heterotrimeric complex model (49). Protein-fold recognition servers identified the mammalian Gα, Gβ, and Gγ subunits as the most compatible templates for the *Arabidopsis* Gα, Gβ, and Gγ subunits, and sequence/structure self-compatibility scores verified the structural integrity of the *Arabidopsis* atomic models. The G-protein complex subunits, built individually, assemble into the heterotrimeric complex in a way that conserves the interactions at the protein-protein interfaces within the mammalian and *Arabidopsis* heterotrimeric complex structures (**Figure 2**). Four residues that form hydrogen bonds between Gγ and Gβ residues in the mammalian structures are invariant in *Arabidopsis* Gγ proteins (AGG1 E_{40}, S_{51}, D_{66}, N_{77}), as are four residues contributing hydrophobic contacts (AGG1 L_{37}, P_{67}, L_{68}, L_{69}). Throughout this review, the amino acid numbers provided are based on *Arabidopsis* G-protein subunits (**Figure 2**).

Extant Plant G Proteins Bear Closest Resemblance to Primordial G Proteins and Most Ancestral-Like Interactions

Amino acid residues that are conserved within all plant Gα subunits can be separated into two groups based on sequence similarity with the mammalian Gα subunits. Conserved plant residues that match the amino acid value in all mammalian Gα subunits are classified as invariant, whereas conserved plant residues are class-specific if they match the amino acid value in at least one, but not all, of the four major Gα classes. Class-distinctive residues are a subset of class-specific residues where the amino acid value in three of the four major Gα classes is conserved, but retains a distinctive amino acid value in the remaining class. Class-distinctive residues were identified for all four of the major Gα classes using mammalian sequences and then were used to predict the sequence of the primordial Gα by evaluating the amino acid value retained in plants and fungi at the class-distinctive positions (B. Temple & A. Jones, unpublished). This approached revealed that the plant amino acid value at class-distinctive positions most frequently matched the conserved value of three classes and not the class-distinctive amino acid value, indicating that the plant Gα descended from Gα zero, the earliest Gα ancestor to extant Gα proteins. (Follow the Supplemental Material link in the online version of this chapter or at **http://www.annualreviews.org/** to view **Supplemental PDF 1**.) It was proposed that Gα evolution along the mammalian lineage consisted of gene duplications coupled with random mutations hitting at different residue positions, but at the same functional regions in different gene copies. Gene duplication along the plant lineage was a relatively rare event, with most plant species possessing a single Gα, a single Gβ, and only 2 Gγ subunits.

Similar to plant Gα proteins, plant Gβ and Gγ subunits contain invariant, class-specific, and plant-specific amino acid residues when compared to the multigene families of mammalian subunits (see **Supplemental PDFs 2** and **3**). Thirty Gγ1 amino acid residues out of a total of 98 residues in *Arabidopsis* AGG1 are highly conserved within the plant family (**Figure 3**), whereas 298 plant Gβ residues out of a total of 377 residues in *Arabidopsis* AGB1 are highly conserved within the plant family (green residues). The residues conserved in the plant Gγ predominately lie at the N and C termini, except for a few residues that contact Gβ in the middle of the protein (**Figures 2** and **3**). Of the 30 residues conserved in the plant Gγ, 3 residues are invariant and 11 show class-specific conservation between the plant family and the 12 subfamilies of mammalian Gγ subunits. The remaining Gγ residues are specific to the plant family. Of the 298 residues conserved in the *Arabidopsis* Gβ, 128 are invariant and 65 show class-specific conservation between the plant family and the 5 subfamilies of mammalian Gβ subunits.

The presence of invariant and class-specific residues in plant Gβ and Gγ subunits suggests that mammalian Gβ and Gγ classes evolved in a manner similar to the evolution of Gα classes from a single primordial G-protein complex whose ancient characteristics are still evident in the extant plant G-protein complex. The combination of gene duplication, the random mutations in functional regions, and the creation of new functions yielded the diverse set of mammalian G proteins evident today. Since the split between the plant and mammalian lineages, higher plants seem to have diverged only in Gγ and probably only by a single gene duplication, which stands in contrast to the divergence seen in the mammalian G-protein complex. What has emerged from these types of analyses is the knowledge that the plant heterotrimeric G-protein complex bears the closest resemblance to the primordial G-protein complex.

The Gα Subunit

Interactions between the Gα subunit and the Gβγ dimer occur in three regions on the Gα subunit, the N-terminal helical region, a small interface at switch I, and an extended interface at switch II. Functional motifs characteristic of a Gα subunit include myristolation and S-acetylation sites, N- and C-terminal receptor binding sites, and three switch regions that sense the nucleotide-bound state. Mutation of the potential myristalization or S-acetylation motifs shifts localization of Gα from the plasma membrane to the cytosol (1), indicating that the plant Gα is plasma membrane delimited by a lipid anchor and that myristylation or S-acetylation alone is insufficient to direct GPA1 to the membrane. The P-loop (G_{45}AGESGKS) for NTP binding, the DxxGQ motif (D_{218}VGGQ) for GTP hydrolysis, and the NKxD motif (N_{287}KFD) for guanine recognition are all conserved in plant Gαs. Mutation of Q222 → L yields a constitutively active (GTPAse dead) variant that has proven invaluable for dissecting the function of the activated Gα (11, 18, 24, 26, 38). The switch I target Arg target (R_{190}) for cholera-toxin-mediated ribosylation is conserved, but the C-terminal cysteine ribosylation target, conserved in Gαi subfamily members, is not present in plants, therefore calling into question the interpretation of the large number of published conclusions based on pertusis toxin sensitivity as a diagnostic for plant G-protein action in various physiologies.

Significant functional divergence has probably occurred within the plant Gα-protein family, as deduced from the sequence dissimilarity occurring in regions of the protein associated with particular functions. For example, the C terminus of most plant subunits shows sequence similarity to mammalian Gα proteins, especially the G(i) and G(o) families. This region is known for imparting receptor and effector specificity (33). The monocot cereals (wheat, rice, and barley) have a plant-specific variation in this critical functional region of the Gα protein that may have led to loss of the associated functionalities (see **Supplemental PDF 4**).

To date, seven noncanonical, extra-large GTP binding proteins (XLG) were identified in plants, three in *Arabidopsis* (29), and four in rice. Six of the seven XLGs range are 800–900 residues in length and one rice XLG is much shorter at 538 residues (see **Supplemental**

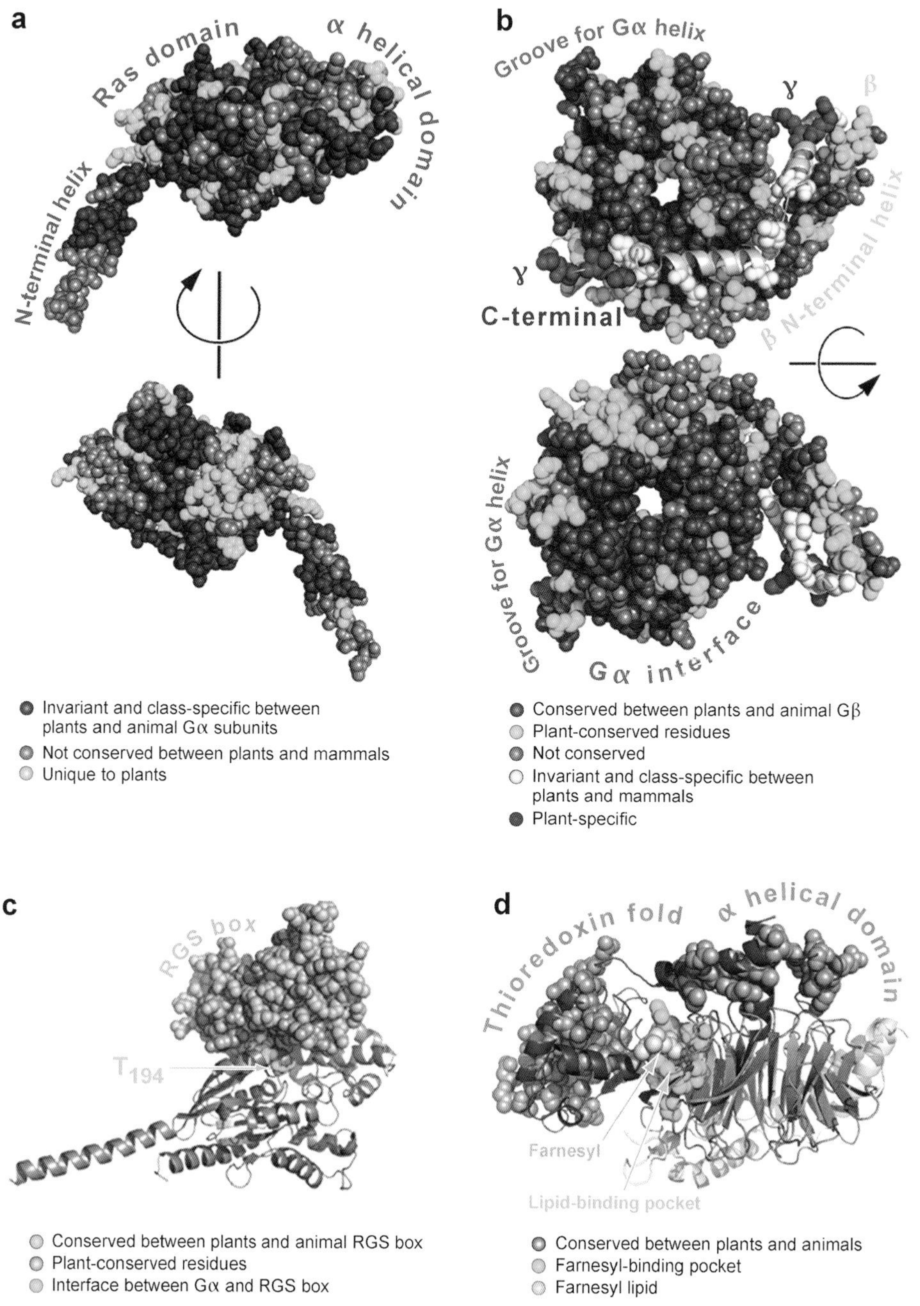

PDF 5). All seven XLGs share significant sequence similiarity over the C-terminal 633 residues, with the last 414 C-terminal residues being recognized as a variant of a Gα domain. Sequence identity within the Gα domain among the seven plant XLGs is 47% on average, but the AtXLG1 Gα domain only shares 29% identity with AtGPA1. Not all functional motifs are equally conserved within the Gα–like domain of the XLGs. The P-loop shows some variation from the canonical form of [A or G]-x-x-x-x-G-K-[S or T] to a form G-x-x-x-x-G-T-S in the XLG in four of the XLGs, including AtXLG1, which binds GTP (29). The NKxD motif is invariant in all XLGs except for one rice XLG protein and AtXLG2, where it has become TKxD. All three switches in the XLGs show significant modification from the canonical form. Switch I has a 15-residue insert in the middle, and shows little sequence conservation to the canonical form. Switch II of the XLGs has a single-residue insert in the middle of a highly conserved region and otherwise shows significant divergence in sequence. Similarly, switch III is quite divergent from the canonical switch III and has undergone both deletions and mutations depending on the particular XLG. Given the significant modifications found in the switch regions, it is not surprising that an interaction between the XLG and AtAGB1, or any Gα effector that binds the switch region, has not yet been detected (L. Ding, & S.M. Assmann, unpublished). In the plant XLGs, these three regions show significant diversification and extra-large specific inserts, leading us to conclude that it is not likely that these extra-large plant proteins interact with the Gβγ dimer in a traditional fashion, if at all.

The Gβ Subunit

At the N terminus of the Gβ subunit is a helical structure that forms a coiled-coil/hydrophobic interaction with the Gγ subunit that is essentially irreversible under nondenaturing conditions. The remaining residues form a seven-bladed propeller structure from repeating units that are called WD40 repeat motifs. The Gα N-terminal helix binds into a groove on the side of the β propeller while the Ras-like and helical domains bind to the top surface of the Gβ propeller (**Figure 3*b***). Besides nucleotide-dependent binding with Gα, Gβ contains conserved interfaces for other elements of the heterotrimeric complex.

Phosducin is one such Gβγ interactor and is involved in sequestering the Gβγ dimer in mammals, thus inhibiting signaling through Gβγ or enhancing signaling through Gα (44). *Arabidopsis* appears to have one phosducin-like gene (At5g14240) that is approximately 50% similar in sequence to animal phosducin (see **Supplemental PDF 6**). Conserved residues are shown in **Figure 3*d***. Plant Gβγ dimers share significant sequence similarity with the conserved phosducin/Gβγ interface in the mammalian complex and with the regions of the mammalian Gβγ that undergo conformational changes on binding

Figure 3

Components of the plant G-protein complex. (*a*) The Gα subunit (AtGPA1) is shown in two views indicating the three main domains. Note that the plant residues form patches on the surface suggestive of an interface between Gα and effector(s). (*b*) Two views of the Gβγ dimer are shown. The pocket that interfaces with the N-terminal α helix of Gα is shown. For the Gγ subunit, the yellow residues that largely form the interface between Gβ and Gγ are invariant and class-specific (see text) between plants and mammals whereas dark blue residues are plant-specific. (*c*) The RGS-box [PDB ID 1AGR (47)] is shown with Gα. A critical residue (T_{194}) is discussed in the text. (*d*) Phosducin [PDB ID IA0R (30)] is docked to the Gβγ dimer. Phosducin has two domains and both interact with Gβγ along an extended interface. Residues that are conserved between plants and animals are indicated. Residues on Gβ that form the farnesyl-binding pocket described in the text are also shown, as is the farnesyl lipid.

phosducin. Phosducin consists of two domains, an N-terminal helical domain and a C-terminal thioredoxin-like domain that bind independently to Gβγ (6, 30). Most of the contact residues between phosducin and Gβγ are conserved to the mammalian residue in the putative plant phosducin. An X-ray crystal structure of the mammalian phosducin/Gβ1γ1 complex showed the farnesyl moiety of an intact Gγ buried in a cavity of the Gβ subunit that was formed by conformational changes upon binding phosducin (30). Burying the farnesyl moiety within the Gβ-propeller structure results in a signal transduction cascade that is turned off by sequestration of Gβγ to the cytosol. Loew et al. (30) identified residues that formed the binding pocket and residues that underwent conformational change upon binding phosducin, which all lie in the C-terminal 35 amino acids of Gβ (**Figure 3*d***). Myung & Garrison (37) mutated these residues along with several of the residues in the region undergoing conformational change and measured reduced activation of PLCβ and type II adenylyl cyclase (37). Sequence similarity comparisons between the plant Gβγ and the mammalian Gβγ suggest the farnesyl binding pocket is present and conserved in plants, and many elements of the mammalian phosducin/Gβγ interface are also present and conserved in plants. Of the six residues identified as forming the farnesyl pocket, four are invariant between plants and mammals (T_{358}, S_{360}, K_{366}, W_{368} in AGB1), one is a conservative substitution ($V \rightarrow I_{344}$ in AGB1), and one is not conserved between plants and mammals ($F \rightarrow N_{364}$ in AGB1). In addition, two residues critical for phosducin-induced conformational change (37) are also invariant between plants and mammals (H_{340}, R_{343} in AGB1). The interface between phosducin and Gβγ is extensive, with many residues throughout the interface invariant or similar between plants and mammals (**Figure 2*b***). Much of this interface overlaps precisely with the Gα-Gβ switch interface. Although an interaction between plant phosducin and plant Gβγ has not yet been shown, a reasonable hypothesis is that the phosducin/Gβγ interaction with induced conformational change and buried farnesyl moiety will be found in plants, and that it is a primordial interaction, existing at the time of the split between mammalian and plant lineages.

The Gγ Subunits

Arabidopsis has two Gγ subunits. Molecular masses for both Gγ subunits are on the order of 11 kDa and both contain a C-terminal CAAX-box for isoprenyl modification and an N-terminal α-helical coiled-coil domain for interactions with Gβ subunits. Rice Gγ1 (RGG1) has a molecular mass of 10.5 kDa and the expected C-terminal CAAX-box. In contrast, rice Gγ2 (RGG2) (24) has a significantly larger molecular weight of 16.9 kDa, with a Gγ subunit-like domain located at the C terminus and an extra 57 amino acid residues at the N terminus. In addition, RGG2 lacks the C-terminal CaaX motif. Mammalian Gβγ dimers have differing abilities to regulate effectors and the specificity for this regulation resides in the structural requisites of the N and C termini. Except for contact residues, plant Gγ proteins show extensive conservation only in these two regions. Finally, Gγ uses a DPLL motif that serves as an important hydrophobic contact to Gβ (**Figures 2*b*** and **3*b***).

THE UNUSUAL PROPERTIES OF THE PLANT G-PROTEIN COMPLEX

Kinetic Properties of Gα

The first demonstration showing GTPAse activity of a plant Gα was with a recombinant tomato Gα subunit (2). More recently, a detailed kinetic analysis found that AtGPA1 has one of the slowest GTPAse activities described ($K_{cat} = 0.12\ s^{-1}$) (55; C. Johnston & F.S. Willard, unpublished data). Moreover, guanine-nucleotide exchange for the plant Gα, AtGPA1 (more specifically the GDP release, which is the limiting step in all previously described, nonplant Gα subunits), is the fastest among all Gα subunits. Fast release of GDP and fast loading of GTP onto AtGPA1, coupled with a poor ability to hydrolyze that nucleotide, indicate that AtGPA1, and most likely all plant Gαs, are in the activated state by default. Indeed, when modeled using the in vitro rate constants, 99% of AtGPA1 is loaded with GTP at steady state. This is in stark contrast to the situation with animal Gα subunits where the amount of GTP-loaded Gα is typically below 10%. Since the limiting step of the guanine-nucleotide cycle in animals is GDP release, GPCRs solved the problem by acting as ligand-regulated GEFs (46). However, in *Arabidopsis*, it may be the GAP activity that limits nucleotide cycling, which raises the question of whether plant cells have 7TM proteins that behave as GEFs. As a corollary, this unusual property suggests that plants have ligand-regulated GAPS. In other words, the central question is whether the regulation of nucleotide cycling in plants is backward to the well-described model for animals?

The structural basis for this unique property is unknown but of high interest as it concerns the mechanism of some fascinating endocrine diseases (19). Few mutations decrease the affinity of GDP to Gα, but mutations in the C terminus reveal a role for these residues in stabilizing the bound nucleotides (42). In particular, a C325A mutation in the Gαi class decreased the affinity of GDP to Gα by tenfold (48). Although this region of plant Gαs is highly conserved, the corresponding position in plants is a threonine.

RGS1, a Potential Ligand-Activated GAP

Paradoxically, it appears that the default state of Gα is the activated GTP-bound form. The discovery of an unusual RGS1 protein in *Arabidopsis* provided the missing piece to this strange puzzle. The 2003 discovery of AtRGS1 by Chen and coworkers (11) introduced a hybrid protein with an RGS-box coupled to a predicted 7TM domain unlike any known RGS protein (11). The 7TM domain of AtRGS1 does not share significant sequence similarity to known plant or animal proteins. Unlike for most animal GPCRs, the third intracellular loop of AtRGS1, which is critical for G-protein coupling in animal GPCRs, is extremely short, thus challenging a GEF functional role for AtRGS1. The hybrid nature of AtRGS1 prompts the obvious questions: Is AtRGS1 a GEF, a GAP, or both? Does AtRGS1 have a ligand and does this ligand regulate one or both activities? AtRGS1 interacts with AtGPA1 in vitro and in vivo (11). GAP activity of the RGS-box, located in the amino acid C-terminal domain, was shown biochemically and the full-length AtRGS1 was shown to behave as an RGS1 in vivo both by biochemical and genetic approaches (11). The 248-amino acid N-terminal half of AtRGS1 strongly predicts 7TM spans, but this has yet to be biochemically confirmed.

The ubiquity of AtRGS1 is still unclear. To date, no obvious nonplant RGS1 homolog has been reported, nor has a full-length RGS1 plant homolog other than *Arabidopsis* been reported, although ESTs from 12 crop species with partial overlap with the AtRGS1 RGS-box, the 7TM, or both are in the Phytome database (**http://www.phytome.org**) at the time of this writing. A plant RGS protein lacking the 7TM domain appears to be encoded in the *Medicago truncatulata* genome (Barrel medic: UniProt ID: Q1T3 × 1_MEDTR; submitted to sequence databases March 2006), and thus it is not clear whether AtRGS1 is unique to a particular group of plants. The *Medicago* RGS-protein homolog shares 60% sequence identity to the RGS-box of AtRGS1.

In animals the RGS domain forms a helical bundle that directly interacts predominately with the three switches in Gα—the regions undergoing the largest conformational change during GTP hydrolysis—and binds with highest affinity to the transition state for GTP hydrolysis (13, 47). The most extensive contacts between RGS4 and Giα1 in switches I and II are predominately conserved between plants and animals (**Figure 3c**). Threonine194 in AtGPA1 is invariant with a critical residue in the mammalian Gα subunits that makes extensive contact with the RGS domains. This threonine is conserved in all mammalian Gα subunits except for the G_s and G_{12} classes, which have not yet been reported to bind RGS domains. Interestingly, T_{194} is invariant in all plant Gα subunits except for rice, wheat, and barley. Three of the seven contacts to this threonine are conserved in the two plant RGS-boxes. Similarly, switch II residues Q_{222} and E_{225} in AtGPA1 are invariant in plants and invariant with the corresponding two residues in mammals that lie at the RGS/Gα interface.

By analogy to animal GPCRs, the 7TM domain of AtRGS1 suggests receptor function. Clues to the identity of the presumed ligand come from detailed studies of mutants lacking AtRGS1 and its cognate heterotrimeric subunits (7, 8, 11, 12). *Arabidopsis* seedlings lacking AtRGS1 have altered sensitivities to both ABA and to D-glucose, two plant hormones that may share substantial signal transduction elements. Altered sensitivity as opposed to a complete lack of perception suggests that loss of AtRGS1 confers a loss in GAP function rather than receptor-GEF function.

Are There Other 7-Transmembrane Proteins in the G-Protein Complex?

In humans, GPCRs constitute a superfamily of 7TM proteins that can typically be subdivided into five subfamilies based on sequence similarities and shared topologies. However, the overall conservation between, and to some extent even within, the subfamilies is low (43). Consequently, divergent GPCRs have been difficult to identify. Divergent GPCRs in anthropoda were predicted using vertebrate GPCR-protein models and an alignment-based algorithm (17, 25, 35), but this approach was not successful with *Arabidopsis* (J. Kim, S.M. Assmann & A.M. Jones, unpublished). Recently, Moriyama and coworkers (36) developed a tool to find highly divergent GPCR candidates using multiple nonalignment approaches (25, 36), retrieving 394 *Arabidopsis* protein sequences (see **Supplemental Table 1**). This list includes all of the 22 previously reported 7TM proteins, suggesting that the candidate list may be near saturation. Specifically, this new bioinformative tool retrieved RGS1 (11), 15 proteins called MLOs (14), and GCR1 (23). Note that among these candidate plant GPCRs, only *Arabidopsis* GCR1 shares similarity, albeit weak, to any known GPCR—specifically, the cyclic AMP receptor (CAR1) found in the slime mold and the Class B Secretin family GPCRs (22, 23). Restricting candidates in this list to those predicted to have only seven membrane spans with the amino terminus located on the extracellular face of the membrane culls the candidate list down to only 54 proteins. This stringent list contains GCR1, AtRGS1, and seven of the 15 MLOs. To summarize, 7TM proteins, potential GPCRs, are an ancient protein type with low complexity in plants (15).

To date, only two of these 7TM proteins, namely GCR1 and AtRGS1, have been shown to physically interact with a plant Gα subunit.

G-PROTEIN-COUPLED SIGNALING IN ANIMALS VS. PLANTS

In metazoans, there are dozens of effectors and a thousand GPCRs. Adenyl cyclase (AC) and phospholipase Cβ (PLCβ) are the most utilized effectors in animals, but plants lack a canonical AC and PLCβ and do not harbor other obvious homologs of animal effectors such as G-protein inwardly-rectifying potassium channels or modifiers such as arrestin and GPCR kinases. Moreover, none of the known plant 7TM proteins share enough sequence homology to animal GPCR to suggest relatedness. Clearly, although both animals and plants contain the nexus heterotrimeric G-protein complex, the signaling pathways of the two kingdoms are otherwise greatly divergent above and below this signaling node.

This was done using both Y2H assays and by immunoprecipitation from plant extracts (11, 39). Note, however, that GCR1 has both AtGPA1-dependent and -independent roles (9, 40).

AtRGS1 has been shown to interact genetically with AtGPA1 in *Arabidopsis* and via genetic complementation in yeast (11). Using Förster resonance energy transfer to detect proximity in vivo, we showed that AtGPA1-CFP donor complexes with an AtRGS1-YFP receiver in a D-glucose-dependent manner (J.P. Taylor & A.M. Jones, unpublished data). Furthermore, overexpression phenotypes for AtRGS require AtGPA1 (A.M. Jones, unpublished), and *rgs1* loss-of-function phenotypes are recapitulated by overexpression of a GTPase-dead Gα, AtGPA$^{(Q222L)}$, suggesting that the sole function of AtRGS1 is to regulate the active state of Gα.

Other Unusual Components of the G-Protein Complex

Besides AtRGS1, and most likely, GCR1, only a few other proteins are suspected to complex with the heterotrimer or its subunits. In some cases, these proteins interact preferentially with the activated state of Gα, but GAP activity for any plant protein other than AtRGS1 has not been shown; consequently, these interactors are not yet designated as G-protein effectors. For example, a cupin-domain protein-designated pirin (PRN1) was proposed to be a Gα interactor based on Y2H results (28). The site of interaction between the two proteins was not localized beyond the notation that the cupin domain of AtPirin1 was not required for the interaction. In addition, no preference was detected for either the active (GTP-bound) or inactive (GDP-bound) state of AtGPA1. Although in vivo interactions have yet to be shown, *prn1* loss-of-function mutations confer some of the phenotypes known of *gpa1* null mutations, albeit epistasis has not yet been determined. The biochemical function of PRN1 is unknown. A second potential AtGPA1 interactor that came from the Y2H screen described above was prephenate dehydratase (PD1), a cytosolic enzyme involved in blue-light regulation of the shikimate pathway. Warpeha and coworkers (53) propose that the G-protein complex contains GCR1, PD1, PRN1, and AtGPA1, and that GCR1 may be the GPCR for blue light and/or ABA. GEF activity for GCR1 has not been demonstrated.

THF1 was found to interact in a Y2H configuration, in vitro, and in vivo in plant cells with the activated form of AtGPA1 (18). THF1 is a protein located on the outer membrane of plastids, including non-amyoplasts of root cells. Loss-of-function *thf1* mutations confer several of the sugar-related phenotypes of *rgs1* and GTPase-deficient AtGPA1 plants, indicating that THF1 genetically interacts with GPA1. Further support comes from the observation that null mutations of *thf1* are epistatic to null mutations of *gpa1*. THF1 has one, possibly two, plastid membrane spans. The cytosolic, C-terminal region of THF1 contains the AtGPA1 interaction interface. THF1 is required for proper membrane formation (51), which prompted the suggestion that THF1 is involved in trafficking the sugar-activated heterotrimeric complex, including AtRGS1.

Perhaps the best-characterized potential effector to date is phospholipase D alpha 1 (PLDα1). Zhao & Wang (56) used in vitro coimmunoprecipitation between recombinant PLDα1 and wild-type and mutated AtGPA1 to conclude that these proteins interact specifically at an interface comprised of a so-called DRY motif on PLDα1. DRY motifs can be found on some 7TM GPCRs in the third internal loop and are important for physical coupling between receptor and Gα. The interaction between PLDα1 and AtGPA1 is nucleotide dependent, with PLDα1 exhibiting a preference for the GDP-bound state and GTP inhibiting the interaction between the two proteins. Interaction between AtGPA1 and PLDα1 was confirmed by isothermal titration calorimetry. Mutation of the DRY motif reduced the affinity approximately 150-fold (34). Finally, PLDα1 also exhibited some GAP activity toward AtGPA1, fulfilling the second criterion for effector function.

A number of other candidate components of the complex have been proposed. However, direct physical interaction between these proteins has not yet been demonstrated. For example, indirect evidence suggests that plant Gα may interact with ion channels (2, 52) and phospholipase C (3). The G$\beta\gamma$ dimer interacts genetically with a Golgi-localized hexose transporter (50).

Why is the G-Protein Repertoire So Simple?: A Speculation

Remarkably, the plant heterotrimeric complex has changed little over the past 1.6 billion years since plants and animals diverged. The fact that there are just one or two possible plant complexes compared to the thousands found in metazoans begs the following question: Why did plants keep it simple? One likely scenario is that evolution of this signaling pathway was originally constrained, but at some point prior to the radiation of the metazoans this constraint was lost. We need only to look at the differences in structure and function of the complex between plants and metazoans to speculate. The original Gα came on the scene as a constitutive activator (always on) of some function, and later regulation of its activated state occurred with the acquisition of a glucose-regulated GAP, a primitive sugar receptor. This bipartite molecule acted on the primitive complex and it was not until this gene split and gave rise to the progenitors of extant animal GPCRs and the opposing cytosolic RGS proteins that this evolutionary constraint was released. This gene may not have occured in the plant lineage (certainly not in *Arabidopsis*) and thus the extant G-protein signal pathway remains ancient.

SUMMARY POINTS

1. Plants have a simple repertoire of canonical heterotrimeric G-protein complexes and associated elements, with most plants having only two possible heterotrimers despite what their possible involvement is in many signal networks.
2. The extant plant heterotrimeric complex approximates the structure of the primordial complex and therefore likely serves the primordial function.
3. The kinetic properties for guanine-nucleotide cycling in the plant heterotrimeric complex suggest that GAP activity, and not GDP dissociation, are rate limiting. This is in opposition to the case of metazoan G proteins.
4. Regulated GAP function in *Arabidopsis* probably occurs via an unusual RGS1 protein.

FUTURE DIRECTIONS FOR PLANT G-PROTEIN RESEARCH AND ITS ANTICIPATED IMPACT ON THE FIELD IN GENERAL

There are several reasons why *Arabidopsis* is a model system for studying signal transduction that utilizes the heterotrimeric G-protein complex: (*a*) Plants, particularly *Arabidopsis*, have few heterotrimeric G-protein complexes, (*b*) genetic manipulation of *Arabidopsis* is fast, and (*c*) loss of function of the key elements known so far in G-protein cycling do not confer developmental defects. The latter point should not be understated as it means that, unlike for so many other gene knockouts, it is still possible to study G-protein signal transduction through a genetic approach because *Arabidopsis* plants lacking G proteins appear normal.

However, if plants are the next model system, we must ask if our knowledge of plant G-protein signaling can be extrapolated to the complex G-protein networks used by metazoans? Indeed, because it appears that the primordial functions of G-protein signaling are found in plants, we can determine what the basic connectivities are and how they occur so as to map these to surfaces of the complex and learn how these surfaces are used in higher-order heterotrimeric complexes.

We urgently need to know answers to the following key questions:

1. What is the entire repertoire of proteins that physically interact with the complex and its dissociated components?
2. What is the precise structure of the plant heterotrimer and Gα and the Gβγ dimer interacting with RGS1 and its respective effectors?
3. What are the roles of 7TM proteins, if any, in plant G-protein signaling?
4. Is nucleotide cycling in plants regulated by receptor GAPs?

ACKNOWLEDGMENTS

Work in A.M.J.'s lab on the *Arabidopsis* G-protein complex is supported by the NIGMS (GM65989-01), the DOE (DE-FG02-05er15671), and the NSF (MCB-02,09711). We thank Drs. S.M. Assmann, J.R. Botella, J-G. Chen, C. Johnston, J. Kim, E. Moriyama, D.P. Siderovski, J. P. Taylor, and F.S. Willard for sharing unpublished data.

LITERATURE CITED

1. Adjobo-Hermans MJW, Goedhart J, Gadella Jr, TWJ. 2006. Plant G protein heterotrimers require dual lipidation motifs of Gα and Gγ and do not dissociate upon activation. *J. Cell Sci.* 119:5087–97
2. Aharon GS, Snedden WA, Blumwald E. 1998. Activation of a plant plasma membrane Ca^{2+} channel by TGα1, a heterotrimeric G protein α-subunit homologue. *FEBS Lett.* 424:17–21
3. Apone F, Alyeshmerni N, Wiens K, Chalmers D, Chrispeels MJ, Colucci G. 2003. The G-protein-coupled receptor GCR1 regulates DNA synthesis through activation of phosphatidylinositol-specific phospholipase C. *Plant Physiol.* 133:571–79
4. Assmann SM. 2002. Heterotrimeric and unconventional GTP binding proteins in plant signaling. *Plant Cell* 14:S355–73
5. Assmann SM. 2004. Plant G proteins, phytohormones, and plasticity: three questions and a speculation. *Science STKE* 264:re20
6. Blundell TL, Burke DF, Chirgadze D, Dhanaraj V, Hyvonen M. 2000. Protein-protein interactions in receptor activation and intracellular signaling. *Biol. Chem.* 381:955–59
7. Chen JG, Gao Y, Jones AM. 2006. Differential roles of Arabidopsis heterotrimeric G-protein subunits in modulating cell division in roots. *Plant Physiol.* 141:887–97
8. Chen JG, Jones AM. 2004. AtRGS1 function in Arabidopsis thaliana. *Methods Enzymol.* 389:338–50
9. Chen JG, Pandey S, Huang J, Alonso JM, Ecker JR, et al. 2004. GCR1 can act independently of heterotrimeric G-protein in response to brassinosteroids and gibberellins in Arabidopsis seed germination. *Plant Physiol.* 135:907–15

10. Chen JG, Ullah H, Temple B, Liang JS, Guo JJ, et al. 2006. RACK1 mediates multiple hormone responsiveness and develpmental processes in Arabidopsis. *J. Exp. Bot.* 57:2697–708
11. Chen JG, Willard FS, Huang J, Liang J, Chasse S, et al. 2003. A seven-transmembrane RGS protein that modulates plant cell proliferation. *Science* 301:1728–31
12. Chen Y, Ji FF, Xie H, Liang JS, Zhang JH. 2006. The regulator of G-protein signaling proteins involved in sugar and abscisic acid signaling in Arabidopsis seed germination. *Plant Physiol.* 140:302–10
13. Coleman DE, Berghuis AM, Lee E, Linder ME, Gilman AG, Sprang SR. 1994. Structures of active conformations of Giα1 and the mechanism of GTP hydrolysis. *Science* 265:1405–12
14. Devoto A, Hartmann HA, Piffanelli P, Elliott C, Simmons C, et al. 2002. Molecular phylogeny and evolution of the plant-specific seven-transmembrane MLO family. *J. Mol. Evol.* 56:77–88
15. Fredriksson R, Schioth HB. 2005. The repertoire of G-protein-coupled receptors in fully sequenced genomes. *Mol. Pharmacol.* 67:1414–25
16. Fujisawa Y, Kato H, Iwasaki Y. 2001. Structure and function of heterotrimeric G proteins in plants. *Plant Cell Physiol.* 42:789–94
17. Hill CA, Fox AN, Pitts RJ, Kent LB, Tan PL, et al. 2002. G protein-coupled receptors in Anopheles gambiae. *Science* 298:176–78
18. Huang J, Taylor JP, Chen JG, Uhrig JF, Schnell DJ, et al. 2006. The plastid protein THYLAKOID FORMATION1 and the plasma membrane G-protein GPA1 interact in a novel sugar-signaling mechanism in Arabidopsis. *Plant Cell* 18:1226–38
19. Iiri T, Herzmark P, Nakamoto JM, van Dop C, Bourne HR. 1994. Rapid GDP release from Gs α in patients with gain and loss of endocrine function. *Nature* 371:164–68
20. Jones AM. 2002. G-protein-coupled signaling in Arabidopsis. *Curr. Opin. Plant Biol.* 5:402–7
21. Jones AM, Assmann SM. 2004. Plants: the latest model system for G-protein research. *EMBO Rep.* 5:572–78
22. Josefsson LG. 1999. Evidence for kinship between diverse G-protein coupled receptors. *Gene* 239:333–40
23. Josefsson LG, Rask L. 1997. Cloning of a putative G-coupled receptor from Arabidopsis thaliana. *Eur. J. Biochem.* 249:415–20
24. Kato C, Mizutani T, Tamaki H, Kumagai H, Kamiya T, et al. 2004. Characterization of heterotrimeric G protein complexes in rice plasma membrane. *Plant J.* 38:320–31
25. Kim J, Moriyama EN, Warr CG, Clyne PJ, Carlson JR. 2000. Identification of novel multi-transmembrane proteins from genomic databases using quasi-periodic structural properties. *Bioinformatics* 16:767–75
26. Komatsu S, Abbasi F, Kobori E, Fujisawa Y, Kato H, Iwasaki Y. 2005. Proteomic analysis of rice embryo: An approach for investigating G protein-regulated proteins. *Proteomics* 5:3932–41
27. Lambright DG, Sondek J, Bohm A, Skiba NP, Hamm HE, Sigler PB. 1996. The 2.0 Å crystal structure of a heterotrimeric G protein. *Nature* 379:311–19
28. Lapik VR, Kaufman LS. 2003. The Arabidopsis cupin domain protein AtPirin1 interacts with the G protein α subunit GPA1 and regulates seed germination and early seedling development. *Plant Cell* 15:1578–90
29. Lee YR, Assmann SM. 1999. Arabidopsis thaliana extra large GTP-binding protein (AtXLG1): a new class of G protein. *Plant Mol. Biol.* 40:55–64

30. Loew A, Ho YK, Blundell TL, Bax B. 1998. Phosducin induces a structural change in transducin β-γ. *Structure* 6:1007–19
31. Mason MG, Botella JR. 2000. Completing the heterodimer: Isolation and characterization of an Arabidopsis thaliana G protein γ-subunit cDNA. *Proc. Natl. Acad. Sci. USA* 97:14784–88
32. Mason MG, Botella JR. 2001. Isolation of a novel G-protein γ -subunit from Arabidopsis thaliana and its interaction with Gβ(1). *Biochim. Biophys. Acta* 1520:147–53
33. Mazzoni MR, Hamm HE. 2004. G-protein organization and signaling. In *Handbook of Cell Signaling*, ed. RA Bradshaw, EA Dennis, pp. 335–41. Boston: Academic
34. Mishra G, Zhang W, Deng F, Zhao J, Wang X. 2006. A bifurcating pathway directs abscisic acid effects on stomatal closure and opening in Arabidopsis. *Science* 312:264–66
35. Moriyama EN, Kim J. 2005. Protein family classification with discriminant function analysis. In *Data Mining the Genomes: 23rd Stadler Genetics Symposium*, ed. JP Gustafson, pp. 121–32. New York: Kluwar Acad./Plenum
36. Moriyama EN, Strope PK, Opiyo SO, Chen Z, Jones AM. 2006. Mining the Arabidopsis thaliana genome for highly-divergent seven transmembrane receptors. *Genome Biol.* 7:R96
37. Myung CS, Garrison JC. 2000. Role of C-terminal domains of the G protein β subunit in the activation of effectors. *Proc. Natl. Acad. Sci. USA* 97:9311–16
38. Oki K, Fujisawa Y, Kato H, Iwasaki Y. 2005. Study of the constitutively active form of the α subunit of rice heterotrimeric G proteins. *Plant Cell Physiol.* 46:381–86
39. Pandey S, Assmann SM. 2004. The Arabidopsis putative G protein-coupled receptor GCR1 interacts with the G protein α subunit GPA1 and regulates abscisic acid signaling. *Plant Cell* 16:1616–32
40. Pandey S, Chen JG, Jones AM, Assmann SM. 2006. G-protein complex mutants are hypersensitive to abscisic acid regulation of germination and postgermination development. *Plant Physiol.* 141:243–56
41. Perfus-Barbeoch L, Jones AM, Assmann SM. 2004. Plant heterotrimeric G protein function: insights from Arabidopsis and rice mutants. *Curr. Opin. Plant Biol.* 7:719–31
42. Posner BA, Mixon MB, Wall MA, Sprang SR, Gilman AG. 1998. The A326S mutant of Giα1 as an approximation of the receptor-bound state. *J. Biol. Chem.* 273:21752–58
43. Schioth HB, Fredriksson R. 2005. The GRAFS classification system of G-protein coupled receptors in comparative perspective. *Gen. Comp. Endocrinol. 5th Int. Symp. Fish Endocrynol.* 142:94–101
44. Schulz R. 2001. The pharmacology of phosducin. *Pharm. Res.* 43:1–10
45. Sondek J, Bohm A, Lambright DG, Hamm HE, Sigler PB. 1996. Crystal structure of a G-protein βγ dimer at 2.1 Å resolution. *Nature* 379:369–74
46. Sprang SR. 1997. G protein mechanisms: Insights from structural analysis. *Annu. Rev. Biochem.* 66:639–78
47. Tesmer JJ, Berman DM, Gilman AG, Sprang SR. 1997. Structure of RGS4 bound to AlF4-activated Gi1: stabilization of the transition state for GTP hydrolysis. *Cell* 89:251–61
48. Thomas TC, Schmidt CJ, Neer EJ. 1993. G-protein α o subunit: mutation of conserved cysteines identifies a subunit contact surface and alters GDP affinity. *Proc. Natl. Acad. Sci. USA* 90:10295–99
49. Ullah H, Chen JG, Temple B, Boyes DC, Alonso JM, et al. 2003. The β subunit of the Arabidopsis G protein negatively regulates auxin-induced cell division and affects multiple developmental processes. *Plant Cell* 15:393–409
50. Wang HX, Perdue T, Weerasinghe R, Taylor JP, Cakmakci NG, et al. 2006. A golgi hexose transporter involved in heterotrimeric G protein-regulated early development in Arabidopsis. *Mol. Biol. Cell.* 17:4257–69

51. Wang Q, Sullivan RW, Kight A, Henry RL, Huang JR, et al. 2004. Deletion of the chloroplast-localized Thylakoid formation1 gene product in Arabidopsis leads to deficient thylakoid formation and variegated leaves. *Plant Physiol.* 136:3594–3604
52. Wang XQ, Ullah H, Jones AM, Assmann SM. 2001. G protein regulation of ion channels and abscisic acid signaling in Arabidopsis guard cells. *Science* 292:2070–72
53. Warpeha KM, Lateef SS, Lapik Y, Anderson M, Lee BS, Kaufman LS. 2006. G-protein-coupled receptor 1, G-protein Gα-subunit 1, and prephenate dehydratase 1 are required for blue light-induced production of phenylalanine in etiolated Arabidopsis. *Plant Physiol.* 140:844–55
54. Wilkie TM, Gilbert DJ, Olsen AS, Chen XN, Amatruda TT, et al. 1992. Evolution of the mammalian G protein α subunit multigene family. *Nat. Genet.* 1:85–91
55. Willard FS, Siderovski DP. 2004. Purification and in vitro functional analysis of the Arabidopsis thaliana regulator of G-protein signaling-1. *Methods Enzymol.* 389:320–38
56. Zhao J, Wang X. 2004. Arabidopsis phospholipase Dα 1 interacts with the heterotrimeric G-protein α-subunit through a motif analogous to the DRY motif in G-protein-coupled receptors. *J. Biol. Chem.* 279:1794–800

Alternative Splicing of Pre-Messenger RNAs in Plants in the Genomic Era

Anireddy S.N. Reddy

Department of Biology and Program in Molecular Plant Biology, Colorado State University, Fort Collins, Colorado 80523; email: reddy@colostate.edu

Annu. Rev. Plant Biol. 2007. 58:267–94

The *Annual Review of Plant Biology* is online at plant.annualreviews.org

This article's doi:
10.1146/annurev.arplant.58.032806.103754

First published online as a Review in Advance on January 12, 2007

1543-5008/07/0602-0267$20.00

Key Words

mRNA metabolism, pre-mRNA splicing, serine/arginine-rich (SR) proteins, splicing factors, stresses

Abstract

Primary transcripts (precursor-mRNAs) with introns can undergo alternative splicing to produce multiple transcripts from a single gene by differential use of splice sites, thereby increasing the transcriptome and proteome complexity within and between cells and tissues. Alternative splicing in plants is largely an unexplored area of gene expression, as this phenomenon used to be considered rare. However, recent genome-wide computational analyses have revealed that alternative splicing in flowering plants is far more prevalent than previously thought. Interestingly, pre-mRNAs of many spliceosomal proteins, especially serine/arginine-rich (SR) proteins, are extensively alternatively spliced. Furthermore, stresses have a dramatic effect on alternative splicing of pre-mRNAs including those that encode many spliceosomal proteins. Although the mechanisms that regulate alternative splicing in plants are largely unknown, several reports strongly suggest a key role for SR proteins in spliceosome assembly and regulated splicing. Recent studies suggest that alternative splicing in plants is an important posttranscriptional regulatory mechanism in modulating gene expression and eventually plant form and function.

Contents

Pre-mRNA splicing: a process that removes introns from precursor-mRNA (also called primary transcript) and joins exons to generate mRNA. This process takes place cotranscriptionally in the spliceosome

INTRODUCTION

The recent sequencing of plant genomes revealed that the coding region of about 80% of nuclear genes is interrupted by non-coding introns (1, 52). The generation of functional messenger RNAs (mRNAs) from these intron-containing precursor-mRNAs (pre-mRNAs) involves excising introns with remarkable precision and joining the exons in pre-mRNAs. This process, termed "pre-mRNA splicing," occurs cotranscriptionally in the spliceosome, a large multicomponent megadalton complex.

Soon after the discovery of exons and introns in adenovirus 2 genes (10, 19), it was proposed that different combinations of exons could be joined together to produce multiple mRNAs from a single gene (32). Although alternative splicing has been studied with some genes as isolated events, the extent of alternative splicing in multicellular eukaryotes was not known until recently. Many recent studies suggest that alternative splicing likely plays an important role in regulating gene expression at the posttranscriptional level. Pre-mRNA splicing imprints information necessary for nucleocytoplasmic transport of spliced mRNA, nonsense mediated mRNA decay, and mRNA localization in the cytoplasm (48). In plants, it is now clear that alternative splicing is prevalent and can generate tremendous transcriptome and proteome complexity (126). As this is an emerging area of study in plants, a critical discussion on whether the alternative splicing postulated from comparisons of genomic and cDNA sequences is real and the relevance of this transcriptome/proteome diversity to plant growth and development is warranted.

In this review, I limit my discussion to alternative splicing in flowering plants, how this affects the transcriptome and proteome diversity, and implications of this complexity. The aspects of plant splicing not covered here, such as spliceosome assembly, sequence requirements for splicing of plant pre-mRNAs, localization, and dynamics of spliceosomal proteins are discussed in detail in other comprehensive reviews (16, 29, 39, 87, 88, 104, 110, 117).

CHARACTERISTICS OF PLANT EXONS AND INTRONS

Global analysis of gene structure has revealed a number of differences in the architecture of plant and animal genes. In *Arabidopsis thaliana* the average gene is ~2.4 kb long with about 5 exons and 4 introns (**Table 1**). The

Table 1 Characteristics of plant and human introns

Organism	Average gene size (kb)	Average # of exons/gene	Average # of introns/gene	Average length of intron	Average length of exon[a]	Nucleotide composition of introns	Nucleotide composition of exons	References
Arabidopsis	2.4	~5	4	173	172	67% AT (42% T, 25% A) 33% GC (20%C, 13% G)	57% AT (29% T, 28% A) 43% GC (23% G, 20%C)	(1, 52, 126, 132)
Rice	4.5	~5[b]	~4[b]	433	193	73% AT 37% GC	48.6% AT 51.4% GC	(126, 132)
Human	27	8.8	7.8	~3000	130	51.9% AT 42.1%GC	49.5% AT 50.5% GC	(67, 132)

[a]This size refers to internal exons only.
[b]Information on the number of exon and introns per gene is from **http://www.tigr.org/tdb/e2k1/osa1/riceInfo/info.shtml.**

highest number of experimentally verified introns in *Arabidopsis* is found in At3g48190, which has 78 introns and encodes a homolog of the human *ATAXIA-TELANGIECTASIA MUTATED* (*ATM*) (1). In humans, the average gene is 28 kb long with 8.8 exons of about 130 nucleotides and 7.8 introns with an average size of ~3000 nucleotides (67). Introns in plant genes are much shorter compared with introns in human genes (**Table 1**). Plant introns are U or UA rich and exons are G rich. The compositional bias for UA or U richness in introns is important for recognition of splice sites and for efficient splicing of introns in plant cells (16, 74, 87, 88, 104, 110, 115, 139). The composition of introns in dicots and monocots also differs. Compared with *Arabidopsis*, rice introns are longer and have higher GC content (**Table 1**), suggesting differences in splicing in these two lineages.

How splicing machinery identifies exonic and intronic regions in plants is poorly understood and a subject of great interest. Four intron-defining splicing signals are important for accurate splicing of pre-mRNAs in metazoans. These include (*a*) a consensus sequence at the 5′ donor site with a conserved GU dinucleotide that is recognized by U1 small nuclear ribonucleoprotein particle (snRNP), (*b*) another consensus sequence at the 3′ acceptor site with a conserved AG dinucleotide that is recognized by U2 auxillary factor 35 (U2AF35), (*c*) a polypyrimidine tract at the 3′ end of the intron, which is recognized by U2AF65, and (*d*) a branch point (CURAY, where R is purine and Y is pyrimidine) recognized by U2 snRNP and located about 17–40 nucleotides upstream of the 3′ splice site (12). Analysis of the 5′ and 3′ splice sites in all introns of *Arabidopsis* and rice indicates that these sites are very similar to humans with some subtle differences in the frequencies of specific nucleotides at specific positions (**Figure 1**). In *Arabidopsis*, noncanonical splice sites occur in only 0.7% of all splice sites (1). Analysis of sequences upstream of the 3′ splice site of all *Arabidopsis* introns indicates that the polyprimidine stretch that is found in animals

Spliceosome: a megadalton macromolecular machinery consisting of five snRNPs and many non-snRNP proteins that catalyze two transesterification reactions during pre-mRNA splicing. It precisely excises introns and joins exons in pre-mRNA to generate mRNA. About 300 proteins are present in the spliceosome

Alternative splicing: a process by which the two or more different mRNAs are produced from a gene. This occurs due to alternate use of 5′ and/or 3′ splice sites by the spliceosome

Transcriptome: all transcribed RNAs and their levels within a cell, tissue or organism. Transcriptome is dynamic and a number of factors (e.g, environmental conditions, stresses, and developmental cues) alter the transcriptome

Proteome: all proteins and their levels within a cell, tissue, or organism. Like the transcriptome, the proteome of a cell is affected by many factors

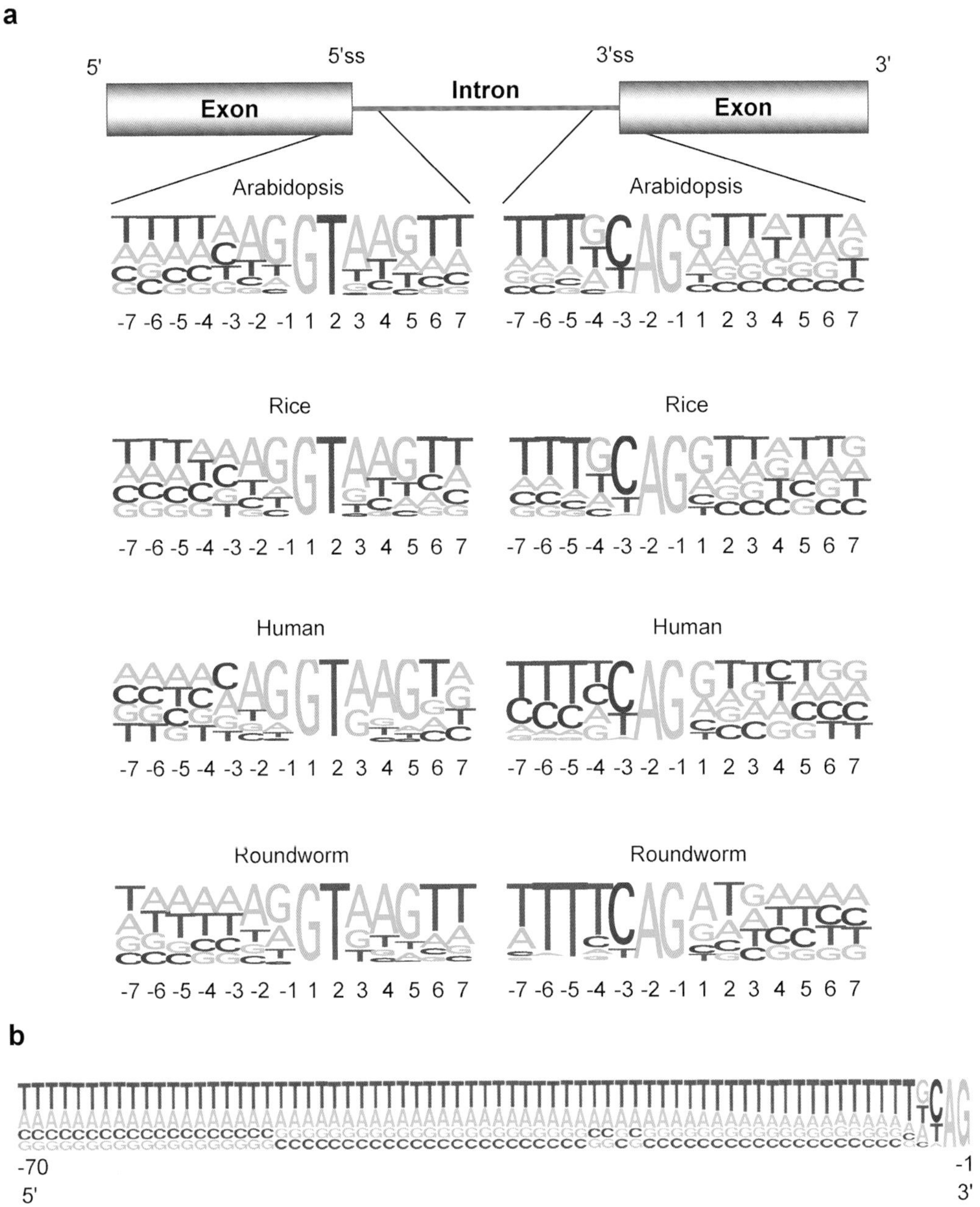

Figure 1

(*a*) The sequence patterns at the 5′ splice site (*left*) and 3′ splice site (*right*) of introns in *Arabidopsis*, rice, human, and roundworm. Top: Schematic diagram showing two exons and an intron. Bottom: The frequencies of A, C, G, and T at each position are represented by the height of the corresponding letter. Fourteen nucleotides (7 from the exon and 7 from the intron) at the 5′ and 3′ splice sites of all genes in these organisms were extracted from the annotated sequences and used to construct this figure. (*b*) The sequence pattern at the 3′ end of the *Arabidopsis* introns. The frequencies of A, C, G, and T at each position are represented by the height of the corresponding letter. The last 70 nucleotides of all introns were retrieved from the annotated sequences and used to generate this figure.

is mostly U in plants (**Figure 1**). Recently, it was shown that U-rich elements can function as a splicing signal or a polypyrimidine tract (115). Furthermore, the branch point sequence (CURAY) is not obvious here. This is probably because of variation in the position of the branch point in different introns. These differences between plant and animal intron architecture and composition suggest that the mechanisms involved in splice site recognition likely differ in these organisms.

SPLICE SITE RECOGNITION

Although the short consensus sequences at the donor, acceptor, and branch sites are necessary for splice site recognition, these sequences alone are not sufficient. Studies with animal systems indicate that other exonic and intronic *cis*-acting regulatory sequences (4 to 18 nucleotides long) that bind *trans*-acting factors influence splice site selection during constitutive and alternative splicing. In animal systems, these regulatory sequences are classified into four groups: exonic splicing enhancers (ESEs), exonic splicing silencers (ESSs), intronic splicing enhancers (ISEs), and intronic splicing suppressors (ISSs). In metazoans, it is established that recognition of exons and introns is achieved by these loosely conserved *cis*-regulatory sequences in exons and introns and their interaction with splicing regulators. Binding of proteins (e.g., SR proteins, discussed below) to these sequences regulates the recruitment of splicing machinery to splice sites. Two models termed exon definition and intron definition have been proposed to explain splice site recognition (see **Figure 2**) (9). In organisms such as humans and other vertebrates where large intron sequences separate fairly small exons, the exon definition model is favored (22, 121). Exons in animal pre-mRNAs contain purine-rich ESEs, which are recognized by regulatory splicing factors (13). In plants, little is known about exonic regulatory sequences. Although plant exons are rich in GC content, specific exonic sequences involved in splice

Splice sites: nucleotide sequences surrounding the exon-intron boundaries that are critical for pre-mRNA splicing

Small nuclear ribonucleoprotein particle (SnRNP): consists of a small nuclear RNA and several proteins

Constitutive splicing: generation of only one type of mRNA from intron-containing genes by using the same set of 5′ and 3′ splice sites

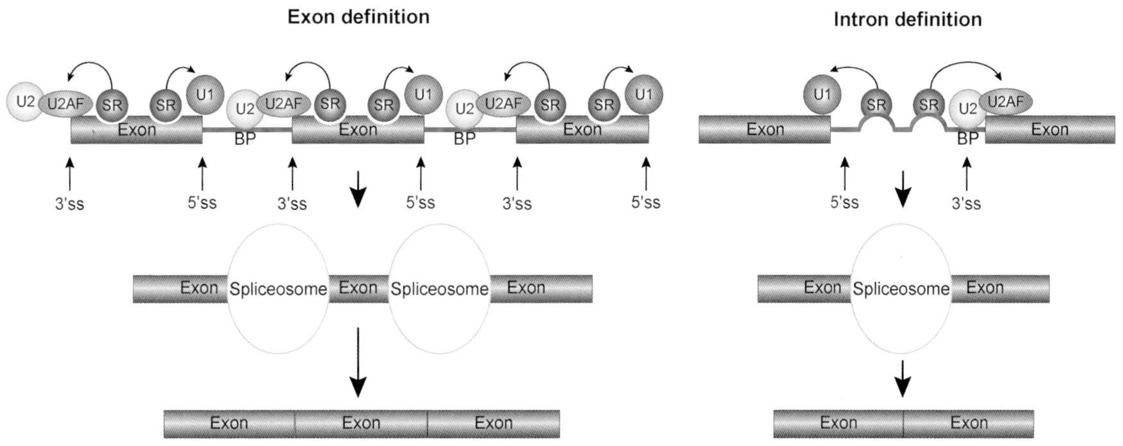

Figure 2

Exon and intron definition models illustrate how splicing machinery accurately recognizes splice sites. The exon definition model argues that splicing machinery assembles on the exon by initially recognizing splice sites around an exon. Splicing regulators such as serine/arginine-rich (SR) and other proteins that bind to ESE recruit U1 snRNP to the 5′ splice site and U2AF to the 3′ splice site, which then recruits U2 snRNP to the branch point. According to the intron definition model, the intronic splicing regulator (ISR) sequences are recognized by the splicing regulators, which then recruit U1 snRNP and U2AF to the 5′ and 3′ splice sites, respectively. Half circles in exons and introns indicate exonic and intronic splicing enhancers, respectively. SR, serine/arginine-rich protein; BP, branch point; U2AF, U2 auxillary factor.

Serine/arginine-rich (SR) proteins: a family of highly conserved phosphoproteins in eukaryotes with modular domain organization. Members of this family contain one or two RRMs at the N terminus and an RS domain at the C terminus. They function as essential splicing factors and also regulate alternative splicing

Splicing factors: non-snRNP proteins required for spliceosome assembly, transesterification reactions, or regulation of constitutive and alternative splicing

site recognition or regulation of splicing have not been identified (17, 18, 69). The intron defintion mode of splice site recognition is thought to occur in organisms with small introns in their genes (122). There is some evidence to support both of these models in plants (15, 18, 64, 72, 92, 93, 110, 116, 130). However, based on the high frequency of intron retention (56% in *Arabidopsis* and 53.5% in rice as compared with 5% in humans) in alternatively spliced transcripts in plants, it has been proposed that splice site recognition in plants occurs predominantly by intron definition (126).

COMPOSTION OF PLANT SPLICEOSOME

The spliceosome, a large RNA-protein machine, consists of 5 snRNAs and nearly 300 proteins. The assembly of the spliceosome is well understood in animals, was previously reviewed (59, 104), and hence is covered only briefly here. In higher eukaryotes there are two types of spliceosomes. The major U2-type spliceosome, which consists of U1, U2, U4, U5, and U6 snRNPs, catalyzes the removal of introns with canonical (GT-AG) splice sites. The minor U12-type spliceosome that contains U11, U12, U4atac, U5, and U6atac snRNPs recognizes a small percentage of introns (<1% in *Arabidopsis* and humans) with noncanonical splice sites. All U snRNPs contain one snRNA and several proteins, some of which are common to all snRNPs. Apart from U snRNPs, the spliceosome contains many non-snRNP proteins [heterogenous nuclear ribonculeoprotein particles (hnRNPs), DExD/H-box, and SR proteins] (125). Because the ancestral eukaryotes (plasmodium, trypanosomes, yeast, and *Caenorhabditis elegans*) lack U11/U12, it is hypothesized that the minor spliceosome was absent in the "first eukaryote" (21). In plants, the presence and functionality of U6atac and U12 were first reported in *Arabidopsis* and additional U snRNAs of the U12 spliceosome were identified using bioinformatics analysis (114, 125). Zhu & Brendel (139) reported a total of 165 U12-type introns in *Arabidopsis*, and both types of introns (U2 and U12) can occur in the same gene. The composition and function of the minor spliceosome in plants appear to be similar to that in humans (84). The fact that the minor spliceosome is present in vertebrates and plants but not in ancestral eukaryotes indicates that the minor spliceosome evolved multiple times during evolution.

In plants, it is not known how many proteins make up the spliceosome. Proteomic analysis of purified animal spliceosomes has resulted in the identification of about 300 distinct proteins in this complex (59). Although plant spliceosomes have not been isolated, analysis of the *Arabidopsis* genome for genes encoding spliceosomal proteins from nonplant systems has revealed the presence of many of the proteins found in metazoans (125). A total of 74 snRNA genes and 395 genes encoding spliceosomal and spliceosome-associated proteins have been identified in this model plant. Most animal spliceosomal proteins in major and minor spliceosomes are conserved in plants (84, 125), suggesting that the composition of plant spliceosomes is most likely similar to that of animals. However, splicing regulators such as SR proteins (discussed below), SR protein kinases, and hnRNP proteins are vastly expanded in plants with many novel proteins, indicating plant-specific mechanisms in splice site recognition and splicing regulation (125). A number of in vivo studies on plant pre-mRNA splicing have suggested differences in sequence requirements between plant and animal splice site recognition (reviewed in 16, 87, 104, 110).

ALTERNATIVE SPLICING IN PLANTS

Alternative splicing generates two or more mRNAs from the same pre-mRNA by using different splice sites. Depending on the combination of alternatively used regions of a gene, a large number of alternatively

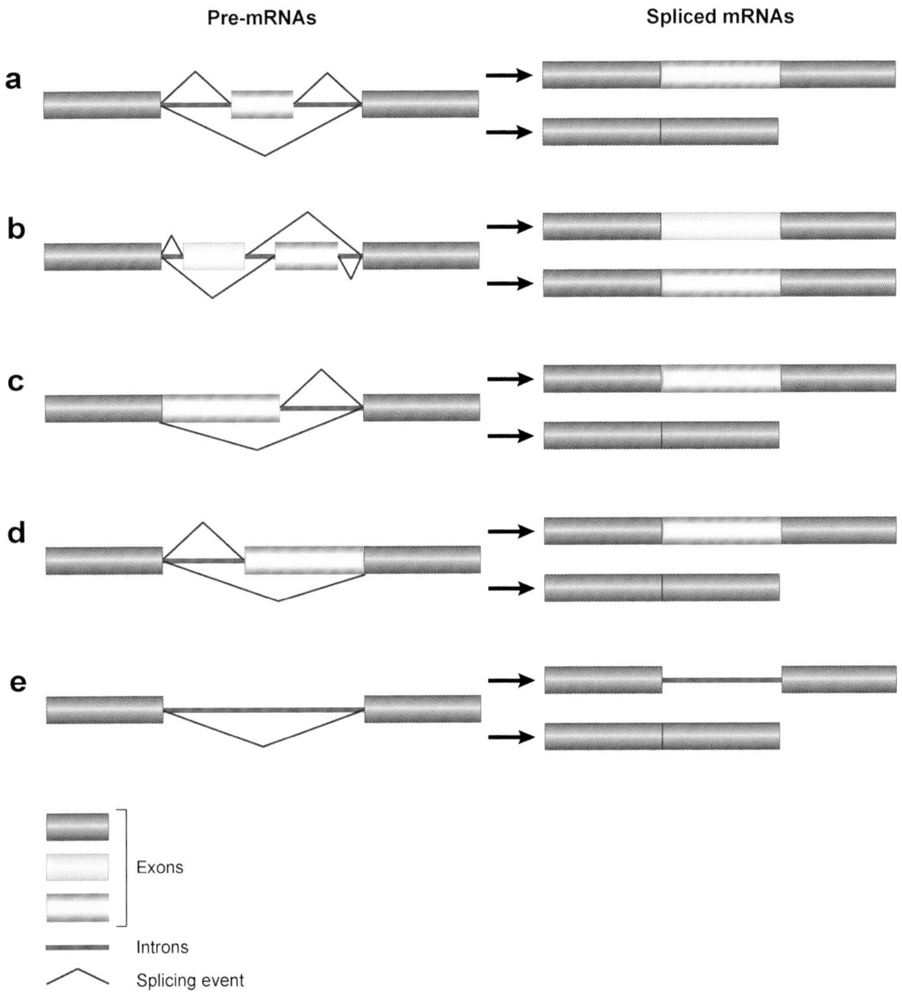

Figure 3

Some common types of alternative splicing. Pre-mRNAs are on the left and spliced mRNAs are on the right. Exons are represented by colored boxes and introns by horizontal lines. Yellow or blue boxes or a horizontal line represent sequences that are either included or excluded from the mRNA. Lines above and below the boxes indicate alternative splicing events. (*a*) Cassette exon. An exon is either included or excluded from the mRNA. (*b*) Mutually exclusive exons. Adjacent exons are spliced in such a way that only one of them is included at a time in the mRNA. (*c*) Alternative 5′ splice site. Different-size mRNAs are produced depending on the use of a proximal or distal 5′ splice site. (*d*) Alterative 3′ splice site. Different-size mRNAs are produced depending on the usage of a proximal or distal 3′ splice site. (*e*) Intron retention. An intron is either retained or excised in the mRNA, resulting in different-size transcripts. Two or more of the above alternative splicing types can occur in a single pre-mRNA to generate multiple mature mRNA from a single gene.

spliced mRNAs can be generated. The most commonly observed alternative splicing types are shown in **Figure 3**. Alternative splicing greatly increases transcriptome diversity and the alternatively spliced transcripts may encode distinct proteins, thus expanding the coding capacity of genes and contributing to the proteome complexity of higher organisms. Alternative splicing affects many aspects of RNA metabolism, including mRNA

Table 2 Database resources on plant spliceosomal proteins and alternative splicing

Database	URL	Species	Reference
Arabidopsis splicing-related genes (ASRG)	**http://www.plantgdb.org/SRGD/ASRG/index.php**	*Arabidopsis*, yeast, and animals	(125)
Alternative splicing in plants (ASIP)	**http://www.plantgdb.org/ASIP/**	*Arabidopsis*, rice	(126)
TIGR-*Arabidopsis* splicing variations	**http://www.tigr.org/tdb/e2k1/ath1/altsplicing/splicing_variations.shtml**	*Arabidopsis*	(44, 45)
TIGR-nonconsensus splice sites	**http://www.tigr.org/tdb/e2k1/ath1/Arabidopsis_nonconsensus_splice_sites.shtml**	*Arabidopsis*	(44, 45)
RARGE-alternative splicing events	**http://rarge.gsc.riken.go.jp/a_splicing/**	*Arabidopsis*	(51)
AtNoPDB-nucleolar protein database	**http://bioinf.scri.sari.ac.uk/cgi-bin/atnopdb/home**	*Arabidopsis* and human	(99)
Plant Alternative Splicing database (PASDB)	**http://pasdb.genomics.org.cn/**	Plants	(138)
ASTRA Alternative Splicing and TRanscription Archives (ASTRA)	**http://alterna.cbrc.jp/**	Plants and animals	(96)
mRNA metabolism proteins	**http://www.albany.edu/faculty/dab/IPGE_HUB.html**	*Arabidopsis*	(8)
Rice alternatively spliced genes	**http://rice.tigr.org/tdb/e2k1/osa1/expression/alt_spliced.info.shtml**	Rice	
Exonic splicing enhancers	**http://www.life.umd.edu/labs/mount/2010-splicing/**	*Arabidopsis*	

degradation through nonsense mediated decay (NMD) and other mechanisms, mRNA recruitment to ribosomes, and translation efficiency. The effects of alternative splicing on proteins include production of protein isoforms with loss or gain of function and altered cellular localization, protein stability, enzyme activity, and/or posttranslational modifications (120).

Historically, alternative splicing in plants was underappreciated primarily because it was considered rare. Until 2001, studies on alternative splicing in plants were limited to a few genes and the extent of alternative splicing in plants was not known (104). One outcome of the large-scale sequencing of genomes and expressed sequence tags (ESTs)/cDNAs is that it allowed global analysis of alternative splicing. The alignment of all human cDNAs/ESTs with genome sequences suggests that up to 60% of human genes are alternatively spliced (94). More recent analysis using splicing-sensitive arrays containing exon junction oligos revealed that 74% of multiexon genes in humans undergo alternative splicing (57). During the past four years, the estimate of alternative splicing in *Arabidopsis* increased from about 5% to 22% (1, 44, 51, 95, 97, 126, 140). This is likely to still be an underestimate as there are still relatively few cDNA sequences in plants relative to that available in humans. **Table 2** provides the URL addresses of publicly available databases on alternative splicing in plants.

To analyze the genome-wide alternative splicing events in *Arabidopsis* and rice, Wang & Brendel (126) aligned the gene sequences with EST/cDNA sequences. This analysis revealed that 4707 *Arabidopsis* genes (22% of total genes) and 6568 rice genes (21.2% of total genes) undergo alternative splicing. Unlike in humans, where exon skipping is more prevalent (**Table 3**), 56% and 53.5% of all alternative splicing events in *Arabidopsis* and rice, respectively, are a result of intron retention, whereas exon skipping accounts for only 8%

Table 3 Extent of alternative splicing in plants and humans

Organism	Total EST/cDNAs used	AS[a]	Alt. 5′SS[c]	Alt. 3′SS[c]	Alt. 5′ and 3′ SS[c]	Exon Skipping[c]	Intron Retention[c]	Reference
Arabidopsis	385,349	21.8	3.3 (10.2)[d]	6.7 (21.9)[d]	0.9 (3.7)[d]	1.8 (8.1)[d]	14.3 (56.1)[d]	(126)
Rice	369,218	21.2	3.2 (11.3)[d]	5.5 (15.1)[d]	1.8 (6.3)[d]	3.2 (13.8)[d]	14.6 (53.5)[d]	(126)
Humans	3.123 million	40–60% (74)[b]	37			58	5	(14, 43, 57, 94)

[a]The percentage of total genes with observed alternative splicing. This estimate is based on alignment of cDNA/ESTs with genome sequence.
[b]The estimate in parenthesis is based on splicing-sensitive microarrays that monitor splicing at every exon-exon junction.
[c]Percentage of total genes that showed the event.
[d]The number in parenthesis indicates the proportion of that event relative to the total number of alternative splicing events.

of alternative splicing events (**Table 3**). Alternative splicing due to alternative 5′ and 3′ splice sites is least prevalent in plants. The differences in the frequencies of alternative splicing events between plants and animals may reflect the differences in pre-mRNA splicing between these organisms. The homologs of 40% of alternatively spliced *Arabidopsis* genes in rice are also alternatively spliced. This conservation of alternative splicing events across phylogenetically diverse dicots and monocots suggests that this process is important. Alternative splicing of 22% of genes in *Arabidopsis* and rice is likely an underestimation as the number of cDNAs/ESTs used in these computational analyses is one tenth of what has been used in humans (~330,000–380,000 in plants as compared with more than 3.1 million in humans) (14, 126). The current estimate of alternative splicing in plants will likely increase as more cDNAs/ESTs become available for such analysis. Furthermore, the use of whole-genome tiling arrays and/or splicing-sensitive chips that contain exon-exon junction probes should reveal the full extent of alternative splicing in plants (57).

In *Arabidopsis*, 78.4% of alternative splicing events are in the coding region. About 50% of alternative splicing events in the coding region have a premature termination codon (PTC) and are the potential targets of nonsense mediated mRNA decay (126). In nonplant systems, most transcripts with a PTC located >50 nucleotides upstream of an exon-exon junction are targets of NMD (75, 118). NMD, which requires exon junction complexes (EJCs), is linked to translation and thought to occur during the "pioneer round" of mRNA translation (90). These splice variants with PTC may be degraded by NMD, which is likely to regulate the abundance of different transcripts. Plants, like nonplant systems, have essential components of the NMD pathway and detect and degrade transcripts with PTC (4, 47). Three proteins (UPF1, UPF2, and UPF3) that are critical components of the NMD pathway exist in plants. It was recently shown that knockout mutants of *UPF1* and *UPF3* in *Arabidopsis* do not degrade mRNAs with PTC (4, 47). In plants, it has been reported that some mRNAs that are degraded due to PTC do not have introns (100), suggesting that more than one mechanism for mRNA surveillance may exist in plants.

A large fraction of splice variants in plants differ only in the noncoding regions [in 5′ untranslated region (UTR) and 3′ UTR] of the transcripts (1, 96, 126), suggesting that alternative splicing may have functions other than generating protein diversity and/or NMD. In *Arabidopsis*, 21.6% of all alternative splicing events occur in the UTRs (15.2% in 5′ UTR and 6.4% in 3′ UTR). These alternative splicing events may have a role in mRNA transport, stability, and/or translational regulation as has been reported in animals (62, 128). Furthermore, 5′ UTR may create a new initiation codon that may or may not be in-frame with the downstream initiation codon and thereby regulate translation (56).

It is not known how many of these splice variants are actually recruited to ribosomes for translation. A transgenic line expressing a tagged ribosomal protein that permits immunoaffinity purification of polysomes should be useful for analyzing ribosomal recruitment of splice variants (133). A similar approach with a tagged ribosomal protein expressing in specific cell types using cell-specific promoters should allow analysis of splice variant recruitment in individual cell types. However, the presence of mRNAs on polysomes does not imply that those transcripts are producing protein through multiple rounds of translation. This is because mRNAs with PTC that meet the criterion for NMD also go through one round of translation. A more conclusive approach to demonstrate that the splice variants are translated is to use isoform-specific antibodies that recognize the predicted proteins.

In plants, the extent of cell-type specific alternative splicing of pre-mRNAs is not known. This is primarily because of our inability to obtain individual cell types. However, recent advances in fluorescent tagging of all root cell types and isolation of individual cell types by fluorescence-activated cell sorting should permit such an analysis (11). Analysis of alternative splicing in different cell types using splicing-sensitive arrays is expected to provide novel information on cell-specific alternative splicing and its role in cell fate.

Alternative Splicing of Pre-mRNAs of Splicing Regulators

Functional categorization of alternatively spliced genes using gene ontology has revealed that genes in all functional categories undergo alternative splicing. However, some functional groups are over-represented (threefold higher than overall alternative splicing), whereas other functional groups are under-represented in both *Arabidopsis* and rice (126). The over-represented gene ontology terms are "nuclear speckle" and "light harvesting complex" (126). It is interesting that most of the nuclear speckle proteins are either involved in pre-mRNA splicing or other aspects of RNA metabolism. Other studies also indicate that genes encoding splicing regulators are extensively alternatively spliced. Many rice SR genes, a conserved family of splicing regulators, generate multiple transcripts (55). Analyses of pre-mRNA splicing of 19 *Arabidopsis* SR genes indicate extensive alternative splicing. Remarkably, about 95 transcripts are produced from only 15 genes, thereby increasing the complexity of the SR gene family transcriptome by sixfold (98). Furthermore, alternative splicing of some SR genes is controlled in a developmental and tissue-specific manner, indicating tight regulation of alternative splicing and leading to differences in the abundance of splice variants in tissues and at developmental stages. Sequence analysis of splice variants revealed that predicted proteins from most of these variants either lacked one or more modular domains due to in-frame translation termination codons or contained additional amino acids at the end (55, 98). Because of the modular nature of various domains in SR proteins, the proteins produced from splice variants may have altered functions. For example, proteins that lack one of the domains may be localized differently or may still interact with some spliceosomal proteins and function as dominant negative regulators. The fact that the alternative splicing type in many genes including SR genes is conserved across phylogenetically divergent organisms strongly suggests a biological role for alternative splicing (50, 126, 138).

Regulation of Alternative Splicing of Spliceosomal Genes by Stresses

Several reports indicate that various biotic (viral and bacterial pathogens) and abiotic stresses (heat, cold, heavy metals) influence alternative splicing of pre-mRNAs (25, 58, 88, 98, 104, 136). Heat stress inhibits splicing of pre-mRNAs of maize polyubiquitin and *hsp70* (20, 46), *Arabidopsis hsp81*, and rice *waxy* gene (68). Exposure of maize seedlings to cadmium increased the level of unspliced, intron-containing transcripts (91). Interestingly, temperature stress (cold and heat) dramatically altered alternative splicing of pre-mRNAs of several SR genes (98). In several cases, new splice variants appeared in response to stress. In addition, some splice variants either increased or decreased by abiotic stresses. These results suggest that stresses alter SR protein expression, which in turn alters the splicing of other pre-mRNAs including SR pre-mRNAs.

A recent study supports this, reporting that cold and other stresses affect alternative splicing profiles of *Arabidopsis* genes (51). It has also been shown that manipulating the expression of SR protein alters the splicing of its own pre-mRNA and other SR genes (60, 79). *STABLIZED 1* (*STA1*), which encodes a stress-induced nuclear protein similar to the human 102 kD U5 snRNP-associated protein, is necessary for pre-mRNA splicing of a cold-induced gene and turnover of unstable mRNAs (73). Expression of one of the U1 snRNP-specific proteins (U1A) or the PRP38-like spliceosomal protein increased salt tolerance in yeast. Overexpression of the PRP38-like protein conferred salt tolerance in *Arabidopsis* plants (30). Proteins implicated in other aspects of mRNA metabolism (e.g., mRNA export, mRNA capping) were also reported to be important in ABA signaling and plant responses to abiotic stresses (30, 38, 49, 63, 76, 129). Together these results indicate that pre-mRNA splicing and mRNA metabolism play important roles in stress responses. How stress alters the alternative splicing pattern of plant SR genes remains to be elucidated. Stress may regulate alternative splicing by altering the (*a*) ratio of splicing factors by changing splicing pattern, (*b*) subcelluar redistribution of splicing regulators (SRs, hnRNPs, proteins kinases, etc.), (*c*) phosphorylation/dephosphorylation status of splicing regulators, and/or (*d*) expression of SR genes.

Stress regulation of alternative splicing of splicing regulators may allow plants to quickly regulate splicing and gene expression of many unrelated genes. Accumulating evidence suggests that stresses change the alternative splicing of splicing regulators. There is also evidence for autoregulation of alternative splicing of SR genes as well as regulation by other SR proteins. Feedback positive and negative autoregulation or regulation of SR pre-mRNAs by other SR proteins may be involved in stress-regulated alternative splicing. There are several examples of autoregulation of alternative splicing of spliceosomal genes (SR30, SRZ33, GRP7) as well as regulation of alternative splicing of other spliceosomal proteins (U1-70K, SR1, AtGRP8) by other SR proteins. The combined effect of these is altered constitutive and alternative splicing of other genes, resulting in altered transcriptome in response to these signals (**Figure 4**). This enables plants to rapidly alter their transcriptome posttranscriptionally in response to changing environmental conditions.

Do Splicing Errors Contribute to Enhanced Transcriptome Complexity?

It is now clear that splice variants are abundant in plants. One reason for this extensive alternative splicing might be that the splice variants have important functions in increasing proteome size and/or regulating the abundance of functional transcripts by regulated unproductive splicing and translation (RUST) (118). However, it is not known if the splice variants are genuine and whether they represent products of regulated splicing with functional implications. The mere presence of a splice variant in tissues does not mean that it has a biological function. An alternative explanation for the observed transcriptome complexity is that pre-mRNA splicing in plants is

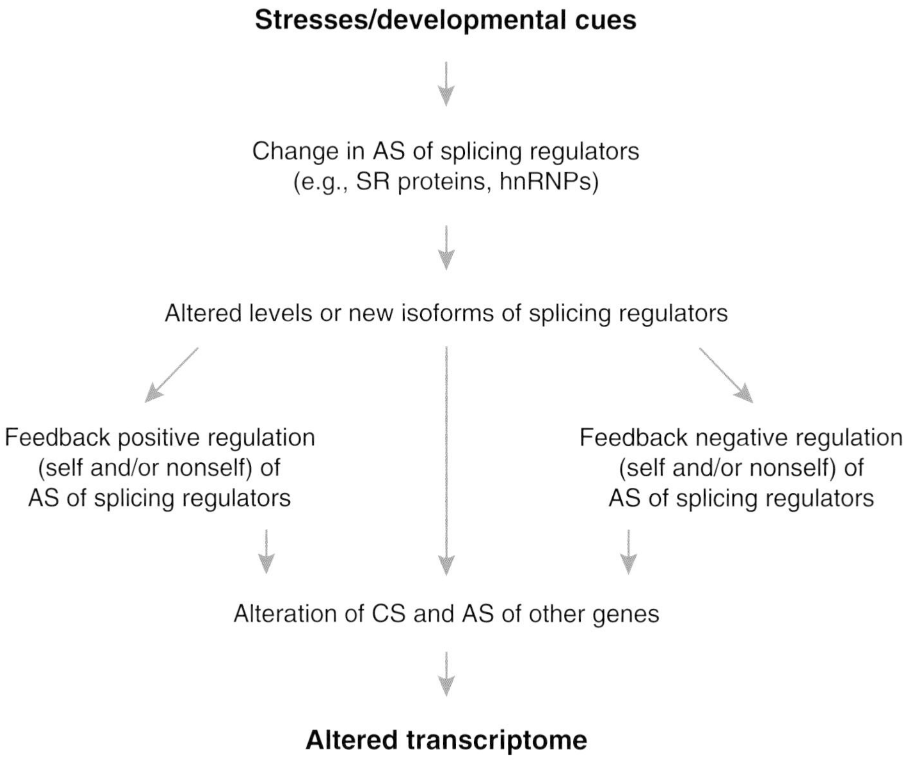

Figure 4

Schematic diagram illustrating how stresses and developmental cues can change the transcriptome by altering the splicing pattern of splicing regulators.

inherently inefficient and the multiple spliced variants are generated due to splicing errors. Although there are no documented cases of splicing errors by the spliceosome (41), it is likely that splicing machinery (like replication machinery) makes occasional splicing errors, resulting in the generation of splice variants. However, a biological role for these cannot be excluded.

Several observations suggest that splice variants may have a biological role. First, if the generation of splice variants was due to splicing errors then one would expect that most intron-containing genes should produce splice variants. However, alternative splicing is predominant in some gene families whereas other gene families with similar size and number of introns do not produce splice variants. Second, in many cases alternative splicing occurs in genes that encode multidomain proteins where splice variants encode proteins that differ in their domain organization and hence are likely to differ in function. Third, numerous studies have provided evidence that alternative splicing in plants is regulated by tissue-specific, developmental cues and stresses (34, 60, 79, 98, 113, 131). Fourth, the position of alternatively spliced introns is conserved across evolutionarily distant plants (Arabidopsis, rice, and ferns) (50, 55, 126). Fifth, splice isoforms with introns are recruited for translation (97). Sixth, many mRNAs with retained introns are not removed by RNA surveillance mechanisms in the cytoplasm. Finally, and most importantly, in a few cases where biological functions of splice variants were investigated, the presence of splice variants, including the intron-retained forms, was necessary for gene functioning (see below). Hence, all splice variants deserve close scrutiny to determine if they have a regulatory role before they are ignored as artifacts. Only detailed functional studies with each splice variant of a gene can resolve this. However, because of the vast number of splice variants, it is a daunting task to address functions of all isoforms. One way to deal with this initially is to narrow the candidates using the conservation of the alternative splicing event in phylogenetically diverse organisms. However, there is a caveat in assuming that only conserved alternative splicing types are meaningful. Alternative splicing events that are not evolutionarily conserved are not necessarily unimportant as they may be specific to one organism and reflect the biology of that organism and/or might have evolved more recently and contributed to the diversification of species.

SERINE/ARGININE-RICH PROTEINS

SR proteins, a family of conserved splicing factors, contain one or two RNA recognition motifs (RRMs) at the N terminus and a C-terminal domain with many repeating arginines and serines [the arginine/serine-rich (RS) domain]. The SR proteins are essential splicing factors required for both constitutive and alternative splicing (13). They bind pre-mRNAs and function as activators or repressors of both constitutive and alternative splicing. The RRM recognizes *cis*-acting sequences in pre-mRNA, and the RS domain functions as a protein-protein interaction module to recruit other proteins of the splicing machinery and contacts the pre-mRNA branch point (111). In addition, RS domains harbor signals for nuclear and subnuclear localization and nucleocytoplasmic shuttling (13). Among eukaryotes, flowering plants have the highest number of SR proteins with a total of 24 in rice and 19 in *Arabidopsis* (50, 55, 103). The differences between plants and animals in intron/exon architecture and plant-specific *cis*-elements involved in pre-mRNA splicing could partly account for this expansion. The presence of novel plant SR proteins and expansion of this family of proteins in plants coupled with increased transcriptome complexity of SR genes and complex interactions with spliceosomal proteins (see below) indicate that these proteins play key roles in plant pre-mRNA splicing.

In metazoans, RS domains of SR proteins are extensively phosphorylated in vivo and this is thought to influence their RNA binding activity and subcellular localization, protein-protein interactions, RNA-protein interactions, and splicing activities (13, 112). Plant SR proteins, like animal SR proteins, are phosphoproteins (79, 80). Phosphoproteomics of *Arabidopsis* proteins has revealed that 13 of the 19 SR proteins are phosphorylated in vivo (24, 28). In plants, three different types of protein kinases (Clk/LAMMER-type, SRPKs, and MAP kinases) phosphorylate one or more SR proteins (28, 36, 109). The interaction of plant SR proteins with SRs and other proteins (e.g., SRp34/SR1, AFC2, CypRS64) is modulated by the phosphorylation status of SR proteins (103). Ectopic expression of a LAMMER-type kinase in *Arabidopsis* altered the splicing pattern of spliceosomal genes (SR30, SR1/SR34, and U1-70K), as well as the expression of numerous other genes, and caused development defects (108).

Interaction studies revealed a complex network of direct interactions among SR proteins and SR protein interactions with other spliceosomal proteins (35, 36, 42, 78, 84, 85) (**Figure 5**). In addition to interactions with other SR proteins, they interacted with three other groups of proteins. These include U1 and U11 snRNP-specific proteins (70K/35K), protein kinases, and cylcophilins. U1 and U11 snRNP play a key role in 5′ splice site recognition in major and minor spliceosomes, respectively (34, 37, 104). Hence, the known interaction of five SR proteins with U1-70K and 11 SRs with U11-35K strongly suggests their involvement in recruiting corresponding snRNPs to the 5′ splice site and/or connecting these snRNPs to other members of the spliceosome. Furthermore, the interaction of several plant-specific SR proteins with U1-70K and U11-35K indicates that early stages of spliceosome formation and/or splice site selection in plants may differ from animals. Interactions with protein kinases

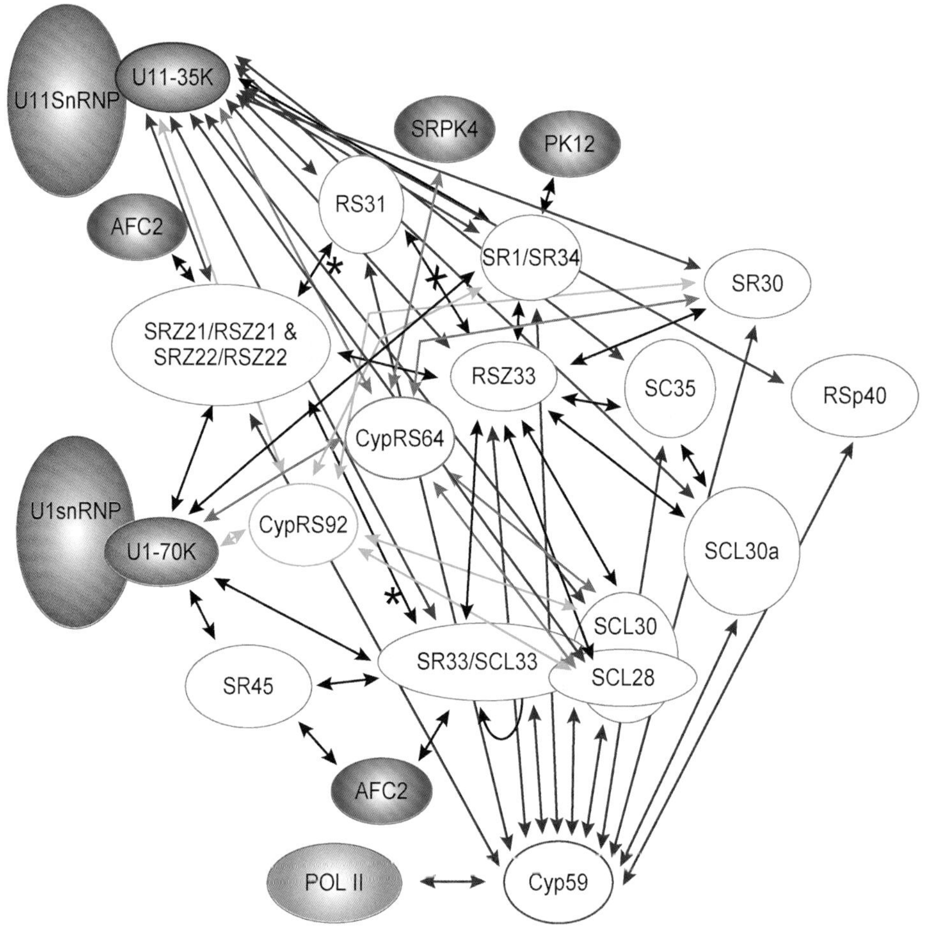

Figure 5

A complex network of interactions among serine/arginine-rich (SR) proteins and between SR proteins and other proteins. Each interaction in the network, indicated by a double-headed arrow, is based on published reports (35, 36, 42, 78, 84, 85, 108). SR proteins are shown in cyan, U1 and U11 small nuclear ribonucleoprotein (snRNP) proteins in red, protein kinases in dark blue, cyclophilin-like proteins in yellow, and RNA polymerase II in pink. SRZ21/RSZ21 and SRZ22/RSZ22 are two different proteins with similar interactions with other proteins. The interaction of SRZ22/RSZ22 was not tested with Cyp59. SR33/SCL33 interacts with itself (shown with an *arrow turning on itself*). Asterisks indicate that SR33/SCL33 and RS31 interact only with SRZ21/RSZ21. RS31 and RSZ33 do not interact (indicated with a *cross*). For clarity, all interactions with Cyp59 are shown with red double-headed arrows, interactions with CypRS64 with orange double-headed arrows, interactions with CypRS92 with cyan double-headed arrows, and all SR interactions with U11-35K with dark blue double-headed arrows. Black double-headed arrows indicate all other interactions. AFC2, a LAMMER-type protein kinase; PK12, a LAMMER-type kinase from tobacco; CypRS, cyclophilin-like protein with RS domain, Cyp59, cyclophilin with multiple domains.

(SRPK, AFC) that phosphorylate SR proteins suggest regulation of activity/function of these proteins by phosphorylation and dephosphorylation. As described above, phosphorylation of some SR proteins by these kinases has been shown in vitro. Cyp59 interacts with 11 SR proteins and also with the C-terminal domain of the largest subunit of RNA polymerase II (42), suggesting that it may be involved in and/or connecting both of these processes. In summary, these complex interactions of SR proteins indicate their importance in splice site recognition and spliceosome assembly. Sixteen of the 19 SRs are localized to the nucleus in a diffuse nucleoplasmic pool and in speckles (2, 26, 82, 83, 106, 123). Heat caused dramatic reorganization of speckles (2), suggesting that redistribution caused by heat may affect splicing.

REGULATION OF CONSTITUTIVE AND ALTERNATIVE SPLICING

Expression studies using Northerns, reverse transcription–polymerase chain reaction (RT-PCR) and promoter-reporter fusions have shown differential expression of SR proteins in different tissues and cell types (27, 35, 36, 60, 70, 79, 81). Most SR proteins showed distinct as well as overlapping expression patterns. Expression data from microarrays indicate that all SR genes are expressed in different tissues at different levels (**http://www.weigelworld.org/resources/microarray/AtGenExpress/**). These data also indicate that SR genes show relatively low expression.

There is no in vitro splicing system derived from plant cells. Hence, plant biologists have relied on three approaches to analyze SR function in pre-mRNA splicing. First, an animal in vitro splicing system was used to demonstrate whether plant SRs function as essential splicing factors. Second, proteins are constitutively overexpressed either stably or transiently and analyzed for alternative splicing of genes. Finally, gene knockout lines are beginning to be used for functional studies. Using the first approach, it has been shown that several plant SR proteins, including some plant-specific ones, can complement the splicing-deficient HeLa cell S100 extract and promote splice site switching in vitro (71, 79–81). Overexpression of SR30 influenced splice site choice and changed the alternative splicing pattern of its own pre-mRNA and that of several other endogenous plant genes (RS31, SR34, U1-70K). Similarly, overexpression of maize ASF/SF2-like proteins in cultured cells also altered 5′ splice site selection (31). Ectopic expression of SR30 greatly reduced the amount of SR1/SR34 full-length mRNA and showed pleiotropic changes in morphology and development, including larger rosette leaves and flowers, four- or five-branched trichomes, reduced apical dominance, and delayed transition from the vegetative to the flowering stage (79). Overexpression of another SR protein (RSZ33) altered the splicing pattern of two other SR genes (SR30 and SR1/SR34) whereas the splicing of SR30 was enhanced and correct splicing of the tenth intron of SR1/SR34 was promoted (60). It was also shown that RSZ33 regulates its own expression.

In addition to splicing changes, *RSZ33* overexpression showed many developmental abnormalities including twin embryos, impaired bilateral symmetry during embryogenesis, abnormal flowers and trichomes, enhanced cell expansion and changes in cell polarity, and reduced pollen germination (60). More recently, Isshiki et al. (55) overexpressed nine rice SR proteins in protoplasts and analyzed the splicing of the *Waxy^b-GUS* fusion gene. They demonstrated that two SR proteins (RSp29 and RSZp23) enhance splicing and alter the 5′ splice site selection. Using domain deletion and swapping experiments, it was shown that the RS domain is essential for splicing and that the RRM1 is required for the enhanced splicing and splice site selection, whereas the RRM2 was dispensable. Overexpression of two other rice SR proteins (RSZ36 or SRp33b) in transgenic rice altered the alternative splicing of other SR protein pre-mRNAs as well as their own. Overexpression of maize ASF/SF2-like genes (ZmSRp31 or ZmSRp32) in cultured cells enhances pre-mRNA splicing of the wheat dwarf virus replication initiator protein (31). A T-DNA insertion mutant of *SR45*, a plant-specific SR protein, showed pleiotropic developmental defects and resulted in dramatic changes in the alternative splicing pattern of pre-mRNAs of other SR genes (3). These studies indicate the role of SR proteins as important regulators of constitutive and alternative splicing of pre-mRNAs. However, overexpression of several rice SR proteins in rice plants showed no obvious phenotypes in plant growth and development (55). A mutant of rice RSZ36 also showed no growth or developmental defects. It is likely that some SR proteins have redundant functions and the lack of one is compensated for by other related SR proteins. In *C. elegans*, inactivation of some single SRs showed no phenotypes whereas inactivation of two or more SR proteins simultaneously caused lethality or developmental defects (77). Because plants have a large number of SR proteins, including several closely related ones, functional redundancy may pose a problem in analyzing their function(s) using a single gene knockout approach. Generating double and triple knockout may be necessary to unravel their functions.

Studies on splicing of U2 and U12 introns in plant cells have shown that the splicing efficiency of plant pre-mRNAs depends on both intronic elements (U or UA content), composition of exons, and/or other potential signals (ISEs/ISSs, ESEs/ESSs) (104). However, specific *cis*-elements that bind to plant SR proteins have not been identified. Some RNA binding proteins that interact specifically with U-rich sequences were identified in plants (66). Overexpression of one of these proteins [U-rich binding protein 1 (UBP1)] enhanced the splicing of poorly spliced U2 introns, suggesting that UBP binding to the U-rich sequences in introns recruits splicing machinery to pre-mRNAs. However, overexpression of UBP1 did not affect the splicing activity of U12 introns (74), suggesting differences between U2 and U12 intron splicing. UBP1 interacts with two plant-specific proteins (UBA1a and UBA2a) that contain one (UBA1) or two (UBA2) RRMs and an acidic domain. Both of these proteins, like UBP, bind oligouridylates. Overexpression of these proteins enhanced steady-state levels of reporter mRNA either in a promoter-dependent (in the case of UBA1a) or promoter-independent (in the case of UBA2a) manner (65, 86). Each of these proteins belongs to a small gene family consisting of three members each. It was proposed that these proteins function as a complex and stabilize mRNA by binding to U-rich sequences. There are a large number of hnRNPs in plants, including some plant-specific ones that might be important for splicing regulation (125). Overexpression of one of these RNA binding proteins (AtGRP7) altered the alternative splicing pattern of its own pre-mRNAs and those of another RNA binding protein (AtGRP8) (119). However, the functions of many hnRNP proteins are unknown.

The importance of the GC content in exons in efficient splicing has also been shown in plants (17, 69). An AG-rich exonic element that is capable of promoting downstream 5′ splice site selection was reported (93). Yoshimura et al. (131) identified a putative *cis*-element (with a high AU content), which is highly conserved in this gene in various plants, upstream of the acceptor site in intron 12. Deleting this element reduced the splicing efficiency of intron 12, confirming the importance of this element in splicing. Using gel shift assays it was shown that a nuclear protein binds to this *cis*-element. However, the identity of this protein is not known. There are other examples in plants where alternative splicing regulates enzyme activity, protein function, or cellular localization of a protein (104). The exon upstream of the U12 intron in *Luminidepends* (*LD*) contains two regions (potential ESEs) that increase splicing (74). It is speculated that SR proteins and/or other RNA binding proteins bind to these elements and promote spliceosome assembly. **Figure 6** shows the potential roles of plant SR and other RNA binding proteins during the early stages of spliceosome formation.

Currently, an NSF 2010 project aimed at identifying splicing signals in plants using computational tools and experimental

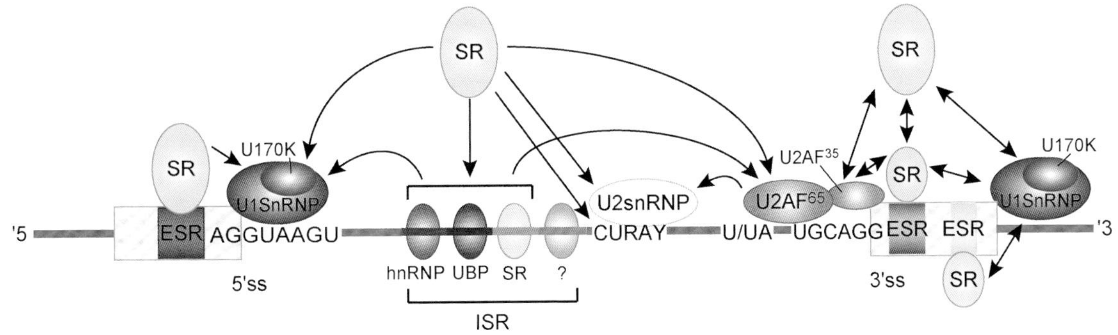

Figure 6

A model illustrating the possible functions of plant SR and other RNA binding proteins during the early stages of spliceosome assembly. White boxes indicate exons and the horizontal lines between and on either side of each exon indicate introns. Consensus sequences at 5′ and 3′ splice sites in plants are from **Figure 1*a***. Based on **Figure 1*b***, the U or U/A region is shown in place of the animal polypyrimidine tract. Colored boxes in exons indicate putative exonic splicing regulators (ESRs). Colored lines in introns indicate intronic splicing regulators (ISRs) such as a U-rich region. *Trans*-factors (SR and other RNA binding proteins) that are likely to interact with exonic and intronic elements are shown. SR proteins can promote the recognition of 5′ and 3′ splice sites by recruiting U1 snRNP, U2AF, and other spliceosomal proteins. In addition, they can bridge the spliceosomal components at 5′ and 3′ splice sites. Some interactions shown here have experimental evidence (e.g., interaction among SR proteins, interactions of several SR proteins with U1-70K and U11-35K, UBP binding to U-rich sequences, etc.), and others are hypothetical interactions. Arrows indicate SR protein-mediated interactions. SR, serine/arginine-rich protein; $U2AF^{65}$, U2 auxillary factor large subunit; $U2AF^{35}$, U2 auxillary factor small subunit; hnRNP, heterogeneous nuclear ribonucleoprotein particle proteins.

verification of the predicted ESE using splicing reporters is underway (**http://www.life.umd.edu/labs/mount/2010-splicing/**). Systematic evolution of ligands by exponential enrichment (SELEX) and genomic SELEX have been successfully used to identify target sequences for the RRMs of metazoan SR proteins and other RNA binding proteins (13, 61). Using genomic SELEX to identify natural binding sequences of plant SR proteins should also be useful in identifying regulatory *cis*-elements. It is likely that at least some *cis*-elements involved in alternative splicing of plant pre-mRNAs are conserved during evolution. With the availability of the complete genome sequence of several phylogenetically diverse plants and the expected completion of genomes of other crop plants together with ESTs/cDNAs (33, 52, 132), it should be possible to use bioinformatics tools to identify the putative exonic and intronic *cis*-regulatory sequences involved in constitutive and regulated splicing. The activity of such predicted elements can then be tested experimentally. In animals, such analyses are already providing some useful insights into regulated splicing (40).

BIOLOGICAL ROLES OF ALTERNATIVE SPLICING

Although several thousand plant genes are now known to produce multiple transcripts (104, 126), the precise functions of most of the splice variants and their encoded proteins are not known. Functional studies conducted with a few splice variants indicate important functions of alternative splicing. In only a few of the following examples proteins encoded by splice variants have been verified using isoform-specific antibodies or amino acid sequence analysis (127, 131).

Photosynthesis

It is interesting that one gene of the over-represented class of alternative splicing genes belongs to photosynthesis. Rubisco activase, a nuclear-encoded chloroplast protein required for the light activation of ribulose 1,5-bisphosphate carboxylase/oxygenase (Rubisco), was one of the first genes to show alternative splicing in dicots and monocots and to generate two proteins that differ only at the carboxyl terminus (107, 124, 127, 137). Although both forms activate Rubisco, only the large isoform is redox regulated (134, 135).

Defense Responses

Some reports indicate that alternative splicing of pre-mRNAs of resistance (R) genes plays an important role in plant defense responses. In tobacco, the N gene, a member of a class of R genes that encode proteins with Toll/interleukin 1 receptor (TIR)—nucleotide binding site (NBS)–leucine-rich repeat (LRR) (TIR-NBS-LRR) domains, confers resistance to tobacco mosaic virus (TMV). The N gene produces two transcripts [a short (N_S) and a long (N_L) transcript] that lack or contain, respectively, an alternative exon in intron 3. The N_S encodes the full-length protein (1144 amino acids) whereas the inclusion of the alternative exon in the N_L shifts the reading frame, resulting in a truncated protein (652 amino acids) that lacks 13 of the 14 LRR regions. TMV infection regulates the splicing pattern of the N gene in such a way that N_L becomes more abundant after 4–8 h of infection (25). Transgenic plants expressing N_S transcript did not show complete resistance to TMV. Interestingly, transgenic lines expressing N_S cDNA containing intron 3 and the 3′ sequence of the N gene or a genomic region capable of producing both N_S and N_L transcripts showed complete resistance, suggesting the importance of regulated alternative splicing of the N gene in disease resistance (25). Alternative splicing was also observed with pre-mRNAs of several other plant disease R genes encoding the TIR-NBS-LRR family of proteins (58). *Arabidopsis RPS4*, another disease R gene that encodes a member of the TIR-NBS-LRR family, produces three transcripts by alternative splicing. In addition to a constitutively spliced transcript, retention of intron 2 and 3 or only intron 3 generates two other transcripts with an in-frame stop codon. Zhang & Gassmann (136) demonstrated that the presence of constitutively as well as alternatively spliced transcripts is necessary for *RPS4* function. Although these reports established the requirement of alternative transcripts, a mechanistic understanding of how these splice variants function is unknown.

Flowering

Alternative processing of *FCA* pre-mRNA controls the developmental switch from the vegetative to the reproductive phase. *FCA*, which encodes a nuclear ABA receptor with an RNA binding domain, produces four transcripts (α, β, γ, and δ) by alternative processing (101, 102). Three (α, β, and γ) of these are generated by differential processing of intron 3. In transcript α, intron 3 is retained, premature cleavage and polyadenylation within intron 3 yields transcript β, and removal of intron 3 generates transcript γ. Alternative splicing of intron 13, resulting in excision of a large intron, yields transcript δ. Only transcript γ encodes the full-length protein, which complements the late flowering phenotype in *fca*. The other three encode truncated proteins. Alternative splicing of the FCA transcript limits the amount of FCA protein, both spatially and temporally, which in turn controls flowering. Quesada et al. (101) demonstrated that *FCA* negatively regulates its own expression by promoting cleavage and polyadenylation within intron 3. This generates a truncated inactive transcript, which limits the production of full-length *FCA* mRNA, and this autoregulation prevents precocious flowering. Alternative processing of intron 3 of *FCA* is conserved between *Arabidopsis* and *Brassica napus* (89), suggesting that alternative splicing is also important in the transition to flowering in other species. Thermal induction of flowering upregulates genes involved in RNA splicing, including 6 SR proteins (6). Overexpression of SR30 that regulates alternative splicing of other SR pre-mRNAs results in delayed flowering (79), implying a role for regulated splicing in flowering.

Grain Quality in Rice

In rice, the amylose quantity affects the seed quality. A high amylose quantity in rice grains is not desirable. The cultivated rice species, which have less amylose content in the endosperm, are derived from wild varieties that have high amylose content. A single mutation at the 5′ splice site of one gene (*Waxy*) that affects the alternative splicing of its pre-mRNA results in the reduced levels of amylose. The *Waxy* (*Wx*) gene encodes a granule-bound starch synthase that is necessary for the synthesis of amylose in endosperm. In one of the two naturally occurring alleles of the *Wx* gene (Wx^a and Wx^b), the Wx^b has a mutation (G to T) in the 5′ splice site in intron 1, causing a tenfold reduction in its expression as compared with Wx^a. It was later shown that this mutation reduces the splicing efficiency of Wx^b pre-mRNA and promotes alternative splicing at cryptic sites in exon 1, resulting in reduced content of amylose in varieties that contain this allele (53). Two mutations (*du-1* and *du-2*) that reduce the splicing of Wx^b without affecting the splicing of Wx^a have been isolated, suggesting that *DU-1* and *DU-2* encode regulators of alternative splicing (54).

EVOLUTION OF ALTERNATIVE SPLICING

Although pre-mRNA splicing occurs in all eukaryotes, alternative splicing is more prevalent in highly evolved multicellular eukaryotes as compared with ancestral unicellular eukaryotes. The number of intron-containing genes and the extent of alternative splicing have increased considerably from ancestral unicellular eukaryotes to complex multicellular organisms. In plants and animals, 80% of genes have introns, with up to 74% of intron-containing genes undergoing alternative splicing. In contrast, the budding yeast has introns in about 4% of its genes (~250 out of the 6000 genes) and only six of these have two introns (7). Alternative splicing in unicellular eukaryotes such as budding yeast is rare or nonexistent (7, 23, 41), suggesting that alternative splicing may have evolved more recently. However, it is also possible that the rarity of alternative splicing in unicellular organisms could be a derived state owing to the loss of alternative splicing machinery.

How did alternative splicing evolve? Comparing splice sites in unicellular eukaryotes that show no alternative splicing with multicellular eukaryotes that show extensive alternative splicing reveals a high degree of plasticity in the 5′ splice sites in multicellular organisms (5). Two theories have been proposed to explain the evolution of alternative splicing (5). According to one theory that is sequence based, mutations in a constitutive splice site result in a suboptimal or weaker site, thereby allowing the splicing machinery to skip the splicing site. The second *trans*-factor-based model invokes the evolution of splicing regulatory factors, which might result in an increase or decrease in the binding of the splicing machinery to constitutive sites. This is expected to release the selective pressure from the splice site, thereby allowing mutations that weaken the splice site. These two theories are not mutually exclusive and both may have contributed to the evolution of alternative splicing. The vast expansion of splicing regulators such as hnRNPs and SR proteins in complex organisms such as humans and plants may have partly contributed to an increase in alternative splicing in these organisms.

CONCLUDING REMARKS

The unexpected prevalence of alternative splicing in plants raises myriad questions. Are the splice variants real? Do they encode protein isoforms that are functionally different? How much of this alternative splicing contributes to proteome diversity? What is the role of splice variants in controlling cellular processes and ultimately plant growth and development? Do splice variants function as regulatory RNAs to control gene expression by different means? What proteins and mechanisms regulate alternative splicing? Answering these questions is a major challenge facing plant biologists. I hope that answers to some of these questions are forthcoming in the next decade. It is well established that introns enhance gene expression (reviewed in 104). Recent analysis of expression data of all *Arabidopsis* genes reveals that highly expressed genes have four or more introns (1), which supports earlier reports that introns enhance gene expression. Thus, a better understanding of pre-mRNA splicing will have implications in optimizing transgene expression in plants. Furthermore, analysis of alternative splicing across phylogenetically divergent as well as closely related species may also help us understand if transcriptome diversity generated by alternative splicing contributed to speciation. It appears that alternative splicing in plants will take center stage and emerge as an important posttranscriptional regulatory mechanism in fine-tuning gene expression.

SUMMARY POINTS

1. Although plant spliceosomes have not been isolated, analysis of the completed sequences of the *Arabidopsis* genome for the presence of known spliceosomal proteins in nonplant systems indicates that the core spliceosomal machinery is conserved between plants and animals. However, specific families, especially splicing regulators, are greatly expanded in plants and some plant splicing proteins do not have homologs in animal systems, suggesting the existence of plant-specific splice site recognition and splicing regulatory mechanisms.
2. Plant and animal introns have similar 5′ and 3′ splice sites. However, plant introns differ from animals in their size, nucleotide composition, branch point sequence, and polypyrimdine tract. These differences indicate that some of the initial events involved in splice site recognition are likely unique to plants.
3. Alternative splicing in flowering plants is much more prevalent than previously thought, with pre-mRNAs of more than 21% of all genes producing multiple transcripts generating considerable transcriptome complexity and likely proteome diversity. Intron retention accounts for more than 50% of all alternative splicing events in plants, but it is not prevalent in animals.
4. The large number of SR proteins in flowering plants, including several plant-specific members, and their structural features, localization, and complex network of interactions among themselves and with other spliceosomal proteins strongly indicate an important role for these proteins in the early stages of spliceosome assembly and regulated splicing.
5. Pre-mRNAs of SR proteins undergo extensive alternative splicing. Because SR proteins are modular and are involved in multiple stages of splicing, it is likely that the splice variants and the encoded proteins are functionally significant.
6. Alternative splicing of plant SR pre-mRNAs is either autoregulated and/or modulated by other SR proteins. This type of feedback regulation is likely important in controlling the levels of SR proteins and splicing of other pre-mRNAs. This offers a quick way

to modulate constitutive and alternative splicing of pre-mRNAs of nonspliceososmal genes in response to signals and thereby alter the transcriptome.

7. Stresses have a dramatic effect on alternative splicing of pre-mRNA SR proteins, which in turn could regulate the splicing of their own and/or other SR pre-mRNAs. It is likely that part of the known stress-induced changes in the transcriptome could be due to changes in pre-mRNA splicing of the affected genes.

ACKNOWLEDGMENTS

I thank Drs. Golovkin, Ali, Goud, Hanumappa, and Prasad for their contributions to splicing research in my laboratory; Dale Richardson for preparing the splice site logos, Justin Anderson for preparing the figures; and Gul Shad Ali, Irene Day, and Saiprasad Goud for constructive comments on the manuscript. Work on pre-mRNA splicing research in my laboratory is supported by a grant from the U.S. Department of Energy (DE-FG02-04ER15556).

LITERATURE CITED

1. Alexandrov NN, Troukhan ME, Brover VV, Tatarinova T, Flavell RB, Feldmann KA. 2006. Features of *Arabidopsis* genes and genome discovered using full-length cDNAs. *Plant Mol. Biol.* 60:69–85
2. Ali GS, Golovkin M, Reddy ASN. 2003. Nuclear localization and in vivo dynamics of a plant-specific serine/arginine-rich protein. *Plant J.* 36:883–93
3. Ali GS, Palusa SG, Golovkin M, Prasad J, Manley JL, Reddy ASN. 2006. A plant-specific splicing factor regulates multiple developmental processes. In review
4. Arciga-Reyes L, Wootton L, Kieffer M, Davies B. 2006. UPF1 is required for nonsense-mediated mRNA decay (NMD) and RNAi in *Arabidopsis*. *Plant J.* 47:480–89
5. Ast G. 2004. How did alternative splicing evolve? *Nat. Rev. Genet.* 5:773–82
6. Balasubramanian S, Sureshkumar S, Lempe J, Weigel D. 2006. Potent induction of *Arabidopsis thaliana* flowering by elevated growth temperature. *PLoS Genet.* 2:e106
7. Barrass JD, Beggs JD. 2003. Splicing goes global. *Trends Genet.* 19:295–98
8. Belostotsky DA, Rose AB. 2005. Plant gene expression in the age of systems biology: Integrating transcriptional and post-transcriptional events. *Trends Plant. Sci.* 10:347–53
9. Berget SM. 1995. Exon recognition in vertebrate splicing. *J. Biol. Chem.* 270:2411–14
10. Berget SM, Moore C, Sharp PA. 1977. Spliced segments at the 5′ terminus of adenovirus 2 late mRNA. *Proc. Natl. Acad. Sci. USA* 74:3171–75
11. Birnbaum K, Jung JW, Wang JY, Lambert GM, Hirst JA, et al. 2005. Cell type-specific expression profiling in plants via cell sorting of protoplasts from fluorescent reporter lines. *Nat. Methods* 2:615–19
12. Black DL. 2003. Mechanisms of alternative premessenger RNA splicing. *Annu. Rev. Biochem.* 72:291–336
13. Bourgeois CF, Lejeune F, Stevenin J. 2004. Broad specificity of SR (serine/arginine) proteins in the regulation of alternative splicing of premessenger RNA. *Prog. Nucleic Acids Res. Mol. Biol.* 78:37–88
14. Brett D, Pospisil H, Valcarcel J, Reich J, Bork P. 2002. Alternative splicing and genome complexity. *Nat. Genet.* 30:29–30

15. Brown JW. 1996. *Arabidopsis* intron mutations and premRNA splicing. *Plant J.* 10:771–80
16. Brown JW, Simpson CG. 1998. Splice site selection in plant premRNA splicing. *Annu. Rev. Plant Physiol. Plant Mol. Biol.* 49:77–95
17. Carle-Urioste JC, Brendel V, Walbot V. 1997. A combinatorial role for exon, intron and splice site sequences in splicing in maize. *Plant J.* 11:1253–63
18. Carle-Urioste JC, Ko CH, Benito MI, Walbot V. 1994. In vivo analysis of intron processing using splicing-dependent reporter gene assays. *Plant Mol. Biol.* 26:1785–95
19. Chow LT, Gelinas RE, Broker TR, Roberts RJ. 1977. An amazing sequence arrangement at the 5′ ends of adenovirus 2 messenger RNA. *Cell* 12:1–8
20. Christensen AH, Sharrock RA, Quail PH. 1992. Maize polyubiquitin genes: Structure, thermal perturbation of expression and transcript splicing, and promoter activity following transfer to protoplasts by electroporation. *Plant Mol. Biol.* 18:675–89
21. Collins L, Penny D. 2005. Complex spliceosomal organization ancestral to extant eukaryotes. *Mol. Biol. Evol.* 22:1053–66
22. Consortium IHGS. 2001. Initial sequencing and analysis of the human genome. *Nature* 409:860–921
23. Davis CA, Grate L, Spingola M, Ares MJ. 2000. Test of intron predictions reveals novel splice sites, alternatively spliced mRNAs and new introns in meiotically regulated genes of yeast. *Nucleic Acids Res.* 28:1700–6
24. de la Fuente van Bentem S, Anrather D, Roitinger E, Djamei A, Hufnagl T, et al. 2006. Phosphoproteomics reveals extensive in vivo phosphorylation of *Arabidopsis* proteins involved in RNA metabolism. *Nucleic Acids Res.* 34:3267–78
25. **Dinesh-Kumar SP, Baker BJ. 2000. Alternatively spliced N resistance gene transcripts: their possible role in tobacco mosaic virus resistance. *Proc. Natl. Acad. Sci. USA* 97:1908–13**

25. Demonstrates that alternatively spliced transcripts of a disease resistance gene in tobacco are necessary to confer disease resistance to tobacco mosaic virus. It is not known if alternatively spliced transcripts are translated.

26. Docquier S, Tillemans V, Deltour R, Motte P. 2004. Nuclear bodies and compartmentalization of premRNA splicing factors in higher plants. *Chromosoma* 112:255–66
27. Fang Y, Hearn S, Spector DL. 2004. Tissue-specific expression and dynamic organization of SR splicing factors in *Arabidopsis*. *Mol. Biol. Cell.* 15:2664–73
28. Feilner T, Hultschig C, Lee J, Meyer S, Immink RGH, et al. 2005. High throughput identification of potential *Arabidopsis* mitogen-activated protein kinases substrates. *Mol. Cell. Proteomics* 4:1558–68
29. Filipowicz W, Gniadkowski M, Klahre U, Liu HX. 1995. Pre-mRNA splicing in plants. In *Pre-mRNA Processing*, ed. AI Lamond, 4:65–77. Georgetown, TX: Landers
30. Forment J, Naranjo MA, Roldan M, Serrano R, Vicente O. 2002. Expression of *Arabidopsis* SR-like splicing proteins confers salt tolerance to yeast and transgenic plants. *Plant J.* 30:511–19
31. Gao H, Gordon-Kamm WJ, Lyznik LA. 2004. ASF/SF2-like maize premRNA splicing factors affect splice site utilization and their transcripts are alternatively spliced. *Gene* 339:25–37
32. Gilbert W. 1978. Why genes in pieces? *Nature* 271:501
33. Goff SA, Ricke D, Lan TH, Presting G, Wang R, et al. 2002. A draft sequence of the rice genome (Oryza sativa L. ssp. japonica). *Science* 296:92–100
34. Golovkin M, Reddy ASN. 1996. Structure and expression of a plant U1 snRNP 70K gene: Alternative splicing of U1 snRNP 70K premRNAs produces two different transcripts. *Plant Cell* 8:1421–35
35. Golovkin M, Reddy ASN. 1998. The plant U1 small nuclear ribonucleoprotein particle 70K protein interacts with two novel serine/arginine-rich proteins. *Plant Cell* 10:1637–48

36. Golovkin M, Reddy ASN. 1999. An SC35-like protein and a novel serine/arginine-rich protein interact with *Arabidopsis* U1-70K protein. *J. Biol. Chem.* 274:36428–38
37. Golovkin M, Reddy ASN. 2003. Expression of U1snRNP 70K antisense transcript using APEATALA3 promoter suppresses the development of petals and stamens. *Plant Physiol.* 132:1884–91
38. Gong Z, Dong CH, Lee H, Zhu J, Xiong L, et al. 2005. A DEAD box RNA helicase is essential for mRNA export and important for development and stress responses in *Arabidopsis*. *Plant Cell* 17:256–67
39. Goodall G, Levy J, Mieszczak M, Filipowicz W. 1990. Plant RNA-binding proteins. *Mol. Biol. Rep.* 14:137
40. Goren A, Ram O, Amit M, Keren H, Lev-Maor G, et al. 2006. Comparative analysis identifies exonic splicing regulatory sequences–The complex definition of enhancers and silencers. *Mol. Cell* 22:769–81
41. Graveley BR. 2001. Alternative splicing: increasing diversity in the proteomic world. *Trends Genet.* 17:100–7
42. Gullerova M, Barta A, Lorkovic ZJ. 2006. AtCyp59 is a multidomain cyclophilin from *Arabidopsis thaliana* that interacts with SR proteins and the C-terminal domain of the RNA polymerase II. *RNA* 12:631–43
43. Gupta S, Zink D, Korn B, Vingron M, Haas SA. 2004. Genome wide identification and classification of alternative splicing based on EST data. *Bioinformatics* 20:2579–85
44. Haas BJ, Delcher AL, Mount SM, Wortman JR, Smith RKJ, et al. 2003. Improving the *Arabidopsis* genome annotation using maximal transcript alignment assemblies. *Nucleic Acids Res.* 31:5654–66
45. Haas BJ, Volfovsky N, Town CD, Troukhan M, Alexandrov N, et al. 2002. Full-length messenger RNA sequences greatly improve genome annotation. *Genome Biol.* 3:RESEARCH0029
46. Hopf N, Plesofsky-Vig N, Brambl R. 1992. The heat shock response of pollen and other tissues of maize. *Plant Mol. Biol.* 19:623–30
47. Hori K, Watanabe Y. 2005. UPF3 suppresses aberrant spliced mRNA in *Arabidopsis*. *Plant J.* 43:530–40
48. Huang YQ, Steitz JA. 2005. SRprises along a messenger's journey. *Mol. Cell* 17:613–15
49. Hugouvieux V, Kwak JM, Schroeder JI. 2001. An mRNA cap binding protein, ABH1, modulates early abscisic acid signal transduction in *Arabidopsis*. *Cell* 106:477–87
50. Iida K, Go M. 2006. Survey of conserved alternative splicing events of mRNAs encoding SR proteins in land plants. *Mol. Biol. Evol.* 23:1085–94
51. Iida K, Seki M, Sakurai T, Satou M, Akiyama K, et al. 2004. Genome-wide analysis of alternative premRNA splicing in *Arabidopsis thaliana* based on full-length cDNA sequences. *Nucleic Acids Res.* 32:5096–103
52. Initiative, Arabidopsis Genome (AGI) 2000. Analysis of the genome sequence of the flowering plant *Arabidopsis thaliana*. *Nature* 408:796–815
53. Isshiki M, Morino K, Nakajima M, Okagaki RJ, Wessler SR, et al. 1998. A naturally occurring functional allele of the rice waxy locus has a GT to TT mutation at the 5′ splice site of the first intron. *Plant J.* 15:133–38
54. Isshiki M, Nakajima M, Satoh H, Shimamoto K. 2000. dull: rice mutants with tissue-specific effects on the splicing of the waxy premRNA. *Plant J.* 23:451–60
55. Isshiki M, Tsumoto A, Shimamoto K. 2006. The serine/arginine-rich protein family in rice plays important roles in constitutive and alternative splicing of premRNA. *Plant Cell* 18:146–58

55. By overexpressing SR proteins either transiently or stably in rice, authors demonstrate that SR proteins regulate alternative splicing of pre-mRNAs of a reporter gene and other SR proteins by changing splice site choice.

56. Jin XP, Turcott E, Englehardt S, Mize GJ, Morris DR. 2003. The two upstream open reading frames of oncogene mdm2 have different translational regulatory properties. *J. Biol. Chem.* 278:25716–21

57. Johnson JM, Castle J, Garrett-Engele P, Kan ZY, Loerch PM, et al. 2003. Genome-wide survey of human alternative premRNA splicing with exon junction microarrays. *Science* 302:2141–44

58. Jordan T, Schornack S, Lahaye T. 2002. Alternative splicing of transcripts encoding Toll-like plant resistance proteins—what's the functional relevance to innate immunity? *Trends Plant Sci.* 7:392–98

59. Jurica MS, Moore MJ. 2003. Pre-mRNA splicing: Awash in a sea of proteins. *Mol. Cell* 12:5–14

60. Kalyna M, Lopato S, Barta A. 2003. Ectopic expression of atRSZ33 reveals its function in splicing and causes pleiotropic changes in development. *Mol. Biol. Cell* 14:3565–77

61. Kim S, Shi H, Lee DK, Lis JT. 2003. Specific SR protein-dependent splicing substrates identified through genomic SELEX. *Nucleic Acids Res.* 31:1955–61

62. Kuersten S, Goodwin EB. 2003. The power of the 3′ UTR: Translational control and development. *Nat. Rev. Genet.* 4:626–37

63. Kuhn JM, Schroeder JI. 2003. Impacts of altered RNA metabolism on abscisic acid signaling. *Curr. Opin. Plant Biol.* 6:463–69

64. Lal S, Choi JH, Shaw JR, Hannah LC. 1999. A splice site mutant of maize activates cryptic splice sites, elicits intron inclusion and exon exclusion, and permits branch point elucidation. *Plant Physiol.* 121:411–18

65. Lambermon MH, Fu Y, Wieczorek Kirk DA, Dupasquier M, Filipowicz W, Lorkovic ZJ. 2002. UBA1 and UBA2, two proteins that interact with UBP1, a multifunctional effector of premRNA maturation in plants. *Mol. Cell. Biol.* 22:4346–57

66. **Lambermon MH, Simpson GG, Wieczorek Kirk DA, Hemmings-Mieszczak M, Klahre U, Filipowicz W. 2000. UBP1, a novel hnRNP-like protein that functions at multiple steps of higher plant nuclear premRNA maturation. *EMBO J.* 19:1638–49**

66. A nuclear protein (UBP1) that binds U-rich intronic sequence and 3′ UTR was identified in tobacco. Overexpression of this protein enhanced removal of introns that are inefficiently processed and enhanced mRNA level from intronless genes.

67. Lander ES, Linton LM, Birren B, Nusbaum C, Zody MC, et al. 2001. Initial sequencing and analysis of the human genome. *Nature* 409:860–921

68. Larkin PD, Park WD. 1999. Transcript accumulation and utilization of alternate and nonconsensus splice sites in rice granule-bound starch synthase are temperature-sensitive and controlled by a single-nucleotide polymorphism. *Plant Mol. Biol.* 40:719–27

69. Latijnhouwers MJ, Pairoba CF, Brendel V, Walbot V, Carle-Urisote JC. 1999. Test of the combinatorial model of intron recognition in a native maize gene. *Plant Mol. Biol.* 41:637–44

70. Lazar G, Goodman HM. 2000. The *Arabidopsis* splicing factor SR1 is regulated by alternative splicing. *Plant Mol. Biol.* 42:571–81

71. Lazar G, Schall T, Maniatis T, Goodman H. 1995. Identification of a plant serine-arginine-rich protein similar to the mammalian splicing factor SF2/ASF. *Proc. Natl. Acad. Sci. USA* 92:7672–76

72. Lazarova GI, Kerckhoffs LH, Brandstadter J, Matsui M, Kendrick RE, et al. 1998. Molecular analysis of PHYA in wild-type and phytochrome A-deficient mutants of tomato. *Plant J.* 14:653–62

73. Lee BH, Kapoor A, Zhu J, Zhu JK. 2006. STABILIZED1, a stress-upregulated nuclear protein, is required for pre-mRNA splicing, mRNA turnover, and stress tolerance in *Arabidopsis*. *Plant Cell* 18:1736–49

74. **Lewandowska D, Simpson CG, Clark GP, Jennings NS, Barciszewska-Pacak M, et al. 2004. Determinants of plant U12-dependent intron splicing efficiency. *Plant Cell* 16:1340–52**

75. Lewis BP, Green RE, Brenner SE. 2003. Evidence for the widespread coupling of alternative splicing and nonsense-mediated mRNA decay in humans. *Proc. Natl. Acad. Sci. USA* 100:189–92

76. Li J, Kinoshita T, Pandey S, Ng CK, Gygi SP, et al. 2002. Modulation of an RNA-binding protein by abscisic-acid-activated protein kinase. *Nature* 418:793–97

77. Longman D, Johnstone IL, Caceres JF. 2000. Functional characterization of SR and SR-related genes in *Caenorhabditis elegans*. *EMBO J.* 19:1625–37

78. Lopato S, Forstner C, Kalyna M, Hilscher J, Langhammer U, et al. 2002. Network of interactions of a novel plant-specific Arg/Ser-rich protein, atRSZ33, with atSC35-like splicing factors. *J. Biol. Chem.* 277:39989–98

79. **Lopato S, Kalyna M, Dorner S, Kobayashi R, Krainer AR, Barta A. 1999. atSRp30, one of two SF2/ASF-like proteins from *Arabidopsis thaliana*, regulates splicing of specific plant genes. *Genes Dev.* 13:987–1001**

80. Lopato S, Mayeda A, Krainer AR, Barta A. 1996. Pre-mRNA splicing in plants: Characterization of Ser/Arg splicing factors. *Proc. Natl. Acad. Sci. USA* 93:3074–79

81. Lopato S, Waigmann E, Barta A. 1996. Characterization of a novel arginine/serine-rich splicing factor in *Arabidopsis*. *Plant Cell* 8:2255–64

82. Lorkovic ZJ, Barta A. 2004. Compartmentalization of the splicing machinery in plant cell nuclei. *Trends Plant Sci.* 9:565–68

83. Lorkovic ZJ, Hilscher J, Barta A. 2004. Use of fluorescent protein tags to study nuclear organization of the spliceosomal machinery in transiently transformed living plant cells. *Mol. Biol. Cell.* 15:3233–43

84. Lorkovic ZJ, Lehner R, Forstner C, Barta A. 2005. Evolutionary conservation of minor U12-type spliceosome between plants and humans. *RNA* 11:1095–1107

85. Lorkovic ZJ, Lopato S, Pexa M, Lehner R, Barta A. 2004. Interactions of *Arabidopsis* RS domain containing cyclophilins with SR proteins and U1 and U11 snRNP-specific proteins suggest their involvement in premRNA splicing. *J. Biol. Chem.* 279:33890–98

86. Lorkovic ZJ, Wieczorek Kirk DA, Klahre U, Hemmings-Mieszczak M, Filipowicz W. 2000. RBP45 and RBP47, two oligouridylate-specific hnRNP-like proteins interacting with poly(A)+ RNA in nuclei of plant cells. *RNA* 6:1610–24

87. Lorkovic ZJ, Wieczorek Kirk DA, Lambermon MH, Filipowicz W. 2000. Pre-mRNA splicing in higher plants. *Trends Plant Sci.* 5:160–67

88. Luehrsen KR, Taha S, Walbot V. 1994. Nuclear premRNA processing in higher plants. *Nucleic Acids Res.* 47:149–93

89. **Macknight R, Duroux M, Laurie R, Dijkwel P, Simpson G, Dean C. 2002. Functional significance of the alternative transcript processing of the *Arabidopsis* floral promoter FCA. *Plant Cell* 14:877–88**

90. Maquat LE. 2004. Nonsense-mediated mRNA decay: splicing, translation and mRNP dynamics. *Nat. Rev. Mol. Cell Biol.* 5:89–99

91. Marrs KA, Walbot V. 1997. Expression and RNA splicing of the maize glutathione S-transferase Bronze2 gene is regulated by cadmium and other stresses. *Plant Physiol.* 113:93–102

92. McCullough AJ, Baynton CE, Schuler MA. 1996. Interactions across exons can influence splice site recognition in plant nuclei. *Plant Cell* 8:2295–307

93. McCullough AJ, Schuler MA. 1997. Intronic and exonic sequences modulate 5′ splice site selection in plant nuclei. *Nucleic Acids Res.* 25:1071–77

74. Authors performed mutational analyses of U12 introns of three different *Arabidopsis* genes and identified the intronic elements that are important for splicing efficiency of U12 introns.

79. First paper to address in vivo roles of an SR protein, showing that ectopic expression of SR30 causes many developmental defects.

89. Shows that alternative processing of the *FCA* transcript regulates the amount of FCA protein and that the level of the FCA protein controls flowering time.

94. Modrek B, Lee C. 2002. A genomic view of alternative splicing. *Nat. Genet.* 30:13–19
95. Nagasaki H, Arita M, Nishizawa T, Suwa M, Gotoh O. 2005. Species-specific variation of alternative splicing and transcriptional initiation in six eukaryotes. *Gene* 364:53–62
96. Nagasaki H, Arita M, Nishizawa T, Suwa M, Gotoh O. 2006. Automated classification of alternative splicing and transcriptional initiation and construction of visual database of classified patterns. *Bioinformatics* 22:1211–16
97. Ner-Gaon H, Halachmi R, Savaldi-Goldstein S, Rubin E, Ophir R, Fluhr R. 2004. Intron retention is a major phenomenon in alternative splicing in *Arabidopsis*. *Plant J.* 39:877–85
98. Palusa SG, Ali GS, Reddy ASN. 2006. Alternative splicing of premRNAs of *Arabidopsis* serine/arginine-rich proteins and its regulation by hormones and stresses. *Plant J.* In press
99. Pendle AF, Clark GP, Boon R, Lewandowska D, Lam YW, et al. 2005. Proteomic analysis of the *Arabidopsis* nucleolus suggests novel nucleolar functions. *Mol. Biol. Cell* 16:260–69
100. Petracek ME, Nuygen T, Thompson WF, Dickey LF. 2000. Premature termination codons destabilize ferredoxin-1 mRNA when ferredoxin-1 is translated. *Plant J.* 21:563–69
101. Quesada V, Macknight R, Dean C, Simpson GG. 2003. Autoregulation of FCA premRNA processing controls *Arabidopsis* flowering time. *EMBO J.* 22:3142–52
102. Razem FA, El-Kereamy A, Abrams SR, Hill RD. 2006. The RNA-binding protein FCA is an abscisic acid receptor. *Nature* 439:290–94
103. Reddy ASN. 2004. Plant serine/arginine-rich proteins and their role in premRNA splicing. *Trends Plant Sci.* 9:541–47
104. Reddy ASN. 2001. Nuclear pre-mRNA splicing in plants. *Crit. Rev. Plant Sci.* 20:523–71
105. Deleted in proof
106. Reddy ASN, Ali GS, Golovkin M. 2004. *Arabidopsis* U1snRNP-70K protein and its interacting proteins: Nuclear localization and in vivo dynamics of a novel plant-specific serine/arginine-rich protein. In *The Nuclear Envelope*, ed. D Evans, J Bryant, pp. 279–95. Oxford: BIOS Sci.
107. Rundle SJ, Zielinski RE. 1991. Organization and expression of two tandemly oriented genes encoding ribulosebisphosphate carboxylase/oxygenase activase in barley. *J. Biol. Chem.* 266:4677–85
108. **Savaldi-Goldstein S, Aviv D, Davydov O, Fluhr R. 2003. Alternative splicing modulation by a LAMMER kinase impinges on developmental and transcriptome expression. *Plant Cell* 15:926–38**
109. Savaldi-Goldstein S, Sessa G, Fluhr R. 2000. The ethylene-inducible PK12 kinase mediates the phosphorylation of SR splicing factors. *Plant J.* 21:91–96
110. Schuler MA. 1998. Plant premRNA splicing. In *A Look Beyond Transcription: Mechanisms Determining mRNA Stability and Translation in Plants*, ed. J Bailey-Serres, DR Gallie, pp. 1–19. Rockville, MD: Am. Soc. Plant Physiol.
111. Shen H, Green MR. 2004. A pathway of sequential arginine-serine-rich domain-splicing signal interactions during mammalian spliceosome assembly. *Mol. Cell* 16:363–73
112. Shen H, Green MR. 2006. RS domains contact splicing signals and promote splicing by a common mechanism in yeast through humans. *Genes Dev.* 20:1755–65
113. Shi H, Xiong L, Stevenson B, Lu T, Zhu JK. 2002. The *Arabidopsis* salt overly sensitive 4 mutants uncover a critical role for vitamin B6 in plant salt tolerance. *Plant Cell* 14:575–88
114. Shukla GC, Padgett RA. 1999. Conservation of functional features of U6atac and U12 snRNAs between vertebrates and higher plants. *RNA* 5:525–38
115. Simpson CG, Jennings SN, Clark GP, Thow G, Brown JW. 2004. Dual functionality of a plant U-rich intronic sequence element. *Plant J.* 37:82–91

108. Authors overexpressed a LAMMER-type protein kinase that phosphorylates SR proteins and demonstrated that it alters alternative splicing of specific endogenous genes and the transcriptome profile and causes many developmental abnormalities.

116. Simpson CG, McQuade C, Lyon J, Brown JW. 1998. Characterization of exon skipping mutants of the COP1 gene from *Arabidopsis*. *Plant J.* 15:125–31
117. Simpson GG, Filipowicz W. 1996. Splicing of precursors to mRNA in higher plants: mechanism, regulation, and subnuclear organization of the spliceosomal machinery. *Plant Mol. Biol.* 32:1–41
118. Soergel DAW, Lareau LF, Brenner SE. 2006. Regulation of gene expression by coupling of alternative splicing and NMD. In *Nonsense-Mediated mRNA Decay*, ed. LE Maquat, pp. 175–96. Austin, TX: Landes Bioscience. **http://www.Eurekah.com**
119. Staiger D, Zecca L, Wieczorek Kirk DA, Apel K, Eckstein L. 2003. The circadian clock regulated RNA-binding protein AtGRP7 autoregulates its expression by influencing alternative splicing of its own premRNA. *Plant J.* 33:361–71
120. Stamm S, Ben-Ari S, Rafalska I, Tang Y, Zhang Z, et al. 2005. Function of alternative splicing. *Gene* 344:1–20
121. Sterner DA, Carlo T, Berget SM. 1996. Architectural limits on split genes. *Proc. Natl. Acad. Sci. USA* 93:15081–85
122. Talerico M, Berget SM. 1994. Intron definition in splicing of small *Drosophila* introns. *Mol. Cell. Biol.* 14:3434–45
123. Tillemans V, Dispa L, Remacle C, Collinge M, Motte P. 2005. Functional distribution and dynamics of *Arabidopsis* SR splicing factors in living plant cells. *Plant J.* 41:567–82
124. To KY, Suen DF, Chen SC. 1999. Molecular characterization of ribulose-1,5-bisphosphate carboxylase/oxygenase activase in rice leaves. *Planta* 209:66–76
125. Wang BB, Brendel V. 2004. The ASRG database: identification and survey of *Arabidopsis thaliana* genes involved in premRNA splicing. *Genome Biol.* 5:R102
126. Wang BB, Brendel V. 2006. Genomewide comparative analysis of alternative splicing in plants. *Proc. Natl. Acad. Sci. USA* 103:7175–80
127. Werneke JM, Chatfield JM, Ogren WL. 1989. Alternative mRNA splicing generates the two polypeptides in spinach and *Arabidopsis*. *Plant Cell* 1:815–25
128. Wilkie GS, Dickson KS, Gray NK. 2003. Regulation of mRNA translation by 5′- and 3′-UTR-binding factors. *Trends Biochem. Sci.* 28:182–88
129. Xiong L, Gong Z, Rock CD, Subramanian S, Guo Y, et al. 2001. Modulation of abscisic acid signal transduction and biosynthesis by an Sm-like protein in *Arabidopsis*. *Dev. Cell* 1:771–81
130. Yi Y, Jack T. 1998. An intragenic suppressor of the *Arabidopsis* floral organ identity mutant apetala3–1 functions by suppressing defects in splicing. *Plant Cell* 10:1465–77
131. Yoshimura K, Yabuta Y, Ishikawa T, Shigeoka S. 2002. Identification of a *cis* element for tissue-specific alternative splicing of chloroplast ascorbate peroxidase premRNA in higher plants. *J. Biol. Chem.* 277:40623–32
132. Yu J, Hu SN, Wang J, Wong GK, Li B, et al. 2002. A draft sequence of the rice genome (Oryza sativa L. ssp. indica). *Science* 296:79–92
133. Zanetti ME, Chang IF, Gong F, Galbraith DW, Bailey-Serres J. 2005. Immunopurification of polyribosomal complexes of *Arabidopsis* for global analysis of gene expression. *Plant Physiol.* 138:624–35
134. Zhang N, Kallis RP, Ewy RG, Portis ARJ. 2002. Light modulation of Rubisco in *Arabidopsis* requires a capacity for redox regulation of the larger Rubisco activase isoform. *Proc. Natl. Acad. Sci. USA* 99:3330–34
135. Zhang N, Portis ARJ. 1999. Mechanism of light regulation of Rubisco: a specific role for the larger Rubisco activase isoform involving reductive activation by thioredoxin-f. *Proc. Natl. Acad. Sci. USA* 96:9438–43

119. Reports that overexpression of an hnRNP protein changes alternative splicing of its own pre-mRNA and pre-mRNA of another hnRNP protein, indicating a function for plant hnRNPs in alternative splicing.

126. Using all available cDNAs/ESTs of *Arabidopsis* and rice and computational tools to perform a global analysis of alternative splicing in plants, authors revealed that a large fraction of genes in flowering plants are alternatively spliced.

127. Using peptide mapping and amino acid sequence analysis, authors show that two splice variants of ribulosebisphosphate carboxylase/oxygenase activase encode two distinct proteins.

131. A conserved sequence element in an intron of chloroplast ascorbate peroxidase gene, which produces four splice variants, was identified. Deletion of this sequence element reduced splicing efficiency.

136. Zhang XC, Gassmann W. 2003. RPS4-mediated disease resistance requires the combined presence of RPS4 transcripts with full-length and truncated open reading frames. *Plant Cell* 15:2333–42
137. Zhang Z, Komatsu S. 2000. Molecular cloning and characterization of cDNAs encoding two isoforms of ribulose-1,5-bisphosphate carboxylase/oxygenase activase in rice (*Oryza sativa* L.). *J. Biochem.* 128:383–89
138. Zhou Y, Zhou C, Ye L, Dong J, Xu H, et al. 2003. Database and analyses of known alternatively spliced genes in plants. *Genomics* 82:584–95
139. Zhu W, Brendel V. 2003. Identification, characterization and molecular phylogeny of U12-dependent introns in the *Arabidopsis thaliana* genome. *Nucleic Acids Res.* 31:4561–72
140. Zhu W, Schlueter SD, Brendel V. 2003. Refined annotation of the *Arabidopsis* genome by complete expressed sequence tag mapping. *Plant Physiol.* 132:469–84

The Production of Unusual Fatty Acids in Transgenic Plants

Johnathan A. Napier

Rothamsted Research, Harpenden, Hertfordshire Al5 2JQ, United Kingdom; email: johnathan.napier@bbsrc.ac.uk

Annu. Rev. Plant Biol. 2007. 58:295–319

The *Annual Review of Plant Biology* is online at plant.annualreviews.org

This article's doi:
10.1146/annurev.arplant.58.032806.103811

1543-5008/07/0602-0295$20.00

Key Words

desaturase, elongase, industrial fatty acids, polyunsaturated fatty acids, seeds, triacylglycerol

Abstract

The ability to genetically engineer plants has facilitated the generation of oilseeds synthesizing non-native fatty acids. Two particular classes of fatty acids are considered in this review. First, so-called industrial fatty acids, which usually contain functional groups such as hydroxyl, epoxy, or acetylenic bonds, and second, very long chain polyunsaturated fatty acids normally found in fish oils and marine microorgansims. For industrial fatty acids, there has been limited progress toward obtaining high-level accumulation of these products in transgenic plants. For very long chain polyunsaturated fatty acids, although they have a much more complex biosynthesis, accumulation of some target fatty acids has been remarkably successful. In this review, we consider the probable factors responsible for these different outcomes, as well as the potential for further optimization of the transgenic production of unusual fatty acids in transgenic plants.

Contents

LA: linoleic acid

ALA: α-linolenic acid

TAG: triacylglycerol

INTRODUCTION

Collectively, higher plants display enormous variation in the fatty acids they synthesize, with current estimates exceeding well over 300 different examples (3). This is in marked contrast to animals, which collectively synthesize a far smaller range of fatty acids (46). However, although the Plant Kingdom contains great diversity, the number of fatty acids that are common to all plants is relatively low and is represented by the saturated fatty acids such as palmitic acid (16:0) and stearic acid (18:0) and unsaturated (i.e., containing a double bond) fatty acids such as oleic acid (18:1Δ9), linoleic acid (LA) (18:2Δ9, 12), and α-linolenic acid (ALA) (18:3Δ9,12,15) (14).[1] Therefore, it is possible to class any other fatty acid as "unusual," with such deviation from the norm due to differences in chain length (i.e., greater than 18 carbons, shorter than 14), position, or number of double bonds (i.e., polyunsaturated), or due to the presence of modifications other than simple double bonds (i.e., hydroxylation, epoxy groups, etc.) (112). However, for the purpose of this review we focus only on a limited subset of the heterogeneity present in plant fatty acid diversity. We examine attempts to use genetic engineering to transfer genes from one plant (or non-plant) species to another, with the aim of accumulating unusual fatty acids in a more agriculturally amenable oilseed (39, 110, 138). We also discuss how such approaches usually result in suboptimal accumulation of the target fatty acid, which speaks volumes about the complexity of plant lipid biochemistry and also our understanding of it (2, 37, 57, 76, 117). In particular, we examine attempts to produce two different generic types of unusual fatty acids in transgenic plants, namely (*a*) industrial fatty acids such as the hydroxylated fatty acid ricinoleic acid and (*b*) very long chain polyunsaturated fatty acids similar to those found in fish oils.

WHY TRY AND MODIFY PLANT OIL COMPOSITION?

Although the higher plants that synthesize unusual fatty acids are dispersed across many different and unrelated taxa, one common factor unites many of them: the unusual fatty acids of interest are usually only found in the seed oils (144). In other words, the unusual fatty acid is only synthesized in developing seed tissues and accumulates in the seed storage lipid in the form of triacylglycerol (TAG) (3, 11). The accumulation of oils in seeds is a trait

[1] A note on fatty acid nomenclature: the length of the fatty acid, in number of carbons is designated by the first number, the number of double bonds by the digit after the colon. The position of the double bond(s) is defined by a Δ symbol, with the subsequent numbers indicating position. Thus, 18:1Δ9 is an eighteen-carbon fatty acid, containing one double bond between the ninth and tenth carbons.

that mankind has been exploiting for millennia, culminating in the selection and generation of a number of oilseed crops that are now grown around the world, mainly for nutritional purposes (60, 75). However, in plants that accumulate unusual fatty acids, most of these are not agronomically-adapted and are therefore unsuitable for modern agriculture, limiting the availability of their seed oils. Although it is possible to use conventional plant breeding techniques to select for lines that might show suitable agronomic performance, such a process is time-consuming and labor intensive (78, 110). The advent of plant genetic engineering over 20 years ago heralded the arrival of a new approach to plant breeding, allowing the direct introduction of the gene(s) encoding any trait of interest (61). Not only did this allow a much more precise incorporation of genetic information into a target plant (compared with the uncontrolled and variable transfer occurring during sexual recombination), but it also circumvented the need for sexual compatibility between two parental lines (78). Thus, it became possible to genetically engineer a plant with genes from any organism, not just a related plant species: This was demonstrated in early experiments by using selectable marker genes derived from bacteria (61). Thus, the potential of introducing novel traits into higher plant crop species allowed researchers to consider the concept of the "designer oilseed," in which traits not present in a current oilseed crop (such as soy, maize, brassicas) could be added by genetic engineering (77, 138). Such designer oilseeds might contain fatty acids that are useful for industrial processes, either as lubricants or as chemical feedstocks (57, 117), or might contain fatty acids of nutritional or beneficial health properties (14, 101, 142). The progress toward generating various designer or genetically improved oilseeds such as these is the focus of this review. In both cases, new pressures on diminishing natural resources make a strong case for their development and deployment.

PLANT LIPID METABOLISM

In addition to the arrival of plant transformation technologies, which facilitated the introduction and expression of any gene of interest, plant lipids have benefited from many years of research on their underlying biochemistry (46, 130, 131). In that respect, it is probably fair to say that our understanding of lipid metabolism in plants is further advanced than equivalent studies in animal or microbial systems. In addition to these pioneering biochemical studies on the synthesis of plant fatty acids, the isolation in the model plant *Arabidopsis thaliana* of genetic mutants defective in aspects of fatty acid metabolism has provided another powerful approach to unraveling the complexities of these pathways (87, 146). In particular, the *fad* (*fatty acid desaturation*) mutant series, which showed alteration to the levels of unsaturated fatty acids, has proved crucial for several reasons (122). First, it demonstrated that alterations to plant fatty acid composition could be mediated by genetic intervention, and that this need not result in severe perturbation to plant form and function (17, 68). Second, the *fad* [and others such as *fab* (151)] mutants confirmed earlier biochemical studies that indicated the role of both the plastid and the endoplasmic reticulum (ER) in plant fatty acid modification (18, 74). This is described as the prokaryotic and eukaryotic pathways, in which fatty acids are synthesized in the plastid (the prokaryotic compartment) and then either further modified in this organelle or exported to the ER for modification in this so-called eukaryotic pathway (73, 122). The presence of these two fatty acid–modifying pathways in most plant tissues and cells also explained (through functional redundancy) the viability of many *Arabidopsis* single *fad* mutants. Finally, the mapping and cloning of the genes disrupted by the *fad* mutations provided the molecular identity of enzymes that had previously proved difficult (if not impossible) to characterize by biochemical purification due to their membrane-bound status

FAD: fatty acid desaturase

(122). Although the soluble plastidial desaturase responsible for converting stearic acid to oleic acid (by introducing a double bond at the Δ9 position) was purified biochemically and protein sequence information was used to identify a candidate cDNA clone (113), limited progress had been made with similar studies on the microsomal (ER) or plastidial membrane-bound desaturases [although a notable exception to this was the purification of a plastidial omega-6 desaturase (111).] The genetic mapping and cloning of the FAD3 locus revealed the identity of this microsomal desaturase (5), and with the similar cloning of other fatty acid desaturase (FAD) genes (55, 68, 88) it became clear that polypeptide sequences of different desaturases from different organelles still shared some highly conserved motifs. In particular, eight histidine residues were conserved in three "histidine boxes" and shown to be absolutely required for enzyme activity via site-directed mutagenesis (6, 114). Subsequent cloning of FAD family orthologs from many different plant species confirmed the presence of these conserved histidine residues (112). Therefore, on the basis of these conserved motifs it became possible to identify candidate desaturase sequences from any suitable DNA sequence. This, in turn, enabled an alternative approach to laborious biochemical purification to identify plant membrane desaturases by mining genomic or cDNA [expressed sequence tag (EST)] databases for candidate genes for subsequent functional characterization via heterologous expression (22). Note that the histidine box motifs are conserved beyond just higher plants, and that this motif has also proved crucial in the identification of many animal and microbial desaturases (83, 125).

MODIFYING THE LIPID COMPOSITION OF PLANTS TO ACCUMULATE INDUSTRIAL FATTY ACIDS

The predominant fatty acids found in most oilseed crops do not, on the whole, have any particularly useful properties in terms of industrial applications, by which we predominantly mean use of these fatty acids as a replacement for petrochemicals (117). Given the growing realization that global petrochemical reserves are finite, the possibility of using transgenic plants to synthesize "oleochemicals" that could serve as petrochemical substitutes is one of considerable interest and significance. For example, the hydroxylated fatty acid ricinoleic acid (12-hydroxy-octadec-*cis*-9-enoic acid; 12-OH 18:1Δ9) can serve as a precursor for chemical conversion to many useful products, ranging from lubricants, emulsifiers, and inks to biodiesel formulation and nylon precursors (57, 69, 110). Ricinoleic acid is the major fatty acid in castor (*Ricinus communis*, from which the fatty acid derives its name), and the storage triacylglycerols of castor accumulate up to 90% ricinoleic acid (69, 145). Nevertheless, castor is not considered an agronomically suitable plant species, not least because of the similar seed-specific accumulation of the potent toxin ricin. Therefore, efforts have been expanded to identify the gene responsible for the synthesis of ricinoleic acid (through the C12 hydroxylation of oleic acid, i.e., hydroxylation at the twelfth carbon of this fatty acid), and the introduction of this new gene into a suitable oilseed crop by genetic engineering.

Biochemical characterization of the synthesis of ricinoleic acid revealed that this reaction was carried out in microsomal fractions and utilized oleic acid that was esterified to membrane phospholipids as its substrate (7, 120). In that respect, this requirement for phosphoglycerolipid substrates was similar to other higher plant fatty acyl modifying activities (119, 120); previous studies had shown that the enzymes equivalent to FAD2 and FAD3 (Δ12-desaturase, responsible for the synthesis of LA; Δ15-desaturase, responsible for the synthesis of ALA) utilized phosphoglycerolipid substrates (123, 124, 132, 133). Such membrane lipid-dependent desaturation is not generally considered to occur in animal or yeast systems, where it is assumed that substrates for fatty acid modification are in the form of acyl-CoAs (129). With castor, although ricinoleic acid accumulates to very high levels in TAGs, almost none is detected in the phospholipid membranes, even though this is the site of synthesis. Thus, it has been hypothesized that castor, in addition to the unusual C12 hydroxylase that is directly responsible for the synthesis of ricinoleic acid, must also contain a very efficient mechanism for channeling this unusual fatty acid from phospholipids to TAG (39, 69, 145).

The first identification of the C12-hydroxylase from castor was obtained by van de Loo et al. (143), who generated an EST sequence database of developing castor seed transcripts from which to identify a suitable candidate hydroxylase. The authors hypothesized that in view of the phospholipid substrate dependence of the C12-hydroxylase and the similar regiospecificity to the FAD2 Δ12 desaturase (i.e., both enzymes carry out reactions on the twelfth carbon of oleic acid), the castor enzyme may represent a diverged or paralogous form of FAD2, and hence show sequence similarity to that latter enzyme (143). Functional characterization of two closely related FAD2-like cDNAs from castor by expression in transgenic tobacco and *Arabidopsis* identified one as a *bona fide* FAD2 ortholog and the other as C12-hydroxylase (designated FAH12) (16, 143). These studies provided several very important insights. First, it demonstrated the feasibility of identifying fatty acid modifying activities by "paralogy," working on the hypothesis that an unusual activity may have resulted from divergence from a more prevalent archetype. Second, the levels of ricinoleic acid achieved in transgenic plants were very low (<1% in tobacco, 17% in *Arabidopsis*), even when the expression of the FAH12 hydroxylase was under the control of strong seed-specific promoters (16, 143). This second observation confirmed the earlier biochemical studies regarding substrate channeling from the site of synthesis to the site of storage (145), and indicated that although the primary biosynthesis of ricinoleic acid might only represent a single gene trait (i.e., FAH12), the accumulation of this fatty acid in TAGs most likely required additional secondary activities (117, 118). This observation also appears to be true for the accumulation of other fatty acids in transgenic plants (72).

The identification of the oleic acid C12-hydroxylase as a paralog of the FAD2 oleic acid Δ12 desaturase facilitated the identification of analogous enzymes from a number of plant species that accumulate unusual industrial fatty acids. For example, a FAD2-related epoxygenase cloned from *Crepis palaestina* generates up to 15% vernolic acid (12,13-epoxy-9-octadecanoic acid) when expressed in the seeds of transgenic *Arabidopsis* (63); this is in contrast to the ~60% vernolic acid found in seed oil of *C. palaestina*. Similarly, a FAD2-related acetylenase identified from *Crepis alpina* directs the synthesis of up to 25% crepenynic acid (9-octadecen-12-ynoic acid) in transgenic *Arabidopsis* (versus ~70% in *C. alpina*) (63). In addition, variant FAD2 activities that generate fatty acids with conjugated double bonds, useful as drying agents in inks and paints, have been identified. FAD2-like sequences isolated from *Morordica charantia* and *Impatiens balsamina* direct the synthesis of the conjugated fatty acids α-eleostearic acid (18:3Δ9*cis*,11*trans*,13*trans*) and α-parinaric acid (18:4Δ9*cis*,11*trans*,13*trans*,15*cis*) in transgenic soybean embryos, although only at a third of the levels observed in the gene-donor species (19). A FAD2-like conjugase identified from *Calendula officinalis* results in the accumulation of low levels of calendic acid (18:3Δ8*trans*,10*trans*,12*cis*) when expressed in yeast (23, 98) and a related enzyme from *Punica granatum* (pomegranate) has also been described (52, 56). More recently, two FAD2-like open reading frames (ORFs) were isolated from *Aleurites fordii* (Tung tree, rich in α-eleostearic acid) (38), one of which was shown by heterologous expression in yeast to encode a conjugase similar to those isolated from *M. charantia* and *I. balsimina*. In particular, the

FAD2-like activities from tung (designated FADX) and pomegranate were shown to be bi-functional, acting as both a (*trans*-) desaturase and a conjugase (38, 56). These observations have led to speculation as to the evolution of these variant forms of FAD2 in terms of the substrate specificity of the enzymes and also the amino acid determinants that define activity (e.g., desaturation vs. "unusual" activity") (38). In the case of the FAH12 hydroxylase, site-directed mutagenesis studies identified four residues that, if altered, can convert the FAD2 desaturase to a C12-hydroxylase, although a reciprocal conversion could only be achieved with the introduction of seven changes into FAH12 (13, 15). It is interesting to note that most of these functional determinant residues are in close proximity to the histidine box motifs (12). A similar conclusion as to the importance of residues in proximity to these motifs was arrived at by the studies on the tung conjugase mentioned above (38).

Another unusual fatty acid that was recently shown to be the product of FAD2-like activities is dimorphecolic acid (9-hydroxy-18:2Δ10*trans*,12*trans*), which contains both a hydroxyl group and conjugated double bonds. Two FAD2-like ESTs were obtained from *Dimorpotheca sinuate* (which typically accumulates ~60% of this unusual fatty acid in its seed oil) (21). One FAD2 variant, designated DsFAD2-1, was found to encode a Δ12*trans* desaturase, whereas the second variant (DsFAD2-2) converted this product into dimorphecolic acid by carrying out hydroxylation at the C9 position and also generating the double bond at the Δ10 position (21). Because the DsFAD2-2 enzyme appeared to have a strong preference for 18:2Δ9,12*trans* (i.e., the product of DsFAd2-1), this indicated the contribution of both activities to the synthesis of this unusual fatty acid. In line with the transgenic accumulation of unusual fatty acids derived from variant FAD2-like activities, very low levels of dimorphecolic acid were detected in transgenic soybean embryos on coexpression of DsFAD2-1 and DsFAD2-2 (21).

OPTIMIZING THE ACCUMULATION OF VARIANT FAD2-DERIVED FATTY ACIDS: WHAT'S THE PROBLEM?

Collectively, the accumulation of hydroxylated, epoxygenated, conjugated, or acetylenated fatty acids in transgenic plants has been relatively disappointing, not least of all because of the high levels that these unusual fatty acids accumulate to in the gene-donor species (57). In addition to the low levels of target fatty acids in transgenic plants, expression of FAD2-like activities also often result in elevated levels of oleic acid (116, 117, 139). This was initially hypothesized to be a result of negative physical interactions and/or titration between the endogenous FAD2 and the related transgene FAD2-like activity, resulting in the decrease in oleate desaturation (116). However, this was formally disproved by the observation that expression of a structurally unrelated C12 epoxygenase cytochrome P450 from Euphorbia resulted in elevated levels of oleic acid as well as low levels of vernolic acid (24). Leaving aside this intriguing example of convergent evolution in the biosynthesis of unusual fatty acids, these data clearly demonstrated that the presumptive inhibition of FAD2 Δ12-desaturase activity was not mediated through a homology-dependent process (24, 117). Given that FAD2 and FAD2-like activities utilize membrane-linked substrates, another reason for impaired synthesis may result from inhibition of endogenous FAD2 activity via the non-native products of the transgene FAD2-like enzymes. In such a scenario, unusual fatty acids such as ricinoleic acid accumulate at their site of synthesis in microsomal phospholipids, resulting in either the inhibition of proximal FAD2 or reducing the flux of oleic acid available to FAD2 for desaturation. Studies on transgenic *Arabidopsis* plants expressing the C12 hydroxylase have revealed the presence of ricinoleic acid in seed phospholipid membranes (20), unlike the situation in castor where almost no trace of this fatty acid is present in phospholipids (145). It therefore seems likely that some form of "editing" activity is required to remove the products of FAD2-like activities, and such editing enzymes are not likely present in plant species that do not accumulate unusual fatty acids (72, 145). Because many of these FAD2-like derived unusual fatty acids contain highly chemically active groups, it would make physiological sense for such fatty acids to be removed from membrane bilayers (8) such as those present in the ER and compartmentalized in storage lipid TAGs (67). There is also evidence indicating that some plant species synthesize acetylenated fatty acids via diverged FAD2 activities in response to fungal infection, presumably utilizing these unusual fatty acids as part of a defense response (25).

An additional consideration in terms of optimizing the synthesis of this class of unusual fatty acid has been investigated in several studies in which competing activities for substrate fatty acids were removed. The accumulation of ricinoleic acid in transgenic *Arabidopsis* seeds was doubled by expressing the castor FAH12 hydroxylase in a *fad2/fae1* mutant background (resulting in significant elevation of the oleic acid content) (121). However, a quite similar increase in the accumulation of ricinoleic acid was observed in the *fad3* mutant, even though this mutation (a disruption of the microsomal Δ15 desaturase) does not alter the levels of oleic acid (121). Thus, there appears to be no direct relationship between the levels of substrate (oleic acid) and FAH12 product (ricinoleic acid), even though both mutations enhance the accumulation of this unusual fatty acid. Another interesting observation is that independent of the background used (WT, *fad2*, *fad3*) maximal levels of ricinoleic acid accumulation in seed lipids never exceeded 20% of total fatty acids, implying the presence of some additional factors that limit accumulation (121).

Analogous Experiments to Define the Substrate

Product relationships have been carried out for the *C. paleastina* Δ12 epoxygenase Cpal2, which utilizes LA for the synthesis of vernolic acid. Coexpression of Cpal2 with the *C. paleastina* FAD2 resulted in a 1.4-fold increase in the level of vernolic acid in *Arabidopsis* seeds to a maximal level of 9.9% of total fatty acid (116). This increase was attributed to the capacity provided by the transgene *C. paleastina* FAD2 to synthesize additional LA. Reciprocal experiments were carried out using the *Arabidopsis* triple mutant *fad3/fad7-1/fad8*, which is devoid of all Δ15-desaturase activity and therefore contains high levels of LA. Expression of Cpal2 in the *fad3/fad7-1/fad8* mutant resulted in a 1.6-fold increase in the accumulation of vernolic acid compared with wild-type background, although the overall levels (3.6%) of the target fatty acid were lower than those previously observed (100).

Perhaps of greater significance, a recent study (156) combined both of these approaches, coexpressing the Cpal2 epoxygenase with the *C. paleastina* FAD2 Δ12-desaturases in the *Arabidopsis fad3/fae1* double mutant (defective in microsomal Δ15 desaturation and also oleic acid elongation, hence containing elevated ALA). This resulted in a large increase in the overall levels of vernolic acid (up to 21% of total seed fatty acids), which represents a significant improvement on previous attempts to synthesize this fatty acid in transgenic plants. Two additional important observations were made in this study. First, the *C. paleastina* FAD2 Δ12-desaturase could be replaced with an equivalent activity from another plant species without reducing the level of vernolic acid (156). Thus, the fact that the elevated levels of 21% were achieved with a "matched pair" of activities from one organism (i.e., *C. palaestina*) was apparently coincidental. Second, coexpression of the Cpal2 epoxygenase and CpFAD2 in transgenic cotton seeds resulted in relatively similar high levels (16.9%) of vernolic acid, with this depending on the presence of the transgene FAD2 Δ12-desaturase, confirming the importance of this additional activity contributing to the synthesis of substrate LA. It is also interesting to note that expression of

VLC-PUFA: very long chain polyunsaturated fatty acids

EPA: eicosapentaenoic acid

DHA: docosahexaenoic acid

ARA: arachidonic acid

the Cpal2 epoxygenase in cotton resulted in only modest increases in oleic acid levels; this would imply that cotton has superior endogenous activities for the editing of unusual fatty acids from phospholipids, relieving the build-up of oleate.

In a situation that appears similar to attempts to produce unusual monounsaturated fatty acids in transgenic plants (137), the collective attempts to produce "industrial" fatty acids in transgenic plants have so far failed to replicate the levels observed in the non-agronomic plants that accumulate them to high levels (57). However, progress is being made toward the optimization of their transgene-mediated synthesis, and a number of key factors are now emerging. These are: (*a*) substrate level and availability, (*b*) editing/removal of the non-native fatty acid from the site of synthesis in phospholipid membranes, and (*c*) efficient incorporation into TAGs. Strategies that might allow further increases to the currently reported levels of ricinoleic or vernolic acid include the identification and coexpression of suitable acyl-exchange editing activities (8, 118, 127, 128) or unusual fatty acid–specific TAG biosynthetic activities such as diacylglycerol:acytransferase (DGAT) (48, 69, 118, 145) in conjunction with the use of RNAi approaches to modulate the endogenous competing activities (e.g., desaturases etc). A very recent study on tung indicated the presence of two discrete forms of DGAT, leading to speculation that one form (DGAT2) may represent a dedicated TAG-biosynthetic activity for unusual fatty acids (115). Alternatively, it may be possible to take unbiased approaches to identifying key factors in host plants that facilitate the accumulation of unusual fatty acids. For example, proteomic approaches have been used to try and identify native plant factors in the microsomal fraction of castor (66). A genetic approach has also been undertaken, in which an *Arabidopsis* line expressing the FAH12 hydroxylase was transformed with a library of castor cDNAs and the resulting population of transgenic plants screened for individual lines that showed elevated accumulation of ricinoleic acid (65). From screening ~4000 lines, three transgenic plants coexpressing a castor cDNA and FAH12 showed a small but significant increase in the target fatty acid. These three cDNAs were found to encode a scaffold-attachment region protein, oleosin, and a phosphatidylethanolamine-binding protein, although their precise role in ricinoleic acid synthesis and accumulation remains to be defined (65).

TRANSGENIC SYNTHESIS OF VERY LONG CHAIN POLYUNSATURATES: MAKING FISH OILS IN PLANTS

Unlike the industrial fatty acids described above, very long chain polyunsaturated fatty acids (VLC-PUFAs, defined as C_{20+} fatty acids containing three or more double bonds) are almost completely absent from higher plants (84, 147). The primary biosynthesizing sources of VLC-PUFAs are marine microbes such as algae, which form the base of an aquatic food web that culminates in the accumulation of these fatty acids in fish oils (33, 89, 149). In addition, some fungi and lower plants have a capacity to synthesize VLC-PUFAs, and animals can convert dietary fatty acids such as LA and ALA to these more complex forms (43, 108, 147). Much interest has recently been focused on the importance of omega-3 VLC-PUFAs such as eicosapentaenoic acid (EPA) (20:5Δ5,8,11,14,17) and docosahexaenoic acid (DHA) (22:6Δ4,7,10,13,16,19) as cardiovascular-protective components of the human diet (26). It is also known that the omega-6 VLC-PUFA arachidonic acid (ARA) (20:4Δ5,8,11,14) and DHA are important for optimal neonatal health and development (30, 84, 149), and because of this these two VLC-PUFAs are now standard additions to infant formula milks.

However, sources of VLC-PUFAs for human health and nutrition are becoming problematic. In omega-3 VLC-PUFAs (sometimes referred to as "oceanic" fatty acids), current sources, in the form of fish stocks, are in severe decline due to decades of overfishing (33, 49, 81). In addition to this non-sustainable exploitation of a natural resource, environmental pollution of the oceans has led to reports of accumulation of toxic compounds such as heavy metals, dioxins, and plasticizers in fish oils (101). In omega-6 ARA, the current predominant source of this VLC-PUFA is via fermentative culture of the filamentous fungi *Morteriella alpina* (152). However, such production systems are expensive to maintain and have limited flexibility to significantly increase capacity. In view of all these factors, there has been considerable interest in the production of VLC-PUFAs in transgenic plants to provide a sustainable, clean, and cheap source of these fatty acids (37, 39, 60, 85, 141).

The biosynthesis of VLC-PUFAs is catalysed by an alternating sequence of fatty acid desaturation and elongation and therefore requires two distinct types of primary biosynthetic activities (desaturases and elongating activities, also known as elongases). Thus, transgenic production of VLC-PUFAs is inherently more complicated than the situation for industrial fatty acids, although both systems have a number of shared problems, as discussed below. The biosynthetic pathway for VLC-PUFAs has been extensively reviewed over the past few years (54, 64, 83, 95, 108) and genes encoding all the primary biosynthetic activities have been isolated from many different organisms. The primary activities required for converting LA and ALA to either ARA, EPA, or DHA are shown in **Figure 1**. The possibility of producing VLC-PUFAs in transgenic plants became clear from the earliest attempts to characterize the activities required for this trait. Δ6-desaturation represents the first committed step on the VLC-PUFA biosynthetic pathway, and expression of a Δ6-desaturase isolated from a cyanobacteria resulted in the accumulation of low levels of γ-linolenic acid (GLA) (18:3Δ6,9,12) and stearidonic acid (SDA) (18:4Δ6,9,12,15) in transgenic tobacco plants (99). Much higher accumulation of GLA and SDA (combined Δ6-desaturated fatty acids ~20% of total) was obtained by expression of a *Borago officinalis* (common borage) Δ6-desaturase (borage being one of the few plant species that can synthesize GLA) (109, 148). Although initial experiments were carried out using a constitutive promoter, subsequent studies have shown very high levels (up to 40%) of GLA and/or SDA as a result of seed-specific expression (103). In the cyanbacterial Δ6-desaturase, low activity was most likely attributed to a requirement for (plastidial) ferredoxin as an electron donor, but a microsomal (ER) location for the transgene product (2, 37, 86). Results from the expression of the borage Δ6-desaturase highlighted a number of points. First, unlike the industrial fatty acids such as ricinoleic, crepynenic, and vernolic acids, GLA could accumulate to levels similar (or greater) to those found in the gene-donor species (28, 41, 45, 51, 96, 155). Second, in a manner similar to the FAD2-like desaturases, the microsomal Δ6-desaturase utilizes phospholipid-linked substrates (34, 135), indicating that the presence of Δ6-desaturated fatty acids is either tolerated in phospholipid membranes or easily recognized by endogenous editing enzymes (104). Third, coexpression of Δ6-desaturases from plant or fungal sources with FAD2 Δ12-desaturases resulted in elevated levels of GLA (53), whereas coexpression with FAD3 Δ15-desaturase resulted in high levels of SDA (40). This latter point highlights the unselective nature of the Δ6-desturase in terms of either omega-6 (LA) or omega-3 (ALA) substrates, and the importance of the fatty acid composition of the transgenic host plant.

GLA: γ-linolenic acid

SDA: stearidonic acid

As shown in **Figure 1**, biosynthesis of the two C_{20} VLC-PUFAs ARA and EPA requires three primary enzyme activities: (*a*) Δ6-desaturase, (*b*) Δ6-elongase, and (*c*) Δ5-desaturase. Several studies have now evaluated the cumulative performance of these activities from different species in transgenic plants, resulting in the accumulation of ARA and EPA to varying levels. Abbadi et al. (1)

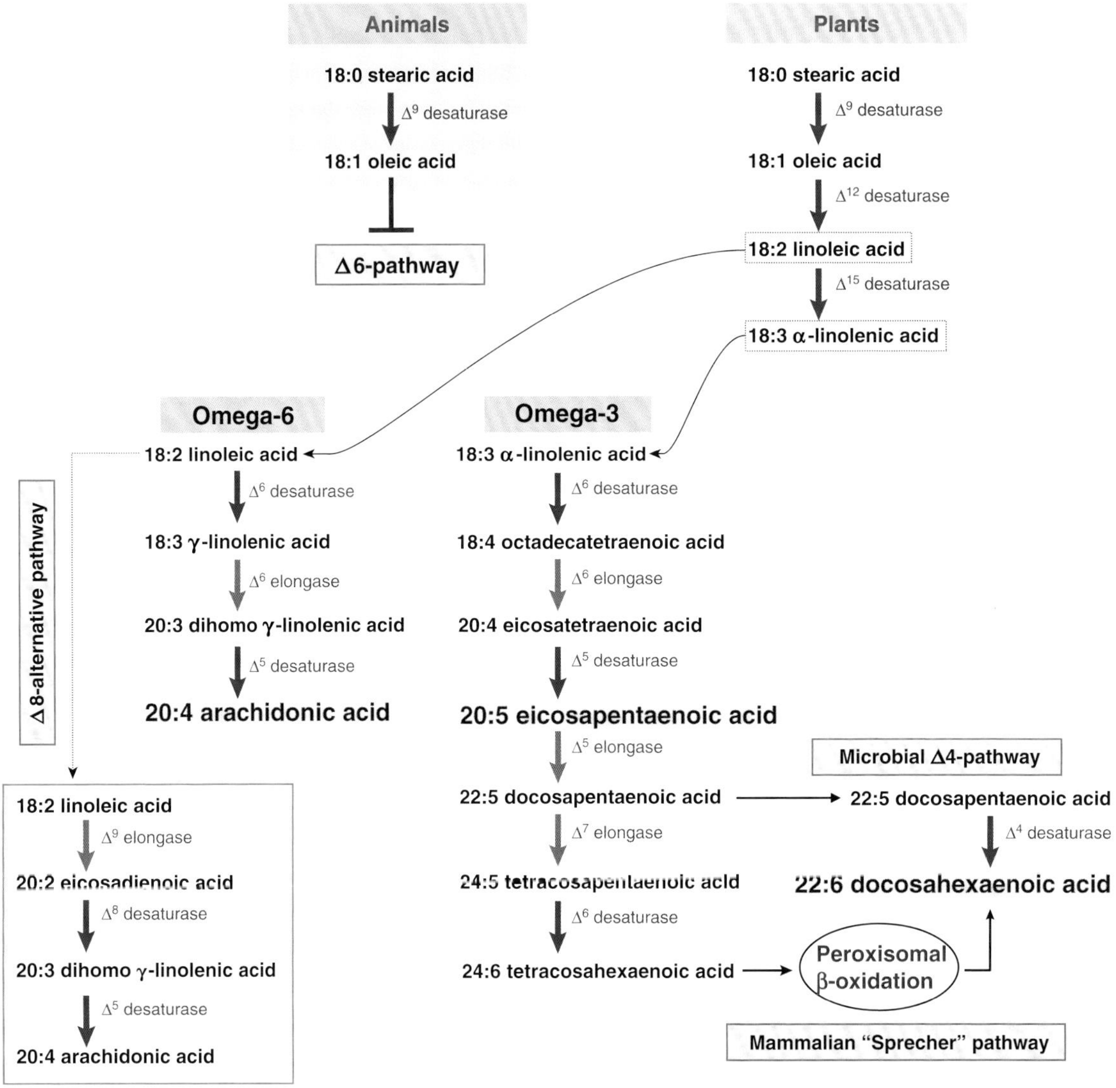

Figure 1

Schematic representation of VLC-PUFA biosynthesis. The various routes for synthesis of arachidonic acid (ARA), eicosapentaenoic acid (EPA), and docosahexaenoic acid (DHA) are shown, as mediated by the key enzyme activities (desaturases, elongases). The predominant Δ6-pathway (via the Δ6-desaturase) is shown, as is the alternative Δ8-pathway. Two routes for DHA synthesis are shown, although only the simpler Δ4-desaturase pathway is described in relation to transgenic plant expression.

coexpressed algal desaturases (36) and a moss elongase (153) in transgenic tobacco and linseed. The rationale for this was that tobacco seeds contain high levels of LA, whereas linseed contains high levels of ALA. As outlined above, this should facilitate the accumulation of ARA in tobacco and EPA in linseed. However, although EPA and ARA were synthesized in transgenic plants, these target fatty acids were present at a very low level (<1%) (1). In contrast, Δ6-desaturation products accumulated at very high levels (~25% of total) (1), indicating that the pathway was stalled after the first committed reaction (Δ6-desaturation)

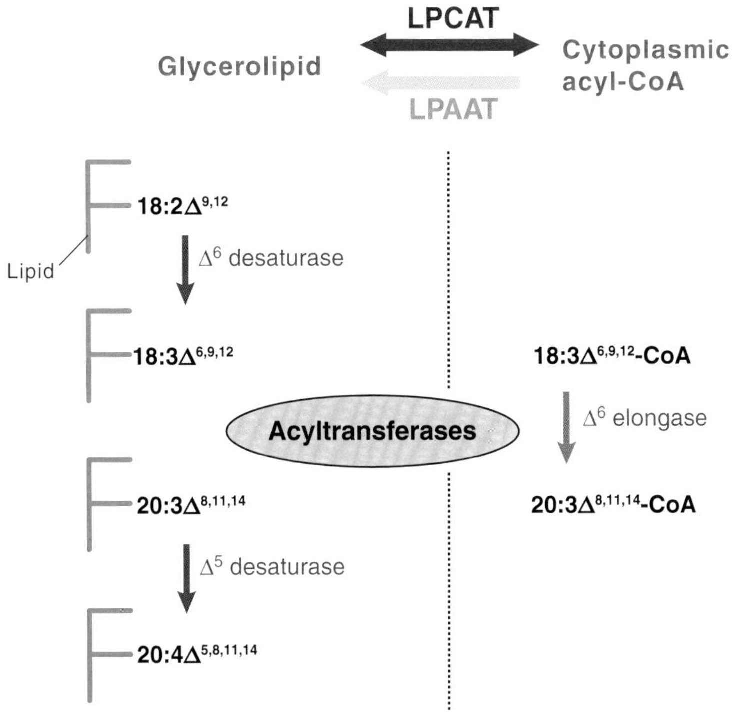

Figure 2

Schematic representation of the substrate-dichotomy bottleneck. Nonanimal fatty acid desaturation uses glycerolipid-linked substrates, whereas fatty acid elongation requires acyl-CoA substrates. The exchange of the fatty acids between phospholipids and the acyl-CoA pool is an enzyme-mediated process (via acyltransferases). Non-native fatty acids (i.e., the products of transgenic VLC-PUFA activities) may not be efficiently exchanged between these two metabolically active pools.

(**Figure 1**). Further detailed biochemical analysis of these transgenic linseed lines indicated several constraints that were preventing the synthesis of EPA. First, there appeared to be limited acyl exchange from the phospholipid site of Δ6-desaturation into the acyl-CoA pool, as determined by an absence of Δ6-desaturated fatty acids such as GLA and SDA in the acyl-CoA pool, but their abundance in phospholipids (1, 33, 34, 85). Fatty acid elongation requires acyl-CoA substrates (64), and it is therefore essential to have efficient acyl exchange between phospholipids and the acyl-CoA pool to maintain a suitable flux of substrates for elongation (1, 85). This requirement represents a generic bottleneck in VLC-PUFA biosynthesis, and we have described this as "substrate-dichotomy" (85) because the two key enzyme activities require different acyl substrates (illustrated in **Figure 2**). The problem of generic substrate dichotomy has been discussed at length by us (84, 85) and others (33, 34, 101, 117, 141), and some potential solutions are outlined below.

In addition to this generic problem, detailed analysis of lipid species present in the transgenic linseed lines also indicated the likelihood of species-specific factors that limited the potential to synthesize EPA in linseed. These implied the presence of a strong acyl-CoA-independent phospholipid:diacylglycerol acyltransferase (PDAT) activity (31), which channeled fatty acids directly from their site of desaturation on phosphatidylcholine (PC) into TAG, compartmentalizing them away from the VLC-PUFA biosynthetic activities (11, 31). It was also hypothesized that linseed lacked endogenous acyl-exchange activities that recognized non-native fatty acids such as GLA and SDA as substrates (33, 85). Although this latter point might be considered a generic bottleneck, other studies have demonstrated the strong influence of host plant species on the successful synthesis of ARA and EPA, presumptively through some native activities that can overcome the substrate dichotomy problem. For example, transgenic soy and *Brassica juncea* (Indian mustard) have been transformed with the same biosynthetic activities as used in tobacco and linseed, but with much greater success, resulting in ~15–20% EPA in seed oils (59, 150). Some of this enhanced performance of the transgenic VLC-PUFA pathway could possibly be ascribed to "superior" forms of the primary biosynthetic activities (50, 51), and in both soy and *B. juncea* additional secondary activities were introduced (such as a FAD2 Δ12-desaturase and ω3-desaturases). However, it is unlikely that these desaturases will directly contribute to enhancing acyl exchange, and it therefore seems more likely that soy and Brassicas contain endogenous acyltransferases with a broader substrate specificity than those found in linseed or tobacco. One unexpected caveat to that is a recent study in transgenic soy in which the three primary biosynthetic activities from *M. alpina* were expressed, resulting in only very low levels of ARA and EPA (27). Whether this represents a specific problem with enzymes from this fungus or some other factor remains unclear.

Obtaining transgenic plants with relatively high levels of EPA allowed the further engineering of these lines for the synthesis of DHA by adding another elongase and desaturase (**Figure 1**) (71, 92, 95, 97). Disappointingly, transgenic DHA levels have been relatively low in the small number of studies published to date, with levels ranging from 0.5% (*Arabidopsis*) (102) to 1.5% (*B. juncea*) (150) to 3.3% (soy) (59). However, it should be remembered that this represents some of the most complex plant genetic engineering yet attempted, with five primary biosynthetic activities and several secondary enhancing ones (e.g., FAD2, ω3-desaturase) encoded by up to nine transgenes on two separate T-DNAs (150). It is also probable that the additional elongation and desaturation steps required to convert EPA to DHA represent an additional potential substrate-dichotomy bottleneck. Moreover, because soy and *B. juncea* efficiently accumulate EPA but not DHA (59, 150), it seems likely that endogenous acyl-exchange activities are unable to mediate this second step (i.e., EPA to DHA) in the same manner in which they are predicted to for the first (i.e., GLA to EPA) (33, 84, 85). In that respect, it may be that even more additional transgenes encoding acyltransferases from VLC-PUFA-accumulating organisms will be required to enhance the accumulation of DHA (see below for further consideration of this topic).

An interesting variation to the ARA and EPA biosynthetic pathway described above is the so-called alternative Δ8-pathway (**Figure 1**), which has been reported in a taxonomically diverse range of organisms (108). Genes encoding the primary activities for this pathway [Δ9-elongase (93), Δ8-desaturase (106), and Δ5-desaturase] were constitutively expressed in transgenic *Arabidopsis* plants, resulting in the synthesis of ARA and EPA to moderate levels (6.6%, 3% of total FAs) (94). Compared with the much lower levels obtained with the "conventional" Δ6-pathway described above (1), and in conjunction with further biochemical analysis, it appears that the alternative Δ8-pathway is potentially more effective in the synthesis of ARA and EPA due to a reduced requirement for acyl exchange between phospholipids and the acyl-CoA pool (**Figure 2**). Specifically, the first committed step in the Δ8-pathway is the C_2 elongation of endogenous LA and ALA present in the acyl-CoA pool (**Figure 1**) (84, 93). Interestingly, we recently observed that these C_{20} elongation products (20:2Δ11,14 and 20:3Δ11,14,17) accumulate to very high levels in the acyl-CoA pool of transgenic *Arabidopsis* lines expressing activities of the alternative Δ8-pathway (106), indicating the inefficient transfer of these non-native fatty acids out of the acyl-CoA pool. It will be important to determine whether this acyl-exchange bottleneck is species-dependent, analogous to the situation in linseed.

ENHANCING THE LEVELS OF VLC-PUFAS IN TRANSGENIC PLANTS

Although remarkable progress has been made in synthesising VLC-PUFAs in transgenic plants, considerable refinements are required to generate a *bona fide* substitute for fish oils. In particular, oceanic oils are not only rich in EPA and/or DHA, they are invariably devoid of omega-6 fatty acids such as GLA and ARA (49, 101, 140, 149). Thus, it is desirable to ensure the conversion of such omega-6 fatty acids to their omega-3 counterparts. Some success has been achieved in this through the use of ω3-desaturases (91, 126, 150), and more recently a fungal bi-functional Δ12, Δ15 desaturase was shown to convert oleic acid to ALA without significant levels of intermediate omega-6 LA (32); a similar activity was recently characterized from the amoeba *Acanthamoeba castellanii* (105). However, the complete ω3-desaturation of omega-6 fatty acids is hampered by the acyl channeling of potential substrates away from metabolically active pools and into TAG (1, 11, 33, 59, 84, 85). In fact, maintaining a continuous flux of substrates through the VLC-PUFA biosynthetic pathway(s) (**Figure 3**) without

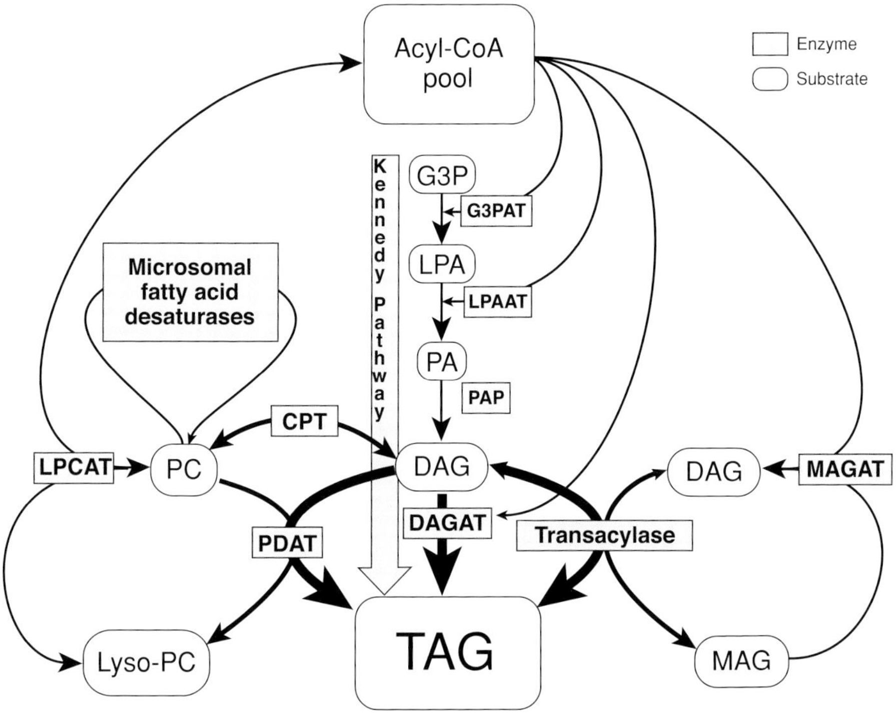

Figure 3

Triacylglycerol biosynthesis and compartmentalization of fatty acids. The dynamic interplay between different enzymes of the Kennedy pathway is shown, as is the contribution of acyl-CoA-independent activities such as phospholipids:diacylglycerol (PDAT). The potentially important activity of acyl-CoA:lyso-phosphatidylcholine acyltransferase (LPCAT) is shown, as is that for diacylglycerol:acyltransferase (DGAT).

significant loss to TAG represents a major challenge. This is obvious even in transgenic plants that accumulate EPA to high levels; for example, transgenic soy seeds also contained high GLA and DHGLA, presumably due to channeling of these fatty acids to TAG (59). Similar loss of potential substrates was also observed in transgenic *B. juncea* (150) despite the fact that both studies utilized transgene ω3-desaturases. Technological "fixes" to such problems are not obvious, not least because it is likely that such channeling represents the sum of multiple different acyl-exchange activities. It is also highly probable that each plant species has a different ratio of these activities, meaning that it is unlikely that any one solution will provide a generic "magic bullet" (82). One possibility may be through the identification of desaturases with preferences for omega-3 substrates, such as have been identified from the Primula and Echium species (42, 107). Expression of a *P. luteola* Δ6-desaturase in transgenic *Arabidopsis* resulted in the synthesis of only SDA, even in the presence of considerable LA substrate (107).

Although the problem of substrate dichotomy appears to be central to lower eukaryotic VLC-PUFA biosynthesis, it may be less so in animals, which are predicted to contain acyl-CoA-dependent desaturases (54, 126, 129). As a slight caveat, it should be noted that while the presence of acyl-CoA-dependent desaturases in, for example, mammals is a biochemical dogma, there is limited direct biochemical evidence to confirm this, not least of all for VLC-PUFA biosynthesis (34). However, it might be predicted that VLC-PUFA biosynthetic desaturases from mammalian sources might demonstrate superior performance in transgenic plants coexpressing acyl-CoA-dependent elongases, although only very limited attempts have been made to test this hypothesis. This is primarily due to concerns that transgenic plants expressing animal genes will have a negative impact on the already sensitive issue of public acceptance of genetically modified (GM) food (78, 81). It is interesting to note that reciprocal studies on the expression of plant genes in transgenic animals have proved successful, particularly in transgenically enabling the biosynthesis of omega-3 fatty acids from dietary omega-6 (62). It has also been shown that overexpression of a Δ6-desaturase can increase endogenous capacity to synthesize EPA in transgenic fish (4). Currently, there are only two published examples of expressing animal desaturases in transgenic plants, the first being the expression of a *Caenorhabditis elegans* (nematode) ω3-desaturase (FAT-1) in *Arabidopsis* (126). Expression of FAT-1 resulted in elevated levels of the endogenous omega-3 fatty acid ALA, as well as the capacity to convert exogenously supplied ARA to EPA (126). Interestingly, the authors of this study suggested that the *C. elegans* FAT-1 desaturase might be a glycerolipid-dependent (rather than acyl-CoA-dependent) desaturase (126). More recently, as part of attempts to produce DHA in transgenic *Arabidopsis*, a bifunctional Δ6, Δ5-desaturase (47) from *Danio rerio* (zebrafish) was used in conjunction with the *C. elegans* elongase (10, 102). Surprisingly, the use of this presumptive acyl-CoA-dependent desaturase did not result in enhanced synthesis of EPA and DHA, although this may simply reflect the low endogenous levels of substrates (LA, ALA) in the acyl-CoA pool (101, 102, 141). However, it also remains to be demonstrated that *D. rerio* desaturase is a *bona fide* acyl-CoA-dependent activity. As an alternative approach to the use of such enzymes, examples have been sought from non-animal sources. An acyl-CoA-dependent Δ6-desaturase was recently identified from the microalga *Ostreococcus tauri* (35), which when coexpressed in yeast with a Δ6-elongase resulted in high levels of C_{20} PUFAs (35). Such results clearly indicate bypassing of the substrate-dichotomy bottleneck (34, 85), although it remains to be demonstrated whether this enzyme will be equally active in transgenic plants.

An additional aspect of VLC-PUFA relating to the C_2 elongation reactions needs to be considered. Microsomal fatty acid elongation occurs as a result of four sequential enzymatic reactions: condensation, ketoreduction, dehydration, and enoyl reduction (64), although transgenic modulation of the elongase is possible through heterologous expression of just the initial condensing enzyme (10, 93, 153). However, it is conceivable that the physical and biochemical interactions between a non-native condensing enzyme and the other three endogenous elongase components may be sub-optimal. Such a situation has been observed in the synthesis of VLC-monounsaturated fatty acids in yeast (90) and was suggested to have an impact on the accumulation of DHA in transgenic *B. juncea* (150). Thus, it may be that for maximal synthesis of VLC-PUFAs additional components of the elongases need to be isolated from suitable EPA- or DHA-accumulating organisms and coexpressed with the transgene condensing enzyme from the same species. Evidence from coexpression of the ketoreductase and condensing enzyme in yeast provides support for this concept (9).

One final contribution to the optimization of VLC-PUFA is likely to be the enzymes involved in acyl exchange and/or lipid channeling. These represent activities that generate the substrate dichotomy, but are also responsible for channeling fatty acids in different lipids, including TAG (11, 58). Because the final aim of producing any unusual fatty acid in transgenic plants is their accumulation in storage triacylgerols, manipulating these traits in favor of the accumulation of VLC-PUFAs is remarkably difficult. As mentioned above, several enzymes (PDAT, DGAT) involved in the biosynthesis of TAG have been suggested to play roles in unusual fatty acid accumulation (39, 57, 117, 145). The acyl-CoA:*lyso*-phosphatidylcholine acyltransferases (LPCATs) have also been suggested as a potential target enzyme to overcome substrate dichotomy (33) through a (reverse) reaction that removes fatty acids from the *sn-2* position of PC to generate an acyl-CoA for renewed exposure to metabolic pathways (134, 136). Thus, such an activity might preferentially remove unusual fatty acids from their site of desaturation and directly incorporate them into the acyl-CoA pool, making them available for transgene-mediated elongation (11, 33). LPCAT genes were recently cloned from non-VLC-PUFA synthesizing mammalian cells (29, 79), although data were only presented for the predominant forward reaction of this enzyme (incorporation of an acyl-CoA into the *sn-2* position of *lyso*-PC) (11, 33, 145). An LPCAT sequence from the VLC-PUFA-accumulating organism *C. elegans* was identified, although a full description of the function of this enzyme has yet to be published (33). In view of the suggestion that *C. elegans* contains glycerolipid desaturases (126), the contribution of this activity to transgenic VLC-PUFA accumulation is awaited with interest. However, the efficacy of heterologous acyltransferases in mediating alterations to seed oil composition has already been demonstrated through the expression of the yeast SLC1 activity in transgenic *B. napus* (154).

SUMMARY

It is now possible to use genetic engineering to modify the fatty acid profiles of plants. For the so-called industrial fatty acids, levels of the target fatty acids such as ricinoleic acid have been relatively low in spite of the fact that this is a relatively simple trait to engineer and that ricinoleic acid accumulates naturally in castor to ~90% of total fatty acids. The FAD2-like family of enzymes that catalyze the synthesis of many industrial fatty acids requires phospholipid-linked substrates. In (transgenic) plants that lack the appropriate "editing" enzymes to remove the products of the FAD2-like enzymes from the phospholipids, accumulation of these unusual fatty acids likely perturbs normal fatty acid biosynthesis and compartmentation. In the case of VLC-PUFAs, levels of target fatty acids are now approaching those found in the native sources (such as marine microbes) (44, 49, 82). There is further scope for enhancing the levels of VLC-PUFA accumulation in transgenic plants, either through the addition of additional acyl-exchange enzymes (to overcome generic bottlenecks in the process) or through the identification of superior (i.e., more efficient) forms of the underpinning primary activities. It should also be remembered that an entirely different (anaerobic) polyketide synthase-like system has been shown to synthesize VLC-PUFAs in some marine bacteria and microbes (80). The genes for this pathway have been identified and expressed in *E. coli* (70), although no data as to performance in transgenic plants is currently available. However, in the case of the accumulation of both industrial fatty acids and VLC-PUFAs, our overall understanding and appreciation of the complexity of plant lipid biochemistry has been greatly expanded through the transgenic studies described in this review. Therefore, not only has biotechnology progress been made, but also much basic knowledge has been generated through these studies.

FUTURE ISSUES

1. What are the various enzymes that mediate acyl exchange in plants that naturally synthesize unusual fatty acids?
2. What transgenic factors will deliver the accumulation to high levels of unusual fatty acids in transgenic plants?
3. Will the accumulation of DHA to high levels in transgenic plants require additional transgene activities? Will it be possible to add such additional genes?
4. Will the general public accept novel genetically modified oils as a substitute for fish oils for use in human nutrition?

ACKNOWLEDGMENTS

Rothamsted Research receives grant-aided support from the Biotechnology and Biological Research Council (BBSRC), United Kingdom. J.A.N. thanks BASF Plant Sciences and the European Union for financial support.

LITERATURE CITED

1. The first demonstration of the use of the Δ6-pathway in transgenic plants to synthesize ARA and EPA in seed oils. It also provides a comprehensive analysis as to the potential bottlenecks in the pathway.

1. **Abbadi A, Domergue F, Bauer J, Napier J, Welti R, et al. 2004. Biosynthesis of very-long-chain polyunsaturated fatty acids in transgenic oilseeds: Constraints on their accumulation. *Plant Cell* 16:2734–48**
2. Abbadi A, Domergue F, Meyer A, Riedel K, Sperling P, et al. 2001. Transgenic oilseeds as sustainable source of nutritionally relevant C_{20} and C_{22} polyunsaturated fatty acids? *Eur. J. Lipid Sci. Technol.* 103:106–13
3. Aitzetmuller K, Matthaus B, Friedrich H. 2003. A new database for seed oil fatty acids—the database SOFA. *Eur. J. Lipid Sci. Technol.* 105:92–103
4. Alimuddin, Yoshizaki G, Kiron V, Satoh S, Takeuchi T. 2005. Enhancement of EPA and DHA biosynthesis by overexpression of Masu salmon Δ6-desaturase-like gene in zebrafish. *Transgenic Res.* 14:159–65
5. Arondel V, Lemieux B, Hwang I, Gibson S, Goodman HM, et al. 1992. Map-based cloning of a gene controlling omega-3 fatty acid desaturation in *Arabidopsis*. *Science* 258:1353–55

6. Avelange-Macherel MH, Macherel D, Wada H, Murata N. 1995. Site-directed mutagenesis of histidine residues in the Δ12 acyl-lipid desaturase of *Synechocystis*. *FEBS Lett.* 361:111–14
7. Bafor M, Smith MA, Jonsson L, Stobart K, Stymne S. 1991. Ricinoleic acid biosynthesis and triacylglycerol assembly in microsomal preparations from developing castor-bean (*Ricinus-communis*) endosperm. *Biochem. J.* 280:507–14
8. Banas A, Johansson I, Stymne S. 1992. Plant microsomal phospholipases exhibit preference for phosphatidylcholine with oxygenated acyl-groups. *Plant Sci.* 84:137–44
9. Beaudoin F, Gable K, Sayanova O, Dunn T, Napier J. 2002. A *Saccharomyces cerevisiae* gene required for heterologous fatty acid elongase activity encodes a microsomal β-ketoreductase. *J. Biol. Chem.* 277:11481–88
10. Beaudoin F, Michaelson L, Hey S, Lewis M, Shewry P, et al. 2000. Heterologous reconstitution in yeast of the polyunsaturated fatty acid biosynthetic pathway. *Proc. Natl. Acad. Sci. USA* 97:6421–26
11. Beaudoin F, Napier JA. 2004. Biosynthesis and compartmentalisation of triacylglycerol in higher plants. In *Lipid Metabolism and Membrane Biogenesis*, ed. G Daum, pp. 267–87. Berlin: Springer-Verlag
12. Broadwater JA, Whittle E, Shanklin J. 2002. Desaturation and hydroxylation—Residues 148 and 324 of *Arabidopsis* FAD2, in addition to substrate chain length, exert a major influence in partitioning of catalytic specificity. *J. Biol. Chem.* 277:15613–20
13. Broun P, Boddupalli S, Somerville C. 1999. A bifunctional oleate 12-hydroxylase: Desaturase from *Lesquerella fendleri*. *Plant J.* 13:201–10
14. Broun P, Gettner S, Somerville C. 1999. Genetic engineering of plant lipids. *Annu. Rev. Nutr.* 19:197–216
15. **Broun P, Shanklin J, Whittle E, Somerville C. 1998. Catalytic plasticity of fatty acid modification enzymes underlying chemical diversity of plant lipids. *Science* 282:1315–17**
16. Broun P, Somerville C. 1997. Accumulation of ricinoleic, lesquerolic, and densipolic acids in seeds of transgenic *Arabidopsis* plants that express a fatty acyl hydroxylase cDNA from castor bean. *Plant Physiol.* 113:933–42
17. Browse J, McConn M, James DJ, Miquel M. 1993. Mutants of *Arabidopsis* deficient in the synthesis of α-linolenate. Biochemical and genetic characterization of the endoplasmic reticulum linoleoyl desaturase. *J. Biol. Chem.* 268:16345–51
18. Browse J, Warwick N, Somerville CR, Slack CR. 1986. Fluxes through the prokaryotic and eukaryotic pathways of lipid synthesis in the '16:3' plant *Arabidopsis thaliana*. *Biochem. J.* 235:25–31
19. **Cahoon E, Carlson T, Ripp K, Schweiger B, Cook G, et al. 1999. Biosynthetic origin of conjugated double bonds: Production of fatty acid components of high-value drying oils in transgenic soybean embryos. *Proc. Natl. Acad. Sci. USA* 96:12935–40**
20. Cahoon E, Dietrich C, Meyer K, Damude H, Dyer J, Kinney A. 2006. Conjugated fatty acids accumulate to high levels in phospholipids of metabolically engineered soybean and *Arabidopsis* seeds. *Phytochemistry* 67:1166–76
21. Cahoon E, Kinney A. 2004. Dimorphecolic acid is synthesized by the coordinate activities of two divergent Δ12-oleic acid desaturases. *J. Biol. Chem.* 279:12495–502
22. Cahoon E, Kinney A. 2005. The production of vegetable oils with novel properties: Using genomic tools to probe and manipulate plant fatty acid metabolism. *Eur. J. Lipid Sci. Technol.* 107:239–43

15. Describes reciprocal mutagenic conversion between the FAD2 desaturase and FAH12 hydroxylase, indicating the close relationship between these two enzymes.

19. Describes the first identification of conjugases, variant FAD2-like activities that generate conjugated double bonds. Expression in transgenic plants indicated the potential of using this approach to synthesize these important industrial fatty acids.

23. Cahoon E, Ripp K, Hall S, Kinney A. 2001. Formation of conjugated Δ8, Δ10-double bonds by Δ12-oleic-acid desaturase-related enzymes—Biosynthetic origin of calendic acid. *J. Biol. Chem.* 276:2637–43
24. Cahoon E, Ripp K, Hall S, McGonigle B. 2002. Transgenic production of epoxy fatty acids by expression of a cytochrome P450 enzyme from *Euphorbia lagascae* seed. *Plant Physiol.* 128:615–24
25. Cahoon E, Schnurr J, Huffman E, Minto R. 2003. Fungal responsive fatty acid acetylenases occur widely in evolutionarily distant plant families. *Plant J.* 34:671–83
26. Calder PC. 2004. *n-3* Fatty acids and cardiovascular disease: evidence explained and mechanisms explored. *Clin. Sci.* 107:1–11
27. Chen R, Matsui K, Ogawa M, Oe M, Ochiai M, et al. 2006. Expression of Δ6, Δ5 desaturase and GLELO elongase genes from Mortierella alpina for production of arachidonic acid in soybean [*Glycine max* (L.) Merrill] seeds. *Plant Sci.* 170:399–406
28. Chen R, Tsuda S, Matsui K, Fukuchi-Mizutani M, Ochiai M, et al. 2005. Production of γ-linolenic acid in *Lotus japonicus* and *Vigna angularis* by expression of the Δ6-fatty-acid desaturase gene isolated from *Mortierella alpina*. *Plant Sci.* 169:599–605
29. Chen X, Hyatt BA, Mucenski ML, Mason RJ, Shannon JM. 2006. Identification and characterization of a lysophosphatidylcholine acyltransferase in alveolar type II cells. *Proc. Natl. Acad. Sci. USA* 103:11724–29
30. Cunnane SC. 2003. Problems with essential fatty acids: time for a new paradigm? *Prog. Lipid Res.* 42:544–68
31. Dahlqvist A, Stahl U, Lenman M, Banas A, Lee M, et al. 2000. Phospholipid: diacylglycerol acyltransferase: An enzyme that catalyzes the acyl-CoA-independent formation of triacylglycerol in yeast and plants. *Proc. Natl. Acad. Sci. USA* 97:6487–92
32. Damude H, Zhang H, Farrall L, Ripp KG, Tomb JF, et al. 2006. Identification of bifunctional Δ12/omega 3 fatty acid desaturases for improving the ratio of omega 3 to omega 6 fatty acids in microbes and plants. *Proc. Natl. Acad. Sci. USA* 103:9446–51
33. Domergue F, Abbadi A, Heinz E. 2005. Relief for fish stocks: oceanic fatty acids in transgenic oilseeds. *Trends Plant Sci.* 10:112–16
34. Domergue F, Abbadi A, Ott C, Zank TK, Zahringer U, et al. 2003. Acyl carriers used as substrates by the desaturases and elongases involved in very long-chain polyunsaturated fatty acids biosynthesis reconstituted in yeast. *J. Biol. Chem.* 278:35115–26
35. Domergue F, Abbadi A, Zahringer U, Moreau H, Heinz E. 2005. In vivo characterization of the first acyl-CoA Δ6-desaturase from a member of the plant kingdom, the microalga *Ostreococcus tauri*. *Biochem. J.* 389:483–90
36. Domergue F, Lerchl J, Zahringer U, Heinz E. 2002. Cloning and functional characterization of *Phaeodactylum tricornutum* front-end desaturases involved in eicosapentaenoic acid biosynthesis. *Eur. J. Biochem.* 269:4105–13
37. Drexler H, Spiekermann P, Meyer A, Domergue F, Zank T, et al. 2003. Metabolic engineering of fatty acids for breeding of new oilseed crops: strategies, problems and first results. *J. Plant Physiol.* 160:779–802
38. Dyer J, Chapital D, Kuan J, Mullen R, Turner C, et al. 2002. Molecular analysis of a bifunctional fatty acid conjugase/desaturase from tung. Implications for the evolution of plant fatty acid diversity. *Plant Physiol.* 130:2027–38
39. Dyer J, Mullen R. 2005. Development and potential of genetically engineered oilseeds. *Seed Sci. Res.* 15:255–67
40. Eckert H, LaVallee B, Schweiger BJ, Kinney AJ, Cahoon EB, Clemente T. 2006. Coexpression of the borage Δ6 desaturase and the *Arabidopsis* Δ15 desaturase results in high accumulation of stearidonic acid in the seeds of transgenic soybean. *Planta* 224:1050–57

41. Garcia-Maroto F, Garrido-Cardenas JA, Rodriguez-Ruiz J, Vilches-Ferron M, Adam AC, et al. 2002. Cloning and molecular characterization of the Δ6-desaturase from two Echium plant species: Production of GLA by heterologous expression in yeast and tobacco. *Lipids* 37:417–26
42. Garcia-Maroto F, Manas-Fernandez A, Garrido-Cardenas JA, Alonso DL. 2006. Substrate specificity of acyl-Δ6-desaturases from Continental versus Macaronesian Echium species. *Phytochemistry* 67:540–44
43. Girke T, Schmidt H, Zahringer U, Reski R, Heinz E. 1998. Identification of a novel Δ6-acyl-group desaturase by targeted gene disruption in *Physcomitrella patens*. *Plant J.* 15:39–48
44. Green A. 2004. From alpha to omega—producing essential fatty acids in plants. *Nat. Biotechnol.* 22:680–82
45. Hamada F, Otani M, Kim SH, Uchida H, Kajikawa M, et al. 2006. Accumulation of γ-linolenic acid by overexpression of a liverwort Δ6 desaturase gene in transgenic rice. *Plant Cell Physiol.* 47(Suppl.):195
46. Harwood JL. 1988. Fatty acid metabolism. *Annu. Rev. Plant Physiol. Plant Mol. Biol.* 39:101–38
47. Hastings N, Agaba M, Tocher D, Leaver M, Dick J, et al. 2001. A vertebrate fatty acid desaturase with Δ5 and Δ6 activities. *Proc. Natl. Acad. Sci. USA* 98:14304–9
48. He X, Turner C, Chen GQ, Lin JT, McKeon TA. 2004. Cloning and characterization of a cDNA encoding diacylglycerol acyltransferase from castor bean. *Lipids* 39:311–18
49. Heinz E. 2006. First breakthroughs in sustainable production of "oceanic fatty acids." *Eur. J. Lipid Sci. Technol.* 108:1–3
50. Hong H, Datla N, MacKenzie SL, Qiu X. 2002. Isolation and characterization of a Δ5-desaturase from *Pythium irregulare* by heterologous expression in *Saccharomyces cerevisiae* and oilseed crops. *Lipids* 37:863–68
51. Hong H, Datla N, Reed DW, Covello PS, MacKenzie SL, Qiu X. 2002. High-level production of γ-linolenic acid in *Brassica juncea* using a Δ6 desaturase from *Pythium irregulare*. *Plant Physiol.* 129:354–62
52. Hornung E, Pernstich C, Feussner I. 2002. Formation of conjugated Δ11, Δ13-double bonds by Δ12-linoleic acid (1,4)-acyl-lipid-desaturase in pomegranate seeds. *Eur. J. Biochem.* 269:4852–59
53. Huang YS, Chaudhary S, Thurmond JM, Bobik EGJ, Yuan L, et al. 1999. Cloning of Δ12- and Δ6-desaturases from *Mortierella alpina* and recombinant production of γ-linolenic acid in *Saccharomyces cerevisiae*. *Lipids* 34:649–59
54. Huang YS, Pereira SL, Leonard AE. 2004. Enzymes for transgenic biosynthesis of long-chain polyunsaturated fatty acids. *Biochimie* 86:793–98
55. Iba K, Gibson S, Nishiuchi T, Fuse T, Nishimura M, et al. 1993. A gene encoding a chloroplast omega-3 fatty acid desaturase complements alterations in fatty acid desaturation and chloroplast copy number of the *fad*7 mutant of *Arabidopsis thaliana*. *J. Biol. Chem.* 268:24099–105
56. Iwabuchi M, Kohno-Murase J, Imamura J. 2003. Δ12-oleate desaturase-related enzymes associated with formation of conjugated trans-Δ11, cis-Δ13 double bonds. *J. Biol. Chem.* 278:4603–10
57. Jaworski J, Cahoon EB. 2003. Industrial oils from transgenic plants. *Curr. Opin. Plant Biol.* 6:178–84
58. Kennedy EP. 1961. Biosynthesis of complex lipids. *Fed. Proc.* 20:934–40
59. Kinney AJ. 2004. Production of very long chain polyunsaturated fatty acids in oilseeds. *Patent WO 2004/071467 A2*

59. Reports the current highest accumulation of EPA and DHA in transgenic plants.

60. Kinney AJ. 2006. Metabolic engineering in plants for human health and nutrition. *Curr. Opin. Plant Biol.* 17:130–38
61. Klee H, Horsch R, Rogers S. 1987. *Agrobacterium*-mediated plant transformation and its further applications to plant biology. *Annu. Rev. Plant Physiol.* 38:467–86
62. Lai L, Kang J, Li R, Wang JD, Witt WT, et al. 2006. Generation of cloned transgenic pigs rich in omega-3 fatty acids. *Nat. Biotech.* 24:435–36
63. **Lee M, Lenman M, Banas A, Bafor M, Singh S, et al. 1998. Identification of non-heme diiron proteins that catalyze triple bond and epoxy group formation. *Science* 280:915–18**
64. Leonard AE, Pereira SL, Sprecher H, Huang YS. 2004. Elongation of long-chain fatty acids. *Prog. Lipid Res.* 43:36–54
65. Lu C, Fulda M, Wallis JG, Browse J. 2006. A high-throughput screen for genes from castor that boost hydroxy fatty acid accumulation in seed oils of transgenic *Arabidopsis*. *Plant J.* 45:847–56
66. Maltman DJ, Simon WJ, Wheeler CH, Dunn MJ, Wait R, Slabas AR. 2002. Proteomic analysis of the endoplasmic reticulum from developing and germinating seed of castor (*Ricinus communis*). *Electrophoresis* 23:626–39
67. Mancha M, Stymne S. 1997. Remodelling of triacylglycerols in microsomal preparations from developing castor bean (Ricinus communis L) endosperm. *Planta* 203:51–57
68. McConn M, Hugly S, Browse J, Somerville C. 1994. A mutation at the *fad8* locus of *Arabidopsis* identifies a second chloroplast omega-3 desaturase. *Plant Physiol.* 106:1609–14
69. McKeon T, Lin JT, Stafford AE. 1999. Biosynthesis of ricinoleate in castor oil. *Adv. Exp. Med. Biol.* 464:37–47
70. Metz JG, Roessler P, Facciotti D, Levering C, Dittrich F, et al. 2001. Production of polyunsaturated fatty acids by polyketide synthases in both prokaryotes and eukaryotes. *Science* 293:290–93
71. Meyer A, Cirpus P, Ott C, Schlecker R, Zahringer U, Heinz E. 2003. Biosynthesis of docosahexaenoic acid in *Euglena gracilis*: Biochemical and molecular evidence for the involvement of a Δ4-fatty acyl group desaturase. *Biochemistry* 42:9779–88
72. Millar AA, Smith MA, Kunst L. 2000. All fatty acids are not equal: Discrimination in plant membrane lipids. *Trends Plant Sci.* 5:95–101
73. Miquel M, Browse J. 1992. *Arabidopsis* mutants deficient in polyunsaturated fatty acid synthesis. Biochemical and genetic characterization of a plant oleoyl-phosphatidylcholine desaturase. *J. Biol. Chem.* 267:1502–9
74. Miquel M, Cassagne C, Browse J. 1998. A new class of *Arabidopsis* mutants with reduced hexadecatrienoic acid fatty acid levels. *Plant Physiol.* 117:923–30
75. Murphy DJ. 1999. Production of novel oils in plants. *Curr. Opin. Plant Biol.* 10:175–80
76. Murphy DJ. 1999. Manipulation of plant oil composition for the production of valuable chemicals. Progress, problems, and prospects. *Adv. Exp. Med. Biol.* 464:21–35
77. Murphy DJ. 1994. Manipulation of lipid metabolism in transgenic plants: Biotechnological goals and biochemical realities. *Biochem. Soc. Trans.* 22:926–31
78. Murphy DJ. 2006. Molecular breeding strategies for the modification of lipid composition. *In Vitro Cell. Dev. Biol. Plant* 42:89–99
79. Nakanishi H, Shindou H, Hishikawa D, Harayama T, Ogasawara R, et al. 2006. Cloning and characterization of mouse lung-type acyl-CoA:lysophosphatidylcholine acyltransferase 1 (LPCAT1). Expression in alveolar type II cells and possible involvement in surfactant production. *J. Biol. Chem.* 281:20140–47

63. First identification of FAD2-like activities capable of directing the synthesis of epoxygenated and acetylenated fatty acids in transgenic plants.

80. Napier JA. 2002. Plumbing the depths of PUFA biosynthesis: a novel polyketide synthase-like pathway from marine organisms. *Trends Plant Sci.* 7:51–54
81. Napier JA. 2007. Transgenic plants as a source of fish oils: Healthy, sustainable and GM. *J. Sci. Food Agric.* 87:8–12
82. Napier JA, Haslam R, Caleron MV, Michaelson LV, Beaudoin F, Sayanova O. 2006. Progress towards the production of very long-chain polyunsaturated fatty acid in transgenic plants: plant metabolic engineering comes of age. *Physiol. Plant.* 126:398–406
83. Napier JA, Michaelson LV, Sayanova O. 2003. The role of cytochrome b_5 fusion desaturases in the synthesis of polyunsaturated fatty acids. *Prostaglandins Leukot. Essent. Fatty Acids* 68:135–43
84. Napier JA, Sayanova O. 2005. The production of very-long-chain PUFA biosynthesis in transgenic plants: towards a sustainable source of fish oils. *Proc. Nutr. Soc.* 64:387–93
85. Napier JA, Sayanova O, Qi BX, Lazarus CM. 2004. Progress toward the production of long-chain polyunsaturated fatty acids in transgenic plants. *Lipids* 39:1067–75
86. Napier JA, Sayanova O, Sperling P, Heinz E. 1999. A growing family of cytochrome b_5 domain fusion proteins. *Trends Plant Sci.* 4:2–4
87. Ohlrogge JB, Browse J, Somerville CR. 1991. The genetics of plant lipids. *Biochim Biophys Acta* 1082:1–26
88. Okuley J, Lightner J, Feldmann K, Yadav N, Lark E, et al. 1994. *Arabidopsis* FAD2 gene encodes the enzyme that is essential for polyunsaturated lipid synthesis. *Plant Cell* 6:147–58
89. Opsahl-Ferstad HG, Rudi H, Ruyter B, Refstie S. 2003. Biotechnological approaches to modify rapeseed oil composition for applications in aquaculture. *Plant Sci.* 165:349–57
90. Paul S, Gable K, Beaudoin F, Cahoon E, Jaworski J, et al. 2006. Members of the *Arabidopsis* FAE1-like 3-ketoacyl-CoA synthase gene family substitute for the Elop proteins of *Saccharomyces cerevisiae*. *J. Biol. Chem.* 281:9018–29
91. Pereira SL, Huang YS, Bobik EG, Kinney AJ, Stecca KL, et al. 2004. A novel omega 3-fatty acid desaturase involved in the biosynthesis of eicosapentaenoic acid. *Biochem. J.* 378:665–71
92. Pereira SL, Leonard AE, Huang YS, Chuang LT, Mukerji P. 2004. Identification of two novel microalgal enzymes involved in the conversion of the omega 3-fatty acid, eicosapentaenoic acid, into docosahexaenoic acid. *Biochem. J.* 384:357–66
93. Qi B, Beaudoin F, Fraser T, Stobart AK, Napier JA, Lazarus CM. 2002. Identification of a cDNA encoding a novel C18-Δ9 polyunsaturated fatty acid-specific elongating activity from the docosahexaenoic acid (DHA)-producing microalga, *Isochrysis galbana*. *FEBS Lett.* 510:159–65
94. Qi B, Fraser T, Mugford S, Dobson G, Sayanova O, et al. 2004. Production of very long chain polyunsaturated omega-3 and omega-6 fatty acids in plants. *Nat. Biotech.* 22:739–45
95. Qiu X. 2003. Biosynthesis of docosahexaenoic acid (DHA, 22:6–4, 7,10,13,16,19): two distinct pathways. *Prostaglandins Leukot. Essent. Fatty Acids* 68:181–86
96. Qiu X, Hong H, Datla N, MacKenzie SL, Taylor DC, Thomas TL. 2002. Expression of borage Δ6 desaturase in *Saccharomyces cerevisiae* and oilseed crops. *Can. J. Bot.* 80:42–49
97. Qiu X, Hong H, MacKenzie SL. 2001. Identification of a Δ4 fatty acid desaturase from *Thraustochytrium* sp involved in the biosynthesis of docosahexanoic acid by heterologous expression in *Saccharomyces cerevisiae* and *Brassica juncea*. *J. Biol. Chem.* 276:31561–66
98. Qiu X, Reed DW, Hong H, MacKenzie SL, Covello PS. 2001. Identification and analysis of a gene from *Calendula officinalis* encoding a fatty acid conjugase. *Plant Physiol.* 125:847–55

94. First demonstration of the feasibility of synthesizing ARA and EPA in transgenic plants, and also a demonstration of the utility of the alternative Δ8-pathway in this process.

99. Reddy AS, Thomas TL. 1996. Expression of a cyanobacterial Δ6-desaturase gene results in γ-linolenic acid production in transgenic plants. *Nat. Biotechnol.* 14:639–42

100. Rezzonico E, Moire L, Delessert S, Poirier Y. 2004. Level of accumulation of epoxy fatty acid in *Arabidopsis thaliana* expressing a linoleic acid Δ12-epoxygenase is influenced by the availability of the substrate linoleic acid. *Theor. Appl. Genet.* 109:1077–82

101. Robert SS. 2006. Production of eicosapentaenoic and docosahexaenoic acid-containing oils in transgenic land plants for human and aquaculture nutrition. *Mar. Biotechnol.* 8:103–9

102. Robert SS, Singh SP, Zhou XR, Petrie JR, Blackburn S, et al. 2005. Metabolic engineering of *Arabidopsis* to produce nutritionally important DHA in seed oil. *Funct. Plant Biol.* 32:473–79

102. Demonstrates the accumulation of DHA in transgenic Arabidopsis, and also the use of animal enzymes in plant metabolic engineering.

103. Sato S, Xing AQ, Ye XG, Schweiger B, Kinney A, et al. 2004. Production of γ-linolenic acid and stearidonic acid in seeds of marker-free transgenic soybean. *Crop Sci.* 44:646–52

104. Sayanova O, Davies GM, Smith MA, Griffiths G, Stobart AK, et al. 1999. Accumulation of Δ6-unsaturated fatty acids in transgenic tobacco plants expressing a Δ6-desaturase from *Borago officinalis*. *J. Exp. Bot.* 50:1647–52

105. Sayanova O, Haslam R, Guschina I, Lloyd D, Christie WW, et al. 2006. A bi-functional desaturase from Acanthamoeba castellanii directs the synthesis of highly unusual n-1 series unsaturated fatty acids. *J. Biol. Chem.* 281(48):36533–41

106. Sayanova O, Haslam R, Qi BX, Lazarus CM, Napier JA. 2006. The alternative pathway C_{20} Δ8-desaturase from the nonphoto synthetic organism *Acanthamoeba castellanii* is an atypical cytochrome b_5-fusion desaturase. *FEBS Lett.* 580:1946–52

107. Sayanova O, Haslam R, Venegas-Caleron M, Napier JA. 2006. Identification of Primula "front-end" desaturases with distinct *n-6* or *n-3* substrate preferences. *Planta.* 224:1269–77

108. Sayanova O, Napier JA. 2004. Eicosapentaenoic acid: biosynthetic routes and the potential for synthesis in transgenic plants. *Phytochemistry* 65:147–58

109. Sayanova O, Smith MA, Lapinskas P, Stobart AK, Dobson G, et al. 1997. Expression of a borage desaturase cDNA containing an N-terminal cytochrome b_5 domain results in the accumulation of high levels of Δ6-desaturated fatty acids in transgenic tobacco. *Proc. Natl. Acad. Sci. USA* 94:4211–16

109. First identification of a VLC-PUFA biosynthetic enzyme and the observation that the desaturases associated with this pathway represented a new class (cytochrome b_5-fusion) of enzyme.

110. Scarth R, Tang J. 2006. Modification of Brassica oil using conventional and transgenic approaches. *Crop Sci.* 46(3):1225–36

111. Schmidt H, Dresselhaus T, Buck F, Heinz E. 1994. Purification and PCR-based cDNA cloning of a plastidial *n-6* desaturase. *Plant Mol. Biol.* 26:631–42

112. Shanklin J, Cahoon EB. 1998. Desaturation and related modifications of fatty acids. *Annu. Rev. Plant Phys.* 49:611–41

113. Shanklin J, Somerville C. 1991. Stearoyl-acyl-carrier-protein desaturase from higher plants is structurally unrelated to the animal and fungal homologs. *Proc. Natl. Acad. Sci. USA* 88:2510–14

114. Shanklin J, Whittle E, Fox BG. 1994. Eight histidine residues are catalytically essential in a membrane-associated iron enzyme, stearoyl-CoA desaturase, and are conserved in alkane hydroxylase and xylene monooxygenase. *Biochemistry* 33:12787–94

114. Describes the identification of conserved histidine residues as being essential for desaturase function, allowing their facile identification on the basis of these motifs.

115. Shockey JM, Gidda SK, Chapital DC, Kuan JC, Dhanoa PK, et al. 2006. Tung tree DGAT1 and DGAT2 have nonredundant functions in triacylglycerol biosynthesis and are localized to different subdomains of the endoplasmic reticulum. *Plant Cell* 18:2294–313

116. Singh S, Thomaeus S, Lee M, Stymne S, Green A. 2001. Transgenic expression of a Δ12-epoxygenase gene in *Arabidopsis* seeds inhibits accumulation of linoleic acid. *Planta* 212:872–79
117. Singh S, Zhou XR, Liu Q, Stymne S, Green AG. 2005. Metabolic engineering of new fatty acids in plants. *Curr. Opin. Plant Biol.* 8:197–203
118. Slabas AR, Hanley Z, Schierer T, Rice D, Turnbull A, et al. 2001. Acyltransferases and their role in the biosynthesis of lipids opportunities for new oils. *J. Plant Physiol.* 158:505–13
119. Smith MA, Cross AR, Jones OT, Griffiths WT, Stymne S, Stobart K. 1990. Electron-transport components of the 1-acyl-2-oleoyl-sn-glycero-3-phosphocholine Δ12-desaturase (Δ12-desaturase) in microsomal preparations from developing safflower (*Carthamus-tinctorius* L) cotyledons. *Biochem. J.* 272:23–29
120. Smith MA, Jonsson L, Stymne S, Stobart K. 1992. Evidence for cytochrome-b_5 as an electron-donor in ricinoleic acid biosynthesis in microsomal preparations from developing castor bean (*Ricinus-communis* L). *Biochem. J.* 287:141–44
121. Smith MA, Moon H, Chowrira G, Kunst L. 2003. Heterologous expression of a fatty acid hydroxylase gene in developing seeds of *Arabidopsis thaliana*. *Planta* 217:507–16
122. Somerville C, Browse J. 1996. Dissecting desaturation: Plants prove advantageous. *Trends Cell Biol.* 6:148–53
123. Sperling P, Heinz E. 1993. Isomeric *sn-1*-octadecenyl and *sn-2*-octadecenyl analogues of lysophosphatidylcholine as substrates for acylation and desaturation by plant microsomal membranes. *Eur. J. Biochem.* 213:965–71
124. Sperling P, Linscheid M, Stocker S, Muhlbach HP, Heinz E. 1993. *In vivo* desaturation of cis-Δ9-monounsaturated to cis-Δ 9,12-diunsaturated alkenylether glycerolipids. *J. Biol. Chem.* 268:26935–40
125. Sperling P, Ternes P, Zank T, Heinz E. 2003. The evolution of desaturases. *Prostaglandins Leukot. Essent. Fatty Acids* 68:73–95
126. Spychalla JP, Kinney AJ, Browse J. 1997. Identification of an animal omega-3 fatty acid desaturase by heterologous expression in *Arabidopsis*. *Proc. Natl. Acad. Sci. USA* 94:1142–47
127. Stahl U, Banas A, Stymne S. 1995. Plant microsomal phospholipid acyl hydrolases have selectivities for uncommon fatty-acids. *Plant Physiol.* 107:953–62
128. Stobart AK, Stymne S. 1985. The interconversion of diacylglycerol and phosphatidylcholine during triacylglycerol production in microsomal preparation of developing cotyledons of safflower (*Carthamus-tinctorius L*). *Biochem. J.* 232:217–21
129. Strittmatter P, Spatz L, Corcoran D, Rogers MJ, Setlow B, et al. 1974. Purification and properties of rat liver microsomal stearyl coenzyme A desaturase. *Proc. Natl. Acad. Sci. USA* 71:4565–69
130. Stumpf PK. 1981. Plants, fatty acids, compartments. *Trends Biochem. Sci.* 6:173–76
131. Stumpf PK. 1984. Plant lipid biosynthesis in 1959 and 1984. *J. Lipid Res.* 25:1508–10
132. Stymne S, Appelqvist LA. 1980. Biosynthesis of linoleate and α-linolenate in homogenates from developing soya bean cotyledons. *Plant Sci. Lett.* 17:287–94
133. Stymne S, Glad G. 1981. Acyl exchange between oleoyl-CoA and phosphatidylcholine in microsomes of developing soya bean cotyledons and its role in fatty-acid desaturation. *Lipids* 16:298–305
134. Stymne S, Stobart AK. 1984. Evidence for the reversibility of the acyl-CoA-lysophosphatidylcholine acyltransferase in microsomal preparations from developing safflower (*Carthamus-tinctorius* L) cotyledons and rat-liver. *Biochem. J.* 223:305–14

135. Stymne S, Stobart AK. 1986. Biosynthesis of γ-linolenic acid in cotyledons and microsomal preparations of the developing seeds of common borage (*Borago officinalis*). *Biochem. J.* 240:385–93
136. Stymne S, Stobart AK, Glad G. 1983. The role of the acyl-CoA pool in the synthesis of poly-unsaturated 18-carbon fatty-acids and triacylglycerol production in the microsomes of developing safflower seeds. *Biochim. Biophys. Acta* 752:198–208
137. Suh M, Schultz D, Ohlrogge J. 2002. What limits production of unusual monoenoic fatty acids in transgenic plants? *Planta* 215:584–95
138. Thelen JJ, Ohlrogge JB. 2002. Metabolic engineering of fatty acid biosynthesis in plants. *Metab. Eng.* 4:12–21
139. Thomaeus S, Carlsson AS, Stymne S. 2001. Distribution of fatty acids in polar and neutral lipids during seed development in *Arabidopsis thaliana* genetically engineered to produce acetylenic, epoxy and hydroxy fatty acids. *Plant Sci.* 161:997–1003
140. Tonon T, Sayanova O, Michaelson LV, Qing R, Harvey D, et al. 2005. Fatty acid desaturases from the microalga *Thalassiosira pseudonana*. *FEBS J.* 272:3401–12
141. Truksa M, Wu G, Vrinten P, Qiu X. 2006. Metabolic engineering of plants to produce very long-chain polyunsaturated fatty acids. *Transgenic Res.* 15:131–37
142. Tucker G. 2003. Nutritional enhancement of plants. *Curr. Opin. Plant Biol.* 14:221–25
143. van de Loo FJ, Broun P, Turner S, Somerville C. 1995. An oleate 12-hydroxylase from *Ricinus communis* L is a fatty acyl desaturase homolog. *Proc. Natl. Acad. Sci. USA* 92:6743–47
144. Voelker T, Kinney AJ. 2001. Variations in the biosynthesis of seed-storage lipids. *Annu. Rev. Plant Phys.* 52:335–61
145. Vogel G, Browse J. 1996. Cholinephosphotransferase and diacylglycerol acyltransferase—Substrate specificities at a key branch point in seed lipid metabolism. *Plant Physiol.* 110:923–31
146. Wallis JG, Browse J. 2002. Mutants of *Arabidopsis* reveal many roles for membrane lipids. *Prog. Lipid Res.* 41:254–78
147. Wallis JG, Watts JL, Browse J. 2002. Polyunsaturated fatty acid synthesis: what will they think of next? *Trends Biochem. Sci.* 27:467–70
148. Whitney HM, Michaelson LV, Sayanova O, Pickett JA, Napier JA. 2003. Functional characterization of two cytochrome b_5-fusion desaturases from *Anemone leveillei*: the unexpected identification of a fatty acid Δ6-desaturase. *Planta* 217:983–92
149. Williams CM, Burdge G. 2006. Long-chain n-3 PUFA: plant v. marine sources. *Proc. Nutr. Soc.* 65:42–50
150. Wu G, Truksa M, Datla N, Vrinten P, Bauer J, et al. 2005. Stepwise engineering to produce high yields of very long-chain polyunsaturated fatty acids in plants. *Nat. Biotechnol.* 23:1013–17

150. Demonstrates high levels of EPA in transgenic plants, and also DHA, through some of the most sophisticated plant genetic engineering to date.

151. Wu J, James DW Jr, Dooner HK, Browse J. 1994. A mutant of *Arabidopsis* deficient in the elongation of palmitic acid. *Plant Physiol.* 106:143–50
152. Yamada H, Shimizu S, Shinmen Y, Akimoto K, Kawashima H. 1992. Production of polyunsaturated fatty acids by microorganisms. *J. Nutr. Sci. Vitaminol.* 1992(Spec. No.):255–58
153. Zank TK, Zahringer U, Beckmann C, Pohnert G, Boland W, et al. 2002. Cloning and functional characterization of an enzyme involved in the elongation of Δ6-polyunsaturated fatty acids from the moss *Physcomitrella patens*. *Plant J.* 31:255–68

154. Zhou XR, Robert S, Singh S, Green A. 2006. Heterologous production of GLA and SDA by expression of an *Echium plantagineum* Δ6-desaturase gene. *Plant Sci.* 170:665–73
155. Zhou XR, Singh S, Liu Q, Green A. 2006. Combined transgenic expression of Δ12-desaturase and Δ12-epoxygenase in high linoleic acid seeds leads to increased accumulation of vernolic acid. *Funct. Plant Biol.* 33:585–92
156. Zou J, Katavic V, Giblin EM, Barton DL, MacKenzie SL, et al. 1997. Modification of seed oil content and acyl composition in the Brassicaceae by expression of a yeast *sn-2* acyltransferase gene. *Plant Cell* 9:909–23

Tetrapyrrole Biosynthesis in Higher Plants

Ryouichi Tanaka and Ayumi Tanaka

Institute of Low Temperature Science, Hokkaido University, N19 W8, Kita-ku, Sapporo 060-0819, Japan; email: rtanaka@lowtem.hokudai.ac.jp, ayumi@lowtem.hokudai.ac.jp

Annu. Rev. Plant Biol. 2007. 58:321–46

The *Annual Review of Plant Biology* is online at plant.annualreviews.org

This article's doi:
10.1146/annurev.arplant.57.032905.105448

First published online as a Review in Advance on January 16, 2007

1543-5008/07/0602-0321$20.00

Key Words

chlorophyll, heme, photosynthesis

Abstract

Tetrapyrroles play vital roles in various biological processes, including photosynthesis and respiration. Higher plants contain four classes of tetrapyrroles, namely, chlorophyll, heme, siroheme, and phytochromobilin. All of the tetrapyrroles are derived from a common biosynthetic pathway. Here we review recent progress in the research of tetrapyrrole biosynthesis from a cellular biological view. The progress consists of biochemical, structural, and genetic analyses, which contribute to our understanding of how the flow and the synthesis of tetrapyrrole molecules are regulated and how the potentially toxic intermediates of tetrapyrrole synthesis are maintained at low levels. We also describe interactions of tetrapyrrole biosynthesis and other cellular processes including the stay-green events, the cell-death program, and the plastid-to-nucleus signal transduction. Finally, we present several reports on attempts for agricultural and horticultural applications in which the tetrapyrrole biosynthesis pathway was genetically modified.

Contents

INTRODUCTION

Higher plants produce four classes of tetrapyrroles, namely, chlorophyll, heme, siroheme, and phytochromobilin. Chlorophyll is a tetrapyrrole macrocycle containing Mg, a phytol chain, and a characteristic fifth ring (**Figure 1**). Chlorophyll serves an essential role in photosynthesis by absorbing light and transfering the light energy or electrons to other molecules. Higher plants have two chlorophyll species, chlorophyll *a* and *b* (**Figure 1**). The methyl group at the C7 position of chlorophyll *a* is replaced by a formyl group in chlorophyll *b*. Heme is another closed macrocycle that contains iron and it plays a vital role in various biological processes including respiration and photosynthesis. Similar to heme, siroheme also contains iron in its closed macrocycle. It is a prosthetic group of nitrite and sulfite reductase that plays central roles in nitrogen and sulfur assimilation, respectively. Phytochromobilin is a linear tetrapyrrole and a chromophore of phytochrome that perceives light and mediates its signal to the nuclei.

The major site of tetrapyrrole biosynthesis in plants occurs in plastids. Only the last few steps of heme biosynthesis are possibly localized in both mitochondria and plastids, a topic that will be discussed in greater detail later. The tetrapyrrole biosynthetic pathway is a multibranched pathway; the steps from glutamate to uroporphyrinogen III are common to the biosynthesis of all of the four classes of tetrapyrroles in higher plants. The next three subsequent steps from uroporphyrinogen III to protoporphyrin IX are common to chlorophyll, heme, and phytochromobilin synthesis. Lastly, the other remaining steps are unique to each branch of the pathway. We categorized those unique steps into the siroheme branch, the heme branch that includes phytochromobilin biosynthesis, the chlorophyll branch, and the chlorophyll cycle that refers to the interconversion steps of chlorophyll *a* and *b* (**Figure 1**). We put the chlorophyll cycle into a separate category because this cycle is regulated by a distinct mechanism from those that govern the other branches. In addition, the chlorophyll cycle has a unique feature distinct from other branches; not only is the chlorophyll cycle a part of the biosynthetic pathway, the latter two reactions of the cycle are also part of the chlorophyll degradation pathway.

In angiosperms, the complexity of the pathway and its intermediates has required many years of effort to identify most of the genes encoding the enzymes for tetrapyrrole biosynthesis (4, 52). These advances have vastly expanded our knowledge since the first cloning of a gene involved in tetrapyrrole synthesis in plants (72). The information gained from these studies has helped to solve the enzyme structures at an atomic resolution, and the structural models have in turn helped our understanding of enzyme chemistry. In addition, a key regulatory component of the pathway, the FLU (TIGRINA-D) protein, was identified (33, 47). Its existence has been indicated since 1974, when von Wettstein and coworkers isolated various mutants impaired with the pathway for tetrapyrrole synthesis (95). It is clear that our knowledge of the biochemical pathway has been greatly enhanced since these pioneering studies. In this review, we summarize the recent findings on tetrapyrrole biosynthesis in higher plants and focus on their cellular biological aspects. In particular, we highlight two major challenges that researchers are currently addressing. The first challenge is to elucidate how plants coordinate multiple steps of the pathway to meet the cellular demands for various tetrapyrroles. The second challenge is to further understand the newly identified interactions between tetrapyrrole biosynthesis and other cellular processes, which include the plastid-to-nucleus signaling pathway and the cellular programs for cell death or leaf senescence. Readers may refer to recent comprehensive reviews on this field for additional information (3, 11, 51, 83, 86, 93, 98, 102).

GluTR: glutamyl-tRNA reductase

GSA: glutamate-1-semialdehyde

ALA: 5-aminolevulinic acid

GSA-AT: glutamate-1-semialdehyde aminotransferase

OVERVIEW OF TETRAPYRROLE BIOSYNTHESIS IN PLANTS

This section gives a brief overview of the enzymatic steps for tetrapyrrole synthesis in plants. A number indicating the position of each step in the entire pathway map for tetrapyrrole biosynthesis precedes the description of each step (**Figure 1**). We reduce the description of enzyme properties and the reactions due to space limitations. Readers may refer to other articles that cover these biochemical properties in more detail (3, 102).

The Common Steps

The common part of the pathway consists of nine enzymatic steps. (**1**) Glutamyl-tRNA synthetase activates glutamate by ligating $tRNA^{Glu}$, a reaction which is common to plastid protein synthesis. (**2**) Glutamyl-tRNA reductase (GluTR) reduces the activated carboxyl group of glutamyl-tRNA to a formyl group to produce glutamate-1-smialdehyde (GSA). (**3**) Subsequently, intermolecular amino-exchange reactions convert GSA into 5-aminolevulinic acid (ALA), which is catalyzed by GSA aminotransferase (GSA-AT). ALA is the universal precursor

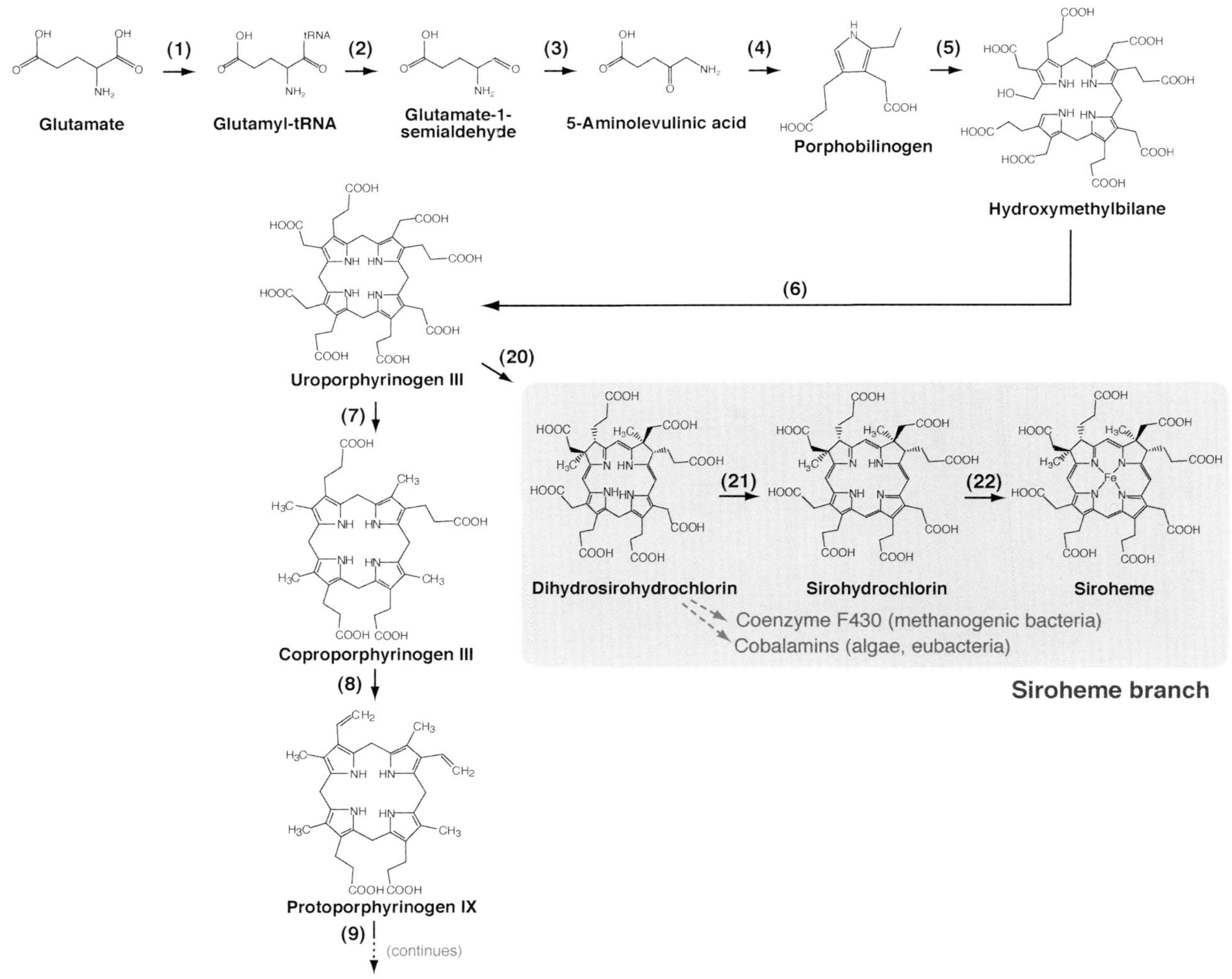

Figure 1

The tetrapyrrole biosynthetic pathway in higher plants. Numbers correspond to the description in the text. The enzymatic steps that do not occur in higher plants are shown with blue dashed lines. The International Union of Pure and Applied Chemistry (IUPAC) numbering scheme for C atoms is shown on the structure of chlorophyll *a*, which is located near the bottom of the figure.

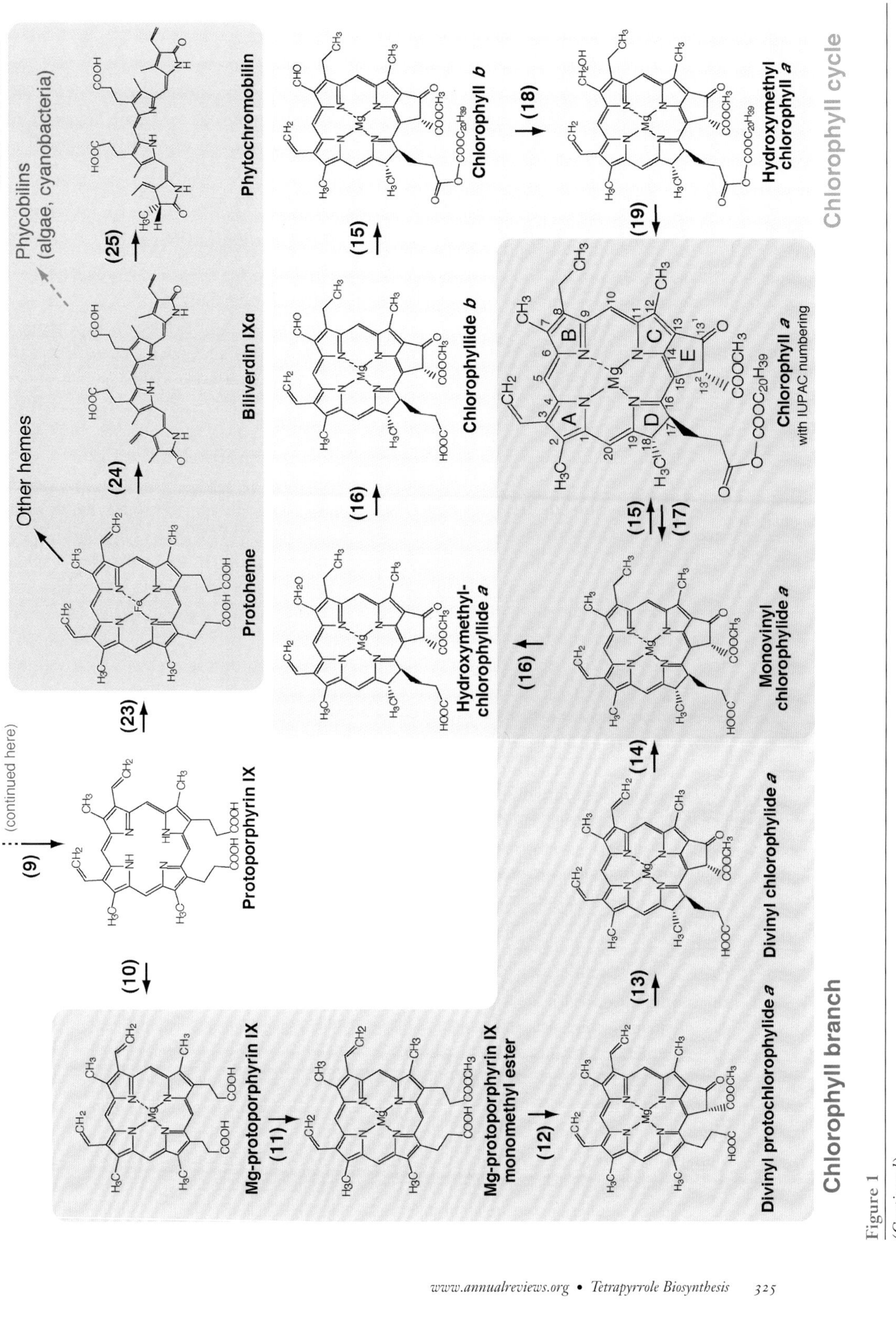

Figure 1
(*Continued*)

PBG: porphobilinogen

CPOX: coproporphyrinogen III oxidase

PPOX: protoporphyrinogen IX oxidase

MgCh: magnesium chelatase

MgMT: Mg-protoporphyrin IX methyltransferase

MgCy: Mg-protoporphyrin IX monomethyl ester cyclase

POR: NADPH: protochlorophyllide oxidoreductase

of tetrapyrrole biosynthesis in organisms. Animals, fungi, yeast, and some bacteria including purple nonsulfur bacteria, synthesize ALA by a distinct route that is a condensation of glycine and succinyl-coenzyme A by ALA synthase. However, plants, algae, and most bacteria including cyanobacteria and archaea synthesize ALA from glutamate (3). Interestingly, the structure of ALA synthase resembles that of GSA-AT, indicating that ALA synthase evolved from GSA-AT. This was concluded because GSA-AT is more ancient on an evolutionary time frame (73a).

(**4**) Two molecules of ALA are then condensed to form a pyrrole molecule, porphobilinogen (PBG), by ALA dehydratase. (**5**) Four molecules of PBG are polymerized by PBG deaminase. As a result, a linear tetrapyrrole, 1-hydroxymethylbillane, is produced. (**6**) Uroporphyrinogen III, the first closed macrocycle in the pathway, is produced from hydroxymethylbillane by uroporphyrinogen III synthase. Note that uroporphyrinogen III and most of the subsequent intermediate molecules of the biosynthetic pathways are photosensitizers and potentially generate singlet oxygen. (**7**) Uroporphyrinogen III is subsequently converted into coproporphyrinogen III in decarboxylation reactions by uroporphyrinogen III decarboxylase. (**8**) Two propinonate groups of coproporphyrinogen III are decarboxylated to form protoporphyrinogen IX by oxygen-dependent coproporhyrinogen III oxidase (CPOX). Although a homolog of bacterial genes encoding oxygen-independent coproporphyrinogen III oxidase exists in the genome of *Arabidopsis thaliana*, there is no reported evidence that this homolog encodes a functional oxygen-independent coproporphyrinogen III oxidase. (**9**) Subsequently, protoporphyrinogen IX oxidase (PPOX) extracts six electrons from protoporphyrinogen IX to form protoporphyrin IX. The plant-type PPOX is a FAD-containing oxidase whose molecular weight is approximately 55 kDa (35, 54). This type of PPOX is often referred to as HemY and found in *Bacillus subtilis*, yeast, and animals. *Escherichia coli* has a unrelated PPOX called HemG whose molecular weight is 15 kDa. Note that cyanobacteria are proposed to contain a currently unidentified type of PPOX because no homologs for *HemY* or *HemG* genes have been identified in the genomes of most cyanobacteria (58).

The Chlorophyll Branch

(**10**) The first step of the chlorophyll branch is the ATP-dependent insertion of the Mg^{2+} ion into protoporphyrin IX, a reaction that is catalyzed by magnesium chelatase (MgCh). This enzyme consists of three subunits, ChlH, ChlI, and ChlD, whose average molecular weights of those subunits are 140, 40, and 70 kDa, respectively. The ChlH subunit is predicted to have the catalytic site, whereas ChlI and ChlD bind to each other to activate ChlH. (102). (**11**) Chlorophyll synthesis subsequently proceeds to the transfer of a methyl group from S-adenosyl-L-methionine to the carboxyl group of the 13-propionate on Mg-protoporphyrin IX by Mg-protoporphyrin IX methyltransferase (MgMT), resulting in the formation of Mg-protoporphyrin IX monomethyl ester. (**12**) In the next reaction, Mg-protoporphyrin IX monomethyl ester cyclase (MgCy) incorporates an atomic oxygen (O) to Mg-protoporphyrin IX, forming 3,8-divinyl protochlorophyllide. In plants, an oxygen-dependent MgCy operates this reaction, whereas purple bacteria and cyanobacteria contain both oxygen-dependent and oxygen-independent MgCy (60). The oxygen-independent MgCy functions to incorporate O from H_2O into Mg protoporphyrin IX monomethyl ester (97b). It is likely that the catalytic subunit of plant MgCy is encoded by the *Crd1* gene (89). However, the involvement of an additional membrane protein and a stromal protein in the MgCy reaction has been suggested (69). (**13**) The D ring of 3,8-divinyl protochlorophyllide is reduced by protochlorophyllide oxidoreductase (POR) to form 3,8-divinyl chlorophyllide. This reaction is absolutely light-dependent in angiosperms because they contain only a light-dependent POR, which belongs to the short-chain dehydrogenase family. Other plants, algae, and cyanobacteria contain both light-dependent and light-independent POR, the latter of which is related to nitrate reductase (for reviews, see 13, 40). Due to the presence of light-independent POR, those organisms can synthesize chlorophyll in darkness. (**14**) The 8-vinyl group of the B ring of this compound is reduced by divinylchlorophyllide reductase (DVR) to form 3-vinyl chlorophyllide *a* (monovinyl chlorophyllide *a*). This enzyme can reduce 3,8-divinyl protochlorophyllide as well; however, the efficiency of this substrate is substantially lower than that of 3,8-divinyl chlorophyllide in vivo and in vitro (N. Nagata, R. Tanaka & A. Tanaka, unpublished results; C.A. Rebeiz, personal communication). Hence, we revise the conventional order of reaction steps illustrated in most previously published articles, and we place the DVR reaction after the POR reaction in the scheme in **Figure 1**. (**15**) In the last step of the chlorophyll branch, 17-propionate on the D ring of monovinyl chlorophyllide *a* is esterified with phytol-pyrophosphate by chlorophyll synthase and results in the formation of chlorophyll *a*.

The Chlorophyll Cycle

(**16**) A portion of the chlorophyll *a* pool is converted to chlorophyll *b*. The key reaction of this conversion is the oxygenation of chlorophyllide *a* to chlorophyllide *b* by the action of a Rieske-type monooxygenase, chlorophyllide *a* oxygenase (CAO). Among organisms that perform oxygenic photosynthesis, only land plants, green algae, and a few groups of cyanobacteria (prochlorophytes) possess chlorophyll *b*. The CAO sequences are contained in all the chlorophyll *b*-synthesizing organisms examined so far. These observations support the hypothesis that the evolution of chlorophyll *b* occurred only once in the evolutionary history of photosynthetic organisms (88; S. Satoh & A. Tanaka, unpublished results). In vitro experiments demonstrate that CAO performs a two-step oxygenation of chlorophyllide *a* as a substrate (59). 7-Hydroxymethyl chlorophyllide *a* is the intermediate of the CAO reaction. The product of CAO is chlorophyllide *b*, which should be subsequently phytylated by chlorophyll synthase to form chlorophyll *b*. (**17**) The phytol chain of chlorophyll *a* can be removed by chlorophyllase, and the resultant chlorophyllide *a* can be used as a substrate of CAO. Hence, plants can increase the amount of chlorophyll *b*, even when the de novo synthesis of chlorophyllide *a* is inhibited (i.e., in darkness). (**18** and **19**) Chlorophyll *b* can be reversibly converted to chlorophyll *a* through 7-hydroxymethyl chlorophyll *a*. These reactions are carried out by chlorophyll *b* reductase (CBR) and 7-hydroxymethyl chlorophyll *a* reductase. Recently, Kusaba et al. iosolated the putative CBR-encoding genes from rice. The predicted structures of those genes indicate that CBR belongs to a short-chain dehydrogenase family (M. Kusaba, unpublished results). It was suggested that 7-Hydroxymethyl chlorophyll *a* reductase was a ferredoxin-dependent enzyme (71); however, the gene encoding this enzyme has not yet been identified.

DVR: divinylchlorophyllide reductase

CAO: chlorophyllide a oxygenase

CBR: chlorophyll b reductase

The Siroheme Branch

(**20**) The first step of the siroheme branch is methylation of uroporphyrinogen III to form dihydrosirohydrochlorin. This reaction is catalyzed by an S-adenosyl-L-methionine-dependent methyltransferase that is encoded by the *UPM1* gene in *A. thaliana* (36). (**21**) This reaction is subsequently followed by an oxidation step that is catalyzed by an oxidase that still remains to be identified. Completion of this oxidation step results in the formation of sirohydrochlorin. (**22**) Subsequently, sirohydrochlorin ferrochelatase inserts Fe^{2+} into sirohydrochlorin to form siroheme. The enzyme responsible for this reaction is encoded by the *SirB* gene in *A. thaliana* (66).

The structure of this enzyme resembles that of protoporphyrin IX ferrochelatase (FeCh) in the heme branch (2).

AGI: a gene locus identifier assigned by the Arabidopsis Genome Initiative

The Heme Branch

(**23**) FeCh inserts Fe^{2+} into protoporphyrin IX to form protoheme (heme *b*). Plants synthesize other types of heme, such as heme *a* and heme *c*. Currently, the biosynthetic steps responsible for synthesizing these heme species have not been elucidated in plants. (**24**) Heme oxygenase catalyzes the oxidation and ring opening of protoheme and yields biliverdin IXα (**25**). This reaction is then followed by the conversion to *3E*-phytochromobilin by phytochromobilin synthase. Isomerization of *3E*-phytochromobilin into the *3Z* isomer occurs before the chromophore is bound to the phytochrome apoprotein. It is not currently known whether this isomerization step is catalyzed by an enzyme or if it proceeds spontaneously (see 86 for a review).

INTRACELLULAR LOCALIZATION OF ENZYMES INVOLVED IN TETRAPYRROLE BIOSYNTHESIS

As described in the introduction, nearly the entire tetrapyrrole pathway is localized in plastids. This localization is not surprising regarding the siroheme and the chlorophyll branch because the end products of these branches function exclusively in plastids. All of the identified steps for phytochromobilin synthesis also localize within plastids (29), even though binding of the chromophore to the apoprotein occurs in cytosol (for a review, see 86). On the contrary, the exact localization of heme biosynthesis is still unknown. Plastid and mitochondrial colocalization of the last three enzymes for heme biosynthesis, namely, CPOX, PPOX, and FeCh, in both plastids and mitochondria have been suggested, but there is still no consensus among researchers regarding this issue.

Recently, Williams et al. (101) reported that *Zea mays* has two isoforms of oxygen-dependent CPOX. With the exception of their putative transit peptide sequences, both of these isoforms are nearly identical on the amino acid level. The authors overexpressed the putative transit sequences of both CPOX isoforms in maize leaf cells with the GFP sequence fused to their C terminus, and found that one of the two fusion proteins localized in plastids, whereas the other localized to mitochondria. Santana et al. (70) reported that one of the two *A. thaliana* genes homologous to known CPOX genes (AGI code: AT4G03205) contains a deletion that causes a frameshift of its open reading frame; therefore, the authors concluded that AT4G03205 was a pseudogene. Furthermore, they showed that another isoform encoded by AT1G03475 exclusively localized in plastids. These results with *A. thaliana* CPOX were consistent with the previous report of Smith et al. (76), which showed that the CPOX activity was localized in plastids of etiolated pea leaves, but not in mitochondria. It is therefore questionable whether plants actually do contain a CPOX isoform that is targeted to mitochondria.

Previously, PPOX activity was detected both in plastids and mitochondria in plants (26, 76). Consistent with these reports, two isoforms of PPOX, namely, PPOX1 and PPOX2, were also found in various plant species including *A. thaliana* (37a, 54) and tobacco (35). In spinach, three forms of PPOX were found (99). Among the spinach isoenzymes, PPOX1 exclusively localizes in plastids. The two remaining isoforms consist of a larger PPOX2 (57 kDa) and a slightly shorter isoform (55 kDa) of PPOX2. It was determined that the slight variation in the two isoforms of PPOX2 is the result of alternative translation initiation from a single mRNA species. Interestingly, the larger and smaller isoforms of PPOX2 were imported into isolated plastids and mitochondria, respectively, in vitro. Supporting these observations, mitochondrial localization for the shorter form of PPOX2 was detected by immunoblotting. Furthermore, immunogold-labeling with PPOX antibodies showed signals in mitochondria. Collectively, these data indicate that PPOX isoforms have dual localizations in both plastids and mitochondria. However, van Lis et al. (92) reported that there is only a single gene encoding PPOX in a green alga, *Chlamydomonas reinhardtii*, and the gene product is exclusively found in plastids.

In contrast to PPOX, there is no strong evidence for localization of FeCh in plant mitochondria. Cornah et al. (8) and Masuda et al. (39) reported that most FeCh activity was associated with plastids. Lister et al. (37b) found that either of the two FeCh isoforms from *A. thaliana* was imported into chloroplasts in vitro. Masuda et al. (39) found that GFP-fusion proteins with either of two isoforms of FeCh from cucumber were targeted to plastids, but not to mitochondria. The specific antibodies against either of the two isoforms of FeCh detected signals only in plastids (39). Furthermore, van Lis et al. (92) also found that there is only a single gene for FeCh, which is targeted to plastids in *C. reinhardtii*. Their study indicated that heme is exclusively produced in plastids and exported to cytosol and mitochondria in *C. reinhardtii*. These data collectively indicate that FeCh is only localized in plastids, but the possibility cannot be ruled out that a fraction of FeCh or an unidentified protein shows minor FeCh activity in mitochondria. If the FeCh protein does not exist in mitochondria, there would not be a biological need for PPOX in this organelle. Further studies are necessary to solve the present discrepancy between the localization of PPOX and FeCh. For example, the analysis of knockout lines for PPOX or FeCh isoforms could be utilized to solve this problem. If the dual-localization hypothesis is true, knockout of the gene encoding a mitochondrial isoform of PPOX or FeCh should have a severe impact on the activities of mitochondrial heme-binding proteins.

The localization of the enzymes involved in tetrapyrrole synthesis within plastids is also a subject of debate. PPOX (99), FeCh (80), MgMT (5), MgCy (89), and CAO (12) are reported to be localized in both envelope and thylakoid membranes (for a review, see 11). Moseley et al. (49) observed that the loss of function of one of the two MgCy isoforms could not be fully compensated by overexpression of the other isoform. Based on this observation, Tottey et al. (89) proposed that chlorophyll-synthesizing enzymes of envelope and thylakoid membranes may supply chlorophyll to different chlorophyll-binding proteins, i.e., the core complexes and the major light-harvesting complexes of PSII (LHCII). The significance of the dual localization of these proteins still remains to be examined.

LHCII: light-harvesting complexes of PSII

REGULATION OF TETRAPYRROLE BIOSYNTHESIS

Overview of the Regulatory Mechanisms

When we consider the biological roles of the tetrapyrrole biosynthesis pathway, it is reasonable to assume that the activity of each branch in the pathway reflects the cellular demand for each end product. At the same time, the activity of the overall tetrapyrrole synthesis may be controlled in response to the demand for a particular tetrapyrrole. For example, to increase the synthesis of chlorophyll, the total tetrapyrrole biosynthetic activity should increase as well as the activity of the chlorophyll branch. If this increase does not occur, the substrate flow to other branches would be sacrificed. Therefore, it is reasonable to assume that plants control the enzymatic activities of the branch points as well as those of the common part of the pathway. Actually, ALA formation was the first regulatory point and it was subjected to multiple regulatory mechanisms. Regulation at this point presents a significant advantage: All of the intermediate molecules of the later steps may generate potentially toxic singlet oxygen under illumination when they accumulate excessively. Genetic modification of the enzyme expression impaired the control mechanisms of tetrapyrrole biosynthesis and it actually resulted in the generation of reactive oxygen species. This topic is discussed below.

Functional studies on mutants of various enzymes of the pathway have greatly contributed to our understanding of the key features of various regulatory mechanisms for tetrapyrrole biosynthesis in plants. We summarize the phenotypes of representative mutants and transgenic plants in which the activity of a certain enzymatic step was impaired. (Follow the Supplemental Material link from the Annual Reviews home page at **http://www.annualreviews.org** to see the **Supplemental Table**.) Interestingly, mutants that were deficient in enzymatic activities of each branch showed different phenotypes from each other. Mutants that had altered ALA formation did not show a necrotic phenotype, but they accumulated lower levels of chlorophyll and heme. Mutants of the latter steps of the common part of the pathway (from uroporphyrinogen III decarboxylase to PPOX) accumulated a substantial amount of intermediate molecules and formed necrotic lesions upon illumination. These results suggested that the levels of the enzymes involved in the steps between uroporphyrinogen III to protoheme are maintained at higher levels in plants. If these enzymes were not maintained at high levels, there would always be a risk of accumulating potentially dangerous intermediate molecules.

Among the mutants of the heme branch, only the FeCh antisense plants showed a necrotic phenotype (61), whereas mutants of the later steps did not show similar necrosis phenotypes (for a review, see 85b). It is possible that the cellular demand for heme biosynthesis might be reflected by the size of the free heme pool. As described below, heme inhibits the activity of GluTR and ALA formation. When heme-binding proteins or phytochrome is actively synthesized, the size of the heme pool may shrink, which may lead to the increased activity for ALA formation, and vice versa.

In contrast to the FeCh mutants, mutants deficient in MgCh did not show a detectable increase in the accumulation of intermediates, nor did they exhibit clear necrotic lesion formation (14, 63). Both MgMT and MgCy mutants showed increased accumulation of the precursor Mg-protoporphyrin IX or its momomethylester (1b, 14). However, the increase in precursor accumulation was so small that most of these mutants did not show obvious necrotic phenotypes, at least under normal growth conditions (1b, 69, 89) (P. E. Jensen & M. Hansson, personal communication). These observations indicate that suppressing the activities of MgCh, MgMT, and MgCy negatively regulated the earlier steps of the pathway, although the mechanism of this feedback control has not been identified. Among these three enzymes, MgCh possibly plays a central role in regulation because only MgCh mutants did not show increased accumulation of the intermediates (63). We should emphasize that this feedback inhibition may not be due to the accumulation of precursors, such as protoporphyrin IX, because protoporphyrin IX accumulation did not lead to suppression of ALA formation in the FeCh mutants (61).

Below we describe the key features of tetrapyrrole regulation in plants in more detail and emphasize the recent findings. We also highlight the regulation of the chlorophyll cycle. The regulation of this cycle is distinct from all other branches.

Regulation of ALA Formation

Among the three steps of ALA synthesis, glutamyl-tRNA reduction is most important in terms of regulation. This step is actually the committed step of tetrapyrrole biosynthesis because the preceding step, which ligates $tRNA^{Glu}$ to glutamate, is shared with protein synthesis in plastids. One of the most obvious observations demonstrating the key role of GluTR is that failure to repress GluTR activities in darkness let to increased accumulation of protochlorophyllide in the *flu* mutant of *A. thaliana* (17). These data indicated that glutamyl-tRNA reduction is the limiting step of chlorophyll biosynthesis.

GluTR is encoded by the small *HemA* gene family in angiosperms. All higher plants examined so far contain at least two *HemA* genes (85a). One type of these genes (*HemA1*) is expressed primarily in photosynthetic tissues, and its expression is induced by light (6, 23, 42, 84, 85a). The expression pattern of *HemA1* is strongly correlated with those of *Lhcb1*, which encodes a light-harvesting complex protein of photosystem II (42, 44). Thus, it is likely that expression of the *HemA1* gene expression reflects the cellular demand for chlorophyll synthesis. In contrast, other *HemA* genes (e.g., cucumber *HemA2* or barley *HemA3*) are either expressed ubiquitously throughout plants, or primarily in nonphotosynthetic tissues, and their expression is not light-inducible (6, 30, 42, 84, 85a). Expression of this second type of *HemA* gene may reflect the demand for heme, siroheme, or phytochromobilin.

Recent findings suggest that controlling the GluTR activity is mainly achieved through feedback regulation by the end products, not through transcriptional regulation. Klaus Apel and coworkers identified an essential component of the feedback regulatory mechanism of GluTR activity by applying elegant genetic approaches. They first identified the *fluorescent (flu)* mutant of *A. thaliana* that accumulated an excessive amount of protochlorophyllide in darkness (47). The mutant was not only deficient in repressing ALA formation in darkness, it also showed an increased ALA-synthesizing capacity in the light (17). They found that the wild-type *Flu* gene encodes a protein of approximately 27-kDa molecular weight that localizes in plastid membranes. Using the yeast two-hybrid system, they demonstrated that FLU directly interacts with GluTR1 encoded by *HemA1*. Interestingly, FLU does not interact with GluTR2, which is encoded by *HemA2* (17, 46). These data suggest that FLU suppresses the GluTR1 activity in response to the excess accumulation of protochlorophyllide and other intermediate molecules of the chlorophyll branch. Moreover, they isolated a second-site mutation (*ulf3*) that suppressed the *flu* phenotype in *A. thaliana* (17). This mutation was mapped to the *Hy1* locus, which encodes heme oxygenase. These data indicate that reduction in the heme oxygenase activity in the *ulf3* mutant led to accumulation of heme. In turn, the accumulation of heme suppressed GluTR activity, and the suppression occurred independent from the action of FLU. These data were the first genetic evidence to demonstrate the feedback regulation of heme on the GluTR activity. This observation is consistent with the reports by Gamini Kannangara and coworkers demonstrating that heme binds GluTR and suppresses its activity in vitro (65, 96). Recently, Strivastava et al. (77) observed that the inhibition of the GluTR activity by heme depended on the presence of an unidentified soluble protein in *C. reinhardtii*. Thus, it is likely that protein-protein interaction is essential in this heme-mediated inhibitory mechanism. It should be emphasized that GluTR1 is subjected to a feedback control from both the chlorophyll branch and the heme branch. This may imply that there is no specific ALA-synthesizing route for heme or chlorophyll, but the ALA-synthesizing enzymes are shared by at least the heme and chlorophyll branches of the pathway. Further investigation focusing on the control mechanism of GluTR activity may provide us with information regarding how plants regulate multiple branches of the pathway simultaneously.

Regulation at the Branching Point of Heme and Chlorophyll Biosynthesis

Activities of the branch point enzymes are possibly regulated by the redox states within chloroplasts. Such a regulation would be reasonable because the redox states within chloroplasts indicate whether the

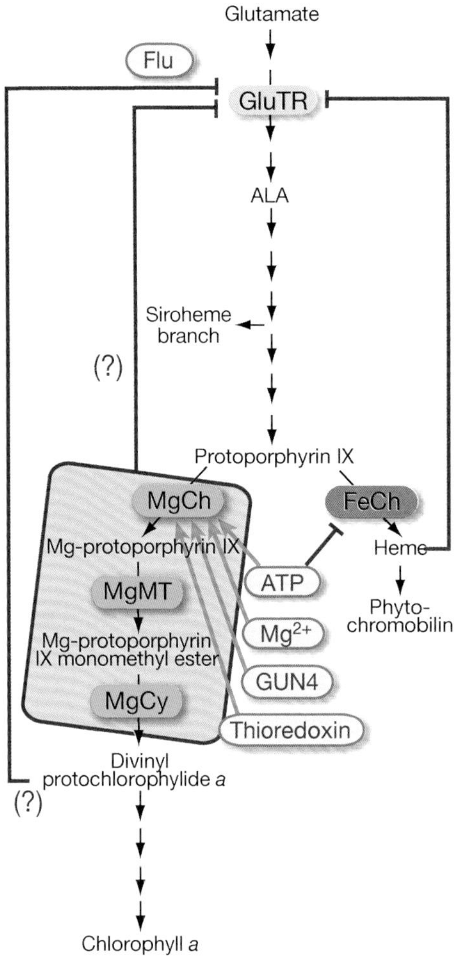

Figure 2

Summary of the current models describing the regulatory mechanisms for tetrapyrrole biosynthesis. Only representative intermediates and enzymes are shown. Green arrows indicate positive regulation, and red arrows indicate negative regulation.

Thioredoxin: a member of the protein disulfide oxidoreductase family. In chloroplast stroma, it activates various proteins by reducing disulfide bonds

photosynthetic activity is excessive or insufficient. In this sense, it is possible that the redox state may reflect the cellular demand for chlorophyll synthesis. In support of this, the ChlI subunit of MgCh was identified as a thioredoxin-target protein (T. Masuda, personal communication) (**Figure 2**). Jensen et al. (27) also provided evidence supporting the redox regulation of the MgCh activity. They found that the ChlH subunit as well as the protein complex formed by ChlI and ChlD subunits of MgCh from a cyanobacterium, *Synechocystis* PCC6803, was inactivated by N-ethylmaleimide that covalently binds thiol groups. Thioredoxin-dependent regulation may account for the observation that MgCh was activated/inactivated in the day/night, respectively (62). Note that GSA-AT, ALA dehydratase, and uroporphyrinogen III decarboxylase were also in the list of potential thioredoxin targets (7).

The ATP/ADP ratio and free Mg^{2+} concentration, which are interdependent in stroma, may also participate in regulating MgCh activity (**Figure 2**). Stromal-free Mg^{2+} concentration increases from 0.5 mM to 2.0 mM upon a shift from darkness to light in spinach chloroplasts (25). Reid & Hunter (67) argued that these changes in Mg^{2+} concentration may activate/inactivate MgCh in vivo in which activity was highly influenced by Mg^{2+} concentration in vitro. The concentration of Mg^{2+}, which is required for the full activation of *Synechocystis* MgCh, is lowered from approximately 6 mM to 2 mM in vitro by the presence of GUN4 protein, a recently identified porphyrin-binding protein that enhances MgCh activity (10). Thus, it is possible that Mg^{2+} concentration regulates the MgCh activity in chloroplasts upon a dark-to-light shift.

Mg^{2+} concentration possibly affects MgCh in a different way; it may influence the localization of MgCh subunits within chloroplasts. Gibson et al. (15) and Nakayama et al. (53) report that when chloroplasts were disrupted in the presence of 5 mM Mg^{2+}, ChlH was associated with membrane. On the contrary, when MgCh was in the presence of 1 mM Mg^{2+}, the protein was found in the stroma fraction. Because PPOX is reported to be localized in the chloroplast membrane (99), it is possible that translocation of ChlH to envelope membrane enables the substrate transfer from PPOX to ChlH.

Another factor that might influence MgCh activity is the ATP/ADP ratios. ATP is necessary to activate MgCh (16, 67), whereas it inhibits FeCh activity (8). In maize chloroplasts, ATP/ADP ratios increased from 1 in dark conditions to 4 in the light (91). Hence, changes in ATP/ADP ratios during day/night shifts may affect the channeling of protoporphyrin IX to the heme or chlorophyll branch (**Figure 2**).

As previously mentioned, GUN4 enhances the activity of MgCh. The *GUN4* gene was originally identified from an *A. thaliana gun4* mutant in which the plastid-to-nucleus signaling pathway was impaired and chlorophyll synthesis was suppressed (32). GUN4 binds both protoporphyrin IX and Mg-protoporphyrin IX and stimulates the activity of MgCh (10, 94). Disruption of the *GUN4* gene reduced the cellular heme levels in *Synechocystis*, suggesting that the protein is also involved in heme biosynthesis (100). It is therefore possible that GUN4 plays a role in the control of the substrate flow into the heme or chlorophyll branch.

Complex Formation of Proteins Involved in Tetrapyrrole Synthesis

For many years, it has been suggested that certain pairs of enzymes involved in tetrapyrrole synthesis form protein complexes to enable the efficient channeling of substrates. Recent progress in the analysis of protein crystal structure for those enzymes presents clear evidence for complex formation.

From the solved crystal structure of GluTR from an archaeon, *Methanopyrus kandleri*, Moser et al. (50) suggested that this protein forms a complex with GSA-AT. Biochemical analysis also supports this model (38, 55). The product of GluTR, glutamate-1-semialdehyde, is likely released from the opposite side of the glutamate recognition site of GluTR, and it immediately enters the active site of GSA-AT (50). This direct channeling may prevent highly reactive aldehyde glutamate-1-semialdehyde from entering aqueous environments.

The formation of a tentative complex of PPOX and FeCh was suggested by Koch et al. (28) and was based on the solved crystal structures of tobacco PPOX (28) and human FeCh (103). They proposed that the product of PPOX, protoporphyrin IX, is directly channeled to the active site of FeCh while the proteins are forming a complex. After the insertion of Fe^{2+} into protoporphyrin IX, FeCh may dissociate from PPOX to release the product (heme) to the membrane. In this way, the accumulation of protoporphyrin IX, which is a highly photosensitizing molecule, would be efficiently prevented. According to this model, PPOX is not always forming a complex with FeCh; thus, MgCh (or, possibly GUN4 protein) should have a chance to access protoporphyrin IX.

MgCh and MgMT are proposed to form a tight complex (18, 74). Shepherd & Hunter (74) reported that the ChlH subunit of *Synechocystis* PCC6803 interacted with MgMT and enhanced its activity. This stimulating effect was not simply due to direct channeling of the substrate; they suggested that ChlH is involved in the reaction chemistry of MgMT.

In 1971, Shlyk proposed an intriguing idea of a mega complex of enzymes that were involved in chlorophyll biosynthesis (75). This idea was based on physiological and biochemical experiments. The hypothetical mega complex was termed the "chlorophyll biosynthetic center." In the future, further improvement of immunochemical and bioimaging techniques may reveal the existence of such a tetrapyrrole biosynthetic center in plastids.

Regulation of the Chlorophyll Cycle

Regulation of the chlorophyll cycle provides a unique example of a regulatory mechanism in which degradation of the key enzyme plays an essential role (**Figure 3**). Plants strictly control the chlorophyll *a* to *b* ratios because each photosynthetic protein requires these pigments in a fixed stoichiometry for the optimal energy transfer. The ratios are regulated through the activity of the chlorophyll cycle, and they always range from 3 (under normal conditions) to 4 (under stressful conditions) in leaves of higher plants. Overexpression of the *CAO* mRNA in plants resulted in only a minor effect on the chlorophyll *a* to *b* ratios, which decreased to 2.7 (64, 81, 82). Even though the *CAO* mRNA levels increased by several-fold in the overexpressors, the CAO protein levels were undetectable in the *CAO* overexpressors as well as in wild type (105). Surprisingly, removing the plant-specific sequence (designated the A domain) from the *CAO* sequence greatly enhanced the accumulation of the CAO protein and resulted in a very low chlorophyll ratio. In this case, the ratios decreased to 2.2, which was never observed with wild type (105). We then overexpressed the GFP-A domain-fusion protein in wild-type *A. thaliana* as well as in a *CAO*-deficient mutant, *chorina1*. The GFP-A domain fusion accumulated to only a trace amount in the wild-type background, whereas it accumulated significantly in the *chorina1* background (105). These results indicate that the A domain suppresses the protein accumulation only when chlorophyll *b* is synthesized, and that the A domain-induced

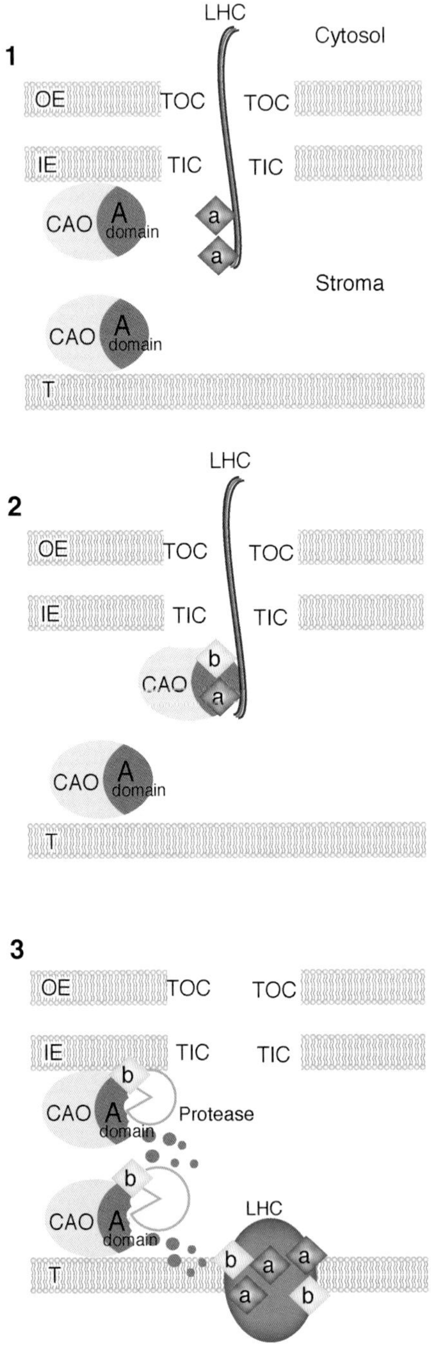

Figure 3

The tentative models showing the role of chlorophyllide *a* oxygenase (CAO) in the transport of light-harvesting complex (LHC) apoproteins, and the control of CAO accumulation. 1. CAO localizes on the inner envelope (IE) and thylakoid (T) membranes. The precursor of the LHC apoprotein is synthesized in cytosol and transported into chloroplasts through translocons in the outer envelope membrane of chloroplasts (TOCs) and translocons in the inner envelope membrane of chloroplasts (TICs). It is hypothesized that the LHC apoprotein binds chlorophyllide *a* during transport [Hoober & Eggink (21)]. 2. CAO on the inner envelope membrane interacts with translocons and LHC, and converts chlorophyllide *a* into chlorophyllide *b*. CAO on the thylakoid membrane may not be involved in the transport of LHC apoprotein. Binding of chlorophyllide *b* to LHC apoprotein may enable the folding and assembling of the apoprotein with chlorophyll *a* and *b* and carotenoid pigments. 3. LHCII is inserted into the thylakoid membrane. Chlorophyll *b* triggers degradation of CAO. The presence of the A domain of CAO is a prerequisite for the degradation of CAO. Currently, it is not clear whether chlorophyll *b* directly interacts with the A domain of CAO to control the stability of CAO.

suppression mechanism can control the accumulation of virtually any protein fused to the A domain.

Coordination of the Entire Pathway for Tetrapyrrole Biosynthesis by Transcriptional Control

As described in the previous sections, fine regulation of the pathway for tetrapyrrole biosynthesis was mainly achieved by modulating enzymatic activities or through protein-protein interactions. Such regulation appears to be compensated by transcriptional control of the genes encoding proteins involved in tetrapyrrole biosynthesis.

Matsumoto et al. (42) performed a transcriptome analysis of *A. thaliana* seedlings using a DNA microarray that covered the majority of genes involved in tetrapyrrole metabolism that were identified at the time of analysis. Their data demonstrated that most of the genes showed synchronized and light-dependent expression patterns. Notably, a group of genes encoding GluTR (HemA1), the ChlH subunit of MgCh, MgCy, and CAO, were regulated by circadian rhythms. These results indicate that transcription of the genes encoding the key enzymes of the pathway was subjected to multiple regulatory mechanisms.

The signal transduction pathway leading to the control of genes for tetrapyrrole biosynthesis is only partly understood. Studies with phytochrome-deficient mutants of *A. thaliana* demonstrated that both phytochrome A and B mediate the light signal to induce *HemA1* gene expression (43, 45). Huq et al. (22) reported that mutants deficient in a phytochrome-interacting transcription factor, PIF1, failed to repress chlorophyll biosynthesis in the dark, resulting in excess accumulation of protochlorophyllide. The authors proposed that PIF1 suppresses expression of the genes involved in chlorophyll biosynthesis in darkness and the phytochrome-mediated light signal releases the negative regulation of PIF1. However, it is currently difficult to anticipate which gene is regulated by PIF1 because there is no report describing that induction of *HemA1* or other genes involved in chlorophyll synthesis led to an increased accumulation of chlorophyll or its intermediates. It is likely that the expression of genes involved in tetrapyrrole biosynthesis is regulated by a complex mechanism comprising many unidentified components.

Recently, Nogaj et al. (56) reported that the protein levels of GluTR are not correlated with the rate of chlorophyll synthesis. These results indicate that modulation of the GluTR activity is more essential in terms of regulating chlorophyll biosynthesis. This report may not immediately deny the importance of transcriptional regulation in tetrapyrrole biosynthesis in a general way. In general, it is thought that when plants attempt to finely modulate the activity of an enzyme, plants should accumulate a sufficient amount of the enzyme. Otherwise, the amount of the enzyme would limit the reaction, and a fine regulation would not be possible. This lack of fine regulation would occur because transcriptional control of protein levels generally takes longer than modulation of an enzyme activity. Thus, it is reasonable to assume that the plants always synthesize a slight surplus of a key enzyme that was actually needed, and activity of the key enzyme would be controlled in order to achieve a fine regulation.

COORDINATED SYNTHESIS AND ASSEMBLY OF CHLOROPHYLL AND APOPROTEINS

When the synthesis of chlorophyll exceeds the accumulation of chlorophyll-binding apoproteins, reactive oxygen species are generated, ultimately leading to cell death. However, when chlorophyll synthesis is not active enough, the amount of fully functional chlorophyll-binding proteins is insufficient to gain the optimal photosynthetic activities (79). Recently, researchers identified various coordination mechanisms for chlorophyll synthesis and apoprotein accumulation in plants.

Chlorophyll and Its Precursors Modulate the Import and Stability of Chlorophyll-Binding Proteins

In chlorophyll-*b*-deficient mutants, the accumulation of LHCII was severely reduced. Kuttkat et al. (31) reported that an analog of chlorophyll *b*, Zn-pheophytin *b*, is required to import LHC apoproteins into etioplasts in vitro. Similarly, Reinbothe et al. (68) found that the import activity of LHC apoproteins into chloroplasts isolated from a CAO-deficient mutant of *A. thaliana* was nearly completely blocked. Furthermore, they demonstrated that CAO interacts with various subunits of translocons present in outer and inner envelope membranes, using a crosslinker, ^{125}I-labeled N-[4-(p-azidosalicyl-amido)butyl]-3 (2-pyridyldithio) propionamide. These observations are consistent with the hypothetical model developed by Hoober & Eggink (21) (**Figure 3**), in which they propose that during the import of LHC apoproteins into chloroplasts, the proteins attach chlorophyllide *a* with specific amino acid residues on the apoproteins. CAO converts the pigments into chlorophyllide *b* on the apoprotein. This conversion helps LHC proteins form a proper conformation, which facilitates stable insertion of the protein-pigment complex into the thylakoid membrane. Evidence supporting this direct interaction was described in *Synechocystis* PCC6803 (104). In their study, they transformed this organism with CAO and/or Lhcb1 genes from *A. thaliana* and found that cotransformation of CAO and Lhcb1 drastically increased the biosynthesis of chlorophyll *b* when it was compared to the transformants that only expressed CAO. These data indicate that presence of LHC protein enhanced the CAO activity.

This hypothesis does not exclude the possiblily of CAO localization on thylakoid membranes. We obtained evidence showing that CAO converts chlorophyll *a* to *b* on thylakoid membranes without coupling to the import process of LHCII (**Figure 3**). We overexpressed the CAO sequence (PhCAO) from a cyanobacterium *Prochlorothrix hollandica* in the *chlorina1* mutant of *A. thaliana* that was deficient in native CAO activity (19). These overexpressing plants produced a large amount of chlorophyll *b*, which was nearly as much as chlorophyll *a*. We found that LHC proteins accumulated to the normal level, even though most of the PhCAO protein localized within the thylakoid membrane (19). These observations demonstrate that localization of CAO in an envelope membrane where translocons should function is not a prerequisite for LHC protein accumulation. In other words, LHC proteins can be transported to thylakoid membranes without the occurrence of import-coupled chlorophyll *b* synthesis on the envelope membrane.

The aforementioned study with PhCAO overexpressing plants suggests that CAO may have another role related to the proper distribution of chlorophyll *b* in photosynthetic machineries, which are composed of many chlorophyll-binding proteins (19). In wild-type plants, proteins in the core antenna complexes only bind chlorophyll *a*, whereas the accessory antenna complexes that are made of LHC proteins bind both chlorophyll *a* and *b*. In the transgenic plants, 40% of the chlorophyll *a* molecules on the core complexes of photosystem I and II were replaced by chlorophyll *b* without a significant loss of photosynthetic activities (19). These observations indicate that the core complexes can bind both chlorophyll species, and that the wild-type CAO plays a role in the selective distribution of chlorophyll *b* in the peripheral antenna complexes. Because the cyanobacterial CAO sequence lacks a region corresponding to the A domain of the CAO sequences of higher plants, we speculate that the A domain takes part in the correct distribution of pigments into photosynthetic machineries.

NEWLY IDENTIFIED INTERACTIONS BETWEEN TETRAPYRROLE BIOSYNTHESIS AND OTHER CELLULAR PROCESSES

Plastid-to-Nucleus Retrograde Signaling

Because the nuclear genome encodes most plastid proteins (24), the nucleus should monitor the plastid conditions to optimize gene expression. It is believed that the plastid sends a signal to the nucleus (referred to as retrograde signaling) in various ways, and Mg-protoporphyrin IX is believed to be one of the plastid signal molecules (57). Compared with wild-type *A. thaliana*, mutations in the genes encoding heme oxygenase, phytochromobilin synthetase, GUN4, and the ChlH subunit of MgCh, whose mutants were designated *gun2*, *3*, *4*, and *5*, mutants, respectively, led to higher *Lhcb1* mRNA levels in the presence of norflurazon. Strand et al. (78) observed a 15-fold increase in the Mg-protoporphyrin IX accumulation by norflurazon treatment in wild type, whereas the levels of the Mg-protoporphyrin IX increased by fivefold in the *gun* mutants. Based on these observations, they proposed that Mg-protoporphyrin IX and Mg-protoporphyrin IX momomethyl ester acted as signaling molecules to repress *Lhcb1* expression in the presence of norflurazon (**Figure 4**). However, there are conflicting results to this model. Two other mutants of MgCh (*cs* and *ch42)* that have defects in one of the two *ChlI* genes did not show the *gun* phenotype, although the MgCh activity was severely impaired (48). Alawady & Grimm (1b) report a decrease in the Mg-protoporphyrin IX levels in tobacco MgMT overexpression lines that did not increase *Lhcb1* expression. Because there appear to be multiple signaling pathways between plastids and the nucleus, and because these pathways likely interact with each other, further dissection of the *gun* phenotypes is necessary to understand the nature of plastid signals.

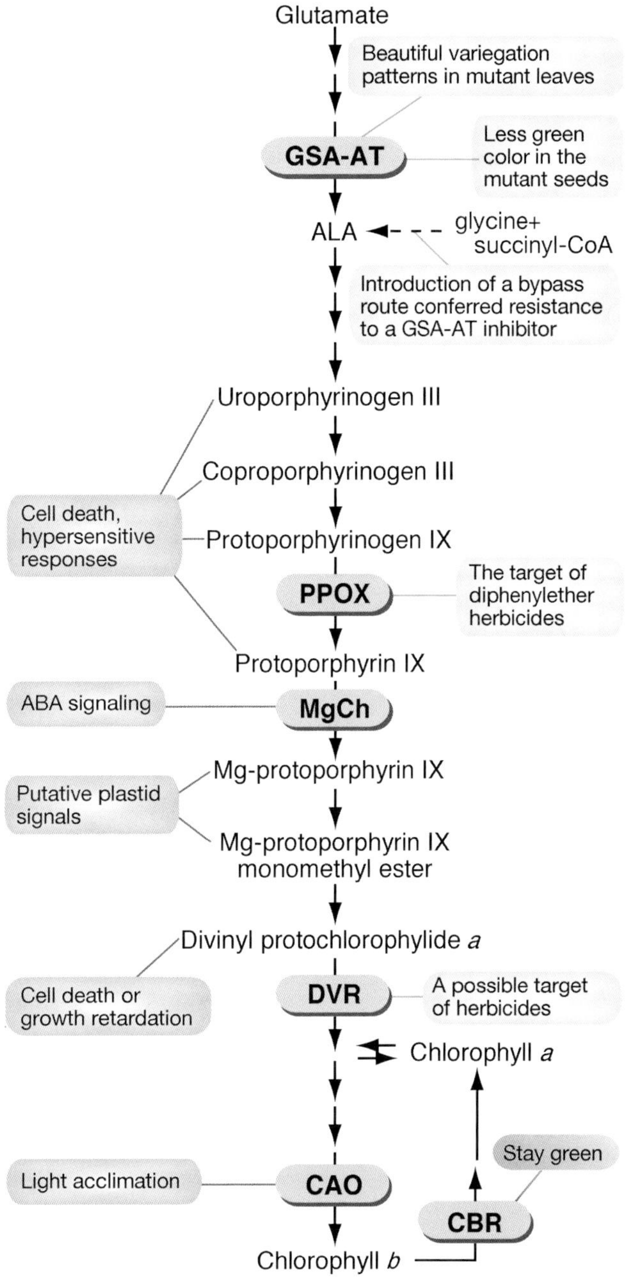

Figure 4

Novel aspects of recent tetrapyrrole research. Red boxes show newly identified interactions of the tetrapyrrole-synthesizing pathway with other biological processes. Blue boxes show examples of agricultural and horticultural applications by genetic manipulation of the enzymes involved in tetrapyrrole biosynthesis.

Nevertheless, these studies imply the possible dependency of nuclear-encoded photosynthesis gene expression on tetrapyrrole biosynthesis.

Hypersensitive responses: one type of the defense responses of plants to pathogen infection. The responses include rapid and local cell death to restrict the spreading of a pathogen

Singlet-Oxygen-Mediating Signaling

Isolating the *flu* mutant was an exciting beginning toward understanding the singlet-oxygen-mediating retrograde-signaling pathway. Here again, an elegant genetic approach was used to isolate the second-site mutation that suppresses the cell-death phenotype of *flu*, which was induced by a combination of protochlorophyllide accumulation in the dark and subsequent illumination (97a). In the successfully isolated mutant, which was designated *executer1*, protochlorophyllide-induced singlet-oxygen formation did not induce cell death. This report was actually the first to demonstrate that the singlet oxygen did not simply induce necrotic cell death, but led to the programmed death of plant cells (**Figure 4**). They revealed that this programmed cell death was clearly distinct from that induced by H_2O_2. At present, it is not clear what the biological role of this cell-death program is. Because tetrapyrrole is the major source of singlet-oxygen generation, it is tempting to speculate that plants use tetrapyrrole molecules for a singlet-oxygen generator with which plants can control the survival of certain cells. Note that the accumulation of intermediate molecules of the common part of tetrapyrrole biosynthesis induced various biological responses such as hypersensitive responses, one of the defense responses of plants to a pathogen (see **Supplemental Table**). These responses were most likely mediated by singlet oxygen. It has been reported that intermediate tetrapyrrole molecules accumulate not only in mutants, but also in wild-type plants under biotic and abiotic stresses (1a, 87). Therefore, it is also likely that these same molecules induce various biological responses in wild-type plants.

Abscisic Acid Signaling

A novel role of ChlH was recently reported by Shen et al. (73b). They identified the ChlH protein as a novel receptor of a phytohormone, abscisic acid (ABA), which regulates plant responses to stressful conditions and various developmental processes. Generally, a hormone receptor binds its substance strongly at a low dissociation constant. The ChlH protein fulfills this requirement: Specifically, it binds Mg-protoporphyrin IX at a low dissociation constant of 32 nM. Genetic evidence also demonstrates the involvement of ChlH in ABA signaling: The ChlH overexpressing plants became hypersensitive to ABA, whereas *chlH*-deficient mutants showed ABA-insensitive phenotypes. The authors demonstrated that the role of ChlH in ABA signaling is apart from its function in chlorophyll biosynthesis or in the plastid-to-nucleus signaling; in other words, the ChlH protein is bifunctional. It is interesting to learn why and how ChlH mediates ABA signals.

Selective Stabilization of LHCII and Appressed Thylakoid Membranes by Chlorophyll *b* Reductase

For many years, it has been known that chlorophyll *b*-deficient mutants have reduced LHCII and grana stacks (appressed regions of thylakoid membranes). It has been implicated that chlorophyll *b* stabilizes LHCII, and LHCII helps form grana lamellae. Recently, a rice mutant deficient in the reduction of chlorophyll *b* was isolated (M. Kusaba, unpublished results) and was designated as *ncy1*. This mutation was mapped to a gene locus encoding a short-chain dehydrogenase. This gene belongs to a small gene family in rice and we show that at least one member of the gene family encodes a functional CBR in vitro (M. Kusaba, H. Ito & A. Tanaka, unpublished results). In *ncy1*, degradation of LHCII and grana lamellae was suppressed during leaf senescence. This observation provides supporting evidence for the role of chlorophyll *b* in stabilizing LHCII, and for the role of LHCII in forming grana lamellae.

AGRICULTURAL AND HORTICULTURAL APPLICATIONS

Transgenic plants in which the tetrapyrrole biosynthetic pathway was genetically modified are potentially useful for agricultural and horticultural applications (**Figure 4**). For example, because tetrapyrrole enzymes are the targets of a number of herbicides, attempts were made to produce transgenic plants tolerant to certain herbicides. Zavgorodnyaya et al. (106) introduced the gene encoding yeast ALA synthase into the tobacco genome, so that the transgenic plants could bypass the glutamate-to-ALA-synthesizing route, resulting in the tolerance of the transgenic plants to a specific inhibitor of GSA-AT, gabaculine.

Because the HemY-type PPOX is the target of the diphenylether herbicides such as acifluorfen or oxyfluorfen (41), there were attempts to produce diphenylether herbicide-tolerant plants by overexpressing the *HemY* gene (34, 99). If identified, overexpression of a cyanobacterial PPOX would lead to further tolerance of transgenic plants because cyanobacteria are insensitive to the diphenylether herbicides.

DVR could be one of the ideal targets of herbicide development. One reason for this is simply because its inhibition causes rapid destruction of cells under high light (52). Another reason is that introducing a yet-unidentified cyanobacterial DVR may yield a bypass of the reaction in the hypothetical overexpressing plants so that the plants would become insensitive to a specific herbicide to plant-type DVR.

To reduce green-color contamination in the commercial *Brassica* seed oil, antisense inhibition of GSA-AT expression was applied to reduce chlorophyll synthesis in the seeds (90). Antisense suppression of GSA-AT may add a commercial value for the leaves of transgenic plants as well, as it sometimes produces a fascinating variegated pattern (20).

A mutation in the CBR gene conferred a partial "stay-green" phenotype to rice (M. Kusaba, unpublished results). Most of the senescent events occurred in this mutant, whereas chlorophylls bound to LHCII were not degraded until cells finally collapsed. The prolonged retention of chlorophylls resulted in the longer conservation of green color in the leaves. It is envisioned that such a stay-green phenotype may add commercial value to plants in the markets. Collectively, the genetic manipulation of the green pigments as well as other tetrapyrroles has a substantial potential for creating agricultural and horticultural value.

SUMMARY POINTS

1. Higher plants produce four classes of tetrapyrroles, namely, chlorophyll, heme, siroheme, and phytochromobilin. Most of the enzymes involved in the tetrapyrrole synthesis pathway have been identified. 7-Hydroxymethyl chlorophyll *a* reductase and dihydrosirohydrochlorin reductase remain to be identified. The existence of unidentified subunits for MgCy has been suggested.
2. Most enzymatic steps of the tetrapyrrole biosynthetic pathway localize within plastids. It is a matter of debate whether the last few steps of heme biosynthesis occur both in plastids and mitochondria, or if they are exclusively restricted to plastids.
3. The tetrapyrrole biosynthetic pathway is regulated via multiple mechanisms. Heme negatively regulates GluTR activity, and protochlorophyllide accumulation also negatively controls the GluTR activity through the action of the FLU protein. The MgCh

activity may be regulated by thioredoxin, GUN4 protein, Mg^{2+} concentration, and ATP/ADP ratios. The chlorophyll cycle is regulated by the abundance of CAO. The accumulation of chlorophyll *b* triggers degradation of CAO. The N-terminal domain of CAO is a prerequisite for the degradation of CAO.

4. Not only does the tetrapyrrole biosynthesis pathway supply the vital molecules that are involved in photosynthesis, respiration, nitrite, and sulfite reduction, and so on, but it is also an integrated part of various cellular processes including gene expression, protein import, and the assembly of essential proteins. Further understanding regarding the regulation and reaction chemistry of enzymes involved in tetrapyrrole synthesis may uncover the indissociable network of multiple biological activities in plants.

ACKNOWLEDGMENTS

We are grateful to Drs. Tatsuru Masuda and Dale Karlson for critical reading of the manuscript. We also thank those who kindly shared unpublished information with us and all of our colleagues for their helpful discussions. We acknowledge financial support by the Grant-in-Aid for Scientific Research (17770027 to R.T.) and the Grant-in-Aid for Creative Scientific Research (17GS0314 to A.T.) from the Ministry of Education, Culture, Sports, Science and Technology of Japan.

LITERATURE CITED

1a. Aarti PD, Tanaka R, Tanaka A. 2006. Effects of oxidative stress on chlorophyll biosynthesis in cucumber (Cucumis sativus) cotyledons. *Physiol. Plant* 128:186–97

1b. Alawady AE, Grimm B. 2005. Tobacco Mg protoporphyrin IX methyltransferase is involved in inverse activation of Mg porphyrin and protoheme synthesis. *Plant J.* 41:282–90

1b. Demonstrates the in vivo correlation of MgMT and MgCh activities.

2. Al-Karadaghi S, Hansson M, Nikonov S, Jonsson B, Hederstedt L. 1997. Crystal structure of ferrochelatase: the terminal enzyme in heme biosynthesis. *Structure* 5:1501–10
3. Beale SI. 1999. Enzymes of chlorophyll biosynthesis. *Photosynth. Res.* 60:43–73
4. Beale SI. 2005. Green genes gleaned. *Trends Plant Sci.* 10:309–12
5. Block MA, Tewari AK, Albrieux C, Marechal E, Joyard J. 2002. The plant S-adenosyl-L-methionine:Mg-protoporphyrin IX methyltransferase is located in both envelope and thylakoid chloroplast membranes. *Eur. J. Biochem.* 269:240–48
6. Bougri O, Grimm B. 1996. Members of a low-copy number gene family encoding glutamyl-tRNA reductase are differentially expressed in barley. *Plant J.* 9:867–78
7. Buchanan BB, Balmer Y. 2005. Redox regulation: a broadening horizon. *Annu. Rev. Plant. Biol.* 56:187–220
8. Cornah JE, Roper JM, Pal Singh D, Smith AG. 2002. Measurement of ferrochelatase activity using a novel assay suggests that plastids are the major site of haem biosynthesis in both photosynthetic and nonphotosynthetic cells of pea (Pisum sativum L.). *Biochem. J.* 362:423–32
9. Deleted in proof
10. **Davison PA, Schubert HL, Reid JD, Iorg CD, Heroux A, et al. 2005. Structural and biochemical characterization of Gun4 suggests a mechanism for its role in chlorophyll biosynthesis. *Biochemistry* 44:7603–12**

10. Presents the crystal structure of the GUN4 protein and also provides a novel insight into the function of GUN4 based on comprehensive kinetic analysis.

11. Eckhardt U, Grimm B, Hortensteiner S. 2004. Recent advances in chlorophyll biosynthesis and breakdown in higher plants. *Plant Mol. Biol.* 56:1–14
12. Eggink LL, LoBrutto R, Brune DC, Brusslan J, Yamasato A, et al. 2004. Synthesis of chlorophyll b: Localization of chlorophyllide a oxygenase and discovery of a stable radical in the catalytic subunit. *BMC Plant Biol.* 4:5
13. Fujita Y, Bauer CE. 2003. The light-independent protochlorophyllide reductase: a nitrogenase-like enzyme catalyzing a key reaction for greening in the dark. In *Chlorophylls and Bilins: Biosynthesis, Synthesis, and Degradation*, ed. KM Kadish, KM Smith, R Guilard, pp. 109–56. New York: Academic
14. Gadjieva R, Axelsson E, Olsson U, Hansson M. 2005. Analysis of gun phenotype in barley magnesium chelatase and Mg-protoporphyrin IX monomethyl ester cyclase mutants. *Plant Physiol. Biochem.* 43:901–8
15. Gibson LC, Marrison JL, Leech RM, Jensen PE, Bassham DC, et al. 1996. A putative Mg chelatase subunit from *Arabidopsis thaliana* cv C24. Sequence and transcript analysis of the gene, import of the protein into chloroplasts, and in situ localization of the transcript and protein. *Plant Physiol.* 111:61–71
16. Gibson LCD, Jensen PE, Hunter CN. 1999. Magnesium chelatase from Rhodobacter sphaeroides: Initial characterization of the enzyme using purified subunits and evidence for a BchI-BchD complex. *Biochem. J.* 337:243–51
17. **Goslings D, Meskauskiene R, Kim C, Lee KP, Nater M, Apel K. 2004. Concurrent interactions of heme and FLU with Glu tRNA reductase (HEMA1), the target of metabolic feedback inhibition of tetrapyrrole biosynthesis, in dark- and light-grown *Arabidopsis* plants. *Plant J.* 40:957–67**

17. Demonstrates the direct interaction of GluTR and the negative feedback regulator, FLU, and provides the first genetic evidence showing the involvement of heme in the feedback regulation of GluTR.

18. Hinchigeri SB, Hundle B, Richards WR. 1997. Demonstration that the BchH protein of Rhodobacter capsulatus activates S-adenosyl-L-methionine:magnesium protoporphyrin IX methyltransferase. *FEBS Lett.* 407:337–42
19. Hirashima M, Satoh S, Tanaka R, Tanaka A. 2006. Pigment shuffling in antenna systems achieved by expressing prokaryotic chlorophyllide a oxygenase in *Arabidopsis*. *J. Biol. Chem.* 281:15385–93
20. Hofgen R, Axelsen KB, Kannangara CG, Schuttke I, Pohlenz HD, et al. 1994. A visible marker for antisense mRNA expression in plants: inhibition of chlorophyll synthesis with a glutamate-1-semialdehyde aminotransferase antisense gene. *Proc. Natl. Acad. Sci. USA* 91:1726–30
21. Hoober J, Eggink L. 2001. A potential role of chlorophylls b and c in assembly of light-harvesting complexes. *FEBS Lett.* 489:1–3
22. Huq E, Al-Sady B, Hudson M, Kim C, Apel K, Quail PH. 2004. Phytochrome-interacting factor 1 is a critical bHLH regulator of chlorophyll biosynthesis. *Science* 305:1937–41
23. Ilag LL, Kumar AM, Soll D. 1994. Light regulation of chlorophyll biosynthesis at the level of 5-aminolevulinate formation in *Arabidopsis*. *Plant Cell* 6:265–75
24. Initiative AG. 2000. Analysis of the genome sequence of the flowering plant *Arabidopsis thaliana*. *Nature* 408:796–815
25. Ishijima S, Uchibori A, Takagi H, Maki R, Ohnishi M. 2003. Light-induced increase in free Mg2+ concentration in spinach chloroplasts: measurement of free Mg2+ by using a fluorescent probe and necessity of stromal alkalinization. *Arch. Biochem. Biophys.* 412:126–32
26. Jacobs JM, Jacobs NJ. 1987. Oxidation of protoporphyrinogen to protoporphyrin, a step in chlorophyll and haem biosynthesis. Purification and partial characterization of the enzyme from barley organelles. *Biochem. J.* 244:219–24

27. Jensen PE, Reid JD, Hunter CN. 2000. Modification of cysteine residues in the ChlI and ChlH subunits of magnesium chelatase results in enzyme inactivation. *Biochem. J.* 352(Pt. 2):435–41
28. Koch M, Breithaupt C, Kiefersauer R, Freigang J, Huber R, Messerschmidt A. 2004. Crystal structure of protoporphyrinogen IX oxidase: a key enzyme in haem and chlorophyll biosynthesis. *EMBO J.* 23:1720–28
29. Kohchi T, Mukougawa K, Frankenberg N, Masuda M, Yokota A, Lagarias JC. 2001. The *Arabidopsis* HY2 gene encodes phytochromobilin synthase, a ferredoxin-dependent biliverdin reductase. *Plant Cell* 13:425–36
30. Kumar AM, Csankovszki G, Soll D. 1996. A second and differentially expressed glutamyl-tRNA reductase gene from *Arabidopsis thaliana*. *Plant Mol. Biol.* 30:419–26
31. Kuttkat A, Edhofer I, Eichacker LA, Paulsen H. 1997. Light-harvesting chlorophyll a/b-binding protein stably inserts into etioplast membranes supplemented with Zn-pheophytin a/b. *J. Biol. Chem.* 272:20451–55
32. Larkin RM, Alonso JM, Ecker JR, Chory J. 2003. GUN4, a regulator of chlorophyll synthesis and intracellular signaling. *Science* 299:902–6
33. Lee KP, Kim C, Lee DW, Apel K. 2003. TIGRINA d, required for regulating the biosynthesis of tetrapyrroles in barley, is an ortholog of the FLU gene of *Arabidopsis thaliana*. *FEBS Lett.* 553:119–24
34. Lermontova I, Grimm B. 2000. Overexpression of plastidic protoporphyrinogen IX oxidase leads to resistance to the diphenyl-ether herbicide acifluorfen. *Plant Physiol.* 122:75–83
35. Lermontova I, Kruse E, Mock HP, Grimm B. 1997. Cloning and characterization of a plastidal and a mitochondrial isoform of tobacco protoporphyrinogen IX oxidase. *Proc. Natl. Acad. Sci. USA* 94:8895–900
36. Leustek T, Smith M, Murillo M, Singh DP, Smith AG, et al. 1997. Siroheme biosynthesis in higher plants. Analysis of an S-adenosyl-L-methionine-dependent uroporphyrinogen III methyltransferase from *Arabidopsis thaliana*. *J. Biol. Chem.* 272:2744–52
37a. Li X, Volrath SL, Nicholl DB, Chilcott CE, Johnson MA, et al. 2003. Development of protoporphyrinogen oxidase as an efficient selection marker for Agrobacterium tumefaciens-mediated transformation of maize. *Plant Physiol.* 133:736–47
37b. Lister R, Chew O, Rudhe C, Lee M-N, Whelan J. 2001. *Arabidopsis thaliana* ferrochelatase-I and -II are not imported into *Arabidopsis* mitochondria. *FEBS Lett.* 506:291–95
38. Luer C, Schauer S, Mobius K, Schulze J, Schubert WD, et al. 2005. Complex formation between glutamyl-tRNA reductase and glutamate-1-semialdehyde 2,1-aminomutase in *Escherichia coli* during the initial reactions of porphyrin biosynthesis. *J. Biol. Chem.* 280:18568–72
39. **Masuda T, Suzuki T, Shimada H, Ohta H, Takamiya K. 2003. Subcellular localization of two types of ferrochelatase in cucumber. *Planta* 217:602–9**

39. Examines the intracellular localization of two isoforms of FeCh using multiple techniques, and concludes that these isoforms primarily localize in plastids. This paper set the stage for reconsidering the localization of tetrapyrrole biosynthesizing enzymes.

40. Masuda T, Takamiya K. 2004. Novel insights into the enzymology, regulation and physiological functions of light-dependent protochlorophyllide oxidoreductase in angiosperms. *Photosynth. Res.* 81:1–29
41. Matringe M, Camadro JM, Labbe P, Scalla R. 1989. Protoporphyrinogen oxidase as a molecular target for diphenyl ether herbicides. *Biochem. J.* 260:231–35
42. Matsumoto F, Obayashi T, Sasaki-Sekimoto Y, Ohta H, Takamiya K, Masuda T. 2004. Gene expression profiling of the tetrapyrrole metabolic pathway in *Arabidopsis* with a mini-array system. *Plant Physiol.* 135:2379–91

43. McCormac AC, Fischer A, Kumar AM, Soll D, Terry MJ. 2001. Regulation of HEMA1 expression by phytochrome and a plastid signal during de-etiolation in *Arabidopsis thaliana*. *Plant J.* 25:549–61
44. McCormac AC, Terry MJ. 2002. Light-signalling pathways leading to the co-ordinated expression of HEMA1 and Lhcb during chloroplast development in *Arabidopsis thaliana*. *Plant J.* 32:549–59
45. McCormac AC, Terry MJ. 2002. Loss of nuclear gene expression during the phytochrome A-mediated far-red block of greening response. *Plant Physiol.* 130:402–14
46. Meskauskiene R, Apel K. 2002. Interaction of FLU, a negative regulator of tetrapyrrole biosynthesis, with the glutamyl-tRNA reductase requires the tetratricopeptide repeat domain of FLU. *FEBS Lett.* 532:27–30
47. Meskauskiene R, Nater M, Goslings D, Kessler F, Op den Camp R, Apel K. 2001. FLU: A negative regulator of chlorophyll biosynthesis in *Arabidopsis thaliana*. *Proc. Natl. Acad. Sci. USA* 98:12826–31
48. Mochizuki N, Brusslan JA, Larkin R, Nagatani A, Chory J. 2001. *Arabidopsis* genomes uncoupled 5 (GUN5) mutant reveals the involvement of Mg-chelatase H subunit in plastid-to-nucleus signal transduction. *Proc. Natl. Acad. Sci. USA* 98:2053–58
49. Moseley JL, Page MD, Alder NP, Eriksson M, Quinn J, et al. 2002. Reciprocal expression of two candidate di-iron enzymes affecting photosystem I and light-harvesting complex accumulation. *Plant Cell* 14:673–88
50. Moser J, Schubert WD, Beier V, Bringemeier I, Jahn D, Heinz DW. 2001. V-shaped structure of glutamyl-tRNA reductase, the first enzyme of tRNA-dependent tetrapyrrole biosynthesis. *EMBO J.* 20:6583–90

50. The solved structure of GluTR beautifully fits with the reported structure of GSA-AT, providing firm evidence for the complex formation of these enzymes.

51. Moulin M, Smith AG. 2005. Regulation of tetrapyrrole biosynthesis in higher plants. *Biochem. Soc. Trans.* 33:737–42
52. Nagata N, Tanaka R, Satoh S, Tanaka A. 2005. Identification of a vinyl reductase gene for chlorophyll synthesis in *Arabidopsis thaliana* and implications for the evolution of Prochlorococcus species. *Plant Cell* 17:233–40
53. Nakayama M, Masuda T, Bando T, Yamagata H, Ohta H, Takamiya KI. 1998. Cloning and expression of the soybean chlH gene encoding a subunit of Mg-chelatase and localization of the Mg2+ concentration-dependent chlH protein within the chloroplast. *Plant Cell Physiol.* 39:275–84
54. Narita S, Tanaka R, Ito T, Okada K, Taketani S, Inokuchi H. 1996. Molecular cloning and characterization of a cDNA that encodes protoporphyrinogen oxidase of *Arabidopsis thaliana*. *Gene* 182:169–75
55. Nogaj LA, Beale SI. 2005. Physical and kinetic interactions between glutamyl-tRNA reductase and glutamate-1-semialdehyde aminotransferase of *Chlamydomonas reinhardtii*. *J. Biol. Chem.* 280:24301–7
56. Nogaj LA, Srivastava A, van Lis R, Beale SI. 2005. Cellular levels of glutamyl-tRNA reductase and glutamate-1-semialdehyde aminotransferase do not control chlorophyll synthesis in *Chlamydomonas reinhardtii*. *Plant Physiol.* 139:389–96

56. Provides a novel insight into the control mechanism of ALA formation. The authors estimate the absolute amounts of GluTR and GSA-AT, and conclude that the protein levels of these enzymes do not control the activity of ALA formation.

57. Nott A, Jung H-S, Koussevitzky S, Chory J. 2006. Plastid-to-nucleus retrograde signaling. *Annu. Rev. Plant. Biol.* 57:739–59
58. Obornik M, Green BR. 2005. Mosaic origin of the heme biosynthesis pathway in photosynthetic eukaryotes. *Mol. Biol. Evol.* 22:2343–53
59. Oster U, Tanaka R, Tanaka A, Rüdiger W. 2000. Cloning and functional expression of the gene encoding the key enzyme for chlorophyll b biosynthesis CAO from *Arabidopsis thaliana*. *Plant J.* 21:305–10

60. Ouchane S, Steunou AS, Picaud M, Astier C. 2004. Aerobic and anaerobic Mg-protoporphyrin monomethyl ester cyclases in purple bacteria: a strategy adopted to bypass the repressive oxygen control system. *J. Biol. Chem.* 279:6385–94
61. Papenbrock J, Mishra S, Mock HP, Kruse E, Schmidt EK, et al. 2001. Impaired expression of the plastidic ferrochelatase by antisense RNA synthesis leads to a necrotic phenotype of transformed tobacco plants. *Plant J.* 28:41–50
62. Papenbrock J, Mock HP, Kruse E, Grimm B. 1999. Expression studies in tetrapyrrole biosynthesis: Inverse maxima of magnesium chelatase and ferrochelatase activity during cyclic photoperiods. *Planta* 208:264–73
63. Papenbrock J, Mock HP, Tanaka R, Kruse E, Grimm B. 2000. Role of magnesium chelatase activity in the early steps of the tetrapyrrole biosynthetic pathway. *Plant Physiol.* 122:1161–69
64. Pattanayak GK, Biswal AK, Reddy VS, Tripathy BC. 2005. Light-dependent regulation of chlorophyll b biosynthesis in chlorophyllide a oxygenase overexpressing tobacco plants. *Biochem. Biophys. Res. Commun.* 326:466–71
65. Pontoppidan B, Kannangara CG. 1994. Purification and partial characterization of barley glutamyl-tRNA(Glu) reductase, the enzyme that directs glutamate to chlorophyll biosynthesis. *Eur. J. Biochem.* 225:529–37
66. Raux-Deery E, Leech HK, Nakrieko KA, McLean KJ, Munro AW, et al. 2005. Identification and characterization of the terminal enzyme of siroheme biosynthesis from *Arabidopsis thaliana*: a plastid-located sirohydrochlorin ferrochelatase containing a 2FE-2S center. *J. Biol. Chem.* 280:4713–21
67. Reid JD, Hunter CN. 2004. Magnesium-dependent ATPase activity and cooperativity of magnesium chelatase from Synechocystis sp. PCC6803. *J. Biol. Chem.* 279:26893–99
68. Reinbothe C, Bartsch S, Eggink LL, Hoober JK, Brusslan J, et al. 2006. A role for chlorophyllide a oxygenase in the regulated import and stabilization of light-harvesting chlorophyll a/b proteins. *Proc. Natl. Acad. Sci. USA* 103:4777–82
69. Rzeznicka K, Walker CJ, Westergren T, Kannangara CG, von Wettstein D, et al. 2005. Xantha-l encodes a membrane subunit of the aerobic Mg-protoporphyrin IX monomethyl ester cyclase involved in chlorophyll biosynthesis. *Proc. Natl. Acad. Sci. USA* 102:5886–91
70. Santana MA, Tan FC, Smith AG. 2002. Molecular characterization of coproporphyrinogen oxidase from Glycine max and *Arabidopsis thaliana*. *Plant Physiol. Biochem.* 40:289–98
71. Scheumann V, Schoch S, Rüdiger W. 1998. Chlorophyll a formation in the chlorophyll b reductase reaction requires reduced ferredoxin. *J. Biol. Chem.* 273:35102–8
72. Schulz R, Steinmuller K, Klaas M, Forreiter C, Rasmussen S, et al. 1989. Nucleotide sequence of a cDNA coding for the NADPH-protochlorophyllide oxidoreductase (PCR) of barley (Hordeum vulgare L.) and its expression in *Escherichia coli*. *Mol. Gen. Genet.* 217:355–61
73a. Schulze JO, Schubert WD, Moser J, Jahn D, Heinz DW. 2006. Evolutionary relationship between initial enzymes of tetrapyrrole biosynthesis. *J. Mol. Biol.* 358:1212–20
73b. Shen YY, Wang XF, Wu FQ, Du SY, Cao Z, et al. 2006. The Mg-chelatase H subunit is an abscisic acid receptor. *Nature* 443:823–26
74. Shepherd M, McLean S, Hunter CN. 2005. Kinetic basis for linking the first two enzymes of chlorophyll biosynthesis. *FEBS J.* 272:4532–39
75. Shlyk AA. 1971. Biosynthesis of chlorophyll b. *Annu. Rev. Plant Physiol.* 22:169–84
76. Smith AG, Marsh O, Elder GH. 1993. Investigation of the subcellular location of the tetrapyrrole-biosynthesis enzyme coproporphyrinogen oxidase in higher plants. *Biochem. J.* 292:503–8

77. Srivastava A, Lake V, Nogaj LA, Mayer SM, Willows RD, Beale SI. 2005. The *Chlamydomonas reinhardtii* gtr gene encoding the tetrapyrrole biosynthetic enzyme glutamyl-trna reductase: structure of the gene and properties of the expressed enzyme. *Plant Mol. Biol.* 58:643–58
78. Strand A, Asami T, Alonso J, Ecker JR, Chory J. 2003. Chloroplast to nucleus communication triggered by accumulation of Mg-protoporphyrinIX. *Nature* 421:79–83
79. Sukenik A, Wyman KD, Bennett J, Falkowski PG. 1987. A novel mechanism for regulating the exitation of photosystem II in a green alga. *Nature* 327:704–7
80. Suzuki T, Masuda T, Singh DP, Tan FC, Tsuchiya T, et al. 2002. Two types of ferrochelatase in photosynthetic and nonphotosynthetic tissues of cucumber. Their difference in phylogeny, gene expression, and localization. *J. Biol. Chem.* 277:4731–37
81. Tanaka R, Koshino Y, Sawa S, Ishiguro S, Okada K, Tanaka A. 2001. Overexpression of chlorophyllide a oxygenase (CAO) enlarges the antenna size of photosystem II in *Arabidopsis thaliana*. *Plant J.* 26:365–73
82. Tanaka R, Tanaka A. 2005. Effects of chlorophyllide a oxygenase overexpression on light acclimation in *Arabidopsis thaliana*. *Photosynth. Res.* 85:327–40
83. Tanaka A, Tanaka R. 2006. Chlorophyll metabolism. *Curr. Opin. Plant Biol.* 9:248–55
84. Tanaka R, Yoshida K, Nakayashiki T, Masuda T, Tsuji H, et al. 1996. Differential expression of two hemA mRNAs encoding glutamyl-tRNA reductase proteins in greening cucumber seedlings. *Plant Physiol.* 110:1223–30
85a. Tanaka R, Yoshida K, Nakayashiki T, Tsuji H, Inokuchi H, et al. 1997. The third member of the hemA gene family encoding glutamyl-tRNA reductase is primarily expressed in roots in Hordeum vulgare. *Photosynth. Res.* 53:161–71
85b. Terry MJ. 1997. Phytochrome chromophore-deficient mutants. *Plant Cell Environ.* 20:740–45
86. Terry MJ, Linley PJ, Kohchi T. 2002. Making light of it: the role of plant haem oxygenases in phytochrome chromophore synthesis. *Biochem. Soc. Trans.* 30:604–9
87. Tewari AK, Tripathy BC. 1998. Temperature-stress-induced impairment of chlorophyll biosynthetic reactions in cucumber and wheat. *Plant Physiol.* 117:851–58
88. Tomitani A, Okada K, Miyashita H, Matthijs H, Ohno T, Tanaka A. 1999. Chlorophyll b and phycobilins in the common ancestor of cyanobacteria and chloroplasts. *Nature* 400:159–62
89. Tottey S, Block MA, Allen M, Westergren T, Albrieux C, et al. 2003. *Arabidopsis* CHL27, located in both envelope and thylakoid membranes, is required for the synthesis of protochlorophyllide. *Proc. Natl. Acad. Sci. USA* 100:16119–24
90. Tsang EW, Yang J, Chang Q, Nowak G, Kolenovsky A, et al. 2003. Chlorophyll reduction in the seed of Brassica napus with a glutamate 1-semialdehyde aminotransferase antisense gene. *Plant Mol. Biol.* 51:191–201
91. Usuda H. 1988. Adenine nucleotide levels, the redox state of the NADP system, and assimilatory force in nonaqueously purified mesophyll chloroplasts from maize leaves under different light intensities. *Plant Physiol.* 88:1461–68
92. van Lis R, Atteia A, Nogaj LA, Beale SI. 2005. Subcellular localization and light-regulated expression of protoporphyrinogen IX oxidase and ferrochelatase in *chlamydomonas reinhardtii*. *Plant Physiol.* 139:1946–58
93. Vavilin DV, Vermaas WF. 2002. Regulation of the tetrapyrrole biosynthetic pathway leading to heme and chlorophyll in plants and cyanobacteria. *Physiol. Plant* 115:9–24
94. Verdecia MA, Larkin RM, Ferrer JL, Riek R, Chory J, Noel JP. 2005. Structure of the Mg-chelatase cofactor GUN4 reveals a novel hand-shaped fold for porphyrin binding. *PLoS Biol.* 3:e151

94. Together with Reference 10, this paper describes the crystal structure of the GUN4 protein.

95. von Wettstein D, Kahn A, Nielsen OF, Gough S. 1974. Genetic regulation of chlorophyll synthesis analyzed with mutants in barley. *Science* 184:800–2
96. Vothknecht UC, Kannangara CG, von Wettstein D. 1998. Barley glutamyl tRNAGlu reductase: Mutations affecting haem inhibition and enzyme activity. *Phytochemistry* 47:513–19
97a. Wagner D, Przybyla D, Op den Camp R, Kim C, Landgraf F, et al. 2004. The genetic basis of singlet oxygen-induced stress responses of *Arabidopsis thaliana*. *Science* 306:1183–85
97b. Walker CJ, Mansfield KE, Smith KM, Castelfranco PA. 1989. Incorporation of atmospheric oxygen into the carbonyl functionality of the protochlorophyllide isocyclic. *Biochem. J.* 257:599–602
98. Warren MJ, Raux E, Schubert HL, Escalante-Semerena JC. 2002. The biosynthesis of adenosylcobalamin (vitamin B12). *Nat. Prod. Rep.* 19:390–412
99. Watanabe N, Che FS, Iwano M, Takayama S, Yoshida S, Isogai A. 2001. Dual targeting of spinach protoporphyrinogen oxidase II to mitochondria and chloroplasts by alternative use of two in-frame initiation codons. *J. Biol. Chem.* 276:20474–81
100. Wilde A, Mikolajczyk S, Alawady A, Lokstein H, Grimm B. 2004. The gun4 gene is essential for cyanobacterial porphyrin metabolism. *FEBS Lett.* 571:119–23
101. Williams P, Hardeman K, Fowler J, Rivin C. 2006. Divergence of duplicated genes in maize: evolution of contrasting targeting information for enzymes in the porphyrin pathway. *Plant J.* 45:727–39
102. Willows RD. 2003. Biosynthesis of chlorophylls from protoporphyrin IX. *Nat. Prod. Rep.* 20:327–41
103. Wu CK, Dailey HA, Rose JP, Burden A, Sellers VM, Wang BC. 2001. The 2.0 A structure of human ferrochelatase, the terminal enzyme of heme biosynthesis. *Nat. Struct. Biol.* 8:156–60
104. Xu H, Vavilin D, Vermaas W. 2001. Chlorophyll b can serve as the major pigment in functional photosystem II complexes of cyanobacteria. *Proc. Natl. Acad. Sci. USA* 98:14168–73
105. Yamasato A, Nagata N, Tanaka R, Tanaka A. 2005. The N-terminal domain of chlorophyllide a oxygenase confers protein instability in response to chlorophyll b accumulation in *Arabidopsis*. *Plant Cell* 17:1585–97
106. Zavgorodnyaya A, Papenbrock J, Grimm B. 1997. Yeast 5-aminolevulinate synthase provides additional chlorophyll precursor in transgenic tobacco. *Plant J.* 12:169–78

105. Describes a novel regulatory mechanism of protein accumulation in plastids in which biosynthesis of a photosynthetic pigment plays an essential role.

RELATED RESOURCES

Roth JR, Lawrence JG, Bobik TA. 1996. Cobalamin (coenzyme B_{12}): synthesis and biological significance. *Annu. Rev. Microbiol.* 50:137–81

Hörtensteiner S. 2006. Chlorophyll degradation during senescence. *Annu. Rev. Plant Biol.* 57:55–77

Suzuki JY, Bollivar DW, Bauer CE. 1997. Genetic analysis of chlorophyll biosynthesis. *Annu. Rev. Genet.* 31:61–89

Plant ATP-Binding Cassette Transporters

Philip A. Rea

Plant Science Institute, Department of Biology, Carolyn Hoff Lynch Biology Laboratory, University of Pennsylvania, Philadelphia, Pennsylvania 19104-6018; email: parea@sas.upenn.edu

Annu. Rev. Plant Biol. 2007. 58:347–75

The *Annual Review of Plant Biology* is online at plant.annualreviews.org

This article's doi:
10.1146/annurev.arplant.57.032905.105406

First published online as a Review in Advance on January 29, 2007

1543-5008/07/0602-0347$20.00

Key Words

subfamilies, organization, functions, localization

Abstract

The ATP-binding cassette (ABC) protein superfamily is one of the largest known, with over 120 members in both *Arabidopsis thaliana* and rice (*Oryza sativa*). Most, but not all, ABC proteins are modularly organized membrane proteins ("ABC transporters") that mediate MgATP-energized transmembrane transport and/or regulate other transporters. The range of processes in which members of the various subclasses of plant ABC transporters have been implicated encompasses polar auxin transport, lipid catabolism, xenobiotic detoxification, disease resistance, and stomatal function. Although it is often possible to predict the likely function of a plant ABC transporter on the basis of its subfamily membership, there are many whose capabilities deviate from what would be predicted from the properties of even their most sequence-related counterparts. When taking account of this and the disparate processes in which the few that have been characterized participate, it is likely that elucidation of the mechanistic basis of any given plant process will necessitate consideration of at least one ABC transporter.

Contents

INTRODUCTION

The capabilities of ATP-binding cassette (ABC) proteins, the majority of which are transporters, are staggering and range from the transport of mineral ions, lipids, and peptides to the regulation of channels and other primary pumps. ABC proteins, in general, and ABC transporters, in particular, are designated as such on the basis of their nucleotide-binding folds (NBFs) or nucleotide-binding domains (NBDs), which share 30–40% sequence identity between family members (38). Each NBF encompasses approximately 200 amino acid residues and contains three defining sequence motifs. These are a Walker A box ($GX_4GK[ST]$) and a Walker B box [(hydrophobic)$_4$[DE] commonly as (hydrophobic)$_4$DEATSALD] (133) separated by approximately 120 amino acid residues, and an ABC signature (alias C) motif ([LIVMFY]S[SG]GX_3[RKA][LIVMYA]-X[LIVFM][AG] commonly as LSSG) situated between the two Walker boxes (**Figure 1**). This is the general case to which there are many exceptions. For instance, in several plant ABC transporters the (hydrophobic)$_4$[DE] Walker B motif is interrupted by hydrophilic residues and in others the canonical [LIVMFY], S, [SG], G, [RKA], [LIVMYA], [LIVMF], and [AG] submotifs of the signature motif frequently contain other residues (75). In defining and delimiting this family it is conservation of the entire NBF with variations, not the possession of any one motif in isolation, which is diagnostic.

Plants are a particularly rich source of ABC proteins. The genome of the model plant *Arabidopsis thaliana* is capable of encoding more than 120 ABC proteins (106) compared with the meager 50–70 ABC proteins encoded by the human genome and the genomes of the fly (*Drosophila melanogaster*) and worm (*Caenorhabditis elegans*). Of the total ABC protein open reading frames (ORFs) identified in *Arabidopsis*, at least 103 likely encode membrane proteins—proteins possessing membrane spans contiguous with one or two NBFs (106). When account is taken of this and the rough equivalence of the total transporter ORF complements of *Arabidopsis* (~700 ORFs) with those of *C. elegans* (~650 ORFs) and *Drosophila* (~600 ORFs), it is evident that *Arabidopsis* allocates nearly twice as many of its transporter genes to ABC transporters as *C. elegans* and *Drosophila* (107).

In this review the *Arabidopsis* ABC transporter superfamily is used as a platform for most considerations. However, this is not to imply that *Arabidopsis* is a plant with a particularly wide range of ABC transporters because it is not. Some 121 ABC protein ORFs have been identified in the next-best genomically characterized plant, rice (*Oryza sativa*), in which almost all of the subfamilies delineated for *Arabidopsis* (106), with the exception of one (the single member subfamily, AOH, below), are represented (27).

Basic Characteristics of ATP-Binding Cassette Transporter-Mediated Transport

There are three salient features of ABC protein-mediated transport in plants (99). First, it is directly energized by MgATP. Transport is energized by MgATP but not by free ATP or nonhydrolyzable ATP analogs such as 5′-(β,γ-imino)triphosphate (AMP-PNP), although other nucleoside triphosphates such as GTP or UTP can, depending on the transporter, partially substitute for ATP. Second, transport is insensitive to the transmembrane H^+ electrochemical potential difference. Elimination of the transmembrane pH gradient and electrical potential that would otherwise be established across plant membranes by primary H^+ pumps through the addition of protonophores or ionophores does not usually inhibit transport. Third, transport is exquisitely sensitive to vanadate. In the presence of vanadate, a metastable analog of orthophosphate, and MgATP, ABC proteins establish a tight complex with nucleoside diphosphates (114, 127, 128). Vanadate thereby arrests the catalytic cycle by substituting for the released phosphate and trapping the other product of hydrolysis, ADP, in the nucleotide-binding site.

Modular Assembly of Core Domains and Motifs

ABC transporters consist of four core structural domains: two transmembrane domains (TMDs) containing multiple (usually 4-6) membrane-spanning α-helices and two NBFs (38) (**Figure 1**). The NBFs couple ATP hydrolysis and ADP release to the energization of transport. The TMDs constitute the pathway for the relay of transport substrates across the lipid bilayer. Although most, if not all, functional ABC transporters consist of a combination of two NBFs and two TMDs organized into an assembly with an internal twofold or pseudotwofold geometry, the core domains may be expressed as separate polypeptides or as multidomain proteins (38). In some ABC transporters, for instance many of those from bacterial sources, the four domains reside on different polypeptides ("quarter molecules"); in others, the domains are fused in various permutations. In the ribose permease system of *Escherichia coli*, the two NBFs are fused; in the mammalian TAP1-TAP2 peptide transporter, the *Schizosaccharomyces pombe* heavy metal tolerance factor 1 (HMT1) and the *Drosophila* pigment transporters (White, Brown, and Scarlet), one TMD is fused with one NBF to form a "half molecule." In the bulk of the transporters from nonplant eukaryotic sources, the four core domains are contiguous on a single polypeptide species ("full molecule") in a "forward" TMD1-NBF1-TMD2-NBF2 or "reverse" NBF1-TMD1-NBF2-TMD2 orientation.

Vanadate: VO_3^-, VO_4^{3-}, or $V_2O_7^{4-}$ anions containing pentavalent vanadium, of which orthovanadate (VO4) is the form that maximally inhibits ABC transporters

HMT1: heavy metal tolerance factor 1

Subfamilies of ATP-Binding Cassette Proteins

Plant ABC proteins can be assigned to 13 subfamilies on the basis of protein size (full, half, or quarter molecule), orientation (forward or reverse), the presence or absence of idiotypic transmembrane and/or linker domains, and overall sequence similarity (106). Full-molecule transporters include:

a)

Subfamily/Domains	Gene name	Localization	Function/Activity
MDR multidrug resistance 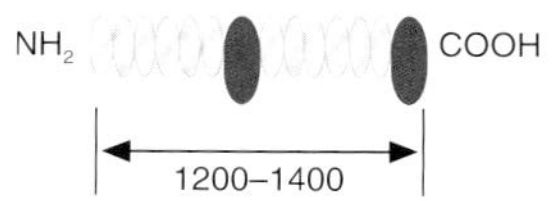 	*AtPGP1* *AtMDR1* *AtPGP4* *BR2* *DWF3* *CjMDR1*	Plasma membrane Plasma membrane Plasma membrane Plasma membrane Plasma membrane Plasma membrane	Auxin export (28, 86, 93) Auxin export (86, 93) Auxin import (123) Auxin export (84) Auxin export (84) Berberine import (117)
MRP multidrug resistance-associated protein 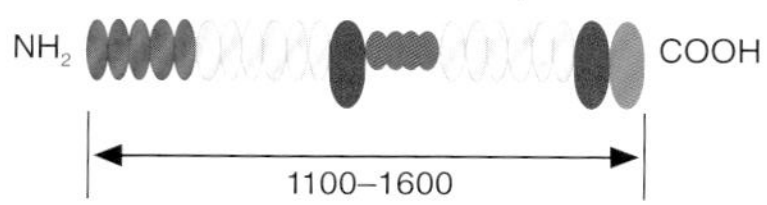 	*AtMRP1* *AtMRP2* *AtMRP3* *AtMRP4* *AtMRP5* *ZmMRP3* *ZmMRP4*	Vacuolar membrane Vacuolar membrane Vacuolar membrane Plasma membrane Plasma membrane Vacuolar membrane Vacuolar membrane	GS-conjugate, glucuronide and folate transport (70, 71, 100) GS-conjugate, glucuronide and chlorophyll catabolite transport (69, 70) GS-conjugate and chlorophyll catabolite transport (126) GS-conjugate and folate transport; stomatal regulation (49, 100) Stomatal regulation (25, 52) Anthocyanin transport (30) Anthocyanin transport (30)
PDR pleiotropic drug resistance 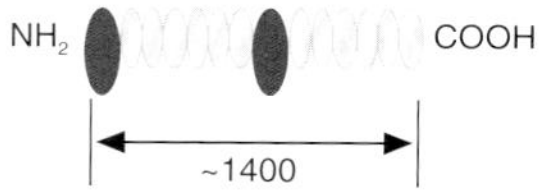 	*AtPDR12* *NpPDR1* *SpTUR2*	Unknown Plasma membrane Plasma membrane	Lead tolerance (60) Sclareolide export and resistance to *Botrytis cinerea* (42, 121) Sclareol export (129)
PMP peroxisomal membrane protein 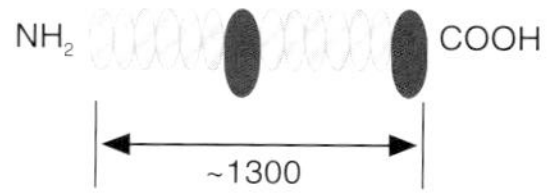 	*AtPMP2*	Peroxisomal membrane	Peroxisomal uptake of IBA, 2,4-DB, LCFAs and/or their LCF acyl-CoAs; JA biosynthesis; seed germination (24, 36, 124, 125, 142)
AOH ABC1 homolog 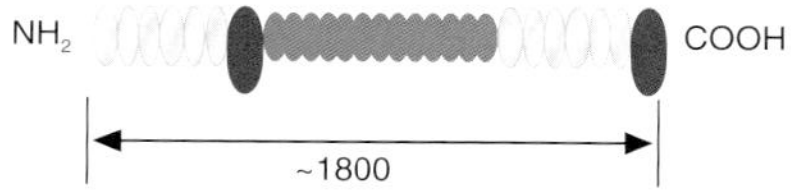 	*AtAOH1*	Unknown	Unknown

Figure 1

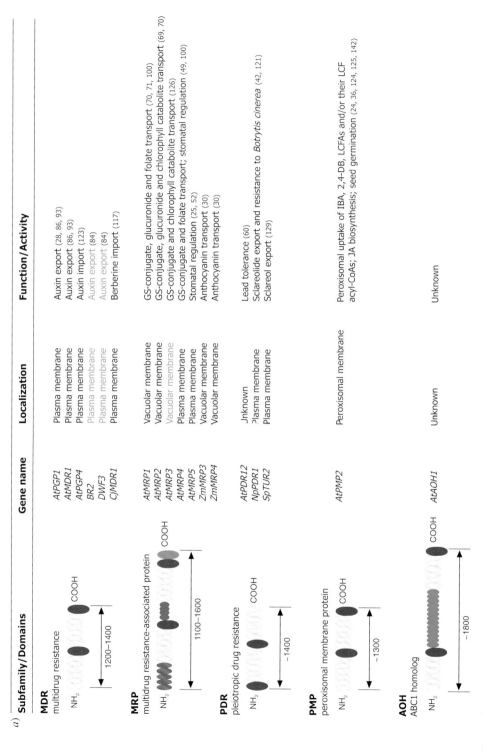

Putative domain organizations of plant ABC transporter subfamilies and examples of representatives whose localizations and/or functions or activities have been determined. The different domains are color-coded as follows: red, NBFs 1 and 2; dark gray, TMD0; light gray, TMDs 1 and 2; blue, linker (L) domain; green, C-terminal extension; yellow, amphipathic N-terminal signal peptide. The gene names listed here are those used in the main body of the text where their aliases, as appropriate, can be found. The localizations and/or functions or activities shown in gray instead of black typeface have been inferred but not demonstrated directly. Numbers delimited by arrows indicate the average number of amino acid residues in the mature translation product. Numbers in parentheses denote the most relevant references.

b) **WBC**

white-brown complex homolog

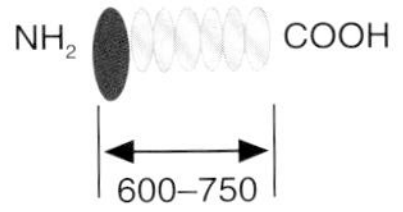

AtWBC12	Plasma membrane	Epidermal wax deposition (94)
AtWBC19	Vacuolar membrane	Kanamycin resistance (78)
GhWBC1	Plasma membrane	Cotton fiber maturation (141)
NtWBC1	Unknown	Lipid export (92)

ATH

ABC2 homolog

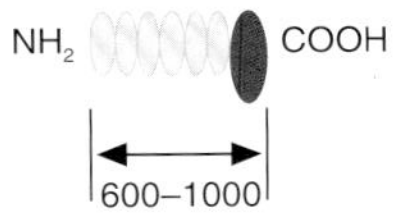

AtATH	Unknown	Unknown
OsATH	Unknown	Unknown

ATM

ABC transporter of the mitochondrion

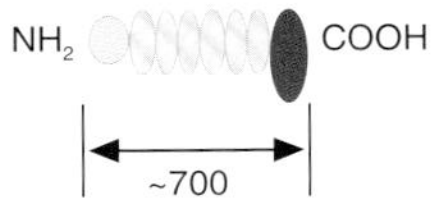

AtATM1	Mitochondria	Mitochondrial export of Fe/S clusters[a] (57)
AtATM2	Punctate inclusions	Unknown[a]
AtATM3	Mitochondria	Mitochondrial export of Fe/S clusters; cadmium and lead tolerance[a] (57, 45)
CrCDS1	Mitochondria	Mitochondrial export of Fe/S clusters; cadmium tolerance (33)

TAP

transporter associated with antigen processing

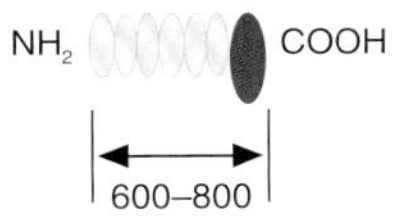

AtTAP1	Punctate inclusions	Unknown[b]
AtTAP2	Punctate inclusions	Unknown[b]

NAP

nonintrinsic ABC protein

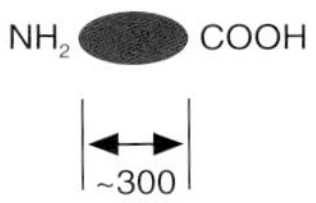

AtNAP1	Plastids	Fe/S cluster maintenance and responses to far-red light (136, 137)
AtNAP6	Plastids	Fe/S cluster maintenance; embryogenesis (137)
AtNAP7	Plastids	Fe/S cluster maintenance; embryogenesis (137)

[a]S. Chen, R. Sánchez-Fernández, A. Dancis & P.A. Rea, unpublished

[b]S. Chen, R. Sánchez-Fernández, M. Peng & P.A. Rea, unpublished

Figure 1

(Continued)

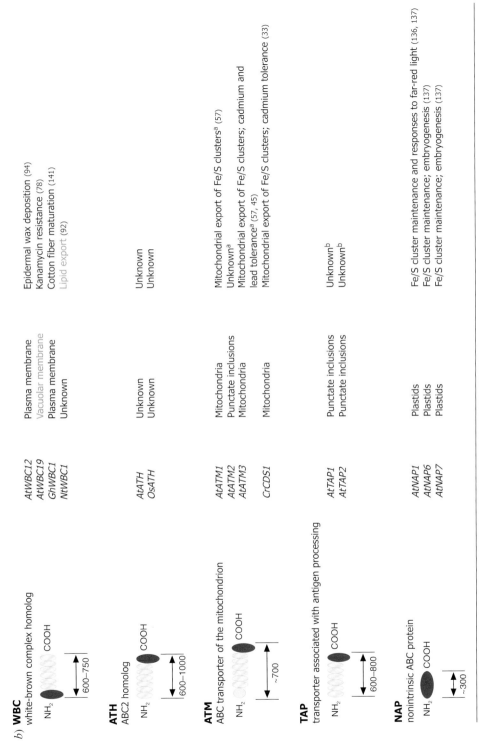

MDR: multidrug resistance homolog

MRP: multidrug resistance-associated protein homolog

PDR: pleiotropic drug resistance homolog

PMP: peroxisomal membrane protein homolog

AOH: ABC1 homolog

WBC: white-brown complex homolog

ATH: ABC2 homolog

ATM: ABC transporter of the mitochondrion homolog

TAP: transporter associated with antigen processing homolog

GCN: yeast GCN20, general control nonrepressible, homolog

SMC: structural maintenance of chromosomes homolog

NAP: nonintrinsic ABC protein

PGP: P-glycoprotein (alias MDR) homolog

the multidrug resistance homologs (MDRs), multidrug resistance-associated protein homologs (MRPs), pleiotropic drug resistance homologs (PDRs), peroxisomal membrane protein homologs (PMPs), and ABC1 homolog (AOH). Half-molecule transporters include: the white-brown complex homologs (WBCs), ABC2 homologs (ATHs), ABC transporter of the mitochondrion homologs (ATMs), and transporter associated with antigenic processing homologs (TAPs). Soluble proteins include: the 2′,5′-oligoadenylate-activated RNase inhibitor homologs (RLIs), yeast GCN 20, general control nonrepressible, homologs (GCNs), and structural maintenance of chromosomes homologs (SMCs) (**Figure 1**). The one category of plant ABC proteins that cannot be categorized in this way is the nonintrinsic ABC proteins (NAPs), a heterogeneous group of soluble or nonintrinsic membrane proteins that do not bear a close resemblance to each other or, with a few exceptions (below), to members of known ABC protein subfamilies from other organisms (**Figure 1**).

The subfamily assignments made in this way are robust with the exception of the ATHs and PMPs. Eleven members of the *Arabidopsis* ATH subfamily fall within the same clade, but five (AtATH8, AtATH9, AtATH10, AtATH12, AtATH13) do not despite their topological equivalence to ABC2. Moreover, none of the outlying AtATHs, apart from AtATH12, which groups within the MDR/MRP/TAP/ATM clade, group within any of the other ABC protein subfamilies. From a re-examination of the annotations made by Sánchez-Fernández et al. (106), and parallel analyses of the rice genome, it is now evident that AtATH8, AtATH9, AtATH10, and AtATH13 are not ABC transporters per se. Instead they are homologs of the yeast *ABC1* gene product (27), a nuclear-encoded protein required for ubiquinone biosynthesis and electron transfer by the mitochondrial bc_1 complex (9, 43). The major factor confounding classification within the plant PMP subfamily is that the one member, AtPMP2, whose PMP-type localization and function is firmly established is not a canonical half-molecule PMP, but instead is a full-molecule transporter (below).

FULL-MOLECULE ATP-BINDING CASSETTE TRANSPORTERS

Multidrug Resistance Homologs

The MDRs (alias P-glycoprotein homologs, PGPs), with a membership of 22 in *Arabidopsis* (106) and 24 in rice (27), represent the second largest ABC protein subfamily and the largest full-molecule ABC transporter subfamily in plants. First identified in mammalian cells because their overexpression confers a multidrug resistance phenotype (31), all MDRs are forward-orientation, full-molecule ABC transporters containing approximately 1200 amino acid residues (**Figure 1**). Many, but not all, of the MDRs from animal sources are plasma membrane efflux pumps competent in the transport of large amphipathic neutral or weakly cationic species (6) and/or the translocation of cationic phospholipids between membrane bilayer leaflets (103). The two most studied plant MDRs are the founding member of the *Arabidopsis* subfamily, AtPGP1 (alias AtMDR1) (22), and AtMDR1 [86, 87; alias AtMDR11 (106) or AtPGP19 (29)].

Both AtPGP1 and AtMDR1 are clearly implicated in auxin transport. Under dim light the hypocotyls of plants ectopically overexpressing *AtPGP1* sense transcripts undergo a 50% greater increase in length compared with wild-type controls, as if they have been treated with low concentrations of exogenous auxin, whereas plants ectopically expressing *AtPGP1* antisense transcripts exhibit 20% less growth than wild-type under the same conditions (118). In agreement with these findings, albeit in a manner indicative of a dominant role for AtMDR1 over AtPGP1, *atmdr1* but not *atpgp1* T-DNA insertion mutants exhibit downward-folded, or epinastic, cotyledons and first true leaves and diminished apical dominance (86).

The hypocotyls of *atmdr1-1* mutants are impaired in the basipetal (apex to base), but not acropetal (base to apex), transport of exogenously applied radiolabeled indole-3-acetic acid (IAA), a phenotype that is simulated by the application of the auxin transport inhibitor 1-*N*-naphthylphthalamic acid (NPA) to wild-type plants (86, 87). This seminal discovery and the finding that *atmdr1 atpgp1* double mutants exhibit a more pronounced mutant morphology associated with even lower rates of basipetal auxin transport versus *atmdr1-1* single mutants, implies that AtMDR1 participates in auxin transport and that AtPGP1 likely has a similar capability but at a level that is only discernible when the activity of AtMDR1 has been diminished sufficiently (86).

Clarifying as it is, this discovery raises some new fundamental questions. Specifically, do AtPGP1 and AtMDR1 directly catalyze transmembrane auxin transport? And, if so, how does this influence the polar translocation of auxin through tissues, for instance from the shoot apex to distal targets, since this process has been attributed to another category of transporters, the putative auxin anion channel (PIN) proteins?

Recent elegant studies go some way to addressing these questions, by confirming that AtMDR1 is a cellular auxin efflux pump while also demonstrating that PIN proteins are not only sufficient for cellular auxin efflux but also primarily responsible for determining the direction of polar auxin translocation by virtue of their polar cellular localization (93). By deploying an estradiol-inducible system for their expression in *Arabidopsis* and tobacco cell suspension cultures and in intact plants it has been established that the PINs and MDRs represent distinct but coexistent auxin efflux proteins. In these studies, PIN-dependent, NPA-inhibitable auxin efflux increases proportionately with an increase in PIN expression in estradiol-inducible cell suspension cultures (93). Similarly, and as would be predicted if the MDRs as well as the PINs contribute to cellular auxin transport, estradiol-induced expression of *AtMDR1* in cell suspensions also elicits net auxin efflux in proportion to the level of expression (93). This is in accord with the results from previous investigations demonstrating diminished efflux of both IAA and the synthetic auxin naphthalene-1-acetic acid (1-NAA) by mesophyll protoplasts isolated from *atpgp1* mutants and accelerated efflux of these same two auxins but not the inactive analog 2-NAA when AtPGP1 is heterologously expressed in yeast or HeLa cells (28).

However, the crucial question that remains to be resolved is how impaired transport by AtPGP1 and/or AtMDR1 compromises basipetal auxin transport by the PINs? If AtMDR1 and AtPGP1 do not interact directly with PIN1, as indicated by the finding that *atmdr1 atpgp1* double mutants are as susceptible as wild-type controls to the perturbing effects of estradiol-induced PIN1 overexpression (93), how do these ABC transporters exert their effects? Is it because, as suggested from the results of computer simulations, polar PIN-mediated auxin translocation via a particular cell column is contingent on a given level of background auxin export by the surrounding tissues, perhaps through the activity of AtMDR1 and/or AtPGP1 (29)? Or is it because the background transmembrane auxin gradients established by AtMDR1 and/or AtPGP1 determine the amenability of PIN to basal localization in the cells making up the translocation pathway? An observation consistent with a mechanism of the latter type is that in *atmdr1 atpgp1* double mutants, and to a lesser extent in *atpgp1* single mutants, PIN1 does not localize to the basal end of hypocotyls cells but is instead associated with randomly distributed punctate cytosolic inclusions (86).

The properties of another AtMDR, AtPGP4 (108) [alias AtMDR4 (106)] further substantiate the concept that plant MDRs, or at least a subset thereof, participate in auxin transport, whereas elucidation of the molecular identities of the *brachytic 2* (*br2*) and *dw3*

PIN: PIN-formed protein, an "auxin anion efflux channel" or "auxin transport facilitator" whose loss of function in mutants confers a pin-like shoot morphology because of impaired polar auxin transport

Glibenclamide: 1-{4-[2-(5-chloro-2-methoxybenzamido)-ethyl]-benzenesulphonyl}-3-cyclohexylurea, a sulfonylurea

dwarfism mutations in maize (*Zea mays*) and sorghum (*Sorghum bicolor*), respectively (84), not only affirms the general applicability of the principle but also illustrates how it has been of utility to plant breeders.

AtPGP4 is a root-specific auxin influx pump (108, 123). *Atpgp4* mutants are impaired in the basipetal redistribution of auxin in the root tip, and exhibit diminished linear and gravitropic root growth, enhanced root hair formation, and decreased sensitivity to NPA (108, 123). Heterologous expression of AtPGP4 in HeLa cells is associated with NPA-reversible IAA uptake (123), and in yeast with enhanced hypersensitivity to the cytotoxic auxin analog 5-fluoroindole (108).

The recessive *br2* mutation of maize is attributable to disruption of a gene encoding an MDR homolog (84). The stems of *br2* mutants, which lack a functional copy of the maize equivalent of AtPGP1, are impaired in the basipetal transport of radiolabeled IAA. This is an important discovery. Although the existence of the *br2* allele, which confers a dwarf phenotype because of shortening of the lower stalk internodes, has been known for more than half a century (61, 120), its molecular basis had eluded definition. A comparable long-known mutation in sorghum, *dw3*, which has also been employed by breeders, has a similar molecular basis (84). Armed with knowledge of the huge impact that plants of short stature have had on the green revolution—a revolution that came about through the introduction of dwarf rice and wheat varieties (44)—one wonders if similar studies of the MDRs from other crop species will prove to be of comparable agronomic significance.

When MDRs were first identified in plants it was thought that they would prove to participate in processes analogous to those in which their mammalian counterparts participate (99). Plants are a rich source of secondary metabolites, many of which, alkaloids primarily, have found widespread use as anticancer agents or precursors for the synthesis of such agents. Against this background, it is surprising how little is known of this aspect of plant MDR function. A notable exception is the *CjMDR1* gene of *Coptis japonica*, a perennial medicinal plant grown in East Asia and Japan (104, 117). In this species, berberine (an intense yellow benzylisoquinoline derivative), is the main alkaloid accumulated. CjMDR1, whose nearest equivalent in *Arabidopsis* is AtPGP4 [alias AtMDR4 (106)], appears to be an endogenous berberine transporter capable of contributing to the net flux of this compound from the root where it is synthesized to the rhizome where it is accumulated. CjMDR1 preferentially localizes to the plasma membrane (117) and the steady-state levels of its transcripts are maximal in the rhizome (138) in immediate proximity to the vascular tissue (117). When heterologously expressed in *Xenopus* öocytes, CjMDR1 promotes the vanadate- and glibenclamide-inhibitable, ATP-dependent uptake of berberine as well as uptake of its precursor reticuline and the quinoline derivative, 4-nitroquinoline *N*-oxide (117). A striking property of CjMDR1, one that it has in common with AtPGP4 (AtMDR4) (123), is that it is a plasma membrane-localized influx pump. This is a characteristic it shares with many prokaryotic ABC transporters but which has yet to be reported for any of the mammalian MDRs.

Much less is known of the MDRs from other plant sources. Although MDR homologs have been cloned from potato (*Solanum tuberosum*) (134) and wheat (*Triticum aestivum*) (110) in screens for genes induced by aluminum and proteins capable of interacting with calmodulin, respectively, the mechanistic basis for why these screens yielded the results they did is not known. It is suspected that induction of the wheat clone *TaMDR1* by aluminum is related to the disruption of calcium homeostasis in that lanthanum, gadolinium, and ruthenium red, three other inhibitors of calcium channels, as well as calcium deprivation also induce expression (110). A similar connection has been proposed for the potato clone, *PMDR1*, based on its interaction with calmodulin in vitro and the known role of the latter in calcium-dependent signal transduction (134).

Multidrug Resistance-Associated Protein Homologs

The second most highly represented subfamily of full-molecule ABC transporters is the MRPs, which number 16 in *Arabidopsis* (106) and 17 in rice (27). Like the MDRs, the MRPs are forward-orientation full molecules. Unlike the MDRs, the MRPs are frequently larger, consisting of at least 1500 amino acid residues, and usually containing three additional subfamily-specific structures: an approximately 200 amino acid residue hydrophobic N-terminal transmembrane domain (TMD0), containing 5 putative transmembrane spans; a linker (L) domain contiguous with NBF1, rich in charged amino acid residues; and a hydrophilic C-terminal extension (**Figure 1**). Two exceptions to this general rule are AtMRP11 and OsMRP12, which lack the first transmembrane domain, TMD0 (113).

One of the original putative *Arabidopsis* MRP ORFs, that for AtMRP15 (106), corresponds to AtMRP9 (55). The genomic sequence of AtMRP15 is more than 99% identical to that of AtMRP9 even in the 5′-UTR, and although the predicted amino acid sequence of the former differs from that of the latter by more than 400 residues, this difference is explicable in terms of three point mutations and a 230-bp insertion within the coding sequence for the N terminus.

Many early studies of MRP-type transport processes in plants focused on the uptake of two model glutathione (GSH) conjugates (GS-conjugates), N-ethylmaleimide-GS (NEM-GS) and S-(2,4-dinitrophenyl)-GS (DNP-GS), and the glutathionated chloroacetanilide herbicide, metolachlor (metolachlor-GS), by isolated vacuoles (74) and vacuolar membrane vesicles purified from plants (67). In this way it was recognized that the transporters responsible bear a close functional resemblance to the GS-conjugate transporting Mg^{2+}-ATPases (GS-X pumps) of mammalian cells and therefore probably belong to the MRP subfamily of ABC transporters (99). Thus, although not precluding the localization of at least some to other membranes (see below), it has come to be assumed that many plant MRPs, like their yeast counterparts, including *Saccharomyces cerevisiae* yeast cadmium factor 1 (YCF1) [see Sidebar: HMT1 in Plants (or the Lack Thereof)], are crucial for the vacuolar sequestration ("excretion storage") of both endogenous compounds and xenobiotics that are susceptible to conjugation with GSH and/or other adducts (reviewed in 99).

YCF1: yeast cadmium factor 1, a vacuolar MRP

HMT1 IN PLANTS (OR THE LACK THEREOF)

In plants, and in some fungi and animals, as exemplified by *S. pombe* and *C. elegans*, respectively, phytochelatins (PCs) bind heavy metals at high affinity and promote their vacuolar sequestration (101, 132). In *S. pombe*, this is attributed to HMT1 (90, 91), a vacuolar membrane-localized forward-orientation half-molecule ABC transporter containing an MRP-like TMD0 and an L sequence contiguous with the NBF. What is perplexing, however, is that MgATP-energized, protonophore-insensitive, vanadate-inhibitable PC uptake by vacuolar membranes from a plant source, oat (*Avena sativa*), has been reported (105), yet plants are devoid of half-molecule, HMT1-like ABC transporters (106). To add further confusion to this issue, the one authentic HMT1 homolog from a nonfungal source, CeHMT1 from *C. elegans*, confers metal tolerance regardless of whether there is upstream PC synthesis (131). Does this mean that in plants, and perhaps in *C. elegans*, the ABC transporter responsible for PC transport has yet to be discovered? The *S. cerevisiae* cadmium factor 1 gene (YCF1), which encodes a vacuolar MRP competent in the transport of both organic GS-conjugates and $Cd \cdot GS_2$ [bis(glutathionato)cadmium] complexes, is pertinent to this issue (65, 66). Although YCF1 itself does not transport PCs, does the topological equivalence of the N-terminal half of YCF1 to HMT1 mean that PC transport in plants is catalyzed by an MRP-type transporter with the facility for recognizing PCs?

Folate: folic acid (pteroyl-L-glutamic acid) anion, vitamin B9

Genetic and biochemical investigations of the maize *Bronze-2* (*bz2*) mutation have helped define the in vivo transport capabilities of plant MRPs. It has been known for some time that the bronze coloration of *bz2* mutants is due to accumulation of the anthocyanin, cyanidin-3-glucoside (C3G) in the cytosol. In wild-type plants anthocyanins are accumulated in the vacuole as purple or red derivatives, but in *bz2* mutants they are restricted to the cytosol where they undergo oxidation and cross-linking to generate brown derivatives. Because they lack the machinery for the cytosolic synthesis of the transport-active anthocyanin derivative C3G-GS (73), it has been suggested that these mutants cannot deploy GS-X pumps for the delivery of these pigments into the vacuole.

In agreement with this speculation, the GS-conjugate of C3G, when synthesized chemically, is transported into isolated vacuolar membrane vesicles at a rate four- to sevenfold greater than that of the model transport substrate DNP-GS by the founder members of the plant MRP subfamily, AtMRP1 and AtMRP2 (70, 71). Furthermore, two maize MRP homologs, ZmMRP3 and ZmMRP4, clearly contribute to the anthocyanin biosynthetic network at the level of vacuolar sequestration (30). ZmMRP3, whose promoter is regulated by the anthocyanin transcription factors B/Pl and R/C1 and specifies a translation product that localizes to the vacuolar membrane, is expressed at high levels in husks, developing tassels and expanding adult leaves, all of which are active in the synthesis of anthocyanin. Transgenic antisense plants lacking a detectable full-length *ZmMRP3* transcript are impaired in pigmentation and assume the *bz2*-like phenotype expected of plants competent in C3G biosynthesis but defective in its vacuolar sequestration. ZmMRP4, a close homolog of ZmMRP3, whose expression is restricted to the aleurone, is required for pigmentation of this tissue (30).

The molecular genetic experiments upon which the concept of MRP-mediated anthocyanin transport across the vacuolar membrane is founded are sound, but there are two interpretative and/or technical issues that complicate identification of the transport-active substrate species in vivo. The first is that all attempts to enzymically glutathionate C3G have failed. Neither Bz2 nor An9, a sequence-divergent petunia type I GST that can complement the Bz2 mutation in planta (1), will conjugate C3G with GSH in vitro under the conditions explored despite their activity toward the model GST substrate 1-chloro-2,4-dinitrobenzene. The second issue is that neither appreciable amounts of glutathionated anthocyanins nor their breakdown products, for instance in the form of sulfur-containing anthocyanin derivatives, have been detected in plant tissues. This may mean that other activities are operative in vivo, activities that are absent in vitro. Alternatively or in addition it may mean that anthocyanin glutathionation, although necessary for transport into the vacuole, yields derivatives that are short-lived in vivo or that at least some GSTs, as exemplified by Bz2 and An9, are primarily responsible for presenting or carrying bound anthocyanin to the vacuolar pump rather than catalyzing the net synthesis of anthocyanin GS-conjugates. Indirect support for the last of these possibilities, the idea that some GSTs act as adaptins for flavonoids, is the finding that purified An9, despite its inability to catalyze the formation of GS-flavonoids under the conditions examined, binds flavonoids with moderate affinity (83).

To date, five unique MRPs, AtMRPs 1–5, have been cloned from *Arabidopsis* and shown to encode functional transporters after heterologous expression in *S. cerevisiae ycf1*Δ strains lacking the yeast vacuolar MRP gene *YCF1* (65, 66; reviewed in 99). All five are capable of transporting GS-conjugates to different degrees in vitro and several are also competent in transporting other amphipathic anions, including glucuronate conjugates, linearized tetrapyrrole catabolites, and the essential vitamin cofactor folate and its derivatives (49, 69–71, 100, 126).

AtMRP2 and AtMRP1 (see below) are the only AtMRPs to have been unequivocally localized to the vacuolar membrane in planta. AtMRP2 is a high-capacity multispecific amphipathic anion pump competent in the transport of a broad range of substrates. Heterologously expressed AtMRP2, like native plant vacuolar membranes (50, 51), is not only competent in the transport of GS-conjugates such as DNP-GS, metolachlor-GS, and C3G-GS, but also linearized tetrapyrroles such as *Brassica napus* nonfluorescent chlorophyll catabolite 1 (Bn-NCC-1), glucuronides such as 17β-estradiol 17-(β-D-glucuronide) ($E_2 17\beta G$), GSH, and oxidized GSH (GSSG) (69). Moreover, DNP-GS stimulates the uptake of $E_2 17\beta G$ and vice versa, and double-label and preloading experiments demonstrate that the two substrates are subject to simultaneous transport by AtMRP2 consequent on their interaction with the transporter from the same (*cis*) side, not the opposite (*trans*) side of the membrane. Evidently, some GS-conjugates and glucuronides reciprocally promote each other's transport (69).

The kinetics of AtMRP2-mediated transport are consistent with a scheme in which: (*a*) $E_2 17\beta G$, DNP-GS, GSH, and Bn-NCC-1 undergo transport via different AtMRP2-dependent pathways; (*b*) $E_2 17\beta G$ and DNP-GS promote each other's transport by binding sites distinct from but tightly coupled to the other's transport pathway (69). The mechanistic basis of these properties is not known, but their existence extends our appreciation of the processes that may converge on transporters of this type. For GS-conjugate and tetrapyrrole transport, the processes that may depend on AtMRP2 and its equivalents in plant species other than *Arabidopsis* may not only include xenobiotic detoxification, antimicrobial compound storage, protection from oxidative stress, and cell pigmentation, but also chlorophyll catabolism (97, 99). For glucuronides, the processes that may depend on AtMRP2 and its equivalents may be broadened to encompass the vacuolar storage of flavonoid antifeedants, UV screening agents, and animal attractants (76).

A property of AtMRP4, and one that it shares with AtMRP1, is its facility for the high-capacity transport of folate (49, 100). In view of the importance of folates in one-carbon metabolism (35), the ability of at least some plant MRPs, like their mammalian counterparts (140), to transport folate may yield insights into how to manipulate plant folate levels. However, if this is to be the case, it will be necessary to reconcile the membrane localization of AtMRP4 with its facility for folate transport. Membrane density fractionation, GFP-tagging, and fluorescence microscopy clearly establish that AtMRP4 localizes to the plasma membrane (49, 100). This implies that if this transporter catalyzes folate transport in the intact plant, the net flux is across the plasma membrane into the apoplast. This begs the question of what physiological roles, if any, apoplastic folates might play. For AtMRP1, which localizes to the vacuolar membrane (48), the key issue is the form or forms in which folate and/or its derivatives are transported. If AtMRP1 indeed participates in the vacuolar storage of folates, it is responsible for delivering only a subfraction of this pool, the monoglutamylated component, into this compartment, or unlike its mammalian counterparts, is able to transport polyglutamates as well as monoglutamates (35, 56, 89, 140).

A fascinating advance in plant MRP research is recognition that their capabilities are probably not restricted to the direct catalysis of primary solute transport. There are members of this subfamily in plants, notably AtMRP4 and AtMRP5, that appear also to have the facility to modulate other transporters. In this regard they share a number of properties with two other animal MRP-like proteins, the cystic fibrosis transmembrane conductance regulator, CFTR, which has chloride channel activity and regulates potassium channels (34), and the sulfonylurea receptor, SUR, a regulatory subunit of the ATP-sensitive potassium channel IKATP (39, 77).

S-type anion channel: slowly activated and deactivated voltage-dependent anion channel implicated in long-term guard cell anion efflux and membrane depolarization during stomatal closure

As an extension of previous investigations of the effects of sulfonylureas, primarily glibenclamide, which completely blocks the activities of CFTR and SUR (34), on stomatal guard cell function (62, 63), it has been demonstrated that *atmrp5-1* T-DNA insertion mutants, unlike their wild-type counterparts, are insensitive to these agents (25). The stomata of wild-type plants, which express high levels of *AtMRP5* (25), are sensitive to glibenclamide-elicited opening in the dark—a phenomenon attributed to inhibition of the calcium-elicited slow anion efflux that ordinarily accompanies closure (62, 63). Moreover, and perhaps more importantly, these effects are not limited to agents, such as sulfonylureas, to which plants are seldom, if ever, exposed. In a better defined, physiologically more meaningful context, the stomata of *atmrp5-1* mutants, in contrast to those of wild-type plants, neither close in response to exogenous calcium or ABA in the light nor open in response to IAA in the dark (52).

One attractive hypothesis that reconciles the properties of AtMRP5 with what is known of stomatal function is based on the negative modulation of slow (S-type) anion channels by this transporter. On the one hand this hypothesis would explain the inability of *atmrp5* stomata to close in response to exogenous calcium and ABA in the light or open in response to IAA in the dark, while retaining the ability to respond to fusicoccin and undergo CO_2-elicited closure (52). On the other hand it is capable of accounting for the diminution of S-type anion channel activity in tobacco guard cell protoplasts ectopically expressing *AtMRP5* (48). Because stomatal closure is triggered by the calcium-dependent activation of S-type anion channels, which upon depolarization activate potassium outward rectifiers (111), the decrease in S-type anion channel activity associated with ectopic expression of *AtMRP5* would be expected to promote sustained stomatal opening and increase drought susceptibility (48).

Less is known of AtMRP4 in this regard except that the effects associated with its loss of function are opposite to those associated with the loss of AtMRP5. Like AtMRP5, AtMRP4 localizes to the plasma membrane and is expressed at high levels in stomata, but unlike their *AtMRP5* equivalents, the stomata of *atmrp4* T-DNA insertion mutants sustain a wider aperture (width to length ratio) than wild-type controls regardless of lighting conditions (49). This phenotype, which is associated with diminished water-use efficiency, is not attributable to decreased sensitivity to ABA. Exogenous ABA inhibits light-elicited stomatal opening similarly in both *atmrp4* mutant and wild-type plants, although the rate of opening in the absence of exogenous ABA is greater in the former, implying that the effects of AtMRP4 are not restricted to absolute changes in stomatal aperture but also apply to the kinetics of opening.

Pleiotropic Drug Resistance Homologs

The third subfamily of full-molecule ABC transporters, the PDRs (**Figure 1**), is encoded by over 15 ORFs in *Arabidopsis* (41, 106, 130) and 23 in rice (12). Yeast PDR5, a reverse-orientation 1511-amino acid protein (5), is the prototype of this family. Considered to be a functional equivalent of the mammalian MDRs, albeit with a reverse orientation and low overall sequence identity, yeast PDR5 is a plasma membrane-localized protein competent in the extrusion of a broad range of anticancer drugs, cyclic peptides, and steroids (54).

Genes encoding bona fide PDR homologs have not been identified in animal and prokaryotic systems, despite their being the most common ABC protein family in yeast (9 out of 29 ORFs) and the fifth most common in *Arabidopsis*. Of the explanations that have been advanced to explain this observation, the most tenable is that this subclass arose from the duplication of a WBC-like reverse-orientation half-molecule ABC transporter (14). Sequence alignments reveal that plant and yeast PDRs and WBCs group in a manner distinct from the other full-molecule and half-molecule ABC transporters (12, 14), indicating a common origin, but that the yeast and plant WBCs group together into a cluster distinct from the yeast and plant PDRs as would be expected if the WBC to PDR transition was a unique event before the divergence of fungi and plants (12, 14).

Four idiotypic motifs containing six or more identical contiguous residues are specific to this subfamily of plant ABC transporters (130). Encompassed by the NBFs, these motifs are deemed "plant PDR signatures." PDR signature 1 (LLLGPP) is immediately N-terminal to and slightly overlaps with the N-terminal Walker A box; signature 2 (GLDSST) starts four residues C-terminal to the N-terminal Walker B box; signature 3 (GLDARAAAIVMR) starts four residues C-terminal to the C-terminal Walker B box; and signature 4 (VCTIHQPSI) starts 86 residues C-terminal to signature 3 (130).

The first plant PDR genes to be identified were *SpTUR2* in the aquaphyte *Spirodela polyrrhiza* (119) and *NpPDR1* (*NpABC1*) in tobacco (*Nicotiana plumbaginifolia*) (42). Initially cloned on the basis of its transcriptional activation by ABA, which in *Spirodela* fronds induces turion (perennating or winter bud) formation, transcription of *SpTUR2* is enhanced by a broad range of stress factors, including cold and salinity, and diminished by kinetins. The tobacco SpTUR2-related PDR5 homolog NpPDR1 encodes a 160 kDa plasma membrane protein whose expression is enhanced by the treatment of *N. plumbaginifolia* cell cultures with sclareolide, a lactone of the antifungal diterpene, sclareol, produced by this plant species (42). Both of these transporters appear to participate in the transport of terpenes and presumably other toxic metabolites elicited by stress. The induction of *NpPDR1* is associated with the energy-dependent exclusion or extrusion of sclareolide by tobacco cell cultures (42), and ectopic expression of *SpTUR2* in *Arabidopsis* confers resistance to sclareol (129). The *Arabidopsis* PDR subfamily member that bears the closest resemblance to SpTUR2 and NpPDR1 is AtPDR12. The steady-state levels of *AtPDR12* transcripts, unlike those for the other *AtPDR*s with the exception of *AtPDR8*, are susceptible to induction by sclareol in seedlings (130). Detailed phylogenetic analyses of the plant PDR subfamily disclose a tightly grouped subcluster consisting of not only SpTUR2, NpPDR1, and NtPDR1 but also AtPDR12 (109, 130).

Brassinolides: class of plant steroids (brassinosteroids) with hormonal activity

Factors that promote the expression of plant PDRs include cycloheximide, brassinolides, herbicides, and high salt for the *AtPDR*s (130); the heavy metals cadmium and zinc, hypoxic stress, jasmonates, auxins, and cytokinins for *OsPDR9* (81) or lead for *AtPDR12* (60); iron starvation for the newly cloned *PDR* from cultivated tobacco (*N. tabacum*), *NtPDR3* (20); and a wide range of microbial elicitors. In the last category is the INF1 elicitin, an elicitor of the late-blight pathogen *Phytophthora infestans*; flagellin, a bacterial proteinaceous hypersensitive reaction elicitor; and yeast extract, a general elicitor in the case of the other member of the *AtPDR12*, *SpTUR2*, *NpPDR1* subcluster, *NtPDR1* (109).

Of this expanded range of factors, lead and iron are of particular interest. With *AtPDR12*, it has been demonstrated that plants constitutively expressing this gene are markedly more tolerant of lead in the growth medium than wild-type controls, which might eventually prove to be of phytoremediative utility (60). For *NtPDR3*, which is probably a component of the iron homeostatic regulatory network because its promoter, like those of other genes activated by iron deprivation (53), contains a *cis*-acting, 16-nucleotide iron deficiency responsive element, IDE1 (20), priority will be to determine if it participates in iron transport. Processes to which NpPDR3 might contribute either directly or indirectly are iron uptake from the soil, perhaps by promoting the uptake of iron chelates, for instance those with organic acids, or the secretion of chelating agents into the rooting medium (12, 20). In addition or alternatively, it may mediate the transport of iron-nicotianamide chelates which are the predominant long-distance transport form of iron within the plant (12).

The participation of PDR-type ABC transporters in plant-pathogen interactions has been a subtext in most considerations of these proteins. Several are induced by some of the elicitors produced by pathogens and a few contribute to the transport of antifungal compounds. But direct evidence to this effect had been lacking until the chance discovery that RNAi-mediated silencing of *NpPDR1* not only confers markedly increased sensitivity to sclareol but also susceptibility to infection by the fungus *Botrytis cinerea* (121). Although *NpPDR1*-silenced plants are increased only marginally in their susceptibility to the bacterium *Pseudomonas syringae* pv tabaci, the same plants consistently undergo spontaneous infection by *B. cinerea*, a fungus that seldom infects wild-type plants (121).

PDRs may also contribute to the transport of signaling molecules or secretion of volatile fragrances (121). For example, the existence of compounds like WAF-1, a sclareol-like labdane diterpene (129) that accumulates in the intercellular space of wounded tobacco leaves and functions as a signal for activation of the expression of wound- and pathogen-elicited defense-related genes (115), is at least consistent with a scheme of this type, although this compound has yet to be tested as a PDR transport substrate. With regard to volatile fragrances it is notable that NpPDR1 is expressed at high levels in the upper parts of petals (121). Although this might mean that this transporter is involved in defense from pathogens deposited on the upper surfaces of plant parts, it might equally signify a role for NpPDR1 in the secretion of monoterpenoid-, sesquiterpernoid-, and/or phenylpropanoid-dominated fragrances for volatilization from these surfaces (21).

Peroxisomal Membrane Protein Homologs

Yeast and animal peroxisomal membranes contain forward-orientation, half-molecule ABC transporters that catalyze the transport of long-chain fatty acids (LCFAs) and/or their acyl-CoA derivatives from the cytosol into the peroxisome (3). The PMP subfamily of yeast contains two members, PXA1 and PXA2, which probably function as a heterodimer (116). The corresponding subfamily of humans contains four members, ABCD1 (alias ALDP), ABCD2 (alias ALDR), ABCD3 (alias PMP70), and ABCD4 (alias PMP69), all of which probably undergo functional homodimerization and/or heterodimerization (3). In yeast, PXA1/PXA2 heterodimers are essential for growth on media containing LCFAs, especially oleic acid, $C_{18:1}$, as the sole carbon source (116). Disruption of either PXA gene diminishes the oxidation of LCFAs to approximately 20% of the wild-type level despite the retention of β-oxidation in detergent lysates of the same mutants (37). In humans, mutation of ALDP is the cause of X-linked adrenoleukodystrophy (82), a progressive, neurological disorder. The primary defect in adrenoleukodystrophy is in the peroxisomal uptake of LCFAs, and LCFA acyl-CoAs associated with the accumulation of LCFAs in plasma and tissues, concomitant with myelin sheath degeneration (3).

It is now clear with the benefit of hindsight that AtPMP2 is the sole member of the PMP subfamily of *Arabidopsis*. AtPMP2, a 1337 amino acid residue forward-orientation full-molecule ABC transporter (**Figure 1**), is not only detectable immunologically as a 150-kDa peroxisomal integral membrane protein (24, 36), but is a protein whose structure appears to be conserved in plants. The rice genome contains two ORFs encoding members of this subfamily, both of which are full-molecule AtPMP2 homologs (27). The half-molecule protein encoded by AtPMP1, by contrast, is a translation product that, despite its resemblance to nonplant PMPs, possesses a putative chloroplast transit peptide not expected of an authentic peroxisomal membrane protein (124).

Three parallel but independent lines of research demonstrate that AtPMP2 is responsible for peroxisomal LCFA uptake. These are the isolation and phenotypic characterization of *Arabidopsis* AtPXA1 (*Arabidopsis thaliana* peroxisomal transporter 1) (142), PED3 (peroxisomal defective 3) (36), and COMATOSE (CTS) AtPMP2 mutants (24).

Arabidopsis pxa1 and *ped3* mutants have two very characteristic phenotypes: (*a*) They are resistant to growth inhibition and the promotion of lateral root initiation by indole-3-butyric acid (IBA) and its analog 2,4-dichlorophenoxybutyric acid (2,4-DB), but not by IAA (36, 142). (*b*) As seedlings they arrest developmentally unless sucrose is added to the growth medium because they are deficient in the catabolism of fatty acids (36, 142). It is now clear that *pxa1* and *ped3* mutants have this pleiotropic phenotype because they lack a functional copy of AtPMP2 and are impaired in the peroxisomal import of fatty acyl-, IBA-, and 2,4-DB-CoA derivatives. As a result, they are unable to β-oxidize stored fatty acids, for the provision of a substrate (succinate) for respiratory energy production, and cannot oxidize IBA or 2,4-DB, to yield the active auxins IAA and 2,4-dichlorophenoxyacetic acid (2,4-D), respectively (36, 142). Allelic *cts* mutants display similar phenotypes including the postgerminal retention of lipid bodies and the maintenance of high levels of triacylglycerol-derived fatty acids and acyl CoAs for the same reason (24).

Further investigations are required, but the effects exerted by AtPMP2 are probably not only because of its role in acyl-CoA transport. Footitt et al. (24) note that the null germination phenotype of strong *cts* alleles is unlikely to be wholly explicable in terms of a deficiency in lipid mobilization because measurements of sucrose in dry *cts* seeds demonstrate levels similar to those of wild-type. This would not be expected if the defect in germination of *cts* mutants were solely attributable to an acute insufficiency of carbohydrates because of impaired reduced carbon entry into and flux via β-oxidation and gluconeogenesis. This contention and the need to further define exactly how AtPMP2 participates in the early steps of germination is reinforced by careful comparisons of the frequencies and patterns of germination of *pxa1* mutants versus *cts-1* and *cts-2* mutants, which reveal a similar phenotype associated with impaired germination despite sustained dry seed sucrose levels in all three cases (124).

Recent work has suggested that AtPMP2 also may be involved in the foliar biosynthesis of jasmonates [jasmonic acid (JA) and its derivatives], lipid-derived signaling compounds implicated in a broad range of plant defense and developmental processes. Basal foliar JA levels are diminished by several-fold and production in response to wounding is markedly attenuated in *cts* mutants (125). Considering that JA biosynthesis is initiated in the chloroplast and terminates in the peroxisome, it has been suggested that AtPMP2 might mediate the transport of a JA biosynthetic intermediate and/or its acyl-CoA derivative into the peroxisome (125). That all known putative AtPMP2 substrates, LCFAs, IBA, 2,4-DB, and JA intermediates, consist of an acyl chain of at least four carbons terminating in a carboxyl group or acyl-CoA substituent is at least consistent with such a scheme (125).

At present, nothing is known of the function of AtPMP1 other than its representation in expressed sequence tag (EST) databases (106). The key questions here are to determine if AtPMP1 indeed localizes to plastid membranes and, if so, whether it has a completely different or perhaps a similar subsidiary role to AtPMP2 in the transport of LCFAs and/or their derivatives.

ATP-Binding Cassette 1 Homologs

The genome of *Arabidopsis* harbors one ORF, AtAOH1, whose translation product bears a close resemblance to mammalian ABC1, a member of the ABCA subfamily (106). As is the case for human and mouse ABC1, AtAOH1 is one of the largest ABC transporter proteins known. It is an 1816-residue protein containing an unusually large putative regulatory domain interrupted by a hydrophobic segment in the central region of the molecule that clearly distinguishes it from other ABC proteins (**Figure 1**). AtAOH1 is the only full-length ABC transporter in *Arabidopsis* for which there is not a yeast homolog or a discernible equivalent in the rice genome (27).

Mutation of human ABC1, which localizes to the plasma membrane and Golgi complex (88) and catalyzes energy-dependent apolipoprotein-mediated cellular lipid efflux (11), is the cause of Tangier disease, a rare autosomal recessive disorder associated with severe high-density lipoprotein deficiency, sterol deposition in macrophages, and systemic atherosclerosis, in humans (59). Nothing is known of the function or localization of AtAOH1, but a logical starting point for its investigation would be to determine if it plays a role in lipid accumulation during seed maturation or lipid mobilization during seed germination, or contributes to lipid mobilization phenomena similar to those associated with ABC1-dependent apoptotic processes in the mouse and worm models (11).

HALF-MOLECULE TRANSPORTERS

White-Brown Complex Homologs

With 29 reverse-orientation half-molecule transporters in *Arabidopsis* (106) and 30 in rice (27) the plant WBC subfamily (**Figure 1**) is many times larger than its equivalents in nonplant eukaryotes with the exception of *Drosophila*, whose genome contains 15 WBC (alias ABCG) genes (15). The yeast genome contains one WBC homolog (ADP1) of unknown function (**http://www.yeastgenome.org/**) (17). The human genome contains five: ABCG1 (alias ABC8), ABCG5, and ABCG8, which participate in the transport of sterols and possibly other lipids (40); ABCG2 (alias BCRP1 for breast cancer resistance protein 1), which contributes to drug, predominantly anthracycline, resistance (19); and ABCG4, whose function is unknown (**http://nutrigene.4t.com/humanabc.htm**).

It was through detailed morphological studies of the wax-secreting epidermal cells of a large collection of *eceriferum* (*cer*), "not wax carrying," *Arabidopsis cer5* mutant lines that the involvement of AtWBC12 in wax secretion was recognized (94). The stems of *cer* mutants have a glossy, bright green appearance because of a decrease in cuticular wax content. In *cer5* mutants, which are mutated in *AtWBC12*, this impairment is accompanied by the accumulation of cytoplasmic laminate inclusions in stem epidermal cells. However, it is not the biosynthesis of wax precursors, predominantly LCFAs and their precursors per se, but instead their delivery to the cell surface that is compromised in *cer5* mutants. The amounts of wax components (alkanes, ketones, and primary and secondary alcohols) on the surfaces of *cer5* mutants are decreased by more than twofold compared with wild-type controls despite a negligible difference in total epidermal (surface plus intracellular) wax content (94). Thus, the transport activity attributed to AtWBC12 is reminiscent of those of mammalian ALDP and the latter's plant equivalent AtPMP2 (24, 36, 142). Indeed, as Pighin et al. (94) note, the trilamellar inclusions seen in *cer5* mutants bear an uncanny resemblance to the inclusions found in mammalian, ALDP mutant, adrenoleukodystrophic cells (2).

AtWBC12-GFP fusions localize to or immediately adjacent to the plasma membrane of epidermal cells as determined by fluorescence microscopy. These data suggest that AtWBC12 contributes to surface wax deposition by forming a homo- or heterodimer through which LCFAs are pumped across the membrane (94). However, this does not exclude other explanations such as the localization of AtWBC12 to a subapical compartment and the delivery of wax components to the cell surface via a vesicular export pathway (112), nor the participation of AtWBC12 in the transport of other chemical species whose transport is needed for the transport and/or stabilization of waxy materials (23). Residual surface wax on the stems of *cer5* mutants and the absence of a discernible phenotype in tissues other than stems and leaves points to the existence of additional AtWBC12-independent wax export mechanisms, perhaps ones mediated by other WBC subfamily members.

The identification of a tobacco WBC subfamily ABC transporter gene, *NtWBC1*, whose transcript levels are highest in stigmata and styles before and during anthesis (92) may be relevant in this context. Lipids are suspected to serve as directional cues for pollen tube growth and penetration into the stigma (135). Based on this and the results of in situ hybridization experiments demonstrating that *NtWBC1* is preferentially expressed in the stigmatic secretory zone and in anthers—a distribution coincident with expression of the gene encoding PEKPIII, a protein that participates in pollination (18)—there is a possibility that NtWBC1 is involved in stigmatic exudate elaboration. The stigmatic secretory zone is the first tissue to come into contact with pollen grains and the one through which the pollen tube grows toward the ovary. A notable feature of NtWBC1, which further implicates it in lipid transactions, is a putative steroidogenic acute regulatory protein signature (START) domain (92), a lipid-binding motif primarily involved in sterol-dependent signal transduction pathways (95). This domain is also recognizable in the *Arabidopsis* At3g55090 (AtWBC16) protein, the nearest equivalent to NtWBC1 from this source, and human ABCG2, which transports lipids (40).

Whether it will prove to have a direct bearing on AtWBC12 and/or NtWBC1 or will yield unexpected insights is not known, but Zhu et al. (141) report the cloning of a WBC subfamily member, *GhWBC1*, from cotton (*Gossypium hirsutum*) whose expression is maximal 5–9 days postanthesis in developing fiber cells. In addition, the expression of *GhWBC1* is weak in the fiber cells of a *li* (*ligon-lintless*) mutant defective in fiber cell elongation, which has led to the suggestion that this putative plasma membrane transporter mediates the translocation of factors necessary for cotton fiber maturation. Of the potential substrates for GhWBC1 that might be necessary for cotton fiber maturation, lipids are candidates in part because of what is known of the transport substrates of several other WBCs and in part because the steady-state levels of transcripts of several genes whose products are involved in lipid biosynthesis—acyl carrier protein, lipid transfer protein, and a putative β-ketoacyl-CoA synthase (64)—are also maximal in rapidly expanding fibers.

In their quest for plant genes that might contribute to detoxification of the explosive 2,4,6-trinitrotoluene (TNT), a potent xenobiotic that heavily contaminates many production facilities and military bases worldwide, Stewart and colleagues conducted systematic cDNA microarray analyses and identified some 52 genes that are upregulated when *Arabidopsis* seedlings are grown on media containing micromolar concentrations of this compound (78). T-DNA insertion mutants for one of these genes, *AtWBC19*, were acquired to assess the role this transporter might play in TNT detoxification. It was during the molecular characterization of these mutants that it was noted that their growth on standard kanamycin selection plates is markedly retarded compared with T-DNA mutants containing the same insertion in other genes. After confirming that *atwbc19* mutants are hypersensitive to the aminoglycoside antibiotic kanamycin it was demonstrated that the wild-type *AtWBC19* gene confers levels of kanamycin resistance comparable to those conferred by conventional bacterial antibiotic resistance markers (ARMs) when expressed in tobacco (*N. tabacum*) (78, 98).

AtWBC19 does not appear to confer resistance to other aminoglycoside antibiotics, for instance gentamycin C or geneticin, that bear a close structural resemblance to kanamycin or to others such as amikacin and hygromycin whose resemblance is more remote (78). As surprising as this is in light of the broad substrate selectivities of many ABC

MDL1 and MDL2, by contrast, localize to the inner mitochondrial membrane, function as homodimers, and, at least in the case of MDL1, promote the mitochondrial export of short-chain, 6–20 amino acid residue, peptides (139). The physiological significance of this yeast peptide export pathway is a mystery because neither *MDL1* nor *MDL2* is essential for cell viability (16). It is appealing to speculate that AtTAPs 1 and/or 2 also catalyze transmembrane peptide translocation, for instance for storage protein mobilization during seed germination and/or for the delivery of pathogen-derived elicitory peptides to the pathogen response machinery.

"SOLUBLE" ATP-BINDING CASSETTE PROTEINS

There are more than 26 ORFs in *Arabidopsis* (106) and 14 in rice (27) encoding ABC proteins lacking contiguous transmembrane spans. Many of these are beyond the scope of this review because they probably encode bona fide soluble proteins. As Sánchez-Fernández et al. (106) detail, these fall into three well-delineated subfamilies: the RLIs, GCNs, and SMCs (see Introduction, above). The remaining so-called "soluble" ABC proteins, the NAPs, on the other hand, although not in a single subfamily, will almost certainly include at least a few that will prove to be the peripheral subunits of transporters. Structurally, the NAPs bear a close resemblance to the ATP-binding peripheral subunits of the many prokaryotic ABC transporters that have been characterized (38). They consist of approximately 300 amino acid residue proteins containing only one NBF (**Figure 1**). That said, quarter-molecule peripheral subunits of this type have yet to be functionally defined in plants. The one NAP, AtNAP1 (alias LAF6), that was initially ascribed a role in plastid protophorphyrin IX transport (80; reviewed in 100) is now known to be a component of the plastid "mobilization of sulfur" (SUF) system responsible for the biogenesis and repair of Fe/S clusters (8, 32, 136). The data available point to the existence of an *E. coli* SufC.SufB.SufD-like (85, 96) plastid AtNAP1.AtNAP6.AtNAP7 complex (136, 137), which, when disrupted, elicits the hyperaccumulation of porphyrins, concomitant with a decrease in chlorophyll content and attenuation of far-red light-regulated gene expression (80). However, this phenotype is not attributable to the abolition of AtNAP1-catalyzed chlorophyll precursor import but instead to the secondary consequences of compromised Fe/S cluster assembly or stability and their effects on steady-state organellar iron levels (136).

CONCLUDING REMARKS

Impressive features of the plant ABC transporter superfamily are its sheer size, the extent to which it is populated by representatives of the MDR, MRP, PDR, WBC, and ATH subfamilies, and its high content of half-molecule transporters compared with other eukaryotic sources. Because several members of the MDR, MRP, PDR, and WBC subfamilies catalyze metabolite, especially secondary metabolite, transport the size of the superfamily as a whole may simply be a reflection of the metabolic versatility of plants (107). This may, in turn, be a factor that has contributed to the high content of half-molecule ABC transporters in plants, of which the WBCs are the majority.

Plant metabolic versatility is manifest at several levels. First, more than 100,000 secondary metabolites have been identified in plants, most of which would be toxic to the cells that produce them if they were not transported across membranes out of the compartments in which they are synthesized (76, 99). Second, although the capacity of green plants for photosynthesis greatly augments their metabolic versatility, this process and its photo-oxidative consequences place even greater demands on the cellular detoxification machinery. Third, plants not only manufacture their own secondary products but also have to contend with those of others (for example, allelochemicals and microbial metabolites). Together, these factors necessitate cellular compartmentation and export mechanisms of wide range and high efficiency.

Although such detoxification mechanisms have considerable biological significance, they also have profound biotechnological implications because the very mechanisms that plants use to protect themselves from their own toxic metabolites and those of others are the mechanisms that must be subverted and/or enhanced if plants are to accumulate secondary compounds or vitamins of high commercial or nutritional value, acquire narrower disease susceptibility profiles, or yield new selectable markers that might meet with less resistance from the public. If what has been learned of anthocyanin, chlorophyll catabolite, and folate transport by the AtMRPs and ZmMRPs, berberine transport by CjMDR1, fatty acyl-CoA transport by AtPMP2, wax transport by AtWBC12, and terpenoid transport by the PDRs applies to the many other plant ABC transporters that remain to be characterized, there is a reasonably high probability that a significant fraction of other compounds of biotechnological interest will also prove to be plant ABC transporter substrates. By the same token, the facility of NpPDR1 and AtWBC19 for conferring resistance to the pathogen *B. cinerea* and the antibiotic kanamycin, respectively, bodes well for the development of new strategies for manipulating plant disease resistance and the introduction of alternate selectable markers for the generation of GM crops. Another potential application of plant ABC transporters would be in the manipulation of such fundamental phenomena as plant stature and water-use efficiency, by using the MDRs and MRPs to alter phytohormone transport and stomatal function.

The closing paragraph of the last *Annual Review* concerned with plant ABC transporters published in 1998 (99) stated that what was written at the time was "no more than a preface to studies of nonchemiosmotic, energy-dependent organic solute transporters in plants. Undoubtedly, investigations of the functional capabilities of other plant ABC transporters. . .whose roles are likely to be equally as significant and varied as those of the MRPs. . .will form the basis of future reviews." This is precisely what has happened. Eight years later it might even be realistic—now that the extent of the plant superfamily is far better defined and the range of processes in which just a few of its members are involved is known—to take these projections one step further and assert that the in-depth characterization of any given plant process will probably necessitate consideration of the part played by at least one ABC transporter.

SUMMARY POINTS

1. ABC transporters, which are highly represented in plants, consist of four core structural modules, two NBFs and two TMDs, and can be classified into subfamilies based on the way these and auxiliary modules are combined and permuted.
2. Most ABC transporters catalyze MgATP-energized, protonophore- and ionophore-insensitive, vanadate-inhibitable transport, but a few in addition to or instead of this appear to regulate channels.
3. The bulk of the known or strongly implicated transport substrates for plant ABC transporters are amphipathic compounds of intermediate complexity. Examples, together with the ABC transporter subfamily or subfamilies shown or suspected to mediate their transport in parenthesis are: natural and synthetic auxins and alkaloids (MDRs); GS-conjugates or endogenous and exogenous toxins, linearized tetrapyrroles, sterol glucuronides, and folates (MRPs); terpenoids (PDRs); LCFAs and/or LCF-acyl CoAs

destined for β-oxidation (PMPs); LCFAs and aminoglycosides (WBCs); and Fe/S clusters and/or their precursors (ATMs).

4. The fundamental plant processes in which ABC transporters are known to participate include polar auxin transport, alkaloid import, tissue pigmentation, vacuolar xenobiotic sequestration, stomatal regulation, disease resistance, lipid catabolism, cuticular wax deposition, antibiotic resistance, assembly of redox-active cytosolic Fe/S proteins, and heavy metal tolerance.

ACKNOWLEDGMENTS

The work from the Rea laboratory was supported by the United States Department of Agriculture National Research Initiative, Department of Energy, and National Science Foundation. The author expresses his appreciation to his many colleagues who either knowingly or unknowingly contributed to the ideas included in this article, and to Annual Reviews for inviting him to write this, his third contribution to the series, the final version of which was much improved thanks to the efforts of an anonymous copy editor and reviewer. For want of space for what is proving to be a research area with a broad range of ramifications, the omission of many fine pieces of work from this review was an inevitability, but one for which the author apologizes. As always, love and thanks are extended to Jenny, Amy and Emily, and Allen and José, all of whom in one way or another made this possible.

LITERATURE CITED

1. Alfenito MR, Souer E, Goodman CD, Buell R, Mol J, et al. 1998. Functional complementation of anthocyanin sequestration in the vacuole by widely divergent glutathione *S*-transferases. *Plant Cell* 10:1135–49
2. Allikmets R, Shroyer NF, Singh N, Seddon J, Lewis RA, et al. 1997. Mutation of the Stargardt disease gene (ABCR) in age-related macular degeneration. *Science* 277:1805–7
3. Almahanu S, Valle D. 2003. Peroxisomal ABC transporters. See Ref. 38a, pp. 497–513
4. Babiychuk E, Fuangthong M, Van Montagu M, Inzé D, Kushnir S. 1997. Efficient gene tagging in *Arabidopsis thaliana* using a gene trap approach. *Proc. Natl. Acad. Sci. USA* 94:12722–27
5. Balzi E, Wang M, Leterme S, Van Dyck L, Goffeau A. 1994. PDR5, a novel yeast multidrug resistance conferring transporter controlled by the transcription regulator PDR1. *J. Biol. Chem.* 269:2206–14
6. Bates SE. 2003. Solving the problem of multidrug resistance ABC transporters in clinical oncology. See Ref. 38a, pp. 359–91
7. Beharry S, Zhong M, Molday RS. 2004. *N*-retinylidenephosphatidylethanolamine is the preferred retinoid substrate for the photoreceptor-specific ABC transporter ABCA4 (ABCR). *J. Biol. Chem.* 279:53972–79
8. Caliebe A, Grimm R, Kaiser G, Lübeck J, Soll J, Heins L. 1997. The chloroplastic protein import machinery contains a Rieske-type iron-sulfur cluster and a mononuclear iron-binding protein. *EMBO J.* 16:7342–50
9. Cardazzo B, Hamel P, Sakamoto W, Wintz H, Dujardin G. 1998. Isolation of an *Arabidopsis thaliana* cDNA by complementation of a yeast abc1 deletion mutant deficient in complex III respiratory activity. *Gene* 221:117–25

10. Cheong N, Madesh M, Gonzales LW, Zhao M, Yu K, et al. 2006. Functional trafficking defects in ATP binding cassette A3 mutants associated with respiratory distress syndrome. *J. Biol. Chem.* 281:9791–800
11. Chimini G, Chambenoit O, Fielding C. 2003. Role of ABC1 in cell turnover and lipid homeostasis. See Ref. 38a, pp. 479–96
12. Crouzet J, Trombik T, Fraysse ÅS, Boutry M. 2006. Organization and function of the plant pleiotropic drug resistance ABC transporter family. *FEBS Lett.* 580:1123–30
13. Csere P, Lill R, Kispal G. 1998. Identification of a human mitochondrial ABC transporter, the functional orthologue of yeast Atm1p. *FEBS Lett.* 441:266–70
14. Dassa E, Bouige P. 2001. The ABC of ABCs: a phylogenetic and functional classification of ABC systems in living organisms. *Res. Microbiol.* 152:211–29
15. Dean M. 2002. *The Human ATP-Binding Cassette (ABC) Transporter Superfamily*. Bethesda, MD: NCBI
16. Dean M, Allikmets R, Gerrard B, Stewart C, Kistler A, et al. 1994. Mapping and sequencing of two yeast genes belonging to the ATP-binding cassette superfamily. *Yeast* 10:377–83
17. DeCottignies A, Goffeau A. 1997. Complete inventory of the yeast ABC proteins. *Nat. Genet.* 15:137–45
18. De Graaf BH, Knuiman BA, Derksen J, Mariani C. 2003. Characterization and localization of the transmitting tissue-specific PELPIII proteins of *Nicotiana tabacum*. *J. Exp. Bot.* 54:55–63
19. Doyle LA, Yang W, Abruzzo LV, Krogmann T, Gao Y, et al. 1998. A multidrug resistance transporter from human MCF-7 breast cancer cells. *Proc. Natl. Acad. Sci. USA* 95:15665–70
20. Ducos E, Fraysse ÅS, Boutry M. 2005. *NtPDR3*, an iron-deficiency inducible ABC transporter in *Nicotiana tabacum*. *FEBS Lett.* 579:6791–95
21. Dudareva N, Pichersky E. 2000. Biochemical and molecular genetic aspects of floral scents. *Plant Physiol.* 122:627–33
22. Dudler R, Hertig C. 1992. Structure of an *mdr*-like gene from *Arabidopsis thaliana*: evolutionary implications. *J. Biol. Chem.* 267:5882–88
23. Flores G. 2004. Wax discovery surprises. *Scientist* 5:20041022–01
24. Footitt S, Slocombe SP, Larner V, Kurup S, Wu YS, et al. 2002. Control of germination and lipid mobilization by COMATOSE, the *Arabidopsis* homolog of human ALDP. *EMBO J.* 21:2912–22
25. Gaedeke N, Klein M, Kolukisaoglu U, Forestier C, Muller A, et al. 2001. The *Arabidopsis thaliana* ABC transporter AtMRP5 controls root development and stomatal movement. *EMBO J.* 20:1875–87
26. Gälweiler L, Guan CH, Muller A, Wisman E, Mendgen K, et al. 1998. Regulation of polar auxin transport by AtPIN1 in *Arabidopsis* vascular tissue. *Science* 282:2226–30
27. Garcia O, Bouige P, Forestier C, Dassa E. 2004. Inventory and comparative analysis of rice and *Arabidopsis* ATP-binding cassette (ABC) systems. *J. Mol. Biol.* 343:249–65
28. Geisler M, Blakeslee JJ, Bouchard R, Lee OR, Vincenzetti V, et al. 2005. Cellular efflux of auxin catalyzed by the *Arabidopsis* MDR/PGP transporter AtPGP1. *Plant J.* 44:179–94
29. Geisler M, Murphy AS. 2006. The ABC of auxin transport: the role of p-glycoproteins in plant development. *FEBS Lett.* 580:1094–102
30. Goodman CD, Casati P, Walbot V. 2004. A multidrug resistance-associated protein involved in anthocyanin transport in *Zea mays*. *Plant Cell* 16:1812–26
31. Gottesman M, Pastan I. 1993. Biochemistry of multidrug resistance mediated by the multidrug transporter. *Annu. Rev. Biochem.* 62:385–427

32. Gray J, Close PS, Briggs SP, Johal GS. 1997. A novel suppressor of cell death in plants encoded by the Lls1 gene of maize. *Cell* 89:25–31
33. Hannikenne M, Motte P, Wu MCS, Wang T, Loppes R, Matagne RF. 2005. A mitochondrial half-size ABC transporter is involved in cadmium tolerance in *Chlamydomonas reinhardtii*. *Plant Cell Environ*. 28:863–73
34. Hanrahan J, Gentzsch M, Riordan JR. 2003. The cystic fibrosis transmembrane conductance regulator (ABCC7). See Ref. 38a, pp. 589–618
35. Hanson AD, Gregory JF. 2002. Synthesis and turnover of folates in plants. *Curr. Opin. Biotechnol*. 5:244–49
36. Hayashi M, Nito K, Takei-Hoshi R, Yagi M, Kondo M, et al. 2002. Ped3p is a peroxisomal ATP-binding cassette transporter that might supply substrates for fatty acid β-oxidation. *Plant Cell Physiol*. 43:1–11
37. Hettema EH, Tabak HF. 2000. Transport of fatty acids and metabolites across the peroxisomal membrane. *Biochem. Biophys. Acta* 1486:18–27
38. Higgins CF. 1992. ABC transporters: from microorganisms to man. *Annu. Rev. Cell Biol*. 8:67–113
38a. Holland BI, Cole SPC, Kuchler K, Higgins CF, eds. 2003. *ABC Proteins: From Bacteria to Man*. London: Academic
39. Inagaki N, Gonoi T, Clement JP, Namba N, Inazawa J, et al. 1995. Reconstitution of IKATP: an inward rectifier subunit plus the sulfonylurea receptor. *Science* 270:1166–70
40. Janvilisri T, Venter H, Shahi S, Reuter G, Balakrishnan L, van Veen HW. 2003. Sterol transport by human breast cancer resistance protein (ABCG2) expressed in *Lactococcus lactis*. *J. Biol. Chem*. 278:20645–51
41. Jasinski M, Ducos E, Martinoia E, Boutry M. 2003. The ATP-binding cassette transporters: structure, function, and gene family comparison between rice and *Arabidopsis*. *Plant Physiol*. 131:1169–77
42. Jasinski M, Stukkens Y, Degand H, Purnelle B, Marchand-Brynaert J, Boutry M. 2001. A plant plasma membrane ATP-binding cassette-type transporter is involved in antifungal terpenoid secretion. *Plant Cell* 13:1095–107
43. Johnson A, Gin P, Marbois BN, Hsieh EJ, Wu M, et al. 2005. *COQ9*, a new gene required for the biosynthesis of coenzyme Q in *Saccharomyces cerevisiae*. *J. Biol. Chem*. 280:31397–404
44. Khush GS. 2001. Green revolution: the way forward. *Nat. Rev. Genet*. 2:815–21
45. Kim DY, Bovet L, Kushnir S, Noh EW, Martinoia E, Lee Y. 2006. AtATM3 is involved in heavy metal resistance in *Arabidopsis*. *Plant Physiol*. 140:1–11
46. Kispal G, Csere P, Prohl C, Lill R. 1999. The mitochondrial proteins Atm1p and Nsf1p are essential for biogenesis of cytosolic Fe/S proteins. *EMBO J*. 18:3981–89
47. Klein I, Sarkadi B, Varadi A. 1999. An inventory of the human ABC proteins. *Biochim. Biophys. Acta* 1461:237–62
48. Klein M, Burla B, Martinoia E. 2006. The multidrug resistance-associated protein (MRP/ABCC) subfamily of ATP-binding cassette transporters in plants. *FEBS Lett*. 580:1112–22
49. Klein M, Geisler M, Suh SJ, Kolukisaoglu HÜ, Azevedo L, et al. 2004. Disruption of *AtMRP4*, a plasma membrane ABCC-type ABC transporter, leads to deregulation of stomatal opening and increased drought susceptibility. *Plant J*. 39:219–36
50. Klein M, Martinoia E, Hoffmann-Thoma G, Veissenbock G. 2000. A membrane potential-dependent, ABC-like transporter mediates the vacuolar uptake of rye flavone glucuronides. *Plant J*. 21:289–304

51. Klein M, Martinoia E, Veissenbock G. 1998. Directly energized uptake of β-estradiol-17-(β-D-glucuronide) in plant vacuoles is strongly stimulated by glutathione conjugates. *J. Biol. Chem.* 273:262–70
52. Klein M, Perfus-Barbeoch L, Frelet A, Gaedeke N, Reinhardt D, et al. 2003. The multidrug resistance ABC transporter AtMRP5 is involved in guard cell hormonal signaling and drought tolerance. *Plant J.* 33:119–29
53. Kobayashi T, Nakayama Y, Itai RN, Nakanishi H, Yoshihara T, et al. 2003. Identification of novel *cis*-acting elements IDE1 and IDE2 of the barley IDS2 gene promoter conferring iron-deficiency-inducible, root-specific expression in heterogeneous tobacco plants. *Plant J.* 36:780–93
54. Kolaczkowski M, Van der Rest M, Cybularz Kolaczkowski A, Soumillion JP, Konings WN, Goffeau A. 1996. Anticancer drugs, ionophoric peptides and steroids as substrates of the yeast multidrug transporter Pdr5p. *J. Biol. Chem.* 271:31543–48
55. Kolukisaoglu HÜ, Bovet L, Klein M, Eggmann T, Geisler M, et al. 2002. Family business: the multidrug-resistance related protein (MRP) ABC transporter genes in *Arabidopsis thaliana*. *Planta* 216:107–19
56. Kruh GD, Zeng H, Rea PA, Liu G, Lee K, Belinsky MG. 2001. MRP subfamily transporters and resistance to anticancer drugs. *J. Bioenerg. Biomembr.* 33:493–500
57. Kushnir S, Babiychuk E, Storozhenko S, Davey MW, Papenbrock J, et al. 2001. A mutation of the mitochondrial ABC transporter Sta1 leads to dwarfism and chlorosis in the *Arabidopsis* mutant *starik*. *Plant Cell* 13:89–100
58. Lankat-Buttgereit B, Tampé R. 2003. The transporter associated with antigen processing (TAP): a peptide transport and loading complex essential for cellular immune response. See Ref. 38a, pp. 533–50
59. Lawn RM, Wade DP, Garvin MR, Wang X, Schwartz K, et al. 1999. The Tangier disease gene product ABC1 controls the cellular apolipoprotein-mediated lipid removal pathway. *J. Clin. Invest.* 104:R25–31
60. Lee M, Lee K, Lee J, Noh EW, Lee Y. 2005. AtPDR12 contributes to lead resistance in *Arabidopsis*. *Plant Physiol.* 138:827–36
61. Leng ER, Vineyard ML. 1951. Dwarf and short plants. *Maize Genetics Coop. Newsl.* 25:31–32
62. Leonhardt N, Marin E, Vavasseur A, Forestier C. 1997. Evidence for the existence of a sulfonylurea receptor-like-protein in plants: modulation of stomatal movements and guard cell potassium channels by sulfonylureas and potassium channel openers. *Proc. Natl. Acad. Sci. USA* 94:14156–61
63. Leonhardt N, Vavasseur A, Forestier C. 1999. ATP-binding cassette modulators control abscisic acid-regulated slow anion channels in guard cells. *Plant Cell* 11:1141–52
64. Li CH, Zhu YQ, Meng YL, Wang JW, Xu KX, et al. 2002. Isolation of genes preferentially expressed in cotton fibers by cDNA filter arrays and RT-PCR. *Plant Sci.* 162:1113–20
65. Li ZS, Lu YP, Zhen RG, Szczypka M, Thiele DJ, Rea PA. 1997. A new pathway for vacuolar cadmium sequestration in *Saccharomyces cerevisiae*: YCF1-catalyzed transport of *bis*(glutathionato)cadmium. *Proc. Natl. Acad. Sci. USA* 94:42–47
66. Li ZS, Szczypka M, Lu YP, Thiele DJ, Rea PA. 1996. The yeast cadmium factor protein (YCF1) is a vacuolar glutathione *S*-conjugate pump. *J. Biol. Chem.* 271:6509–17
67. Li ZS, Zhao Y, Rea PA. 1995. Magnesium adenosine 5′-triphosphate-energized transport of glutathione *S*-conjugates by plant vacuolar membrane vesicles. *Plant Physiol.* 107:1257–68
68. Lill R, Kispal G. 2003. ABC transporters in mitochondria. See Ref. 38a, pp. 515–31

69. Liu GS, Sánchez-Fernández R, Li ZS, Rea PA. 2001. Enhanced multispecificity of *Arabidopsis* vacuolar multidrug resistance-associated protein-type ATP-binding cassette transporter, AtMRP2. *J. Biol. Chem.* 276:8648–56
70. Lu YP, Li ZS, Drozdowicz YM, Hörtensteiner S, Martinoia E, Rea PA. 1998. AtMRP2, an *Arabidopsis* ATP-binding cassette transporter able to transport glutathione *S*-conjugates and chlorophyll catabolites: functional comparison with AtMRP1. *Plant Cell* 10:267–82
71. Lu YP, Li ZS, Rea PA. 1997. *AtMRP1* gene of *Arabidopsis* encodes a glutathione *S*-conjugate pump: isolation and functional definition of a plant ABC-binding cassette transporter gene. *Proc. Natl. Acad. Sci. USA* 84:8243–48
72. Maathuis FJM, Filatov V, Krijger GC, Herzyk P, Axelsen KB, et al. 2003. Transcriptome analysis of *Arabidopsis thaliana* cation transport. *Plant J.* 35:675–92
73. Marrs KA, Alfenito MR, Lloyd AM, Walbot V. 1995. A glutathione *S*-transferase involved in vacuolar transfer encoded by the maize gene *Bronze-2*. *Nature* 375:397–400
74. Martinoia E, Grill E, Tommasini R, Kreuz K, Amrhein N. 1993. ATP-dependent glutathione *S*-conjugate "export" pumps in the vacuolar membrane of plants. *Nature* 364:247–49
75. Martinoia E, Klein M, Geisler M, Bovet L, Forestier C, et al. 2002. Multifunctionality of plant ABC transporters: more than just detoxifiers. *Planta* 214:345–55
76. Martinoia E, Klein M, Geisler M, Sánchez-Fernández R, Rea PA. 2000. Vacuolar transport of secondary metabolites and xenobiotics. In *Vacuolar Compartments. Annual Plant Reviews*, ed. DG Robinson, JC Rogers, 5:221–53. Sheffield, UK: Sheffield Academic
77. Matsuo M, Ueda K, Ryder T, Ashcroft F. 2003. The sulfonylurea receptor: an ABCC transporter that acts as an ion channel regulator. See Ref. 38a, pp. 551–75
78. Mentewab A, Stewart CN. 2005. Overexpression of an *Arabidopsis thaliana* ABC transporter confers kanamycin resistance to transgenic plants. *Nat. Biotechnol.* 23:1177–80
79. Mitsuhashi N, Miki T, Senbongi H, Yokoi N, Yano H, et al. 2000. MTABC3, a novel mitochondrial ATP-binding cassette protein involved in iron homeostasis. *J. Biol. Chem.* 275:17536–40
80. Møller SG, Kunkel T, Chua NH. 2001. A plastidic ABC protein involved in intercompartmental communication of light signaling. *Genes Dev.* 15:90–103
81. Moons A. 2003. *Ospdr9*, which encodes a PDR-type ABC transporter, is induced by heavy metals, hypoxic stress and redox perturbations. *FEBS Lett.* 553:370–76
82. Moser HW, Smith KD, Watkins PA, Powers J, Moser AB. 2001. X-linked adrenoleukodystrophy. In *The Metabolic Basis of Inherited Disease*, ed. CR Scriver, AL Beaudet, D Valle, WS Sly, pp. 3257–302. New York: McGraw-Hill. 8th ed.
83. Mueller LA, Goodman CD, Silady RA, Walbot V. 2000. AN9, a petunia glutathione *S*-transferase required for anthocyanin sequestration, is a flavonoid-binding protein. *Plant Physiol.* 123:1561–70
84. Multani DS, Briggs SP, Chamberlain MA, Blakesee JJ, Murphy AS, Johal GS. 2003. Loss of an MDR transporter in compact stalks of maize *br2* and sorghum *dw3* mutants. *Science* 302:81–84
85. Nachin L, Loiseau L, Expert D, Barras F. 2003. SufC: an unorthodox cytoplasmic ABC/ATPase required for [Fe-S] biogenesis under oxidative stress. *EMBO J.* 22:427–37
86. Noh B, Bandyopadhyay A, Peer WA, Spalding EP, Murphy AS. 2003. Enhanced gravi- and phototropism in plant *mdr* mutants mislocalizing the auxin efflux protein PIN1. *Nature* 423:999–1002
87. Noh B, Murphy AS, Spalding EP. 2001. *Multidrug resistance*-like genes of *Arabidopsis* required for auxin transport and auxin-mediated development. *Plant Cell* 13:2441–54

88. Oram JF. 2000. Tangier disease and ABCA1. *Biochim. Biophys. Acta* 1529:321–30
89. Orsomando G, de la Garza RD, Green BJ, Peng MS, Rea PA, et al. 2005. Plant γ-glutamyl hydrolases and folate polyglutamates. Characterization, compartmentation and co-occurrence in vacuoles. *J. Biol. Chem.* 280:28877–84
90. Ortiz DF, Kreppel L, Speiser DM, Scheel G, McDonald G, Ow D. 1992. Heavy metal tolerance in the fission yeast requires an ATP-binding cassette-type membrane transporter. *EMBO J.* 11:3492–99
91. Ortiz DF, Ruscitti T, McCue KF, Ow DW. 1995. Transport of metal-binding peptides by HMT1, a fission yeast ABC-type vacuolar membrane protein. *J. Biol. Chem.* 270:4721–28
92. Otsu CT, da Silva I, de Molfetta JB, da Silva LR, de Almeida-Engler J, et al. 2004. *NtWBC1*, an ABC tranpsporter gene specifically expressed in tobacco reproductive organs. *J. Exp. Bot.* 55:1643–54
93. Petrášek J, Mravec J, Bouchard R, Blakeslee JJ, Abas M, et al. 2006. PIN proteins perform a rate-limiting function in cellular auxin efflux. *Nature* 312:914–18
94. Pighin JA, Zheng H, Balakshin LJ, Goodman IP, Western TL, et al. 2004. Plant cuticular lipid export requires an ABC transporter. *Science* 306:702–4
95. Ponting RE, Aravind L. 1999. START: a lipid-binding domain in StAR, HD-ZIP and signaling proteins. *Trends Biochem. Sci.* 24:130–32
96. Rangachari K, Davis CT, Eccleston JF, Hirst EMA, Saldanha JW, et al. 2002. SufC hydrolyzes ATP and interacts with SufB from *Thermotoga maritima*. *FEBS Lett.* 74:225–28
97. Rea PA. 1999. MRP subfamily ABC transporters from plants and yeast. *J. Exp. Bot.* 50:895–913
98. Rea PA. 2005. A farewell to bacterial ARMs? *Nat. Biotechnol.* 23:1085–87
99. Rea PA, Li ZS, Lu YP, Drozdowicz YM, Martinoia E. 1998. From vacuolar GS-X pumps to multispecific ABC transporters. *Annu. Rev. Plant Physiol. Plant Mol. Biol.* 49:727–60
100. Rea PA, Sánchez-Fernández R, Chen S, Peng M, Klein M, et al. 2003. The plant ATP transporter superfamily: the functions of a few and identities of many. See Ref. 38a, pp. 335–55
101. Rea PA, Vatamaniuk OK, Rigden DJ. 2004. Weeds, worms, and more. Papain's long-lost cousin, phytochelatin synthase. *Plant Physiol.* 136:2463–74
102. Reits EAJ, Vos JC, Gromme M, Neefjes J. 2000. The major substrates for TAP in vivo and from newly synthesized proteins. *Nature* 404:774–78
103. Ruetz S, Gros P. 1994. Phosphatidylcholine translocase: a physiological role for the *mdr2* gene. *Cell* 77:1071–81
104. Sakai K, Shitan N, Sato F, Ueda K, Yasaki K. 2002. Characterization of berberine transport into *Coptis japonica* cells and the involvement of ABC protein. *J. Exp. Bot.* 53:1879–86
105. Salt DE, Rauser WE. 1995. MgATP-dependent transport of phytochelatins across the tonoplast of oat roots. *Plant Physiol.* 107:1293–301
106. Sánchez-Fernández R, Davies TGE, Coleman JOD, Rea PA. 2001. The *Arabidopsis thaliana* ABC protein superfamily, a complete inventory. *J. Biol. Chem.* 276:30231–44
107. Sánchez-Fernández R, Davies TGE, Coleman JOD, Rea PA. 2001. Do plants have more genes than humans? Yes, when it comes to ABC proteins. *Trends Plant Sci.* 6:347–48
108. Santelia D, Vincenzetti V, Bovet L, Fukao Y, Dchtig P, et al. 2005. MDR-like ABC transporter AtPGP4 involved in auxin-mediated lateral root and root hair development. *FEBS Lett.* 579:5399–406
109. Sasabe M, Toyoda K, Shiraishi T, Inagaki Y, Ichinose Y. 2002. cDNA cloning and characterization of tobacco ABC transporter: *NtPDR1* is a novel elicitor-responsive gene. *FEBS Lett.* 518:164–68

110. Sasaki T, Ezaki B, Matsumoto H. 2002. A gene encoding multidrug resistance (MDR)-like protein is induced in aluminum and inhibitors of calcium flux in wheat. *Plant Cell Physiol.* 43:177–85
111. Schroeder JI, Allen GJ, Hugouvieux V, Kwak JM, Waner D. 2001. Guard cell signal transduction. *Annu. Rev. Plant Physiol. Mol. Biol.* 52:627–58
112. Schulz B, Frommer WB. 2004. A plant ABC transporter takes the lotus seat. *Science* 306:622–24
113. Schulz B, Kolukisaoglu Ü. 2006. Genomics of plant ABC transporters: the alphabet of photosynthetic life forms or just holes in membranes? *FEBS Lett.* 580:1010–16
114. Senior AE, Al-Shawi MK, Urbatsch IL. 1995. The catalytic cycle of P-glycoprotein. *FEBS Lett.* 377:285–89
115. Seo S, Seto H, Koshino H, Yoshida S, Ohashi Y. 2003. A diterpene as an endogenous signal for the activation of defense responses to infection with *Tobacco mosaic virus* and wounding in tobacco. *Plant Cell* 15:863–73
116. Shani N, Valle D. 1996. A *Saccharomyces cerevisiae* homolog of the human adrenoleukodystrophy transporter is a heterodimer of two half ATP-binding cassette transporters. *Proc. Natl. Acad. Sci. USA* 93:11901–6
117. Shitan N, Bazin I, Dan K, Obata K, Kigawarm K, et al. 2003. Involvement of CjMDR1, a plant multidrug-resistance-type ATP-binding cassette protein, in alkaloid transport in *Coptis japonica*. *Proc. Natl. Acad. Sci. USA* 100:751–56
118. Sidler M, Hassa P, Hasan S, Ringli C, Dudler R. 1998. Involvement of an ABC transporter in a developmental pathway regulating hypocotyl cell elongation in the light. *Plant Cell* 10:1623–36
119. Smart CC, Fleming AJ. 1996. Hormonal and environmental regulation of a plant PDR5-like ABC transporter. *J. Biol. Chem.* 271:19351–57
120. Stein OL. 1955. Rates of leaf initiation in two mutants of *Zea mays*, dwarf-1 and brachytic-2. *Am. J Bot.* 42:885–92
121. Stukkens Y, Bultreys A, Grec S, Trombik T, Vanham D, Boutry M. 2005. NpPDR1, a pleiotropic drug resistance-type ATP-binding cassette transporter from *Nicotiana plumbaginifolia*, plays a major role in plant pathogen defense. *Plant Physiol.* 139:341–52
122. Sun H, Molday RS, Nathans J. 1999. Retinal stimulates ATP hydrolysis by purified and reconstituted ABCR, the photoreceptor-specific ATP-binding cassette transporter responsible for Stargardt disease. *J. Biol. Chem.* 274:8269–81
123. Terasaka K, Blakeslee JJ, Titapiwatanakun B, Peer WA, Bandyopadhyaay A, et al. 2005. PGP4, an ATP-binding cassette p-glycoprotein, catalyzes auxin transport in *Arabidopsis thaliana* roots. *Plant Cell* 17:2922–39
124. Theodoulou FL, Holdsworth M, Baker A. 2006. Peroxisomal ABC transporters. *FEBS Lett.* 580:1139–55
125. Theodoulou FL, Job K, Slocombe SP, Footitt S, Holdsworth M, et al. 2005. Jasmonic acid levels are reduced in COMATOSE ATP-binding cassette transporter mutants. Implications for transport of jasmonate precursors into peroxisomes. *Plant Physiol.* 137:835–40
126. Tommasini R, Vogt E, Fromenteau M, Hörtensteiner S, Matile P, et al. 1998. An ABC-transporter of *Arabidopsis thaliana* has both glutathione *S*-conjugate and chlorophyll catabolite transport activity. *Plant J.* 13:773–80
127. Urbatsch I, Sankaran B, Bhagat S, Senior AE. 1995. Both P-glycoprotein nucleotide-binding sites are catalytically active. *J. Biol. Chem.* 270:26956–61
128. Urbatsch I, Sankaran B, Weber J, Senior AE. 1995. P-glycoprotein is stably inhibited by vanadate-induced trapping of nucleotide at a single catalytic site. *J. Biol. Chem.* 270:19383–90

129. van den Brûle S, Muller A, Fleming AJ, Smart CC. 2002. The ABC transporter SpTUR2 confers resistance to the antifungal diterpene sclareol. *Plant J.* 30:649–62
130. van den Brûle S, Smart C. 2002. The plant PDR family of ABC transporters. *Planta* 216:95–106
131. Vatamaniuk OK, Bucher EA, Sundaram MV, Rea PA. 2005. CeHMT-1, a putative phytochelatin transporter, is required for cadmium tolerance in *Caenorhabditis elegans*. *J. Biol. Chem.* 280:23684–90
132. Vögeli-Lange R, Wagner GJ. 1990. Subcellular localization of cadmium and cadmium-binding peptides in tobacco leaves. Implications of a transport function for cadmium-binding peptides. *Plant Physiol.* 92:1068–93
133. Walker JE, Saraste M, Runswick MJ, Gay NJ. 1982. Distantly related sequences in α- and β-subunits of ATP synthase, myosin, kinases and other ATP-requiring enzymes and a common nucleotide binding fold. *EMBO J.* 1:945–51
134. Wang W, Takezawa D, Pooviah BW. 1997. A potato cDNA encodes a homologue of mammalian multidrug resistance P-glycoprotein. *Plant Mol. Biol.* 31:683–87
135. Wolters-Arts M, Lush WM, Mariani C. 1998. Lipids are required for directional pollen tube growth. *Nature* 392:818–21
136. Xu XM, Adams S, Chua NH, Møller SG. 2005. AtNAP1 represents an atypical SufB protein in *Arabidopsis* plastids. *J. Biol. Chem.* 280:6648–54
137. Xu XM, Møller SG. 2004. AtNAP7 is a plastidic SufC-like ATP-binding cassette/ATPase essential for *Arabidopsis* embryogenesis. *Proc. Natl. Acad. Sci. USA* 101:9143–48
138. Yasaki K, Shitan N, Takamatsu H, Ueda K, Sato F. 2001. A novel *Coptis japonica* multidrug resistant protein preferentially expressed in the alkaloid accumulating rhizome. *J. Exp. Bot.* 52:877–79
139. Young L, Leonhard K, Tatsuta T, Trowsdale J, Langer T. 2001. Role of the ABC transporter Mdl1 in peptide export from mitochondria. *Science* 219:2135–38
140. Zeng H, Chen ZS, Belinsky MG, Rea PA, Kruh GD. 2001. Transport of methotrexate (MTX) and folates by multidrug resistance protein (MRP) 3 and MRP1: effect of polyglutamylation on MTX transport. *Cancer Res.* 61:7225–32
141. Zhu YQ, Xu KX, Luo B, Wang JW, Chen XY. 2003. An ATP-binding cassette transporter GhWBC1 from elongating cotton fibers. *Plant Physiol.* 133:580–88
142. Zolman BK, Silva ID, Bartel B. 2001. The *Arabidopsis pxa1* mutant is defective in an ATP-binding cassette transporter-like protein required for peroxisomal fatty acid β-oxidation. *Plant Physiol.* 127:1266–78

Genetic and Epigenetic Mechanisms for Gene Expression and Phenotypic Variation in Plant Polyploids

Z. Jeffrey Chen

Department of Molecular Cell and Developmental Biology and Institute for Cellular and Molecular Biology, University of Texas, Austin, Texas 78712;
email: zjchen@mail.utexas.edu

Annu. Rev. Plant Biol. 2007. 58:377–406

The *Annual Review of Plant Biology* is online at plant.annualreviews.org

This article's doi:
10.1146/annurev.arplant.58.032806.103835

First published online as a Review in Advance on February 5, 2007

1543-5008/07/0602-0377$20.00

Key Words

polyploidy, nonadditive gene expression, epigenetic regulation, RNA interference, evolution

Abstract

Polyploidy, or whole-genome duplication (WGD), is an important genomic feature for all eukaryotes, especially many plants and some animals. The common occurrence of polyploidy suggests an evolutionary advantage of having multiple sets of genetic material for adaptive evolution. However, increased gene and genome dosages in autopolyploids (duplications of a single genome) and allopolyploids (combinations of two or more divergent genomes) often cause genome instabilities, chromosome imbalances, regulatory incompatibilities, and reproductive failures. Therefore, new allopolyploids must establish a compatible relationship between alien cytoplasm and nuclei and between two divergent genomes, leading to rapid changes in genome structure, gene expression, and developmental traits such as fertility, inbreeding, apomixis, flowering time, and hybrid vigor. Although the underlying mechanisms for these changes are poorly understood, some themes are emerging. There is compelling evidence that changes in DNA sequence, *cis*- and *trans*-acting effects, chromatin modifications, RNA-mediated pathways, and regulatory networks modulate differential expression of homoeologous genes and phenotypic variation that may facilitate adaptive evolution in polyploid plants and domestication in crops.

Contents

Polyploid: an individual or cell that has more than two sets of chromosomes

Allopolyploid: a polyploid that contains two or more sets of genetically distinct chromosomes, usually by hybridization between different species

Autopolyploid: a polyploid created by the multiplication of one basic set of chromosomes

INTRODUCTION

Polyploids can be classified into allopolyploids and autopolyploids based on the origins and levels of ploidy (25, 49, 135) (**Figure 1**). An autopolyploid results from doubling a diploid genome (**Figure 1*a***). An allopolyploid is formed by the combination of two or more sets of distinct genomes). Mechanisms include interspecific hybridization followed by chromosome doubling (**Figure 1*b***), fertilization of unreduced gametes between two diploid species (**Figure 1*c***), or interspecific hybridization between two autotetraploids (**Figure 1*d***). The two identical chromosomes (red or blue) within a species are homologous, while the chromosomes derived from different species (red and blue) are orthologous but become homoeologous within an allotetraploid. In an allopolyploid, only bivalents are formed because meiotic pairing occurs between homologous chromosomes (**Figure 1*b*,*c*,*d***). If the homoeologous chromosomes have some segments that are homologous (**Figure 1*e***), pairing may occur between the homoeologous chromosomes, resulting in the formation of multivalents and segmental allotetraploids (135). Here we consider allopolyploids and amphidiploids or disomic allopolyploids to be synonyms. Strictly speaking, only bivalents are formed in the amphidiploids and disomic allopolyploids, whereas multivalents may be formed in the allopolyploids or segmental allopolyploids.

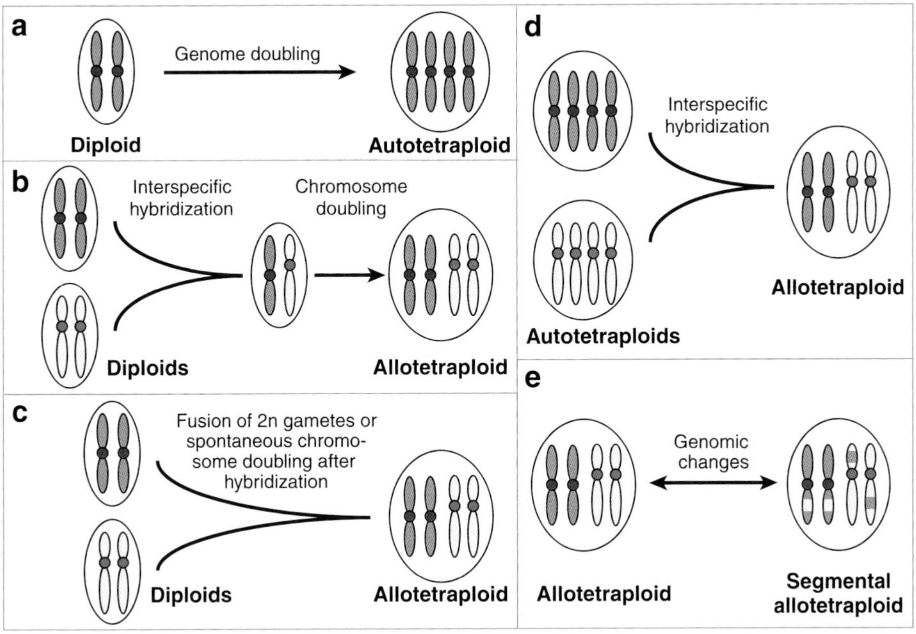

Figure 1

Illustration of auto- and allopolyploids. For simplicity, only one pair of homologous chromosomes (either in red or blue) is shown in a diploid. (*a*) Formation of an autotetraploid by doubling a basic set of chromosomes. A triploid (not shown) can be formed by hybridization between a diploid and an autotetraploid. (*b*) Formation of an allotetraploid by interspecific hybridization followed by chromosome doubling. (*c*) Formation of an allopolyploid by fusion of unreduced gametes in two diploid species. (*d*) Formation of an allotetraploid by interspecific hybridization between two autotetraploid species. (*e*) An allotetraploid (*left*) may become a segmental allotetraploid (*right*) if homoeologous chromosomes contain some homologous chromosomal segments.

In addition to polyploidy, some plant and animal species exist as intraspecific and interspecific hybrids (96, 124). Many plants that transmit as diploids are actually paleopolyploids (ancient polyploids), which are derived from at least one event of whole-genome duplication (WGD) followed by massive gene loss and genomic reorganization through a process known as diploidization (152). *Arabidopsis* (14, 17, 144), rice (154), and maize (45) are good examples of diploidized paleopolyploids. An estimated 30–70% of plant species are of polyploid origin (93, 152). That estimate is as high as 100% if paleopolyploids are included (152).

Polyploidy is a fundamental but relatively underexplored biological process. It is widespread but little is known about how duplicate genes and genomes function in the early stages of hybridization, and how the duplicate genes maintain and diverge functions during plant evolution and crop domestication. Many polyploids are ancient, and their exact progenitors are often unknown. Resynthesized polyploids with known progenitors are excellent materials for dissecting gene expression and genomic changes in early stages and comparisons with older polyploids (28, 79, 132, 149). In addition to the parental phenotypes, polyploids give rise to phenotypes that are intermediate between the two parents and to novel phenotypes that are absent in or exceed features of the contributing parents (77, 82, 119), suggesting nonadditive gene expression. Some traits, such as increasing levels of drought tolerance, apomixis, pest

Ploidy: the number of basic chromosome sets

Homologous: genes or structures that share a common evolutionary ancestor

Orthologous: chromosomes or genes in different species that have evolved from the same ancestor

Homoeologous: chromosomes or genes in related species that are derived from the same ancestor and coexist in an allopolyploid

Segmental allopolyploid: An allopolyploid that contains some homologous segments between homoeologous chromosomes, in which meiotic pairing occurs between homologous chromosomes as well as homoeologous chromosomes

resistance, flowering-time variation, and organ size, may allow polyploids to enter new niches or improve their fitness. Indeed, polyploids may survive better than their diploid progenitors in harsh environments, such as high altitudes and latitudes and cold climates, whereas both diploids and polyploids often thrive and cohabit in mild conditions (49, 135). Moreover, polyploidy is a means of permanent fixation of hybrid vigor and dosage regulation, which may be why many crops (e.g., wheat, cotton, oats, canola, potato, peanuts, sugarcane, coffee, and strawberry) are of polyploid origin (63, 93). Thus, polyploidy has been studied in the context of evolution, genetics, breeding, and molecular biology (25, 29, 50, 76, 79, 86, 95, 111, 132, 149).

Interspecific hybridization and allopolyploidization occur frequently in plant taxa including *Brassica* (133), *Gossypium* (150), *Senecio* (1), *Spartina* (10), *Tragopogon* (141), and *Triticum* (40, 130). Furthermore, hybrids can be formed between different genera including *Triticum* (wheat)-*Secale* (rye) (62), *Triticum* (wheat)-*Hordeum* (barley) (106), *Zea* (maize)-*Avena* (oat) (123), and *Zea* (maize)-*Tripsacum* (gamma grass) (58). Some allopolyploids (e.g., *Tragopogon miscellus* and *T. mirus*) were produced in natural conditions as recently as ~80 years ago, and new *Tragopogon* allotetraploids appear to form every year (141). In contrast, polyploids are rarer in animals than in plants (90, 107). Interspecific hybrids occur in vertebrates and mammals (e.g., a mule is a hybrid between a horse and a donkey), but they cannot produce offspring probably because of genomic incompatibility and/or imbalance in imprinting and sex chromosome dosage (107, 112). Polyspermy (fertilization of more than one sperm into one ovum) causes human triploids in 1–3% of conceptions, and the triploid fetuses are aborted (100). An isolated case of a tetraploid South American rodent (*Tympanoctomy barrerae*) is still debatable (44, 139). Except for endopolyploidy (a diploid

Figure 2

Hypotheses to explain changes in genome structure and function in polyploids. (*a*) Genetic and epigenetic changes in polyploid genomes. Genetic changes include sequence mutations and chromosomal instabilities (deletion/insertion, translocation, transposition, etc.). Epigenetic changes may occur at transcriptional and post-transcriptional levels. Transcriptional regulation is associated with the formation of heterochromatin (*black segments in red chromosomes*) and euchromatin (*irregularly-shaped long arms of blue chromosomes*), leading to gene silencing or activation. Chromatin modifications [through multiple mechanisms including RNA interference (RNAi)] may transcriptionally silence or activate mobile elements, protein-coding genes, and rRNA genes that are responsible for novel variation in allopolyploids. Post-transcriptional regulation involves RNAi, RNA processing, and stability. AA: diploid genome; AAAA or BBBB: autotetraploid; AABB: allotetraploid. Blue segments in red chromosomes or red segments in blue chromosomes in an allotetraploid indicate nonreciprocal exchanges between homoeologous chromosomes. White segments in the red chromosomes indicate a deletion. (*b*) A dosage-dependent (12) or "rheostat" (102) model suggests additive effects of gene expression in a diploid (*upper panel*) and tetraploid (*lower panel*). The duplicate loci in the tetraploid provide additional levels of controls for gene expression. The levels of gene expression are shown as "off" (0) and by gray arrows (1, 2, and 3), and the maximum level (2 in diploid and 4 in tetraploid) is shown by green arrows. This model may explain dominant effects of hybrid vigor but does not explain overdominant performance or novel variation in allopolyploids. (*c*) A regulatory compatibility model suggests that regulatory factors (proteins) produced from orthologous genes generate incompatible heterologous products (*heterodimers between red squares and blue circles*). Alternatively, heterologous proteins may perform better than homologs (*homodimers between red squares or blue circles*), which may explain overdominant effects and hybrid vigor. (*d*) Hypotheses for testing additive and nonadditive gene regulation in *Arabidopsis* allotetraploids. The null hypothesis (Ho) is that gene expression levels in an allotetraploid (Allo) are equal to the sum of two progenitors, *A. thaliana* autotetraploid (At4) and *A. arenosa* (Aa). Typical seedling leaves in At4, Aa, and Allo are shown. Ha: alternative hypothesis.

individual with cells containing more than 2 C amount of DNA in their nuclei) in some cell types (37), aneuploid and polyploid cells in animals and humans are often associated with malignant cell proliferation or carcinogenesis (136).

MECHANISMS FOR GENE EXPRESSION DIVERGENCE BETWEEN ALLOPOLYPLOIDS AND THEIR PARENTS

Several mechanisms may affect the fate of orthologous and homoeologous genes in polyploids (25, 29, 76, 79, 86, 95, 111, 132, 149) (**Figure 2**). First, the majority of homoeologous genes are coexpressed. Second, some duplicate genes are lost, mutate, or diverge (due to genetic changes). The half-life of an active paralogous gene that becomes mutated or lost is estimated to be 2–7 million years (87). Third, epigenetic changes may reprogram gene expression and developmental patterns of new allopolyploids. The impact of these mechanisms on various polyploids can be very different. For example, sequence elimination is predominately observed in wheat and *Tragopogon* allopolyploids (40, 130, 141); chromosomal translocations and transposition (insertion of a DNA fragment into homoeologous chromosomes) are common in *Brassica* allopolyploids (133); and changes in gene expression appear to be a major consequence in *Arabidopsis* and cotton

Paleopolyploid: an ancient polyploid that undergoes chromosomal rearrangements, gene loss, and mutations and eventually becomes diploidized such that it behaves cytogenetically as a diploid

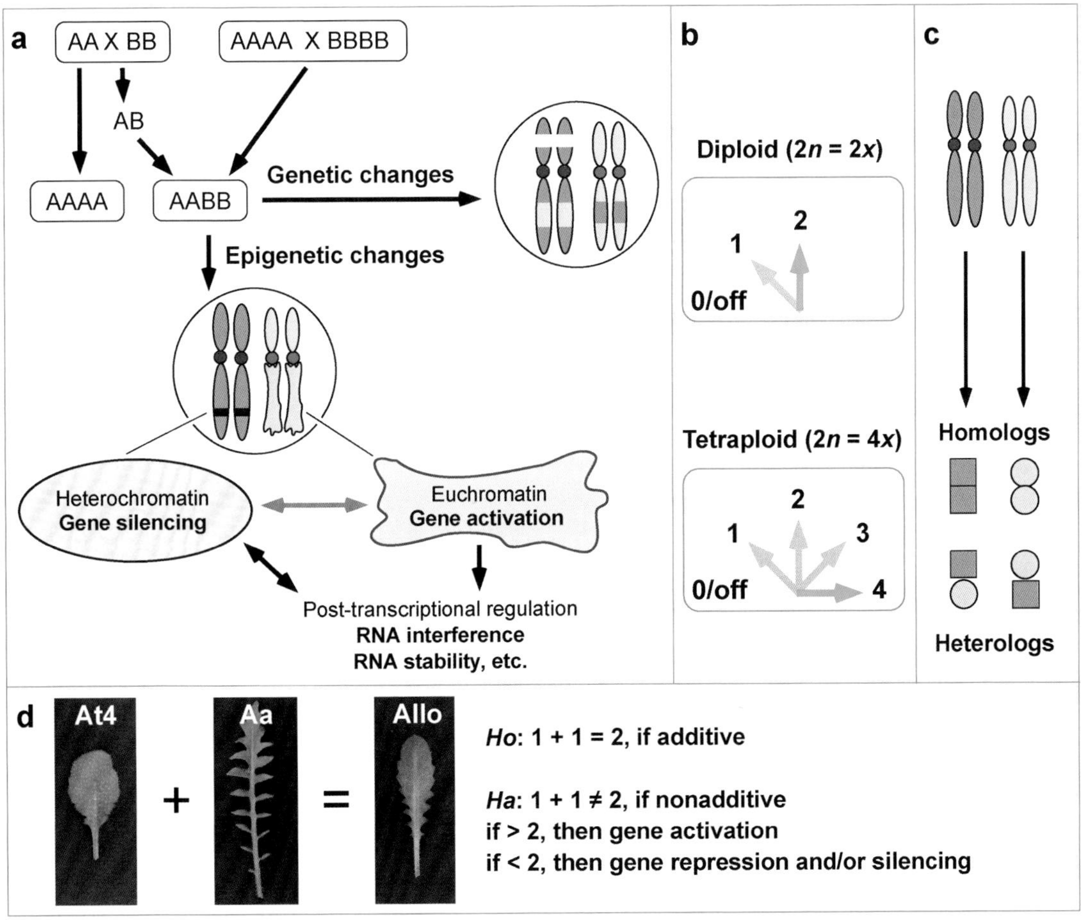

allopolyploids (2, 73, 147, 148). Moreover, genetic and epigenetic changes may be interrelated (25); reactivating transposons by chromatin modifications or RNA-mediated pathways may lead to chromosomal breakages and rearrangements. Eliminating DNA sequences (regulatory and/or coding regions) may alter dosage-dependent gene regulation and chromatin structure.

Nonadditive gene expression: levels of gene expression in an auto- or allopolyploid are not equal to the sum of progenitors, suggesting gene activation including dominance, overdominance, or gene repression

Apomixis: only one parent (usually female) contributes genes to the offspring

Aneuploid: an individual in which the chromosome number is not an exact multiple of the typical haploid set for that species

Paralogous: two or more genes in the same species that share a single ancestral origin

Epigenetic: heritable changes in gene expression without changes in primary DNA sequences

Genomic shock: release of genome-wide chromatin constraints of gene expression including activation of transposons in response to genomic, environmental, and physiological changes

Mutations, Sequence Eliminations, and Chromosomal Rearrangements (Genetic Changes)

Elimination of chromosome- or genome-specific sequences may occur during polyploid formation (**Figure 2*a***). The stochastic changes of duplicate genes may promote polyploid speciation, which is supported by studies in *Brassica* (133), wheat (40, 130), and *Tragopogon* (141). In the resynthesized allopolyploids, loss of parental fragments and/or the appearance of novel fragments are commonly observed. Rapid sequence elimination in the resynthesized allopolyploid wheat may account for a relatively high amount (~14%) of genome- or chromosomal-specific DNA sequences (40, 130), suggesting that differential elimination of genome-specific sequences facilitates pairing between homologous chromosomes but not homoeologous chromosomes.

Changes in DNA sequence may contribute to the loss of duplicate gene expression and function. Indeed, many isozyme loci are lost during polyploidization, such as chlorophyll a/b binding protein genes in *Polystichum munitum*, leucine aminopeptidase loci in tetraploid *Chenopodium*, and phosphoglucose isomerase loci in homosporous fern and *Clarkia* (48, 116). Estimates indicate that in the salmonid and cyprinid fish, the loss of duplicate isozyme loci can be as high as 35–65%, suggesting that loss of duplicate gene function is common after polyploidization, which occurred 50 million years ago (Mya) in this lineage (41). In *Tragopogon*, 9 of 10 genes that display expression differences are associated with changes in allelic DNA sequence (141). However, loss of gene function may also suggest an epigenetic cause (see below).

Epigenetic Regulation of Orthologous Loci

Genetic mutations can explain the cause of gene loss over evolutionary time, but many silencing phenomena may be epigenetically controlled, especially in the early stages of polyploid formation. When two different genomes are combined into a single cell, they must respond to the consequences of genome duplication, especially duplicate copies of genes with similar or redundant functions. Increased gene or genome dosage may induce disease syndromes and abnormal development (7, 38). Thus, the expression of orthologous genes must be reprogrammed through epigenetic mechanisms (**Figure 2*a***) in the early process of polyploidization. This resembles the "genomic shock" phenomenon proposed by McClintock (98). The genomic shock occurs rapidly in interspecific hybrids and allopolyploids, resulting in demethylation of retroelements (92), relaxation of imprinting genes (20, 68, 145), and silencing and activation of homoeologous genes (2, 26, 69, 73, 91, 146–148), including rRNA genes subjected to nucleolar dominance (expression of rRNA genes from only one progenitor in an interspecific hybrid or allopolyploid) (117, 122). Epigenetic changes, which are potentially reversible, provide an effective and flexible means for a polyploid cell to respond to polyploidy or genomic shock. Moreover, gene silencing or activation that is initially epigenetic and reversible could be one step toward a genetically fixed and irreversible state. Epigenetic silencing may also accelerate sequence mutation rates of the affected genes, as observed in repeat-induced point mutations in duplicated genes of *Neurospora crassa* (128).

Mechanisms for epigenetic regulation of homoeologous genes in the allopolyploids are reminiscent of those for X-chromosome inactivation (75), gametic imprinting (143), paramutation (21, 121, 134), and homology-dependent gene silencing (11, 67, 94). However, ploidy-dependent gene regulation has some unique features. First, epigenetic interactions are established among four alleles of two homoeologous loci in allotetraploids compared with two alleles of one locus in a diploid. Second, homoeologous genes from different parental origins may be up- or down-regulated in a chromosomal domain (73, 147), which is different from dosage compensation that often refers to concerted or unidirectional changes in gene expression. Third, at least some epigenetic silencing phenomena in allopolyploids are stochastically established and require multiple generations (24, 104, 148), probably because of the complex process of sorting out chromosome pairing in the allopolyploids. Fourth, pairing occurs mainly between homologous chromosomes, but occasionally between homoeologous chromosomes in allopolyploids (**Figure 1*e***), which may affect gene expression. Finally, because of divergence of regulatory sequences between the progenitors and of heterologous proteins produced in the allopolyploids, *cis*- and *trans*-acting effects on homoeologous genes (146, 151) in various biological pathways constitute a major mode of gene regulation in the allopolyploids.

Additive and Nonadditive Expression of Orthologous Loci

Many genes display dosage dependency and are expressed additively in aneuploids and polyploids (53). If the levels of gene expression and phenotypic variation in the progenitors are additive, they would have the mid-parent values (MPVs) in the polyploids; that is, one plus one is equal to 2 (**Figure 2*d***). If the gene expression is nonadditive (different from the MPV), the values would be larger than 2 or smaller than 2. The former would suggest gene activation including overdominance, whereas the latter would suggest repression and/or silencing. One model to explain additive gene regulation is that there are extra control settings or rheostat potentials (four levels) in a tetraploid compared with two levels in a diploid (**Figure 2*b***). In *Arabidopsis* and *Brassica*, the alleles in the *FLOWERING LOCUS C* (*FLC*) loci display additive effects on flowering time (102). In *Arabidopsis* allotetraploids, the expression of *A. thaliana* and *A. arenosa FLC* loci is additive, giving rise to a late flowering phenotype (146). Furthermore, up to ~90% of the transcriptome is expressed additively in resynthesized *Arabidopsis* allotetraploids (147). Although odd and even dosage effects may vary in a ploidy series (see below), coexpression and coevolution of orthologs and paralogs suggest that a selective advantage is obtained from dosage dependency (15, 47).

Similar numbers of genes (5–11%) are differentially expressed between the parents and resynthesized allotetraploids in stable allotetraploids and five selfing generations (148), and a slightly low number of genes (~2.5%) are differentially expressed between a natural allotetraploid, *A. suecica*, and its assumed progenitors (73). Most gene expression changes observed in early generations are maintained in the late generations and natural allotetraploids, suggesting that rapid and stochastic changes in the resynthesized allotetraploids are responsible for adaptive evolution (148).

Furthermore, allopolyploidy may induce regulatory incompatibilities as well as selective advantage by combining heterologous protein products (**Figure 2*c***). For instance, there is evidence that some protein heterodimers may not function as well as homodimers or vice versa (115, 118). Thus, a silencing strategy could balance regulatory incompatibility and the advantages of having multiple copies of orthologous genes or gene products (e.g., transcriptional factors) spontaneously produced in an allopolyploid cell. Alternatively, novel interactions between heterologous protein products may provide a molecular basis for hybrid vigor and novel adaptation.

X-chromosome inactivation: during mammalian development, the repression of one of the two X chromosomes in the somatic cells of females as a method of dosage compensation

Gametic imprinting: the expression of a gene depends on its parental origin in the offspring

Paramutation: heritable changes in gene expression due to allelic interactions

Homology-dependent gene silencing: repression of gene expression by homologous DNA sequences such as transgenes or endogenous loci

INSTANTANEOUS SPECIATION AND A MODEL FOR POLYPLOIDY STUDIES

New species are often gradually formed because of geographical and ecological separations from an ancestral species (49). However, new species are believed to have arisen suddenly via polyploidization in plants and some animals, including vertebrates such as amphibians and lizards (18, 49, 93, 112). For example, *Arabidopsis suecica* ($2n = 4x = 26$) is a natural allotetraploid formed 12,000 to 1.5 Mya (64, 71, 125). The two progenitor species, *A. thaliana* and *Arabidopsis arenosa* (108, 123), split ~6 Mya (71), similar to the distance between humans and chimpanzees (~6.3 million years) (114). Despite this distance, *A. thaliana* autotetraploid ($2n = 4x = 20$) and *A. arenosa* tetraploid ($2n = 4x = 32$) can hybridize to produce *A. suecica*-like plants ($2n = 4x = 26$) (**Figure 3*a*,*b*,*c***). *A. arenosa* is thought to be an autotetraploid (31), but sequencing analysis sugegsts that it is not a pure autotetraploid (146) (L. Tian, J. Wang & Z.J. Chen, unpublished). The resynthesized allotetraploids are meiotically stable (30, 147) and contain five pairs of *A. thaliana* chromosomes and eight pairs of *A. arenosa* chromosomes (28, 30, 147) (**Figure 3*c***). Compared with resynthesized *Brassica* and wheat allopolyploids that undergo rapid changes in chromosomal structure and DNA sequences (40, 133), the frequency of aneuploids and chromosome abnormalities in *Arabidopsis* resynthesized allotetraploids is relatively low (30).

The nascent allotetraploids (F_1 individuals) are genetically identical (**Figure 3*a***) and showed subtle phenotypic variation. Some variation among F_1 individuls may derive from heterozygosity of the outcrossing tetraploid *A. arenosa* parent, whereas other variation may result from interactions between *A. arenosa* and the different genotypes of *A. thaliana* used in interspecific hybridizations. The degree of variation depends on parental genotypes used in the interspecific hybridization. For example, the seed set is higher in the nascent allotetraploids (F_1) between *A. arenosa* and *A. thaliana* C24 or *Ler* ecotype than those between *A. arenosa* and *A. thaliana* Columbia (20, 31), indicating genotypic effects on interspecific hybridization. Hybridization was successful only in the crosses using *A. thaliana* as a maternal parent and *A. arenosa* as a pollen donor (31) (J. Wang, L. Tian & Z.J. Chen, unpublished), probably because *A. arenosa* is outcrossing and self-incompatible. Most F_1 individuals and selfing progeny in late generations resemble the *A. arenosa* parent and *A. suecica* (31, 91, 147) (**Figure 3*c***), although diverse phenotypes are observed in segregating populations (F_2-F_3) (**Figure 3*b***). Therefore, *A. arenosa* appears to be phenotypically dominant over *A. thaliana* in the allotetraploids (28, 147).

The allotetraploids obtained from selfing the F_1s show stable karyoptes in the fifth generation (**Figure 3*c***), but exhibit a wide range

Figure 3

Production of *Arabidopsis* allotetraploids. (*a*) Schematic karyotypes of F_1 nascent allotetraploids and their progenitors. Seedlings of the two progenitors, *A. thaliana* autotetraploid (At4) and *A. arenosa* tetraploid (Aa), are shown. (*b*) Phenotypic variation was infrequently observed in the F_1 allotetraploids but was very common in the segregating populations (F_2) (*A–D*). *A* and *D*: *A. arenosa*-like, *B*: *A. thaliana*-like, and *C*: intermediate. (*c*) Independent lineages of allotetraploids are produced by selfing multiple F_1 lines. Allopolyploids self-pollinate spontaneously (or can be manually self-pollinated), and progressive inbreeding occurs in advanced generations. Karyotypes in a meiotic cell of a resynthesized allotetraploid (Allo733 in the fifth generation) show 5 pairs of *A. thaliana* centromeres (At Cen, *red*) and 8 pairs of *A. arenosa* centromeres (Aa Cen, *green*) (147) (reproduced with kind copyright permission of the Genetics Society of America). Seedling of Allo733 is shown. All plants were photographed when they were 4–5 weeks old. The size bars indicate 3 cm.

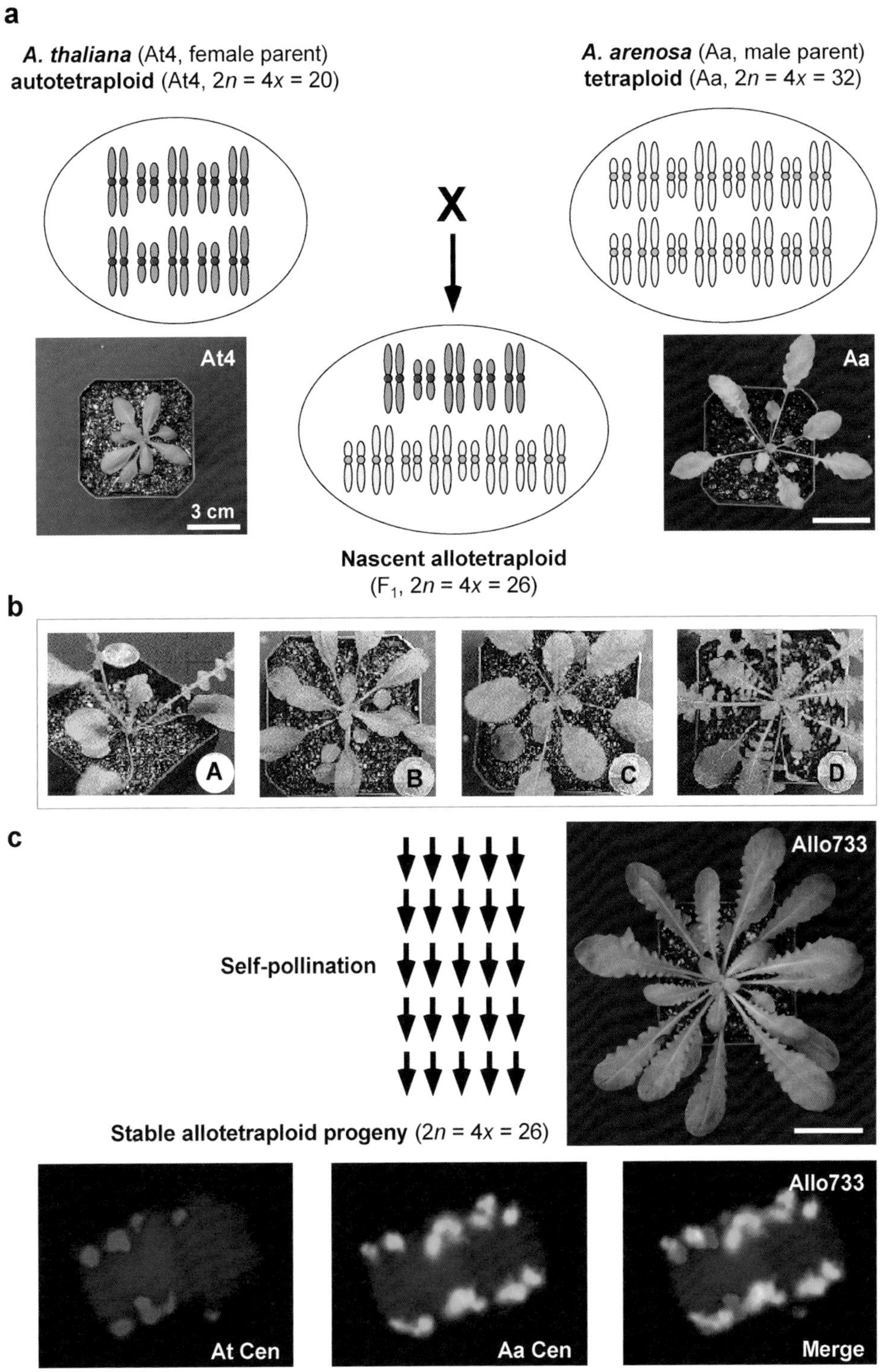
a
A. thaliana (At4, female parent)
autotetraploid (At4, 2n = 4x = 20)
A. arenosa (Aa, male parent)
tetraploid (Aa, 2n = 4x = 32)
X
At4
3 cm
Aa
Nascent allotetraploid
(F1, 2n = 4x = 26)
b
A
B
C
D
c
Self-pollination
Allo733
Stable allotetraploid progeny (2n = 4x = 26)
At Cen
Aa Cen
Allo733
Merge

Amphiplasty: alterations of chromosomal morphology in interspecific hybrids or amphidiploids

of variants, some of which are absent in either parent (transgression) (**Figure 4*a***). Moreover, the allotetraploids display hybrid vigor: larger rosettes, more leaves, longer and wider leaves, and taller plants than the parents. The fertility rate of the plants in the selfing progeny varies from one lineage to another (30, 31). The overall level of fertility improves after each generation of selfing, suggesting that genome incompatibility and gene expression divergence between the progenitors are gradually overcome (148).

The flower colors varied from pink (like *A. arenosa*) in the early generation (F_1) to a mixture of pink and white flowers in the intermediate generations (S2–4) and white in the late generation (S5). During selfing (S3), there is a low frequency of mixed white and pink flowers in the same flower branch (**Figure 4*b***, bottom), which is transient and mosaic (derived from the same zygote). The appearance of varigation within the same flower branch suggests rapid changes in the expression of genes involved in anthocyanin synthesis pathways probably via epigenetic regulation.

NONADDITIVE GENE REGULATION AND TRANSCRIPTOME DOMINANCE IN ALLOPOLYPLOIDS

In 1928, Navashin coined the term "amphiplasty" to describe chromosomal changes in interspecific hybrids of *Crepis* (110). He defined "differential amphiplasty" as specific changes in a few chromosomes (disappearance of satellites or secondary constrictions) and "general amphiplasty" as the overall changes in chromosomal morphology (shortening, thickening, or lengthening of chromosomes) from one species in the interspecific hybrids or amphidiploids. Changes in chromosomal morphology might also affect gene expression. Indeed, following the pioneering work of Navishin & McClintock (97, 110), several contemporary researchers demonstrated that differential amphiplasty is synonymous to nucleolar dominance (117, 122). The disappearance of the secondary constrictions is caused by silencing of rDNA loci in those chromosomes (117). Nucleolar dominance is observed in *Drosophila* interspecific hybrids and *Xenopus*, *Arabidopsis*, *Brassica*, and wheat allopolyploids (117, 122). The dominance is reversible and developmentally regulated and is controlled by chromatin modifications involving DNA methylation and histone acetylation (26, 27). Blocking histone acetylation or DNA methylation derepresses the silenced rRNA genes subjected to nucleolar dominance. Both DNA methylation and histone hypoacetylation reinforce the formation of the "inactive" chromatin state, resulting in gene silencing (72). The silencing of rDNA chromatin requires at least one histone deacetylase (AtHDA6) that is localized in nucleoli (35).

General amphiplasty may be similar to the effects of genomic shock (98). Combining two genomes in a "new" polyploid cell

→

Figure 4

(*a*) Phenotypic variation of the plants after five generations of selfing. The plants include two parents, *A. thaliana* autotetraploid (At4) and *A. arenosa* (Aa), six allotetraploids (S–1 to S–6), and a natural allotetraploid, *A. suecica* (As). Allotetraploids have the same chromosome number (**Figure 3*c*** and data not shown) but display phenotypic and flowering-time variation. Some plants (S5–5 and S5–6) resemble *A. suecica*, whereas others (S5–1 to S5–4) show novel phenotypes. Differential expression of parental genes may contribute to the phenotypic variation of *Arabidopsis* allotetraploids. (*b*) Evidence for epigenetic silencing in resynthesized allopolyploids. *A. suecica* (As, *white flower*) is the naturally occurring allotetraploid derived from *A. thaliana* (not shown) and *A. arenosa* (Aa), which have white and pink flowers, respectively. The flower colors in the third generation of resynthesized allotetraploids (S3–1, 2, and 3) segregate from all white (S3–1) to all pink (S3–3). Variegated flower colors in the same inflorescence branch (S3–2) indicate epigenetic regulation of genes controlling flower color pathways. The same-size bars are 50 mm in (*a*), except for At4, and 5 mm in (*b*), except for As.

a
At4
Aa
S5-1
50 mm
50 mm
S5-2
S5-3
S5-4
50 mm
S5-5
S5-6
As
50 mm
b
S3-1
S3-2
S3-3
5 mm
As
S3-2
Aa
5 mm

may generate the genomic shock and release some constraints imposed on unstable elements locked in a junk yard (e.g., transposable elements in heterochromatin). Little is known about the consequences of general amphiplasty or genomic shock on interspecific hybrids or allopolyploids that have balanced pairs of chromosomes.

To begin to test this, Wang et al. (144) studied transcriptome divergence in *Arabidopsis* allotetraploids and their progenitors. First, they compared gene expression differences between the two progenitors using the spotted oligo-gene microarrays designed from ~26,000 annotated genes that share a high percentage of sequence identities between *A. thaliana* and *A. arenosa*. Most of the oligos can cross-hybridize with both *A. thaliana* and *A. arenosa* genes (74). More than 15% of the transcriptome is differentially expressed between *A. thaliana* and *A. arenosa* that diverged ~6 Mya. Approximately 2,100 genes (8%) are more abundantly expressed in *A. thaliana* than in *A. arenosa*, whereas 1,818 genes (7%) are expressed at higher levels in *A. arenosa* than in *A. thaliana*. Second, Wang et al. (144) compared mRNA abundance in an allotetraploid with the mid parental value (MPV: an equal mixture of RNAs from two parents). If the genes from two progenitors are additively expressed (**Figure 2*d***), their cumulative expression levels in the allotetraploid are equal to MPV. Nonadditive expression suggests that at least one of the homoeologous genes is up- or downregulated. There may also be instances in which silencing of a locus is compensated by increased expression of its homoeologous locus, which cannot be detected in this comparison. Wang et al. (146) found that 2,011 genes (~8%) are nonadditively expressed in two independently derived allotetraploids using a common variance analysis, and up to ~38% genes using a per-gene variance analysis. Interestingly, ~68% of the genes that are nonadditively expressed in the allotetraploids are differentially expressed between the two parents, suggesting that the genes with species-specific expression patrterns are subjected to expression changes in the allopolyploids. Remarkably, among the nonadditively expressed genes, more than 65% of the genes are downregulated in the allotetraploids. Among them, >94% of the genes that are expressed at higher levels in *A. thaliana* than in *A. arenosa* are downregulated in the allotetraploids. These data indicate that the genes with *A. thaliana* expression patterns tend to be repressed, whereas the genes with *A. arenosa* expression patterns are transcriptionally dominant in the allotetraploids, coincident with the phenotypic and nucleolar dominance of *A. arenosa* in the allotetraploids (117, 147) (**Figures 3** and **4**).

Interestingly, similar levels of transcriptional changes were observed in maize diploid hybrids (9.8%) (140) and polyploid taxa of *Senecio* (5%) (60), wheat (7.7%) (59), and cotton (5%) (2). The high percentage of gene expression changes in the *Tragopogon* allopolyploids (~17.5%) (141) is partly associated with a high level of polymorphism (~11%) within populations between the two parents and a moderate amount of variation (>2.5%) among allopolyploid populations. These numbers are also similar to those observed in interspecific hybrids of *Drosophila* (103, 151), suggesting that the levels of transcriptional changes induced by hybridization may be fairly consistent even across plant and animal kingdoms. Transcriptome dominance is also observed in an analysis of ~210,000 expressed sequence tags (ESTs) derived from an ovular cDNA library of tetraploid cotton (*Gossypium hirsutum* L.) (153). The upland cotton was formed by ancient interspecific hybridization between AA and DD genome species (150). AA subgenome ESTs of all functional classifications including cell cycle control and transcription factor activity were selectively enriched in *G. hirsutum* L., a result consistent with the production of long lint fibers in AA genome species. Therefore, transcriptome dominance is likely a general consequence of hybridization effects on gene expression in interspecific hybrids and allopolyploids.

The number of genes displaying expression changes in *A. thaliana* autopolyploids is much smaller than that in the allotetraploids (147). In yeast, Galitski et al. (43) found that 10 genes are induced and seven genes are reduced in response to an increase in ploidy levels (haploid, diploid, triploid, and tetraploid). The cell size increases with increasing ploidy levels, which is correlated with repression of G1 cyclins (*Cln1* and *Pcl1*). *FLO11*, a gene important to the invasiveness of the yeast cells, is repressed with increasing ploidy levels. The reduction of *FLO11* expression in cells of higher ploidy is correlated with diminished invasion, suggesting a role of ploidy-dependent gene regulation in adaptive evolution. Collectively, the data indicate that genome doubling has smaller effects on gene expression changes than intergenomic hybridization.

What factors affect transcriptome dominance in the allopolyploids? Is the gene repression controlled by widespread chromatin modifications or a few "key" regulatory genes? Over time, the progenitor species may have evolved to possess species-specific gene expression patterns. Modulation of the species-specific expression of these genes may determine the outcome of transcriptional and posttranscriptional competition between the two parental genomes in their offspring. Changes in chromatin landscape on repressed genes may result from concerted modifications of many genes in one species, perhaps by a mechanism similar to that for nucleolar dominance (117). Alternatively, expression changes in a few regulatory genes such as transcription factors and microRNAs may induce *trans*-acting effects on many downstream pathways (25). For example, a Myb transcription factor gene is responsible for hybrid-induced incompatibilities in *Drosophila* interspecific hybrids (8). Also, a single miRNA can regulate hundreds of genes involved in the transition from one developmental stage to another (39).

The role of chromatin modifications in silencing or activating protein-coding genes in allopolyploids has been demonstrated in several recent studies (70, 73, 91, 104, 146, 148). Silenced genes can be reactivated by aza-dC (73), a chemical inhibitor of DNA methylation, or by downregulation of the genes encoding DNA methyltransferases using RNA interference (RNAi) (148). Treating allotetraploids with aza-dC generates pleiotropic effects on natural and synthetic allotetraploids including reactivation of mobile elements (91). Reactivation of transposons is also observed in the synthetic allotetraploids (92). The above data suggest that two species may have possessed different levels of chromatin modifications for many genes that display species-specific expression patterns. Perturbation of chromatin structure may have occurred during the formation of interspecific hybrids or allopolyploids, leading to the changes in gene expression.

Factors other than chromatin modifications may also be responsible for genome-wide nonadditive gene regulation. Nonadditively expressed genes are randomly distributed along the chromosomes (147). Within a small chromosomal region in which *TCP3* and *RFP* genes are located, *A. thaliana TCP3* is expressed, whereas *A. arenosa TCP3* is silenced (73). For *RFP*, *A. thaliana RFP* is repressed, whereas *A. arenosa RFP* is expressed. Interestingly, the neighboring genes located between *TCP3* and *RFP* loci are coexpressed. The above data are reminiscent of the silenced rRNA genes that are restricted in the rDNA loci (81). Furthermore, in the *met1*-RNAi *A. suecica* lines, several silenced genes are not reactivated (148). These data argue that widespread chromatin remodeling does not explain nonadditive regulation for all genes, but support the notion that each gene is regulated through interactions among homoeologous loci such as paramutation-like phenomena observed in *A. thaliana* tetraploids (104).

ALLOPOLYPLOIDY AND HETEROSIS

Genome-wide nonadditive gene regulation observed in the allotetraploids correlates with expression divergence between the parents. Thus, hybrids derived from distantly related species may induce a high level of gene expression changes in a nonadditive fashion, providing molecular bases of hybrid vigor (13) and phenotypic variation in the allotetraploid progeny (31). Hybrid vigor refers to the performance of an F_1 hybrid higher than MPV or the best parent. The genetic basis for heterosis is predicted to be associated with dominant complementation of slightly deleterious recessives (dominance model) (19, 66) or overdominant gene action in which genes have greater expression in heterozygous conditions (overdominance model) (32, 36). According to the dominance model, highest performance should be observed when all dominant favorable genes from both parents are in homozygous conditions. The overdominance model suggests that heterosis should reach its peak at the maximum levels of heterozygosity and dissipate when approaching homozygosity. Moreover, overdominance is accompanied by nonallelic or epistatic interactions, and epistasis is involved in most QTLs associated with inbreeding depression and heterosis in corn (137) and rice (84). Comparing genome-wide gene expression data with phenotypic traits (QTLs) may provide new insights into the role of gene expression changes in various biological pathways that give rise to hybrid vigor.

The gene expression changes observed in maize diploid hybrids (6, 140) and genome-wide transcriptome dominance in *Arabidopsis* allotetraploids (147, 148) support both dominance and overdominance models. Many genes in energy, metabolism, cellular biogenesis, and plant hormonal regulation are upregulated in the allotetraploids (147), which may contribute to the hybrid vigor observed in the allotetraploids. Although the underlying mechanisms are unknown, one possibility is modulation of a few key regulators in the allotetraploids that may control downstream genes in various biological pathways (146, 147) such as photosynthesis and metabolism (Z. Ni & Z.J. Chen, unpublished).

Alternatively, *cis*- and *trans*-acting effects involving regulatory sequence changes (see below), chromatin modifications, and RNA-mediated pathways (25) (**Figure 2*a***) may explain dominance, overdominance, and epistasis. The interactions between the diverged orthologous protein products may determine repression or activation of progenitors' genes in allopolyploids (**Figure 2*c***) of *Arabidopsis* (148), cotton (2), *Senecio* (60), and wheat (59, 69), interspecific hybrids (151) in *Drosophila*, intraspecific diploid hybrids in maize (6, 54, 140), and sex-dependent gene regulation in *Drosophila* (46, 120). These mechanisms are not mutually exclusive, and the diverged protein-protein and protein-DNA interactions in allopolyploids may trigger repression of the protein-coding genes and rDNA loci derived from one progenitor (e.g., *A. thaliana*) via chromatin modifications (26, 73, 146, 148) or novel expression patterns leading to hybrid vigor (**Figure 2*c***).

CIS- AND *TRANS*-ACTING EFFECTS ON TRANSCRIPTIONAL REGULATION IN ALLOPOLYPLOIDS

Stable allopolyploids provide an excellent system for testing *cis*- and *trans*-acting effects because a common set of protein factors is present in the same allotetraploid cells. After the unification of the distinct genomes, differences in *cis*- and *trans*-regulation contribute to changes in the expression of orthologs that become homoeologous pairs in the allopolyploid or interspecific nucleus (146, 151). *Cis*-regulatory divergence directly acts on single genes or localized chromatin domains such as promoters or enhancers and may result in asymmetric accumulation of homoeologous transcripts in allpolyploids.

There is evidence for *cis*- and *trans*-effects on orthologous or homoeologous genes in the allotetraploids (146) and interspecific hybrids (33, 151). Differential expression of progenitors' genes in *Arabidopsis* allopolyploids (73, 148) and interspecific hybrids (33), *Drosophila* interspecific hybrids (151), and maize diploid hybrids (138) is mainly caused by *cis*-regulatory changes. Progenitor-specific differences in expression in the same cells are most likely due to allelic or epigenetic differences. In contrast, expression divergence due to alterations in *trans*-regulatory hierarchies should result in two kinds of expression changes. The first is a difference in the sum of homoeologous mRNAs compared with the mid-value of the two parents or nonadditive gene expression. Indeed, the divergently expressed orthologs comprise ~68% of the genes that were expressed in a nonadditive fashion in two allotetraploids (146, 147), implicating *trans*-acting effects. The second is a change in the ratio of homoeolog-encoded mRNAs in an allopolyploid compared with the ratio of the two orthologs in a 1:1 mixture of the parental mRNAs (147). Such a difference would demonstrate a regulatory interaction between the parental genomes (for example, the failure of an interspecific heterodimer to activate transcription) (25) (**Figure 2*c***). A regulatory hierarchy (12) model suggests that *trans*-regulatory differences predominate in allopolyploids (25).

The species-specific expression patterns observed in *Arabidopsis* allotetraploids (147) may result from sequence divergence at regulatory elements during the ~6 million years that separate the parental species. *Cis*- and *trans*-acting regulation and epigenetic modifications of homoeologous genes may change regulatory interactions in a biological pathway (**Figure 5*a***). This has been demonstrated in a subset of genes controlling flowering-time variation. *A. arenosa* and *A. thaliana* (L*er*) diverged in flowering habits probably because of selective adaptation to cold and warm climates (108, 123), respectively. Natural variation of flowering time is largely controlled by two epistatically acting loci, namely, *FRIGIDA* (*FRI*) (65) and *FLOWERING LOCUS C* (*FLC*) (102, 131). *FRI* upregulates *FLC* expression that represses flowering in *A. thaliana*. *A. thaliana* has a nonfunctional *AtFRI* (65), whereas *A. arenosa FRI* (*AaFRI*) is functional (146). Compared with *A. thaliana* (*AtFLC*), *A. arenosa FLC* (*AaFLC*) loci possess deletions in the promoter and first intron that are important to *cis*-regulation of *FLC* expression. In resynthesized allotetraploids, *AaFRI* complements nonfunctional *AtFRI* and interacts in *trans* with *AtFLC*, making the synthetic allotetraploids winter annual in a dosage-dependent manner. *AaFRI* acts on *AtFLC* in *trans* and on *AaFLC* in *cis* because *A. thaliana FRI* is nonfunctional. The different effects of *AaFRI* on *AtFLC* and *AaFLC* loci are likely dependent on the sequence divergence in their *cis*-regulatory elements (e.g., deletions in the promoter and first intron). *AtFLC* and *AaFLC* upregulation is mediated by H3-K9 acetylation and H3-K4 methylation, suggesting a role of FRI in locus-specific chromatin modifications (146).

Although our model (**Figure 5*a***) simplifies the flowering pathway that involves >80 genes (80), it offers one explanation of the fate of orthologous genes involved in biological pathways during allopolyploidization. Many orthologous genes might have diverged in their *cis*-regulatory elements that confer strong or weak, dominant or recessive alleles, tissue-specific expression, and/or developmental regulation. The regulatory networks may be reset by *cis*- and *trans*-acting effects via chromatin modification immediately after allopolyploidization. Over generations, genetic and epigenetic changes are subject to selection and adaptation, and additional genes (e.g., MAF, a FLC-MAF family member of MADS-box genes, in *A. suecica*) (146) may be activated for allopolyploids to occupy an environmental niche. A similar mechanism may be responsible for the functional diversification of orthologous genes in developmental regulation of gene expression, a phenomenon known as subfunctionalization of duplicate genes (88).

Flowering time directly affects plant reproduction and adaptation. Therefore, sequence evolution and epigenetic regulation play interactive and pervasive roles in mediating the regulatory incompatibilities between divergent genomes, leading to natural variation and selective adaptation during allopolyploid evolution.

POST-TRANSCRIPTIONAL REGULATION IN ALLOPOLYPLOIDS

Some gene expression variation observed in polyploids may be controlled at the level of post-transcriptional regulation (70). Silencing a duplicate copy of homoeologous RNA in polyploids may be part of an RNA-mediated pathway similar to cosuppression (67, 94) or RNAi (42). Silencing of transgenes is correlated with transgene dosage in *Drosophila* (113) and ploidy levels in *Arabidopsis* (105). Activation of Wis 2–1A retrotransposon in the newly synthesized wheat allotetraploids drives the readout transcripts from adjacent sequences including the antisense or sense strands of known genes, leading to the silencing or activation of respective corresponding genes (70). RNA-mediated silencing of duplicate genes in polyploids is a developmental strategy. Production of progenitor-dependent RNA transcripts may be associated with

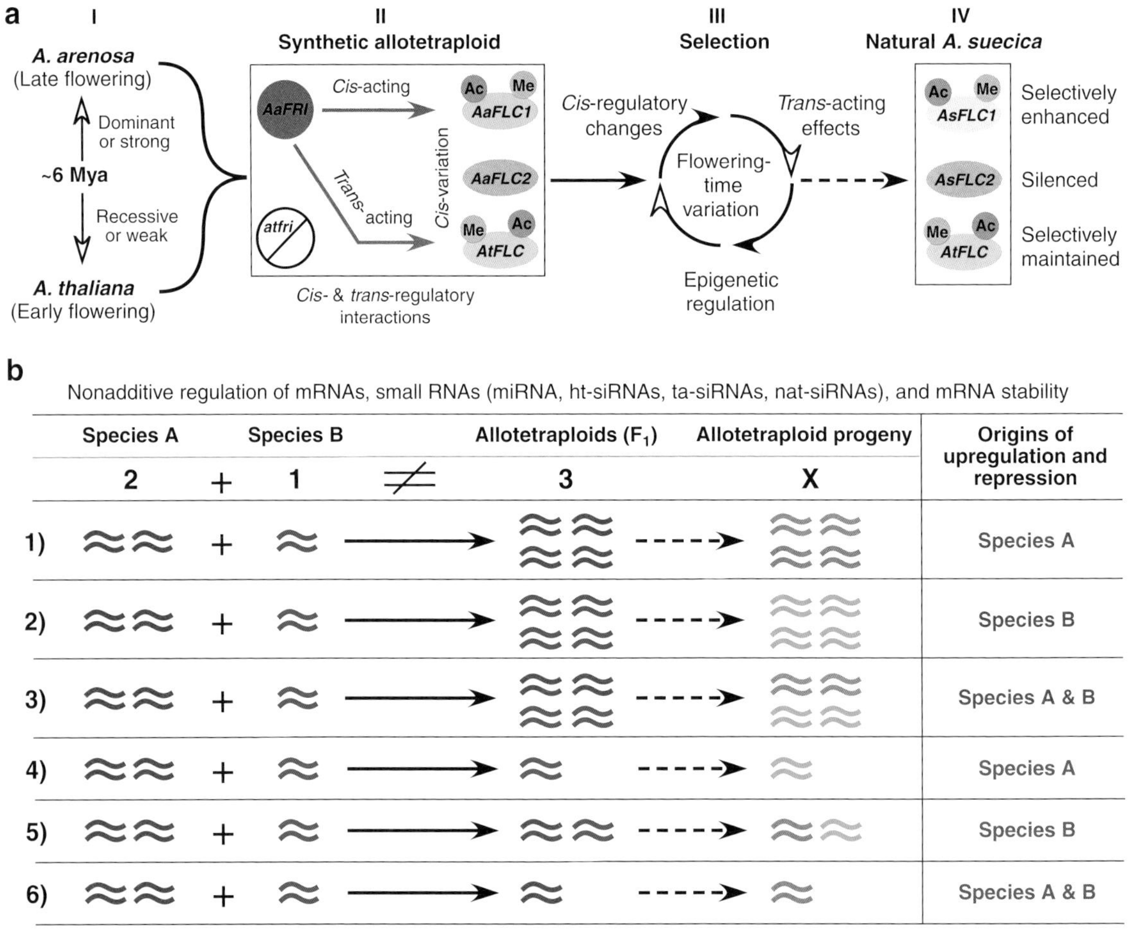

mRNA accumulation and stability during growth and development (55). For example, a subset of genes involved in mRNA stability displayed expression variation in allotetraploids (E. Kim & Z.J. Chen, unpublished). *CCR4*, a gene involved in RNA stability and degradation in yeast and animals (34), is differentially accumulated in leaves and flower buds, suggesting a role of RNA stability in transcript accumulation in allopolyploids.

Over time, species may have adapted to spatial and temporal regulation of RNA transcripts including mRNAs, small RNAs, and additional noncoding RNA transcripts that could accumulate nonadditively in the allotetraploids (**Figure 5*b***). The small RNAs may serve as negative regulators for the expression of target genes originating from two parents. First, the loci encoding miRNAs and siRNAs may diverge during the evolution of the progenitors as observed in the *FRI* and *FLC* loci (146). Sequence divergence in promoter regions and *cis*-acting elements leads to the expression variation when these loci are present in the same cell nuclei. Alternatively, differential expression of *trans*-acting factors may cause gene expression changes in the allopolyploids, as predicted (25). Second, antiviral RNAi genes involved in the biogenesis of small RNAs such as dicers, Argonaute, and RNA-dependent RNA

Figure 5

Models for transcriptional and post-transcriptional regulation of nonadditive gene expression in allopolyploids. (*a*) *Cis*- and *trans*-acting effects on transcriptional regulation of nonadditive gene expression in allotetraploids. I. During evolution, changes in *cis*-regulatory elements may confer "strong/dominant" and "weak/recessive" loci. II. Combination of these loci in the same allotetraploid cells induces *cis*- and *trans*-acting effects. Because *A. thaliana* FRI (AtFRI) is nonfunctional, *A. arenosa* FRI (AaFRI) *trans*-activates *A. thaliana FLC* (*AtFLC*) and *cis*-regulates one of the *A. arenosa FLC* (*AaFLC1*) loci, leading to overdominance or very late flowering in resynthesized allotetraploids. The *cis*-regulation and *trans*-activation are maintained by histone acetylation (Ac in a *small purple circle*) and methylation (Me in a *small orange circle*). Functional *AaFRI* is in a large red circle, and nonfunctional *A. thaliana FRI* (*atfrc*) is in a white small circle with a slash line. The straight and inclined red arrows indicate *cis*- and *trans*-acting effects, respectively, by AaFRI. *FLC* loci are in oval circles. Blue and green indicates active *AtFLC1* and *AaFLC1*, respectively, and gray indicates silenced *AaFLC2*. The differences in the expression of three *FLC* loci may reflect *cis*-regulatory variation (146). III. *Cis*- and *trans*-acting interactions and epigenetic regulation of *FRI* and *FLC* loci may facilitate selection for flowering-time variation in resynthesized allotetraploids (indicated by a circular wheel). IV. In *A. suecica*, additional selection forces may enhance the expression of *AaFLC1* (color change from *dark green* in II to *yellow green* in IV) and related genes such as *MAFs* (146), which facilitate adaptation to cold climate. This simple model may be generalized to explain a mechanism for regulatory interactions between divergent loci in other pathways in the allotetraploids (see text for details). (*b*) Nonadditive accumulation of mRNAs and small RNAs in allotetraploids (null hypothesis H_0: $2 + 1 = 3$). There is evidence that 15–40% of the genes are expressed differently between two closely related species of *Arabidopsis* (147), and a subset of those genes are subjected to nonadditive expression in the allotetraploids. Up- or downregulation of mRNAs may be caused by activation of genes originating in parent A, B, or both. We exclude the situation in which the expression of one locus is compensated by its corresponding homoeologous locus. If the nonadditively accumulated transcripts are small RNAs, such as microRNAs (miRNAs), heterochromatic and transposon-related siRNAs (ht-siRNAs), *trans*-acting siRNAs (ta-siRNAs), and natural anti-sense (nat-siRNAs), they typically negatively regulate the expression of their target genes. The red and blue wave lines indicate RNA transcripts or small RNAs produced in species A and B, respectively. We hypothetically assign two RNA molecules for species A and one RNA molecule for species B. Note that the number of RNAs in the nascent allotetraploids (F_1) is not equal to 3 ($2 + 1$), suggesting nonadditive regulation (dominance and overdominance if larger than 3 or repression if smaller than 3). Possible changes (*dashed arrows*) in RNA composition and accumulation (column X) may occur in self-pollinating progeny, which are indicated by orange and green wave lines, respectively. According to the hypothetical outcome, the origins of small RNA upregulation (*red* in 1–3) and repression (*blue* in 4–6) are shown in the last column.

polymerases (9, 155) generally diverge faster than other proteins during evolution (106). Combining two divergent proteins in the allotetraploids may alter enzymatic activity and specificity. As a result, different pools of small RNAs could accumulate in the allopolyploids. Third, natural sense and anti-sense transcripts and other read-through transcripts may participate in defense mechanisms (16) and may be accumulated differently in the progenitors. These transcripts affect the expression of neighboring loci as well as other loci via *trans*-acting effects. Fourth, mRNA transcript abundance is different in the progenitors. Although the coding sequences are very similar among *Arabidopsis* and its related species, sequences at the noncoding regions (5′- and 3′-ends) diverge relatively rapidly (L. Tian, M. Ha & Z.J. Chen, unpublished), which may affect processing and stability of RNA transcripts (55). Finally, each species is differentiated by the presence or absence of species-specific repetitive DNA sequences, including transposons that may affect the chromatin structure and expression of their neighboring genes. Differences in DNA replication and perturbation of chromatin structures among different species may induce the release of transposons and aberrant RNA transcripts that cause "genomic shock" and many downstream effects, as previously predicted (25, 98).

Neofunctionalization: gain of a novel function (or expression pattern) from one copy of the gene duplicate

Subfuncationalization: functional (or expression) divergence of gene duplicate from the ancestral gene (e.g., tissue-specific expression of gene duplicate)

DEVELOPMENTAL REGULATION OF ORTHOLOGOUS AND HOMOEOLOGOUS GENES

During polyploid evolution, both copies of orthologous genes may remain if dosage effects are advantageous (142), or one copy of the gene duplicate may evolve a novel function via neofunctionalization (89). Alternatively, both copies may diverge their functions or expression patterns in different organs or tissues via subfunctionalization (88). Indeed, silenced rRNA genes in vegetative tissues are reactivated during flower development (27). In a survey of 40 genes in cotton, 10 genes (25%) display unequal expression in allotetraploids and exhibit organ-specific expression patterns (2). For 5 genes, the A-subgenome loci are expressed higher than the D-homoeologous loci, whereas for the other 4 genes, the D-subgenome loci are expressed higher than the A-homoeologous loci. For some homoeologous gene pairs, one locus (e.g., *AdhA*) is silenced in one organ, whereas the other locus is silenced in another organ. This silencing scheme is genotype-independent and occurs in both synthetic and natural cotton allotetraploids (3), suggesting rapid subfunctionalization of duplicate genes and stable maintenance of tissue-specific expression patterns during evolution.

Although the mechanisms for developmental control of the expression of orthologous genes are unclear, developmental regulation of orthologous genes immediately after allopolyploid formation suggests that duplicate genes provide genetic robustness against null mutations (52) and dosage-dependent selective advantage (15, 142). Moreover, immediate divergence in the expression of orthologous genes in allopolyploids provides a virtually inexhaustible reservoir for generating genetic variation and phenotypic diversification, which facilitates natural selection and adaptive evolution.

ODD AND EVEN DOSAGE EFFECTS ON GENE REGULATION IN POLYPLOIDS

Dosage-dependent gene regulation shows odd and even effects, which may affect additive and nonadditive gene regulation in polyploids. Using B-A chromosome translocation lines in maize, Birchler and his colleagues (53) generated a series of lines with different doses of A chromosomes that could be used to measure gene expression in response to changes in chromosome dosage. Gene expression levels are generally positively correlated with the dosage of the genes or chromosomes in these lines. However, the expression levels of ~10% genes are either reduced or negatively correlated with odd chromosome dosages (e.g., one, three, and five). One possibility is that dosage-dependent gene regulation is associated with chromosome pairing because one or more copies of chromosomes in odd dosages cannot pair properly.

The odd and even effects on gene regulation are also observed in the study of transgene expression in diploid and triploid hybrids derived from the crosses of diploid or tetraploid plants with a diploid strain containing a single copy of a transgenic resistance gene in an active state (105). The expression of the transgene is reduced in the triploids compared with the diploid hybrids, leading to the loss of the resistant phenotype at various stages of seedling development in some individuals. The reduction of gene expression was reversible under selective tissue culture conditions. This type of suppression was observed for a single-copy insert in the absence of other *trans*-acting copies of the transgene and is therefore different from homology-dependent gene silencing. An increase in ploidy or chromosome dosage can give rise to epigenetic gene silencing, generating stochastic variations in gene expression patterns. Although the expression of the transgene in a haploid or a pentaploid was not studied, odd ploidy may result in a new type of epigenetic repression. The expression of the transgene is repressed only in the triploids in which one set of chromosomes is likely not paired or improperly paired.

Ploidy-dependent gene regulation suggests a sensing mechanism for gene dosage and DNA content via chromosome pairing. Although somatic pairing has not been documented in plants, such transient pairing has been observed in humans, *Drosophila*, and yeast (101). Homologous pairing has been implicated in transvection, position-effect variegation, and transgene gene silencing (5, 61, 113), all of which involve alterations in gene expression.

PARAMUTATION-LIKE EFFECTS IN POLYPLOIDS

Paramutation is the result of heritable changes in gene expression that occur upon interaction between alleles (21). The phenomenon was first discovered in plants and later found in many other organisms including mammals (mouse and human) (21, 121, 134). The paramutagenic allele induces the change in the expression state of the paramutable allele. A paramutation-like phenomenon was also discovered in the tetraploid plants containing active and inactive transgene alleles of hygromycin phosphotransferase (HPT) (104). Active alleles that are *trans*-inactivated by their silenced counterparts are observed in tetraploid but not in diploid plants, and this occurred only in progeny resulting from self-fertilization of plants heterozygous for the active and inactive HPT allele. The occurrence of transgene paramutation only in tetraploid plants indicates that active and inactive alleles go through meiosis together. This led to the hypothesis of pairing-based *trans*-inactivation. This predication is consistent with observations in tetraploid tomato, where the frequency of paramutation of a specific paramutagenic allele at the *sulfurea* locus is different between diploid, triploid, and tetraploid plants and depends on the ratio of paramutagenic to paramutable alleles (57). This suggests a counting mechanism for polyploidy-dependent paramutation, which may be similar to that for X-chromosome inactivation (75).

A paramutation-like phenomenon occurs in the progeny of genetic crosses between heterozygotes and between heterozygotes and wild-type mice independent of gender combination (121). The phenomenon is speculated to be associated with aberrant RNAs resulting from the paramutagenic allele that are packaged in sperm and cause paramutation upon transmission to the next generation. Indeed, paramutation depends on a RNA-dependent RNA polymerase; the *rdr101* mutation prevents paramutation in maize (4). However, paramutation in *Arabidopsis* tetraploids is probably not associated with RNA because *trans*-activation does not occur in the F_1 generation (104). Moreover, crosses of decrease in the DNA methylation (*ddm1*) mutant with a paramutable tetraploid do not change paramutation phenotypes in the F_1 or F_2 but do in the F_3 family, which is consistent with the gradual loss of DNA methylation by *ddm1*. The data suggest that methylation occurs later, and is speculated to occur during physical contact of the epialleles during meiosis and after the silencing is established (104). Alternatively, sorting out pairing between homologous and homoeologous chromosomes in polyploids may require a few more rounds of meiosis.

Many paramutation phenomena are associated with repeated sequences (21, 134). Multicopy genes or repetitive intergenic regions are a major trigger for the formation of silenced chromatin. Repeated sequences, whether inverted or tandem, can give rise to the production of dsRNA, an important trigger for RNA silencing as well as heterochromatin formation (85). In addition, repetitive sequences are also able to associate physically with their homologs in nonmeiotic cells (134). It is conceivable that different repeat sequences originating in the progenitors may trigger abnormal siRNA production and heterochromation formation that are responsible for paramutation-like or other epigenetic phenomena in allopolyploids.

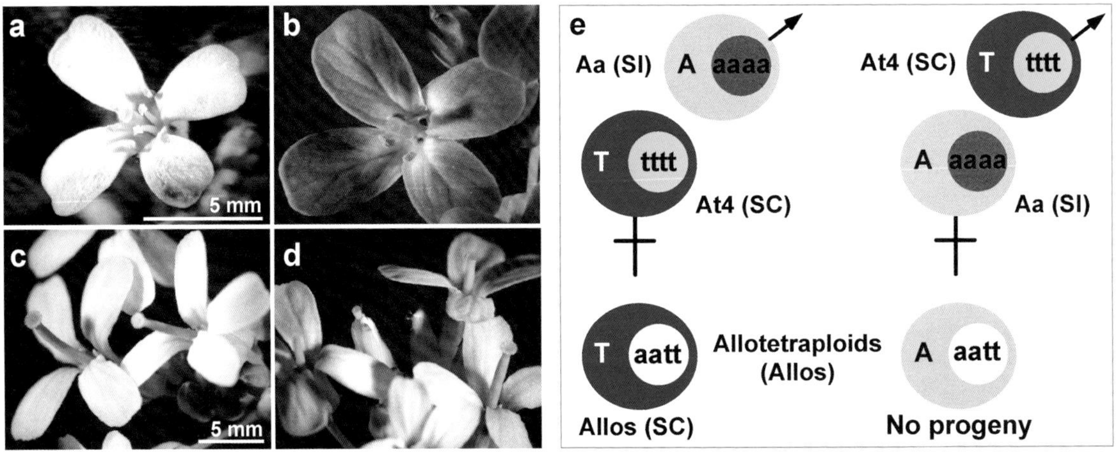

Figure 6

Switching of self-incompatibility in resynthesized *Arabidopsis* allotetraploids. (*a*) Flower morphology of self-compatible *A. suecica*. (*b*) Self-incompatible *A. arenosa* and (*c* and *d*) two self-compatible synthetic allotetraploid lines. The altered flower organs (elongated stigma and relatively short stamens) in the synthetic allotetraploids (*c* and *d*) may contribute to a low fertility without manual pollination. Size bars (5 mm) are the same in (*a*) and (*b*) or in (*c*) and (*d*). (*e*) Diagram of cytoplasmic and nuclear interactions in resynthesized allotetraploids (Allos, $2n = 4x = 26$). T or A and tt or aa denote cytoplasm and nuclear genomes, respectively. The cytoplasmic-nuclear genotypes for autotetraploid *A. thaliana* (At4) are Ttttt, tetraploid *A. arenosa* (Aa), Aaaaa; and allotetraploids, and either Taatt or Aaatt in reciprocal crosses. Blue and gray indicate *A. thaliana* and *A. arenosa* cytoplasms, whereas green, red, and yellow indicate nuclear genomes of *A. thaliana*, *A. arenosa*, and allotetraploids, respectively. *A. thaliana* is self-compatible (SC), whereas *A. arenosa* is self-incompatible (SI). The resynthesized allotetraploids (Allos) are SC. Fertile seeds are readily obtained from the cross using *A. thaliana* as a maternal parent and *A. arenosa* as a pollen donor, and the reciprocal cross is usually unsuccessful (31).

FERTILITY, SELF-INCOMPATIBILITY, AND CYTOPLASMIC-NUCLEAR INTERACTIONS IN ALLOPOLYPLOIDS

Incompatibility between alien cytoplasmic and nuclear genomes and between alien nuclear genomes is believed to be a barrier leading to reproductive isolation, speciation, and developmental abnormalities in vertebrates and plants (18, 77, 78, 82, 112). Breaking down this barrier is essential in forming a new polyploid species (78). Seed fertility may be controlled by a few genes or many genetic loci. Three imprinted genes, *PHERES1*, *MEIDOS*, and *MEDEA*, are silenced in allotetraploids in a dosage-dependent manner (68). Disrupting maternal imprinting of *AtPHERES1* and paternal imprinting of MEDEA may reduce seed viability in the allopolyploids. Imbalance of paternally and maternally imprinted genes in the endosperm may also cause reproductive failures (20).

Another factor affecting seed fertility is the breeding system. Self-incompatibility is a mechanism for preventing inbreeding in many plant species (22). The sporophytic self-incompatibility system in the family Brassicaceae has been used as a model system to study mating system evolution in plants (99, 108). *A. arenosa* is an outbreeder (self-incompatible), and *A. thaliana* is an inbreeding plant (self-compatible). They diverged from the same ancestor ~6 Mya (67, 108). However, the natural allotetraploid *A. suecica* and the resynthesized allopolyploids are self-compatible (**Figure 6**), suggesting that the mating system switches immediately following polyploidization. The loss of self-incompatibility in the first generation of allotetraploids is not caused by the segregation of S-alleles in the allotetraploids because all possible alleles are present. The data suggest rapid epigenetic changes in the expression of the genes important to self-incompatibility, probably including the well-characterized loci encoding S-locus receptor kinase (SRK) and S-locus cysteine-rich (SCR) proteins in Brassicaceae (109, 127).

In many cases, polyploidization converts self-incompatible diploids into self-compatible tetraploids in *Nicotiana* and *Solanum* with a gametophytic system, and some allopolyploids become self-compatible regardless of the mating types of their parents (18, 49, 82). Selfing in the allopolyploids may have an advantage for adapting new allopolyploids because of increased levels of heterozygosity. Inbreeding depression (1/18 or ~5% of homozygosity in one selfing generation) in allopolyploids is relatively low compared with that in a diploid (50%). An

ADAPTIVE EVOLUTION AND EXPRESSION DIVERGENCE BETWEEN DUPLICATE GENES

Theoretical prediction suggests that one copy of a gene duplicate would become lost by accumulation of deleterious mutations over an evolutionary timescale (87). Evidently, many duplicate genes are retained during evolution, and the redundancy conferred by duplicate genes may facilitate species adaptation (107) and genetic robustness (52) against changes in environmental conditions and developmental programs. Gene expression analyses indicate that duplicate genes offer genetic robustness against null mutations in yeast (52) and tend to experience expression divergence during development and to evolve faster between *Drosophila* species and within yeast species than single-copy genes (51, 83). Using all duplicate gene pairs derived from a recent WGD (14, 17, 144) and gene expression microarrays in *A. thaliana* (126), Ha et al. (56) found that expression divergence between gene duplicates is significantly higher in response to external stresses than to internal developmental changes. Rapid divergence between gene duplicates in response to abiotic and biotic stresses may facilitate subfunctionalization (88), neofunctionalization (89), and the evolution of an adaptive mechanism to environmental changes (98, 129). A relatively slow rate of expression divergence between the duplicates may provide dosage-dependent selective advantage (15, 142) and enable organisms to fine-tune complex regulatory networks. Orthologous and homoeologous genes in allopolyploids may have similar evolutionary fates.

extreme form of reproductive modification is apomixis that is commonly associated with polyploidy (18, 49). Resynthesized allopolyploids may be released from reproductive failure if they are capable of vegetative or seed apomixis.

The interactions between cytoplasm-nuclear and nuclear-nuclear genomes in the allopolyploids may induce genomic shock (98) and general amphiplasty (110) that are manifested by differential accumulation of transcripts originating from divergent species, leading to transcriptome dominance and activation or silencing of one or both homoeologous loci through genetic and/or epigenetic mechanisms. As a consequence, allopolyploids display hybrid vigor, flowering-time variation, inbreeding, apomixis, and selective advantage. Over time, orthologous or homoeologous loci in the allopolyploids may diverge their functions via neofunctionalization and subfunctionalization (88, 89), as predicted for the paralogous loci (see Sidebar: Adaptive Evolution and Expression Divergence Between Duplicate Genes).

In polyploid populations of separate origin, one population may lose function from one copy of an orthologous gene, while a second population may lose function from a second copy of this ortholog. This "reciprocal silencing" of duplicated genes in polyploid genomes would ultimately lead to hybrid lethality (8), promoting reproductive isolation and the origin of new species. Following this model, the stochastic silencing and subfunctionalization of orthologous genes in different lineages of allopolyploids (148) may play a major role in the origin of new species. Together, these mutually inclusive mechanisms may contribute significantly to the adaptation potential, domestication, and evolution of polyploid plants.

SUMMARY POINTS

1. Allopolyploidy induces a wider range of gene expression changes than autopolyploidy, suggesting hybridization has greater effects on gene expression and phenotypic variation than genome doubling.
2. Genetic changes occur immediately after interspecific hybridization or allopolyploidization. These include genome-specific sequence elimination, as observed in *Triticum* and *Tragopogon* allopolyploids and chromosomal rearrangements in *Brassica* allotetraploids. *Arabidopsis* and cotton (*Gossypium*) allotetraploids show very low levels of genomic changes.
3. Approximately 15–43% of the transcriptome diverged in expression between *A. thaliana* and *A. arenosa* (split ~6 Mya). The genes that display species-specific expression patterns are preferentially subject to nonadditive regulation in the resynthesized allotetraploids, indicating modulation of transcriptome divergence between the two progenitors.
4. The genes that are highly expressed in *A. thaliana* are repressed in the resynthesized allotetraploids, which is correlated with the nucleolar dominance and overall suppression of *A. thaliana* phenotypes in resynthesized and natural allotetraploids.
5. Some genes that display nonadditive expression are developmentally regulated, which may lead to subfunctionalization of these homoeologous genes in the allopolyploids.
6. Establishment of gene expression patterns in the resynthesized allotetraploids is usually rapid and stochastic and in some cases may require several rounds of meiosis (by selfing).

7. Most nonadditively expressed genes in the resynthesized allotetraploids are present in both parents, suggesting an epigenetic cause and *cis*- and *trans*-acting effects on nonadditive gene expression.
8. Sequence divergence in regulatory elements between the progenitors may cause genetic and epigenetic effects on expression of the genes involved in a biological pathway. *Trans*-acting differences exert their effects on *cis*-regulatory elements via epigenetic mechanisms, indicating interrelated regulatory interactions between genetic and epigenetic effects.
9. Reprogramming gene expression patterns in allopolyploids via chromatin modifications and RNA-mediated pathways may cause changes in adaptive traits such as flowering-time variation, self-incompatibility, and hybrid vigor.
10. Although the majority of data are documented from studies of resynthesized allotetraploids, similar changes in genomic organization and gene expression may occur in the natural allopolyploids and facilitate polyploid evolution and speciation.

FUTURE ISSUES

Addressing the following questions is essential to illuminate new insights into molecular and evolutionary impact on polyploid formation and speciation.

1. How and why are genetic dominance and nonadditive expression established rapidly in resynthesized allopolyploids, and how are these molecular processes altered or maintained in natural allopolyploids?
2. How do changes in gene expression contribute to intricate regulatory hierarchies and complex biological pathways leading to hybrid vigor and adaptive traits in allopolyploids?
3. Why are ploidy changes more deleterious in animals than in plants, and why are polyploid cells often carcinogenetic in animals?

ACKNOWLEDGMENTS

I thank Jianlin Wang, Hyeon-See Lee, Lu Tian, Zhongfu Ni, Misook Ha, Letricia Nogueira, Eun-Deok Kim, Jinsuk Lee, and Suk-Hwan Yang for their contributions to the gene expression data, and Donald Levin, Andrew Woodward, and two anonymous reviewers for critical reading and suggestions to improve the manuscript. I apologize for not citing many enlightening reviews and papers published in this exciting field owing to space limitations. The work was supported by the grants from the National Science Foundation (MCB0608602, DBI0501712 and DBI0624077) and the National Institutes of Health (GM067015).

LITERATURE CITED

1. Abbott RJ, Lowe AJ. 2004. Origins, establishment and evolution of new polyploid species: Senecio cambrensis and S. eboracensis in the British Isles. *Biol. J. Linn. Soc.* 82:467–74

2. Examines tissue-specific expression of 40 protein-coding genes in allotetraploid cotton, suggesting subfunctionalization for 10 homoeologous genes.

2. **Adams KL, Cronn R, Percifield R, Wendel JF. 2003. Genes duplicated by polyploidy show unequal contributions to the transcriptome and organ-specific reciprocal silencing. *Proc. Natl. Acad. Sci. USA* 100:4649–54**
3. Adams KL, Percifield R, Wendel JF. 2004. Organ-specific silencing of duplicated genes in a newly synthesized cotton allotetraploid. *Genetics* 168:2217–26
4. Alleman M, Sidorenko L, McGinnis K, Seshadri V, Dorweiler JE, et al. 2006. An RNA-dependent RNA polymerase is required for paramutation in maize. *Nature* 442:295–98
5. Aramayo R, Metzenberg RL. 1996. Meiotic transvection in fungi. *Cell* 86:103–13
6. Auger DL, Gray AD, Ream TS, Kato A, Coe EH Jr, Birchler JA. 2005. Nonadditive gene expression in diploid and triploid hybrids of maize. *Genetics* 169:389–97
7. Bailey JA, Gu Z, Clark RA, Reinert K, Samonte RV, et al. 2002. Recent segmental duplications in the human genome. *Science* 297:1003–7
8. Barbash DA, Siino DF, Tarone AM, Roote J. 2003. A rapidly evolving MYB-related protein causes species isolation in *Drosophila*. *Proc. Natl. Acad. Sci. USA* 100:5302–7
9. Bartel B, Bartel DP. 2003. MicroRNAs: At the root of plant development? *Plant Physiol.* 132:709–17
10. Baumel A, Ainouche ML, Bayer RJ, Ainouche AK, Misset MT. 2002. Molecular phylogeny of hybridizing species from the genus Spartina Schreb. (Poaceae). *Mol. Phylogenet. Evol.* 22:303–14
11. Bender J, Fink GR. 1995. Epigenetic control of an endogenous gene family is revealed by a novel blue fluorescent mutant of Arabidopsis. *Cell* 83:725–34
12. Birchler JA. 2001. Dosage-dependent gene regulation in multicellular eukaryotes: Implications for dosage compensation, aneuploid syndromes, and quantitative traits. *Dev. Biol.* 234:275–88
13. Birchler JA, Auger DL, Riddle NC. 2003. In search of the molecular basis of heterosis. *Plant Cell* 15:2236–39
14. Blanc G, Hokamp K, Wolfe KH. 2003. A recent polyploidy superimposed on older large scale duplications in the *Arabidopsis* genome. *Genome Res.* 13:137–44
15. Bomblies K, Doebley JF. 2006. Pleiotropic effects of the duplicate maize FLORICAULA/LEAFY genes zfl1 and zfl2 on traits under selection during maize domestication. *Genetics* 172:519–31
16. Borsani O, Zhu J, Verslues PE, Sunkar R, Zhu JK. 2005. Endogenous siRNAs derived from a pair of natural cis-antisense transcripts regulate salt tolerance in Arabidopsis. *Cell* 123:1279–91
17. Bowers JE, Chapman BA, Rong J, Paterson AH. 2003. Unravelling angiosperm genome evolution by phylogenetic analysis of chromosomal duplication events. *Nature* 422:433–38
18. Briggs D, Walters SM. 1997. *Plant Variation and Evolution*. Cambridge, UK: Cambridge Univ. Press. 512 pp.
19. Bruce AB. 1910. The Mendelian theory of heredity and the augmentation of vigor. *Science* 32:627–28
20. Bushell C, Spielman M, Scott RJ. 2003. The basis of natural and artificial postzygotic hybridization barriers in Arabidopsis species. *Plant Cell* 15:1430–42
21. Chandler VL, Stam M. 2004. Chromatin conversations: mechanisms and implications of paramutation. *Nat. Rev. Genet.* 5:532–44
22. Charlesworth D, Wright SI. 2001. Breeding systems and genome evolution. *Curr. Opin. Genet. Dev.* 11:685–90
23. Deleted in proof

24. Chen ZJ, Comai L, Pikaard CS. 1998. Gene dosage and stochastic effects determine the severity and direction of uniparental ribosomal RNA gene silencing (nucleolar dominance) in *Arabidopsis* allopolyploids. *Proc. Natl. Acad. Sci. USA* 95:14891–96
25. Chen ZJ, Ni Z. 2006. Mechanisms of genomic rearrangements and gene expression changes in plant polyploids. *BioEssays* 28:240–52
26. Chen ZJ, Pikaard CS. 1997. Epigenetic silencing of RNA polymerase I transcription: a role for DNA methylation and histone modification in nucleolar dominance. *Genes Dev.* 11:2124–36
27. Chen ZJ, Pikaard CS. 1997. Transcriptional analysis of nucleolar dominance in polyploid plants: biased expression/silencing of progenitor rRNA genes is developmentally regulated in *Brassica*. *Proc. Natl. Acad. Sci. USA* 94:3442–47
28. Chen ZJ, Wang JL, Tian L, Lee HS, Wang JYJ, et al. 2004. The development of an *Arabidopsis* model system for genome-wide analysis of polyploidy effects. *Biol. J. Linn. Soc.* 82:689–700
29. Comai L. 2005. The advantages and disadvantages of being polyploid. *Nat. Rev. Genet.* 6:836–46
30. Comai L, Tyagi AP, Lysak MA. 2003. FISH analysis of meiosis in *Arabidopsis* allopolyploids. *Chromosome Res.* 11:217–26
31. Comai L, Tyagi AP, Winter K, Holmes-Davis R, Reynolds SH, et al. 2000. Phenotypic instability and rapid gene silencing in newly formed *Arabidopsis* allotetraploids. *Plant Cell* 12:1551–68
32. Crow JF. 1948. Alternative hypothesis of hybrid vigor. *Genetics* 33:477–87
33. de Meaux J, Pop A, Mitchell-Olds T. 2006. Cis-regulatory evolution of Chalcone-synthase expression in the genus *Arabidopsis*. *Genetics* 174:2182–202
34. Denis CL, Chen J. 2003. The CCR4-NOT complex plays diverse roles in mRNA metabolism. *Prog. Nucleic Acid Res. Mol. Biol.* 73:221–50
35. Earley K, Lawrence RJ, Pontes O, Reuther R, Enciso AJ, et al. 2006. Erasure of histone acetylation by *Arabidopsis* HDA6 mediates large-scale gene silencing in nucleolar dominance. *Genes Dev.* 20:1283–93
36. East EM. 1936. Heterosis. *Genetics* 21:375–97
37. Edgar BA, Orr-Weaver TL. 2001. Endoreplication cell cycles: more for less. *Cell* 105:297–306
38. Emanuel BS, Shaikh TH. 2002. Segmental duplications: an 'expanding' role in genomic instability and disease. *Nat. Rev. Genet.* 2:791–800
39. Farh KK, Grimson A, Jan C, Lewis BP, Johnston WK, et al. 2005. The widespread impact of mammalian MicroRNAs on mRNA repression and evolution. *Science* 310:1817–21
40. Feldman M, Liu B, Segal G, Abbo S, Levy AA, Vega JM. 1997. Rapid elimination of low-copy DNA sequences in polyploid wheat: a possible mechanism for differentiation of homoeologous chromosomes. *Genetics* 147:1381–87
41. Ferris SD, Whitt GS. 1976. Loss of duplicate gene expression after polyploidisation. *Nature* 265:258–60
42. Fire A, Xu S, Montgomery MK, Kostas SA, Driver SE, Mello CC. 1998. Potent and specific genetic interference by double-stranded RNA in *Caenorhabditis elegans*. *Nature* 391:806–11
43. Galitski T, Saldanha AJ, Styles CA, Lander ES, Fink GR. 1999. Ploidy regulation of gene expression. *Science* 285:251–54
44. Gallardo MH, Bickham JW, Honeycutt RL, Ojeda RA, Kohler N. 1999. Discovery of tetraploidy in a mammal. *Nature* 401:341

26. Discovery of molecular evidence for epigenetic regulation of rRNA genes subjected to nucleolar dominance (silencing of rRNA genes originating from one progenitor) in *Brassica* allopolyploids, which has provided mechanistic insights into understanding the expression of orthologous genes in allopolyploids.

31. The first extensive survey of phenotypic and gene expression changes in resynthesized *Arabidopsis* allotetraploids.

45. Gaut BS. 2001. Patterns of chromosomal duplication in maize and their implications for comparative maps of the grasses. *Genome Res.* 11:55–66
46. Gibson G, Riley-Berger R, Harshman L, Kopp A, Vacha S, et al. 2004. Extensive sex-specific nonadditivity of gene expression in *Drosophila melanogaster*. *Genetics* 167:1791–99
47. Gottlieb LD. 1982. Conservation and duplication of isozymes in plants. *Science* 216:373–80
48. Gottlieb LD, Ford VS. 1997. A recently silenced, duplicate *PgiC* locus in *Clarkia*. *Mol. Biol. Evol.* 14:125–32
49. Grant V. 1981. *Plant Speciation*. New York: Columbia Univ. Press. 563 pp.
50. Grant-Downton RT, Dickinson HG. 2006. Epigenetics and its implications for plant biology 2. The 'epigenetic epiphany': epigenetics, evolution and beyond. *Ann. Bot.* 97:11–27
51. Gu Z, Rifkin SA, White KP, Li WH. 2004. Duplicate genes increase gene expression diversity within and between species. *Nat. Genet.* 36:577–79
52. Gu Z, Steinmetz LM, Gu X, Scharfe C, Davis RW, Li W-H. 2003. Role of duplicate genes in genetic robustness against null mutations. *Nature* 421:63–66
53. Guo M, Davis D, Birchler JA. 1996. Dosage effects on gene expression in a maize ploidy series. *Genetics* 142:1349–55

53. An extensive survey of gene expression variation in response to dosage changes in maize.

54. Guo M, Rupe MA, Zinselmeier C, Habben J, Bowen BA, Smith OS. 2004. Allelic variation of gene expression in maize hybrids. *Plant Cell* 16:1707–16
55. Gutierrez RA, Ewing RM, Cherry JM, Green PJ. 2002. Identification of unstable transcripts in *Arabidopsis* by cDNA microarray analysis: rapid decay is associated with a group of touch- and specific clock-controlled genes. *Proc. Natl. Acad. Sci. USA* 99:11513–18
56. Ha M, Li WH, Chen ZJ. 2006. External factors accelerate expression divergence between duplicate genes. *Trends Genet.* doi:10.1016/j.tig.2007.02.005
57. Hagemann R, Berg W. 1978. Paramutation at the *sulfurea* locus of *Lycopersicon esculentum* Mill. VII. Determination of the time of occurence of paramutation by the quantitative evaluation of the variegation. *Theor. Appl. Genet.* 53:709–19
58. Harlan JR, De Wet JM, Naik SM, Lambert RJ. 1970. Chromosome pairing within genomes in maize-Tripsacum hybrids. *Science* 167:1247–48
59. He P, Friebe BR, Gill BS, Zhou JM. 2003. Allopolyploidy alters gene expression in the highly stable hexaploid wheat. *Plant Mol. Biol.* 52:401–14
60. Hegarty MJ, Jones JM, Wilson ID, Barker GL, Coghill JA, et al. 2005. Development of anonymous cDNA microarrays to study changes to the Senecio floral transcriptome during hybrid speciation. *Mol. Ecol.* 14:2493–510
61. Henikoff S, Comai L. 1998. *Trans*-sensing effects: the ups and downs of being together. *Cell* 93:329–32
62. Heslop-Harrison JS. 1990. Gene expression and parental dominance in hybrid plants. *Development Suppl.* 1990:21–28
63. Hilu KW. 1993. Polyploidy and the evolution of domesticated plants. *Am. J. Bot.* 80:2521–28
64. Jakobsson M, Hagenblad J, Tavare S, Sall T, Hallden C, et al. 2006. A unique recent origin of the allotetraploid species *Arabidopsis* suecica: Evidence from nuclear DNA markers. *Mol. Biol. Evol.* 23:1217–31
65. Johanson U, West J, Lister C, Michaels S, Amasino R, Dean C. 2000. Molecular analysis of FRIGIDA, a major determinant of natural variation in *Arabidopsis* flowering time. *Science* 290:344–47
66. Jones DF. 1917. Dominance of linked factors as a means of accounting for heterosis. *Genetics* 2:466–79

67. Jorgensen R. 1990. Altered gene expression in plants due to trans interactions between homologous genes. *Trends Biotechnol.* 8:340–44
68. Josefsson C, Dilkes B, Comai L. 2006. Parent-dependent loss of gene silencing during interspecies hybridization. *Curr. Biol.* 16:1322–28
69. Kashkush K, Feldman M, Levy AA. 2002. Gene loss, silencing and activation in a newly synthesized wheat allotetraploid. *Genetics* 160:1651–59
70. **Kashkush K, Feldman M, Levy AA. 2003. Transcriptional activation of retrotransposons alters the expression of adjacent genes in wheat. *Nat. Genet.* 33:102–6**
71. Koch MA, Haubold B, Mitchell-Olds T. 2000. Comparative evolutionary analysis of chalcone synthase and alcohol dehydrogenase loci in *Arabidopsis*, *Arabis*, and related genera (Brassicaceae). *Mol. Biol. Evol.* 17:1483–98
72. Lawrence RJ, Earley K, Pontes O, Silva M, Chen ZJ, et al. 2004. A concerted DNA methylation/histone methylation switch regulates rRNA gene dosage control and nucleolar dominance. *Mol. Cell* 13:599–609
73. Lee HS, Chen ZJ. 2001. Protein-coding genes are epigenetically regulated in *Arabidopsis* polyploids. *Proc. Natl. Acad. Sci. USA* 98:6753–58
74. Lee HS, Wang JL, Tian L, Jiang HM, Black MA, et al. 2004. Sensitivity of 70-mer oligonucleotides and cDNAs for microarray analysis of gene expression in *Arabidopsis* and its related species. *Plant Biotechnol. J.* 2:45–57
75. Lee JT, Jaenisch R. 1997. The (epi)genetic control of mammalian X-chromosome inactivation. *Curr. Opin. Genet. Dev.* 7:274–80
76. Leitch IL, Bennett MD. 1997. Polyploidy in angiosperms. *Trends Plant Sci.* 2:470–76
77. Levin DA. 1983. Polyploidy and novelity in flowering plants. *Am. Nat.* 122:1–25
78. Levin DA. 2003. The cytoplasmic factor in plant speciation. *Syst. Bot.* 28:5–11
79. Levy AA, Feldman M. 2002. The impact of polyploidy on grass genome evolution. *Plant Physiol.* 130:1587–93
80. Levy YY, Dean C. 1998. The transition to flowering. *Plant Cell* 10:1973–90
81. Lewis MS, Pikaard CS. 2001. Restricted chromosomal silencing in nucleolar dominance. *Proc. Natl. Acad. Sci. USA* 98:14536–40
82. Lewis WH. 1980. *Polyploidy: Biological Relevance*. New York: Plenum. 583 pp.
83. Li WH, Yang J, Gu X. 2005. Expression divergence between duplicate genes. *Trends Genet.* 21:602–7
84. Li ZK, Luo LJ, Mei HW, Wang DL, Shu QY, et al. 2001. Overdominant epistatic loci are the primary genetic basis of inbreeding depression and heterosis in rice. I. Biomass and grain yield. *Genetics* 158:1737–53
85. Lippman Z, Martienssen R. 2004. The role of RNA interference in heterochromatic silencing. *Nature* 431:364–70
86. Liu B, Wendel J. 2002. Non-Mendelian phenomenon in allopolyploid genome evolution. *Curr. Genomics* 3:489–505
87. Lynch M, Conery JS. 2000. The evolutionary fate and consequences of duplicate genes. *Science* 290:1151–55
88. Lynch M, Force A. 2000. The probability of duplicate gene preservation by subfunctionalization. *Genetics* 154:459–73
89. Lynch M, O'Hely M, Walsh B, Force A. 2001. The probability of preservation of a newly arisen gene duplicate. *Genetics* 159:1789–804
90. Mable BK. 2004. 'Why polyploidy is rarer in animals than in plants': Myths and mechanisms. *Biol. J. Linn. Soc.* 82:453–66

70. First report of retrotransposon activation and its effects on the expression of adjacent genes in wheat allopolyploids.

91. Documented changes in DNA methylation and transposons in resynthesized *Arabidopsis* allotetraploids.

104. Paramutation-like phenomenon was observed in *Arabidopsis* tetraploids.

110. Discovery of progenitor-dependent chromosome morphological changes in *Crepis* interspecific hybrids.

91. **Madlung A, Masuelli RW, Watson B, Reynolds SH, Davison J, Comai L. 2002. Remodeling of DNA methylation and phenotypic and transcriptional changes in synthetic *Arabidopsis* allotetraploids. *Plant Physiol.* 129:733–46**
92. Madlung A, Tyagi AP, Watson B, Jiang H, Kagochi T, et al. 2005. Genomic changes in synthetic *Arabidopsis* polyploids. *Plant J.* 41:221–30
93. Masterson J. 1994. Stomatal size in fossil plants: evidence for polyploidy in majority of angiosperms. *Science* 264:421–24
94. Matzke MA, Matzke AJM. 1995. How and why do plants inactivate homologous (trans)genes. *Plant Physiol.* 107:679–85
95. Matzke MA, Scheid OM, Matzke AJ. 1999. Rapid structural and epigenetic changes in polyploid and aneuploid genomes. *BioEssays* 21:761–67
96. Mavarez J, Salazar CA, Bermingham E, Salcedo C, Jiggins CD, Linares M. 2006. Speciation by hybridization in Heliconius butterflies. *Nature* 441:868–71
97. McClintock B. 1934. The relationship of a particular chromosomal element to the development of the nucleoli in *Zea mays*. *Z. Zellforsch. Mikrosk. Anat.* 21:294–328
98. McClintock B. 1984. The significance of responses of the genome to challenge. *Science* 226:792–801
99. McCubbin AG, Kao T. 2000. Molecular recognition and response in pollen and pistill interactions. *Annu. Rev. Cell Dev. Biol.* 16:333–64
100. McFadden DE, Kwong LC, Yam IY, Langlois S. 1993. Parental origin of triploidy in human fetuses: evidence for genomic imprinting. *Hum. Genet.* 92:465–69
101. McKee BD. 2004. Homologous pairing and chromosome dynamics in meiosis and mitosis. *Biochim. Biophys. Acta* 1677:165–80
102. Michaels SD, Amasino RM. 1999. FLOWERING LOCUS C encodes a novel MADS domain protein that acts as a repressor of flowering. *Plant Cell* 11:949–56
103. Michalak P, Noor MA. 2003. Genome-wide patterns of expression in *Drosophila* pure species and hybrid males. *Mol. Biol. Evol.* 20:1070–76
104. **Mittelsten Scheid O, Afsar K, Paszkowski J. 2003. Formation of stable epialleles and their paramutation-like interaction in tetraploid *Arabidopsis thaliana*. *Nat. Genet.* 34:450–54**
105. Mittelsten Scheid O, Jakovleva L, Afsar K, Maluszynska J, Paszkowski J. 1996. A change of ploidy can modify epigenetic silencing. *Proc. Natl. Acad. Sci. USA* 93:7114–19
106. Molnar-Lang M, Linc G, Logojan A, Sutka J. 2000. Production and meiotic pairing behaviour of new hybrids of winter wheat (*Triticum aestivum*) x winter barley (*Hordeum vulgare*). *Genome* 43:1045–54
107. Muller HJ. 1925. Why polyploidy is rarer in animals than in plants. *Am. Nat.* 59:346–53
108. Nasrallah JB. 2000. Cell-cell signalling in the self-incompatibility response. *Curr. Opin. Plant Biol.* 3:368–73
109. Nasrallah ME, Liu P, Nasrallah JB. 2002. Generation of self-incompatible *Arabidopsis thaliana* by transfer of two S locus genes from *A. lyrata*. *Science* 297:247–49
110. **Navashin M. 1934. Chromosomal alterations caused by hybridization and their bearing upon certain general genetic problems. *Cytologia* 6:169–203**
111. Osborn TC, Pires JC, Birchler JA, Auger DL, Chen ZJ, et al. 2003. Understanding mechanisms of novel gene expression in polyploids. *Trends Genet.* 19:141–47
112. Otto SP, Whitton J. 2000. Polyploid incidence and evolution. *Annu. Rev. Genet.* 34:401–37
113. Pal-Bhadra M, Bhadra U, Birchler JA. 2002. RNAi related mechanisms affect both transcriptional and posttranscriptional transgene silencing in *Drosophila*. *Mol. Cell* 9:315–27

114. Patterson N, Richter DJ, Gnerre S, Lander ES, Reich D. 2006. Genetic evidence for complex speciation of humans and chimpanzees. *Nature* 441:1103–8
115. Phillips JP, Tainer JA, Getzoff ED, Boulianne GL, Kirby K, Hilliker AJ. 1995. Subunit-destabilizing mutations in *Drosophila* copper/zinc superoxide dismutase: neuropathology and a model of dimer dysequilibrium. *Proc. Natl. Acad. Sci. USA* 92:8574–78
116. Pichersky E, Soltis D, Soltis P. 1990. Defective chlorophyll a/b-binding protein genes in the genome of a homosporous fern. *Proc. Natl. Acad. Sci. USA* 87:195–99
117. Pikaard CS. 1999. Nucleolar dominance and silencing of transcription. *Trends Plant Sci.* 4:478–83
118. Prudhomme M, Mejean V, Martin B, Claverys JP. 1991. Mismatch repair genes of Streptococcus pneumoniae: HexA confers a mutator phenotype in *Escherichia coli* by negative complementation. *J. Bacteriol.* 173:7196–203
119. Ramsey J, Schemske DW. 1998. Pathways, mechanisms, and rates of polyploid formation in flowering plants. *Annu. Rev. Ecol. Syst.* 29:467–501
120. Ranz JM, Castillo-Davis CI, Meiklejohn CD, Hartl DL. 2003. Sex-dependent gene expression and evolution of the *Drosophila* transcriptome. *Science* 300:1742–45
121. Rassoulzadegan M, Grandjean V, Gounon P, Vincent S, Gillot I, Cuzin F. 2006. RNA-mediated non-Mendelian inheritance of an epigenetic change in the mouse. *Nature* 441:469–74
122. Reeder RH. 1985. Mechanisms of nucleolar dominance in animals and plants. *J. Cell Biol.* 101:2013–16
123. Riera-Lizarazu O, Rines HW, Phillips RL. 1996. Cytological and molecular characterization of oat x maize partial hybrids. *Theor. Appl. Genet.* 93:123–35
124. Rieseberg LH, Raymond O, Rosenthal DM, Lai Z, Livingstone K, et al. 2003. Major ecological transitions in wild sunflowers facilitated by hybridization. *Science* 301:1211–16
125. Sall T, Jakobsson M, Lind-Hallden C, Hallden C. 2003. Chloroplast DNA indicates a single origin of the allotetraploid *Arabidopsis* suecica. *J. Evol. Biol.* 16:1019–29
126. Schmid M, Davison TS, Henz SR, Pape UJ, Demar M, et al. 2005. A gene expression map of *Arabidopsis* thaliana development. *Nat. Genet.* 37:501–6
127. Schopfer CR, Nasrallah ME, Nasrallah JB. 1999. The male determinant of self-incompatibility in *Brassica*. *Science* 286:1697–700
128. Selker EU. 1990. Premeiotic instability of repeated sequences in *Neurospora crassa*. *Annu. Rev. Genet.* 24:579–613
129. Seoighe C, Gehring C. 2004. Genome duplication led to highly selective expansion of the *Arabidopsis* thaliana proteome. *Trends Genet.* 20:461–64
130. Shaked H, Kashkush K, Ozkan H, Feldman M, Levy AA. 2001. Sequence elimination and cytosine methylation are rapid and reproducible responses of the genome to wide hybridization and allopolyploidy in wheat. *Plant Cell* 13:1749–59
131. Sheldon CC, Burn JE, Perez PP, Metzger J, Edwards JA, et al. 1999. The FLF MADS box gene: a repressor of flowering in *Arabidopsis* regulated by vernalization and methylation. *Plant Cell* 11:445–58
132. Soltis DE, Soltis PS, Tate JA. 2003. Advances in the study of polyploidy since plant speciation. *New Phytol.* 161:173–91
133. Song K, Lu P, Tang K, Osborn TC. 1995. Rapid genome change in synthetic polyploids of *Brassica* and its implications for polyploid evolution. *Proc. Natl. Acad. Sci. USA* 92:7719–23
134. Stam M, Mittelsten Scheid O. 2005. Paramutation: an encounter leaving a lasting impression. *Trends Plant Sci.* 10:283–90

133. Discovery of rapid DNA sequence changes in synthetic *Brassica* allotetraploids, which has renewed the interests in molecular studies on plant polyploids.

135. Stebbins GL. 1971. *Chromosomal Evolution in Higher Plants*. London: Edward Arnold. 216 pp.
136. Storchova Z, Pellman D. 2004. From polyploidy to aneuploidy, genome instability and cancer. *Nat. Rev. Mol. Cell Biol.* 5:45–54
137. Stuber CW, Lincoln SE, Wolff DW, Helentjaris T, Lander ES. 1992. Identification of genetic factors contributing to heterosis in a hybrid from two elite maize inbred lines using molecular markers. *Genetics* 132:823–39
138. Stupar RM, Springer NM. 2006. Cis-transcriptional variation in maize inbred lines B73 and Mo17 leads to additive expression patterns in the F1 hybrid. *Genetics* 173:2199–210
139. Svartman M, Stone G, Stanyon R. 2005. Molecular cytogenetics discards polyploidy in mammals. *Genomics* 85:425–30
140. Swanson-Wagner RA, Jia Y, DeCook R, Borsuk LA, Nettleton D, Schnable PS. 2006. All possible modes of gene action are observed in a global comparison of gene expression in a maize F1 hybrid and its inbred parents. *Proc. Natl. Acad. Sci. USA* 103:6805–10

140. Microarray analysis of gene expression changes in maize hybrids suggests nonadditive gene regulation, as observed in *Arabidopsis* allopolyploids.

141. Tate JA, Ni Z, Scheen AC, Koh J, Gilbert CA, et al. 2006. Evolution and expression of homeologous loci in *Tragopogon miscellus* (Asteraceae), a recent and reciprocally formed allopolyploid. *Genetics* 173:1599–611

141. An extensive survey of gene expression changes in recent and reciprocally formed *Tragopogon* allotetraploids in natural populations.

142. Thomas BC, Pedersen B, Freeling M. 2006. Following tetraploidy in an *Arabidopsis* ancestor, genes were removed preferentially from one homeolog leaving clusters enriched in dose-sensitive genes. *Genome Res.* 16:934–46
143. Tilghman SM. 1999. The sins of the fathers and mothers: Genomic imprinting in mammalian development. *Cell* 96:185–93
144. Vision TJ, Brown DG, Tanksley SD. 2000. The origins of genomic duplications in *Arabidopsis*. *Science* 290:2114–17
145. Vrana PB, Fossella JA, Matteson P, del Rio T, O'Neill MJ, Tilghman SM. 2000. Genetic and epigenetic incompatibilities underlie hybrid dysgenesis in *Peromyscus*. *Nat. Genet.* 25:120–24
146. Wang J, Tian L, Lee HS, Chen ZJ. 2006. Nonadditive regulation of FRI and FLC loci mediates flowering-time variation in *Arabidopsis* allopolyploids. *Genetics* 173:965–74

146. Evidence for *cis*- and *trans*-acting effects on gene expression of regulatory pathways in allopolyploids.

147. Wang J, Tian L, Lee HS, Wei NE, Jiang H, et al. 2006. Genomewide nonadditive gene regulation in *Arabidopsis* allotetraploids. *Genetics* 172:507–17

147. The first genome-wide analysis of gene expression reveals transcriptome dominance and nonadditive gene regulation in *Arabidopsis* allotetraploids.

148. Wang J, Tian L, Madlung A, Lee HS, Chen M, et al. 2004. Stochastic and epigenetic changes of gene expression in *Arabidopsis* polyploids. *Genetics* 167:1961–73
149. Wendel JF. 2000. Genome evolution in polyploids. *Plant Mol. Biol.* 42:225–49
150. Wendel JF, Cronn RC. 2003. Polyploidy and the evolutionary history of cotton. *Adv. Agron.* 78:139–86
151. Wittkopp PJ, Haerum BK, Clark AG. 2004. Evolutionary changes in cis and trans gene regulation. *Nature* 430:85–88
152. Wolfe KH. 2001. Yesterday's polyploidization and the mystery of diploidization. *Nat. Rev. Genet.* 2:333–41
153. Yang SS, Cheung F, Lee JJ, Ha M, Wei NE, et al. 2006. Accumulation of genome-specific transcripts, transcription factors and phytohormonal regulators during early stages of fiber cell development in allotetraploid cotton. *Plant J.* 47:761–75

153. EST analysis reveals transcription dominance of AA-subgenome in allotetraploid cotton (AADD).

154. Yu J, Wang J, Lin W, Li S, Li H, et al. 2005. The genomes of Oryza sativa: A history of duplications. *PLoS Biol.* 3:e38
155. Zamore PD, Haley B. 2005. Ribo-gnome: the big world of small RNAs. *Science* 309:1519–24

Tracheary Element Differentiation

Simon Turner, Patrick Gallois, and David Brown

University of Manchester, Faculty of Life Sciences, Manchester M13 9PT, United Kingdom; email: simon.turner@manchester.ac.uk

Annu. Rev. Plant Biol. 2007. 58:407–33

The *Annual Review of Plant Biology* is online at plant.annualreviews.org

This article's doi:
10.1146/annurev.arplant.57.032905.105236

1543-5008/07/0602-0407$20.00

Key Words

xylem, secondary cell wall, programmed cell death, microtubules

Abstract

Tracheary elements (TEs) are cells in the xylem that are highly specialized for transporting water and solutes up the plant. TEs undergo a very well-defined process of differentiation that involves specification, enlargement, patterned cell wall deposition, programmed cell death and cell wall removal. This process is coordinated such that adjacent TEs are joined together to form a continuous network. Expression studies on model systems as diverse as trees and cell cultures have contributed to providing a flood of candidate genes with potential roles in TE differentiation. Analysis of some of these genes has yielded important information on processes such as patterned secondary cell wall deposition. The current challenge is to continue this functional analysis and to use these data and build an integrated model of TE development.

Contents

INTRODUCTION

The formation of xylem cells is one of the most intensely studied aspects of cell differentiation in plants. Part of this interest has been driven by the economic importance of xylem as a major constituent of wood and forage crops. Xylem is composed of a number of cell types, but the highly specialized and easily identifiable water-conducting cells, known as tracheids or vessels, have been most intensively studied. Collectively, these two cell types are frequently referred to as tracheary elements (TEs) and have been used to study many different aspects of plant cell differentiation, including developmentally regulated cell death, cell polarity, patterned secondary cell wall (SCW) deposition, and the role of the cytoskeleton (**Figures 1** and **2**). It is the differentiation of TEs that forms the basis of this review.

Tracheids: water-conducting cells that are similar to vessels but with no perforation plate

Figure 1 shows a summary of TE differentiation illustrating the different stages of vessel development referred to in this review. TEs are only one of several xylem cell types. They are easily identified by their characteristic patterned SCW, which is confined to the lateral sides of the cell and excluded from the ends. This SCW is essential to resist the negative pressures generated in the xylem during transpiration (**Figure 3*a***). The final stages of differentiation are characterized by programmed cell death (PCD), a process that removes the cell contents and leaves the cell empty, resulting in what has been described as a "functional corpse." Mature TEs are connected end-to-end by the perforation plate and form a tube specialized for unimpeded water flow, a process for which coordinated cell-to-cell communication between adjacent cells is clearly vital (**Figures 1** and **2*b***).

The intimate association of vessels with other cells, such as xylem parenchyma and fibers, makes it difficult to study TEs in isolation (**Figures 2** and **3**). However, there is an increasing body of evidence demonstrating that surrounding cells contribute to TE differentiation (see below).

An increasingly large number of *Arabidopsis* mutants have been described with alterations in vascular tissue patterning. These mutants exhibit a variety of phenotypes

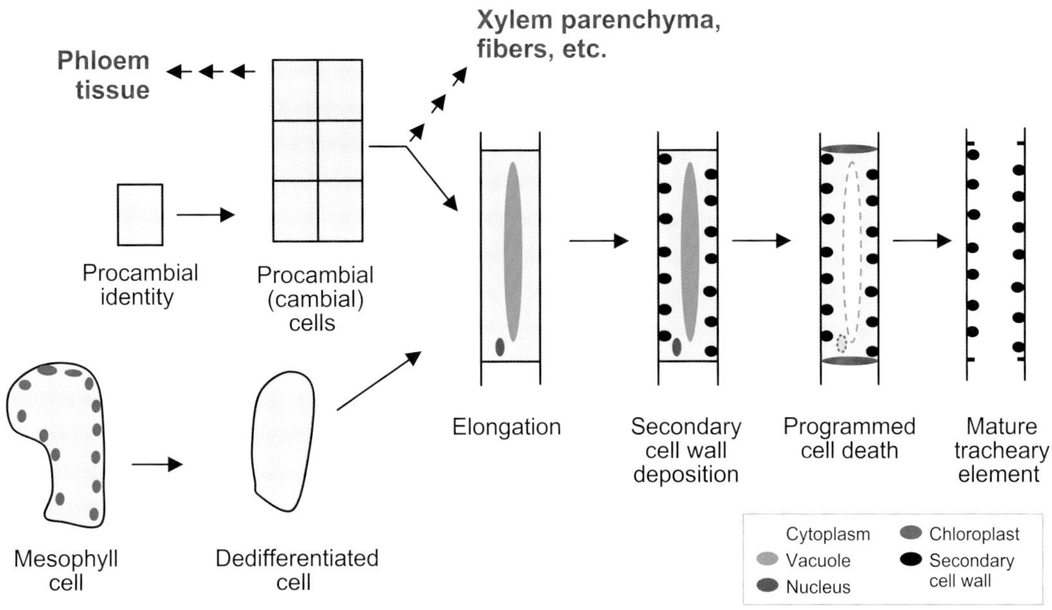

Figure 1

Tracheary element (TE) differentiation. The first sign of primary vascular development is the identification of procambial cells. These cells then divide and differentiate to become both the xylem and phloem. In secondary growth the equivalent cambial cells continue to divide and differentiate over many years. During xylem development procambial/cambial cells form various cell types in addition to TEs. In Zinnia mesophyll cell culture or following wounding, differentiated cells can transdifferentiate to form TEs.

including changes to the spacing and organization of veins of the leaf, or to the organization of xylem and phloem within the vascular bundle. These patterning mutants have been reviewed in this series (129) and recently elsewhere (19, 112) and are not the focus of this review unless they have a clear impact on xylem differentiation.

One distinctive characteristic of recent research on xylem differentiation is the use of widely disparate model systems (**Figure 3**). An ornamental flower, a model weed, and different trees species have all made important contributions to recent progress in the field. Recent research in these systems is characterized by a dramatic increase in data from expression studies and molecular genetic analysis, and these recent developments form the basis of this review.

MODEL SYSTEMS FOR THE STUDY OF XYLEM DIFFERENTIATION

Zinnia Cell Culture

The formation of TEs from isolated leaf mesophyll cells provides an excellent system for studying xylem differentiation (**Figures 1** and **3*b***). No other system offers the ability to regulate differentiation in such a coordinated manner, resulting in the synchronous transdifferentiation of a high percentage (up to 80%) of cells into TEs. Consequently, this enables the study of TE differentiation with minimal interference from other cells types.

Large-scale expression analysis based on cDNA restriction fragment length polymorphisms (RFLPs) identified many genes whose expression alters during Zinnia TE

Vessels: water-conducting cells that are connected together in files via the perforation plate to form a continuous network

TE: tracheary element

SCW: secondary cell wall

Programmed cell death (PCD): a predictable form of cell death. In the xylem, cell contents are removed leaving a hollow conduit for water transport.

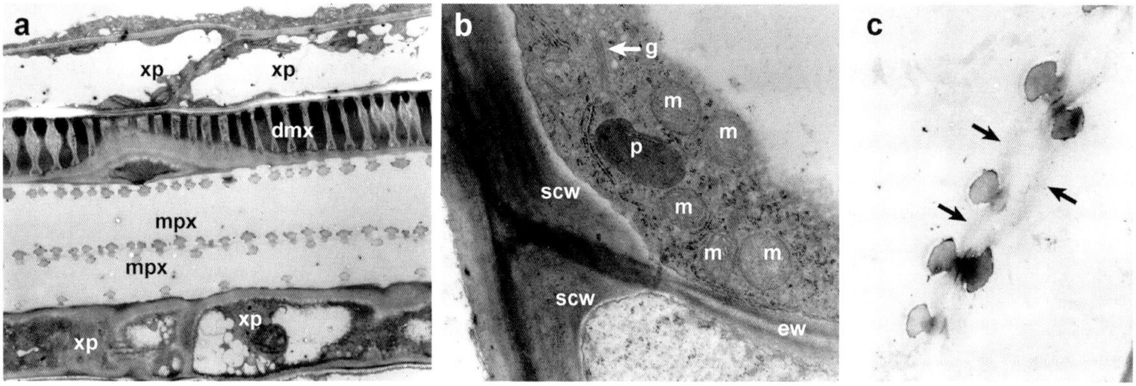

Figure 2

Stages of tracheary element (TE) formation in *Arabidopsis*. (*a*) Stem vascular bundle showing early mature protoxylem (mpx) that is completely differentiated and has lost the cell contents, and developing metaxylem (dmx) that still retains cytoplasm. In xylem tissue, TEs are intimately associated with xylem parenchyma (xp) that retain their contents and form nonpatterned secondary cell walls (SCWs). (*b*) Close-up view of two adjacent TEs. The end wall (ew) will eventually be lost to form the perforation plate that connects the two cells. The lower cell is undergoing cell death and the cytoplasm is degraded. The upper cell shows a very active cytoplasm characteristic of a cell undergoing SCW formation and contains numerous mitochondria (m), plastids (p), golgi (g), and ribosomes as well as extensive endoplasmic reticulum. (*c*) Close-up view of the cell wall between two adjacent mature protoxylem cells. The SCW is clearly visible, and the arrows mark the remnants of the primary cell wall that remains following programmed cell death.

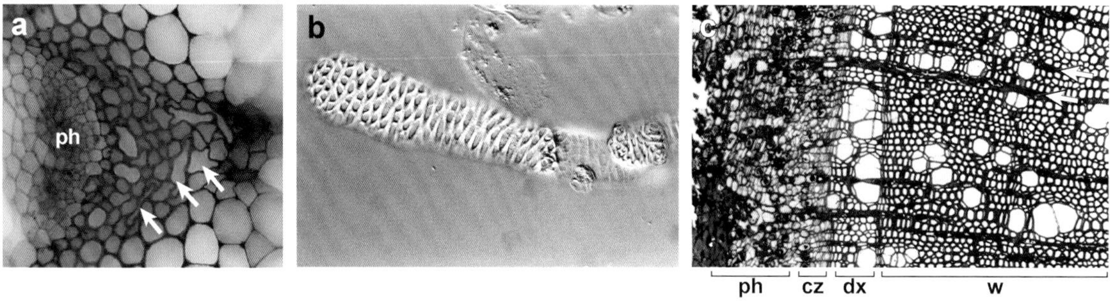

Figure 3

Model systems for studying xylem development. (*a*) Transverse section through a stem vascular bundle from the *Arabidopsis irx3* mutant. Secondary cell walls (SCWs) of the xylem are stained blue and are located opposite the phloem (ph). Metaxylem vessels that have collapsed as a result of a defective secondary cell are marked with arrows. (*b*) Zinnia tracheary elements (TEs) formed as the result of mesophyll cell transdifferentiation. The large cell on the left is fully mature and has a distinct SCW, whereas the adjacent cell on the right just started SCW deposition. The cell above, which contains some chloroplasts and exhibits no signs of TE differentiation, may still contribute to the formation of TEs. (*c*) Wood formation in developing birch showing a transverse section through the birch cambium and the region of wood formation. The cambial initials in the cambial zone (cz) divide to generate new cells that form xylem (dx) and phloem (ph). The distance from the cambial zone is directly related to the stage of xylem differentiation. The wood-forming tissue from the previous year's growth (w) shows the distribution of very large vessels surrounded by numerous fiber cells. The whole tissue is interspersed with ray cells, which are marked by arrows.

differentiation (78). However, the most comprehensive study used microarrays composed of 8000 Zinnia cDNA clones. More than 500 genes were identified whose expression varied more than eightfold during xylem cell differentiation. Hierarchical clustering of these genes identified a number of groupings that exhibit distinctive expression patterns during xylem development (24). Such distinctive patterns can only be easily detected if the differentiation of cells into TEs occurs in a highly synchronized manner and are a testament to the excellence of the Zinnia system.

Although both transient assays and stable transformation may offer some means of carrying out functional analysis of Zinnia genes, rapid progress toward the functional analyis of these genes is being made by analyzing homologous genes from species more amenable to molecular genetic approaches.

Wood Formation in Trees

The study of xylem differentiation in trees is driven by the economic importance of wood. However, other factors, such as the highly organized nature of xylem differentiation during secondary growth, make trees an attractive model. Adjacent files of cells represent a developmental series from undifferentiated cambial initials through to mature xylem (**Figure 3*c***).In Poplar, the ordered arrangement of the xylem and the relatively large size of trees have been exploited to isolate mRNA from sections cut at different stages of xylem formation (45, 110). Recent detailed analysis of the cambial zone has the resolution to distinguish gene expression within the cambial initials from the xylem mother cells and provides some of the most detailed analysis of gene expression during the early stages of xylem formation in any species (110).

The analysis of xylem formation in wood is complicated by the presence of diverse cell types, including vessels, fibers, and ray cells (**Figure 3*c***). However, many of the genes that exhibit alterations in expression during xylem differentiation in trees have clear homologs in *Arabidopsis*, which is likely to facilitate their functional analysis.

Arabidopsis

Arabidopsis has a clear role in the functional analysis of genes identified in both Zinnia and trees. Extensive gene expression studies of xylem development in *Arabidopsis* have already been described (62). In the stem, primary vascular tissue development represents a developmental series that has been exploited to identify genes that may be involved in SCW formation in the xylem (see below) (15, 28). Other investigators have applied weight and auxin to increase secondary growth in the stem and compared tissues undergoing primary growth with those undergoing secondary growth as a means of identifying genes specifically involved in secondary xylem formation (59, 88). The many different cell types in the *Arabidopsis* stem add complexity to the data analysis, a point illustrated by the observation that the expression of more than 5000 genes, around 20% of the genome, is altered during stem development (28). However, detailed bioinformatic analysis has identified candidates for many aspects of xylem differentiation, particularly SCW formation and its regulation (28).

The problems of tissue complexity have been overcome in one study where hypocotyls were separated into xylem, cambium/phloem, and the nonvascular outer tissues, and used to identify genes preferentially upregulated in one of these tissues (132). In one of the few cell type–specific studies in *Arabidopsis*, cell sorting was used to examine gene expression in cell layers of the developing root. This study included the stele that contains both xylem and phloem (9). Cross-referencing expression data from stems with stele-specific expression in the root is an excellent way to screen candidate genes (15, 28). The wealth of publicly accessible expression data available from expression studies in *Arabidopsis* represents a powerful asset that can be used to support custom array data (15). One study (93) relied

Cambial: part of the cambium

Cambium: lateral meristem located between the xylem and phloem that is responsible for secondary growth

Procambial: part of the procambium

IAA: indole-3-acetic acid

solely on 486 essentially randomly selected, publicly available data sets to identify genes likely involved in SCW synthesis.

The use of *Arabidopsis* in genetic analysis is well documented. A focused genetic screen was carried out to identify ectopic expression of a xylem-specific marker gene. Numerous mutants were identified that are likely to define repressors of xylem differentiation (106). An important breakthrough was the development of a TE cell culture system in *Arabidopsis*. Although not as efficient as the Zinnia system (see above), *Arabidopsis* cell suspensions can be induced to form up to 30% TEs within a 96-h culture period in inductive conditions (62, 87).

FACTORS AFFECTING EARLY XYLEM DIFFERENTIATION

Auxin

The role of auxin in the differentiation of vascular tissue both during normal development and wounding is well documented (104). Recent research has greatly increased the understanding of both auxin transport and auxin signaling (see 65 for recent review). Mutants with impaired auxin signaling have clearly identified a role for auxin in the early stages of vascular patterning during the period when procambial cells are formed (reviewed in 19, 112). Studies of secondary growth in pine demonstrate that the highest levels of auxin are found in the cambium, which is consistent with the role of auxin in maintaining cambial cell identity (123).

The role of auxin in promoting procambial cells to differentiate into xylem is unclear. Auxin is an absolute requirement in the Zinnia system, where its addition during a specific 10-min window is sufficient to induce differentiation (77). Several genes involved in auxin signaling are expressed during TE differentiation in Zinnia (24, 78). The complexity of auxin signaling during xylem development is illustrated by studies in Poplar on the Auxin (AUX)/indole-3-acetic acid (IAA) gene family, which is an important component of the auxin signaling pathway. At least eight members of the AUX/IAA gene family are expressed during secondary growth in Poplar (82). The expression of some genes is specifically correlated with high levels of auxin in the dividing cells of the cambium and xylem mother cells, where auxin may be required to maintain a population of dividing cells. In contrast, at least one of the IAA genes was expressed in mature xylem, where it presumably has a role in the later stages of xylem development and must respond to much lower levels of auxin (82). One suggestion is that auxin may act as a morphogen to define the fate of different tissues. In this model auxin concentrations are translated into positional information via the IAA genes, and the expression of individual IAA genes is regulated by different concentrations of auxin. Consequently, expression of IAA genes in the cambial meristem where auxin concentrations are high would contribute to maintaining a population of cambial cells, whereas other IAA genes that are expressed at lower concentrations of auxin facilitate the later stages of xylem development (8).

Both genetic and inhibitor studies have identified a clear role for polar auxin transport in developing vascular networks and this work was recently reviewed elsewhere (19, 112). In the Zinnia system, xylogenesis may be blocked by auxin efflux inhibitors, which appear to activate metabolism of intracellular auxin, resulting in it degrading more rapidly. Consequently, the concentration of intracellular auxin is lower in these treated cells and this is presumed to be the cause of the low rates of differentiation (130).

A 10-min exposure to both auxin and cytokinin can be sufficient to induce TE differentiation (77); however, exposure to either auxin or cytokinin alone has no effect. This suggests that high levels of both exogenous auxin and cytokinin are required for a commitment step, after which point they are no longer required for TE formation. At least 68 genes are upregulated within 30 min of induction by auxin and cytokinin, and some of these genes were previously implicated in auxin signaling. Of the genes upregulated, a substantial proportion are upregulated by auxin or cytokinin alone (24, 78). However, many genes are upregulated by these hormones, whether or not they are involved in xylem development. Therefore functional analysis of the upregulated genes is essential to establish their roles, if any, in xylem differentiation. One method of narrowing the list of candidates is to cross-reference the Zinnia expression data with that from trees where no application of hormones is required (24).

Cytokinin

The *Arabidopsis woodenleg* (*wol*) mutants contain fewer cell files in the vascular cylinder of the primary root and all the cells differentiate as xylem (108). *wol* is allelic with *cre1*, which is a cytokinin receptor (50, 72). Cytokinin acts via a signaling pathway similar to the bacterial two-component relay systems, with members of the *Arabidopsis* Response Regulator (ARR) family acting downstream of the receptor. Overexpression of one *ARR* gene (*ARR22*) results in the vascular cylinder of the primary root being composed exclusively of xylem, phencopying the *wol* mutants (57). A recent study overexpressed a cytokinin oxidase gene to deplete cytokinin specifically in the procambium (71). The resulting plants also phenocopy the *wol* mutation and provide very good evidence for an essential role for cytokinin in maintaining procambial cells and preventing their differentiation into xylem. In Zinnia cells, cytokinin is essential for differentiation (see above), which seems at odds with its role in maintaining procambial cell identity. However, it is possible that in the Zinnia system, cytokinin promotes mesophyll cell de-differentiation prior to transdifferentiation into TEs.

Brassinosteroids

Although a great deal of attention has been focused on the role of brassinosteroids (BRs) in regulating cell expansion (23), several independent experiments suggest BRs are also involved in the regulation of xylem development. BRs have been detected in developing pine cambium, and more direct evidence has come from the Zinnia system. The addition of Uniconazole, a known inhibitor of BR and gibberellin biosynthesis, to the Zinnia systems blocks the transdifferentiation of TEs in a manner that can be overcome by the addition of exogenous BRs but not gibberellin (54). Uniconazole appears to block the later stages of TE differentiation and prevents the expression of genes associated with PCD and SCW deposition (127). Consistent with the idea that BRs are required for the later stages of TE differentiation is the identification of five different BRs that accumulate both within the cells and in the TE culture medium when TEs are differentiating (128).

Procambium: meristematic cells that generate both the xylem and phloem of the primary vascular tissues

Independent confirmation of a role for BRs in xylem development came from the study of the known *Arabidopsis* BR biosynthesis mutants. *CPD*, *DWF7*, and *DET2* encode a hydroxylase, sterol desaturase, and reductase, respectively, and are all essential for BR biosynthesis (22, 84, 114). *cpd* and *dwarf7* have fewer vascular bundles, with each vascular bundle having comparable amounts of phloem to the wild type, but reduced xylem (22, 114), whereas *det2* has more phloem cells per vascular bundle at the expense of the xylem (18). *bri1* encodes a membrane-bound kinase believed to be part of the plasma membrane BR receptor (66). *bri1* mutants exhibit a similar phenotype to the biosynthetic mutants with an increase in phloem relative to xylem, whereas plants overexpressing *BRI1* have increased amounts of xylem (18). Three other proteins (BRL1, 2, and 3) have been identified based on their similarity to BRI1. Both BRL1 and BRL3, but not BRL2, bind to BRs and may also function as BR receptors (18). *brl1* mutants exhibit a similar phenotype to *bri1*, and the double mutant exhibits an enhanced vascular defect. A triple mutant of *bri1*, *brl1*, and *brl3* has greatly reduced vascular tissue, with both fewer xylem and phloem cells.

These results suggest that in addition to promoting xylem development at the expense of phloem, BRs also promote cell divisions in the procambial cells to provide the precursors for vascular cells (18).

Both auxin (see above) and BRs control a number of similar processes, such as cell elongation and vascular development. The induction of several auxin-inducible IAA genes by BRs (83) suggests a common link in the signaling of these two hormones; however, how these two hormones act together to regulate vascular development remains unclear. One important outcome of the discovery of the role of BRs in xylem differentiation is the development of an *Arabidopsis* cell culture system for TE differentiation in which the addition of BRs to the medium is essential (62, 87).

Xylogen

It is well established that one of most important factors in obtaining a high frequency of TE differentiation in the Zinnia cell culture system is a high cell density in the starting culture. Analysis of Zinnia cells immobilized in a thin sheet of agarose demonstrates that TEs typically form in clusters of cells, suggesting that differentiation of TEs tends to promote differentiation of neighboring cells (80). Furthermore, media isolated from cultures that have already undergone TE differentiation (called conditioned media) can be used to induce TE formation in cultures with a low starting density (80). Both an oligosaccharide (97) and a sulphated pentapeptide (Phytosulfokine) (75) have been suggested as compounds that accumulate and promote TE differentiation. Recently, a compound called xylogen was biochemically purified from differentiating Zinnia TEs (81). The purified protein (ZeXYP1) contains an N-glycosylation site, a signal peptide, and a putative glycosylphosphatidylinositol (GPI) anchor, which are all characteristics of Arabinogalactan proteins (AGPs) (81). Xylogen synthesized heterologously in tobacco cells can promote TE differentiation in the Zinnia cell system. Furthermore, a double mutant combination of two *Arabidopsis* homologs (*AtXYP1* and *AtXYP2*) resulted in leaf veins with a simpler, poorly coordinated pattern and xylem vessels that develop unconnected to the network. Interestingly, xylogen preferentially accumulates at only one end of the cell. Xylogen does not appear to be an essential determinate for xylem development because TEs still differentiate in the *AtXYP1/AtXYP2* double mutant. Taken together, these data suggests that xylogen coordinates the deposition of vascular tissue and that its polar secretion contributes to the continuity of the vascular network by promoting differentiation of the adjacent cells (81).

CLE Peptides

Clavata3-like/ESR (CLE) genes encode small peptides that appear to act as extracellular signaling molecules. The most widely studied member is CLV3, which functions together with the CLV1/CLV2 receptor complex as part of a pathway essential for regulating the organization of the *Arabidopsis* shoot apical meristem. There are at least 25 members of the CLE family in *Arabidopsis*, including several expressed in the vascular tissue. Recently, Ito and colleagues (53) used a bioassay for Zinnia TE differentiation to isolate a 12–amino acid peptide that inhibited TE differentiation. Comparison with the cDNA sequence revealed that the peptide was part of a larger precursor that exhibited homology to CLE genes. Only CLE peptides containing the specific 12–amino acid sequence could prevent TE differentiation and they had no effect on the root apical meristem. The study defines two distinct signaling pathways for CLE genes products: the regulation of apical meristem activity and the control of TE differentiation (53).

HD-Zip Genes and MicroRNAs

Analysis of the homeodomain leucine zipper (HD-Zip) proteins illustrates the

Table 1 Multiple functions of HD-ZIP genes[1]

Arabidopsis gene no.	Gene name	*Arabidopsis* Mutant	Loss of function Phenotype	Zinnia homolog	Overexpression Phenotype
AT4G32880	*ATHB8*	—	Not apparent	ZeHB-10	Overexpression produces weakly radialized vascular bundles
AT2G34710	*ATHB14*	*Phabulosa* (*phb*)	Radialized leaves and vascular tissue	—	—
AT1G30490	*ATHB9*	*Phavoluta* (*phv*)	Radialized leaves and vascular tissue	—	—
AT5G60690		*Revoluta/ interfascicular fiberless* (*rev/ifl*)	Loss of interfascicular fibers, alterations in auxin transport, meristem initiation	ZeHB-11 ZeHB-12	Overexpression produces weakly radialized vascular bundles
AT1G52150	*ATHB15*	*Corona* (*cna*)	Decreased vascular tissue development, increased apical meristem cell numbers	ZeHB-13	

[1]Further information on genes and phenotypes can be found in References 37, 91, 94, 111, and references therein.

complexities involved in relating gene expression patterns in developing xylem to gene function. This family of putative transcription factors has been divided into four classes (I-IV) (111). In *Arabidopsis*, class III is composed of five members (**Table 1**). Mainly due to expression studies in *Arabidopsis* and Zinnia, the role of class III HD-ZIP genes during vascular development has been the focus of much attention (3, 56, 89, 90, 111).

Analysis of the role of the HD-Zip III genes in vascular development is complicated by the complex interactions between family members, and the pleiotropic nature of the mutant phenotypes, which include alteration in the size of the apical meristem and changes in organ polarity (**Table 1**). Promoter-GUS studies using *ATHB8* (an HD-Zip III) show that it is an early marker for procambial development (3). Overexpression of *ATHB8* causes a proliferation of xylem and precocious initiation of secondary growth; however, loss of function mutations in *ATHB8* cause no obvious vascular phenotype (4). The absence of a vascular phenotype in the *ATHB8* knockout cannot be explained by redundancy among the HD-ZIP III group of genes (4) because comprehensive genetic analysis of insertional mutants in all five members of the HD-ZIP group does not support this idea (94). The proliferation of xylem caused by overexpression of *ATHB8* may result from the activation of genes normally regulated by members of the HD-ZIP II class. The rice HD-Zip III, *OsHOX1*, is expressed early in vascular development and appears to enhance the rate at which procambial cells differentiate into xylem (107).

MicroRNA (miRNA): 20–25 nucleotide RNA derived from a stem-loop region of a longer transcript complementary to a gene sequence

In contrast to results with *ATHB8*, down-regulation of *ATHB15* either by mutation (94) or by antisense (58) gives a clear phenotype and results in plants with increased vascular tissue. Overexpression of this gene leads to smaller vascular bundles, consistent with the role of *ATHB15* as a negative regulator of procambial cell specification or proliferation. Recently, three Zinnia class III HD-Zip genes expresed during TE formation were overexpressed in *Arabidopsis* (91). This study included mutant forms that should be resistant to microRNA (miRNA)-mediated cleavage (see below). Overexpression of the mutated *ZeHB10* and *ZeHB12* resulted in much

Figure 4

Microtubules (MTs) in developing xylem. MTs were visualized using yellow fluorescent protein (YFP) attached to an MT-binding protein and exhibit the characteristic banding pattern that marks sites of secondary cell wall deposition. (Photograph courtesy of Raymond Wightman.)

MT: microtubule

DCB: 2,6-dichlorobenzonitrile

before SCW deposition is complete and that MTs may only be required to pattern the early stages of SCW deposition (98).

The debate on the role of MTs in SCW deposition often focuses on the relationship between MTs and cellulose deposition (7). It is clear from numerous studies, however, that bands of cortical MTs in developing TEs mark not only the site of cellulose deposition in SCW, but also the site of deposition of other cell wall components such as lignin, hemicellulose, and proteins (46). Both the cellulose synthase complex and hemicelluloses are likely transported in Golgi vesicles, which must be targeted to specific regions of the cell membrane. Furthermore, other cell wall components, such as lignin, are incorporated into the wall in a localized manner (**Figure 5**). Consequently, bands of cortical MTs in TEs mark the sites for vesicle transport for a number of components, making it a unique example of MT-targeted vesicle fusion.

Inhibition of cellulose deposition by 2,6-dichlorobenzonitrile (DCB) is reported to give wider SCW bands, which do not protrude as far into the cell. These cells are also reported to have a dispersed pattern of lignin deposition (117). These observations support a mechanism in which the patterned deposition of cellulose directs the assembly of lignin to the same regions. Although this is an attractive idea, it is hard to reconcile with genetic studies. Mutants that appear to lack cellulose in the SCW, such as *irx3*, exhibit more uneven but distinctly narrower SCW thickening in contrast to the broader banding seen in DCB-treated cells (41, 122). An alternative hypothesis was presented by Hogetsu (46), who observed the pitted xylem vessels of pea. In these cells MTs and microfibrils are parallel to one another. However, under the pit, where no SCW deposition occurs, the MTs tended to be randomly orientated, with the exception of a distinct accumulation of MTs around the border of the pits. Based on these observations, Hogetsu (46) suggested that the plasma membrane is partitioned into two distinct domains. Domains under SCW thickenings are preferential sites for inserting vesicles containing hemicellulose and cellulose synthase rosettes. Bundles of MTs are postulated to determine and maintain the boundaries between plasma membrane domains (46) (**Figure 5**). If colchicine is used to depolymerize the MTs, the SCW thickenings become rough and poorly defined (87), consistent with the idea that MTs might maintain two different domains in the plasma membrane (46) (**Figure 5**).

Adding MT-stabilizing drugs, such as taxol, to differentiating TEs does not prevent bands of MTs from forming (30). Furthermore, whether the bands are orientated in the longitudinal or transverse plane appears to reflect the organization of MTs prior to TE differentiation. These observations

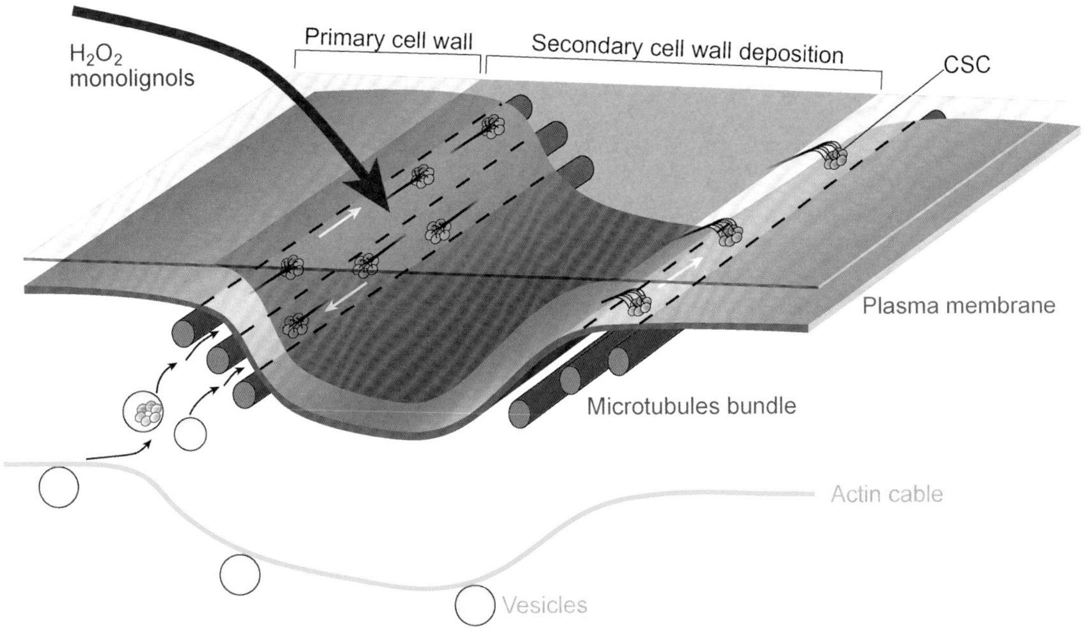

Figure 5

A model for secondary cell wall deposition. The actin cytoskeleton directs the delivery of vesicles containing hemicellulose and the cellulose synthase complex to the plasma membrane. Microtubule (MT) bundles mark the sites of vesicle insertion and maintain plasma membrane partitions, ensuring that the cellulose synthase complex moves parallel to the MTs, coaligning the MTs and the cellulose microfibrils. Adjacent cells deliver components that contribute to the synthesis of lignin in the secondary cell wall thickenings.

suggest that the bands of MTs seen in differentiating TEs result from the bundling of existing cortical MTs (29). MT dynamics has been studied extensively in expanding cells (26). However, few studies address how and where bands of MTs form in developing TEs, although they represent a well-defined system from which to study cytoskeleton organization and dynamics. This point is illustrated by the *wilted-dwarf* mutant of tomato. During normal vessel development, SCW deposition must be excluded from the ends of the cell where the perforation plate forms. This reflects the polarity of the cytoskeleton and consequently restricts the banding of MTs to the lateral wall and excludes them from the wall ends. In mutant plants, the thickening, and presumably the MTs, cover the ends of the cell and partially occlude the perforation plate, thus causing the wilted phenotype (1). Understanding this mutant or the identification of other mutants with altered patterns of MT banding will likely provide insight into the regulation of cell polarity and how this polarity helps determine MT dynamics in developing xylem and other plant cells.

MT bundling proteins, such as MAP65, have been shown to cross-link MTs by forming cross-bridges between them (21). A MAP65 recently isolated from Zinnia (ZeMAP65-1) is upregulated during Zinnia TE differentiation and is expressed in xylem cells of Zinnia vascular bundles. ZeMAP65-1 is associated with bundles of MTs in differentiating Zinnia TEs. When overexpressed in *Arabidopsis* suspension cells it causes bundling of cortical MTs in a manner reminiscent of developing TEs (74). However, expression of ZeMAP65-1 cannot be the complete explanation because in situ experiments

Metaxylem: later-forming xylem; distinguished by larger pitted cells with reticulate SCW patterning

Protoxylem: early-forming xylem; distinguished by spiral (helical) or annular SCW patterning

35S: highly expressed cauliflower mosaic virus RNA

suggest that the gene is strongly expressed in developing xylem parenchyma, which do not exhibit MT banding or patterned SCW deposition (74).

Live cell imaging of MTs has confirmed previous observations that initial transverse bands of MTs branch to form the more complex patterns seen in reticulate pitted metaxylem vessels (87). Although MT behavior and dynamics may have partially explained MT patterning, recent work suggests that transcriptional changes are also important in MT patterning in developing TEs. Kubo and coworkers (62) used xylogenic *Arabidopsis* cultures to identify a family of seven NAC transcription factors, including VND6 and VND7, expressed during TE differentiation. Overexpression of *VND6* and *VND7* resulted in ectopic vessels (see above). Most surprisingly, in Cauliflower mosaic virus strong promoter *35S:VND6* plants both ectopic and root TEs exhibited reticulate or pitted patterns of cell wall thickening characteristic of metaxylem. In contrast, the TEs of *35S:VND7* plants exhibited spiral or annular patterns of SCW thickenings characteristic of protoxylem. Overexpression of the VND6 protein fused to the SRDX strong repression domain driven by the 35S promoter resulted in the repression of metaxylem development in the root, although protoxylem was normal. The converse was found when VND7 was used. These results demonstrate that these two transcription factors independently regulate aspects of protoxylem or metaxylem development (62). It will be intriguing to compare the downstream targets of these two transcription factors and determine how they differ.

Actin

Early reports suggested that actin filaments exhibited a similar pattern to MTs in developing Zinnia TEs. Furthermore, during TE differentiation in Zinnia, disrupting actin with cytochalasin B prevents the reorientation of MTs from a predominantly longitudinal to a transverse orientation that occurs, suggesting a role for actin in reorientating MTs (60, 61). However, studies with intact *Arabidopsis* roots and in trees suggest that actin cables run mostly longitudinally during TE differentiation, which suggests a more probable role in vesicle transport of cell wall components (20, 34).

An MT-independent role for actin microfibrils in the SCW is suggested by the *Arabidopsis* mutants *fra3*, *fra4*, and *fra7*, which are caused by mutation in a GTPase, inositol polyphosphate 5-phosphotase, and phosphoinositide phosphostase, respectively. All these mutants have thinner SCWs and altered patterns of actin cables, but normal patterns of MTs (48, 134, 135). The relationship between the actin defect and decreased SCW deposition is unclear, but it is likely that phoshoinositides regulate the actin cytoskeleton and the resulting cell wall defect may be due to reduced vesicle trafficking of cell wall components (**Figure 5**).

SECONDARY CELL WALL BIOSYNTHESIS

Cellulosc

Studies using *Arabidopsis* mutants have demonstrated that three different cellulose synthase (CESA) proteins [IRREGULAR XYLEM (IRX)1, IRX3, and IRX5] are required for cellulose synthesis in the SCW (118). This organization appears conserved between *Arabidopsis* rice and various tree species. IRX1, IRX3, and IRX5 initially localize within the cell and bands of cortical MTs precede their localization to the plasma membrane. At later stages of TE development, all three proteins colocalize at the plasma membrane with bands of cortical MTs (34), which is consistent with much earlier work that used electron microscopy to localize the cellulose synthase complexes to regions of SCW thickenings (109). The absence of any one of IRX1, IRX3, or IRX5 results in the remaining subunits being retained within the cell, presumably as a result of the inability to form an intact cellulose synthase complex. However, the organization of cortical MTs appears unaffected (34), suggesting that the cellulose synthase complex does not have a role in stabilizing MTs during SCW deposition.

The endoglucanase KORRIGAN is also required for cellulose synthesis in the SCW. It does not appear to be part of the cellulose synthase complex and its role in cellulose synthesis remains unclear (115). Other candidates for genes involved in SCW synthesis have come from microarray studies. Using microarray data, a number of genes were identified that exhibited good coexpression with *IRX1*, *IRX3*, and *IRX5* (15, 93). Mutations in several of these genes gave an *irx* phenotype characteristic of a SCW mutant. However, cell wall analysis revealed that only one mutation in the *COBRA-LIKE4* (*CBL4*) gene exhibited a specific reduction in cellulose (15). Mutation in the rice homolog of *CBL4* had previously been shown to cause a cellulose-deficient phenotype (67). *COBRA* was previously described as a gene required for cellulose microfibril orientation during anisotrophic growth (101, 102), but the analysis described above suggests a wider role for this gene family in cellulose deposition.

Two additional *fra* mutants have provided some insight into how MTs might control the orientation of cellulose deposition. *fra2* is caused by a mutation in a katenin-like gene that encodes a protein with MT severing properties. *fra2* plants exhibit both altered cell shape and MT organization in a wide variety of cell types. However, in the SCW of fibers, cellulose microfibril orientation is clearly altered in a way that reflects altered MT organization, providing one of the best pieces of evidence to support a role of MTs in controlling the orientation of cellulose microfibrils (16, 17). In contrast, *fra1* mutants exhibit normal patterns of MTs, but cellulose microfibril orientation is altered, which results in plants with much weaker cell walls. The *fra1* phenotype is caused by a mutation in a kinesin-like protein and it remains unclear how this might alter cellulose deposition (133). One possibility is that the kinesin is involved in the transport of cell wall components that contribute to cellulose orientation.

Lignin

Lignin is a complex phenolic polymer that is essential for SCW structure and is formed from the oxidative cross-linking of monolignols. Lignin biosynthesis has been revised many times in recent years and was recently reviewed in this series (10); therefore, it is not covered in detail here. In cases where the genes encoding individual steps in lignin biosynthesis are unknown, candidate genes have been identified from microarray data (28). Lignification has identified an intriguing relationship between vessels and the surrounding xylem parenchyma. Evidence from Zinnia TEs suggests that lignification proceeds after cell death. Studies suggest that TEs can use monolignols or dilignols supplied by other cells or added to the media, and these can be incorporated into the cell walls of TEs that have already undergone cell death (47, 120). Similarly, Barcelo (6) looked for the site of $H_2 0_2$ production that is presumed to be required for lignin polymerization in developing Zinnia TEs. $H_2 0_2$ was generated at the plasma membrane of both vessels and thin-walled cells that are not undergoing lignification. Barcelo concluded that the latter cells provide the $H_2 0_2$ necessary for lignification in developing TEs (6). These results also imply that the restriction of lignin to sites of SCW thickening in vessels is determined by the localization of enzymes that polymerize lignin cross-linking. This is consistent with a recent report of a peroxidase from Zinnia specifically localized to the SCW (105).

Hemicellulose

Xylan is the predominant hemicellulose of SCWs (reviewed in 27) and consequently one of the most abundant naturally occurring polymers. Surprisingly, no gene for xylan synthesis has been identified, although the characterization of three xylan-deficient mutants generated using reverse genetics has identified three glycosyltransferases that are good candidates for xylan biosynthetic enzymes (15).

Glucomannans are also found in *Arabidopsis* SCWs (42), and the enzyme that synthesizes the backbone of this polymer was recently identified. The enzyme has broad substrate specificity and is a member of the Cellulose synthase-like A (*CslA*) gene family (25, 68).

It is assumed that hemicellulose is synthesized in the Golgi and transported to the cell wall in Golgi-derived vesicles. Identifying more of the enzymes involved in hemicellulose synthesis should clarify the relationship between the sites of hemicellulose synthesis and its localized deposition in the SCW of xylem vessels.

Cell Wall Modification During Cell Death

During cell death, the end walls connecting adjacent cells are removed to form the perforation plate while areas of the lateral primary cell wall not covered by the SCW are extensively modified (**Figures 1** and **2**). This is particularly apparent in protoxylem, in which the regions of SCW deposition are well spaced. In protoxylem, vessels face a particular problem in maintaining the integrity of the vessel for water transpor, while being passively stretched by the elongation of the growing plant. During cell death many of the components of the primary cell wall are digested, with only the cellulose microfibrils and an electron-dense material remaining (**Figure 2*c***). Immunological data suggest that glycine-rich proteins (GRPs) are abundant components of this electron-dense material and link together regions of SCW thickening as well as the SCW of adjacent protoxylem (103). It is suggested that these GRPs are highly specialized to act as load-bearing components that stabilize the hydrolyzed primary cell wall. In doing so they maintain the integrity of the protoxylem vessels for transporting water even as the vessels become stretched during plant growth (96).

Regulation of Secondary Cell Wall Deposition

The regulation of cell wall deposition and particularly lignin deposition by transcription factors has received considerable attention. MYB genes in particular have been implicated in the regulation of lignin biosynthesis. MYB genes from *Arabidopsis* and *Antirrhinum* have been shown to control phenylpropanoid biosynthesis (11, 116). Similarly, a MYB from pine is sufficient to direct lignification in tobacco (92). More recently, a Eucalyptus MYB (*EgMYB2*) that alters lignin deposition in tobacco was identified. However, *EgMYB2*-overexpressing plants generally exhibit much thicker SCWs, suggesting that *EgMYB2* regulates many aspects of SCW synthesis, not just lignification (35). In addition to the two NAC transcription factors described above (*VND6* and 7) (62), another NAC transcription factor appears to have a role in regulating SCW deposition. *NST1* falls within a different NAC subfamily to the VND genes, but results in ectopic SCW when overexpressed (79). In the epidermis, *NST1* overexpression results in cell walls that are patterned and are reminiscent of TEs, whereas ectopic SCWs in mesophyll and other cell types is not patterned, suggesting that *NST1* regulates SCW deposition independently of cell wall patterning. Interestingly, repressing these genes results in the anthers failing to dehisce as a result of cells in the endothecium which surround the pollen grains failing to undergo secondary thickening (79). A similar phenotype was also described for mutations in a MYB gene (113). This highlights an important point: Many features of TE differentiation, such as patterned cell wall deposition, may be shared by other cell types and may be regulated by common mechanisms.

Programmed Cell Death

PCD is an active area of research in plants. Several molecular components have been identified over the past few years, but there is as yet no coherent picture. It appears that plants have evolved PCD mechanisms, but it is not clear what these mechanisms have in common with animal PCD, and, consequently, it is inappropriate to refer to plant PCD as plant apoptosis. Cell death of the TE has long been recognized as a prime example of developmental PCD in plants. It occurs among healthy cells in a predictable pattern that indicates regulation by a developmental program. It is an active process whereby a TE cell up-regulates genes that trigger its destruction. New protein synthesis is required, because inhibiting translation using cyclohexamide blocks cell death of TEs in the Zinnia system (63).

The most striking feature of xylem cell death is the collapse of the vacuole that coincides with the digestion of the nucleus (31, 38). Vacuolar collapse activates or releases hydrolytic enzymes, including proteases (32), DNAses (52) and RNAses (64), into the cell, some of which may cause its destruction. Vacuolar collapse, without accumulation of these enzymes is not sufficient for PCD as it does not induce digestion of the nucleus in non-TE cells in Zinnia cultures (85). In addition, the clearest proof for the essential nature of the hydrolytic enzymes during PCD is demonstrated by the introduction of an antisense for the DNase *ZEN1* gene that suppresses nucleus digestion (52). After the accumulation step, the enzymatic activity of these hydrolytic enzymes dramatically increases at the time of vacuolar collapse, possibly because of their release from the vacuole. In support of this, a vacuolar localization was demonstrated for the protease XCP1 (32). By contrast, there is at least one RNase that accumulates in the ER of TEs and not in the vacuole (64). In fact, some of the hydrolytic enzymes detected may be cytosolic and activated by the acidification of the cytosol that results from vacuolar collapse. Hydrolytic enzymes associated with TE PCD have also been detected in the proteome of maize xylem sap (2) and this may support the idea that some of them are involved in defense and not PCD, as has been suggested for the xylem protease XCP1 (32). Of the several xylem-specific proteases detected, none have yet been shown to be required for the cell death process.

An important and recent development in our understanding of the role of proteases in plant PCD has been the demonstration of the importance of caspase-like proteases in this process (reviewed in 100). There are two main classes of plant caspase-like protease: metacaspases and proteases with caspase-like activity. The plant proteases most related to caspases are the metacaspases, which do not have caspase activity. One antisense study showed a metacaspase to be required for the cell death of the suspensor in embryogenic cell cultures of Norway spruce (14). It is interesting to note that only one of the nine *Arabidopsis* metacaspases is upregulated during the late stages of TE differentiation and this member of the gene family could therefore be one mediator of the PCD process (**Figure 6**). Few plant proteases with caspase-like activity have been identified thus far (100). One of them, the protease vacuolar processing enzyme (VPE), which has caspase 1 activity, is relevant to TE differentiation. Silencing of VPEs suppresses vacuole collapse in TMV-infected leaves. VPE may therefore be required to activate various vacuolar proteins involved in the disintegration of vacuoles that occurs during plant PCD (44). One member of the VPE family in *Arabidopsis* is upregulated during TE formation and could therefore be involved in vacuole collapse (**Figure 6**). However, the absence of any effect following the addition of caspase inhibitors (ICE inhibitors, general inhibitors) in the Zinnia system has been reported (31, 76), but the details have not been published. This suggests that the TE PCD may differ from many plant PCD systems that are clearly sensitive to caspase inhibitors (100). In this context, a study

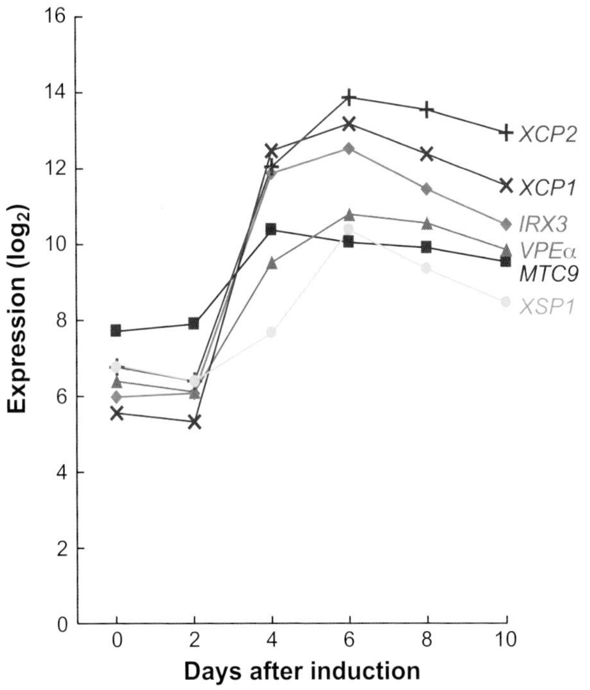

Figure 6

The expression patterns of programmed cell death (PCD)-associated genes during TE differentiation. Expression data from *Arabidopsis* cell suspension cultures (62) were used to examine transcript abundance of numerous putative PCD-related genes. The *IRX3* gene is included as a marker for the later stages of TE differentiation. The genes analyzed include those believed to encode vacuolar processing enzymes (VPE), metacaspases, xylem endopeptidases gene families, and genes involved in autophagy (32, 43, 119, 124). Only genes showing upregulation consistent with a role in PCD are shown. *MTC9*, metacaspase 9; *VPEα*, vacuolar processing enzyme; *XCP*, xylem cysteine proteases.

using the full range of caspase inhibitors available would prove very informative, as would knockout and overexpression studies of metacaspases and caspase-like proteases.

It has been suggested that TE cell death is a form of autophagy (126). There is little evidence to support this. On the contrary, autophagy may limit PCD, as shown in TMV-induced cell death (69). In addition, the upregulation of a suite of genes is a marker for autophagy, but in our microarray analysis, none of the 13 autophagy genes from *Arabidopsis* are upregulated at any time in xylem differentiation (**Figure 6**). This suggests that autophagy is not a component of PCD during TE differentiation.

Another important development in plant PCD has been the realization of a role for mitochondria. In particular, the release of cytochrome C from mitochondria is an early marker of cell death despite the fact that this does not appear to activate PCD, as described in animal models (5). One study showed that TE mitochondria are involved in the PCD process and cytochrome C is released before the vacuole ruptures (131). This suggests that vacuole collapse is not a trigger of TE PCD but more likely a final execution stage. Vacuole collapse is such a dramatic and visible event that it has attracted a lot of attention, possibly to the detriment of earlier PCD events that remain to be characterized.

Despite a wealth of information on the demise of the TE, it is striking that virtually nothing is known of the induction pathway for TE PCD at the biochemical or molecular level. Some studies have suggested that calcium (39) and nitric oxide (33), two known PCD regulators in plants, may be involved. However, little is known about which genes are regulating the process with such contained tissue localization. Is the trigger for PCD cell autonomous? Is PCD an integral part of the TE differentiation program or is it a general cell death module that is activated at the end of TE differentiation? The mutant *gpx* shows that PCD can occur without the completion of SCW synthesis and it is conceivable that in other mutants, secondary cell wall deposition could occur in the absence of PCD (79, 121). It is striking that so far there are no TE differentiation mutants in which only the PCD step is absent or has been delayed. Such mutants would represent an important breakthrough and their identification is likely to require specific genetic screens.

CONCLUSIONS

The variety of experimentally amenable models and the economic importance of wood have led to intensive study of TE differentiation. Consequently, a lot of data have recently been generated on this subject. This is particularly true of transcriptional data. To maximize the utility of these data it will be necessary to integrate microarray data from diverse species such as Poplar, Zinnia, and *Arabidopsis*. It is clearly not possible to assign a function to a gene simply based on expression analysis. Consequently, one of the biggest challenges is to analyze these data and select target genes for further functional analysis. More proteomic data and metabolomics data will likely be available shortly; a future goal will be to integrate these data to develop a real systems model for TE differentiation. Although such a model is still some way off, TE differentiation is well ahead of the study of many plant cell types and appears on course to pioneer this approach.

SUMMARY POINTS

1. A large amount of transcriptomic analysis, including recent comprehensive microarray data from *Arabidopsis*, has led to the identification of many genes whose expression is altered during TE differentiation.
2. Functional analysis of many genes expressed during TE differentiation is still needed.
3. Initiation of TE differentiation is regulated by complex interactions of the plant growth regulators auxin, cytokinin, and BR.
4. Xylogen, an AGP that promotes TE differentiation and that exhibits a polar secretion, appears to contribute to the continuity of the xylem network.
5. Transcriptional regulation plays a role in various aspects of TE differentiation. In some cases, such as the HD-ZIP III genes where the corresponding mutants have pleiotropic phenotypes, it is difficult to separate their effects on TE differentiation from those of other aspects of plant development.
6. The cytoskeleton and MTs in particular appear to determine the site of SCW synthesis and are confined to the lateral walls and excluded from the end wall in a way that reflects the highly polarized organization of the cell.
7. The analysis of two transcription factors VND6 and VND7 uncovered a role for transcription in determining patterns of cell wall deposition.
8. Although large increases in hydrolytic enzymes, vacuole collapse, and cell wall digestion characterize the later stages of PCD during TE differentiation, factors triggering early events during PCD have not been identified.

FUTURE ISSUES

1. Several plant growth regulators such as auxin, BRs, and cytokinin influence TE differentiation, and understanding the network that integrates these signals is essential.
2. The identification of an AGP (xylogen) and a CLE peptide, which regulate TE differentiation, appears to define new signaling pathways. Which other components are involved in these pathways and how these components function remain to be determined.

3. Many putative transcriptional regulators are expressed during TE differentiation. Identification of their downstream targets and how they integrate to form a transcriptional network is a major challenge.
4. A proper understanding of TE differentiation will not be complete without more comprehensive proteomic and metabolomic data, which will need to be integrated with the transcriptomic data.
5. Recent work suggests that different patterns of transcription lead to altered patterns of SCW deposition, and a better understanding of how these transcriptional changes translate to alteration in the cytoskeleton is needed.
6. Identifying mutants that better separate the different stages of TE development, such as mutants that fail to undergo PCD, will greatly help with functional analysis of genes expressed during TE differentiation.
7. Mutation in several genes that are upregulated in TE differentiation results in very pleiotropic phenotypes. Dissecting the role of these genes in vascular development from their role in other aspects of plant development requires further analysis.

LITERATURE CITED

1. Alldridge NA. 1964. Anomolous vessel elements in the *wilty-dwarf* tomato. *Bot. Gaz.* 125:138–42
2. Alvarez S, Goodger JQ, Marsh EL, Chen S, Asirvatham VS, Schachtman DP. 2006. Characterization of the maize xylem sap proteome. *J. Proteome Res.* 5:963–72
3. Baima S, Nobili F, Sessa G, Lucchetti S, Ruberti I, Morelli G. 1995. The expression of the Athb-8 homeobox gene is restricted to provascular cells in *Arabidopsis thaliana*. *Development* 121:4171–82
4. Baima S, Possenti M, Matteucci A, Wisman E, Altamura MM, et al. 2001. The *Arabidopsis* ATHB-8 HD-zip protein acts as a differentiation-promoting transcription factor of the vascular meristems. *Plant Physiol.* 126:643–55
5. Balk J, Chew SK, Leaver CJ, McCabe PF. 2003. The intermembrane space of plant mitochondria contains a DNase activity that may be involved in programmed cell death. *Plant J.* 34:573–83
6. Barcelo AR. 2005. Xylem parenchyma cells deliver the H2O2 necessary for lignification in differentiating xylem vessels. *Planta* 220:747–56
7. Baskin TI. 2001. On the alignment of cellulose microfibrils by cortical microtubules: a review and a model. *Protoplasma* 215:150–71
8. Bhalerao RP, Bennett MJ. 2003. The case for morphogens in plants. *Nat. Cell Biol.* 5:939–43
9. Birnbaum K, Shasha DE, Wang JY, Jung JW, Lambert GM, et al. 2003. A gene expression map of the *Arabidopsis* root. *Science* 302:1956–60
10. Boerjan W, Ralph J, Baucher M. 2003. Lignin biosynthesis. *Annu. Rev. Plant Biol.* 54:519–46
11. Borevitz JO, Xia YJ, Blount J, Dixon RA, Lamb C. 2000. Activation tagging identifies a conserved MYB regulator of phenylpropanoid biosynthesis. *Plant Cell* 12:2383–93

12. Bourquin V, Nishikubo N, Abe H, Brumer H, Denman S, et al. 2002. Xyloglucan endotransglycosylases have a function during the formation of secondary cell walls of vascular tissues. *Plant Cell* 14:3073–88
13. Bowman JL. 2004. Class III HD-Zip gene regulation, the golden fleece of ARGONAUTE activity? *BioEssays* 26:938–42
14. Bozhkov PV, Filonova LH, Suarez MF. 2005. Programmed cell death in plant embryogenesis. *Curr. Top. Dev. Biol.* 67:135–79
15. **Brown DM, Zeef LAH, Ellis J, Goodacre R, Turner SR. 2005. Identification of novel genes in *Arabidopsis* involved in secondary cell wall formation using expression profiling and reverse genetics. *Plant Cell* 17:2281–95**

15. Both custom and publicly available data were used to identify genes upregulated during secondary cell wall formation and to identify novel components required for cellulose and xylan biosynthesis.

16. Burk DH, Liu B, Zhong RQ, Morrison WH, Ye ZH. 2001. A katanin-like protein regulates normal cell wall biosynthesis and cell elongation. *Plant Cell* 13:807–27
17. Burk DH, Ye ZH. 2002. Alteration of oriented deposition of cellulose microfibrils by mutation of a katanin-like microtubule-severing protein. *Plant Cell* 14:2145–60
18. Cano-Delgado A, Yin YH, Yu C, Vafeados D, Mora-Garcia S, et al. 2004. BRL1 and BRL3 are novel brassinosteroid receptors that function in vascular differentiation in *Arabidopsis*. *Development* 131:5341–51
19. Carlsbecker A, Helariutta Y. 2005. Phloem and xylem specification: pieces of the puzzle emerge. *Curr. Opin. Plant Biol.* 8:512–17
20. Chaffey N, Barlow P, Barnett J. 2000. A cytoskeletal basis for wood formation in angiosperm trees: the involvement of microfilaments. *Planta* 210:890–96
21. Chan J, Jensen CG, Jensen LCW, Bush M, Lloyd CW. 1999. The 65-kDa carrot microtubule-associated protein forms regularly arranged filamentous cross-bridges between microtubules. *Proc. Natl. Acad. Sci. USA* 96:14931-36
22. Choe SW, Noguchi T, Fujioka S, Takatsuto S, Tissier CP, et al. 1999. The *Arabidopsis* dwf7/ste1 mutant is defective in the Delta(7) sterol C-5 desaturation step leading to brassinosteroid biosynthesis. *Plant Cell* 11:207–21
23. Clouse SD, Sasse JM. 1998. Brassinosteroids: essential regulators of plant growth and development. *Annu. Rev. Plant Physiol. Plant Molec. Biol.* 49:427–51
24. **Demura T, Tashiro G, Horiguchi G, Kishimoto N, Kubo M, et al. 2002. Visualization by comprehensive microarray analysis of gene expression programs during transdifferentiation of mesophyll cells into xylem cells. *Proc. Natl. Acad. Sci. USA* 99:15794–99**

24. Utilized a large cDNA array to examine transcriptional changes during TE differentiation in Zinnia.

25. Dhugga KS, Barreiro R, Whitten B, Stecca K, Hazebroek J, et al. 2004. Guar seed beta-mannan synthase is a member of the cellulose synthase super gene family. *Science* 303:363–66
26. Dixit R, Cyr R. 2004. The cortical microtubule array: from dynamics to organization. *Plant Cell* 16:2546–52
27. Ebringerova A, Heinze T. 2000. Xylan and xylan derivatives—biopolymers with valuable properties, 1—Naturally occurring xylans structures, procedures and properties. *Macromol. Rapid Commun.* 21:542–56
28. Ehlting J, Mattheus N, Aeschliman DS, Li EY, Hamberger B, et al. 2005. Global transcript profiling of primary stems from *Arabidopsis thaliana* identifies candidate genes for missing links in lignin biosynthesis and transcriptional regulators of fiber differentiation. *Plant J.* 42:618–40
29. Falconer MM, Seagull RW. 1985. Xylogenesis in tissue-culture—Taxol effects on microtubule reorientation and lateral association in differentiating cells. *Protoplasma* 128:157–66

30. Falconer MM, Seagull RW. 1986. Xylogenesis in tissue-culture. 2. Microtubules, cell-shape and secondary wall patterns. *Protoplasma* 133:140–48
31. Fukuda H. 1996. Xylogenesis: initiation, progression, and cell death. *Annu. Rev. Plant Physiol. Plant Molec. Biol.* 47:299–325
32. Funk V, Kositsup B, Zhao C, Beers EP. 2002. The *Arabidopsis* xylem peptidase XCP1 is a tracheary element vacuolar protein that may be a papain ortholog. *Plant Physiol.* 128:84–94
33. Gabaldon C, Ros LVG, Pedreno MA, Barcelo AR. 2005. Nitric oxide production by the differentiating xylem of *Zinnia elegans*. *New Phytol.* 165:121–30
34. Gardiner JC, Taylor NG, Turner SR. 2003. Control of cellulose synthase complex localization in developing xylem. *Plant Cell* 15:1740–48
35. Goicoechea M, Lacombe E, Legay S, Mihaljevic S, Rech P, et al. 2005. EgMYB2, a new transcriptional activator from Eucalyptus xylem, regulates secondary cell wall formation and lignin biosynthesis. *Plant J.* 43:553–67
36. Gray-Mitsumune M, Mellerowicz EJ, Abe H, Schrader J, Winzell A, et al. 2004. Expansins abundant in secondary xylem belong to subgroup a of the alpha-expansin gene family (1 w). *Plant Physiol.* 135:1552–64
37. Green KA, Prigge MJ, Katzman RB, Clark SE. 2005. CORONA, a member of the class III homeodomain leucine zipper gene family in arabidopsis, regulates stem cell specification and organogenesis. *Plant Cell* 17:691–704
38. Groover A, DeWitt N, Heidel A, Jones A. 1997. Programmed cell death of plant tracheary elements: differentiating in vitro. *Protoplasma* 196:197–211
39. Groover A, Jones AM. 1999. Tracheary element differentiation uses a novel mechanism coordinating programmed cell death and secondary cell wall synthesis. *Plant Physiol.* 119:375–84
40. Gustafson AM, Allen E, Givan S, Smith D, Carrington JC, Kasschau KD. 2005. ASRP: the *Arabidopsis* Small RNA Project Database. *Nucleic Acids Res.* 33:D637–40
41. Ha MA, MacKinnon IM, Sturcova A, Apperley DC, McCann MC, et al. 2002. Structure of cellulose-deficient secondary cell walls from the irx3 mutant of *Arabidopsis thaliana*. *Phytochemistry* 61:7–14
42. Handford MG, Baldwin TC, Goubet F, Prime TA, Miles J, et al. 2003. Localisation and characterization of cell wall mannan polysaccharides in *Arabidopsis thaliana*. *Planta* 218:27–36
43. Hara-Nishimura I, Hatsugai N, Nakaune S, Kuroyanagi M, Nishimura M. 2005. Vacuolar processing enzyme: an executor of plant cell death. *Curr. Opin. Plant Biol.* 8:404–8
44. Hatsugai N, Kuroyanagi M, Yamada K, Meshi T, Tsuda S, et al. 2004. A plant vacuolar protease, VPE, mediates virus-induced hypersensitive cell death. *Science* 305:855–58
45. Hertzberg M, Aspeborg H, Schrader J, Andersson A, Erlandsson R, et al. 2001. A transcriptional roadmap to wood formation. *Proc. Natl. Acad. Sci. USA* 98:14732–37
46. Hogetsu T. 1991. Mechanism for formation of the secondary wall thickening in tracheary elements—microtubules and microfibrils of tracheary elements of *Pisum-Sativum* L and *Commelina-Communis* L and the effects of Amiprophosmethyl. *Planta* 185:190–200
47. Hosokawa M, Suzuki S, Umezawa T, Sato Y. 2001. Progress of lignification mediated by intercellular transportation of monolignols during tracheary element differentiation of isolated Zinnia mesophyll cells. *Plant Cell Physiol.* 42:959–68
48. Hu Y, Zhong RQ, Morrison WH, Ye ZH. 2003. The *Arabidopsis* RHD3 gene is required for cell wall biosynthesis and actin organization. *Planta* 217:912–21
49. Im KH, Cosgrove DJ, Jones AM. 2000. Subcellular localization of expansin mRNA in xylem cells. *Plant Physiol.* 123:463–70

50. Inoue T, Higuchi M, Hashimoto Y, Seki M, Kobayashi M, et al. 2001. Identification of CRE1 as a cytokinin receptor from *Arabidopsis*. *Nature* 409:1060–63
51. Israelsson M, Sundberg B, Moritz T. 2005. Tissue-specific localization of gibberellins and expression of gibberellin-biosynthetic and signaling genes in wood-forming tissues in aspen. *Plant J.* 44:494–504
52. Ito J, Fukuda H. 2002. ZEN1 is a key enzyme in the degradation of nuclear DNA during programmed cell death of tracheary elements. *Plant Cell* 14:3201–11
53. **Ito Y, Nakanomyo I, Motose H, Iwamoto K, Sawa S, et al. 2006. Dodeca-CLE peptides as suppressors of plant stem cell differentiation. *Science* 313:842–45**

53. Biochemical purification identified a 12 amino acid active component of a CLE peptide precursor defining a novel pathway that prevents TE differentiation.

54. Iwasaki T, Shibaoka H. 1991. Brassinosteroids act as regulators of tracheary-element differentiation in isolated Zinnia mesophyll-cells. *Plant Cell Physiol.* 32:1007–14
55. Jones-Rhoades MW, Bartel DP, Bartel B. 2006. MicroRNAs and their regulatory roles in plants. *Annu. Rev. Plant Biol.* 57:19–53
56. Kang J, Dengler N. 2002. Cell cycling frequency and expression of the homeobox gene ATHB-8 during leaf vein development in *Arabidopsis*. *Planta* 216:212–19
57. Kiba T, Aoki K, Sakakibara H, Mizuno T. 2004. *Arabidopsis* response regulator, ARR22, ectopic expression of which results in phenotypes similar to the wol cytokinin-receptor mutant. *Plant Cell Physiol.* 45:1063–77
58. Kim J, Jung JH, Reyes JL, Kim YS, Kim SY, et al. 2005. microRNA-directed cleavage of ATHB15 mRNA regulates vascular development in *Arabidopsis* inflorescence stems. *Plant J.* 42:84–94
59. Ko JH, Han KH. 2004. *Arabidopsis* whole-transcriptome profiling defines the features of coordinated regulations that occur during secondary growth. *Plant Mol. Biol.* 55:433–53
60. Kobayashi H, Fukuda H, Shibaoka H. 1987. Reorganization of actin-filaments associated with the differentiation of tracheary elements in Zinnia mesophyll-cells. *Protoplasma* 138:69–71
61. Kobayashi H, Fukuda H, Shibaoka H. 1988. Interrelation between the spatial disposition of actin-filaments and microtubules during the differentiation of tracheary elements in cultured Zinnia cells. *Protoplasma* 143:29–37
62. Kubo M, Udagawa M, Nishikubo N, Horiguchi G, Yamaguchi M, et al. 2005. Transcription switches for protoxylem and metaxylem vessel formation. *Genes Dev.* 19:1855–60
63. Kuriyama H. 1999. Loss of tonoplast integrity programmed in tracheary element differentiation. *Plant Physiol.* 121:763–74
64. Lehmann K, Hause B, Altmann D, Kock M. 2001. Tomato ribonuclease LX with the functional endoplasmic reticulum retention motif HDEF is expressed during programmed cell death processes, including xylem differentiation, germination, and senescence. *Plant Physiol.* 127:436–49
65. Leyser O. 2006. Dynamic integration of auxin transport and signaling. *Curr. Biol.* 16:R424–33
66. Li JM, Chory J. 1997. A putative leucine-rich repeat receptor kinase involved in brassinosteroid signal transduction. *Cell* 90:929–38
67. Li YH, Qian O, Zhou YH, Yan MX, Sun L, et al. 2003. BRITTLE CULM1, which encodes a COBRA-like protein, affects the mechanical properties of rice plants. *Plant Cell* 15:2020–31
68. Liepman AH, Wilkerson CG, Keegstra K. 2005. Expression of cellulose synthase-like (Csl) genes in insect cells reveals that CslA family members encode mannan synthases. *Proc. Natl. Acad. Sci. USA* 102:2221–26

69. Liu Y, Schiff M, Czymmek K, Talloczy Z, Levine B, Dinesh-Kumar SP. 2005. Autophagy regulates programmed cell death during the plant innate immune response. *Cell* 121:567–77

70. Lu SF, Sun YH, Shi R, Clark C, Li LG, Chiang VL. 2005. Novel and mechanical stress-responsive microRNAs in *Populus trichocarpa* that are absent from *Arabidopsis*. *Plant Cell* 17:2186–203

71. Mahonen AP, Bishopp A, Higuchi M, Nieminen KM, Kinoshita K, et al. 2006. Cytokinin signaling and its inhibitor AHP6 regulate cell fate during vascular development. *Science* 311:94–98

72. Mahonen AP, Bonke M, Kauppinen L, Riikonen M, Benfey PN, Helariutta Y. 2000. A novel two-component hybrid molecule regulates vascular morphogenesis of the *Arabidopsis* root. *Genes Dev.* 14:2938–43

73. Mallory AC, Bartel DP, Bartel B. 2005. MicroRNA-directed regulation of *Arabidopsis* AUXIN RESPONSE FACTOR17 is essential for proper development and modulates expression of early auxin response genes. *Plant Cell* 17:1360–75

74. Mao GJ, Buschmann H, Doonan JH, Lloyd CW. 2006. The role of MAP65-1 in microtubule bundling during Zinnia tracheary element formation. *J. Cell Sci.* 119:753–58

75. Matsubayashi Y, Takagi L, Omura N, Morita A, Sakagami Y. 1999. The endogenous sulfated pentapeptide phytosulfokine-alpha stimulates tracheary element differentiation of isolated mesophyll cells of zinnia. *Plant Physiol.* 120:1043–48

76. McCann MC, Stacey NJ, Roberts K. 2000. Targeted cell death in xylogenesis. In *Programmed Cell Death in Animals and Plants*, pp. 193–201. Oxford: Scientific Publishers Ltd.

77. Milioni D, Sado PE, Stacey NJ, Domingo C, Roberts K, McCann MC. 2001. Differential expression of cell-wall-related genes during the formation of tracheary elements in the Zinnia mesophyll cell system. *Plant Mol. Biol.* 47:221–38

78. Milioni D, Sado PE, Stacey NJ, Roberts K, McCann MC. 2002. Early gene expression associated with the commitment and differentiation of a plant tracheary element is revealed by cDNA-amplified fragment length polymorphism analysis. *Plant Cell* 14:2813–24

79. Mitsuda N, Seki M, Shinozaki K, Ohme-Takagi M. 2005. The NAC transcription factors NST1 and NST2 of *Arabidopsis* regulate secondary wall thickenings and are required for anther dehiscence. *Plant Cell* 17:2993–3006

80. Motose H, Fukuda H, Sugiyama M. 2001. Involvement of local intercellular communication in the differentiation of zinnia mesophyll cells into tracheary elements. *Planta* 213:121–31

81. Motose H, Sugiyama M, Fukuda H. 2004. A proteoglycan mediates inductive interaction during plant vascular development. *Nature* 429:873–78

82. Moyle R, Schrader J, Stenberg A, Olsson O, Saxena S, et al. 2002. Environmental and auxin regulation of wood formation involves members of the Aux/IAA gene family in hybrid Aspen. *Plant J.* 31:675–85

83. Nakamura A, Higuchi K, Goda H, Fujiwara MT, Sawa S, et al. 2003. Brassinolide induces IAA5, IAA19, and DR5, a synthetic auxin response element in arabidopsis, implying a cross talk point of brassinosteroid and auxin signaling. *Plant Physiol.* 133:1843–53

84. Noguchi T, Fujioka S, Takatsuto S, Sakurai A, Yoshida S, et al. 1999. *Arabidopsis* det2 is defective in the conversion of (24R)-24-methylcholest-4-En-3-One to (24R)-24-methyl-5 alpha-cholestan-3-one in brassinosteroid biosynthesis. *Plant Physiol.* 120:833–39

71. Identifies a novel component of the cytokinin signaling pathway and clearly demonstrates a role for cytokinin in maintaining procambial cell identity and suppressing xylem formation.

78. A highly synchronous Zinnia system was used to identify genes expression changes within 30 min of adding inductive media that starts TE differentiation.

81. Identifies and demonstrates the polar secretion of an AGP that promotes TE differentiation and appears to be required to maintain a continuous network of TEs.

85. Obara K, Kuriyama H, Fukuda H. 2001. Direct evidence of active and rapid nuclear degradation triggered by vacuole rupture during programmed cell death in Zinnia. *Plant Physiol.* 125:615–26
86. Oda Y, Hasezawa S. 2006. Cytoskeletal organization during xylem cell differentiation. *Int. J. Plant Sci.* 119:167–77
87. Oda Y, Mimura T, Hasezawa S. 2005. Regulation of secondary cell wall development by cortical microtubules during tracheary element differentiation in *Arabidopsis* cell suspensions. *Plant Physiol.* 137:1027–36
88. Oh S, Park S, Han KH. 2003. Transcriptional regulation of secondary growth in *Arabidopsis thaliana*. *J. Exp. Bot.* 54:2709–22
89. Ohashi-Ito K, Demura T, Fukuda H. 2002. Promotion of transcript accumulation of novel Zinnia immature xylem-specific HD-Zip III homeobox genes by brassinosteroids. *Plant Cell Physiol.* 43:1146–53
90. Ohashi-Ito K, Fukuda H. 2003. HD-Zip III homeobox genes that include a novel member, ZeHB/13 (Zinnia)-ATHB-15 (*Arabidopsis*), are involved in procambium and xylem cell differentiation. *Plant Cell Physiol.* 44:1350–58
91. Ohashi-Ito K, Kubo M, Demura T, Fukuda H. 2005. Class III homeodomain leucine-zipper proteins regulate xylem cell differentiation. *Plant Cell Physiol.* 46:1646–56
92. Patzlaff A, McInnis S, Courtenay A, Surman C, Newman LJ, et al. 2003. Characterisation of a pine MYB that regulates lignification. *Plant J.* 36:743–54
93. Persson S, Wei HR, Milne J, Page GP, Somerville CR. 2005. Identification of genes required for cellulose synthesis by regression analysis of public microarray data sets. *Proc. Natl. Acad. Sci. USA* 102:8633–38
94. Prigge MJ, Otsuga D, Alonso JM, Ecker JR, Drews GN, Clark SE. 2005. Class III homeodomain-leucine zipper gene family members have overlapping, antagonistic, and distinct roles in *Arabidopsis* development. *Plant Cell* 17:61–76
95. Rhoades MW, Reinhart BJ, Lim LP, Burge CB, Bartel B, Bartel DP. 2002. Prediction of plant microRNA targets. *Cell* 110:513–20
96. Ringli C, Keller B, Ryser U. 2001. Glycine-rich proteins as structural components of plant cell walls. *Cell. Mol. Life Sci.* 58:1430–41
97. Roberts AW, Donovan SG, Haigler CH. 1997. A secreted factor induces cell expansion and formation of metaxylem-like tracheary elements in xylogenic suspension cultures of Zinnia. *Plant Physiol.* 115:683–92
98. Roberts AW, Frost AO, Roberts EM, Haigler CH. 2004. Roles of microtubules and cellulose microfibril assembly in the localization of secondary-cell-wall deposition in developing tracheary elements. *Protoplasma* 224:217–29
99. Roberts AW, Uhnak KS. 1998. Tip growth in xylogenic suspension cultures of *Zinnia elegans* L.: implications for the relationship between cell shape and secondary-cell-wall pattern in tracheary elements. *Protoplasma* 204:103–13
100. Rotari VI, Rui H, Gallois P. 2005. Death by proteases in plants: whodunit. *Physiol. Plant* 123:376–85
101. Roudier F, Fernandez AG, Fujita M, Himmelspach R, Borner GHH, et al. 2005. COBRA, an *Arabidopsis* extracellular glycosyl-phosphatidyl inositol-anchored protein, specifically controls highly anisotropic expansion through its involvement in cellulose microfibril orientation. *Plant Cell* 17:1749–63
102. Roudier F, Schindelman G, DeSalle R, Benfey PN. 2002. The COBRA family of putative GPI-anchored proteins in *Arabidopsis*. A new fellowship in expansion. *Plant Physiol.* 130:538–48

87. Microtubule dynamics in *Arabidopsis* cell cultures differentiating into TEs were examined using a fluorescent MT report showing that bands of TEs split into at least two distinct groups either side of the growing secondary cell wall.

103. Ryser U, Schorderet M, Zhao GF, Studer D, Ruel K, et al. 1997. Structural cell-wall proteins in protoxylem development: evidence for a repair process mediated by a glycine-rich protein. *Plant J.* 12:97–111
104. Sachs T. 1991. Cell polarity and tissue patterning in plants. *Development* 91(Suppl.):83–93
105. Sato Y, Demura T, Yamawaki K, Inoue Y, Sato S, et al. 2006. Isolation and characterization of a novel peroxidase gene ZPO-C whose expression and function are closely associated with lignification during tracheary element differentiation. *Plant Cell Physiol.* 47:493–503
106. Sawa S, Demura T, Horiguchi G, Kubo M, Fukuda H. 2005. The ATE genes are responsible for repression of trans differentiation into xylem cells in *Arabidopsis*. *Plant Physiol.* 137:141–48
107. Scarpella E, Rueb S, Boot KJM, Hoge JHC, Meijer AH. 2000. A role for the rice homeobox gene Oshox1 in provascular cell fate commitment. *Development* 127:3655–69
108. Scheres B, Dilaurenzio L, Willemsen V, Hauser MT, Janmaat K, et al. 1995. Mutations affecting the radial organization of the *Arabidopsis* root display specific defects throughout the embryonic axis. *Development* 121:53–62
109. Schneider B, Herth W. 1986. Distribution of plasma-membrane rosettes and kinetics of cellulose formation in xylem development of higher-plants. *Protoplasma* 131:142–52
110. Schrader J, Nilsson J, Mellerowicz E, Berglund A, Nilsson P, et al. 2004. A high-resolution transcript profile across the wood-forming meristem of poplar identifies potential regulators of cambial stem cell identity. *Plant Cell* 16:2278–92
111. Sessa G, Steindler C, Morelli G, Ruberti I. 1998. The *Arabidopsis* Athb-8, -9 and -14 genes are members of a small gene family coding for highly related HD-ZIP proteins. *Plant Mol. Biol.* 38:609–22
112. Sieburth LE, Deyholos MK. 2006. Vascular development: the long and winding road. *Curr. Opin. Plant Biol.* 9:48–54
113. Steiner-Lange S, Unte US, Eckstein L, Yang CY, Wilson ZA, et al. 2003. Disruption of *Arabidopsis thaliana* MYB26 results in male sterility due to nondehiscent anthers. *Plant J.* 34:519–28
114. Szekeres M, Nemeth K, Koncz-Kalman Z, Mathur J, Kauschmann A, et al. 1996. Brassinosteroids rescue the deficiency of CYP90, a cytochrome P450, controlling cell elongation and de-etiolation in arabidopsis. *Cell* 85:171–82
115. Szyjanowicz PMJ, McKinnon I, Taylor NG, Gardiner J, Jarvis MC, Turner SR. 2004. The irregular xylem 2 mutant is an allele of korrigan that affects the secondary cell wall of *Arabidopsis thaliana*. *Plant J.* 37:730–40
116. Tamagnone L, Merida A, Parr A, Mackay S, Culianez-Macia FA, et al. 1998. The AmMYB308 and AmMYB330 transcription factors from antirrhinum regulate phenylpropanoid and lignin biosynthesis in transgenic tobacco. *Plant Cell* 10:135–54
117. Taylor JG, Owen TP, Koonce LT, Haigler CH. 1992. Dispersed lignin in tracheary elements treated with cellulose synthesis inhibitors provides evidence that molecules of the secondary cell-wall mediate wall patterning. *Plant J.* 2:959–70
118. Taylor NG, Howells RM, Huttly AK, Vickers K, Turner SR. 2003. Interactions among three distinct CesA proteins essential for cellulose synthesis. *Proc. Natl. Acad. Sci. USA* 100:1450–55
119. Thompson AR, Doelling JH, Suttangkakul A, Vierstra RD. 2005. Autophagic nutrient recycling in *Arabidopsis* directed by the ATG8 and ATG12 conjugation pathways. *Plant Physiol.* 138:2097–110
120. Tokunaga N, Sakakibara N, Umezawa T, Ito Y, Fukuda H, Sato Y. 2005. Involvement of extracellular dilignols in lignification during tracheary element differentiation of isolated Zinnia mesophyll cells. *Plant Cell Physiol.* 46:224–32

110. Exploits the advantages of studying xylem differentiation using wood to generate a very high-resolution profile of gene expression across the poplar cambium.

121. Turner SR, Hall M. 2000. The gapped xylem mutant identifies a common regulatory step in secondary cell wall deposition. *Plant J.* 24:477–88
122. Turner SR, Somerville CR. 1997. Collapsed xylem phenotype of *Arabidopsis* identifies mutants deficient in cellulose deposition in the secondary cell wall. *Plant Cell* 9:689–701
123. Uggla C, Moritz T, Sandberg G, Sundberg B. 1996. Auxin as a positional signal in pattern formation in plants. *Proc. Natl. Acad. Sci. USA* 93:9282–86
124. Vercammen D, van de Cotte B, De Jaeger G, Eeckhout D, Casteels P, et al. 2004. Type II metacaspases Atmc4 and Atmc9 of *Arabidopsis thaliana* cleave substrates after arginine and lysine. *J. Biol. Chem.* 279:45329–36
125. Wang JW, Wang LJ, Mao YB, Cai WJ, Xue HW, Chen XY. 2005. Control of root cap formation by microRNA-targeted auxin response factors in *Arabidopsis*. *Plant Cell* 17:2204–16
126. Weir IE, Maddumage R, Allan AC, Ferguson IB. 2005. Flow cytometric analysis of tracheary element differentiation in *Zinnia elegans* cells. *Cytometry Part A* 68A:81–91
127. Yamamoto R, Demura T, Fukuda H. 1997. Brassinosteroids induce entry into the final stage of tracheary element differentiation in cultured Zinnia cells. *Plant Cell Physiol.* 38:980–83
128. Yamamoto R, Fujioka S, Demura T, Takatsuto S, Yoshida S, Fukuda H. 2001. Brassinosteroid levels increase drastically prior to morphogenesis of tracheary elements. *Plant Physiol.* 125:556–63
129. Ye ZH. 2002. Vascular tissue differentiation and pattern formation in plants. *Annu. Rev. Plant Biol.* 53:183–202
130. Yoshida S, Kuriyama H, Fukuda H. 2005. Inhibition of transdifferentiation into tracheary elements by polar auxin transport inhibitors through intracellular auxin depletion. *Plant Cell Physiol.* 46:2019–28
131. Yu XH, Perdue TD, Heimer YM, Jones AM. 2002. Mitochondrial involvement in tracheary element programmed cell death. *Cell Death Differ.* 9:189–98
132. Zhao CS, Craig JC, Petzold HE, Dickerman AW, Beers EP. 2005. The xylem and phloem transcriptomes from secondary tissues of the *Arabidopsis* root-hypocotyl. *Plant Physiol.* 138:803–18
133. Zhong RQ, Burk DH, Morrison WH, Ye ZH. 2002. A kinesin-like protein is essential for oriented deposition of cellulose microfibrils and cell wall strength. *Plant Cell* 14:3101–17
134. Zhong RQ, Burk DH, Morrison WH, Ye ZH. 2004. FRAGILE FIBER3, an *Arabidopsis* gene encoding a type II inositol polyphosphate 5-phosphatase, is required for secondary wall synthesis and actin organization in fiber cells. *Plant Cell* 16:3242–59
135. Zhong RQ, Burk DH, Nairn CJ, Wood-Jones A, Morrison WH, Ye ZH. 2005. Mutation of SAC1, an *Arabidopsis* SAC domain phosphoinositide phosphatase, causes alterations in cell morphogenesis, cell wall synthesis, and actin organization. *Plant Cell* 17:1449–66

RELATED RESOURCES

Ehrhardt DW, Shaw SL. 2006. Microtubule dynamics and organization in the plant cortical array. *Annu. Rev. Plant Biol.* 57:859–75

Hussey PJ, Ketelaar T, Deeks MJ. 2006. Control of the actin cytoskeleton in plant cell growth. *Annu. Rev. Plant Biol.* 57:109–25

Jansson S, Tuskan G, Douglas CJ. 2007. *Populus*: a model system for plant biology. *Annu. Rev. Plant Biol.* 58:435–58

Populus: A Model System for Plant Biology

Stefan Jansson[1] and Carl J. Douglas[2]

[1]Department of Plant Physiology, Umeå Plant Science Center, Umeå University, SE-901 87 Umeå, Sweden; email: stefan.jansson@plantphys.umu.se

[2]Department of Botany, University of British Columbia, Vancouver BC V6T 1Z4, Canada; email: cdouglas@interchange.ubc.ca

Annu. Rev. Plant Biol. 2007. 58:435–58

The *Annual Review of Plant Biology* is online at plant.annualreviews.org

This article's doi:
10.1146/annurev.arplant.58.032806.103956

First published online as a Review in Advance on February 5, 2007

1543-5008/07/0602-0435$20.00

Key Words

genomics, wood formation, seasonality, flowering, biotic interactions, natural variation

Abstract

With the completion of the *Populus trichocarpa* genome sequence and the development of various genetic, genomic, and biochemical tools, *Populus* now offers many possibilities to study questions that cannot be as easily addressed in *Arabidopsis* and rice, the two prime model systems of plant biology and genomics. Tree-specific traits such as wood formation, long-term perennial growth, and seasonality are obvious areas of research, but research in other areas such as control of flowering, biotic interactions, and evolution of adaptive traits is enriched by adding a tree to the suite of model systems. Furthermore, the reproductive biology of *Populus* (a dioeceous wind-pollinated long-lived tree) offers both new possibilities and challenges in the study and analysis of natural genetic and phenotypic variation. The relatively close phylogenetic relationship of *Populus* to *Arabidopsis* in the Eurosid clade of Eudicotyledonous plants aids in comparative functional studies and comparative genomics, and has the potential to greatly facilitate studies on genome and gene family evolution in eudicots.

Contents

INTRODUCTION: WHY DO WE NEED A TREE MODEL SYSTEM?

During the last two decades, researchers have been astonished to see the very high level of conservation of biological function in all living organisms. For example, a long list of genes known to cause human heritable diseases have very close homologs in the genome of *Arabidopsis thaliana* (3). Nevertheless, the striking differences in appearance and physiology of different species show that a single model system cannot be used to answer all biological questions. In plants, *Arabidopsis* has been adopted as the prime model system and an impressive number of tools and techniques are now available to understand gene function in this plant. *Arabidopsis* was chosen as a model species for obvious reasons: small physical size, rapid generation time, straightforward genetics, high fecundity, and small genome size. However, *Arabidopsis* is, in many respects, an unusual plant. As an almost obligate inbreeder, heterozygosity has been reduced to a minimum. Although this greatly facilitates functional studies, most plant species have different reproductive strategies, making *Arabidopsis* a genetic extreme. The very accelerated life cycle of *Arabidopsis* also makes many traits that are essential in many (or most) plants unimportant in *Arabidopsis*. Two obvious examples are wood formation and seasonality of growth.

Physiologically and genetically, in many respects trees represent an opposite extreme to *Arabidopsis* in the spectrum of land plants, with long life spans and generation times, and woody perennial growth habits. However, trees do not form a monophylogenetic group but are found among many higher plant genera and families, and have arisen multiple times during land plant evolution. *Populus* is found in the angiosperm Euroside I clade together with *Arabidopsis*. Thus, *Arabidopsis* is more related to *Populus* than to the vast majority of other dicot taxa including those with trees, not to mention monocots like rice or gymnosperm trees such as conifers, lineages that separated from the eudicots long before the radiation of eudicot families 100–120 million years ago (mya) (**Figure 1**). Thus, *Populus*, although relatively closely related to *Arabidopsis*, offers a new model system to study an expanded repertoire of biological processes that better represent the breadth of plant biology.

The development of *Populus* as a model system for tree and woody perennial plant biology has been largely driven by the rapid development of genomic and molecular biology resources for this genus, as discussed below, culminating in the completion of a draft sequence of the *Populus trichocarpa* (black cottonwood) genome (106). This information will facilitate studies on the comparative biology of *Arabidopsis* and *Populus* as representatives of two angiosperm extremes,

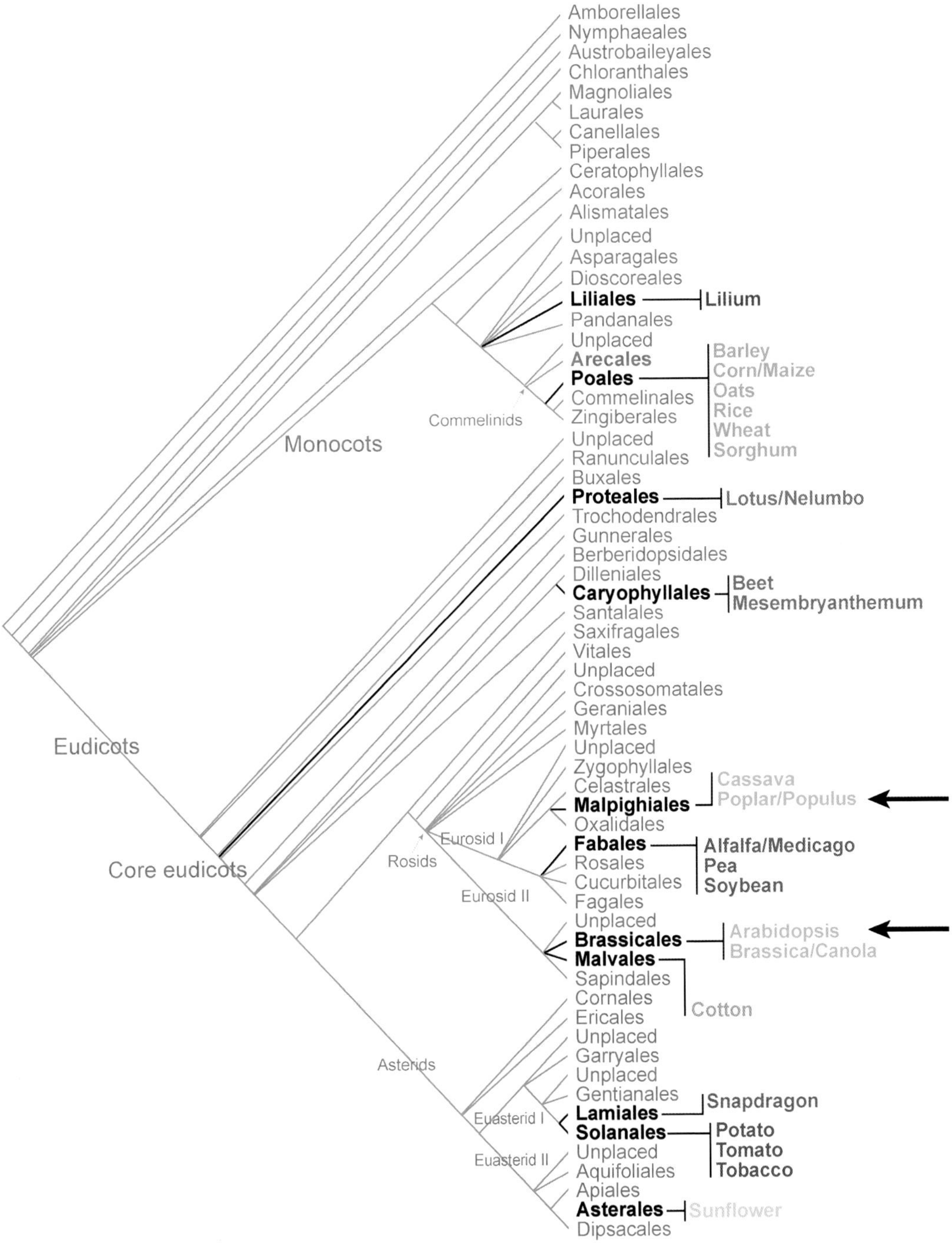

Figure 1

Angiosperm phylogeny showing the Eurosid clade containing *Populus* and *Arabidopsis*, relative to other species with significant sequence information (highlighted in color). Data and images from P.F. Stevens, Angiosperm Phylogeny Web site: **http://www.mobot.org/MOBOT/research/APweb/** (Version 7, May 2006)

Leaf senescence: the yellowing of leaves; occurs in the autumn in trees

Whole-genome shotgun sequencing: random sequencing of genomic libraries at high redundancy followed by computer-aided sequence contig assembly

BAC: bacterial artificial chromosome

enabling discovery of mechanisms conserved among eudicots. Perhaps even more importantly, such studies should reveal how different lineages, subject to different selection pressures, have used common cellular machinery in different ways. This is illustrated by recent studies on the CONSTANS(CO)/FLOWERING LOCUS T (FT) regulatory module. In *Arabidopsis*, this pathway regulates the photoperiod-dependent induction of flowering (55). In *Populus*, however, this module regulates not only flowering but also bud set in the late season (9), a process absent in *Arabidopsis*. Obviously, in the *Populus* lineage, the signal provided by the photoperiod-dependent oscillations of *CO* and *FT* mRNAs has been co-opted to regulate two different developmental pathways, whereas *Arabidopsis* only uses one of these. This example illustrates the power of comparative biology to provide important insights into the evolution of signal transduction pathways, insights that would be less obvious given a single model. These kinds of studies can be used to better understand the marvelous phenotypic and functional diversification of higher plants. They also shed light on the evolution of traits that has enabled plants to colonize most terrestrial ecosystems, making the existence of heterotrophic organisms like humans not only possible but also beautiful.

WHAT TOOLS DO WE HAVE TO STUDY *POPULUS* BIOLOGY?

DNA Sequence and Physical Map

The major sets of tools and resources that have propelled *Populus* into the set of model plants are the genomics resources. The first set of expressed sequence tags (ESTs), 5692 sequences from libraries made from wood-forming tissues, was published in 1998 (97). Since then, numerous groups have contributed to EST sequencing efforts, using multiple species or hybrids, organs, tissues, and treatments, with 376,565 *Populus* sequences in the National Center for Biotechnology Information (NCBI) EST database as of August 2006. These ESTs represent sequences not only from an expanded effort to capture expressed genes in wood-forming tissues (20, 82, 96), but also includes sequences from, for example, abiotic stress treatments (11, 69), plants and cell cultures subjected to biotic stresses such as herbivory, elicitor treatment, and systemic wounding (17, 82), and leaves undergoing autumnal leaf senescence (6). From an analysis of 102,019 ESTs, Sterky et al. (96) identified 24,644 unique sequences, concluding that the high degree of similarity between *Arabidopsis* and *Populus* genes would facilitate comparative genomics approaches. Furthermore, these ESTs, as well as those provided by other investigators and a set of over 4000 full-length cDNA sequences (106) greatly facilitated annotation of the genome, and provide relative expression data ("digital northern") for a large subset of *Populus* genes (96).

For sequencing of the genome, a *P. trichocarpa* female individual ("Nisqually-1"), previously used in breeding programs, was chosen (106). The strategy employed whole-genome shotgun sequencing of small insert libraries, generating 4.2×10^9 high-quality (Phred $>$ 20) nucleotides, or a depth of coverage of approximately 8.5 X of the 480 Mb *Populus* genome (106). In parallel, a physical map derived from a 50,000-clone Nisqually-1 bacterial artificial chromosome (BAC) library was constructed by restriction enzyme fragment fingerprinting (51a) (**http://www.bcgsc.ca/platform/mapping/data**). Including paired-end reads [BAC end sequences (BES)] in the data set aided large-scale assembly of the genome into 2447 major scaffolds (410 Mb), and integration of the physical map with the genome sequence. Finally, polymorphic microsatellite genetic markers anchored 335 Mb of genome sequence to the 19 *Populus* linkage groups. Annotation of the genome assembly was carried out using a diversity of ab initio gene calling programs, aided by *Populus* EST and full-length cDNA sequences, and the *Arabidopsis* genome sequence, together with manual annotation for selected gene families, yielding 45,555 promoted gene models (**http://genome.jgi-psf.org/Poptr1_1/Poptr1_1.home.html**), the largest for any completely sequenced plant genome to date.

DNA Microarrays, Proteomics, and Metabolomics

Functional genomics tools such as DNA microarrays have been developed in parallel with EST and genome sequencing. These include a 25,000-element cDNA array developed in Sweden (**http://www.populus.db.umu.se/project.html**), a 27,000-element PICME cDNA array developed in France and available for purchase (**http://www.picme.at/**), and a 15,400-element Treenomix cDNA array developed in Canada (82). In addition, subsequent to release of the *Populus* genome sequence, *Populus* short oligomer-based full-genome arrays or array probes have been or are being commercially developed by Affymetix, Inc., NimbleGen, Inc., and Operon, Inc. A publicly accessible database containing *Populus* DNA microarray data was recently opened (93; **http://www.upscbase.db.umu.se**).

Generation of proteomic data is in its infancy in *Populus*, but annotation of 45,555 gene models from the genome sequence should add efficiency of in silico protein predictions from mass spectrometry data. Du et al. (25) generated a data set of 244 proteins that accumulate during *P. tomentosa* wood formation, Renaut et al. (83) used proteomics to identify proteins associated with cold acclimation in the stem, and Plomion et al. (78a) cataloged more than 300 proteins expressed in different tissues and in response to drought, largely by LC-MS/MS of spots excised from two-dimensional protein gels. Similarly, although the complex biochemistry of *Populus* provides interesting opportunities for comparing biochemical phenotypes with genotype, environment, and development, published metabolomic studies are few. However, Andersson-Gunnerås et al. (2) monitored global changes in metabolites—by gas chromatography/time-of-flight mass spectrometry—and transcript abundance during tension wood formation, allowing models for reprogramming of carbohydrate metabolism to be developed. Morreel et al. (67) used a genetical metabolomics approach to identify quantitative trait loci (QTL) that control flux into the complex set of flavonoids in a *Populus* pedigree. This work illustrates the potential for profiling natural variation in chemical defense compounds to identify regulatory loci. Finally, metabolic profiling of transgenic *Populus* lines altered in lignin composition revealed unexpected changes in carbohydrates and other metabolites relative to wild-type lines (84), demonstrating the potential for metabolic profiling to reveal biochemical phenotypes and discriminate between closely related individuals.

QTL: quantitative trait locus/loci

Genetic Tools and Transformation

Genetic resources have been developed in *Populus* over decades, both in the way of F_1 and F_2 populations, often derived from interspecific crosses and backcrosses, and collections of genetically diverse wild genotypes. Examples include family 331, a three-generation pedigree derived from a *P. trichocarpa* female parent and a *P. deltoides* male parent (28), and F_1 family 545 (99), derived from a cross between *P. trichocarpa* Nisqually-1 (clone 383–2499) and a *P. deltoides* male parent. These and other similar populations have allowed genetic analysis of Mendelian and quantitative trait loci.

Genetic transformation, along with the ability to perform reverse genetic analyses, is key to functional studies. Certain, but not all, *Populus* genotypes can be subject to *Agrobacterium*-mediated transformation and overexpression or RNAi-mediated downregulation of target genes, as reviewed by Busov et al. (14). Naylor et al. (70) recently demonstrated that viral-induced

Salicoid duplication: a whole genome duplication event that occurred near the emergence of *Salix* and *Populus* lineages 60–65 mya

MAPK: mitogen-activated protein kinase

gene silencing (VIGS) might be possible in *Populus* using PopMV as a vector. Despite logistical challenges in working with transgenic trees that require large amounts of greenhouse or field space, activation tagging approaches have been successfully employed in *Populus* (15) for gene identification based on mutant phenotype, and generation of insertionally mutagenized populations hold great promise in the future (14).

WHAT CAN WE DO WITH A SECOND DICOT GENOME SEQUENCE?

The availability of a second eudicot genome sequence has the potential to modify the *Arabidopsis*-centric view of plant biology. Genomic studies in *Populus* can give insights into plant genome evolution, gene family structure, and certain important biological processes poorly developed in *Arabidopsis*. Some of these processes are discussed below. In terms of genome evolution, analysis of both ESTs (95) and the whole genome sequence (106) has revealed a rather recent whole genome duplication (salicoid duplication) that appears to have coincided with the emergence of the *Populus* and *Salix* genera. Most interestingly, the rate of gene evolution seems to be much slower in *Populus* than in *Arabidopsis*, since the molecular clock places the time of gene duplication at 8–13 mya, in conflict with fossil evidence for *Populus* 60–65 mya (106). The greatly decelerated molecular clock in *Populus* is perhaps a consequence of its long generation time and contribution of "ancient gametes" from long-lived *Populus* genotypes that can clonally propagate (106), and suggests that *Populus* genome reorganization following the whole genome duplication is a dynamic process that is still continuing. Gene loss after the salicoid genome duplication has been smaller than gene loss following a previous whole genome duplication shared by *Arabidopsis* and *Populus*, so a single-copy gene in *Arabidopsis* is typically represented by two copies in *Populus*. Most segments of the *Populus* genome mapped to 1 of 19 linkage groups have a paralagous segment located on one or more other linkage groups, highlighting the highly duplicated nature of the genome (**Figure 2**). However, some gene families like F-box proteins have been much more expanded in *Arabidopsis* relative to *Populus*, and gene family expansions by tandem duplication may be less common than in *Arabidopsis* (106).

By examining differences in gene family structure between *Arabidopsis* and *Populus*, eudicot gene family evolution and its consequences for plant function can now be investigated in more detail. Differences in retention and loss of duplicated gene family members in the two species will help to reveal genes important in common eudicot functions, as well as genes that may have specialized roles in the two lineages, studies that would be impossible without a second genome to compare to the *Arabidopsis* reference genome. "Classical" gene families like those encoding expansins (89), MADS-box (58), Lhc (CAB) proteins (53), and mitogen-activated protein kinases (MAPKs) (36) have been reanalyzed incorporating genome data, and large and complex families like carbohydrate-active enzymes (CAZymes) have been attacked (29). Not surprisingly, newly arisen gene duplicates may have different evolutionary rates (102). Many duplicate genes in the *Populus* genome have undergone subfunctionalization in expression, rationalizing their retention (106), and promotor sequences have been analyzed by phylogenetic footprinting to yield information about redundancy/neofunctionalization/subfunctionalization (19). The *Populus* genome has also enabled comparative analysis of less well-studied elements such as microRNAs (miRNAs) and their potential targets at the whole genome level. Although the number of miRNA families is conserved in *Populus* and *Arabidopsis*, the number of miRNAs and miRNA targets is almost twofold higher in *Populus* (106), and it has been suggested that one function of an

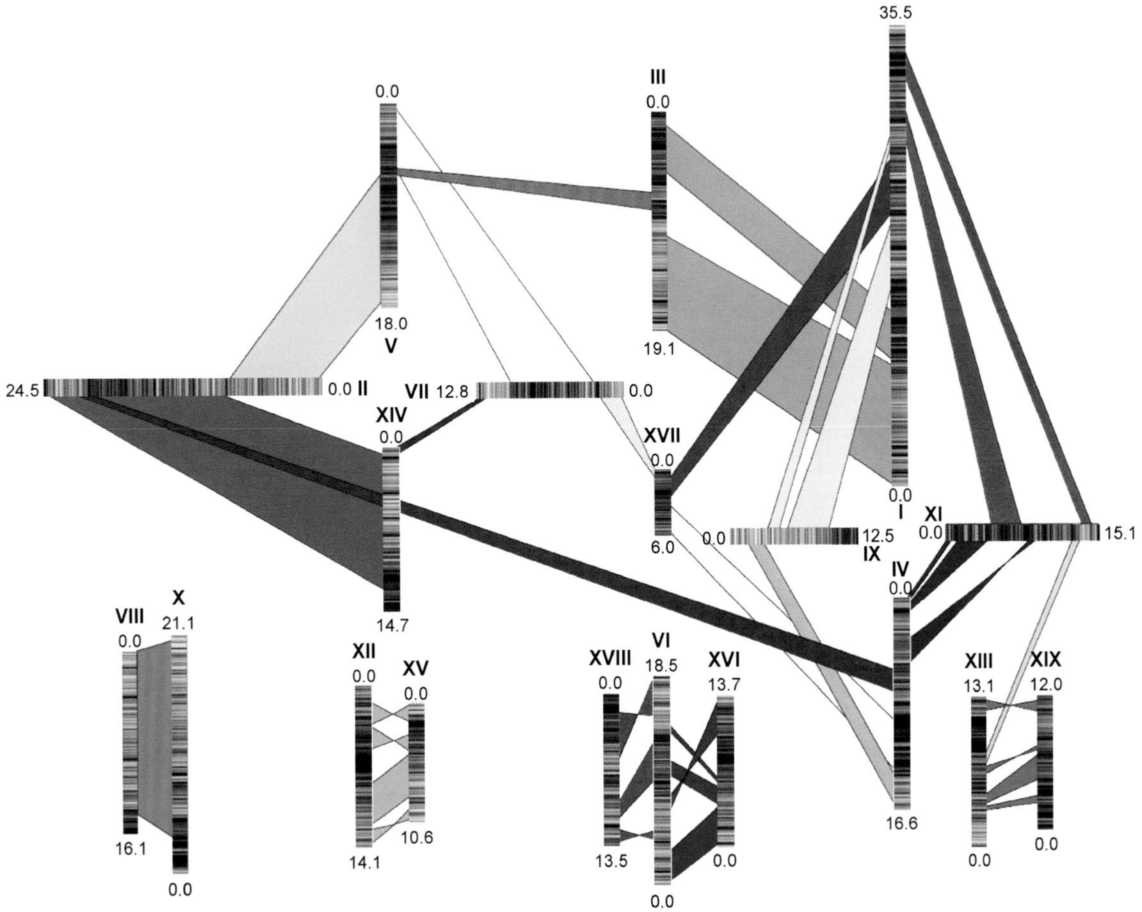

Figure 2

The 19 linkage groups (LGs) of *Populus trichocarpa*. Regions connected by colored bands are homologous and retained since the salicoid genome duplication. Genes on LG annotated as those with high (*green*) and low (*red*) expressed sequence tag (EST) support. Three LG pairs have been retained almost intact after the duplication; LG VI is a fusion between LG XVI and XVIII. The remaining 10 LGs have undergone more extensive, and complex, reorganization, as described by Tuskan et al. (106).

expanded miRNA repertoire in *Populus* is related to wood development (60).

WHAT CAN WE STUDY BETTER IN *POPULUS* THAN IN *ARABIDOPSIS*?

Although annuals have been essential in understanding plant development, responses to biotic and abiotic stimuli, and trait evolution, the biology of *Populus* provides an opportunity to study processes for which herbaceous models are less well suited. Below, we summarize five such general areas of investigation.

Wood Development

One defining characteristic of vascular land plants is the ability to undergo secondary growth. The evolution of the secondary growth habit in the eudicots, and aspects of the regulation of vascular cambium and secondary xylem formation, was recently reviewed by Groover (34). It is clear that

Secondary growth and cambium: radial growth brought about by the vascular cambium, a cylindrical meristem in stems

Complementary DNA (cDNA)-AFLP: amplified fragment length polymorphism

GA: gibberellic acid

secondary growth has evolved multiple independent times in eudicot lineages, and even *Arabidopsis* can be induced to undergo secondary growth in certain conditions (16). Although this strongly suggests conservation of at least some key gene functions, secondary growth and the seasonal regulation of vascular cambium activity and wood formation are still defining characteristics of the life history of trees. *Populus* has provided excellent material for experiments on xylem development and functioning of the vascular cambium (reviewed in 64) and exciting new experimental avenues are now available.

Formation and functioning of the vascular cambium. The transition from primary to secondary growth takes place near the shoot apex; secondary growth is evident within the first three internodes of the uppermost internode that can be isolated. Prassinos et al. (80) identified 271 differentially regulated transcript-derived fragments (TDFs) and corresponding genes by complementary DNA-amplified fragment length polymorphism (cDNA-AFLP), many of which were differentially regulated in the transition to secondary growth, and van Raemdonck et al. (107) identified a similar number. Unsurprisingly, genes involved in secondary wall formation and lignin biosynthesis were prominent in both sets, but several intriguing transcription factors and other potential signaling intermediates were associated with transition to secondary growth. Among these candidates, *PtaRHE1* has an apparent cell-type-specific expression pattern in ray cell initials of the developing vascular cambium (107).

Once the vascular cambium is established, it can remain active for decades or centuries, and is likely to be governed by internal regulatory circuitry that may be analogous to that in apical meristems (34). A key breakthrough in the field was microarray-based expression profiling over the course of wood development after isolation of cells at specific developmental stages by tangential cryo-sectioning (40). Although other techniques for isolation of cells at discrete stages of wood development, such as staged sampling following reactivation of the cambium in the spring (42) and laser capture microdissection (71), also hold promise, this approach gave the first glimpse into global expression patterns that accompany wood development, revealing, for example, the coordinated expression of putative cyclin genes involved in cell cycle control in the zone of cell division within the vascular cambium. Capitalizing on rapid advances in the development of microarray resources, Schrader et al. (91) used a much larger cDNA microarray and refined sampling strategies to probe patterns of gene expression at higher resolution, specifically in the cambial zone. Of particular note is the evidence, based on coordinated patterns of gene expression, of distinct subdomains within the 60-μm cambial zone, including phloem mother cells, cambial stem cells, and xylem mother cells. From a functional point of view, analysis of gene expression patterns at this resolution revealed spatially distinct expression patterns of homologs encoding known regulators of the apical meristem function, such as receptor-like kinases of the CLAVATA class (PttCLV1 and PttRLK3), and regulatory transcription factors (WUS, ANT) (**Figure 3**). Based on the opposing expression patterns of *PttCLV1* (phloem side) and *PttRLK3* (xylem side), an attractive hypothesis is that regulatory circuits controlled by these receptors independently control the formation of secondary xylem and phloem, allowing for flexibility in the relative amounts of these tissues formed in accordance with environment and seasonality (91) (**Figure 3**). Based on their expression patterns, *WUS*-like *PttHB3* and *PttANT* may also be part of feedback regulatory circuits controlling cell proliferation at the vascular cambium.

Secondary xylem development. Of particular interest in wood formation is the regulation of fiber cell length, a trait of commercial interest in papermaking. Eriksson et al. (26) showed that gibberellins (GAs) play a key role

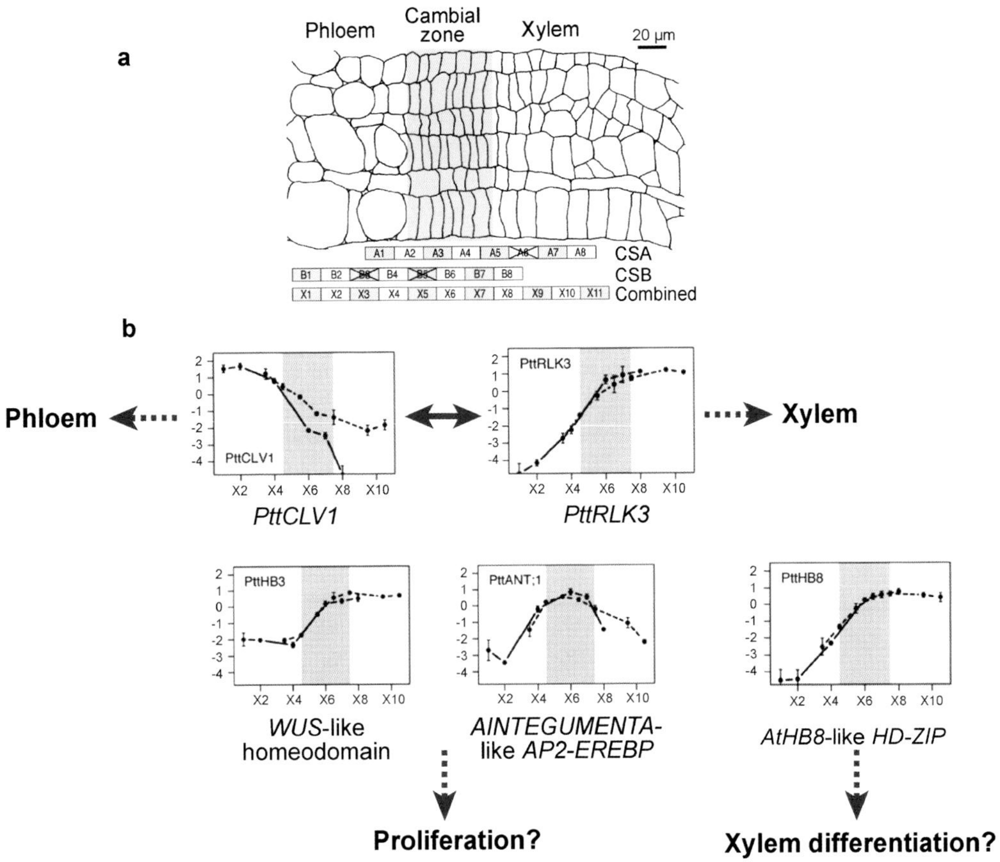

Figure 3

Model for control of xylem and phloem differentiation and cell proliferation at the *Populus* vascular cambium, based on global gene expression profiling across the vascular cambium. Data taken and images modified from Schrader et al. (91), copyright American Society of Plant Biologists. (*a*) Tangential cyro-sectioning strategy across the cambial zone (shown in *gray*), showing tissue used for RNA isolation. (*b*) Microarray expression profiles (in relative units) for selected receptor-like kinase (RLK) and transcription factor genes. Expression patterns suggest antagonistic action of RLKs PttCLV1 and PttRLK3 in phloem and xylem differentiation, and possible roles for *Populus WUS*-like and *AINTEGUMANTA*-like transcription factors in cell proliferation at the cambium, and a *Populus AtHB-8*-like transcription factor in secondary xylem differentiation.

in this process by overexpressing an *Arabidopsis* GA20 oxidase gene in transgenic hybrid aspen; 20-fold higher levels of bioactive GA_1 and GA_4 gave significantly enhanced height growth and xylem fiber length. Endogenous GA_1 and GA_4 levels peak sharply in the zone of xylem cell expansion and elongation immediately adjacent to the vascular cambium (45, 46). This supports a model whereby GA_1 and GA_4 are key regulators of cell expansion and elongation during secondary xylem formation. Consistent with a model of GA-controlled cell expansion and elongation, a set of genes showing highest expression in the zone of cell expansion was selectively upregulated in the GA_1/GA_4-enhanced trees relative to wild-type trees (45). This suggests the existence of regulons involved in the control of cell expansion that are directly or indirectly controlled by GA levels. Targets of such

CesA: cellulose synthase A subunit

Secondary cell wall: formed outside the primary wall in specialized cells; rich in lignin, cellulose, and hemicellulose

AGP: arabinogalactan protein

PCD: programmed cell death

regulons might include proteins such as expansins, which are important in cell wall loosening. For example, a specific ∂-expansin gene family member (*PttEXP1*) is expressed in the cell expansion zone of secondary xylem and may be involved in intrusive growth of xylem fibers (33).

Secondary wall formation is a hugely important process in forest ecosystems, because the bulk of the biomass of trees is cellulose and encrusting lignin in secondary walls. It is now possible to catalog the complete sets of genes encoding lignin biosynthetic enzymes and cellulose synthase (CesA) subunits, revealing the biosynthetic "toolboxes" for these processes. The *Populus* lignin biosynthetic toolbox (106) is roughly twice as big as that of *Arabidopsis*, and the higher number of genes encoding key cytochrome P450-mediated enzymes required for hydroxylation of the phenylpropanoid phenolic ring at the 4′, 3′, and 5′ positions suggests a greater need for maintaining flux into monolignol biosynthesis during the massive commitment to secondary wall biosynthesis. Alternatively, it could reflect evolutionary pressure for subfunctionalization, as apparent for duplicated cinnamate-4-hydroxylase (*C4H*) genes (61). *Populus* contains 18 apparent *CesA* genes (22), relative to the 10 known in *Arabidopsis*. Five of these genes are homologs of genes encoding *Arabidopsis* CesA4, CesA7, and CesA8 subunits, which are dedicated to cellulose biosynthesis in secondary walls (103). Expression data support specific roles for *Populus* CesA proteins in secondary wall biosynthesis during wood formation (for example, 50), and recent work suggests that dedication of specific CesA subunits to cellulose biosynthesis in secondary cell walls occurred well before divergence of angiosperms and gymnosperms (68).

More functional information is available for lignin biosynthesis, reviewed recently by Boerjan et al. (8) and Li et al. (59). Catalyzed in part by the importance of *Populus* as a commercial plantation species, multiple transgenic lines with modified lignin characteristics have been generated by misexpression of lignin biosynthetic genes. A thorough study of the pulping characteristics of wood from field-tested lines showed the potential to generate modified lignin trees with superior wood quality (78).

Functioning and regulation of the CesA rosette complex in the plasma membrane require numerous other proteins such as COBRA, KORRIGAN, and sucrose synthase (23, 88), and coexpression analyses in *Arabidopsis* have highlighted potentially important players such as COBRA-like 4 (COBL4), germin-like, chitinase-like 2 (CTL2), and fasciclin-like arabinogalactan (AGP) proteins in cellulose secretion during secondary wall formation (for example, 12). Homologs of these genes are present in *Populus* (49, 106) and are often coregulated with genes encoding secondary wall-specific *CesA* genes (40). Several studies have also highlighted the potentially important role of fasciclin-like AGPs in secondary wall formation (for example, 77).

The final step in wood development and tracheary element differentiation, programmed cell death (PCD), has mainly been investigated in Zinnia, but PCD in the context of wood formation can been studied in *Populus*. Moreau et al. (66) isolated populations of cells enriched for those undergoing PCD and those still undergoing secondary wall formation, and used EST analysis and cDNA microarray expression profiling to identify novel proteases, and potential regulatory proteins and mechanisms involved in PCD, opening the door to further functional studies.

The mechanisms regulating secondary xylem formation are not yet clear. Key transcriptional regulators of meristem and vascular differentiation known or inferred from *Arabidopsis* studies include KNOTTED-like homeodomain proteins such as SHOOT-MERISTEMLESS and BREVIPEDICELLUS and class III HD-ZIP proteins such as AtHB-8 and REVOLUTA (34), as well as MYB proteins (85). Homologs of genes encoding these proteins are found in the *Populus* genome and expression of some, such as class III *HD-ZIP* (54), *KNOX* (34a), and *MYB* genes (51), are correlated with secondary xylem formation in *Populus*. Using the 15.4K Treenomix cDNA *Populus* microarray (82), we identified a set of KNOTTED-like and MYB transcription factors common to *Populus* and *Arabidopsis* whose expression is highly correlated secondary wall formation in both species and in some cases with spruce (M. Friedman, L. Johnson, E. Li, M. Ellis & C. Douglas, unpublished). Interestingly, as in *Arabidopsis*, there is evidence of miRNA regulation of *Populus* HD-ZIP proteins (54), and novel *Populus* miRNAs downregulated during wood formation may play additional important regulatory roles in this process (60). This and other data strongly suggest that subtle changes in regulatory networks common to herbaceous plants and trees, rather than entirely new pathways, are responsible for the elaboration of the wood-forming habit of trees. Functional approaches performed in parallel in *Populus* and *Arabidopsis* now have the potential to unravel differential use of common regulatory networks.

Seasonality/Phenology

Trees in temperate climates must not only be able to adapt to the seasonal changes that restrict their growth, but they must also be able to withstand often severe winter conditions. As a typical deciduous tree, the yearly activity of a *Populus* tree in a climate of alternating summer and winter seasons is depicted in **Figure 4**.

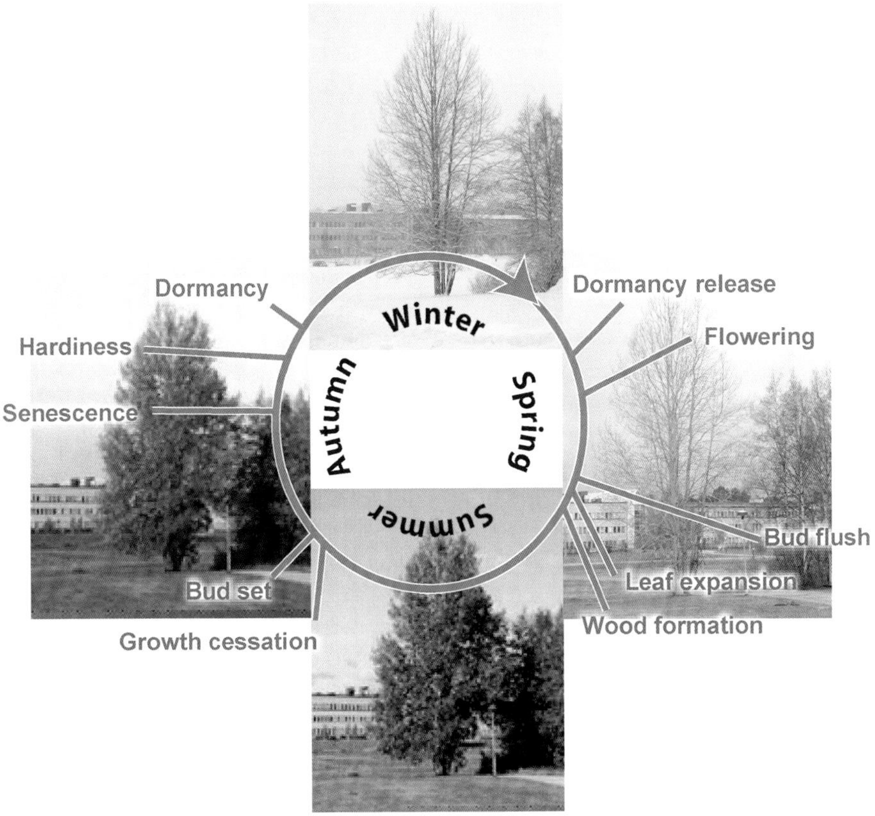

Figure 4

The annual cycle of a *Populus* tree growing at Umeå University, Umeå, Sweden.

The ability to anticipate harsh winter conditions is a highly adaptive trait, and the shorter days in autumn are the most important input signal. Photoreceptors detect shortening photoperiods, and both phytochrome A and at least one of the two *Populus* phytochrome B genes (*phyB2*) have roles in this process (73). As a first event in the preparation for winter, trees cease growth and set dormant buds. Time of bud set is easy to score phenotypically, is under very strong genetic control, and has been a model trait studied from the perspective of natural variation within *Populus* species and pedigrees (for example, 28). Chilling temperatures do not seem not to influence timing of bud set, but trees that have not set buds when the temperature drops below zero will stop growing and rapidly set buds (V. Luquez, D. Hall, B. Albrectsen, J. Karlsson, P. Ingvarsson, S. Jansson, submitted). Recent studies have shown that bud set is under control of the CO/FT regulatory module in *Populus*. Expression of the *Populus CO* gene is subject to strong diurnal regulation, peaking in the evening, and if *CO* mRNA accumulates before dawn, it activates the FT regulatory protein, which represses bud set. As days get shorter, the CO protein may only accumulate after dawn and is not able to induce the repressor FT, and bud set occurs (9).

Later in the autumn, leaf senescence starts and chlorophyll and the vast majority of the leaf proteins are degraded. Again, day length seems to be the main trigger in *Populus*. Not all trees start senescence according to the calendar but an aspen that has been studied in detail over many years always starts chlorophyll degradation around September 11 in northern Sweden (52). Leaf senescence has been studied in many systems, but the autumn leaf provides at attractive model; the process seems to be similar to leaf senescence in annual plants, but it is triggered in a different way. Key cellular and transcriptional events in autumn senescence in *Populus* have been described (52). Autumn senescence has clear adaptive value, because remobilization of nitrogen, in particular, is important in forest ecosystems where nitrogen is often a limiting factor for plant growth. However, perhaps even more important are dormancy and associated cold hardiness that develop in overwintering aboveground parts. Using microarray expression profiling of wood-forming tissues, Schrader et al. (90) identified marker genes for cambial dormancy. Endodormancy refers to an initial stage in the development of full dormancy in which presentation of favorable conditions does not result in resumption of growth. Studies in *Populus* have shown that once endodormancy has been induced by short day length, extended periods of chilling temperatures are needed to condition trees to break dormancy (86) and put the tree in a "standby" position, waiting for spring conditions that are permissive for growth. The tools now available in *Populus* make it an excellent system to study the regulatory circuitry underlying these events, which likely involves abscisic acid and a signaling pathway including *PtABI3*, a *Populus* ortholog of the *Arabidopsis ABI3* gene (87).

In the buds formed in the autumn, leaf development is arrested at an early stage. Timing of bud flush in the spring is under genetic control; when dormancy is broken, a tree of a given genotype has a requirement for a certain temperature sum for bud flush. Considerable variation in timing of bud flush often (27) exists between populations from different latitudes. In a full-grown tree, at bud flush in the spring, all leaves develop in parallel so that at a given date, all leaves of the tree are of the same age. Gene expression studies have shown that the environmental influence on gene expression during onset of leaf development in the spring can largely be separated from developmental and leaf-age-dependent influences (110) in *Populus*, studies that would not be possible in annuals. Gene expression in leaves up to one month after bud flush seems to be mainly determined by a developmental program, but later in the summer environmental factors are much more important, making *Populus* leaves an attractive system to study environmental influences on gene expression (A. Sjödin, K. Wissel & S. Jansson, unpublished).

Flowering

Many trees such as *Populus* have extended juvenile phases of several years before reproductive maturity and flowering (reviewed in 13). Thus, studies of flowering time in *Populus* add another dimension to the picture of regulation flowering in angiosperms that could not be obtained in annuals alone. As in annuals, trees are under selective pressure to coordinate timing of annual flowering with other trees of the same species to maximize reproductive outcome. However, the juvenility to maturity transition can be better studied in a tree such as *Populus*, which lies at an extreme relative to herbaceous weedy plants such as *Arabidopsis*. It is possible that the herbaceous growth habit is a derived state in evolution (94), so herbs can be considered mutant trees that flower during their first year of growth, making typical traits found in trees, such as woodiness, unnecessary.

Populus species flower early in the season, often before bud burst. The overwintering buds are either vegetative or reproductive so the decision if, or how much, to flower next year is taken in the summer, before bud set. The *FT* gene is a regulator of FT originally identified in *Arabidopsis* (55). Unlike *Arabidopsis*, *Populus* contains two *FT* genes (*FT1* and *FT2*; 91% amino acid identity) (41). Remarkably, transgenic *Populus* trees where ectopic *FT1* expression is driven by a constitutive promoter can flower in tissue culture six weeks after transformation (**Figure 5**) (9), and those ectopically expressing *FT2* flower within one year (41). In wild-type trees, *FT* expression increases as trees get older or reach reproductive maturity (9, 41); therefore, *FT* is also a regulator of maturity. In *Arabidopsis*, chromatin structure is altered by the *EARLY BOLTING IN SHORT DAYS* (*EBS*) gene, making FT susceptible to activation by CO, and by analogy it has been hypothesized that a gradual modification of chromatin structure as a *Populus* individual gets older could eventually make FT accessible for activation by CO (9). In contrast to the single flowers induced by overexpression of the meristem identity gene *LEAFY* (109), *FT* overexpressors form catkins that look very similar to those found in wild-type plants (9, 41).

In contrast to hermaphroditic plants such as *Arabidopsis* with perfect flowers, *Populus* species are dioeceous and produce male and female flowers on different individuals. Gender is genetically determined, although hermaphroditic flowers have been reported in many *Populus* species, especially on female trees (for example, 98). New opportunities to understand the molecular mechanisms underlying sex determination in plants have therefore been opened up with the genome sequence of a dioeceous plant. For example, it may be possible to identify gender-specific markers, as those found in *Salix* (1), and gene expression profiling could reveal gender-biased gene expression pointing to sex-determining loci. Several other unresolved matters remain regarding the control of flowering in trees such as *Populus*, including the competence of a meristem to alternate between vegetative and reproductive growth, and the mode of action of environmental factors such as light and temperature, that determine to what extent a tree will flower the following year.

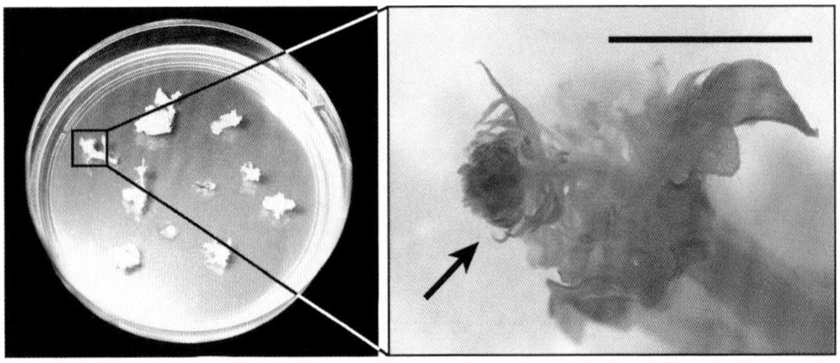

Figure 5

Populus in tissue culture flowering six weeks after transformation with the *FLOWERING LOCUS T1 (FT1)* gene. This illustrates the conservation of flowing time regulation in annuals such as *Arabidopsis* and trees such as *Populus*, as well as how generation times of tress can be drastically reduced allowing for efficient tree breeding. Copyright *Science*, redrawn from Reference 9.

Interactions with Other Organisms

Phenolic glycosides (PGs): phenolic molecules with sugars attached by O-glycosidic bonds

LRR: leucine-rich repeat

Phytochemistry and chemical ecology. With a life span of decades, *Populus* faces challenges and opportunities with respect to interactions with other organisms that are distinct from those of annual plants. *Populus* must defend itself against herbivores and pathogens year after year in order to survive. Furthermore, with an extensive root system that must maintain efficiency in water and nutrient uptake over similar time periods, the selective pressure to develop beneficial interactions with soil microflora may be greater than for short-lived organisms.

The complexity of *Populus* chemical ecology is likely to greatly exceed that of *Arabidopsis* or domesticated crop plants. An important part of this repertoire revolves around phenolic metabolism. Phenolic glycosides (such as salicin), condensed tannins, and other flavonoids may constitute up to 35% of foliar dry weight in some *Populus* species (104), and the recent annotation of the complete set of *Populus* shikimate/phenylpropanoid genes involved in biosynthesis of these (104) lays the groundwork for further studies in chemical defense, which have great quantitative variation (38). Arimura et al. (5) showed that forest tent caterpillars feeding on *Populus* leaves (but not mechanical wounding) induced a diversity of local and systemic volatile terpenoid emissions, suggesting that terpenoid chemical diversity in *Populus* may have a role for potential tri-trophic plant-herbivore-parasitoid interactions.

Pathogens. The number of *Populus* nucleotide binding site (NBS) – leucine-rich repeat (LRR) *R* resistance genes (399) is about double the number in *Arabidopsis* and appears to have increased both due to gene retention following genome duplication and by tandem gene duplication (106). Pathogen-induced signaling pathways appear conserved, consistent with reports of rapid transcriptional activation of phenylpropanoid in *Populus* by elicitor treatment (21), and the activation pattern of MAPKs (35). However, a diversification of *R*-gene-based pathogen surveillance mechanisms appears to have occurred in the *Populus* lineage relative to *Arabidopsis* and rice (106), and it will be interesting to determine if this trait is common to perennials.

One of the major fungal pathogens of the *Populus* species is the fungal rust pathogen *Melampsora* sp., and studies on the genetics of this interaction go back at least two decades (for example, 79). Recent studies have revealed Mendelian loci (e.g., *MXC3* and *MER*) (99, 111) and QTL for quantitative and qualitative resistance (48). Fine mapping of *R* gene-containing loci combined with BAC physical map and genome sequence information should make positional cloning of genetically identified *R* genes in *Populus* feasible (57, 111). Future studies on these and other host-pathogen systems in *Populus* [for example, involving *Septoria* (72)] provide fertile ground for biologists interested in plant-pathogen interactions. Sequencing of the *Melampsora larici-populina* genome by the Joint Genomics Institute (**http://www.jgi.doe.gov/sequencing/why/CSP2006/poplarrust.html**) will greatly enhance research in this area.

Beneficial interactions. Unlike *Arabidopsis*, associations with beneficial microorganisms are well described in *Populus*. *Populus* is a host for ectomycorrhizal and abuscular mycorrhizal fungal symbionts and is the first species with a complete genome sequence capable of undergoing such symbiotic relationships. With the completion or impending completion of the genome sequences of two important *Populus* mycrorrhizal fungi, *Laccaria bicolor* (ectomycorrhizal) and *Glomus intraradices* (arbuscular mycorrhizal) (56), interesting possibilities for investigating the molecular, physiological, and environmental basis for these interactions are possible. QTLs controlling *Populus*-mycorrhizal fungal interactions have been identified (101), a high-affinity ammonium transporter (PttAMT1.2) whose expression correlates with successful ectomycorrhizal association has been found (92), and infection-induced aquaporins involved in water transport during ectomycorrhizal associations have been identified (63).

While still in its infancy, studies on the colonization of internal *Populus* tissues by potentially beneficial bacterial endophytes are intriguing. More than 50 different fungal endophytes have been isolated from a single *Populus* tree (B. Albrectsen, M. Wedin, S. Jansson, J. Karlsson, unpublished) and whole genome shotgun sequences, derived from DNA libraries made from surface-sterilized material included sequences from 22 bacterial genera, some which had thousands of sequence reads (106), suggesting that these may represent endophytes rather than contaminating surface bacteria. Experimental evidence for efficient bacterial endophyte colonization of *Populus* exists (30), and Doty et al. (24) even identified the nitrogen-fixing bacterium *Rhizobium tropici* as an endophyte of *Populus*.

Herbivory. As an ecologically dominant tree, *Populus* is subject to herbivory by a variety of animals. Although browsing of *Populus* by mammals is particularly interesting from a chemical ecology point of view (32) and offers a system that will not soon be tackled by *Arabidopsis* biologists, insect herbivory in *Populus* has received the most attention. Phenolic products are likely involved in defending *Populus* against insect herbivory. Although genetically determined variation in phenolic glycoside levels plays a major role in insect performance as herbivores on *Populus* (74), rapid induction of *dihydroflavonol reductase* (*DFR*) gene expression and condensed tannin accumulation in response to herbivory suggest that these flavonoid-derived compounds are also important (76). Potential protein-based chemical defenses in *Populus*, such as expression of genes encoding polyphenol oxidase (PPO) and Kunitz trypsin inhibitors, have been studied at the molecular level (for example, 39), and overexpression of *PPO* in transgenic *Populus* increased insect mortality and decreased fitness (108). Microarray studies have confirmed and extended these findings, and highlighted the importance of octadeonoid and ethylene signaling in mediating defense responses (82).

Natural Variation

Trees typically have higher levels of genetic diversity and lower levels of genetic differentiation between populations than other plants (37). As a wind-pollinated obligate outbreeder, *Populus* may have even higher variation than some other trees. Each year, a female *Populus* can produce tens of millions of seeds that could potentially have thousands of different fathers. Seeds of many *Populus* species have hairs (hence the name cottonwood), and can be dispersed for many kilometers by the wind (summarized by G.C. Wyckoff and J.C. Zasada at **http://www.nsl.fs.fed.us/wpsm/Populus.pdf**). A high level of diversity and heterozygosity was demonstrated early in *Populus* using isozyme and random-amplified polymorphic DNA (RAPD) markers (for example, 7), but the genomic tools now available (for example, 31, 43) make it possible to dissect natural variation in more detail. Sequencing and physical map construction of the highly heterozygous *P. trichocarpa* Nisqually-1 genotype reveal extensive haplotype diversity in this single individual (2.6 single nucleotide polymorphisms or small indels per kilobase) (106), and haplotype-specific BAC clones from the Nisqually-1 *Populus* physical map differ by indels up to 10 kb in size (51a).

Populations of different *Populus* species seem to have different levels of nucleotide diversity. For example, *P. tremula* is more variable (on average, nucleotide diversity is 0.01 in genes) (43) than *P. trichocarpa* (on average, 0.0018) (31). Connecting phenotypic variation to genotypic variation in *Populus* using traditional QTL analysis presents challenges given its long generation time and the years required to generate segregating populations, as well as the logistics of maintaining large populations in the field. The first QTLs

Endophytes: fungi and bacteria that live internally in association with plants

RAPD: random-amplified polymorphic DNA

Association mapping: process of finding a genetic marker that associates with a trait in a large population

identified in *Populus*, for growth, form, and bud phenology (10, 28), were mapped in pedigrees established 15 to 20 years earlier. *Populus* pedigrees have subsequently been used to map QTLs for numerous other traits, for example, osmotic potential (105), drought response (100), and response to elevated CO_2 (81), in addition to the uses mentioned above. Completion of the genome sequence anchored to genetic and physical maps, combined with other genomic tools, will facilitate analysis of candidate genes underlying such *Populus* QTLs.

As an alternative to QTL analysis, association mapping shows great promise for exploiting natural variation to understand phenotypic diversity. The efficient mixing of alleles in outbreeding species ensures that those that give the highest fitness will accumulate in a population at a given site. In contrast to inbreeders like *Arabidopsis*, false positives resulting from population structure (4) are therefore less of a problem. However, the prospect of making genome-wide scans for association is hampered by the low linkage disequilibrium (LD). Even in *P. trichocarpa*, where LD extends longer than in *P. tremula* (31), scoring of close to one million evenly distributed markers may be necessary to have a reasonable chance of finding association to a crucial polymorphism. If candidate genes in a QTL region can be analyzed using association mapping, it should be possible to relatively rapidly identify SNPs or indels linked to phenotypic variation. By way of example, this approach has been used to show that two nonsynonymous SNPs in the *phyB2* gene together explain about 10% of the variation in critical photoperiod for bud set (44, P. Ingvarsson and S. Jansson, unpublished).

Dioecy adds another level of complexity to natural variation in *Populus*. Selection can act differently on males and females, leading to variation between males and females in traits such as growth rate (reviewed in 65). Thus, there is potential for linkage of traits to sex-determining loci not present in most other plant species. Finally, intriguing questions regarding the potential role of epigenetic mechanisms in generating additional variation in a perennial plant species with long generation times can be addressed in *Populus*.

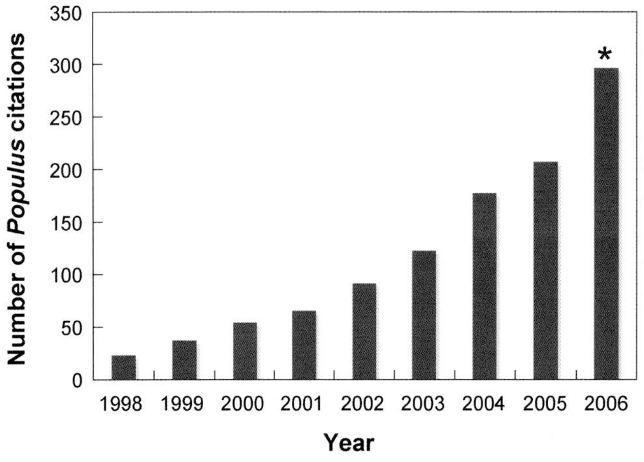

Figure 6

Number of citations in National Center for Biotechnology Information (NCBI) Medline containing the term "*Populus*," from 1998 to 2006. *Data for 2006 is partial (through November 2006).

FOR THE FUTURE

In addition to the areas we have reviewed here, there are many others for which *Populus* provides a good model system to understand plant biology. For example, *Populus* could be used for studies on long-distance transport of water and nutrients, studies on systemic signaling (5, 76), studies on seasonal nitrogen cycling and nitrogen storage as part of the perennial growth habit (18), studies on adaptation to stressful environments (for example, 11), and studies on the impact of predicted future increased CO_2 levels on trees and forest ecosystems (for example, 81). Comparative genomics and comparative functional studies capitalizing on the similarities and differences between *Arabidopsis* and *Populus* are likely to be fruitful for studies on basic plant processes and the evolution of plant growth and development. The high levels of natural variation in wild populations of *Populus*, and the almost infinite number of recessive alleles in these populations, can be capitalized on for both basic and applied research purposes, potentially contributing toward domestication of *Populus* (7) and the understanding of the evolution of adaptive traits. In addition to its traditional uses as a species for wood and fiber, *Populus*, as a model and plantation tree, offers opportunities for research into woody plants as sources of ligno-cellulosic feedstocks for biofuels (for example, 75). For these reasons, *Populus* will likely be a model system that will grow in importance for basic and applied studies, as it has over the past several years. One measure of the increasing interest in *Populus* as a model system is the nearly tenfold increase in NCBI Medline *Populus* citations per year from 1998 (the year of publication of the first *Populus* EST data set, marking the beginning of *Populus* genomics) (97) to 2006 (the year of publication of the *Populus* genome sequence) (106) (**Figure 6**). We predict that this number will continue to increase, catalyzed by the availability of the *Populus* genome sequence, the availability of *Populus* genomic and molecular tools, and the large number of interesting biological questions that can be addressed in this plant.

SUMMARY POINTS

1. Sequencing of the *Populus* genome revealed that its genome has undergone a rather recent genome duplication/polyploidization (60–65 mya; 8–12 mya by the molecular clock), which has shaped gene family organization in the genus.
2. A suite of critical genomic and molecular tools (genome sequence, EST collections, DNA microarrays, transformation protocols, etc.) have been developed for *Populus*.
3. *Populus* is a close relative of *Arabidopsis*, so gene functions are often conserved.
4. *Populus* provides opportunities to study important plant processes absent or poorly developed in *Arabidopsis*, such as wood formation and autumn senescence, as well as traits like flowering, sex determination, and biotic interactions from a comparative point of view.
5. *Populus* exhibits an unusual amount of natural variation that now can be explored with respect to adaptation, evolution, and biotechnology.

ACKNOWLEDGMENTS

The authors thank numerous members of the *Populus* research community for their ideas and contributions to *Populus* genomics and biology, and apologize for their inability to include all such work in this review due to space limitations. They would like to particularly thank Gerald Tuskan for his tireless efforts on behalf of *Populus* genomics and genetics, for advice on this review, and for generating **Figure 1**. Bo Segerman is acknowledged for providing **Figure 2**. The authors' work in *Populus* is funded by grants from the Knut and Alice Wallenberg Foundation, the Swedish Foundation for Strategic Research, the Swedish Research Council, Kempestiftelserna, and the European Commission (contract No. QLK5-CT-2002-00953 (POPYOMICS) (S.J.), and by Genome Canada and the Province of British Columbia (to the Treenomix project), and a Natural Sciences and Engineering Research Council of Canada (NSERC) Discovery Grant (C.J.D.).

LITERATURE CITED

1. Alstrom-Rapaport C, Lascoux M, Wang YC, Roberts G, Tuskan GA. 1998. Identification of a RAPD marker linked to sex determination in the basket willow (*Salix viminalis* L.). *J. Hered.* 89:44–49
2. Andersson-Gunnerås S, Mellerowicz EJ, Love J, Segerman B, Ohmiya Y, et al. 2006. Biosynthesis of cellulose-enriched tension wood in *Populus*: global analysis of transcripts and metabolites identifies biochemical and developmental regulators in secondary wall biosynthesis. *Plant J.* 45:144–65
3. *Arabidopsis* Genome Initiative. 2000. Analysis of the genome sequence of the flowering plant *Arabidopsis thaliana*. *Nature* 408:796–815
4. Aranzana MJ, Kim S, Zhao KY, Bakker E, Horton M, et al. 2005. Genome-wide association mapping in *Arabidopsis* identifies previously known flowering time and pathogen resistance genes. *PLoS Genet.* 1:531–39
5. **Arimura G, Huber DPW, Bohlmann J. 2004. Forest tent caterpillars (*Malacosoma disstria*) induce local and systemic diurnal emissions of terpenoid volatiles in hybrid poplar (*Populus trichocarpa* x *deltoides*): cDNA cloning, functional characterization, and patterns of gene expression of (−)-germacrene D synthase, PtdTPS1. *Plant J.* 37:603–16**
6. Bhalerao R, Keskitalo J, Sterky F, Erlandsson R, Bjorkbacka H, et al. 2003. Gene expression in autumn leaves. *Plant Physiol.* 131:430–42
7. Boerjan W. 2005. Biotechnology and the domestication of forest trees. *Curr. Opin. Biotechnol.* 16:159–66
8. Boerjan W, Ralph J, Baucher M. 2003. Lignin biosynthesis. *Annu. Rev. Plant Biol.* 54:519–46
9. **Bohlenius H, Huang T, Charbonnel-Campaa L, Brunner AM, Jansson S, et al. 2006. CO/FT regulatory module controls timing of flowering and seasonal growth cessation in trees. *Science* 312:1040–43**
10. **Bradshaw HD, Stettler RF. 1995. Molecular-genetics of growth and development in *Populus*. 4. Mapping QTLs with large effects on growth, form, and phenology traits in a forest tree. *Genetics* 139:963–73**
11. Brosche M, Vinocur B, Alatalo ER, Lamminmaki A, Teichmann T, et al. 2005. Gene expression and metabolite profiling of *Populus euphratica* growing in the Negev desert. *Genome Biol.* 6:R101
12. Brown DM, Zeef LAH, Ellis J, Goodacre R, Turner SR. 2005. Identification of novel genes in *Arabidopsis* involved in secondary cell wall formation using expression profiling and reverse genetics. *Plant Cell* 17:2281–95
13. Brunner AM, Nilsson O. 2004. Revisiting tree maturation and floral initiation in the poplar functional genomics era. *New Phytol.* 164:43–51
14. Busov VB, Brunner AM, Meilan R, Filichkin S, Ganio L, et al. 2005. Genetic transformation: a powerful tool for dissection of adaptive traits in trees. *New Phytol.* 167:9–18
15. Busov VB, Meilan R, Pearce DW, Ma C, Rood SB, Strauss SH. 2003. Activation tagging of a dominant gibberellin catabolism gene (GA 2-oxidase) from poplar that regulates tree stature. *Plant Physiol.* 132:1283–91
16. Chaffey N, Cholewa E, Regan S, Sundberg B. 2002. Secondary xylem development in *Arabidopsis*: a model for wood formation. *Physiol. Plant* 114:594–600
17. Christopher ME, Miranda M, Major IT, Constabel CP. 2004. Gene expression profiling of systemically wound-induced defenses in hybrid poplar. *Planta* 219:936–47

5. Novel biochemical analysis reveals importance of terpenoid emissions in response to insect herbivory.

9. An example of the power of *Arabidopsis/Populus* comparative biology.

10. First demonstration of QTL mapping in *Populus* using an interspecific hybrid population.

18. Cooke JE, Weih M. 2005. Nitrogen storage and seasonal nitrogen cycling in *Populus*: bridging molecular physiology and ecophysiology. *New Phytol.* 167:19–30
19. De Bodt S, Theissen G, Van de Peer Y. 2006. Promoter analysis of MADS-box genes in eudicots through phylogenetic footprinting. *Mol. Biol. Evol.* 23:1293–303
20. Dejardin A, Leple JC, Lesage-Descauses MC, Costa G, Pilate G. 2004. Expressed sequence tags from poplar wood tissues–a comparative analysis from multiple libraries. *Plant Biol.* 6:55–64
21. de Sa MM, Subramaniam R, Williams FE, Douglas CJ. 1992. Rapid activation of phenylpropanoid metabolism in elicitor-treated hybrid poplar (*Populus trichocarpa* Torr. & Gray x *Populus deltoides* Marsh) suspension-cultured cells. *Plant Physiol.* 98:728–37
22. Djerbi S, Lindskog M, Arvestad L, Sterky F, Teeri TT. 2005. The genome sequence of black cottonwood (*Populus trichocarpa*) reveals 18 conserved cellulose synthase (*CesA*) genes. *Planta* 221:739–46
23. Doblin MS, Kurek I, Jacob-Wilk D, Delmer DP. 2002. Cellulose biosynthesis in plants: from genes to rosettes. *Plant Cell Physiol.* 43:1407–20
24. Doty SL, Dosher MR, Singleton GL, Moore AL, Van Aken B, et al. 2005. Identification of an endophytic *Rhizobium* in stems of *Populus*. *Symbiosis* 39:27–35
25. Du J, Xie HL, Zhang DQ, He XQ, Wang MJ, et al. 2006. Regeneration of the secondary vascular system in poplar as a novel system to investigate gene expression by a proteomic approach. *Proteomics* 6:881–95
26. Eriksson ME, Israelsson M, Olsson O, Moritz T. 2000. Increased gibberellin biosynthesis in transgenic trees promotes growth, biomass production and xylem fiber length. *Nat. Biotechnol.* 18:784–88
27. Farmer RE Jr, Reinholt RW. 1986. Genetic variation in dormancy relations of balsam poplar along a latitudinal transect in northwestern Ontario. *Silvae Genet.* 35:38–42
28. Frewen BE, Chen TH, Howe GT, Davis J, Rohde A, et al. 2000. Quantitative trait loci and candidate gene mapping of bud set and bud flush in *Populus*. *Genetics* 154:837–45
29. Geisler-Lee J, Geisler M, Coutinho PM, Segerman B, Nishikubo N, et al. 2006. Poplar carbohydrate-active enzymes. Gene identification and expression analyses. *Plant Physiol.* 140:946–62
30. Germaine K, Keogh E, Garcia-Cabellos G, Borremans B, van der Lelie D, et al. 2004. Colonisation of poplar trees by gfp expressing bacterial endophytes. *FEMS Microbiol. Ecol.* 48:109–18
31. Gilchrist EJ, Haughn GW, Ying CC, Otto SP, Zhuang J, et al. 2006. Use of Ecotilling as an efficient SNP discovery tool to survey genetic variation in wild populations of *Populus trichocarpa*. *Mol. Ecol.* 15:1367–78
32. Gill RMA. 1992. A review of damage by mammals in north temperate forests. 3. Impact on trees and forests. *Forestry* 65:363–88
33. Gray-Mitsumune M, Mellerowicz EJ, Abe H, Schrader J, Winzell A, et al. 2004. Expansins abundant in secondary xylem belong to subgroup A of the alpha-expansin gene family. *Plant Physiol.* 135:1552–64
34. Groover AT. 2005. What genes make a tree a tree? *Trends Plant Sci.* 10:210–14

34a. Groover AT, Mansfield SD, DiFazio SP, Dupper G, Fontana JR, et al. 2006. The *Populus* homeodomox gene *ARBORKNOX1* reveals overlapping mechanisms regulating the shoot apical meristem and the vascular cambium. *Plant Mol. Biol.* 61:917–32

35. Hamel LP, Miles GP, Samuel MA, Ellis BE, Seguin A, Beaudoin N. 2005. Activation of stress-responsive mitogen-activated protein kinase pathways in hybrid poplar (*Populus trichocarpa* x *Populus deltoides*). *Tree Physiol.* 25:277–88

36. Hamel LP, Nicole MC, Sritubtim S, Morency MJ, Ellis M, et al. 2006. Ancient signals: Comparative genomics of plant MAPK and MAPKK gene families. *Trends Plant Sci.* 11:192–98
37. Hamrick JL, Godt MJW, Sherman-Broyles SL. 1992. Factors influencing levels of genetic diversity in woody plant species. *New For.* 6:95–124
38. Harding SA, Jiang H, Jeong ML, Casado FL, Lin HW, Tsai CJ. 2005. Functional genomics analysis of foliar condensed tannin and phenolic glycoside regulation in natural cottonwood hybrids. *Tree Physiol.* 25:1475–86
39. Haruta M, Major IT, Christopher ME, Patton JJ, Constabel CP. 2001. A Kunitz trypsin inhibitor gene family from trembling aspen (*Populus tremuloides* Michx.): cloning, functional expression, and induction by wounding and herbivory. *Plant Mol. Biol.* 46:347–59
40. **Hertzberg M, Aspeborg H, Schrader J, Andersson A, Erlandsson R, et al. 2001. A transcriptional roadmap to wood formation. *Proc. Natl. Acad. Sci. USA* 98:14732–37**

40. First *Populus* DNA microarray used for transcript profiling around the vascular cambium. A landmark paper for *Populus* functional genomics.

41. Hsu CY, Liu Y, Luthe DS, Yuceer C. 2006. Poplar FT2 Shortens the juvenile phase and promotes seasonal flowering. *Plant Cell* 18:1846–61
42. Iliev I, Savidge R. 1999. Proteolytic activity in relation to seasonal cambial growth and xylogenesis in *Pinus banksiana*. *Phytochemistry* 50:953–60
43. Ingvarsson PK. 2005. Nucleotide polymorphism and linkage disequilibrium within and among natural populations of European aspen (*Populus tremula* L., Salicaceae). *Genetics* 169:945–53
44. **Ingvarsson PK, Garcia MV, Hall D, Luquez V, Jansson S. 2006. Clinal variation in *phyB2*, a candidate gene for day-length-induced growth cessation and bud set, across a latitudinal gradient in European aspen (*Populus tremula*). *Genetics* 172:1845–53**

44. Demonstrates the power of association mapping in *Populus*.

45. Israelsson M, Eriksson ME, Hertzberg M, Aspeborg H, Nilsson P, Moritz T. 2003. Changes in gene expression in the wood-forming tissue of transgenic hybrid aspen with increased secondary growth. *Plant Mol. Biol.* 52:893–903
46. Israelsson M, Sundberg B, Moritz T. 2005. Tissue-specific localization of gibberellins and expression of gibberellin-biosynthetic and signaling genes in wood-forming tissues in aspen. *Plant J.* 44:494–504
47. Jelinski DE, Cheliak WM. 1992. Genetic diversity and spatial subdivision of *Populus tremuloides* (Salicaceae) in a heterogeneous landscape. *Am. J. Bot.* 79:728–36
48. Jorge V, Dowkiw A, Faivre-Rampant P, Bastien C. 2005. Genetic architecture of qualitative and quantitative *Melampsora larici-populina* leaf rust resistance in hybrid poplar: genetic mapping and QTL detection. *New Phytol.* 167:113–27
49. Joshi CP, Bhandari S, Ranjan P, Kalluri UC, Liang X, et al. 2004. Genomics of cellulose biosynthesis in poplars. *New Phytol.* 164:53–61
50. Kalluri UC, Joshi CP. 2004. Differential expression patterns of two cellulose synthase genes are associated with primary and secondary cell wall development in aspen trees. *Planta* 220:47–55
51. Karpinska B, Karlsson M, Srivastava M, Stenberg A, Schrader J, et al. 2004. MYB transcription factors are differentially expressed and regulated during secondary vascular tissue development in hybrid aspen. *Plant Mol. Biol.* 56:25570

51a. Kelleher C, Chiu R, Shin H, Bosdet IE, Martin I, et al. 2007. A physical map of the highly heterozygous *Populus* genome: integration with the genome sequence and genetic map and analysis of haplotype variation. *Plant J.* In press

52. Keskitalo J, Bergquist G, Gardestrom P, Jansson S. 2005. A cellular timetable of autumn senescence. *Plant Physiol.* 139:1635–48

53. Klimmek F, Sjodin A, Noutsos C, Leister D, Jansson S. 2006. Abundantly and rarely expressed Lhc protein genes exhibit distinct regulation patterns in plants. *Plant Physiol.* 140:793–804
54. Ko JH, Prassinos C, Han KH. 2006. Developmental and seasonal expression of PtaHB1, a *Populus* gene encoding a class III HD-Zip protein, is closely associated with secondary growth and inversely correlated with the level of microRNA (miR166). *New Phytol.* 169:469–78
55. Koornneef M, Alonso-Blanco C, Peeters AJ, Soppe W. 1998. Genetic control of flowering time in *Arabidopsis*. *Annu. Rev. Plant Physiol. Plant Mol. Biol.* 49:345–70
56. Lammers P, Tuskan GA, DiFazio SP, Podila GK, Martin F. 2004. Mycorrhizal symbionts of *Populus* to be sequenced by the United States Department of Energy's Joint Genome Institute. *Mycorrhiza* 14:63–64
57. Lescot M, Rombauts S, Zhang J, Aubourg S, Mathe C, et al. 2004. Annotation of a 95-kb *Populus deltoides* genomic sequence reveals a disease resistance gene cluster and novel class I and class II transposable elements. *Theor. Appl. Genet.* 109:10–22
58. Leseberg CH, Li A, Kang H, Duvall M, Mao L. 2006. Genome-wide analysis of the MADS-box gene family in *Populus trichocarpa*. *Gene* 378:84–94
59. Li LG, Lu SF, Chiang V. 2006. A genomic and molecular view of wood formation. *Crit. Rev. Plant Sci.* 25:215–33
60. Lu SF, Sun YH, Shi R, Clark C, Li LG, Chiang VL. 2005. Novel and mechanical stress-responsive MicroRNAs in *Populus trichocarpa* that are absent from *Arabidopsis*. *Plant Cell* 17:2186–203
61. Lu SF, Zhou YH, Li LG, Chiang VL. 2006. Distinct roles of cinnamate 4-hydroxylase genes in *Populus*. *Plant Cell Physiol.* 47:905–14
62. Deleted in proof
63. Marjanovic Z, Uehlein N, Kaldenhoff R, Zwiazek JJ, Weiss M, et al. 2005. Aquaporins in poplar: what a difference a symbiont makes! *Planta* 222:258–68
64. Mellerowicz EJ, Baucher M, Sundberg B, Boerjan W. 2001. Unravelling cell wall formation in the woody dicot stem. *Plant Mol. Biol.* 47:239–74
65. Mitton JB, Grant MC. 1996. Genetic variation and the natural history of quaking aspen. *BioScience* 46:25–31
66. Moreau C, Aksenov N, Lorenzo MG, Segerman B, Funk C, et al. 2005. A genomic approach to investigate developmental cell death in woody tissues of *Populus* trees. *Genome Biol.* 6:R34
67. Morreel K, Goeminne G, Storme V, Sterck L, Ralph J, et al. 2006. Genetical metabolomics of flavonoid biosynthesis in *Populus*: a case study. *Plant J.* 47:224–37
68. Nairn CJ, Haselkorn T. 2005. Three loblolly pine *CesA* genes expressed in developing xylem are orthologous to secondary cell wall *CesA* genes of angiosperms. *New Phytol.* 166:907–15
69. Nanjo T, Futamura N, Nishiguchi M, Igasaki T, Shinozaki K, Shinohara K. 2004. Characterization of full-length enriched expressed sequence tags of stress-treated poplar leaves. *Plant Cell Physiol.* 45:1738–48
70. Naylor M, Reeves J, Cooper JI, Edwards ML, Wang H. 2005. Construction and properties of a gene-silencing vector based on Poplar mosaic virus (genus *Carlavirus*). *J. Virol. Methods* 124:27–36
71. Nelson T, Tausta SL, Gandotra N, Liu T. 2006. Laser microdissection of plant tissue: What you see is what you get. *Annu. Rev. Plant Biol.* 57:181–201
72. Newcombe G, Ostry N. 2001. Recessive resistance to *Septoria* stem canker of hybrid poplar. *Phytopathology* 91:1081–84

73. Olsen JE, Junttila O. 2002. Far red end-of-day treatment restores wild type-like plant length in hybrid aspen overexpressing phytochrome A. *Physiol. Plant* 115:448–57

74. Osier TL, Lindroth RL. 2006. Genotype and environment determine allocation to and costs of resistance in quaking aspen. *Oecologia* 148:293–303

75. Pan XJ, Gilkes N, Kadla J, Pye K, Saka S, et al. 2006. Bioconversion of hybrid poplar to ethanol and coproducts using an organosolv fractionation process: optimization of process yields. *Biotechnol. Bioeng.* 94:851–61

76. Peters DJ, Constabel CP. 2002. Molecular analysis of herbivore-induced condensed tannin synthesis: cloning and expression of dihydroflavonol reductase from trembling aspen (*Populus tremuloides*). *Plant J.* 32:701–12

77. Pilate G, Dejardin A, Laurans F, Leple JC. 2004. Tension wood as a model for functional genomics of wood formation. *New Phytol.* 164:63–72

78. Pilate G, Guiney E, Holt K, Petit-Conil M, Lapierre C, et al. 2002. Field and pulping performances of transgenic trees with altered lignification. *Nat. Biotechnol.* 20:607–12

78. Summary of large-scale multiyear field tests of lignin-modified trees from multiple European collaborators and convincing demonstration of the feasibility of biotechnological approaches for *Populus* improvement.

78a. Plomion C, Lalanne C, Clavero S, Meddour H, Kohler A, et al. 2006. Mapping the proteome of poplar and application to the discovery of drought-stress responsive proteins. *Proteomics* 6:6509–27

79. Prakash CP, Heather WA. 1985. Inheritance of resistance to races of *Melampsora medusae* in *Populus deltoides*. *Silvae Genet.* 35:74–77

80. Prassinos C, Ko JH, Yang J, Han KH. 2005. Transcriptome profiling of vertical stem segments provides insights into the genetic regulation of secondary growth in hybrid aspen trees. *Plant Cell Physiol.* 46:1213–25

81. Rae AM, Ferris R, Tallis MJ, Taylor G. 2006. Elucidating genomic regions determining enhanced leaf growth and delayed senescence in elevated CO_2. *Plant Cell Environ.* 29:1730–41

82. Ralph S, Oddy C, Cooper D, Yueh H, Jancsik S, et al. 2006. Genomics of hybrid poplar (*Populus trichocarpa x deltoides*) interacting with forest tent caterpillars (*Malacosoma disstria*): normalized and full-length cDNA libraries, expressed sequence tags, and a cDNA microarray for the study of insect-induced defences in poplar. *Mol. Ecol.* 15:1275–97

82. Generation of large EST data set and full-length cDNA libraries and production of 15.4-K micorarray used for first genome-wide analysis of transcriptional response to insect herbivory in *Populus*.

83. Renaut J, Lutts S, Hoffmann L, Hausman JF. 2004. Responses of poplar to chilling temperatures: proteomic and physiological aspects. *Plant Biol.* 6:81–90

84. Robinson AR, Gheneim R, Kozak RA, Ellis DD, Mansfield SD. 2005. The potential of metabolite profiling as a selection tool for genotype discrimination in *Populus*. *J. Exp. Bot.* 56:2807–19

85. Rogers LA, Campbell MM. 2004. The genetic control of lignin deposition during plant growth and development. *New Phytol.* 164:17–30

86. Rohde A, Howe G, Olsen J, Moritz T, van Montagu M, et al. 2000. Molecular aspects of bud dormancy in trees. In *Molecular Biology of Woody Plants*, ed. S Jain, S Minocha, pp. 89–134. Dordrecht: Kluwer Acad.

87. Rohde A, Prinsen E, De Rycke R, Engler G, van Montagu M, Boerjan W. 2002. PtABI3 impinges on the growth and differentiation of embryonic leaves during bud set in poplar. *Plant Cell* 14:1885–901

88. Roudier F, Fernandez AG, Fujita M, Himmelspach R, Borner GH, et al. 2005. COBRA, an *Arabidopsis* extracellular glycosyl-phosphatidyl inositol-anchored protein, specifically controls highly anisotropic expansion through its involvement in cellulose microfibril orientation. *Plant Cell* 17:1749–63

89. Sampedro J, Carey RE, Cosgrove DJ. 2006. Genome histories clarify evolution of the expansin superfamily: new insights from the poplar genome and pine ESTs. *J. Plant Res.* 119:11–21
90. Schrader J, Moyle R, Bhalerao R, Hertzberg M, Lundeberg J, et al. 2004. Cambial meristem dormancy in trees involves extensive remodelling of the transcriptome. *Plant J.* 40:173–87
91. **Schrader J, Nilsson J, Mellerowicz E, Berglund A, Nilsson P, et al. 2004. A high-resolution transcript profile across the wood-forming meristem of poplar identifies potential regulators of cambial stem cell identity. *Plant Cell* 16:2278–92**
92. Selle A, Willmann M, Grunze N, Gessler A, Weiss M, Nehls U. 2005. The high-affinity poplar ammonium importer PttAMT1.2 and its role in ectomycorrhizal symbiosis. *New Phytol.* 168:697–706
93. Sjödin A, Bylesjö M, Skogström O, Eriksson D, Nilsson P, et al. 2006. UPSC-BASE - *Populus* transcriptomics online. *Plant J.* 48:806–17
94. Sporne KR. 1980. A reinvestigation of character correlations among dicotyledons. *New Phytol.* 85:419–49
95. Sterck L, Rombauts S, Jansson S, Sterky F, Rouze P, Van de Peer Y. 2005. EST data suggest that poplar is an ancient polyploid. *New Phytol.* 167:165–70
96. **Sterky F, Bhalerao RR, Unneberg P, Segerman B, Nilsson P, et al. 2004. A *Populus* EST resource for plant functional genomics. *Proc. Natl. Acad. Sci. USA* 101:13951–56**
97. Sterky F, Regan S, Karlsson J, Hertzberg M, Rohde A, et al. 1998. Gene discovery in the wood-forming tissues of poplar: analysis of 5,692 expressed sequence tags. *Proc. Natl. Acad. Sci. USA* 95:13330–35
98. Stettler R. 1971. Variation in sex expression of black cottonwood and related hybrids. *Silvae Genet.* 20:42–46
99. Stirling B, Newcombe G, Vrebalov J, Bosdet I, Bradshaw HD. 2001. Suppressed recombination around the *MXC3* locus, a major gene for resistance to poplar leaf rust. *Theor. Appl. Genet.* 103:1129–37
100. Street NR, Skogström O, Sjödin A, Tucker J, Rodríguez-Acosta M, et al. 2006. The genetics and genomics of drought response in *Populus*. *Plant J.* 48:321–41
101. Tagu D, Bastien C, Faivre-Rampant P, Garbaye J, Vion P, et al. 2005. Genetic analysis of phenotypic variation for ectomycorrhiza formation in an interspecific F1 poplar full-sib family. *Mycorrhiza* 15:87–91
102. Talyzina NM, Ingvarsson PK. 2006. Molecular evolution of a small gene family of wound inducible Kunitz trypsin inhibitors in *Populus*. *J. Mol. Evol.* 63:108–19
103. Taylor NG, Howells RM, Huttly AK, Vickers K, Turner SR. 2003. Interactions among three distinct CesA proteins essential for cellulose synthesis. *Proc. Natl. Acad. Sci. USA* 100:1450–55
104. Tsai CJ, Harding SA, Tschaplinski TJ, Lindroth RL, Yuan YN. 2006. Genome-wide analysis of the structural genes regulating defense phenylpropanoid metabolism in *Populus*. *New Phytol.* 172:47–62
105. Tschaplinski TJ, Tuskan GA, Sewell MM, Gebre GM, Todd DE, Pendley CD. 2006. Phenotypic variation and quantitative trait locus identification for osmotic potential in an interspecific hybrid inbred F2 poplar pedigree grown in contrasting environments. *Tree Physiol.* 26:595–604
106. **Tuskan GA, DiFazio S, Jansson S, Bohlmann J, Grigoriev I, et al. 2006. The genome of black cottonwood, *Populus trichocarpa* (Torr & Gray ex. Brayshaw). *Science* 313:1596–604**

91. A fine-scale dissection of gene expression in the vascular cambium reveals regulatory circuits.

96. Description and documentation of *Populus* genomics tools: Generation and analysis of a 100-K EST data set, use of the EST data set for digital expression profiling in *Populus*, and production of a 25-K cDNA microarray.

106. The *Populus* paper! Generation and analysis of the genome sequence, data on genome duplication, comparative genomics, and gene family evolution, among other things.

107. van Raemdonck D, Pesquet E, Cloquet S, Beeckman H, Boerjan W, et al. 2005. Molecular changes associated with the setting up of secondary growth in aspen. *J. Exp. Bot.* 56:2211–27
108. Wang JH, Constabel CP. 2004. Polyphenol oxidase overexpression in transgenic *Populus* enhances resistance to herbivory by forest tent caterpillar (*Malacosoma disstria*). *Planta* 220:87–96
109. Weigel D, Nilsson O. 1995. A developmental switch sufficient for flower initiation in diverse plants. *Nature* 377:495–500
110. Wissel K, Pettersson F, Berglund A, Jansson S. 2003. What affects mRNA levels in leaves of field-grown aspen? A study of developmental and environmental influences. *Plant Physiol.* 133:1190–97
111. Yin TM, DiFazio SP, Gunter LE, Jawdy SS, Boerjan W, Tuskan GA. 2004. Genetic and physical mapping of *Melampsora* rust resistance genes in *Populus* and characterization of linkage disequilibrium and flanking genomic sequence. *New Phytol.* 164:95–105

Related Resources

Brunner AM, Busov VB, Strauss SH. 2004. Poplar genome sequence: functional genomics in an ecologically dominant plant species. *Trends Plant Sci.* 9:49–56

Cronk QCB. 2005. Plant eco-devo: the potential of poplar as a model organism. *New Phytol.* 166:39–48

Taylor G. 2002. *Populus*: *Arabidopsis* for forestry. Do we need a model tree? *Ann. Bot.* 90:681–89

Oxidative Modifications to Cellular Components in Plants

Ian M. Møller,[1,*] Poul Erik Jensen,[2]
and Andreas Hansson[2]

[1]Department of Agricultural Sciences, [2]Department of Plant Biology, Faculty of Life Sciences, University of Copenhagen, DK-1871 Frederiksberg C, Denmark;
email: imm@life.ku.dk, peje@life.ku.dk

Annu. Rev. Plant Biol. 2007. 58:459–81

The *Annual Review of Plant Biology* is online at plant.annualreviews.org

This article's doi:
10.1146/annurev.arplant.58.032806.103946

First published online as a Review in Advance on February 8, 2007

1543-5008/07/0602-0459$20.00

*Present address: Department of Genetics and Biotechnology, Faculty of Agricultural Sciences, University of Aarhus, DK-8830 Tjele, Denmark;
email: ian.max.moller@agrisci.dk

Key Words

polyunsaturated fatty acids, proteins, reactive nitrogen species, reactive oxygen species, signal transduction, stress

Abstract

Reactive oxygen species (ROS) and reactive nitrogen species (RNS) are produced in many places in living cells and at an increased rate during biotic or abiotic stress. ROS and RNS participate in signal transduction, but also modify cellular components and cause damage. We first look at the most common ROS and their properties. We then consider the ways in which the cell can regulate their production and removal. We critically assess current knowledge about modifications of polyunsaturated fatty acids (PUFAs), DNA, carbohydrates, and proteins and illustrate this knowledge with case stories wherever possible. Some oxidative breakdown products, e.g., from PUFA, can cause secondary damage. Other oxidation products are secondary signaling molecules. We consider the fate of the modified components, the energetic costs to the cell of replacing such components, as well as strategies to minimize transfer of oxidatively damaged components to the next generation.

Contents

INTRODUCTION

ROS: reactive oxygen species

RNS: reactive nitrogen species

The production of reactive oxygen species (ROS), such as superoxide ($O_2^{\bullet-}$), hydrogen peroxide (H_2O_2), hydroxyl radicals ($HO^{\bullet}$), and singlet oxygen ($^1O_2$), is an unavoidable consequence of aerobic metabolism (41). Reactive nitrogen species (RNS), such as the hormone nitric oxide ($NO^{\bullet}$) and peroxinitrite ($ONOO^-$), are also formed. In some cases, ROS are produced in plants as products of mainstream enzymatic reactions, e.g., glycolate oxidase in the peroxisomes during photorespiration. In other cases, ROS production appears to be an unavoidable accident, e.g., the $O_2^{\bullet-}$ produced by the mitochondrial electron transport chain (ETC). Hypoxia and anoxia are special cases where most of the ROS production occurs during the reoxygenation phase (12). ROS can also be produced to perform certain tasks, e.g., production of $O_2^{\bullet-}$ on the outer surface of the plasma membrane by the NADPH oxidase following pathogen recognition and in a variety of other processes (105). It is generally observed that ROS production and ROS-induced damage increase during abiotic and biotic stress and ROS are important signaling molecules (5, 33, 36, 67).

The emphasis on the role of ROS in signaling has led to the suggestion that the expression "oxidative stress" should be replaced by "oxidative signaling" (33). However, being reactive molecules, ROS oxidize all types of cellular components. To some extent the cell has learned to live with that and some modifications can be used in metabolic regulation, e.g., the oxidation of protein cysteines. In other cases, the modification prevents the molecule from performing its original function, i.e., it is damaging. Typically, more damage is observed under stress conditions when the ROS levels are increased (**Figure 1**). At the same time the oxidized products can be important secondary signaling molecules, and in such cases damage and signaling are two sides of the same story (**Figure 1**).

In this review we focus on the specific oxidative modifications caused by ROS without losing sight of ROS signaling. First, we briefly summarize the current knowledge about ROS turnover in plant cells. We then consider the type and extent of oxidative modifications in plant cells. Finally, we look at the possible repair processes and estimate the energetic cost of replacing modified components. Throughout the review we compare and

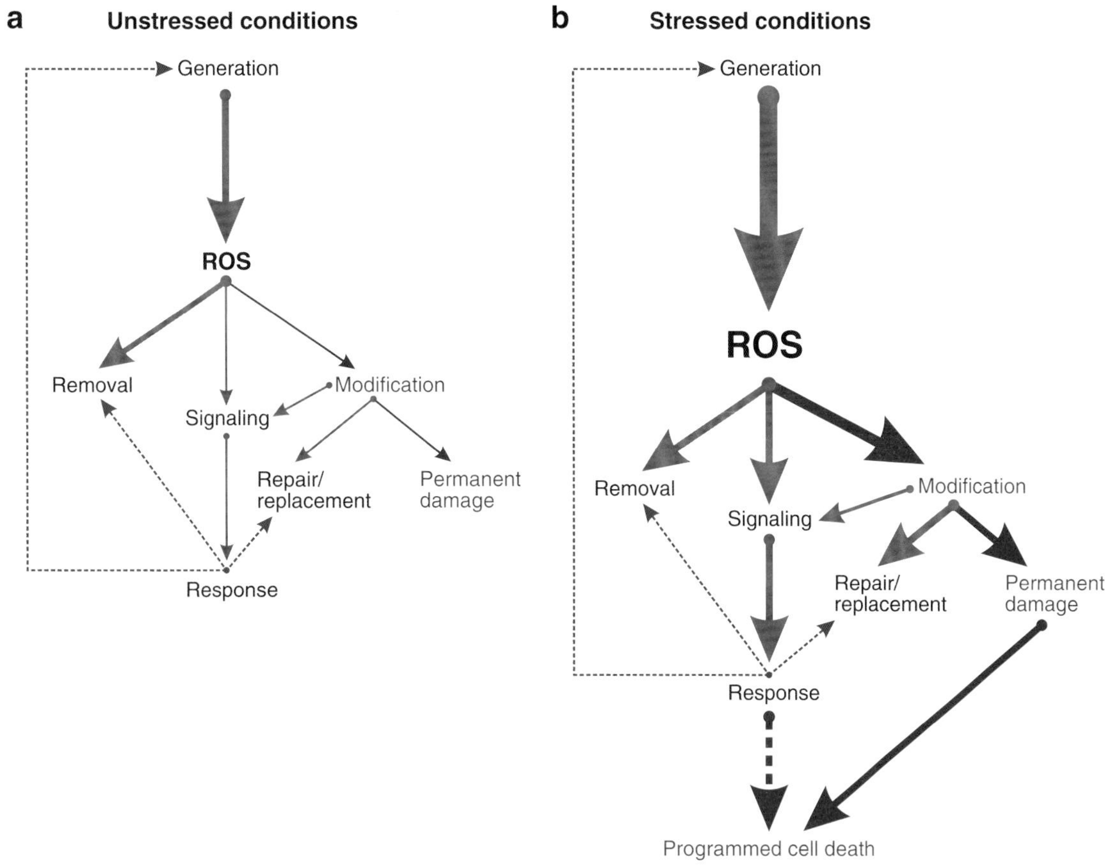

Figure 1

The relationship between reactive oxygen species (ROS) production, removal, modification, signaling, and damage in plant cells under (*a*) unstressed and (*b*) stressed conditions. The arrow from "Modification" to "Signaling" indicates that some of the modified molecules are secondary signal molecules.

contrast our knowledge about oxidative modifications in plants with the greater knowledge gained from experiments on fungi and animals.

ROS PRODUCTION AND REMOVAL

The response of the plant and its components to ROS depends on the ROS in question as well as on its concentration, site of production, and interaction with other stress molecules. It also depends on the developmental stage and prehistory of the plant cell (e.g., previous stress encounters) (36). In this section we look at the types of ROS produced and the mechanisms for regulating the rate of production. In the following section we look at processes removing ROS and then analyze the modifications and/or damage that cellular components can sustain when interacting with ROS.

ROS Production

In green plant parts in the light, the chloroplasts and peroxisomes (through photorespiration) are the main ROS producers (32).

$NO^{\bullet}$: nitric oxide (or nitrogen monoxide), a volatile signaling molecule, is an RNS and a radical

$ONOO^{-}$: peroxinitrite, an RNS formed by the condensation of $O_2^{\bullet-}$ and $NO^{\bullet}$. It is very reactive, especially in the protonated state.

ETC: electron transport chain

$^1O_2$: singlet oxygen, an ROS that is oxygen in a higher energy state, is mainly produced in the chloroplasts

PSI and II: photosystem I and II

$O_2^{\bullet-}$: superoxide, an ROS that is both a radical and an anion

SOD: superoxide dismutase

$HO^{\bullet}$: hydroxyl radical, the most reactive of all the common ROS

In nongreen plant parts or in darkness, the mitochondria appear to be the main ROS producers (64, 68). The chloroplasts produce $^1O_2$ at photosystem II (PSII) and $O_2^{\bullet-}$ at photosystem I (PSI) (6) and PSII (84) as byproducts. The mitochondria produce $O_2^{\bullet-}$ at complexes I and III, also as byproducts. An estimated 1–5% of the oxygen consumption of isolated mitochondria results in ROS production (68). The peroxisomes produce $O_2^{\bullet-}$ and H_2O_2 in several key metabolic reactions (26). And, finally, the NADPH oxidase in the plasma membrane produces $O_2^{\bullet-}$, which participates in several physiological processes (105).

$O_2^{\bullet-}$ can be converted into H_2O_2, e.g., by superoxide dismutase (SOD), and H_2O_2 can give rise to $HO^{\bullet}$ through the Fenton reaction, which is catalyzed mainly by free transition metal ions. The different ROS have very different properties (**Table 1**). H_2O_2 is relatively stable and its concentration in plant tissues is in the micromolar to low millimolar range, probably depending on the compartment (14, 41, 86). The other ROS have very short half-lives and are probably present at very low concentrations. They also have different reactivities: Whereas $HO^{\bullet}$ reacts rapidly with all types of cellular components, $O_2^{\bullet-}$ reacts primarily with protein Fe-S centers and $^1O_2$ is particularly reactive with conjugated double bonds as found in polyunsaturated fatty acids (PUFAs). This means that they leave different footprints in the cell in the form of different oxidatively modified components.

H_2O_2 is already an established messenger in bacteria, yeast, and mammals where transcription factors are sensors (56) (see also below), as they may also be in plants (66). Its relative stability and ability to cross membranes possibly through aquaporins (9, 10) makes H_2O_2 a good messenger molecule. However, all the ROS forms can, in principle, act as messengers either directly or by using an oxidized product as a secondary messenger. Using a secondary messenger, e.g., transcription factors, depends on the distance the ROS molecule travels before reacting with cellular components (**Figure 2**; see also **Table 1**).

Table 1 Important ROS and RNS in plant tissues and their basic properties[a]

Property	Singlet oxygen ($^1O_2$)	Superoxide ($O_2^{\bullet-}$)	Hydrogen peroxide (H_2O_2)	Hydroxyl radical ($HO^{\bullet}$)	Peroxynitrite ($ONOO^-$)[e]
Half-life[b]	1 μs	1 μs	1 ms	1 ns	?
Distance traveled[c]	30 nm	30 nm	1 μm	1 nm	?
Cellular concentration	?	?	μM-mM[f]	?	?
Reacts with					
Lipids	PUFA	Hardly	Hardly	Rapidly	Yes
DNA	Mainly guanine	No	No	Rapidly	Especially guanine
Carbohydrates[d]	No	No	No	Rapidly	?
Proteins	Trp, His, Tyr, Met Cys[g]	Fe-S centers	Cysteines	Rapidly[h]	Tyr, Trp, Phe, Met

[a]Where no source is indicated the information is from Halliwell & Gutteridge (41).

[b]In biological systems (9, 85, 99).

[c]Distance traveled in one half-life if the diffusion coefficient is assumed to be 10^{-9} m^2s^{-1}.

[d]Reference 34.

[e]Peroxynitrite is included instead of $NO^{\bullet}$. It is formed by the condensation of $NO^{\bullet}$ with $O_2^{\bullet-}$, and it is the protonated form (ONOOH) that is most reactive (41).

[f]References 14, 41, 86.

[g]Reference 22.

[h]Order of reactivity for all the amino acids given by Xu & Chance (114), the most reactive being Cys, Met, Trp, Tyr, and Phe.

The presence of relatively high concentrations of H_2O_2 in plant tissues (14, 41, 86) has consequences. To limit the inadvertent formation of the very reactive $HO^•$ in the Fenton reaction, the cell must keep the concentration of free metal ions (mainly Fe, Cu, and Mn) extremely low by binding to metallochaperone molecules. This means that there is much less than one free copper ion in the average unstressed yeast cell (31, 87).

Nitrate reductase and the mitochondrial ETC appear to be major sources of $NO^•$ in plant cells (38, 83). Since $O_2^{•-}$ is also produced in all mitochondria (68), the reactive fusion product $ONOO^-$ will probably also be formed there.

In addition to producing ROS themselves, plants may also be exposed to anthropogenic ROS (and RNS) in the environment. Thus, ozone (O_3), sulfur dioxide (SO_2), and nitrogen dioxide (NO_2) are all major air pollutants that can cause oxidative modifications to many cellular components, and can even induce programmed cell death (41, 81).

Regulating ROS Production

In mitochondria, ROS production is generally caused by an over-reduction of the ETC. The mitochondria have several strategies for preventing or limiting such an over-reduction (68, 70). The stimulation of proton leakage in potato mitochondria by $O_2^{•-}$, probably through a stimulation of the uncoupling protein, provides an interesting feedback mechanism for lowering the reduction level of the mitochondrial ETC, thus preventing further $O_2^{•-}$ production (18).

The photosynthetic apparatus is constantly challenged by ROS generation during the light reactions of photosynthesis. Changes in light intensity and other environmental stresses can result in the absorption of more light energy than can be utilized for photosynthesis and plants have several mechanisms for dissipating this excess and preventing over-reduction of the photosynthetic ETC (52, 77).

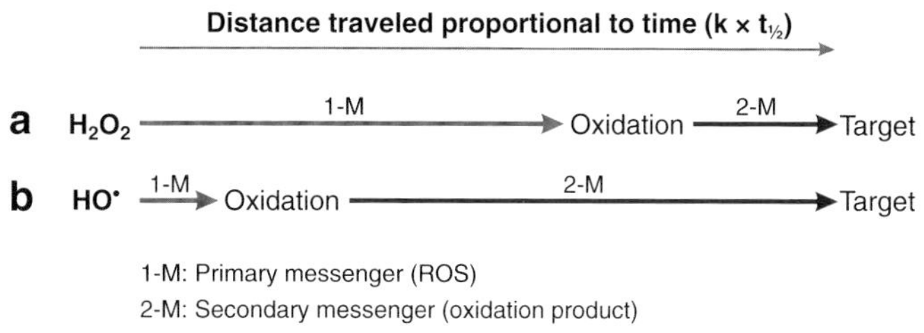

Figure 2

Signal transduction involving reactive oxygen species (ROS) as a primary messenger and a secondary messenger that is an oxidation product of a cellular component. The ultimate target is a gene or a group of genes. (*a*) The relatively long-lived H_2O_2 brings the signal close to the target, where it oxidizes an H_2O_2 sensor, which could be a transcription factor (56). (*b*) A short-lived ROS, in this case $HO^•$, reacts with a cellular component ($HO^•$ sensor) close to the site of production. The oxidation product, the second messenger, brings the signal to the gene(s) or perhaps to other transcription factors.

Removal of ROS

The plant cell and its organelles—peroxisomes (26), chloroplasts (6), and mitochondria (68, 70, 75)—contain multiple enzymes or enzyme systems for removing ROS (5, 13, 67). The reducing power for all of these systems (except SOD and catalase, which use internal oxidation/reduction of their substrate) derives directly or indirectly from NADPH or ferridoxin. It is outside the scope of this review to discuss their properties in detail, but **Table 2** contains an overview.

PUFA: polyunsaturated fatty acid

Several of the ROS-removing systems mentioned in **Table 2** use the reversible oxidation of peptide- or protein-bound cysteines to reduce ROS, thus illustrating the high reactivity of that amino acid with ROS (see below).

The balance between ROS production and the activities of these ROS-removing systems determines the type and concentration of ROS present and thus to what extent signaling and/or damage will occur.

MODIFICATIONS TO POLYUNSATURATED FATTY ACIDS

The polyunsaturated fatty acids (PUFAs) linoleic acid (18:2) and linolenic acid (18:3)

Table 2 ROS-removal mechanisms in plant cells[a]

Mechanisms	Removes (product)	Cellular location[b]	Useful references
SOD	$O_2^{\bullet -}$ (H_2O_2)	Chl, Cyt, Mit, Per	(26, 67)
Catalase	H_2O_2 (H_2O)	Mit?, Per	(26, 67, 68)
Peroxidases	H_2O_2 (H_2O)	Many locations	(82)
Ascorbate/glutathione cycle	H_2O_2 (H_2O)	Chl, Cyt?, Mit, Per	(26, 67, 70, 78)
Glutathione peroxidases	H_2O_2 (H_2O)	Chl, Cyt, ER, Mit	(67, 88)
	Lipid hydroperoxides		
	Other hydroperoxides		
Peroxiredoxin system	H_2O_2 (H_2O)	Chl, Cyt, Mit, Nucl	(27, 67)
	Alkyl hydroperoxides		
	Peroxinitrite		
Thioredoxin system[c]	H_2O_2 (H_2O)[c]	Chl, Cyt, Mit	(13, 67)
Glutaredoxin system[c]	H_2O_2 (H_2O)[c]	Chl, Cyt, Mit, Sec	(13, 67, 91)
	Hydroperoxides		
Carotenes and tocopherol	$^1O_2$ (O_2)	Chl	(6)

[a]Most of these are enzymes or enzyme systems. A good general reference is Mittler et al. (67).
[b]Chl, chloroplasts (plastids); Cyt, cytosol; ER, endoplasmic reticulum; Mit, mitochondria; Nucl, nucleus; Per, peroxisomes; Sec, secretory pathway.
[c]These systems are mainly involved in regulating the sulfhydryl/disulfide ratio in proteins. Their importance in removing H_2O_2 is uncertain.

are the major fatty acids in the plant membrane galactolipids (thylakoid membrane) and phospholipids (all other membranes). PUFAs are particularly susceptible to attack by $^1O_2$ and $HO^\bullet$, giving rise to complex mixtures of lipid hydroperoxides (**Figure 3**) (41, 73). Extensive PUFA peroxidation decreases the fluidity of the membrane, increases leakiness, and causes secondary damage to membrane proteins (see below) (40). Several aldehydes, e.g., 4-hydroxy-2-nonenal (HNE) and malondialdehyde (MDA), as well as hydroxyl and keto fatty acids, are formed as a result of PUFA peroxidation (**Figure 3**). The aldehyde breakdown products can form conjugates with DNA and proteins (see below). Aldehydes formed in the mitochondria may be involved in causing cytoplasmic male sterility in maize because a restorer gene in this species encodes a mitochondrial aldehyde dehydrogenase (62, 69).

HNE: 4-hydroxy-2-nonenal

MDA: malondialdehyde

Radiolysis: here, the formation of ROS (mainly $HO^\bullet$) from water by a pulse of high-energy electrons

In animals, oxidized fatty acids are selectively released by certain phospholipases, but such enzymes have not yet been identified in plants. In plant cells, some of the PUFA oxidation products function as secondary messengers either directly or after enzymatic modification (**Figure 3**) (73).

Glutathione peroxidases comprise an enzyme family using glutathione to reduce H_2O_2, lipid hydroperoxides, and other hydroperoxides (**Table 2**) (41). One member of this family is the phospholipid-hydroperoxide glutathione peroxidase, which can act directly on lipid hydroperoxide without the need to release the hydroperoxy fatty acid (88).

Membrane lipids appear to decrease initial radiolytic oxidative modifications (mostly by $HO^\bullet$) to membrane proteins (25). This is energetically advantageous to the cell because a damaged lipid molecule or a PUFA is cheaper to remove and replace than a protein molecule (see below).

Case Story

Lipid peroxidation in the *flu* mutant. To assess the role of $^1O_2$ relative to those of

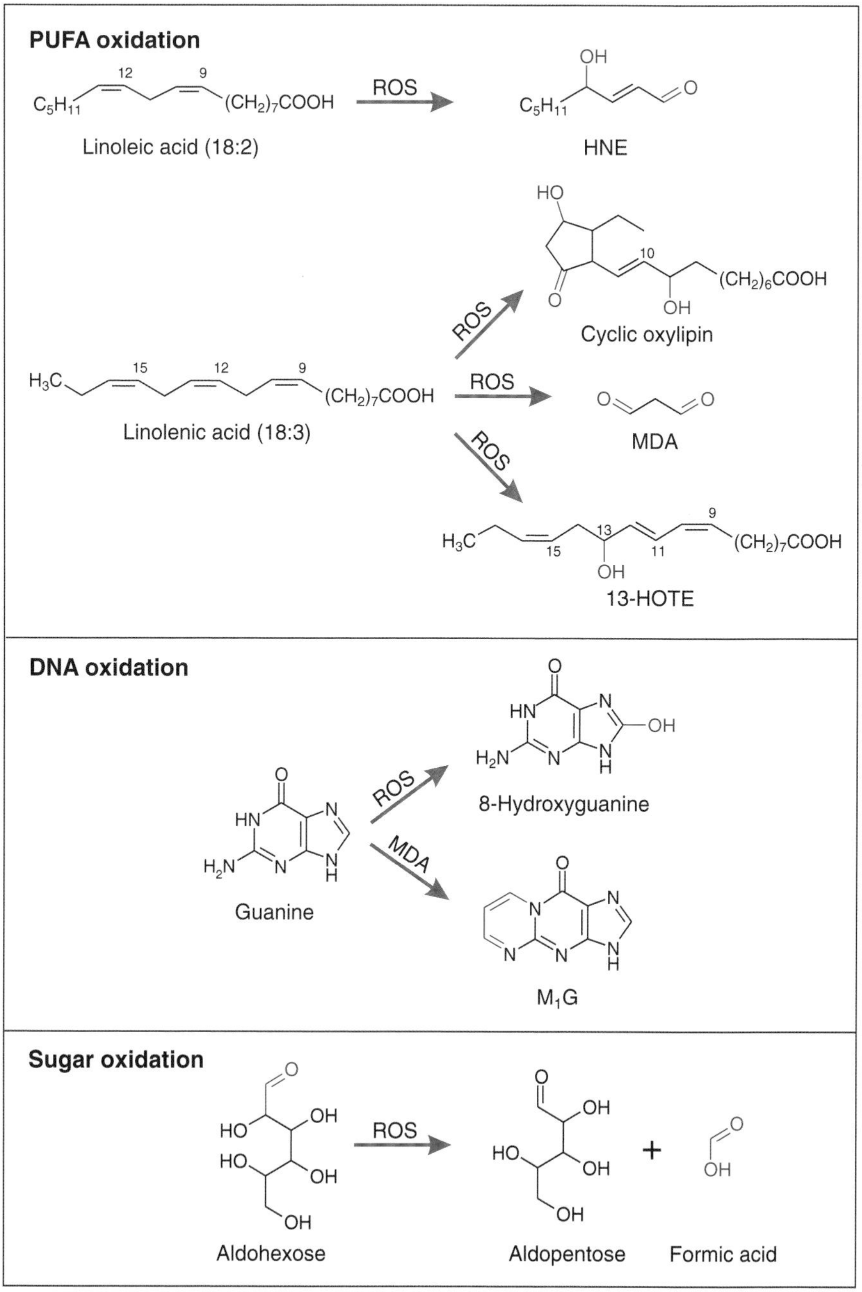

Figure 3

Commonly observed oxidative modifications of polyunsaturated fatty acids (PUFAs), DNA, and carbohydrates. In plant cells, 4-hydroxy-2-nonenal (HNE) and malondialdehyde (MDA) are formed mainly by oxidation of linoleic acid (18:2) and linolenic acid (18:3), respectively (41). The 13-HOTE and cyclic oxylipins are only examples of the wide range of similar products arising from peroxidation of linolenic acid (73). Similar products would be created by oxidation of linoleic acid. Both PUFA and guanine are drawn as free although in the cell they would normally be oxidized while attached to a lipid and DNA, respectively. Modifications caused by reactive oxygen species (ROS) are highlighted in red. The group oxidized in the sugar (aldohexose) is shown in blue. M_1G, pyrimido[1,2-a]-purin-10(3H)-one.

other ROS, op den Camp et al. (80) took advantage of the Arabidopsis *flu* mutant. In darkness, this mutant accumulates a specific chlorophyll precursor, protochlorophyllide, in its chloroplasts. Because protochlorophyllide is a photosensitizer that specifically generates $^1O_2$, this system provided highly localized production of $^1O_2$ without generating other oxidizing species. The results clearly indicate that the chloroplasts generated $^1O_2$

HOTE: hydroxyoctadecatrienoic acid

selectively and rapidly activated distinct sets of nuclear genes that were different from those induced by $O_2^{\bullet-}/H_2O_2$. At the same time a transient and selective accumulation of a stereospecific isomer of free hydroxy-octadecatrieonic acid (HOTE) (**Figure 3**), which is an oxidation product of linolenic acid, was observed. The observation of stereospecificity was interpreted to be indicative of an enzymatic (lipoxygenase-mediated) oxidation rather than direct chemical oxidation by $^1O_2$. The authors concluded that $^1O_2$ does not primarily act as a toxin but rather as a signal activating several stress-response pathways (80). This conclusion was later supported by the identification of EXECUTER1, a hitherto unknown nucleus-encoded chloroplastic protein. Although the double mutant *flu/executer1* accumulates the same amount of free protochlorophyllide generating the same amount of $^1O_2$ as the *flu* mutant, the double mutant did not initiate any stress response (110).

The oxidation of linolenic acid reported by op den Camp et al. (80) presumably occurred while it was still part of the membrane galactolipids and the oxidized product(s) was subsequently released by enzymatic action. Thus, it cannot be excluded that a lipase or perhaps enzymes degrading other peroxidation products generated the stereospecificity bias observed (only one stereoisomer of HOTE was detected).

The primary oxidation event in the above study would probably take place within 0.1–0.3 μm of the site of $^1O_2$ generation as this is as far as $^1O_2$ can diffuse in the few microseconds it exists (**Table 1**). The HOTE formed is a possible secondary messenger (73).

MODIFICATIONS TO DNA

DNA can be modified by ROS in many different ways, mainly on the nucleotide bases. $HO^\bullet$ is the most reactive, $^1O_2$ primarily attacks guanine, and H_2O_2 and $O_2^{\bullet-}$ do not react at all (112). 8-Hydroxyguanine is the most commonly observed modification (**Figure 3**). Given that mtDNA and ctDNA are close to major sites of ROS production and that these DNA forms have no histones and no chromatin structure, one might expect high rates of modification and equally high repair rates and this appears to be the case for mammalian mitochondria (104). No comparable information is available for plant cells. However, the presence of multiple copies of mtDNA and ctDNA enables the cell to select against negative mutations.

ROS damage to both mtDNA and nDNA is not completely random and mutation clusters at hot spots have been observed (41). However, no gene has been identified as being particularly susceptible to ROS damage.

In addition to direct DNA oxidation, ROS can also indirectly modify DNA. A common type of damage involves conjugation of the PUFA breakdown product MDA with guanine, which creates an extra ring (**Figure 3**) (47). In addition to mutations, oxidative DNA modifications can lead to changes in the methylation of cytosines, which is important for regulating gene expression (40).

A number of mechanisms are available for repairing DNA damage both in the nucleus and in the mitochondria. These include direct reversal of the damage, replacement of the base, and replacement of the whole nucleotide (55, 106). An accumulation of DNA damage is observed in several human diseases characterized by ROS accumulation, implying that the repair mechanisms are overwhelmed (112).

MODIFICATIONS TO CARBOHYDRATES

$HO^\bullet$ reacts with free carbohydrates, such as sugars, and polyols (100). Transgenic tobacco accumulating mannitol in the chloroplasts shows increased resistance to oxidative stress possibly because mannitol removes $HO^\bullet$ before it reacts with more vital cellular components (97). The oxidation of sugars with $HO^\bullet$ often releases formic acid as the main breakdown product (44). This may be the long-sought-after source of substrate for the enigmatic enzyme formate dehydrogenase (17, 42, 50).

Plant cell wall polysaccharides are susceptible to oxidative scission mediated by $HO^•$ in vitro under physiologically relevant conditions (34). Cell extension during elongation growth in plants is mediated by auxin, which promotes apoplastic ROS production. Cell wall–bound peroxidases use the ROS to generate $HO^•$ close to the site of scission (95). Support that $HO^•$ actually acts as a wall-loosening agent in vivo comes from chemical features, which are diagnostic of an $HO^•$ attack, that can be found in cell wall polysaccharides of ripening fruit (35). These oxidative modifications are not damaging as they are not detrimental to the modified component or the cell.

MODIFICATIONS TO PROTEINS

Protein oxidation is defined here as covalent modification of a protein induced by ROS, RNS, or byproducts of oxidative stress. Most types of protein oxidations are essentially irreversible whereas a few involving sulfur-containing amino acids are reversible (37). Protein oxidation is widespread and often used as a diagnostic marker for oxidative stress. The most common oxidative protein modifications are shown in **Figure 4**. A number of excellent reviews have been written about protein oxidation in mainly mammalian tissues (see 21, 22, 25, 41, and 96 as well as **http://www.medicine.uiowa.edu/FRRB/VirtualSchool/Virtual.html**). Protein oxidation in plants has been reviewed for mitochondria (45, 70, 71), but until now not for the whole plant cell.

Sulfur-Containing Amino Acids

Cys and Met are quite reactive, especially with $^1O_2$ and $HO^•$ (**Table 1**). The oxidation of thiol to disulfide (R_1-S-S-R_2, cystine) (**Figure 4**), which can be caused by several different ROS, is a very important metabolic redox regulation mechanism. Intra- or intermolecular disulfide bonds can be formed between cysteine side chains and the reduced form can be regenerated by the thioredoxin (Trx) or glutaredoxin systems (**Table 2**). A large number of potential Trx-regulated proteins have been identified in different cellular compartments including chloroplasts and mitochondria (13). The further oxidation of cysteine via cysteine sulfenic acid (R-SOH) to cysteine sulfinic acid (R-SO_2H) (**Figure 4**) is also enzymatically reversible and probably involved in signaling pathways (11). The highest level of cysteine oxidation, cysteic acid (R-SO_3H) (**Figure 4**), appears to be irreversible and damaging (37).

Trx: thioredoxin

Cysteine can also form mixed disulfides primarily with the cysteine-containing tripeptide glutathione and this might serve to protect the cysteine group against further oxidation. The mixed disulfides are re-reduced by glutaredoxins again involving glutathione (37). Glutathione can also be used to remove H_2O_2 (**Table 2**) and is the main nonprotein thiol in the cell (78).

Oxidation of methionine to methionine sulfoxide (**Figure 4**) is another reversible modification. An interesting example is the small heat shock protein in chloroplasts, which is inactivated by methionine sulfoxidation, but reactivated by reduction catalyzed by the enzyme peptide methionine sulfoxide reductase using Trx as the reductant (39). In Arabidopsis, a null mutation in a gene encoding a cytosolic isoform of the enzyme showed increased ROS content, lipid peroxidation, and protein oxidation at the end of a long night, which was clearly stressful to the plant (8).

It has been suggested that some peripheral methionine residues act as endogenous antioxidants protecting the active site and other sensitive domains in the protein while helping to remove ROS (58). It is quite likely that reversible methionine sulfoxidation will turn out to be an important regulatory mechanism (101). Further oxidation of Met to the sulphone (**Figure 4**) appears to be irreversible and damaging to the protein.

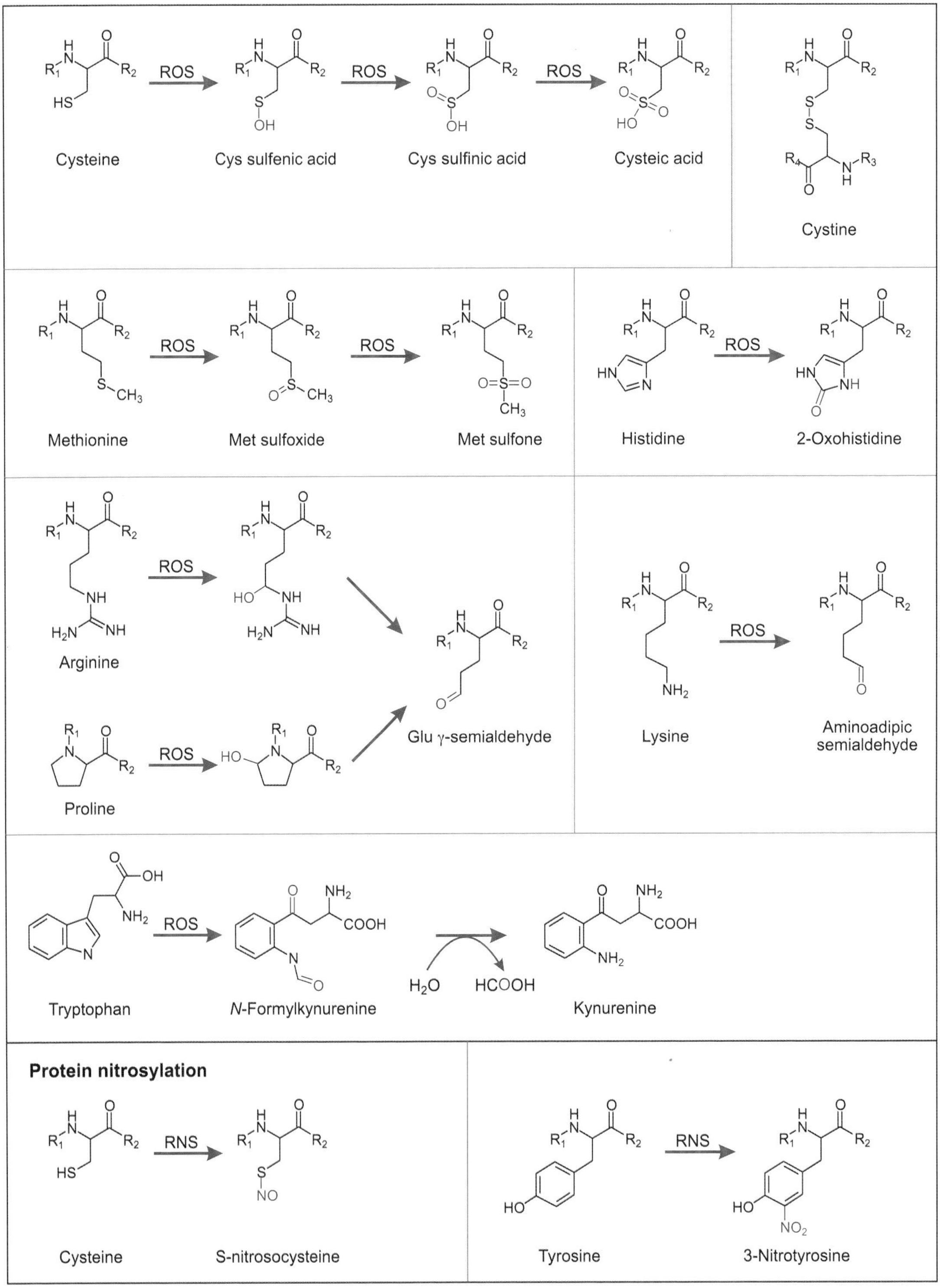
ROS
Cysteine
Cys sulfenic acid
Cys sulfinic acid
Cysteic acid
Cystine
Methionine
Met sulfoxide
Met sulfone
Histidine
2-Oxohistidine
Arginine
Proline
Glu γ-semialdehyde
Lysine
Aminoadipic semialdehyde
Tryptophan
N-Formylkynurenine
H2O
HCOOH
Kynurenine
Protein nitrosylation
RNS
Cysteine
S-nitrosocysteine
Tyrosine
3-Nitrotyrosine

Carbonylation

Apart from the reactions involving the sulfur-containing amino acids, carbonylation is the most commonly occurring oxidative protein modification. There are no indications that carbonylation is reversible (96). The oxidation of a number of protein amino acids—particularly Arg, His, Lys, Pro, Thr, and Trp—give free carbonyl groups (**Figure 4**). The detection of carbonyl groups is relatively simple, e.g., by conjugation with dinitrophenylhydrazine followed by antibody detection of the conjugate (59). Protein carbonylation has therefore been studied extensively.

In mammalian cells, protein carbonylation is found at a basic level of 1 nmol/mg protein increasing to 8 nmol/mg in samples from diseased tissue. This is, on average, equivalent to 0.05–0.40 carbonyl group per 50 kDa protein molecule, although these values may be overestimates (25). Thus, in stressed tissues a significant proportion of all proteins are carbonylated. An analysis of plant proteins gave an estimated 4 nmol carbonyl groups per milligram of protein (76, 89). Values as high as 60–700 nmol/mg have also been reported (107), but these are likely serious overestimates.

An in vivo treatment of pea plants with Cd^{2+} (89) or a strong in vitro oxidative treatment (76) raised the carbonylation level from 4 to 5.6 and 34 nmol/mg or about 0.3 and 1.7 carbonyl groups per 50 kDa protein molecule, respectively. In Arabidopsis, protein carbonylation in total protein extracts increased during the vegetative phase, but decreased sharply at the start of the reproductive phase, staying relatively low until and during senescence. It was suggested that a low degree of protein carbonylation during the reproductive phase could be part of a strategy to limit the transfer of oxidatively damaged components to the offspring (49) (see also below).

In wheat leaves, protein carbonylation was higher in the mitochondria than in chloroplasts and peroxisomes (7). This indicates that the mitochondria are more susceptible to oxidative damage and/or the removal of modified proteins is less efficient in the mitochondria. A number of carbonylated proteins have been identified in the mitochondrial matrix, 20 of which were probably carbonylated in vivo and a further 31 were oxidized by an in vitro treatment with $HO^\bullet$ (53) (follow the Supplemental Material link from the Annual Reviews home page at **http://www.annualreviews.org** to see **Supplemental Table 1**). Many of these oxidized proteins are Krebs cycle enzymes, other redox-active enzymes, or chaperones, or are involved in ROS detoxification (SOD). None of them are located immediately adjacent to the primary sites of ROS production in the ETC, so migration of ROS from the ETC to the matrix is implied.

The level of protein carbonylation in isolated pea leaf peroxisomes rose from 6.9 to 16.3 nmol per mg of peroxisomal protein as a result of Cd^{2+} treatment of the intact plant. These values are higher than for whole plant extracts from control and Cd^{2+}-treated plants, respectively, which could be the result of a higher local ROS concentration in the peroxisomes (89).

As mentioned above, Nguyen & Donaldson (76) exposed extracted proteins from castor bean peroxisomes to $HO^\bullet$ and monitored protein carbonylation and

Figure 4

Commonly observed oxidative modifications of protein amino acids. With the exception of Trp, which is shown as the free amino acid for convenience, all the amino acids are shown as part of a polypeptide chain. However, the names shown are those of free amino acids (i.e., proline rather than prolyl). Modifications caused by reactive oxygen species (ROS) or reactive nitrogen species (RNS) are highlighted in red. Cysteic acid is also known as cysteine sulfonic acid.

enzyme activities. The degree of carbonylation increased strongly over the first 90 min without a detectable change in the overall protein pattern. This indicates that the oxidative treatment did not lead to chain breakage or degradation. In response to the oxidative treatment, the activities of malate synthase, isocitrate lyase, catalase, and malate dehydrogenase decreased by 50–100%. Catalase was the protein with the most carbonylation and its activity was reduced by 80%. Carbonylated catalase has also been identified in other investigations (53, 89) (**Supplemental Table 1**).

Sweetlove et al. (102) identified shortened products of several proteins in mitochondria isolated from Arabidopsis cell cultures exposed to H_2O_2. Presumably, these smaller forms had been formed by oxidative breakage of peptide bonds in the proteins. This is surprising given that an in vitro treatment of peroxisomal proteins with $HO^\bullet$ did not cause substantial chain breakage, as judged by one-dimensional gel electrophoresis (76). The greater separation power of two-dimensional gel electrophoresis employed by Sweetlove et al. (102) likely made it possible to detect relatively small amounts of breakdown products.

In dry Arabidopsis seeds most of the carbonylation was found in the storage proteins, but carbonylation of a number of other proteins increased strongly during seed germination. The oxidized proteins derived from several cellular compartments including the cytosol, chloroplasts, and mitochondria (48) (**Supplemental Table 1**).

Tryptophan Oxidation

Tryptophan oxidation is another apparently irreversible protein modification (96). Mass spectrometric analysis of mitochondrial proteins permitted the identification of numerous proteins containing oxidized Trp in the form of N-formylkynurenine (**Figure 4**) as well as the specific site(s) of modification (72). With the exception of the oxidation-sensitive aconitase (a Krebs cycle enzyme that contains an Fe-S center), all of these proteins were either redox active or subunits in redox-active enzyme complexes (including the ETC). The same site was modified in (*a*) several adjacent spots containing the P-protein of the glycine decarboxylase complex, (*b*) two different isoforms of the mitochondrial processing peptidase in complex III, and (*c*) the same Trp residues in Mn-SOD in both rice and potato mitochondria. This indicates that Trp oxidation is not a random process. It is possible that exposed Trp residues act as intramolecular antioxidants, as suggested for Met residues (58).

Nitrosylation

The covalent attachment of the messenger $NO^\bullet$ to cysteine thiol (**Figure 4**) is a post-translational modification that potentially regulates the function of proteins. Protein (and glutathione) thiols can react with $NO^\bullet$ derivatives to produce a range of products including disulfides, sulfenic, sulfinic and sulfonic acids, as well as S-nitrosothiols (19).

The most common $NO^\bullet$ derivative is peroxynitrite ($ONOO^-$), which is formed through the condensation reaction of $NO^\bullet$ with $O_2^{\bullet-}$ ($NO^\bullet + O_2^{\bullet-} \rightarrow ONOO^-$) (**Table 1**) (41). In this way, $NO^\bullet$ is intimately linked with ROS, but $NO^\bullet$ does not cause S-nitrosylation on its own; it seems to involve S-nitrosothiols (20). However, peroxynitrite can cause depletion of SH groups and other antoxidants, oxidation of lipids, deamination of DNA bases, nitration of aromatic amino acid residues in proteins (**Figure 4**), and oxidation of methionine to its sulfoxide (41).

In a proteomic study of proteins from Arabidopsis cell suspension cultures or leaves treated with an $NO^\bullet$-donor or gaseous $NO^\bullet$, respectively, 63 proteins from cell cultures and 52 proteins from leaves that represent candidates for S-nitrosylation were identified. Among these were stress-related, redox-related, signaling/regulating, structural, photosynthetic, and metabolic proteins, indicating that $NO^{\bullet}$ is involved in the regulation of all of these processes (61) (**Supplemental Table 1**).

Interaction with Products of PUFA Oxidation

In vitro treatment of isolated mitochondria or enzyme complexes with HNE, one of the products of PUFA peroxidation (**Figure 3**), causes an inhibition of the decarboxylating enzyme complexes, e.g., pyruvate dehydrogenase, in mitochondria, probably by forming an adduct with their lipoamide moiety (65, 102, 103). HNE also reacts with the regulatory Cys (and other amino acids) in the alternative oxidase, thereby inhibiting its activity (111).

In vivo treatments of Arabidopsis cell cultures or pea plants, which increase the production of $O_2^{\bullet -}$ (and consequently other ROS), also increase the amount of PUFA breakdown as measured by MDA formation. At the same time, the amount and number of proteins conjugated with HNE increase dramatically while glycine oxidation by isolated mitochondria is inhibited (103, 111).

What Makes a Protein Susceptible to Oxidative Damage?

Cherry et al. (15) attempted to answer the question using the method of directed evolution. A fungal peroxidase was subjected to multiple rounds of directed evolution and an enzyme that had 100 times the oxidative stability (measured as the half-life at 40°C, pH 10.5 in the presence of 0.2 mM H_2O_2) of the wild-type enzyme was obtained. Changing three oxidizable amino acids near the active site heme group only improved the stability fivefold. The other changes required to increase the stability a further 20-fold changed the overall structure of the active site and probably reduced the accessibility of susceptible groups in the interior of the protein to H_2O_2 (15).

Case Stories of Protein Oxidation

Here we outline several case stories to illustrate the interactions between ROS and proteins.

Ribulose-1,5-bisphosphate carboxylase/oxygenase. Rubisco is located in the stroma of chloroplasts where it catalyzes the primary reactions of CO_2 assimilation and photorespiration. ROS trigger Rubisco degradation and the large subunit of Rubisco is site-specifically cleaved into five major fragments in light-treated leaf discs under chilling temperatures (74). In this case, the fragmentation was completely inhibited by n-propyl gallate or 1,2-dihydrobenzene-3,5-disulfonic acid, suggesting the involvement of $HO^{\bullet}$ and $O_2^{\bullet -}$, respectively. The fragmentation was stimulated by $FeSO_4$ and suppressed by an iron-specific chelator (deferoxamine), indicating a role for free Fe-ions in the generation of ROS. The results indicate that in situ formation of $HO^{\bullet}$ via the Fenton reaction between iron and H_2O_2 in vivo causes the fragmentation of the large subunit of Rubisco.

Transition metal ions like Fe^{2+} and Cu^{+} bound to proteins can generate ROS that can potentially cleave the polypeptide backbone. The ROS generated probably only diffuse relatively short distances from the generation site, suggesting that proximity is a very important parameter for the cleavage reaction of proteins. Therefore, it is likely that metals bound to the protein prior to damage are the source of ROS. An in vitro study with the Rubisco large subunit identified specific cleavage at six residues (Gly, Ala, Asp) located within a 1.2-nm radius around a metal-binding site. They all had the $C_{\alpha}H$ group of their peptide backbone completely or partially exposed to the bound metal (63).

Rubisco: ribulose-1,5-bisphosphate carboxylase/oxygenase

Photosystem I. PSI is degraded in intact leaves under chilling-light conditions, resulting in destruction of the Fe-S centers and detectable degradation of the PSI-B reaction center subunit (94). The data indicate the involvement of ROS, probably $HO^{\bullet}$ produced by the Fenton reaction between photoreduced Fe-S centers and H_2O_2. PSI photodestruction results in the release of free iron from the damaged Fe-S centers. Iron ions leaking from the thylakoid membrane into the stroma of the chloroplast could lead to the generation of $HO^{\bullet}$, which can attack stromal proteins such as Rubisco.

A similar case is the release of Fe^{2+} from damaged Fe-S centers in mitochondrial aconitase in mammalian cells, e.g., as a result of heat stress, and the subsequent collateral damage (43). Aconitase in plant mitochondria is also sensitive to oxidative damage (70, 72) (**Supplemental Table 1**).

D1 protein (PSII). Under illumination D1 has the highest turnover rate of all thylakoid membrane proteins. The cause of D1 fragmentation is likely $^1O_2$ produced by the reaction center chlorophyll of PSII (P680), because in vitro treatment with $^1O_2$-generating substances results in a similar fragmentation pattern of the D1 protein, as observed under in vivo photoinhibitory conditions (79). However, it is not the $^1O_2$, per se, that causes cleavage of the peptide backbone, because the fragmentation is not observed in a mutant lacking an FtsH protease. Rather, the accumulation of oxidized amino acids in the D1 protein triggers a conformational change that renders the D1 protein susceptible to the FtsH protease (98).

Superoxide dismutase. Two groups of proteins must be able to function in the presence of ROS: (*a*) ROS-detoxifying enzymes and (*b*) enzymes that produce ROS as part of their normal catalytic cycle. SOD, three forms of which are found in different compartments, qualifies on both of these counts. This raises several questions: Are these enzymes more resistant to ROS-induced damage? And, if yes, how is this resistance achieved? Does oxidation of these enzymes lead to activation, which would be an interesting auto-regulatory mechanism? Do these enzymes have a more rapid turnover than other proteins? Carbonylation of several ROS-metabolizing enzymes including SOD and catalase has been identified (**Supplemental Table 1**).

Mammalian Cu,Zn-SOD found in the cytosol and the mitochondrial intermembrane space is inactivated by oxidation of the Cu^{2+} His ligand in the catalytic site (54). In a proteomic study of the role of SOD in neurodegenerative diseases, several spots were identified as Cu,Zn-SOD isoform 1. One spot contained a carbonylated form, which could be caused by His oxidation (**Figure 4**), but the amino acid modified was not identified. Another spot contained a cysteine sulfonic acid (**Figure 4**), a modification in vitro caused by high concentrations of H_2O_2, the SOD reaction product (16).

The inactivation of human Mn-SOD, which removes $O_2^{\bullet -}$ in the mitochondrial matrix, by nitration of a tyrosine (**Figure 4**) near the Mn in the active site has been reported (115).

Transcription factors. The transcription factor PerR in *Bacillus subtilis* contains two His residues coordinating bound Fe^{2+}. Upon exposure to low levels (<10 μM) of H_2O_2 one or both of the His are oxidized presumably by $HO^{\bullet}$ generated by a Fenton reaction involving the bound Fe^{2+}. The oxidized transcription factor derepresses the PerP regulon, encoding enzymes acting to detoxify peroxides. The His oxidation is probably irreversible and leads to degradation of the protein. Thus, the transcription factor is sacrificed to help minimize damage from Fenton reactions elsewhere in the cell (56). Several other transcription factors have been identified in bacteria, yeast, and mammals, which use the reversible oxidation of cysteines to sense H_2O_2 (56).

WHAT HAPPENS TO OXIDIZED PROTEINS?

It is generally thought that oxidized proteins are degraded relatively rapidly, probably because a change in conformation exposes more hydrophobic residues, which are recognized by proteases. However, massive protein damage can lead to the formation of protein aggregates that apparently cannot be degraded (23, 25) or are degraded by autophagy (113).

A number of proteases are found in all the major cellular compartments: cytosol (109), chloroplast (1, 93), and mitochondria (46, 71). It is unlikely that specific proteases are dedicated to the removal of oxidatively modified proteins because there are so many types of oxidative modifications. It is more likely that the proteases recognize proteins with a more open conformation. However, an unfolded conformation does not appear to be a requirement for degradation (90). We need to know more about the function and specificity of the proteases before we can understand the turnover of oxidized proteins.

Peptides deriving from mitochondrial matrix proteins have been found outside the mitochondria (51). Such peptides or even individual oxidized amino acids released from oxidized proteins are second messenger candidates.

THE COST OF ROS-INDUCED MODIFICATIONS OF CELLULAR COMPONENTS

The fate of oxidized proteins can be graded by the energetic cost to the cell (**Table 3**). The cheapest method, one used extensively in metabolic regulation, is to reverse the modification, i.e., reduce the oxidized sulfur-containing amino acid side chain back to its original form. At the second level, the oxidatively modified polypeptide is removed, degraded, and replaced by a new polypeptide. Presumably, the degradation goes to the amino acid level, where the vast majority of unmodified amino acids are recycled, whereas the modified amino acids are further metabolized. If a damaged polypeptide is part of a protein complex, the complex might dissociate, the damaged peptide be replaced, and the complex reassembled. The most expensive solution (short of programmed cell death) is to remove the entire organelle by autophagy, e.g., mitochondria by mitophagy (57).

Table 3 Estimated costs of replacing damaged cellular components[a]

Component	Size	Cost[e]
One amino acid molecule	100–200 Da	10^{f}
One PUFA molecule[b]	250–300 Da	4×10^1
One phospholipid molecule (with 2 PUFAs)	700–800 Da	10^2
One average protein molecule[c]	50,000 Da	2×10^3
One average protein molecule imported across a membrane	50,000 Da	10^4
One mitochondrion[d]	1 pg protein	2×10^{10}–10^{11}

[a]All values are very approximate.
[b]This is an estimate of the cost of synthesizing the PUFA from acetyl-CoA and does not consider the potential cost of excising and degrading the oxidized PUFA.
[c]An estimate of the cost of synthesizing the protein from amino acids. The cost of RNA turnover, e.g., mRNA, has not been considered, but it would probably not change the estimate significantly.
[d]Calculated assuming that 1 mg mitochondrial protein (consisting of 50 kDa proteins) has a matrix volume of 1 μl and a mitochondrion is a 2 μm cylinder with a diameter of 0.5 μm. The cost of replacing the ca 25% (w/w) phospholipids (28) is about half the cost. The cost of DNA and RNA turnover has not been considered, but it would probably not change the estimate significantly.
[e]In ATP equivalents/molecule or unit.
[f]This is only to give an order of magnitude.

The replacement for oxidatively damaged proteins will often have to cross membranes and/or be inserted into membranes. It is difficult to estimate the total energy consumption of this translocation and insertion, but it appears to be costly. Estimates of 10^3–10^4 ATP per protein molecule translocated have been obtained for secretion of bacterial proteins using the secretory pathway (29, 60) and translocation of proteins that use the chloroplastic twin arginine (cpTat) pathway (3).

An Arabidopsis null mutation in a gene encoding a cytosolic peptide methionine sulfoxide reductase (see above) showed increased ROS content, lipid peroxidation, and (curiously) protein carbonylation at the end of a long night. This coincided with increased protein turnover and respiration. It was interpreted to mean that increased protein turnover caused by oxidative damage was so extensive that it required increased respiration to produce the necessary extra ATP (8). Let us therefore estimate the energetic costs of replacing all the proteins in a plant mitochondrion. The cost of synthesizing and degrading 1 mg mitochondrial protein containing 20 nmol protein with an average size of 50 kDa is 20–200 μmol ATP (see **Table 3**). Assuming that the mitochondria are oxidizing Krebs cycle intermediates such as malate at 100 nmol oxygen $min^{-1}mg$ $protein^{-1}$ (typical rates under in vitro conditions), they will produce around 400 nmol ATP $min^{-1}mg$ $protein^{-1}$. It will therefore take approximately 1–10 h for the mitochondria to produce enough ATP to replace all their proteins.

What is the half-life of mitochondrial proteins? Estimates range from 20 h to >100 h for mammalian mitochondria (see 71 and references therein). Thus, if we assume that the average lifespan is 50 h in plant mitochondria, then mitochondrial protein turnover will consume 2–20% of all the ATP the mitochondria can make working at their maximum rate. However, this may be an underestimate because mitochondria generally do not respire at their maximum rate in situ and because the activity of energy-wasteful enzymes in the mitochondrial ETC, such as the alternative oxidase, can lower the ATP yield (e.g., 30). We must also keep in mind that the relative contribution of oxidative modifications to total protein turnover remains to be established. A significant part of cellular protein turnover is regulated to suit metabolic flexibility and that will also contribute to mitochondrial protein turnover.

HOW TO PREVENT THE TRANSFER OF OXIDATIVE DAMAGE TO THE NEXT GENERATION

It is important to prevent, or at least limit, the transfer of damaged components, especially DNA and DNA-containing organelles, to the next generation. Several very different strategies have been described for plants, mammals, and yeast.

The biochemical activity is very high in the stamen where pollen are produced and the mitochondria appear to be working close to their maximal capacity because most mutations in genes encoding mitochondrial proteins lead to dramatically lowered pollen pro duction (e.g., 92, 108). High activity generally means high ROS production, although direct evidence for this in stamen is lacking. However, mitochondria (and plastids) are generally maternally inherited in plants. Therefore, organelles derive from the relatively quiescent egg cell, thus minimizing oxidative damage to the fertilized egg cell (4).

Another possible strategy has been proposed for mammalian cells where a small, biochemically relatively inactive subpopulation of "breeding" mitochondria in each cell is dividing while the biochemically active, and therefore relatively damaged, mitochondria do not divide (57). Finally, a third strategy has been observed in yeast where carbonylated proteins, regardless of subcellular localization, are prevented from entering the daughter cell by an as-yet-unknown mechanism (2).

SUMMARY POINTS

1. In each compartment of the plant cell, ROS formation and removal are tightly regulated.
2. Specifically, the formation of $HO^{\bullet}$ from H_2O_2 in the Fenton reaction is limited by keeping the concentration of free metal ions extremely low.
3. Oxidation of PUFA generates many products, some of which are secondary signaling molecules in plants whereas others, including MDA and HNE, can form adducts and damage DNA and proteins.
4. Nitrosylation and reversible oxidation of cysteine and methionine serve regulatory purposes.
5. Many other types of protein oxidations, e.g., carbonylation, are irreversible and probably mainly damaging. There is a relatively high level of protein carbonylation in plant tissues, which increases during stress. Many well-known proteins are oxidized, but little is known about the percentage of the molecules modified or the effect of the modifications on the properties of the proteins. Oxidized proteins are normally degraded rapidly by a number of proteases present in all plant cell compartments.
6. The cost of replacing all the proteins in a plant mitochondrion is estimated to be 2–20% of the ATP produced by the mitochondrion.
7. Yeast, mammals, and plants have mechanisms for minimizing the transfer of oxidatively modified components, especially DNA, to the next generation.
8. Many oxidative modification products are involved in metabolic regulation and signal transduction. Thus, oxidative damage and signaling are often two sides of the same story.

ACKNOWLEDGMENTS

We are grateful to Drs. Mats X. Andersson, G. Patrick Bienert, Axel Brennicke, Christine Finnie, Thomas P. Jahn, David M. Logan, Allan G. Rasmusson, Tinna Stevnsner, and Hans Thordal-Christensen for helpful comments. Funding from The Danish Veterinary and Agricultural Research Council (23-03-0105) and the Danish Natural Science Research Council (272-05-0360) to P.E.J. is gratefully acknowledged.

LITERATURE CITED

1. Adam Z, Rudella A, van Wijk KJ. 2006. Recent advances in the study of Clp, FtsH and other proteases located in chloroplasts. *Curr. Opin. Plant Biol.* 9:234–40
2. **Aguilaniu H, Gustafsson L, Rigoulet M, Nyström T. 2003. Asymmetric inheritance of oxidatively damaged proteins during cytokinesis. *Science* 299:1751–53**
3. Alder NN, Theg SM. 2003. Energetics of protein transport across biological membranes: a study of the thylakoid ΔpH-dependent/cpTat pathway. *Cell* 112:231–42
4. Allen JF. 1996. Separate sexes and the mitochondrial theory of ageing. *J. Theor. Biol.* 180:135–40
5. Apel K, Hirt H. 2004. Reactive oxygen species: Metabolism, oxidative stress and signal transduction. *Annu. Rev. Plant Biol.* 55:373–99

2. Carbonylated proteins are shown to be prevented from entering the yeast daughter cell by an unknown mechanism.

6. Asada K. 2006. Production and scavenging of reactive oxygen species in chloroplasts and their functions. *Plant Physiol.* 141:391–96
7. Bartoli CG, Gomez F, Martinez DE, Guiamet JJ. 2004. Mitochondria are the main target for oxidative damage in leaves of wheat (*Triticum aestivum* L.). *J. Exp. Bot.* 55:1663–69
8. Bechtold U, Murphy DJ, Mullineaux PM. 2004. Arabidopsis peptide methionine sulfoxide reductase2 prevents cellular oxidative damage in long nights. *Plant Cell* 16:908–19
9. Bienert GP, Møller ALB, Kristiansen KA, Schulz A, Møller IM, et al. 2007. Specific aquaporins facilitate the diffusion of hydrogen peroxide across membranes. *J. Biol. Chem.* 282:1183–92
10. Bienert GP, Schjoerring JK, Jahn TP. 2006. Membrane transport of hydrogen peroxide. *Biochim Biophys Acta* 1758:994–1003
11. Biteau B, Labarre J, Toledano MB. 2003. ATP-dependent reduction of cysteine-sulphinic acid by *S. cerevisiae* sulphiredoxin. *Nature* 425:980–84
12. Blokhina O, Virolainen E, Fagerstedt KV. 2003. Antioxidants, oxidative damage and oxygen deprivation stress: A review. *Ann. Bot.* 91:179–94
13. Buchanan BB, Balmer Y. 2005. Redox regulation: A broadening horizon. *Annu. Rev. Plant Biol.* 56:187–20
14. Cheeseman JM. 2006. Hydrogen peroxide concentrations in leaves under natural conditions. *J. Exp. Bot.* 57:2435–44
15. Cherry JR, Lamsa MH, Schneider P, Vind J, Svendsen A, et al. 1999. Directed evolution of a fungal peroxidase. *Nat. Biotechnol.* 17:379–84
16. Choi J, Rees HD, Weintraub ST, Levey AI, Chin L-S, Li L. 2005. Oxidative modifications and aggregation of Cu,Zn-superoxide dismutase associated with Alzheimer and Parkinson diseases. *J. Biol. Chem.* 280:11648–55
17. Colas des Francs-Small C, Ambard-Bretteville F, Small ID, Remy R. 1993. Identification of a major soluble protein in mitochondria from nonphotosynthetic tissues as NAD-dependent formate dehydrogenase. *Plant Physiol.* 102:1171–77
18. Considine MJ, Goodman M, Echtay KS, Laloi M, Whelan J, et al. 2003. Superoxide stimulates a proton leak in potato mitochondria that is related to the activity of the uncoupling protein. *J. Biol. Chem.* 278:22298–302
19. Costa NJ, Dahm CC, Hurrell F, Taylor ER, Murphy MP. 2003. Interactions of mitochondrial thiols with nitric oxide. *Antioxid. Redox Signal.* 5:291–305
20. Dahm CC, Moore K, Murphy MP. 2006. Persistent S-nitrosation of Complex I and other mitochondrial membrane proteins by S-nitrosothiols but not nitric oxide or peroxynitrite. Implications for the interaction of nitric oxide with mitochondria. *J. Biol. Chem.* 281:10056–65
21. Dalle-Donne I, Scaloni A, Giustarini D, Cavarra E, Tell G, et al. 2005. Proteins as biomarkers of oxidative/nitrosative stress in disease: The contribution of redox proteomics. *Mass Spectrom. Rev.* 24:55–99
22. Davies MJ. 2004. Reactive species formed on proteins exposed to singlet oxygen. *Photochem. Photobiol. Sci.* 3:17–25
23. Davies MJ, Shringarpure R. 2006. Preferential degradation of oxidized proteins by the 20S proteasome may be inhibited in aging and in inflammatory neuromuscular diseases. *Neurology* 66:S93–96
24. Davletova S, Rizhsky L, Liang HJ, Shengqiang Z, Oliver DJ, et al. 2005. Cytosolic ascorbate peroxidase 1 is a central component of the reactive oxygen gene network of Arabidopsis. *Plant Cell* 17:268–81
25. Dean RT, Fu SL, Stocker R, Davies MJ. 1997. Biochemistry and pathology of radical-mediated protein oxidation. *Biochem. J.* 324:1–18

26. del Rio LA, Sandalio LM, Corpas FJ, Palma JM, Barroso JB. 2006. Reactive oxygen species and reactive nitrogen species in peroxisomes. Production, scavenging, and role in cell signaling. *Plant Physiol.* 141:330–35
27. Dietz K-J, Jacob S, Oelze M-L, Laxa M, Tognetti V, et al. 2006. The function of peroxiredoxins in plant organelle redox metabolism. *J. Exp. Bot.* 57:1697–709
28. Douce R. 1985. *Mitochondria in Higher Plants: Structure, Function, and Biogenesis*. Orlando, FL: Academic
29. Driessen AJM. 1992. Precursor protein translocation by the *Escherichia coli* translocase is directed by the protonmotive force. *EMBO J.* 11:847–53
30. Escobar MA, Geisler DA, Rasmusson AG. 2006. Reorganization of the alternative pathways of the Arabidopsis respiratory chain by nitrogen supply: Opposing effects of ammonium and nitrate. *Plant J.* 45:775–88
31. Finney LA, O'Halloran TV. 2003. Transition metal speciation in the cell: Insights from the chemistry of metal ion receptors. *Science* 300:931–36
32. Foyer CH, Noctor G. 2003. Redox sensing and signalling associated with reactive oxygen in chloroplasts, peroxisomes and mitochondria. *Physiol. Plant.* 119:355–64
33. Foyer CH, Noctor G. 2005. Oxidant and antioxidant signalling in plants: A re-evaluation of the concept of oxidative stress in a physiological context. *Plant Cell Environ.* 28:1056–1071
34. Fry SC. 1998. Oxidative scission of plant cell wall polysaccharides by ascorbate-induced hydroxyl radicals. *Biochem J.* 332:507–15
35. Fry SC. 2004. Primary cell wall metabolism: Tracking the careers of wall polymers in living plant cells. *New Phytol.* 161:641–75
36. Gechev TS, Van Breusegem F, Stone JM, Denev I, Laloi C. 2006. Reactive oxygen species as signals that modulate plant stress responses and programmed cell death. *BioEssays.* 28:1091–101
37. Ghezzi P, Bonetto V. 2003. Redox proteomics: Identification of oxidatively, modified proteins. *Proteomics* 3:1145–53
38. Gupta KJ, Stoimenova M, Kaiser WM. 2005. In higher plants, only root mitochondria, but not leaf mitochondria reduce nitrite to NO, in vitro and in situ. *J. Exp. Bot.* 56:2601–9
39. Gustavsson N, Kokke BP, Härndahl U, Silow M, Bechtold U, et al. 2002. A peptide methionine sulfoxide reductase highly expressed in photosynthetic tissue in *Arabidopsis thaliana* can protect the chaperone-like activity of a chloroplast-localized small heat shock protein. *Plant J.* 29:545–53
40. Halliwell B. 2006. Reactive species and antioxidants. redox biology is a fundamental theme of aerobic life. *Plant Physiol.* 141:312–22
41. **Halliwell B, Gutteridge JMC, eds. 1999. *Free Radicals in Biology and Medicine*. Oxford: Oxford Univ. Press. 3rd ed.**
42. Igamberdiev AU, Bykova NV, Kleczkowski LA. 1999. Origins and metabolism of formate in higher plants. *Plant Physiol. Biochem.* 37:503–13
43. Ilangovan G, Venkatakrishnan CD, Bratasz A, Osinbowale S, Cardounel AJ, et al. 2006. Heat shock-induced attenuation of hydroxyl radical generation and mitochondrial aconitase activity in cardiac H9c2 cells. *Am. J. Physiol. Cell Physiol.* 290:C313–24
44. Isbell HS, Frush HL, Martin ET. 1973. Reactions of carbohydrates with hydroperoxides. 1. Oxidation of aldoses with sodium peroxide. *Carbohydr. Res.* 26:287–95
45. Ito J, Heazlewood JL, Millar AH. 2007. The plant mitochondrial proteome and the challenge of defining the post-translational modifications responsible for signalling and stress effects on respiratory functions. *Physiol. Plant.* 129:207–24

41. This is the ROS handbook. The fourth edition came out in December 2006.

46. Janska H. 2005. ATP-dependent proteases in plant mitochondria: What do we know about them today? *Physiol. Plant.* 123:399–405
47. Jeong Y-C, Nakamura J, Upton PB, Swenberg JA. 2005. Pyrimido[1,2-α]-purin-10(3H)-one, M_1G, is less prone to artifact than base oxidation. *Nucleic Acids Res.* 33:6426–34
48. Job C, Rajjou L, Lovigny Y, Belghazi M, Job D. 2005. Patterns of protein oxidation in Arabidopsis seeds and during germination. *Plant Physiol.* 138:790–802
49. Johansson E, Olsson O, Nyström T. 2004. Progression and specificity of protein oxidation in the life cycle of Arabidopsis thaliana. *J. Biol. Chem.* 279:22204–8
50. Juszczuk IM, Bykova NV, Møller IM. 2007. Protein phosphorylation in plant mitochondria. *Physiol. Plant.* 129:90–113
51. Kambacheld M, Augustin S, Tatsuta T, Müller S, Langer T. 2005. Role of the novel metallopeptidase MoP112 and saccharolysin for the complete degradation of proteins residing in different subcompartments of mitochondria. *J. Biol. Chem.* 280:20132–39
52. Krieger-Liszkay A, Trebst A. 2006. Tocopherol is the scavenger of singlet oxygen produced by the triplet states of chlorophyll in the PSII reaction centre. *J. Exp. Bot.* 57:1677–84
53. Kristensen BK, Askerlund P, Bykova NV, Egsgaard H, Møller IM. 2004. Identification of oxidised proteins in the matrix of rice leaf mitochondria by immunoprecipitation and two-dimensional liquid chromatography-tandem mass spectrometry. *Phytochemistry* 65:1839–51
54. Kurahashi T, Miyazaki A, Suwan S, Isobe M. 2001. Extensive investigations on oxidized amino acid residues in H_2O_2-treated Cu,Zn-SOD protein with LC-ESI-Q-TOF-MS, MS/MS for the determination of the copper-binding site. *J. Am. Chem. Soc.* 123:9268–78
55. Larsen NB, Rasmussen M, Rasmussen LJ. 2005. Nuclear and mitochondrial DNA repair: Similar pathways? *Mitochondrion* 5:89–108
56. Lee JW, Helmann JD. 2006. The PerR transcription factor senses H_2O_2 by metal-catalysed histidine oxidation. *Nature* 440:363–67
57. Lemasters JJ. 2005. Selective mitochondrial autophagy, or mitophagy, as a targeted defense against oxidative stress, mitochondrial dysfunction, and aging. *Rejuvenation Res.* 8:3–5
58. Levine RL, Mosoni L, Berlett BS, Stadtman ER. 1996. Methionine residues as endogenous antioxidants in proteins. *Proc. Natl. Acad Sci. USA* 93:15036–40
59. Levine RL, Willams JA, Stadtman ER, Shacter E. 1994. Carbonyl assays for determination of oxidatively modified proteins. *Methods Enzymol.* 233:346–57
60. Lill R, Cunningham K, Brundage LA, Ito K, Oliver D,Wickner W. 1989. SecA protein hydrolyzes ATP and is an essential component of the protein translocation ATPase of *Escherichia coli*. *EMBO J.* 8:961–66
61. Lindermayr C, Saalbach G, Durner J. 2005. Proteomic identification of S-nitrosylated proteins in Arabidopsis. *Plant Physiol.* 137:921–30
62. Liu F, Cui XQ, Horner HT, Weiner H, Schnable PS. 2001. Mitochondrial aldehyde dehydrogenase activity is required for male fertility in maize. *Plant Cell* 12:1063–78
63. Luo S, Ishida H, Makino A, Mae T. 2002. Fe^{2+}-catalyzed site-specific cleavage of the large subunit of ribulose 1,5-bisphosphate carboxylase close to the active site. *J. Biol. Chem.* 277:12382–87
64. Maxwell DP, Wang Y, McIntosh L. 1999. The alternative oxidase lowers mitochondrial reactive oxygen production in plant cells. *Proc. Natl. Acad. Sci. USA* 96:8271–76

53. A number of mitochondrial proteins are identified as being carbonylated in vivo or in vitro. The paper demonstrates the power of combining immunoprecipitation with two-dimensional HPLC linked online to a mass spectrometer.

56. A bacterial transcription factor senses H_2O_2 by oxidizing a His group which binds Fe^{2+}. The oxidized form of the protein activates a group of genes encoding peroxide-detoxifying enzymes.

61. More than 100 nitrosylated proteins are identified in Arabidopsis cell cultures and leaves by a biotin switch method. This provides us with a list of pathways potentially regulated by NO•.

64. The alternative oxidase is demonstrated to have an important function in regulating ROS production in plant mitochondria and plant cells.

65. Millar AH, Leaver CJ. 2000. The cytotoxic lipid peroxidation product, 4-hydroxy-2-nonenal, specifically inhibits decarboxylating dehydrogenases in the matrix of plant mitochondria. *FEBS Lett.* 481:117–21
66. Miller G, Mittler R. 2006. Could heat shock transcription factors function as hydrogen peroxide sensors in plants? *Ann. Bot.* 98:279–88
67. Mittler R, Vanderauwera S, Gollery M, Van Breusegem F. 2004. Reactive oxygen gene network of plants. *Trends Plant Sci.* 9:490–98
68. Møller IM. 2001. Plant mitochondria and oxidative stress. Electron transport, NADPH turnover and metabolism of reactive oxygen species. *Annu. Rev. Plant Physiol. Plant Mol. Biol.* 52:561–91
69. Møller IM. 2001. A more general mechanism of cytoplasmic male sterility? *Trends Plant Sci.* 6:560
70. Møller IM. 2007. Mitochondrial electron transport and oxidative stress. In *Annual Plant Reviews: Plant Mitochondria*, ed. DC Logan, pp. 185–211. Oxford, Blackwell
71. Møller IM, Kristensen BK. 2004. Protein oxidation in plant mitochondria as a stress indicator. *Photochem. Photobiol. Sci.* 3:730–35
72. Møller IM, Kristensen BK. 2006. Protein oxidation in plant mitochondria detected as oxidized tryptophan. *Free Radic. Biol. Med.* 40:430–35
73. Mueller MJ. 2004. Archetype signals in plants: The phytoprostanes. *Curr. Opin. Plant Biol.* 7:441–48
74. Nakano R, Ishida H, Makino A, Mae T. 2006. In vivo fragmentation of the large subunit of ribulose-1,5-bisphosphate carboxylase by reactive oxygen species in an intact leaf of cucumber under chilling-light conditions. *Plant Cell Physiol.* 47:270–76
75. Navrot N, Rouhier N, Gelhaye E, Jacquot J-P. 2007. ROS generation and antioxidant systems in plant mitochondria. *Physiol. Plant.* 129:185–95
76. Nguyen AT, Donaldson RP. 2005. Metal-catalyzed oxidation induces carbonylation of peroxisomal proteins and loss of enzymatic activities. *Arch. Biochem. Biophys.* 439:25–31
77. Niyogi KK. 2000. Safety valves for photosynthesis. *Curr. Opin. Plant Biol.* 3:455–60
78. Noctor G, Foyer CH. 1998. Ascorbate and glutathione: Keeping active oxygen under control. *Annu. Rev. Plant Physiol. Plant Mol. Biol.* 49:249–79
79. Okada K, Ikeuchi M, Yamamoto N, Ono T, Miyao M. 1996. Selective and specific cleavage of the D1 and D2 proteins of photosystem II by exposure to singlet oxygen: factors responsible for the cleavage of proteins. *Biochim. Biophys. Acta* 1274:73–79
80. op den Camp RG, Przybyla D, Ochsenbein C, Laloi C, Kim C, et al. 2003. Rapid induction of distinct stress responses after the release of singlet oxygen in Arabidopsis. *Plant Cell* 15:2320–32
81. Overmyer K, Brosché M, Kangasjärvi J. 2003. Reactive oxygen species and hormonal control of cell death. *Trends Plant Sci.* 8:335–42
82. Passardi F, Cosio C, Penel C, Dunand C. 2005. Peroxidases have more functions than a Swiss army knife. *Plant Cell Rep.* 24:255–65
83. Planchet E, Gupta KJ, Sonoda M, Kaiser WM. 2005. Nitric oxide emission from tobacco leaves and cell suspensions: rate limiting factors and evidence for the involvement of mitochondrial electron transport. *Plant J.* 41:732–43
84. Pospisil P, Arato A, Krieger-Liszkay A, Rutherford AW. 2004. Hydroxyl radical generation by photosystem II. *Biochemistry* 43:6783–92
85. Pryor WA. 1986. Oxy-radicals and related species: Their formation, lifetimes, and reactions. *Annu. Rev. Physiol.* 48:657–67
86. Puntarulo A, Sanchez RA, Boveris A. 1988. Hydrogen peroxide metabolism in soybean embryonic axes at the onset of germination. *Plant Physiol.* 86:626–30

87. Rae TD, Schmidt PJ, Pufahl RA, Culotta VC, O'Halloran TV. 1999. Undetectable intracellular free copper: The requirement of a copper chaperone for superoxide dismutase. *Science* 284:805–8
88. Rodriguez Milla MA, Maurer A, Huete AR, Gustafson JP. 2003. Glutathione peroxidase genes in Arabidopsis are ubiquitous and regulated by abiotic stresses through diverse signaling pathways. *Plant J.* 36:602–15
89. Romero-Puertas MC, Palma JM, Gómez M, del Rio LA, Sandalio LM. 2002. Cadmium causes the oxidative modification of proteins in pea plants. *Plant Cell Environ.* 25:677–86
90. Röttgers K, Zufall N, Guiard B, Voos W. 2002. The ClpB homolog Hsp78 is required for the efficient degradation of proteins in the mitochondrial matrix. *J. Biol. Chem.* 277:45829–37
91. Rouhier N, Gelhaye E, Jacquot J-P. 2004. Plant glutaredoxins: still mysterious reducing systems. *Cell. Mol. Life Sci.* 61:1266–77
92. Sabar M, de Paepe R, de Kouchkovsky Y. 2000. Complex I impairment, respiratory compensations, and photosynthetic decrease in nuclear and mitochondrial male sterile mutants of *Nicotiana sylvestris*. *Plant Physiol.* 124:1239–49
93. Sakamoto W. 2006. Protein degradation machineries in plastids. *Annu. Rev. Plant Biol.* 57:599–621
94. Scheller HV, Haldrup A. 2005. Photoinhibition of photosystem I. *Planta* 221:5–8
95. Schopfer P, Liszkay A, Bechtold M, Frahry G, Wagner A. 2002. Evidence that hydroxyl radicals mediate auxin-induced extension growth. *Planta* 214:821–28
96. Shacter E. 2000. Quantification and significance of protein oxidation in biological samples. *Drug Metab. Rev.* 32:307–26
97. Shen B, Jensen RG, Bohnert HJ. 1997. Increased resistance to oxidative stress in transgenic plants by targeting mannitol biosynthesis to chloroplasts. *Plant Physiol.* 113:1177–83
98. Silva P, Thompson E, Bailey S, Kruse O, Mullineaux CW, et al. 2003. FtsH is involved in the early stages of repair of photosystem II in *Synechocystis* sp PCC 6803. *Plant Cell* 15:2152–64
99. Skovsen E, Snyder JW, Lambert JDC, Ogilby PR. 2005. Lifetime and diffusion of singlet oxygen in a cell. *Phys. Chem. Lett. B* 109:8570–73
100. Smirnoff N, Cumbes QJ. 1989. Hydroxyl radical scavenging activity of compatible solutes. *Phytochemistry* 28:1057–60
101. Sundby C, Härndahl U, Gustavsson N, Åhrman E, Murphy DJ. 2005. Conserved methionines in chloroplasts. *Biochim. Biophys. Acta* 1703:191–202
102. Sweetlove LJ, Heazlewood JL, Herald V, Holtzapffel R, Day DA, et al. 2002. The impact of oxidative stress on Arabidopsis mitochondria. *Plant J.* 32:891–904
103. Taylor NL, Day DA, Millar AH. 2002. Environmental stress causes oxidative damage to plant mitochondria leading to inhibition of glycine decarboxylase. *J. Biol. Chem.* 277:42663–68
104. Thorslund T, Sunesen M, Bohr VA, Stevnsner T. 2002. Repair of 8-oxoG is slower in endogenous nuclear genes than in mitochondrial DNA and is without strand bias. *DNA Repair* 20:1–13
105. Torres MA, Dangl JL. 2005. Fuctions of the respiratory burst oxidase in biotic interactions, abiotic stress and development. *Curr. Opin. Plant Biol.* 8:397–403
106. Tuteja N, Singh MB, Misra MK, Bhalla PL, Tuteja N. 2001. Molecular methanisms of DNA damage and repair: Progress in plants. *Crit. Rev. Biochem. Mol. Biol.* 36:337–97
107. Vanacker H, Sandalio LM, Jimenez A, Palma JM, Corpas FJ, et al. 2006. Roles for redox regulation in leaf senescence of pea plants grown on different sources of nitrogen nutrition. *J. Exp. Bot.* 57:1735–45

108. Vedel F, Lalanne E, Sabar M, Chetrit P, de Paepe R. 1999. The mitochondrial respiratory chain and ATP synthase complexes: Composition, structure and mutational studies. *Plant Physiol. Biochem.* 37:629–43
109. Vierstra RD. 2003. The ubiquitin/26S proteasome pathway, the complex last chapter in the life of many plant proteins. *Trends Plant Sci.* 8:135–42
110. Wagner D, Przybyla D, op den Camp R, Kim C, Landgraf F, et al. 2004. The genetic basis of singlet oxygen-induced stress responses of *Arabidopsis thaliana*. *Science* 306:1183–85
111. Winger AM, Millar AH, Day DA. 2005. Sensitivity of plant mitochondrial terminal oxidases to the lipid peroxidation product 4-hydroxy-2-nonenal (HNE). *Biochem. J.* 387:865–70
112. Wiseman H, Halliwell B. 1996. Damage to DNA by reactive oxygen and nitrogen species: Role in inflammatory disease and progression to cancer. *Biochem. J.* 313:17–29
113. Xiong Y, Contento AL, Nguyen PQ, Bassham DC. 2007. Degradation of oxidized proteins by autophagy during oxidative stress in Arabidopsis. *Plant Physiol.* 143:291–99
114. Xu GZ, Chance MR. 2005. Radiolytic modification and reactivity of amino acid residues serving as structural probes for protein footprinting. *Anal. Chem.* 77:4549–55
115. Yamakura F, Taka H, Fujimura T, Murayama K. 1998. Inactivation of human manganese-superoxide dismutase by peroxynitrite is caused by exclusive nitration of tyrosine 34 to 3-nitrotyrosine. *J. Biol. Chem.* 273:14085–89

Cumulative Indexes

Contributing Authors, Volumes 48–58

Chapter Titles, Volumes 48–58

Genetics and Molecular Biology

Cell Differentiation

Tissue, Organ, and Whole Plant Events

Methods